Sustainable Global Resources of Seaweeds Volume 1

Ambati Ranga Rao • Gokare A. Ravishankar
Editors

Sustainable Global Resources of Seaweeds Volume 1

Bioresources, Cultivation, Trade and Multifarious Applications

Springer

Editors
Ambati Ranga Rao
Department of Biotechnology
Vignan's Foundation of Science
Technology and Research
Deemed to be University
Vadlamudi, Guntur, Andhra Pradesh, India

Gokare A. Ravishankar
Dr. C. D. Sagar Centre for Life Sciences
Dayananda Sagar College of Engineering
Dayananda Sagar Institutions
Bengaluru, Karnataka, India

ISBN 978-3-030-91957-3 ISBN 978-3-030-91955-9 (eBook)
https://doi.org/10.1007/978-3-030-91955-9

This Springer imprint is published by the registered company Springer Nature Switzerland AG
The registered company address is: Gewerbestrasse 11, 6330 Cham, Switzerland

Professor V. Subramanian

This volume is dedicated to Prof. Velusamy Sivasubramanian, a doyen in the field of applied phycology.
He has established a private company Phycospectrum Environmental Research Center (PERC) in Chennai, India, where he is serving as the founding director. He continues to develop processes using algal forms for mitigation of pollution in industrial effluents and for environmental clean-up. He has been a keen supporter of research undertaken by the editors, and we are grateful to him for the same.

He was born in the year 1955, in Udumalpet, Tamil Nadu, India. He did his schooling in Gandhi Kala Nilayam, Karattur,

Udumalpet district. He obtained BSc and MSc in Botany in Madras university in the year 1975 and 1977, respectively. He obtained his PhD degree in the year 1983 in Botany from Madras University, under the supervision of Dr V.N Rajarao. He has spent more than 25 years in teaching and in conducting research on algal technologies.

He realized the potentials of microalgal forms in converting the pollutants such as hazardous chemicals, heavy metals, industrial wastes, and wastewaters to acceptable levels for discharge to environment. World's first phyco-remediation plant was set up by him at SNAP (SNAP Natural and Alginate Products) alginate industry at Ranipet, Tamil Nadu. This was followed by his contributions to the development of several large-scale industrial treatment plants adopting algae-based technologies.

He implemented algae-based treatment technology in industries in India, like SAR Chandra, Kakinada; Perfetti Van melle, Chennai; Ultralmarine and Pigments, Ranipet; Thirumalai Chemicals, Ranipet; STAHL India, Ranipet; Hindustan Coca Cola, Ahmedabad; Arvind Mills, Ahmedabad; Tata Steel, Jamshedpur. He has also put up petrochemical effluent treatment plants using algal forms at Pacific Rubiales Industry, Colombia. Recently, he was instrumental in setting up Iran's first commercial algal cultivation facility at Chabahar. Now another project was undertaken for Astral Industrial Corporation in Sicily, Cyprus.

He has guided 20 students for PhD and has published more than 100 research articles in national and international journals. He has served as the core committee expert on second-generation biofuels at the Ministry of New and Renewable Energy (MNRE), Government of India. Research program monitoring committees of Council of Scientific and Industrial Research, Department of Biotechnology (DBT), New Delhi. He was a national committee member for CO_2 sequestration research of Department of Science and Technology; DBT–Indian Oil Corporation, Faridabad. He is the editorial board member of Seaweed Research and Utilization Journal and assistant editor of Phykos of Phycological Society of India; International Journal of Engineering, Science, and Technology, and Indian Hydrobiology.
We wish success in his ecofriendly endeavors to pollution abatement for protecting the health of our planet.

Preface

Upon realizing the need for a comprehensive treatise on seaweed cultivation and their utilization for food and nutritional security, we the editors present to you two volumes on seaweeds which are elegantly published by Springer Nature. The demand for increasing the food production for the growing world population, from nearly 8.0 billion presently, which is expected to reach 9.9 billion by 2050, has been a daunting task. Sustainable development goals of the UN for achieving zero hunger through enhanced food production and distribution also focus on dietary diversification of the world population to meet the food, nutritional, and nutraceutical needs. Moreover, the approaches to the enhancement in food production are also limited by the scarcity of water needed for agriculture. In this context, the utilization of seaweed-biodiversity provides tremendous opportunities to employ them to produce large quantities of quality biomass for food applications, using marine water resources. Furthermore, they can be cultivated without the dependence on fertilizers and other agrochemical inputs which are otherwise needed in agricultural farming systems.

Seaweeds have been consumed, traditionally as food, in the eastern countries for centuries. Their consumption is also rapidly expanding to other regions of the world. The seaweed recipes are gaining popularity owing to their culinary properties and health attributes. Seaweed constituents, such as agar and carrageenan, are used in food processing. Presently, seaweeds are realizing applications in the health and wellness sectors. The growing global demand for seaweeds has resulted in the development of farming methods for large-scale production of biomass and international trade. Food safety and standards issues for seaweed products are gaining prominence for local and global needs. The researches to unravel the hidden potentials of seaweed for health and therapeutic applications is in full swing.

Because of the utility value and economic implication of the seaweeds and their products for food, health, pharmaceutical, cosmeceutical, and a myriad of uses, their biodiversity is being increasingly explored for a wide variety of applications.

Here, we have made our best efforts to compile a vast body of knowledge by leveraging the experience of the experts in the field of seaweed science and technology for the benefit of all stakeholders. The material contained here will cater to a vast audience including students, teachers, scientists, food and health experts, technologists, policymakers, and environmentalists. It should also address the professionals in the food, agriculture, health, pharmaceutical, cosmeceutical, environmental, and several emerging technologies for sustainable utilization of seaweeds and their biodiversity for the benefit of the global population.

The above-mentioned aspects are presented in two volumes, and the brief description of each of the volumes provides an insight into their broad contents.

Volume I and II is contributed by 122 and 127 authors, respectively, from 21 countries.

Volume I deals with seaweed bioresources, cultivation, trade, and multifarious applications:

The seaweed farming in various parts of the world with a focus on Asia, Europe, Australia, and South America has been detailed by the authors from the respective parts of the world. They have also presented state-of-the-art technologies and product development strategies, including the quality of the produce and market potentials. The ecological implications of seaweed cultivation concerning the industrial aquaculture scenario have been presented. Seaweeds as biofertilizers, aquaculture feeds, livestock feeds, and agricultural applications have been detailed. Industrial adoption of seaweeds for bioremediation, wastewater treatment, and bioenergy generation are dealt which have environmental applications with far-reaching implications. The biorefinery approach to Valorization of the process with complete fractionation of constituents for value addition has been dealt with.

Volume II deals with seaweeds for food, pharmaceutical, and health applications:

The seaweed as food with nutritional advantages has been detailed. The use of several edible forms such as *Monostroma*, *Caulerpa*, *Palmaria*, *Gracillaria*, *Porphyra*, *Laminaria*, *Fucus*, *Undaria*, etc., used routinely in various recipes as sea vegetables, salads, soups, meat analogs have been dealt extensively. The use of seaweeds as sources of nutraceuticals has been explained in detail, and the products in the international market, as well as their trade, have been reviewed. The occurrence of pigments, such as fucoxanthin, phlorotannins, phycocyanin, phycoerythrin, Beta-carotene, and several more, are presented in detail. In addition to these, the fatty acids, vitamins, and minerals have been described. These aforementioned seaweed constituents have innumerable health applications. The current understanding of their bioactive properties and pharmacological actions – including their bioefficacy, safety, and toxicity aspects — is described. Antidiabetic, antioxidant, antiobesity, cardioprotective, pre and probiotic properties with augmentation of gut microbiome, antiviral effects, and many more direct benefits to health and disease management through seaweed-based products are described. The advancement of science and technology of seaweeds has brought to light the utility of seaweed as a rich source of nutraceuticals and also as a source of micronutrients. Global trade of seaweed-based foods and quality considerations have been brought out elegantly, which will govern the future expansion of the seaweed industry. Their use in cosmeceuticals is fast expanding with a high degree of value addition to the herbal formulations. The seaweed ingredients are also of value in extending the shelf life of food products. Biosynthetic and genomic studies to genetically modify seaweeds for enhanced productivity have been brought to light.

The collective efforts of scientists from all over the world represent global perspectives on the topics. The authors have put in their best in presenting you with their own research experience in handling various systems, thereby offering practical solutions to the industrial exploitation of seaweed resources. The editors expect a wide range of readership of these volumes which fills in the demand for up-to-date knowledge in the utilization of seaweed biodiversity for commercial exploitation in a sustainable manner.

Guntur, AP, India
Bangalore, India

Ambati Ranga Rao
Gokare A. Ravishankar

Acknowledgments

We the editors of this volume wish to place on record the encouragement received from worldwide supporters interested in seaweeds science, technology, and trade. Most importantly, the spontaneity in acceptance to contribute articles by the authors has resulted in timely publication as scheduled. We are grateful to the authors for their cooperation, despite the pandemic time, by keeping up the commitment to bringing to the table a vast body of knowledge and information on seaweeds for the benefit of all stakeholders. We are highly indebted to the authors for their scholarly inputs provided to the international audience.

We wholeheartedly appreciate the efforts of Springer Nature who has done a marvelous job in collation, compilation, and publication with due diligence in bringing out these volumes. Our special thanks are due to Daniel Falatko and his team for their total dedication in bringing out this volume in such an elegant manner.

Editors are thankful to the support of their families during the pandemic for the cooperation and encouragement received to complete this task.

A.R.R thanks his wife, Deepika; daughter, Jesvisree; parents, Venkateswaralu and Tulasidevi; brothers; sisters in law; sisters; and brothers-in-law.

G.A.R thanks his wife, Shyla; son, Prashanth; daughter-in-law, Vasudha; and daughter, Apoorva.

A.R.R is indebted to Dr. L. Rathaiah, Chairman; Mr. L. Sri Krishnadevarayalu, Vice Chairman (Member of Parliament); Prof. Dr. K. Ramamurthy Naidu, Chancellor; Dr. M.Y.S. Prasad, Vice-Chancellor; Dr. Kavi Kishor, Scientific Advisor; and Dr. Madhusudhan Rao, Director, Engineering and Management, Dean Academics, Dean R&D, and Head, Biotechnology Department, Vignan's Foundation for Science, Technology and Research University for providing facility and support to fulfill this additional assignment.

G.A.R. expresses his gratitude to Dr. Premachandra Sagar, Vice Chairman, Dayananda Sagar Institutions, and Pro-Chancellor of Dayananda Sagar University, Bengaluru, for granting permission to take this additional responsibility and for his unstinted support.

G.A.R. offers his special thanks to the Council of Scientific and Industrial Research (CSIR) for including him as an expert in CSIR Mission on Seaweeds – a multi-institutional project on seaweeds cultivation and utilization as an economic activity supported by the Government of India under the "Atmanirbhar Bharath" program.

Editors are thankful to the institutions of the Government of India, the Department of Biotechnology and the Department of Science and Technology, for their support of our research project on algae-based foods at CSIR-Central Food Technological Research Institute (CFTRI), Mysuru, India.

Ambati Ranga Rao
Gokare A. Ravishankar

Contents

About the Editors

Ambati Ranga Rao is a senior scientist and associate professor in the Department of Biotechnology at Vignan's Foundation for Science, Technology, and Research (Deemed to be University), Andhra Pradesh, India. He is involved in both teaching cum research for graduate and undergraduate students. Recently, he is listed as the Top 2% World's Scientist reported by Stanford University, USA.

He holds bachelor's and master's degree from Acharya Nagarjuna University, Andhra Pradesh, India, and PhD degree from the University of Mysore. He started his research career in 2004 as a research assistant at the Department of Plant Cell Biotechnology, Council of Scientific and Industrial Research (CSIR)–Central Food Technological Research Institute (CFTRI), Mysuru, India, under the supervision of Dr. G.A. Ravishankar and Dr. R. Sarada. He was awarded Senior Research Fellow of Indian Council of Medical Research (ICMR), New Delhi, in the year 2007. His PhD work at CFTRI focused on the production of astaxanthin from cultured green alga *Haematococcus pluvialis* and its biological activities.

He worked extensively on process optimization of algal biomass production, mass culture of various algal species in raceway ponds and photobioreactors and downstream processing of algal metabolites, and evaluation of their possible nutraceutical applications in *in vitro* and *in vivo* models. Furthermore, Dr. Ranga Rao was involved in a project on "Studies on field cultivation and harvesting of seaweeds *Porphyra*, *Enteromorpha*, *Eucheuma* and their use in processed foods".

He worked as lead scientist in Algal Technologies, Carot Labs Pvt. Ltd, India; postdoctoral research associate in Laboratory of Algal Research and Biotechnology, Arizona State University, USA, under the supervision of Prof. Milton Sommerfeld and Prof. Qiang Hu; visiting assistant professor in Food Science and Technology Program, Beijing Normal University and Hong Kong Baptist University, United International College, China, under the supervision of Prof. Bo Lei; and visiting senior research fellow (associate professor grade) in Institute of Ocean and Earth Sciences, University of Malaya, Malaysia, under the guidance of Prof. Phang Siew-Moi.

He is the author of 50 peer-reviewed publications, 60 international/national conferences/symposia/invited talks/FDPs/workshops/STC, and 28 chapters in books. His research citations exceed 3300 with h-index (20) and i10-index (26) as Google Scholar. He has delivered lectures at international/national conferences/symposia in the United States, Canada, Brazil, China, Malaysia, Indonesia, and Oman.

He has edited three books (CRC Press and Academic Press, USA), as coeditor, namely, *Handbook of Algal Technologies and Phytochemicals: Volume I Food, Health, and Nutraceutical Applications*; *Handbook of Algal Technologies and Phytochemicals: Volume II Phycoremediation, Biofuels, and Global Biomass Production*; and *Global Perspectives on Astaxanthin: from Industrial Production to Food, Health, and Pharmaceutical Application*.

He was selected for the Junior Scientist of the Year Award (2015) by National Environmental Science Academy, New Delhi, India; honored TWAS-Young Affiliate (2014–2018) by Regional Office of South East Asia and the Pacific Chinese Academy of Sciences (CAS), China; received Young Scientist Award (2014) at the World Food Science Congress by International Union of Food Science and Technology (IUFoST), Canada; Carl Storm International Diversity Fellowship Award (2010) by Gordon Research Conferences, USA.

He is a lifetime member of the Society of Applied Biotechnology, India; National Environmental Science Academy, India; and Asia PGPR Society of Sustainable Agriculture, USA and Association of food scientists and technologists of India.

He is an associate fellow of Andhra Pradesh Akademi of Sciences (2019), Government of Andhra Pradesh, India, and also a fellow of the Society of Applied Biotechnology (2013), India.

He has received research grants and travel grant fellowship as both international and national awards, under Young Scientist schemes. He is also serving as an editorial board member, guest editor for special issues, and reviewer for reputed international and national journals.

Gokare A. Ravishankar professor of biotechnology, is presently the vice president of R&D in Life Sciences and Biotechnology at Dayananda Sagar Institutions, Bengaluru, India. Earlier, he had a distinguished research career of over 30 years at the Central Food Technological Research Institute (CFTRI), Mysore, and in the institutions of government of India. He served as chairman of the board of studies in biotechnology at the Visvesvaraya Technological University, Belgavi, and academic council member of Dayananda Sagar University. He has also been a member of the boards of eight universities. He served as visiting professor to universities in Japan, Korea, Taiwan, and Russia.

He is an internationally recognized expert in the areas of food science and technology, plant biotechnology, algal biotechnology, food biotechnology and postharvest technologies, plant secondary metabolites, functional foods, herbal products, genetic engineering, and biofuels.

He holds a master's and a PhD degree from Maharaja Sayajirao University of Baroda. He mentored over 40 PhD students, 62 master's students, 7 postdocs, and 8 international guest scientists and authored over 260 peer-reviewed research papers in international and national journals, 50 reviews, 55 patents in India and abroad, and edited 5 books, with an h-index of 70 and over 22,000 citations. He has presented over 220 lectures in various scientific meetings in India and abroad, including visits to about 30 countries.

He has received international honors as a Fellow of the International Academy of Food Science and Technology (Canada) and Institute of Food Technologists (USA), and is a Certified Food Scientist of USA.

He was honored as a fellow of several organizations in India, namely, the National Academy of Sciences, National Academy of Agricultural Sciences, Association of Microbiologists of India, Society of Agricultural Biochemists, Society of Applied Biotechnologists, Indian Botanical Society, Biotechnology Research Society of India, and the Association of Food Scientists and Technologists of India. He is also an elected member of the Plant Tissue Culture Association of India.

Dr. Ravishankar received several coveted honors and awards as follows: Young Scientist award (Botany) by the then Prime Minister of India in 1992; National Technology Day Award of Government of India in 2003; Laljee Goodhoo Smarak Nidhi Award for food biotechnology R&D of industrial relevance; and the prestigious, Professor V. Subramanyan Food Industrial Achievement Award; Professor S.S. Katiyar Endowment Lecture Award in New Biology by Indian Science Congress; Professor Vyas Memorial Award of Association of Microbiologists of India; Professor V.N. Raja Rao Endowment Lecture Award in Applied Botany, University of Madras; Lifetime Achievement Award by the Society of Applied Biotechnologists, Dr. Diwaker Patel Memorial Award by Anand Agricultural University, Anand; Prof. C.S. Paulose Memorial Oration Award by Society for Biotechnologists of India; Prof. Gadgil Memorial Lecture Award from Plant Tissue Culture Association of India.

He has held honorary positions as President in the Society of Biological Chemists, Mysore Chapter and President of Association of Microbiologists of India, and Mysore and Bangalore Chapters. He is a lifetime member of the Nutrition Society of India and several biotechnology societies including the Society for Biotechnologists of India, Biotechnology Research Society of India, International Coffee Genome Network, American Society of Plant Biologists, and Global Harmonization Initiative.

He is a consultant to World Bank projects in the domains of postharvest technologies, plant biotechnologies for value addition to crop plants, and food biotechnologies. He has also served as advisor and resource person in international conferences, seminars, workshops, and short courses, and has convened national and international seminars in biology, biotechnology, and food science and technology. He is an associate editor and reviewer in a large number of reputed research journals.

Contributors

B. Aiyamperumal Centre of Advanced Study in Marine Biology, Faculty of Marine Sciences, Annamalai University, Parangipettai, Chidambaram, India

Ambati Ranga Rao Department of Biotechnology, Vignan`s Foundation of Science, Technology and Research (Deemed to be University), Vadlamudi, Guntur, Andhra Pradesh, India

P. Anantharaman Centre of Advanced Study in Marine Biology, Faculty of Marine Sciences, Annamalai University, Parangipettai, Chidambaram, India

Glacio S. Araújo Federal Institute of Education, Science and Technology of Ceará – IFCE, Aracati, Ceará, Brazil

K. Arunkumar Phycoscience Laboratory, Department of Plant Science Central University of Kerala, Kasaragod, Kerala, India

Oce Astuti Faculty of Fishery and Marine Sciences, Halu Oleo University, Kendari, Indonesia

Kiril Bahcevandziev Agricultural College of Coimbra (ESAC/IPC), Research Centre for Natural Resources Environment and Society (CERNAS), Institute of Applied Research (IIA), Coimbra, Portugal

Bahtiar Faculty of Fishery and Marine Sciences, Halu Oleo University, Kendari, Indonesia

Abdul Muis Balubi Faculty of Fishery and Marine Sciences, Halu Oleo University, Kendari, Indonesia

Budiyanto Faculty of Fishery and Marine Sciences, Halu Oleo University, Kendari, Indonesia

Elisa Capuzzo Centre for the Environment, Fisheries and Aquaculture Science (Cefas), Weymouth, UK

Josefina Plaza Cazón Centro de Investigación y Desarrollo en Fermentaciones Industriales, CINDEFI (CCT La Plata-CONICET, UNLP), La Plata, Argentina

Facultad de Ciencias Exactas, La Plata, Argentina

Ana Cerar Algen, algal technology centre, llc, Ljubljana, Slovenia

Dinesh Chandra Department of Biological Sciences, CBS&H, G.B. Pant University of Agriculture and Technology, Almora, Uttarakhand, India

GIC Chamtola, Almora, Uttarakhand, India

Anong Chirapart Department of Fishery Biology, Faculty of Fisheries, Kasetsart University, Bangkok, Thailand

Bruno Cognie Université de Nantes, MMS, Nantes, France

João Cotas University of Coimbra, MARE – Marine and Environmental Sciences Centre, Department of Life Sciences, Coimbra, Portugal

Priscilla Decottignies Université de Nantes, MMS, Nantes, France

Charu Deepika Institute of Molecular Bioscience, The University of Queensland, Brisbane, QLD, Australia

Dhara Dixit Department of Earth and Environmental Science, Krantiguru Shyamji Krishna Verma Kachchh University, Bhuj, Gujarat, India

Edgardo R. Donati Centro de Investigación y Desarrollo en Fermentaciones Industriales, CINDEFI (CCT La Plata-CONICET, UNLP), La Plata, Argentina

Facultad de Ciencias Exactas, La Plata, Argentina

Justine Dumay Université de Nantes, MMS, Nantes, France

Martha Bonnet Dunbar CTAQUA, Aquaculture Technology Centre, El Puerto de Santa María (Cádiz), Cádiz, Spain

ICMAN-CSIC, Institute of Marine Sciences of Andalusia, Spanish National Research Council, Puerto Real (Cádiz), Cádiz, Spain

Islam Mahmoud El-Manaway Science Faculty, Botany Department, Suez Canal University, Ismailia, Egypt

Sunita Facknath Department of Agricultural and Food Science, Faculty of Agriculture, University of Mauritius, Reduit, Mauritius

Joël Fleurence Université de Nantes, MMS, Nantes, France

Felicia Fonseca Centro de Investigação de Montanha (CIMO), Instituto Politécnico de Bragança, Bragança, Portugal

Sara García-Poza University of Coimbra, MARE – Marine and Environmental Sciences Centre, Department of Life Sciences, Coimbra, Portugal

M. Garcia-Vaquero Section of Food and Nutrition, School of Agriculture and Food Science, University College Dublin, National University of Ireland, Belfield, Dublin, Ireland

Marta Castilla Gavilan Université de Nantes, MMS, Nantes, France

Thiyam General Department of Biological Sciences, CBS&H, G.B. Pant University of Agriculture and Technology, Almora, Uttarakhand, India

Ana M. M. Gonçalves University of Coimbra, MARE – Marine and Environmental Sciences Centre, Department of Life Sciences, Coimbra, Portugal

Department of Biology and CESAM, University of Aveiro, Aveiro, Portugal

Nguyen Cam Ha Institute of Biotechnology, Vietnam Academy of Science and Technology (VAST), Hanoi, Vietnam

Graduate University of Science and Technology, VAST, Hanoi, Vietnam

Ben Hankamer Institute of Molecular Bioscience, The University of Queensland, Brisbane, QLD, Australia

Deepthi Hebbale Energy & Wetlands Research Group, Centre for Ecological Sciences, Bangalore, India

Centre for Sustainable Technologies, Bangalore, India

Dang Diem Hong Institute of Biotechnology, Vietnam Academy of Science and Technology (VAST), Hanoi, Vietnam

Graduate University of Science and Technology, VAST, Hanoi, Vietnam

ThuyLoi University, Hanoi, Vietnam

Eun Kyoung Hwang Aquaculture Industry Research Division, East Sea Fisheries Research Institute, National Institute of Fisheries Science, Gangneung, Republic of Korea

Marta Illera-Vives Instituto de Biodiversidad Agraria y Desarrollo Rural (IBADER), Universidad de Santiago de Compostela, Lugo, Spain

Muhammad Sajjad Iqbal Department of Botany, University of Gujrat, Gujrat, Pakistan

Wardha Jalil Faculty of Fisheries, Dayanu Iksanuddin University, Baubau, Southeast Sulawesi, Indonesia

Pannaga P. Jutur Omics of Algae Group, Industrial Biotechnology, International Centre for Genetic Engineering and Biotechnology, Aruna Asaf Ali Marg, New Delhi, India

Mukul S. Kareya Omics of Algae Group, Industrial Biotechnology, International Centre for Genetic Engineering and Biotechnology, Aruna Asaf Ali Marg, New Delhi, India

Ma'ruf Kasim Faculty of Fishery and Marine Sciences, Halu Oleo University, Kendari, Indonesia

Saleha Khan Department of Fisheries Management, Bangladesh Agricultural University, Mymensingh, Bangladesh

Stefan Kraan The Seaweed Company Blue Turtle Ltd., Mausrevagh, Kilcoona, Headford, Ireland

Bhanooduth Lalljee Department of Agricultural and Food Science, Faculty of Agriculture, University of Mauritius, Reduit, Mauritius

Yoran Le Strat Université de Nantes, MMS, Nantes, France

S. W. Lee Faculty of Agro Based Industry, Universiti Malaysia Kelantan Campus Jeli, Jeli, Kelantan, Malaysia

Khanjanapaj Lewmanomont Faculty of Fisheries, Kasetsart University, Bangkok, Thailand

Adolfo López-Fabal Departamento de Producíón Vegetal y Proyectos de Ingeniería, Escuela Politécnica Superior de Ingeniería, Universidad de Santiago de Compostela, Lugo, Spain

M. Elvira López-Mosquera Instituto de Biodiversidad Agraria y Desarrollo Rural (IBADER), Universidad de Santiago de Compostela, Lugo, Spain

S. Mahesh Centre of Advanced Study in Marine Biology, Faculty of Marine Sciences, Annamalai University, Parangipettai, Chidambaram, India

Khalid M. Mahrose Animal and Poultry Production Department, Faculty of Technology and Development, Zagazig University, Zagazig, Egypt

Erik Malta CTAQUA, Aquaculture Technology Centre, El Puerto de Santa María (Cádiz), Cádiz, Spain

João Carlos Marques University of Coimbra, MARE – Marine and Environmental Sciences Centre, Department of Life Sciences, Coimbra, Portugal

I. K. Martina Faculty of Agro Based Industry, Universiti Malaysia Kelantan Campus Jeli, Jeli, Kelantan, Malaysia

Anju Mehra Omics of Algae Group, Industrial Biotechnology, International Centre for Genetic Engineering and Biotechnology, Aruna Asaf Ali Marg, New Delhi, India

Izabela Michalak Wrocław University of Science and Technology, Faculty of Chemistry, Department of Advanced Material Technologies, Wrocław, Poland

Rok Mihelic University of Ljubljana, Biotechnical faculty, Agricultural department, Ljubljana, Slovenia

Ankita Mishra Department of Crop Physiology, Institute of Agriculture, Visva-Bharati, Bolpur, West Bengal, India

M. Mohd Faculty of Agro Based Industry, Universiti Malaysia Kelantan Campus Jeli, Jeli, Kelantan, Malaysia

Navid Moheimani Algae R&D Centre, Harry Butler Institute, Environmental and Conservation Sciences, Murdoch University, Murdoch, WA, Australia

Arijit Sankar Mondal Department of Microbiology, Guru Nanak Institute of Pharmaceutical Science and Technology (Life Science), Kolkata, India

Sananda Mondal Department of Crop Physiology, Institute of Agriculture, Visva-Bharati, Bolpur, West Bengal, India

Tiago Morais Lusalgae, Lda, Incubadora de Empresas da Figueira da Foz, Figueira da Foz, Portugal

University of Coimbra, MARE – Marine and Environmental Sciences Centre, Department of Life Sciences, Coimbra, Portugal

Michèle Morançais Université de Nantes, MMS, Nantes, France

Wellem Muskita Faculty of Fishery and Marine Sciences, Halu Oleo University, Kendari, Indonesia

Nadeem Nazurally Department of Agricultural and Food Science, Faculty of Agriculture, University of Mauritius, Reduit, Mauritius

Swaleha Hudaa Neetoo Department of Agricultural and Food Science, Faculty of Agriculture, University of Mauritius, Reduit, Mauritius

Asha A. Nesamma Omics of Algae Group, Industrial Biotechnology, International Centre for Genetic Engineering and Biotechnology, Aruna Asaf Ali Marg, New Delhi, India

A. Noorjahan Centre of Advanced Study in Marine Biology, Faculty of Marine Sciences, Annamalai University, Parangipettai, Chidambaram, India

Nurdiana Faculty of Fishery and Marine Sciences, Halu Oleo University, Kendari, Indonesia

Onu La Ola Faculty of Fishery and Marine Sciences, Halu Oleo University, Kendari, Indonesia

Diana Pacheco University of Coimbra, MARE – Marine and Environmental Sciences Centre, Department of Life Sciences, Coimbra, Portugal

Ruma Pal Phycology Laboratory, Department of Botany, Centre of Advanced Study (CAS), University of Calcutta, Kolkata, West Bengal, India

Debasish Panda Department of Crop Physiology, Institute of Agriculture, Visva-Bharati, Bolpur, West Bengal, India

Utama Kurnia Pangerang Faculty of Fishery and Marine Sciences, Halu Oleo University, Kendari, Indonesia

Chan Sun Park Department of Marine and Fisheries Resources, Mokpo National University, Muan, Republic of Korea

Leonel Pereira University of Coimbra, MARE – Marine and Environmental Sciences Centre, Department of Life Sciences, Coimbra, Portugal

Reza Rabiei Department of Natural Resources Researches, Agriculture and Natural Resources Research and Education Center, Agricultural Research, Education and Extension Organization (AREEO), Bandar Abbas, Iran

Abdul Rahman Faculty of Fishery and Marine Sciences, Halu Oleo University, Kendari, Indonesia

T. V. Ramachandra Energy & Wetlands Research Group, Centre for Ecological Sciences, Bangalore, India

Centre for Sustainable Technologies, Bangalore, India

Centre for infrastructure, Sustainable Transportation and Urban Planning [CiSTUP], Indian Institute of Science, Bangalore, India

Sarah Hamdy Rashedy Marine Environment Division, Hydrobiology Lab, National Institute of Oceanography and Fishers, Alexandria, Egypt

Gokare A. Ravishankar Dr. C. D. Sagar Centre for Life Sciences, Dayananda Sagar College of Engineering, Dayananda Sagar Institutions, Bangalore, Karnataka, India

Robert Reinhardt Algen, algal technology centre, llc, Ljubljana, Slovenia

Jinnath Rehana Ritu Department of Fisheries Management, Bangladesh Agricultural University, Mymensingh, Bangladesh

A. S. Roselina Faculty of Entrepreneurship and Business, Universiti Malaysia Kelantan City Campus, Pengkalan Chepa, Kelantan, Malaysia

T. Antony Roseline Department of Botany, Alagappa Government Arts College Karaikudi, Tamil Nadu, India

Rapeeporn Ruangchuay Department of Technology and Industry, Faculty of Science and Technology, Prince of Songkla University, Meaung, Pattani, Thailand

Ruslaini Faculty of Fishery and Marine Sciences, Halu Oleo University, Kendari, Indonesia

Gour Gopal Satpati Department of Botany, Bangabasi Evening College, University of Calcutta, Kolkata, West Bengal, India

Sarban Sengupta Phycology Laboratory, Department of Botany, Centre of Advanced Study (CAS), University of Calcutta, Kolkata, West Bengal, India

Nowrin Akter Shaika Department of Fisheries Management, Bangladesh Agricultural University, Mymensingh, Bangladesh

Ritu Sharma Department of Microbiology, Guru Nanak Institute of Pharmaceutical Science and Technology (Life Science), Kolkata, India

José W. A. Silva Federal Institute of Education, Science and Technology of Ceará – IFCE, Aracati, Ceará, Brazil

Jelveh Sohrabipour Department of Natural Resources Researches, Agriculture and Natural Resources Research and Education Center, Agricultural Research, Education and Extension Organization (AREEO), Bandar Abbas, Iran

A. B. Soro TEAGASC Food Research Centre, Ashtown, Dublin, Ireland

Tiago Sousa University of Coimbra, Marine and Environmental Sciences Centre (MARE), Department of Life Sciences, Coimbra, Portugal

B. K. Tiwari TEAGASC Food Research Centre, Ashtown, Dublin, Ireland

Nitin Trivedi DBT-ICT Centre for Energy Biosciences, Institute of Chemical Technology, Matunga, Mumbai, India

Vincent Turpin Université de Nantes, MMS, Nantes, France

V. R. Umashree Department of Plant Science, Central University of Kerala, Kasaragod, Kerala, India

Brian von Herzen Climate Foundation, Woodford, QLD, Australia

W. Wendy Department of Basic Knowledge and Entrepreneurship, Center of Fundamental and Continuing Education, Universiti Malaysia Terengganu, Kuala Nerus, Terengganu, Malaysia

Juliane Wolf Institute of Molecular Bioscience, The University of Queensland, Brisbane, QLD, Australia

X. Zhu TEAGASC Food Research Centre, Ashtown, Dublin, Ireland

School of Biosystems and Food Engineering, University College Dublin, National University of Ireland, Belfield, Dublin, Ireland

Maja Berden Zrimec Algen, algal technology centre, llc, Ljubljana, Slovenia

Part I
Seaweeds Bioresources, Ecology, Biology, Composition, Cultivation and Quality Considerations for Trade

Chapter 1
The Ecology and Physiology of Seaweeds: An Overview

Islam Mahmoud El-Manaway and Sarah Hamdy Rashedy

Abbreviations

CCA Crustose Calcareous Algae
DIN Dissolved Inorganic Nitrogen
DOC Dissolved Organic Carbon
GBR Great Barrier Reef
UV Ultraviolet Radiation

1.1 Introduction

Seaweeds are a collective term used for benthic marine macroalgae that are generally visible to the naked eye. They occupy various ecological niches, including shallow and deep coral reefs, deep inter-reef areas, sandy bottoms, seagrass beds, mangroves roots, and rocky intertidal zones in all coastline areas of the world. They can be found in almost all aquatic environments, from marine to brackish and freshwater, and from the tropical islands near the equator to Polar Regions. The number of species described by taxonomists is increasing worldwide but most likely there are several thousand species of seaweed attained from Algaebase (Guiry and Guiry 2021) which is the most accurate evaluation. Macroalgae have a significant ecological role in the marine ecosystem. They produce oxygen and they are a carbon dioxide sink (Raven et al. 2011). Moreover, many species provide protective habitats for a wide range of flora and fauna to preserve the coastal community (Omer et al. 2021), they also provide a direct source of food for many animals (Makkar et al. 2016). Finally, macroalgae are of extreme importance in protecting coastal shores,

I. M. El-Manaway
Science Faculty, Botany Department, Suez Canal University, Ismailia, Egypt

S. H. Rashedy (✉)
National Institute of Oceanography and Fisheries, NIOF, Cairo, Egypt
e-mail: sarahamdy.niof@gmail.com

A. Ranga Rao, G. A. Ravishankar (eds.), *Sustainable Global Resources of Seaweeds Volume 1*, https://doi.org/10.1007/978-3-030-91955-9_1

by dissipating wave energy and capturing sediments and nutrients (Hurd et al. 2014a, b). Seaweeds respond to various climatic and physicochemical factors. Survival, growth, and reproduction of their dependence on and vary with numerous key environmental variables such as temperature, salinity, hydrodynamics and wave exposure, nutrients, carbon dioxide, and pH (Harley et al. 2012). These factors form latitudinal patterns of algal distribution (Ramos et al. 2019). The interactions of these parameters influence both the presence and abundance of individual taxa. Major changes in abiotic factors take place along spatial and temporal declivity. Large changes in temperature, light availability, and seasonality are observed; along the coastline, steep gradients in abiotic factors exist stretching from the intertidal to the subtidal zone; but even on very small scales, the abiotic environment of seaweeds may change dramatically, e.g. within algal mats (Hurd et al. 2014a, b).

1.2 Seaweed Taxonomic Groups

Seaweeds broadly comprise species included in three phyla: Rhodophyta (red algae), Ochrophyta (brown algae, class Phaeophyceae) and Chlorophyta (green algae, classes Bryopsidophyceae, Chlorophyceae, Dasycladophyceae, Prasinophyceae, and Ulvophyceae). Red and brown algae are almost exclusively marine, whilst green algae are also common in freshwater (rivers and lakes), and even in terrestrial (rocks, walls, houses, and tree bark in damp places) situations. The main criteria used to identify the different phyla are photosynthetic pigments, storage food products, the cell wall components, the fine structure of the cell, and flagella (Sahoo and Seckbach 2015). Macroalgae classification has been revised in recent years, based on DNA sequence data. Although the classification systems have evolved over the centuries, it is generally agreed as follows:

1.2.1 Phylum Chlorophyta

It includes green species that have chlorophylls *a, b* with several carotenoids (Fig. 1.1). All green macroalgae are classified in a common class, called Ulvophyceae which are a very diverse group including about 1927 species (Guiry and Guiry 2021), and distributed in all seas of the world.

1.2.2 Phylum Ochrophyta

It includes brown species that have chlorophyll a and c, and dominated by fucoxanthin carotenoid that gives them the brownish color (Fig. 1.1). Almost all species are seaweeds, and are of few centimeters up to 50 m in length. They are utilized as food

Fig. 1.1 Three groups of seaweeds collected from the Red Sea by S.H. Rashedy and identified by I.M. El-Manawy. Green algae in **a**, **b**, **c**. (**a**) *Caulerpa racemosa* (Forsskal) J. Agardh, (**b**) *Halimeda opuntia* (Linnaeus) Lamouroux, (**c**) *Codium tomentosum* Stackhouse. Brown seaweed in **d**, **e**, **f**, (**d**) *Sargassum aquifolium* (Turner) **c**. Agardh, (**e**) *Padina boergesenii* Alender & Kraft, (**f**) *Dictyota dichotoma* (Hudson) Lamouroux. Red seaweed in **g**, **h**, **i**. (**g**) *Ganonema farinosum* (Lamouroux) Fan&Yung Wang. (**h**) *Amphiroa anceps* (Lamarck) Decaisne. (**i**) *Actinotrichia fragilis* (Forsskal) Børgesen

products, in cosmetics, and as fertilizers. Alginate is a component of their cell walls, used as emulsifiers, anticoagulants, and in the production of textile and rubber (Pereira and Cotas 2020).

1.2.3 Phylum Rhodophyta

It includes species with brilliant red color due to dominance of phycoerythrin and phycocyanin over chlorophyll *a*, and d, β-carotene, and a number of xanthophylls (Fig. 1.1). Their cell walls contain colloidal components, agar, and carrageenan, which are important for industrial and microbial products. Their extracts are known to have antimicrobial, antiviral, and anticancer activities (Hmani et al. 2021; Lee et al. 2021) (Fig. 1.1)

1.3 Functional Groups of Seaweeds

Seaweeds play vital ecological roles on both the coral reefs and seashores, and their roles have been found to be related to different morphological and structural aspects of the thallus. Thus, a functional-form model has been used to estimate the photosynthetic, nutrient uptake, and structural aspects of seaweeds in relation to grazer (Littler et al. 1983). Outer and inner seaweed structure will impact photosynthetic efficiency, ability to absorb nutrients, and resistance to predation. For example, a thick, tough fucoid such as *Fucus* will be less likely to be eaten. But, due to its thick wall, *Fucus* will show a lower photosynthetic and nutrient uptake efficiency than thin bladed seaweeds like *Ulva* (Wiencke and Bischof 2012). In tropical habitats, macroalgae range from small, structurally simple, filamentous turfs, a few millimeters high, or heavily calcified crustose forms, to large leathery macrophytes, such as *Sargassum*, up to several meters tall. Given this diversity, different macroalgae should be assumed to respond in qualitatively different ways to ecological parameters and the stressors associated with human activities as well as climate changes (Ateweberhan et al. 2005). As an alternative to taxonomic groups, macroalgae can be considered in terms of three functional groupings based on plant attributes and ecological characteristics (El-Manawy 2008). Plant attributes are plant size, toughness, photosynthetic ability, and growth. Ecological characteristics include grazing resistance, physiological adaptation, etc. The three main classes are: (I) algal turfs, (II) upright macroalgae (fleshy and calcified), and (III) crustose calcareous algae. Each category includes several 'functional groups'. This approach is considered more useful by ecologists, because it reflects both physiological traits and the ecological role of algae, whereas ecological roles are not well correlated with taxonomic groupings (El-Manawy 2008).

1.4 Ecological Roles of Seaweeds on Coral Reefs

Macroalgae provide vital ecological functions such as primary production construction, contribution to nitrogen fixation, and cementation of reef framework, facilitation of coral settlement, and creation of habitats for other reef species. The abundance of macroalgae on the coral reefs can be recognized as a cause of coral reef degradation. Species lists in reef plant communities contain from 100 to 250 red, green, and brown taxa. This number may double or triple on fringing coral reefs, neritic waters have higher levels of nutrients (Dawes 1998). Brown macroalgae are only represented by a few species (˃30), some of the most common genera being *Sargassum*, *Padina*, *Dictyota*, and *Turbinaria* (Kuffner et al. 2008). Green macroalgae are dominated by calcified species, whereas red algae are primarily crustose forms. Fleshy and turf-forming seaweeds are less abundant on coral reefs due to extensive grazing but are known to be critical primary producers (Dubinsky and Stambler 2011).

Fig. 1.2 Seaweeds associated with coral reef in the Red Sea, Hurghada Reef, Egypt. (**a**) *Sargassum aquifolium*. (**b**) *Padina boergesenii*. (**c**) *Turbinaria turbinata*

Regarding studies on seaweeds associated with coral reefs in the Red sea, El-Manawy (2008) studied spatial variation in cover and biomass of macroalgae on the fringing reefs of Hurghada, Egypt. Species composition and abundance significantly varied in relation to reef health. Abundant of the upright leathery assemblage of *Padina* with *Sargassum*, *Turbinaria* and *Hormophysa* prevailed the northern reefs (Fig. 1.2).

1.4.1 Contribution to Primary Production

A large proportion of the primary production on a coral reef is contributed by benthic algae, particularly by algal turfs (Dubinsky and Stambler 2011). The primary production of coral reef seaweeds varies according to their morphology, as noted in the functional group. Available research from the Great Barrier Reef (GBR) indicates that primary production by fleshy macroalgae and crustose algae is also important (Dean et al. 2015). The organic matter produced enters the reef food web by several pathways. Many algae are directly consumed by herbivorous fishes, crabs, sea urchins, and mesograzers, while dissolved organic carbon released by the algae into the water enters the microbial food web (Diaz-Pulido and Barron 2020). Some organic matter is exported as detritus by currents and tides to adjacent habitats such as seagrass meadows, mangroves, and the deeper, inter-reef seafloor.

1.4.2 Nitrogen Fixation and Nutrient Retention

In algal turf communities, a group of filamentous cyanobacteria are found on the sand bottom and fix significant amounts of atmospheric nitrogen to sustain their growth independent of dissolved nutrients (Heil et al. 2004). As dissolved inorganic nitrogen (DIN) compounds, their levels in reef waters are very low, greater than 0.4–1.0 μML^{-1}; ammonium-N is maximally followed by nitrate-N and then nitrite-N. The nitrogen cycle is mostly a biological process, with all stages occurring on coral reefs. Due to the rapid growth rates of blue-green algae and intense grazing on turf communities, the organic nitrogen fixed in algal tissue rapidly enters the food web and becomes available for other primary producers (Diaz-Pulido and McCook 2003).

1.4.3 Facilitation of Coral Settlement

The settlement of invertebrates is a fundamental process simulating the structure of marine communities and supports the ability of benthic ecosystems to recuperate from disruption. It is well documented that specific species of crustose calcareous algae are the preferred settlement substrate for numerous coral species. As a result, any disturbance to the CCA-coral settlement association is a concern with recruitment success. There is some proof that explains that ocean acidification (Doropoulos et al. 2012), elevated sea temperatures (Webster et al. 2012), reduced grazing (Birrell et al. 2008), and poor water quality (Negri and Hoogenboom 2011) change the ecological interactions required for optimal coral settlement onto CCA.

1.4.4 Reef Degradation

Seaweeds may also cause reef degradation in ecological phase shifts, where abundant reef-building corals are replaced by abundant fleshy macroalgae (Diaz-Pulido and McCook 2008). Reductions in herbivores due to overfishing and increases in nutrient inputs have been shown to cause increases in fleshy macroalgal abundance, leading to coral overgrowth by algae and, ultimately, reef degradation (El-Manawy 2008). Many adversaries, such as coral bleaching, crown-of-thorns starfish outbreaks, extreme low tides, outbreaks of coral diseases, and storm damage (specifically tropical cyclones) often lead directly to coral mortality. The dead coral skeletons are then rapidly colonized by diverse algal communities (Halford et al. 2004). A reef community dominated by abundant, high-biomass algal turfs or larger, fleshy macroalgae may lead to overgrowth, smothering and/or shading of corals, the exclusion of coral recruitment, and increases in pathogens, resulting in an alternate stable state, with decreased ecological, economic and aesthetic value (Smith et al. 2006).

1.5 The Spatial and Temporal Natural Patterns of Seaweeds

Macroalgae species exhibit populations whose distribution along coastal rocky shores is not uniform, either in space or time. These results from complex ecological processes, such as succession patterns, where different species have different recruitment, growth, and mortality rates (Cervin et al. 2005). Naturally, a variety of different patches of macroalgae species can be observed. These groups differ in a coastline, both spatially and temporally, according to the existence or absence (composition) of different species. For example, different seaweed species display vertical patterns of distribution, from the uppermost to the lowermost tide levels, giving different zones of species or zonation patterns. This is because different species have different adaptive responses to several physicals (e.g., emersion or exposure to the atmosphere), chemical (e.g., salinity), and biotic (e.g., competition, grazing) factors, which can influence the different locations on the shore (Hurd et al. 2014a, b).

1.6 Factors Affecting the Diversity, Distribution and Abundance of Seaweeds

Seaweeds respond to various climatic and physicochemical factors. Survival, growth, and reproduction of their dependence on and vary with numerous key environmental variables such as temperature, salinity, hydrodynamics and wave exposure, nutrients, carbon dioxide, and pH (Harley et al. 2012). These factors form latitudinal patterns of algal distribution (Ramos et al. 2019). The interactions of these parameters impact both the occurrence and abundance of individual taxa.

1.6.1 Biotic and Abiotic Factors

The distribution and abundance of seaweeds depend on biotic and abiotic factors that differ spatially and temporally. Biotic factors include recruitment, mortality, dispersal, competition, and herbivory. Abiotic factors include the available resources, such as light, carbon dioxide, mineral nutrients, substrate, and the physical parameters, such as wave action, aerial exposure, and temperature. All these aspects and their interactions are of particular importance since they are all likely to be altered in space and time (Hu and Fraser 2016).

Light (photosynthetically active radiation) and temperature are the most important natural parameters ruling the development of macroalgal communities. Without light, photosynthesis is not possible and temperature, as for all other organisms, determines the performance of seaweeds at the fundamental levels of enzymatic processes and metabolic function (Harley et al. 2012). Although seaweeds are generally well adapted to their thermal environment, any deviation from the optimum temperature range (daily and seasonal), particularly in situations of environmental stress, contributes to variation (Chung et al. 2011). It may also have a major effect on seaweed survival, reducing and delaying growth and leading to an increase in mortality, which may culminate in species 'loss. The same is true for salinity, where a deviation from the optimum range may lead to differential development of some species relative to others (Harley et al. 2012).

Nutrients are very important for seaweeds growth, they are usually present at low concentrations in marine waters unaffected by anthropogenic inputs, which together with grazing pressure (e.g., herbivorous fish), preserves the density of seaweeds at balanced levels (Diaz-Pulido and McCook 2008). Schaffelke et al. (2005) reported that nutrient increase enhances macroalgal growth and potentially abundance. The Sea urchins are universal herbivores feeding on attached algae found in marine ecosystems ranging from shallow subtidal to depths greater than 100 m. They can survive primarily on detrital seaweeds produced in the shallow photic zone, capturing these organic materials and, therefore, regulate the community structure in shallow algal habitats (Whippo et al. 2011).

Other aspects related to hydrodynamics, such as water flow, currents, waves, and tides, may also influence species distributional patterns because of hydrodynamically driven processes such as recruitment and detachment (Thomsen and Wernberg 2005). Thalli of intertidal macroalgae are size-limited in habitats with heavy wave action, and regardless of their morphology, their maximum size may be limited by water velocities that occur on exposed coasts. In small individuals, flexibility allows the plant to reorient and reconfigure in the flow, assuming a streamlined shape and reducing the applied hydrodynamic force. In large individuals, flexibility allows fronds to go with the flow; a strategy that can allow the plant to minimize hydrodynamic forces (Denny and Gaylord 2002). The combination of these different structural properties of macroalgae with the hydrodynamics and other varied natural factors found on a shore irresponsible for differences in macroalgal communities, giving to intertidal rocky shores the typical vertical distribution of seaweeds.

1.6.2 The Global Climate Changes

Fluctuations in global climate and its variability have an impact on biological, ecological, and socioeconomic systems (Price et al. 2011). The influences of increased atmospheric carbon dioxide, elevated sea temperatures, increasing sea level, and increasing UV radiation can alter the distribution, productivity, and community composition of seaweed (Sunny 2017).

1.6.2.1 Changing in Ocean Temperatures

Sea surface temperature is an important physical attribute of the world's oceans. As the oceans absorb more heat, sea surface temperature increases, and the ocean circulation patterns that transport warm and cold water around the globe change (Huang et al. 2016). Changes in sea surface temperature can alter marine ecosystems in several ways. For example, variations in ocean temperature can affect what species of plants, animals, and microbes are present in a location, alter migration and breeding patterns, and threaten sensitive ocean life such as corals (FAO 2018).

Temperature controls the performance of enzymatic processes and metabolic function of seaweeds (Trincone 2017). Although macroalgae are well adapted to their thermal environment, the environmental change could cause cellular and sub cellular damage which could slow growth (Hoegh-Guldberg et al. 2007).

1.6.2.2 Changing in Ocean Circulation

Changing in ocean circulation causes moderate shifts in species composition and function, particularly of turfs and upright macroalgae. These changes may be sudden or abrupt, depending on the nature of the circulation changes (Diaz-pulido et al. 2007). Algal dispersal is dependent on ocean currents, algal distributions, and ecological functions (e.g., productivity, nitrogen fixation) which are sensitive to changes in water temperature and water quality (Millar 2007). Upright algae, such as *Sargassum* and *Halimeda*, are less homogeneous in distribution than turfs or CCA, and hence may be more sensitive to changes in dispersal by water movements. For example, *Sargassum* spp. distributions are restricted to inshore reefs and therefore changes in ocean circulation could affect populations of these algae (Diaz-pulido et al. 2007).

1.6.2.3 Changes in Ocean Acidity

Carbon dioxide (CO_2) is a critical element for photosynthesis. In terrestrial environments, an excess of CO_2 may be considered as a fertilizer, exciting photosynthesis, but in marine systems, where bicarbonate ions (HCO_3^-) rather than CO_2 are used by most macroalgae as a photosynthetic substrate, other problems may arise because of excessive concentrations in seawater (Branch et al. 2013). However, the increase of carbon dioxide concentration could drive deviations in the chemistry of seawater carbonate and lead to a reduction in pH, the response of different marine plants to this change will vary. They show a variety of essential functional mechanisms affected by carbonate chemistry (e.g., dissolution and calcification rates, growth rates, development, and survival) that allow their responses to varying so broadly. This means it is difficult to predict how marine ecosystems will respond to ocean acidification (Kroeker et al. 2010).

Although all benthic macroalgae will be exposed to changes in pH, CO_2, and calcium carbonate saturation, this will be particularly important for crustose, and upright calcareous macroalgae and will be also potential for changes in the availability of nutrients under reduced pH (Kleypas et al. 2006). The sensitivity of all algal groups is expected due to interactions between the effects of pH and CO_2 enhancement of photosynthesis. However, calcified algae are particularly sensitive to ocean acidification. For example, in the GBR, a decrease in pH from 8 to 7.5 reduced calcification dramatically for the alga *Halimeda tuna* (Sinutok et al. 2012). Reduction of pH may also decrease calcification of *Amphiroa* and *Corallina*. CCA is the algal group that highly sensitive to reductions of carbonate saturation state. Minor changes in pH (from 8.1 to 7.8) reduced calcification by as much as 21 percent for a coral reef community that included CCA (Cornwall et al. 2017).

1.6.2.4 Changes in Light and Ultraviolet Radiation

Ultraviolet (UV) radiation is likely to continue to increase, due to the effects of ozone depletion, and UV levels are already high in tropical regions (McKenzie et al. 2011). The most common impacts on macroalgae include direct damage to the photosynthetic apparatus (Kataria et al. 2014), DNA (Van De Poll et al. 2003), reproductive tissues (Agrawal 2009), and reduction of nutrient uptake (Courtial et al. 2018). All these effects may lead to community changes, due to shifts in relative abundance (Lotze et al. 2002). The vulnerability of algal turfs and upright macroalgae as a whole is moderate since there is potential for adaptation to increased UV radiation and the impacts are likely to be restricted to shallow-water assemblages. The vulnerability of CCA as a group is likely to be low to moderate.

1.6.2.5 Sea Level Rise

Sea level rise due to thermal expansion of the oceans and the melting of glaciers and ice sheets is occurring at a rate of one to two millimeters per year. By 2100, the global sea level is projected to be 310 ± 30 mm higher than in 1990 (IPCC 2019). The direct effect of the increase in sea level is to increase the depth of water. Sea level rise will reduce the light to seaweeds throughout their depth range and may limit plant photosynthesis (Harley et al. 2012). Within all three algal groups, different taxa will have very different colonization and dispersal potentials, resulting in highly variable responses to the increase in the available substrate with sea-level rise (Harley et al. 2006). The assumed scenario of a sea level rise is slow relative to the life spans of most algal turfs, upright macroalgae, and CCA, and thus high rates of colonization, growth, and reproduction will reduce the vulnerability of all macroalgal groups to sea-level rise.

1.7 Role of Seaweeds in Blue Carbon Sequestration

Marine habitats are highly productive ecosystems that contribute to Blue Carbon sequestration (Ortega et al. 2019). Fourqurean et al. 2012 pointed that, seagrass meadows, salt marshes, and mangrove trees have complex root systems that restore great quantities of carbon in soft sediments within their habitat. While most seaweeds lack root systems and grow on rocky shores where sediment accretion does not occur and they do not accumulate carbon-rich sediments, so they have been neglected in Blue Carbon assessments, however, calculations submitted that 25% of exported seaweed carbon is sequestered in long-term reservoirs, such as coastal sediments and the deep sea (Krause-Jensen et al. 2018).

Seaweeds lack lignin (complex molecules that are key structural materials in woody plants) in their supporting structures and therefore the material within them is more easily remineralized (the breakdown of organic material to its simplest inorganic forms) than terrestrial material. However, there is a dissolved organic fraction of carbon (DOC) which is exuded constantly from both living seaweed and from seaweed detritus. If this carbon is transported to below the 'carbon sequestration horizon' (areas of 1000 m or deeper) then it is stored for periods of time significant enough to be considered sequestered. Some biomarker studies have found certain compounds in seaweeds have properties that make them less labile and more refractory (resistant to being metabolized) than other components (Trevathan-Tackett et al. 2015). For example, the sulfated polysaccharide fucoidan has been shown to persist in sediments, and in dissolved form fucoidan is less available as a substrate for bacteria to consume (Barrón and Duarte 2015). Carbohydrates and phenols (a type of organic compound) have also been used as biomarkers to show the seaweed's contribution to sediment carbon (Abdullah et al. 2017).

1.8 Conclusions and Future Perspectives

Seaweeds are of ecological importance because they form structure and habitat that provides shelter, food, and oxygen for thousands of marine organisms, such as fish, sea urchins, and crustaceans, and protect our coasts by reducing wave action, and storm surges. They also support commercial fisheries, are used in foods, cosmetics and medicines. Furthermore, they possess excellent survival strategies to withstand the many environmental stresses that they are exposed to. It's important for us to monitor them so we can understand and manage these vital resources sustainably.

References

Abdullah MI, Fredriksen S, Christie H (2017) The impact of the kelp (*Laminaria hyperborea*) forest on the organic matter content in sediment of the west coast of Norway. Mar Biol Res 13(2):151–160

Agrawal SC (2009) Factors affecting spore germination in algae – review. Folia Microbiol 54(4):273–302

Ateweberhan M, Bruggemann JH, Breeman AM (2005) Seasonal dynamics of *Sargassum ilicifolium* (Phaeophyta) on a shallow reef flat in the southern Red Sea (Eritrea). Mar Ecol Prog Ser 292:159–271

Barrón C, Duarte CM (2015) Dissolved organic carbon pools and export from the coastal ocean. Glob Biogeochem Cycles 29(10):1725–1738. https://doi.org/10.1002/2014GB005056

Birrell CL, McCook LJ, Willis BL, Diaz-Pulido GA (2008) Effects of benthic algae on the replenishment of corals and the implications for the resilience of coral reefs. Oceanogr Mar Biol Annu Rev 46:25–63

Branch TA, De Joseph BM, Ray LJ, Wagner CA (2013) Impacts of ocean acidification on marine seafood. Trends Ecol Evol 28(3):178–186. https://doi.org/10.1016/j.tree.2012.10.001

Cervin GA, Berg P, Jenkins SR (2005) Small-scale disturbance in a stable canopy dominated community: implications for macro-algal recruitment and growth. Mar Ecol Prog Ser 305:31–40

Cornwall CE, Comeau S, McCulloch MT (2017) Coralline algae elevate pH at the site of calcification under ocean acidification. Glob Change Biol 23:4245–4256. https://doi.org/10.1111/gcb.13673

Courtial L, Planas Bielsa V, Houlbrèque F, Ferrier-Pagès C (2018) Effects of ultraviolet radiation and nutrient level on the physiological response and organic matter release of the scleractinian coral *Pocillopora damicornis* following thermal stress. PLoS One 13(10):e0205261. https://doi.org/10.1371/journal.pone.0205261

Chung, IK, Beardall, J, Mehta, S et al. (2011) Using marine macroalgae for carbon sequestration: a critical appraisal. J Appl Phycol 23, 877–886 (2011). https://doi.org/10.1007/s10811-010-9604-9

Dawes JC (1998) Marine botany, 2nd ed. ISBN: 978–0–471-19208-4, 496 Pages

Dean AG, Steneck RS, Tager D, Pandolfi MJ (2015) Distribution, abundance and diversity of crustose coralline algae on the great barrier reef. Coral Reefs 34:581–594. https://doi.org/10.1007/s00338-015-1263-5

Denny MW, Gaylord B (2002) The mechanics of wave-swept algae. J Exp Biol 205:1355–1362

Diaz-Pulido G, Barron C (2020) CO2 enrichment stimulates dissolved organic carbon release in coral reef macroalgae. J Phycol 56:1039–1052. https://doi.org/10.1111/jpy.13002

Diaz-Pulido G, McCook LJ (2003) Relative roles of herbivory and nutrients in the recruitment of coral reef seaweeds. Ecology 84:2026–2033

Diaz-Pulido G, McCook LJ (2008) Macroalgae (seaweeds) in China. In: The state of the great barrier Reef on-line. Great Barrier Reef Marine Park Authority, Townsville

Diaz-pulido G, McCook LJ, AWD L, Lotze HK, Raven JA, Schaffelke B, Smith JE, Steneck RS (2007) Vulnerability of macroalgae of the great barrier Reef to climate change. In: Marshall P, Johnson J (eds) Climate Change and the Great Barrier Reef. Great Barrier Reef Marine Park Authority, Townsville, pp 153–192

Doropoulos C, Ward S, Diaz-Pulido G, Hoegh-Guldberg O, Mumby PJ (2012) Ocean acidification reduces coral recruitment by disrupting intimate larval-algal settlement interactions. Ecol Lett 15:338–346

Dubinsky Z, Stambler N (2011) Coral reefs: an ecosystem in transition. https://doi.org/10.1007/978-94-007-0114-4_17

El-Manawy IM (2008) The spatial variability of macroalgal communities and their functional groupings on the fringing reefs of Ghardaqah. Egypt J Phycol 9:55–69

FAO (2018) Deep-ocean climate change impacts on habitat, fish and fisheries, by Lisa Levin, Maria Baker, and Anthony Thompson (eds). FAO Fisheries and Aquaculture Technical Paper No. 638. Rome, FAO. 186 pp. Licence: CC BY-NC-SA 3.0 IGO

Fourqurean JW et al (2012) Seagrass ecosystems as a globally significant carbon stock. Nat Geosci 5:505–509

Guiry MD, Guiry GM (2021) AlgaeBase. World-wide electronic publication. National University of Ireland, Galway. https://www.algaebase.org; searched on 30 March 2021

Halford A, Cheal AJ, Ryan D, Williams DM (2004) Resilience to large-scale disturbance in coral and fish assemblages on the great barrier Reef. Ecology 85:1892–1905

Harley CDG, Hughes AR, Hultgren KM, Miner BG, Sorte CJB et al (2006) The impacts of climate change in coastal marine systems. Ecol Lett 9:228–241

Harley CDG, Anderson KM, Demes KW, Jorve JP, Kordas RL, Coyle TA, Graham MH (2012) Effects of climate change on global seaweed communities. J Phycol 48(5):1064–1078

Heil CA, Chaston K, Jones A, Bird P, Longstaff B, Costanzo S, Dennison WC (2004) Benthic microalgae in coral reef sediments of the southern great barrier Reef, Australia. Coral Reefs 23:336–343

Hmani I, Ktari L, Ismail A et al (2021) Assessment of the antioxidant and antibacterial properties of red algae (Rhodophyta) from the north coast of Tunisia. Euro Mediterr J Environ Integr 6:13. https://doi.org/10.1007/s41207-020-00222-7

Hoegh-Guldberg O, Mumby PJ, Hooten AJ, Steneck RS, Greenfield P et al (2007) Coral reefs under rapid climate change and ocean acidification. Science 318:1737–1742

Hu Z, Fraser C (2016) Seaweed phylogeography. Springer Netherlands, Dordrecht, p 395. https://doi.org/10.1007/978-94-017-7534-2

Huang B, Thorne PW, Smith TM, Liu W, Lawrimore J, Banzon VF, Zhang H, Peterson TC, Menne M (2016) Further exploring and quantifying uncertainties for extended Reconstructed Sea Surface Temperature (ERSST) version 4 (v4). J Clim 29(9):3119–3142

Hurd CL, Harrison PJ, Biscoff K, Lobban CS (2014a) Seaweed ecology and physiology. Cambridge University Press, Cambridge

Hurd C, Harrison P, Bischof K, Lobban C (2014b) Physico-chemical factors as environmental stressors in seaweed biology. In: Seaweed ecology and physiology. Cambridge University Press, Cambridge, pp 294–348. https://doi.org/10.1017/CBO9781139192637.008

IPCC (Intergovernmental Panel on climate Change) (2019) Climate Change 2019, Summary for Policymakers. In: *IPCC Special Report on the Ocean and Cryosphere in a Changing Climate* [H.- O. Pörtner, D.C. Roberts, V. Masson-Delmotte, P. Zhai, M. Tignor, E. Poloczanska, K. Mintenbeck, M. Nicolai, A. Okem, J. Petzold, B. Rama, N. Weyer (eds.)]. In press. https://report.ipcc.ch/srocc/pdf/SROCC_SPM_Approved.pdf

Kataria S, Jajoo A, Guruprasad KN (2014) Impact of increasing ultraviolet-B (UV-B) radiation on photosynthetic processes. J Photochem Photobiol B Biol 137:55–66

Kleypas JA, Feely RA, Fabry VJ, Langdon C, Sabine CL, Robbins LL (2006) Impacts of ocean acidification on coral reefs and other marine calcifiers, A guide for future research. NSF, NOAA, and the U.S. Geological Survey, St. Petersburg, Florida

Krause Jensen D et al (2018) Sequestration of macroalgal carbon: the elephant in the blue carbon room. Biol Lett 14:20180236

Kroeker KJ, Kordas RL, Crim RN, Singh GG (2010) Meta-analysis reveals negative yet variable effects of ocean acidification on marine organisms. Ecol Lett 13:1419–1434

Kuffner IB, Andersson AJ, Jokel PL, Rogers KS (2008) Decreased abundance of crustose coralline algae due to ocean acidification. Nat Geosci 1:114–117

Lee H-G, Lu Y-A, Je J-G, Jayawardena TU, Kang M C, Lee S H, Kim T H, Lee D-S, Lee J-M, Yim M-J et al (2021) Effects of ethanol extracts from *Grateloupia elliptica*, a red seaweed, and its chlorophyll derivative on 3T3-L1 adipocytes: suppression of lipid accumulation through downregulation of adipogenic protein expression. Mar Drugs 19:91. https://doi.org/10.3390/md19020091

Littler MM, Taylor PR, Littler DS (1983) Algal resistance to herbivory on a Caribbean barrier reef. Coral Reefs 2:111–118

Lotze HK, Worm B, Molis M, Wahl M (2002) Effects of UV radiation and consumers on recruitment and succession of a marine macrobenthic community. Mar Ecol Prog Ser 243:57–66

Makkar HPS, Tran G, Heuzé V, Giger-Reverdin S, Lessire M, Lebas F, Ankers P (2016) Seaweeds for livestock diets: a review. Anim Feed Sci Technol 212:1–17

McKenzie RL, Aucamp PJ, Bais AF, Björn LO, Ilyas M, Madronich S (2011) Ozone depletion and climate change: impacts on UV radiation. Photochem Photobiol Sci 10(2):182–198. https://doi.org/10.1039/c0pp90034f

Millar AJK (2007) The Flindersian and Peronian provinces, algae of Australia series. In: Algae of Australia: introduction. Australian Biological Resources Study CSIRO Publishing, Victoria

Negri AP, Hoogenboom MO (2011) Water contamination reduces the tolerance of coral larvae to thermal stress. PLoS One 6:e19703

Omer MY, Abd El-Wakeil KF, Hussein HNM, Rashedy SH (2021) Invertebrate assemblages associated with seaweeds at different locations in the Red Sea, Egypt. Egypt J Aquat Biol Fish 25(1):407–421

Ortega A, Geraldi NR, Alam I et al (2019) Important contribution of macroalgae to oceanic carbon sequestration. Nat Geosci 12:748–754. https://doi.org/10.1038/s41561-019-0421-8

Pereira L, Cotas J (2020) Alginates – a general looking. In: Pereira L (ed) Alginates – recent uses of this natural polymer. Open access peer-reviewed edited book. Intech Open, London, pp 1–16. https://doi.org/10.5772/intechopen.88381

Price NN, Hamilton SL, Tootell JS, Smith JE (2011) Species specific consequences of ocean acidification for the calcareous tropical green algae Halimeda. Mar Ecol Prog Ser 440:67–78

Ramos AB, Felices RBC, Leon NTE, Muñoz AA, Muñoz CA (2019) Selection criteria for high-value biomass: seasonal and morphological variation of polyphenolic content and antioxidant capacity in two brown macroalgae. J Appl Phycol 31:653–664

Raven JA, Giordano M, Beardall J, Marberly SC (2011) Algal and aquatic plant carbon concentrating mechanisms in relation to environmental change. Photosynth Res 109:281–296

Sahoo D, Seckbach J (2015) The algae world. Springer, Dordrecht Press, p 598. https://doi.org/10.1007/978-94-017-7321-8

Schaffelke B, Mellors J, Duke NC (2005) Water quality in the great barrier Reef region: responses of mangrove, seagrass and macroalgal communities. Mar Pollut Bull 51:279–296

Sinutok S, Hill RM, Doblin A, Ku M, Ralph PJ (2012) Microenvironmental changes support evidence of photosynthesis and calcification inhibition in *Halimeda* under ocean acidification and warming. Singapore123Coral Reefs 31:1201–1213. https://doi.org/10.1007/s00338-012-0952-6

Smith JE, Shaw M, Edwards RA, Obura D, Pantos O, Sala E, Sandin SA, Smriga S, Hatay M, Rohwer FL (2006) Indirect effects of algae on coral: algae-mediated, microbe induced coral mortality. Ecol Lett 9:835–845

Sunny AR (2017) A review on effect of global climate change on seaweed and seagrass. Int J Fish Aquat Stud 5(6):19–22

Thomsen MS, Wernberg T (2005) Miniview: what affects the forces required to break or dislodge macroalgae? Eur J Phycol 40:139–148

Trevathan-Tackett SM, Kelleway J, Macreadie PI, Beardall J, Ralph P, Bellgrove A (2015) Comparison of marine macrophytes for their contributions to blue carbon sequestration. Ecology 96(11):3043–3057

Trincone A (2017) Enzymatic processes in marine biotechnology. Mar Drugs 15(4):93. https://doi.org/10.3390/md15040093

Van De Poll WH, Bischof KAI, Buma AGJ, Breeman AM (2003) Habitat related variation in UV tolerance of tropical marine red macrophytes is not temperature dependent. Physiol Plant 118:74–83

Webster NS, Soo R, Cobb R, Negri AP (2012) Elevated seawater temperature causes a microbial shif on crustose coralline algae with implications for the recruitment of coral larvae. ISME J 5:759–770

Whippo R, Low A, Britton-Simmons K (2011) Effects of the Red Sea urchin on benthic invertebrate communities, a link to spatial subsidies. In Pollock NW (ed) Diving for science. Proceedings of the American Academy of underwater sciences 30th symposium, Dauphin Island, AL, AAUS

Wiencke C, Bischof K (eds) (2012) Seaweed biology. Springer, Berlin, Heidelberg, p 514. https://doi.org/10.1007/978-3-642-28451-9

Chapter 2
Potential Products from Macroalgae: An Overview

Charu Deepika, Gokare A. Ravishankar, and Ambati Ranga Rao

Abbreviations

ABE	Acetone-Butanol-Ethanol fermentation
APC	Allophycocyanin
C/N	Carbon/Nitrogen
CAGR	Compound Annual Growth Rate
CFIA	Canadian Food Inspection Agency
Chl	Chlorophyll
CO_2	Carbon-dioxide
DALYs	Disability-Adjusted Life-Years
DHA	Docosahexaenoic acid
EAA/NEAA	Essential Amino Acid and Non-Essential Amino Acid
EPA	Eicosapentaenoic acid
EU	European Union
FAO	Food and Agriculture Organization
FDA	Food and Drug Administration
H_2S	Hydrogen Sulfide
LCA	Life-Cycle Assessment
LHCs	Light-Harvesting Complexes
NH_3	Ammonia
OECD	The Organization Economic Co-operation and Development

C. Deepika
Institute of Molecular Bioscience, The University of Queensland, Brisbane, QLD, Australia

G. A. Ravishankar
Dr. C. D. Sagar Centre for Life Sciences, Dayananda Sagar College of Engineering, Dayananda Sagar Institutions, Bangalore, Karnataka, India

A. R. Rao (✉)
Department of Biotechnology, Vignan's Foundation of Science, Technology and Research (Deemed to be University), Vadlamudi, Guntur, Andhra Pradesh, India
e-mail: arangarao99@gmail.com

A. Ranga Rao, G. A. Ravishankar (eds.), *Sustainable Global Resources of Seaweeds Volume 1*, https://doi.org/10.1007/978-3-030-91955-9_2

PC	Phycocyanin,
PE	Phycoerythrin
PS-I	Photosystem-I
PS-II	Photosystem-II
PUFA	Poly-Unsaturated Fatty Acid
RNS	Reactive Nitrogen Species
ROS	Reactive Oxygen Species
TEA	Techno-Economic Analysis
UV	Ultraviolet

2.1 Introduction

Energy is the key driver of the global economy. Increasing population and lifestyle has enhanced the demands needing approximately 70% more food (FAO), 50% more fuel (OECD), 50% more freshwater in by next 25 years. (Outlook 2010). The resource limitations have increased at an alarming rate over these issues has prompted several nations to pass legislations to sustainable solutions coupled to industrialization (Baghel et al. 2020; Liu et al. 2020a). These rapidly increasing demands are imposing an upward pressure upon our planetary boundaries which include stratospheric ozone depletion, loss of biosphere integrity, chemical pollution, climate change, ocean acidification, atmospheric aerosol loading, and freshwater consumption (Whiteman et al. 2013; Steffen et al. 2015). Consequently, there is now an increasing need for renewable production systems capable of producing food, fuel, biomaterials and fresh water in a CO_2 neutral approach. Solar driven macroalgae biotechnologies offer promising solutions to address these challenges.

Macroalgae (seaweeds) are macroscopic in structure, naturally available, and they can be cultivated in the vast area of low tide seashore. Seaweeds utilize natural nutrients available in the sea for their growth (Fernand et al. 2017). Seaweeds of various forms are available in large quantities throughout the coastal areas in various parts of the world such as Japan, Philippines, Malaysia, Singapore, Thailand, United States of America, Australia, India and most of the European countries (Hughes 2014; Torres et al. 2019; García-Poza et al. 2020). Nearly eight million tons of fresh seaweeds are being harvested worldwide every year. The total annual production is estimated to be USD 6 billion, of which food products for human consumption contributes USD 5 billion. The growth of several seaweed species occurs along the coastline in Australia, with 36,735 km composing more than 9000 species (Frost and Cullen 1997). Seaweeds are found abundantly in the south-east and west coastal areas of India with a coastline of about 7500 km comprising about 271 genera and 1153 species (Sahayaraj et al. 2014).

Macroalgae are rich in pigments such as phycobiliproteins, chlorophylls, and carotenoids (Young and Britton 2012). Their morphology commonly exhibits a hold fast, stipe and blade (Spalding et al. 2019). They appear as red, green and brown based on their external color and classified as *Rhodophyceae*, *Chlorophyceae*, and

Phaeophyceae, respectively (Spalding et al. 2019). Macroalgal consumption dates back to 500 B.C. However their large-scale production has received attention in the last few decades. They are being recognised for their major biochemical compounds such as, carbohydrates, proteins, minerals, and phytochemicals (Ortega et al. 2019; Lourenço-Lopes et al. 2020; Machado et al. 2020). Carbohydrates present in seaweeds are polysaccharides such as agar, alginate, fucoidan, laminarin, ulvan, cellulose and carrageenan (Deepika 2017a; Gomez et al. 2020; Mohammed et al. 2020). Phycobiliproteins are pigment proteins, can be used in various applications (Bikker et al. 2016). Minerals such as micro and macro elements are present abundantly in seaweeds (Spalding et al. 2019; Lourenço-Lopes et al. 2020). Nearly 83% of seaweeds are used for human consumption while the leftover biomass is used as fertilizers, animal feed additives (Leandro et al. 2020; Shama et al. 2019). Some of the commercially used macroalgae species are shown in Fig. 2.1.

The utilization of seaweed as an alternate energy feedstock has been studied extensively., for biodiesel, bioethanol, and biohydrogen gas production (Ortega et al. 2019). Macroalgae have high photon conversion efficiencies, they synthesize and accumulate large quantities of neutral lipids and carbohydrates along with other valuable co-products, from abundant and inexpensive raw materials. They can be cultivated in saline/coastal seawater (Torres et al. 2019; Baghel et al. 2020). They can utilize growth nutrients such as nitrogen and phosphorous from a variety of wastewater sources, thus providing sustainable bioremediation of wastewater for economic benefit (Baghel et al. 2020). They can also couple CO_2-neutral fuel production with CO_2 sequestration from other power industries (Brennan and Owende 2010; Bikker et al. 2016). Compared to other advanced feedstocks for the production on cellulosic ethanol, basic research on algal genomics is more advanced and gaining increased momentum (Gomez et al. 2020). Already, we have a range of seaweed-based products in the market, evidently reflecting consumer demands and industrial opportunities (Fig. 2.2).

2.2 Macroalgae as a Source of Biofuels

Global warming has put tremendous pressure on the renewable sources of energy. The major factors concerning algal biofuel sector are the issues of sustainability, technological hurdles, and economic considerations (Hessami et al. 2019). As no biofuel is carbon neutral in the present technological and industrial scenario, significant fossil fuel input is needed for cultivating, processing and extracting oil, which might offset the positive aspects of algal biofuel (Laurens and Nelson 2020).

Although macroalgae are good at lipid accumulation and easier to be pyrolyzed at lower pyrolysis temperatures, the use of macroalgae-based bio-oils are not used as fuel due to higher nitrogen content. Biodiesel is traditionally produced by first extracting and purifying oil from raw material, after which the oil is treated by transesterification (Balu et al. 2020). The studies performed on biodiesel production from *Gracilariopsis longissima* and *Chaetomorpha linum* showed that the

Fig. 2.1 Macroalgae species used in algal biorefineries. From top right to bottom left – *Gracilaria refugium, Padina australis, Pelvitia canaliculate, Laminaria hyperbosrea, Ascophylum nodosum, Alaria esculenta, Eucheuma cottonii, Sargassum cristefolium, Turbinaria decurrens, Ulva lactuca, Palmaria palmata, Caulerpa racemose* (Fernandez-García et al. 2011)

consumption of natural resources was approximately 30 times higher in processing the macroalgal biomass due to the need of chemicals in the oil extraction from macroalgae and due to high energy needs in recovering the solvents (Balu et al. 2020).

Ulva lactuca was used as first test species to determine the yields of bio-butanol from macroalgae. Acetone–Butanol–Ethanol (ABE) fermentation process for bio-butanol production using *Clostridium beijerinckii* yielded 0.35 g ABE per gram of sugar consumed (Lee et al. 2009). Bioethanol is produced from macroalgae by the

Fig. 2.2 Commercial macroalgae products in food, nutraceuticals, cosmetics, biofertilizer and livestock feed sectors. *(Created in photoshop based on the commercial catalogues available-Source from Google)*

microbiological process of fermentation with the help of micro-organisms, mainly yeasts and bacteria (Campbell et al. 2020; Nguyen et al. 2020). Glucan is converted into ethanol, but usually the glucan content of macroalgae is low (22–25% in *Chaetomorpha linum*). The achievement of economically sustainable recovery of bioethanol requires not only glucan, but also non-glucan carbohydrates, like alginate, mannitol, agar and carrageenan, which are converted into ethanol (Dahiya et al. 2020).

Promising results were obtained from the anaerobic fermentation of *Saccharina japonica* with *Enterobacter* sp. JMP3, a bacterium capable of degrading mannitol for conversion to ethanol (Ji et al. 2016). The amount of ethanol produced depended on the redox potential of the carbon source (Ramachandra and Hebbale 2020). Compared to glucose, mannitol has a more favourable redox state for production of high amount of ethanol. For the maximal ethanol production from brown algae, a co-fermentation of laminarin and mannitol; and mannitol & alginate have been investigated (Suutari et al. 2015; Mohammad et al. 2019). A study on the two large kelps, *Macrocystis pyrifera* and *Durvillaea antarctica,* with similar elemental composition, showed that either species or their mixture can produce high methane concentrations in a two-phase anaerobic digestion system, which consisted of an anaerobic sequencing batch reactor and an up flow anaerobic filter (Khan et al. 2017). The average biogas production rate was 181.4 mL g^{-1}dw day^{-1} for *M. pyrifera*. The methane content in the biogas was approximately of 60–70%, and CO_2

content approximately 18% ($CH_4/CO_2 > 1$), which is considered adequate for energy recovery.

2.3 Macroalgae as Source for Food, Nutraceuticals and Therapeutics

In the past few decades, there were several studies performed on composition of macroalgae for food applications (Campbell et al. 2020; Laurens and Nelson 2020; Soares et al. 2020). Macroalgae-derived food additives are commonly used in the preparation of fast foods (Gomez et al. 2020). Macroalgae are rich in resistant protein and dietary fiber. Macroalgae-derived food flavours, colours, and nutrients continue to attract considerable commercial attention due to their rich elemental composition (Gomez et al. 2020).

The Asian countries such as China, Japan and Korea have been the large macroalgae consumers for many centuries. The macroalgae species such as *Gracilaria edulis, Gracilaria corticata, Gracilaria chilensis, Laminaria sp., Undaria pinnatifida, Eucheuma spinosum, Eucheuma cottonii* and *Sargassum wightii* were regularly employed as food for human consumption, at several centuries. Macroalgal species contain rich mineral content compared to higher plants and animal products (Miyashita et al. 2020). Reported values of nutritional, mineral and polyunsaturated fatty acids composition of seaweeds are presented in Tables 2.1, 2.2, and 2.3. They were excellent source of vitamins A, B, C and K e.g. 3 g of dried *Pyropia yezoensis* produces 2 μg of Vitamin B_{12}. Similarly, *Ulva sp.* are well-known to have high iron, potassium, magnesium and sodium content and are considered as valuable alternatives for plant-based food due to their enhanced levels of vitamins, proteins (e.g. *Grateloupia filicina),* minerals (e.g. *Palmaria palmata* is rich source of potassium and iodine) and dietary fibers (e.g. *Sargassum fusiforme)* (Miyashita et al. 2020).

Food should basically cater to nutrition and also promotion of health. Nutraceuticals are defined as a substance that may be considered as food which also provides medical and health benefits, encompassing prevention and treatment of disease (Liu et al. 2020a; Miyashita et al. 2020). Nutraceuticals may be individual concentrates or mixtures of polysaccharides, carbohydrates, phytochemicals, lipids, amino acids, indoles, vitamins, probiotics and minerals (Miyashita et al. 2020). Recent years have witnessed a great boom in the market size of nutraceuticals estimated to be USD 150 billion in which macroalgae-derived dietary supplements contributes to 37% (Miyashita et al. 2020).

Macroalgae are promising candidates for the development of therapeutics as they produce different compounds that exhibit a range of bioactivities that include antioxidant, anti-coagulant, anti-cancer, anti-viral, anti-allergic, anti-adhesive, antiangiogenic and anti-inflammatory actions (Deepika 2016; Liu et al. 2020b). The ratio of essential amino acid and non-essential amino acid (EAA/NEAA) is higher in red macroalgae (0.98–1.02) followed by green macroalgae (0.72–0.97) and

Table 2.1 Reported values of nutritional and mineral composition of seaweeds[a]

Seaweeds	Nutritional composition (%)					Mineral content (g/100 g)				
	Protein	Carbohydrates	Dietary fiber	Ash	Lipids	Sodium	Potassium	Phosphorus	Calcium	Magnesium
Caulerpa lentillifera	10–13	38–59	33	24–37	0.86–1.11	8.91	0.7–1.14	1.0	0.78–1.8	0.63–1.65
C. racemosa	17.8–18.4	33–41	64.9	7–19	9.8	2.57	0.31	0.029	1852	0.38–1.61
Ulva lactuca	10–25	36–43	29–55	12.9	0.6–1.6	–	–	0.14	0.84	–
Ulva rigida	18–19	43–56	38–41	28.6	0.9–2.0	1.59	1.56	0.21	0.52	2.09
Chondrus crispus	11–21	56–68	10–34	21	1.0–3.0	1.2–4.27	1.35–3.18	0.13	0.42–1.12	0.60–0.73
Agarophyton chilense	13.7	66.1	–	18.9	1.3	5.46	3.41	–	0.40	0.56
Palmaria palmate	8–35	46–55	29–46	12–37	0.7–3	1.5–2.5	7.0–9.0	0.23	0.56–1.20	0.17–0.61
Neopyropia tenera	28–47	44.3	12–35	8–21	0.7–1.3	3.62	3.5	–	0.39	0.56
Porphyra umbilicalis	29–39	43	29–35	12	0.3	0.94	2.03	0.23	0.33	0.37
Neopyropia yezoensis	31–44	44.4	30–59	7.8	2.1	0.57	2.4	–	0.44	0.65
Fucus vesiculosus	3–14	46.8	45–59	14–30	1.9	2.45–5.46	2.5–4.32	0.31	0.72–0.93	0.67–0.99
Himanthalia elongate	5–15	44–61	33–37	27–36	0.5–1.1	4.1	8.25	0.24	0.72	0.43
Laminaria digitata	8–15	48	37	38	1.0	3.31	0.01–0.08	–	1.0	0.65
Saccharina Japonica	7–8	51.9	10–41	27–33	1.0–1.9	2.53–3.26	4.35–5.95	0.15–0.30	0.22–0.91	0.55–0.75
Saccharina latissima	6–6.26	52–61	30	34.78	0.5–1.1	2.62	4.33	0.16	0.81	0.71
Sargassum fusiforme	11.6	30.6	17–69	19.77	1.4	–	–	–	1.86	0.68
Undaria pinnatifida	12–23	45–51	16–51	26–40	1.05–45	1.60–7.0	5.5–6.81	0.23–0.45	0.68–1.38	0.40–0.68

Morais et al. (2020)

[a]*Values varied upon culture conditions*

Table 2.2 Reported values of vitamin composition of seaweeds[#]

	Vitamin content (mg/100 g)								
Seaweeds	A	B_1	B_2	B_3	B_5	B6	B_8	C	E
Caulerpa lentillifera	–	0.05	0.0	1.09	–	–	–	1.00	2.22
Codium fragile	0.52	0.22	0.55	–	–	–	–	<0.22	–
Ulva lactuca	0.01	<0.02	0.55	98	–	6	–	<0.24	–
Ulva rigida	9581	0.47	0.19	<0.5	1.70	<0.1	0.01	9.42	19.40
Palmaria palmata	1.59	0.07–0.15	0.51–1.91	1.89	–	8.99	–	6.34–34.5	2.2–13.9
Porphyra umbilicalis	3.65	0.14	0.36	–	–	–	–	4.21	–
Fucus vesiculosus	0.30–7	0.02	0.03	–	–	–	–	14.12	–
Himanthalia elongata	0.07	0.02	0.02	–	–	–	–	28.56	–
Laminaria digitata	–	1.25	0.13	61.2		6.41	6.41	35.5	3.43
Laminaria ochroleuca	0.042	0.05	0.21	–	–	–	–	0.35	–
Saccharina latissimi	0.04	0.05	0.21	–	–	–	–	0.35	1.6
Undaria pinnatifida	0.04–0.22	0.17–0.30	0.23–1.4	2.56	–	0.18	–	5.29	1.4–2.5

Morais et al. (2020)
[#]*Values varied upon culture conditions*

Table 2.3 Reported values of polyunsaturated fatty acids (PUFAs) contents of seaweed[a]

	Total fatty acid content (%)					
Seaweed	Saturated	Monounsaturated	PUFAs	ω6 PUFAs	ω3 PUFAs	ω6:ω3 Ratio
Himanthalia elongata	39.06	22.75	38.16	15.08	18.7	0.81
Laminaria ochroleuca	33.82	19.23	46.94	20.99	25.08	0.83
Undaria pinnatifida	20.39	10.5	69.11	22.1	44.7	0.49
Palmaria spp.	60.48	10.67	28.86	2.14	25.52	0.13
Porphyra spp.	64.95	18.91	16.1	7.97	7.2	1.21

Sanchez-Machado et al. (2020), MacArtain et al. (2007)
[a]*Values varied upon culture conditions*

brown macroalgae (0.73), more specifically *A. nodosum* (1.00–1.06) and *Gracilaria corticata* (1.47–1.74) (Ganesan et al. 2019). The macroalgae-derived fucoxanthin (*Fucus vesiculosus*) demonstrated increased insulin sensitiveness, reduced blood glucose level and exhibited anti-obesity effect in diabetic animal models (Kwak

2014). Fucoidan is another macroalgae-derived polysaccharide which is non-toxic, biodegradable and biocompatible were found to enhance cell proliferation suppression by macrophage activation (induce apoptosis) in the case of human colon cancer cells (Caro and Pozo 2015).

2.4 Macroalgae as a Source of Pigments and their Applications

Photosynthesis in macroalgae, microalgae and cyanobacteria, are similar to higher plants. The photosynthetic process occurs in chloroplasts where the highly organized light-harvesting complexes (LHCs) that include chlorophylls, carotenoids, and phycobiliproteins harvest the visible light spectrum (PAR; 400–700 nm) (Grossman et al. 1995).

Macroalgal pigments are extensively used in various industrial products such as pharmaceutical, ink, cosmetic as dyes in food and textile industry (Begum et al. 2016). These pigments being the natural source of colour and derived from edible sources are great replacements for harmful chemical dyes used in the manufacture of commercial products (Begum et al. 2016). Pigments are not only attractive but are valuable to human health (act as anti-oxidant, anti-cancer, anti-inflammatory, anti-obesity, anti-angiogenic and neuroprotective agents) (Manivasagan et al. 2018). Antioxidant properties of these pigments help in scavenging of reactive oxygen species (ROS) and reactive nitrogen species (RNS) which play a vital role in causing various diseases and metabolic disorders such as cancer, diabetes, cataract, aging, and arthritis (Leong et al. 2018).

2.4.1 Chlorophylls

Chlorophylls (*Chl*) are the most abundant natural pigment that give green algae (Chlorophyta) their green colour and are essential for the photosynthetic process, whereby macroalgae derive their energy for metabolism and reproduction. Chemically, chlorophylls can be defined as cyclic tetrapyrroles which aid in light-harvesting and trapping during photosynthesis (Table 2.4). The possible applications of chlorophylls for food and health is shown in Table 2.4.

Table 2.4 Chlorophyll pigments from seaweeds and their applications

Chlorophyll pigments	Seaweeds	Chemical Structure	Chemical formula	Color	Applications
Chl a	*Ulva rigida,* *Ulva pertusa, Ulva lactuca, U.reticulate*	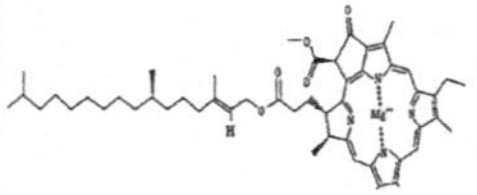 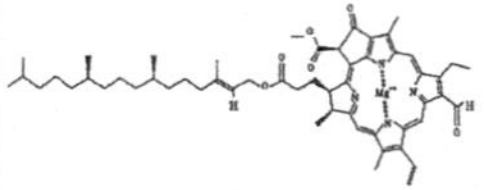	$C_{55}H_{72}MgN_4O_5$	Blue/green	Food colorant in beverages, fruit juices, pasta, dairy products, soups, sweeter preparations Cosmetics (wound healing property and sun creams),textile dye Health supplement, coatings (protective films),
Chl b	*Halimeda macroloba, Codium fragile*	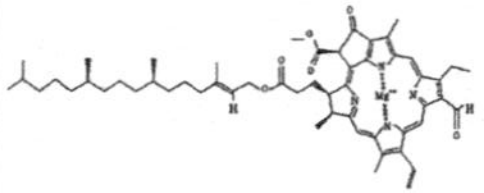	$C_{55}H_{70}MgN_4O_6$	Green/yellow	
Chl c	*Caulerpa racemosa, Ulva fasciata*	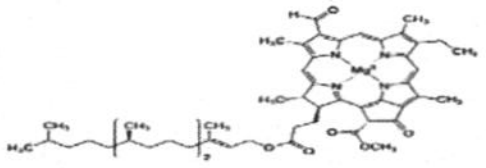	$C_{35}H_{32}N_4O_5$	Yellow	–
Chl d	*Monostroma latissimum, Ulva sp.*	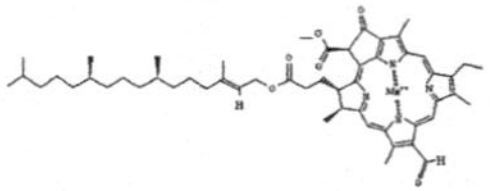	$C_{54}H_{70}MgO_6N_4$	Green	
Chl f	*Colpomenia sinuosa, Hormosira banksii*	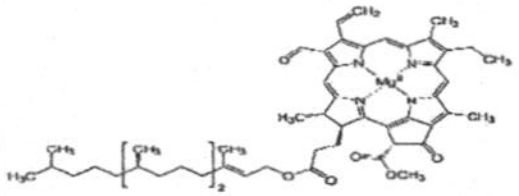	$C_{55}H_{70}O_6N_4Mg$	Green/yellow	

Fernandes et al. (2007), Koca et al. (2007), Danesi et al. (2002), Schoefs (2003)

2.4.2 Carotenoids

Carotenoids are unsaturated hydrocarbons belonging to the tetraterpene group (40-carbon atom terpenoids). There are around 700 types of carotenoids reported and 30 of these have known to have a role in photosynthesis (Young and Britton 2012). Carotenoids are categorized into two classes – xanthophylls and carotenes. Structurally, astaxanthin has 13 double bonds which makes it a potent anti-oxidant (Henríquez et al. 2016; Ambati et al. 2019). Some carotenoids are specific for certain macroalgae species and can be used as chemotaxonomic markers (Guedes et al. 2011; Rammuni et al. 2019). The carotenoids such as β-carotene, lutein and zeaxanthin absorb light apart from the absorption spectrum of chlorophyll molecules and also offers protection against the harmful effect of extreme light intensity by dissipating excess energy absorbed (Guedes et al. 2011). The red algae species of the family *Corallinaceae* (generally known as corallinales) are the highest carotenoid producers (Koizumi et al. 2018). Besides their role in light harvesting, carotenoids also contribute to stabilize the structure and aid in the function of photosynthetic complexes by quenching chlorophyll triplet states and scavenging reactive oxygen species. Their common feature is the presence of coupled double-bond systems, which determine the carotenoids ability to absorb light in a visible range and their characteristic red-orange (Henríquez et al. 2016). Fucoxanthin is the major carotenoid pigment with >10% of total carotenoids produced, in most of the Phaeophycean brown algae, for example., *Ectocarpus siliculosus* (Sáez et al. 2015), *Hijikia fusiformis* (Yan et al. 1999), *Laminaria japonica* (Chen et al. 2018), *Laminaria digitata* (Fertah et al. 2017), *Ascophyllum nodosum* and *Laminaria hyperborean* (Miyashita et al. 2020). Lutein cannot be synthesized by humans and has a protective role against macular degeneration of the eye therefore is an important dietary supplement (E161b in the European Union) (Krinsky et al. 2003). Tables 2.5 and 2.6 showed carotenoid pigments are used in various applications.

There is increasing demand for natural colorants from sustainable sources have the great economical value in different industries such as food, cosmetics, pharmaceuticals, textile, and printing industries. Out of many known carotenoids, only 40 are produced commercially, some of which includes β-carotene and astaxanthin, and, to a lesser extent, lutein, zeaxanthin and lycopene. The major pigments of commercial interest are β-carotene produced using *Dunaliella salina* (14% dry weight) (Pulz and Gross 2004) and astaxanthin from *Haematococcus pluvialis* (3% dry weight) (Ambati et al. 2014) and can be potentially replaced by the red seaweeds such as *Hizikiafusiformis*, *Undaria pinnatifida*, and *Sargassum fulvellum*. The largest astaxanthin consumer is the salmon feed industry (FDA approved 1987). Human nutraceuticals have also expanded as a new market for astaxanthin (FDA approved 1999) owing to its potent antioxidant properties (Table 2.7).

Table 2.5 Carotenoids from seaweeds and their applications

Carotenoid pigments	Seaweeds	Chemical structure	Chemical formula	Colour	Applications
Astaxanthin	*Bossiella californica, B. orbignyana*	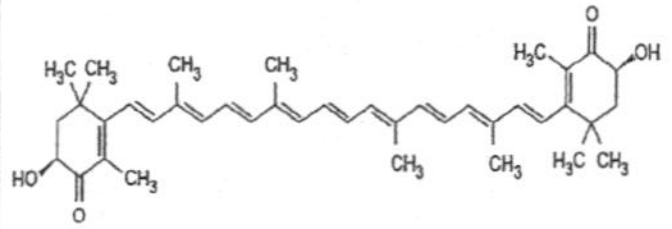	$C_{40}H_{52}O_4$	Red	Aquaculture feed (Salmon pigmentation), nutraceutical, food colour
Fucoxanthin	*Eisenia bicyclis, Fucus distichus*	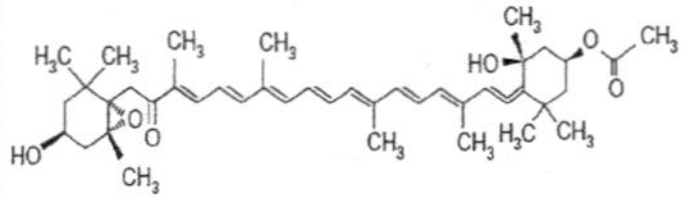	$C_{42}H_{58}O_6$	Olive green	Antioxidant agent, Skin-whitening creams, Antiaging pills
Zeaxanthin	*Sargassum muricum, S. wightii*	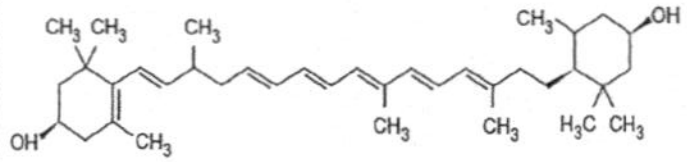	$C_{40}H_{56}O_2$	Yellow	Dietary supplement (eye)
β-Cryptoxanthin	*Laminaria japonica, L. religosa*	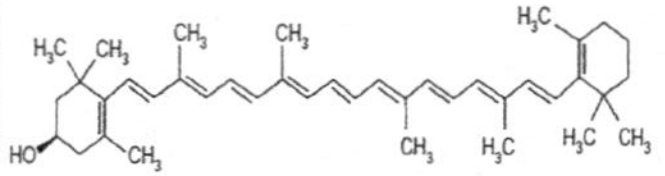	$C_{40}H_{56}O$	Yellow orange	Feed additive, food colour
Lutein	*Petalonia binghamiae*	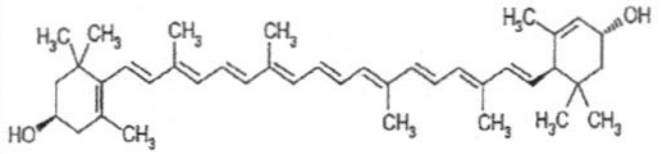	$C_{40}H_{56}O_2$	Yellowish-red	Dietary supplement (eye)
Canthaxanthin	*Hizika fusiformis, Undaria pinnatifida*	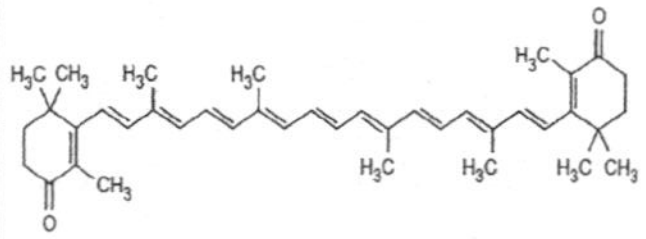	$C_{40}H_{52}O_2$	Orange red	Poultry feed (egg yolk and broiler pigmentation)
Myxoxanthophyll	*Sargassum fulvellum, Fucus distichus*	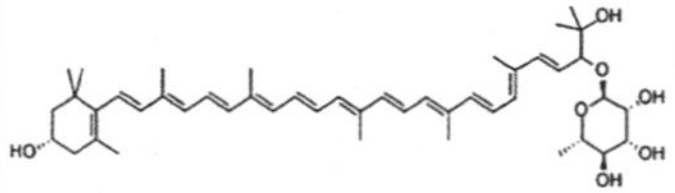	$C_{46}H_{66}O_8$	Bright red	Chemosystematic studies

Carotenoid pigments	Seaweeds	Chemical structure	Chemical formula	Colour	Applications
Echinenone	*Calliarthron tuberculosum, Corallina officinalis*	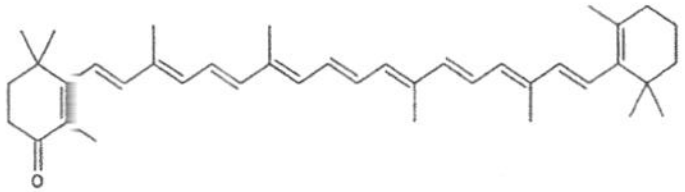	$C_{40}H_{54}O$	Brownish-red	Food colour, Nutraceutical supplement
ζ-Carotene	*Jania tenella, Ectocarpus siliculosus, Laminaria digitata, Ascophyllum nodosum, Laminaria hyperborean*	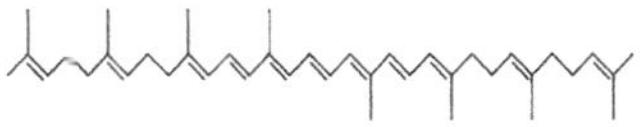	$C_{40}H_{56}$	Light-yellow	Food colour agent
β-Carotene		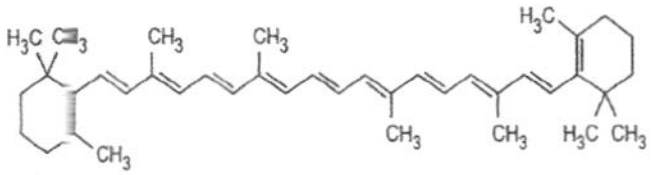	$C_{40}H_{56}$	Orange	Food colour a(margarine and juices), feed additive (fertility, cattle), nutraceutical
γ-Carotene		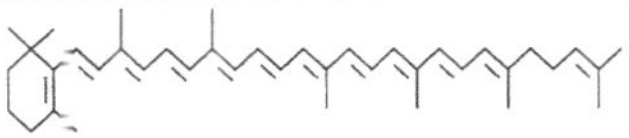	$C_{40}H_{56}$	Yellowish-orange	Food colour agent
Lycopene	*Corallina officinalis, Mastocarpus stellatus*	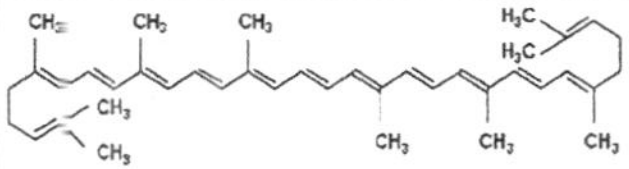	$C_{40}H_{56}$	Red	Food colour, Nutraceutical supplement

Table 2.6 Health benefits of seaweed pigments

Category	Applications	Pigment type
Food	Colorant in food and beverages	β-carotene
	Antioxidant additives	Fucoxanthin
Nutraceuticals and functional foods	Antioxidant potential in vivo	Zeaxanthin, β-carotene, lutein
	Anti-diabetic and anti-obesity property via suppression of insulin levels, accumulation of adipose tissue, and controlling of hyperglycemia	Fucoxanthin
	Blood-pressure lowering properties and lowers stroke risk factors	Fucoxanthin
	Anti-cancer activities via anti-neoplastic effects and ability to inhibit the growth of cancer cells and induce timely and dose depended apoptosis in human breast cancer cells	Fucoxanthin
	Antioxidant, immune modulatory, anti-angiogenic, anti-malarial activites, and anti-inflammatory effects	Astaxanthin; Fucoxanthin
Pet foods	Improve appearance of pet foods while imparting biological effects such as anti-oxidant, and immune enhancing properties	Astaxanthin
Animal feed	Enhanced the color of animal food products such as poultry (eggs, meat colour), seafood (salmon, trout, crustaceans, and shrimp, and krill)	Astaxanthin

Aryee et al. (2018), Pangestuti and Siahaan (2018), Namvar et al. (2014), Brown et al. (2014), Déléris et al. (2016)

2.4.3 *Phycobiliproteins*

Phycobilisomes are major photosynthetic accessory, giant subcellular complexes, composed of many phycobiliproteins joined together by linker proteins, forming a major part of the light harvesting complex (Table 2.8) (Glazer 1988). Phycobiliproteins are made up of multi-chain holoproteins composed of apoproteins with covalently bound phycobilins (open-chain tetrapyrroles) via thioether bonds (cysteine residue at C-3 position). They are produced as a part of the bilin biosynthesis with the catalytic regulation of lyases enzymes (Kannaujiya et al. 2020). The wide spectral composition of phycobiliproteins is contributed by the assembly of mainly three subunits such as phycoerythrin (PE), phycocyanin (PC), and allophycocyanin (APC) (Kannaujiya et al. 2020). Commercially, phycobiliproteins are broadly classified into two categories – phycocyanin and phycoerythrin based on the colour (Singh and Ahluwalia 2013). They are auto fluorescent and water-soluble proteins found in macroalgae (Yeremenko 2004). In photosynthetic macroalgae such as *Furcellaria lumbricalis*, they constitute ~40% dry weight of the cell and the phycobiliprotein concentration may rise up to ~50% in low light conditions(Gomez et al. 2020). Based on the light absorption and fluorescence capability, phycobiliproteins are classified into six groups, namely, allophycocyanin

Table 2.7 Market analysis of commercial algae-based pigments

Commercial pigments	Colour	Production strain	Production cost	Selling price	Market size	CAGR (%)
Astaxanthin	Red	*Haematococcus pluvialis*	USD 552.0/kg	USD 2500–7000/kg	40 USD million	2.3
β-carotene	Yellow	*Dunaliella salina*	USD 105.0/kg	USD 790/kg	200 USD million (2019)	3.3
C-Phycocyanin	Dark Blue	*Arthrospira platensis*	USD 46.0/kg	USD 548/kg	114.8 USD million (2022)	4.7
Fucoxanthin	Orange	*Phaeodactylum*	–	USD 180–42,000/kg	95 USD million (2019)	2.6
Phycoerythrin	Pinkish Red	*Prophyridium, Spirulina*	–	USD 500–50,000/kg	~60 USD million (2019)	–
Lutein	Yellowish red	*Scenedesmus, Muriellopsis, Chlorella*	–	USD 910–15,000/kg	~3 USD million (2019)	3.6
Canthaxanthin	Orange red	*Chlorella, Dunaliella, Scenedesmus*	–	USD 100–500/kg	~1 USD million (2019)	3.7

Hu (2019), Jacob-Lopes et al. (2019)

B, allophycocyanin, C-phycocyanin, phycoerythrocyanin, C-phycoerythrin, and R-phycoerythrin (Li et al. 2020).

There are several reports of production of phycobiliproteins in macroalgae such as – *Porphyra umbilicalis,* (13.3%), *Gelidium corneum* (13.6%), *Gelidium longissima* (10%), and *Corallium rubrum* (5.2%) (Vega et al. 2020).

Phycobiliproteins are widely used in foods therapeutics and cosmetics (Ambati et al. 2019; Ambati et al. 2014). Their characteristic fluorescent properties that make them useful as fluorescent dyes and markers (Rammuni et al. 2019). Phycocyanin (bright blue) has a bright blue colour and is considered more versatile, although it is heat and light sensitive. It is used as a natural colourant in ice cream, soft drinks, candies, chewing gum, desserts, cake decorations, icings and frostings, milk shakes as well as lipsticks and eyeliners (de Amarante et al. 2020). Fluorescent phycobiliproteins are used in fluorescent microscopy, flow cytometry, fluorescence-activated cell sorting, diagnostics, immunolabeling and immunohistochemistry (Richa et al. 2011). Phycobiliproteins also exhibit different levels of therapeutic properties such as anti-oxidant, anti-inflammatory, neuroprotective, anti-angiogenic,

Table 2.8 Phycobiliproteins from seaweeds and their applications

Phycobilin pigments	Seaweeds	Chemical structure of chromophores	Colour	Applications
Allophycocyanin	*Cyanidioschyzon merolae, Corallina officinalis*	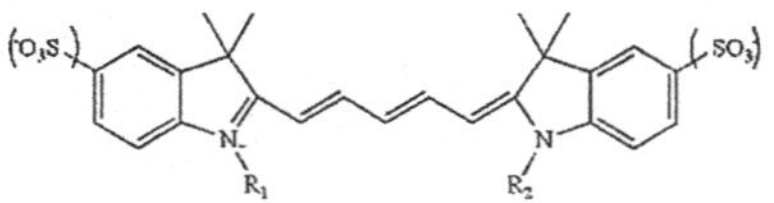	Bright blue	Immunoassays such as FACS, flow cytometry
C-Phycocyanin	*Atractophora hypnoides, Mastocarpus stellatus*	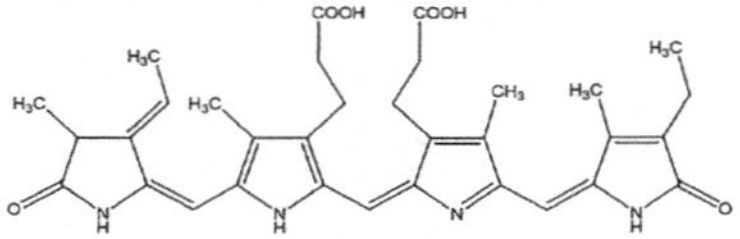	Dark blue	Food colour, Anticancer agent, TR-FRET assays
Phycoerythrocyanin	*Corallina officinalis, Vertebrata simulans*	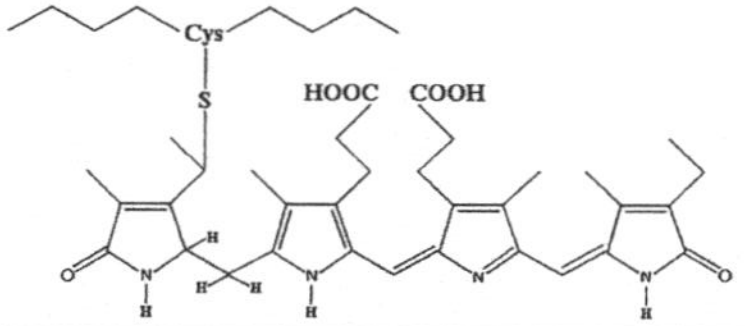	Orange	Fluorescence cell sorting, Cosmetics (Sun cream, Lipsticks, Eyeshadow)
C-Phycoerythrin	*Chondrus crispus, Irish moss, Mastocarpus stellatus*	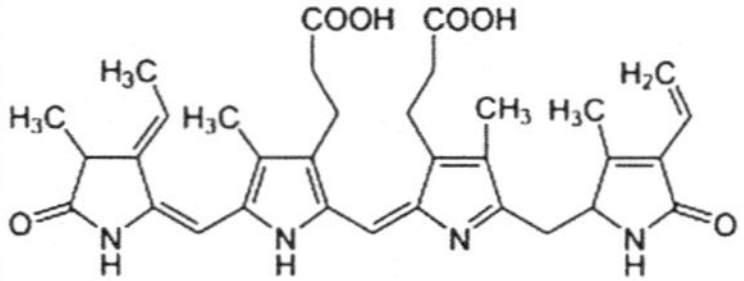	Reddish pink	Protein marker for gel electrophoresis, Dairy products (ice creams, milkshakes)
R-Phycoerythrin	*Chondrus crispus, Mastocarpus stellatus, Vanvoorstia bennettiana*	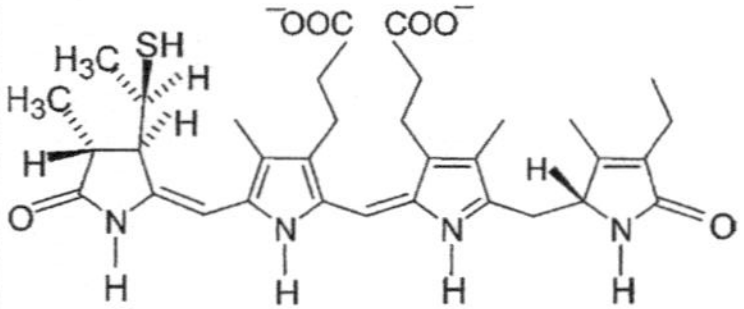	Red	Free radical scavengers in nutraceuticals, Diagnostic immunolabelling

cardioprotective, hepatoprotective, immuno-modulatory, anti-viral, and anti-tumor activity (Manirafasha et al. 2016).

2.5 Cosmetics

The term 'cosmetics' is defined as the product that is only used for topological application on the skin (epidermis) and does not alter the physiological conditions (Lourenço-Lopes et al. 2020). They are formulations that contain active ingredients, such as phytochemicals, essential oils, vitamins, antioxidants, etc. Skin damage is mostly caused due to harmful UV radiation, temperature variations, salinity variations, environmental pollutants like smoke and dust (Bedoux et al. 2014). As described earlier, macroalgal extracts are rich in primary and secondary metabolites, their composition includes a wide range of proteins, amino acids, lipids, polysaccharides, phytochemicals, phycocolloids and phlorotannins (Leandro et al. 2020). The cosmetic industry has a special interest on the secondary metabolites such as phenolic compounds, pigments, sterols, vitamins, and other bioactive compounds. Among them the phlorotannin's are the key ingredient for macroalgae-based cosmetics development. Phlorotannins consist of oligomers or polymers of phloroglucinol which are known to be accumulated in special vesicles called physodes of the cells. They are only found in the brown algae (Phaeophyceae). More specifically the extracts from *Saccharina thunbergii, S. fusiforme, Ishige okamurae, Fucus vesiculosus, Undaria pinnatifida* and *Saccharina japonica* are recognized as good sources of phlorotannins (Campbell et al. 2020; Liu et al. 2020a, b; Soares et al. 2020). Many formulations of cosmetic products use these pigments as anti-aging agents. These constituents act as excellent antioxidants which inhibit the action of free radical to promote Matrix Metallo Proteinase (MMP) activity which damage cell membrane and collagen in skin epidermis preventing the occurrence of freckles and wrinkles (Bedoux et al. 2014). However, the hydrocolloids, which account for 40% of the world's hydrocolloid market are also widely used in cosmetic industry (Provide reference).

Furthermore, The macroalgal extracts are regularly used in the development of skin-whitening and anti-aging creams. The most popularly adapted macroalgae species in both cosmetic and food industry are *Petalonia binghamiae, Scytosiphon lomentaria, Undaria pinnatifida, Piropya dentata, Codium fragile*, and *Umbraulva japonica*. There are a range of studies that demonstrate the effectiveness of secondary metabolites to act against UV damage and prevention of wrinkles (Table 2.9). Topical application of fucoidan from the extracts of *Laminaria sp.* has been shown to have anti-aging activity by increasing hydration and elasticity of cells by stimulating the heparin growth factors within the cells. Whitonyl® is the first macroalgae-based commercial product, derived from *Palmaria palmata*, used for anti-aging and skin-whitening applications (Pimentel et al. 2018). The *Rhodella sp.* (red algae) rich in photosynthetic pigments (Phycobiliproteins – Allophycocyanin, phycoerythrin, phycocyanin) is highly employed in cosmetic development to act as an anti-oxidant,

Table 2.9 Phlorotannins from seaweeds and their applications

Phlorotannins	Chemical Structure	Chemical formula	Applications
Phloroglucinol	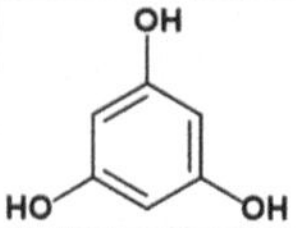	$C_6H_6O_3$	Anti-aging, anti-oxidant
Eckol	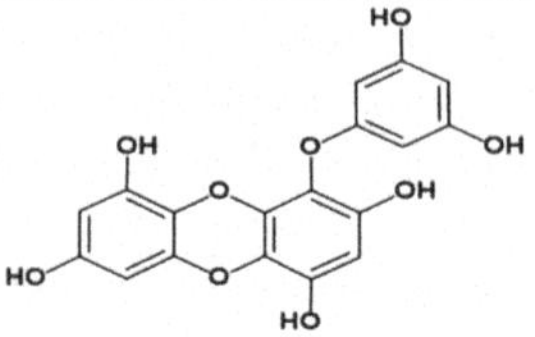	$C_{18}H_{12}O_9$	Skin moisturization and protection
Fucophloroethol A	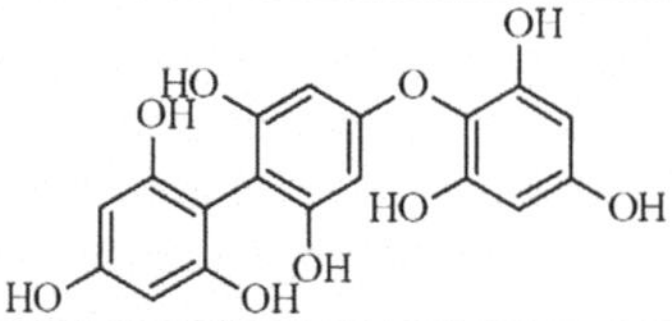	$C_{18}H_{14}O_9$	Anti-elastase, anti-oxidant, skin-whitening
Tetrafucol A	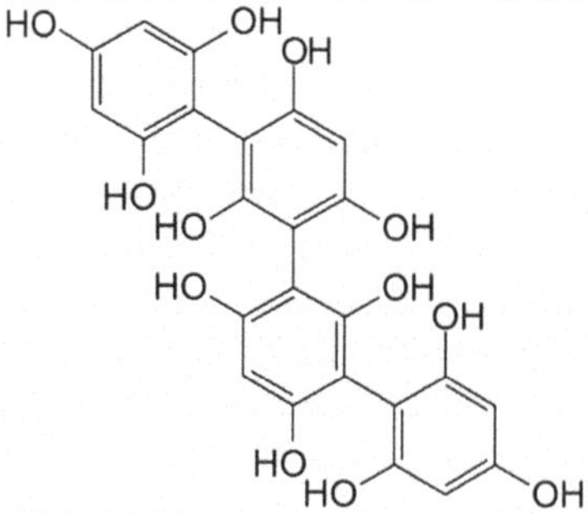	$C_{24}H_{18}O_{12}$	Anti-aging, anti-wrinkling (hyaluronidase Inhibition), lipid peroxidation inhibition
Dieckol	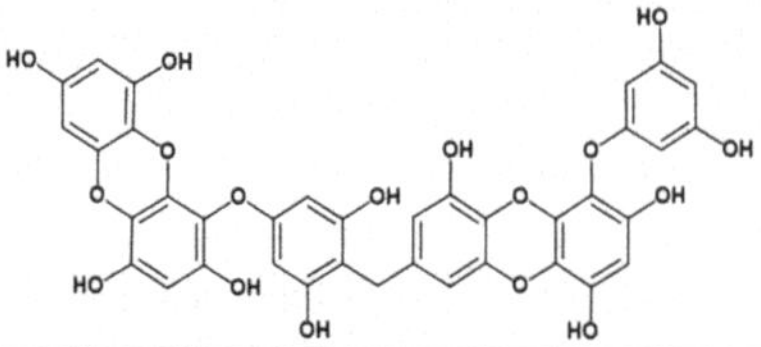	$C_{36}H_{22}O_{18}$	Hair growth

anti-inflammatory, colorant and radical scavenging agent (Chakraborty et al. 2009; Fiedor and Burda 2014; Deepika et al. 2016). The extracts of *Chondrus crispus, Undaria pinnatifida, Vertebrata lanosa* and *P. umbilicalis* are mostly used for skin moisturization and conditioning (Patel et al. 2020). In future it is possible to find a wide variety of products from topical application starting from slimming creams to perfumes, shampoos, sunscreens, and bath salts.

Table 2.10 Phycocolloids from seaweeds and its applications

Phycocolloids	Chemical Structure	Chemical formula	Applications
Agar	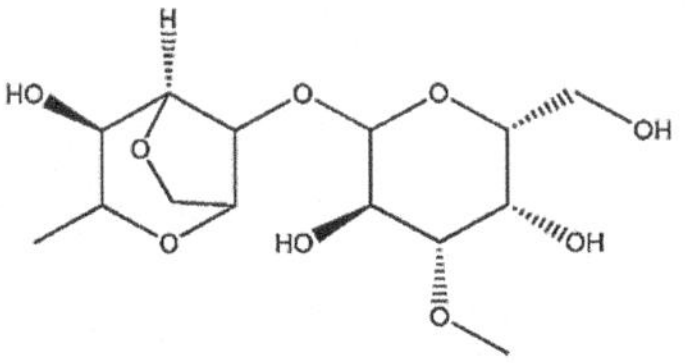	$C_{14}H_{24}O_9$	Bakery products, clarifier in wine breweries, stabilizers for canned food. Impression materials in dentistry, intestinal regulator, excipient in pills, microsphere and beads, growth media, gel electrophoresis, bulk laxative. Textile and paper making aids in making ultra-thin separating films.
Alginate	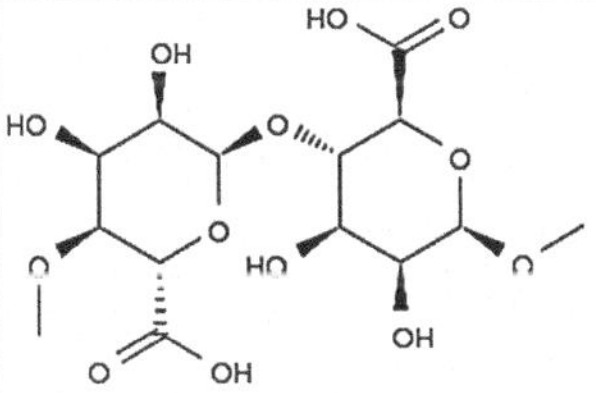	$C_{14}H_{22}O_{13}$	Thickeners in ice-cream. Hydrogel in biomedicine, including wound healing, drug delivery and tissue engineering applications. Wastewater purification (heavy metal sorption)
Carrageenan	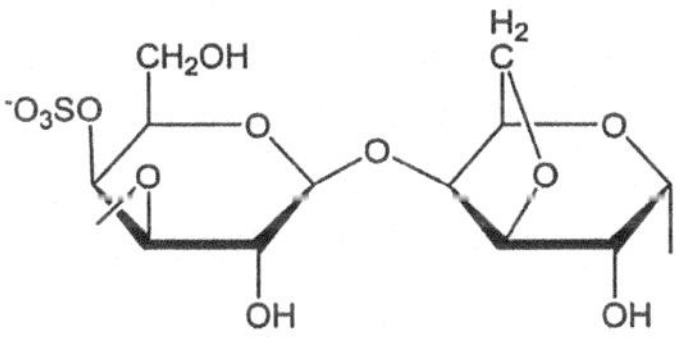	$C_{24}H_{36}O_{25}S_2^{-2}$	Puddings, vegan alternative to gelatin, processed meat, desserts, cream thickener. Thickeners for toothpaste, cosmetic, Inhibitor of Papilloma, dengue and herpes virus, beads for controlled released system, lotion and cream, suspending agents in antacid, eye drops.

2.6 Phycocolloids

Polysaccharides derived from different sources, macroalgal hydrocolloids, or phycocolloids are highly used in the industries. Phycocolloids refer to those polysaccharides extracted from macroalgae, mostly brown and red algae which are prominent producers of three phycocolloids namely, agar, alginate and carrageenan (Gomez et al. 2020; Puspita et al. 2020). These hydrocolloids (water-based colloids) have many applications in food, pharmaceutical, and cosmetic industries as gelling agents, thickeners or stabilizers and emulsifying agents (Table 2.10).

The market size of hydrocolloids is estimated to be USD 10.2 billion by 2027 (4.4% CAGR) and is expected to reach 3879 million tonnes production by 2027 (Newswire 2020). Food and beverages sector alone contributed to 43.7% share of total phycocolloids market (USD 5.4 billion) in 2018. In terms of volume, the global market for hydrocolloids is estimated at 2.2 million metric tons in 2018 (Newswire 2020).

Macroalgal-derived phycocolloids are high molecular weight, structural polysaccharides, found in the cell wall of marine algae that typically form colloidal solutions (Glicksman 2020). Polysaccharides used as thickeners, gelling agents and stabilizers for suspensions and emulsions in diverse industries, including the food, biotechnological, paint, textile and biomedicine (Deepika 2017a, Glicksman 2020). In recent days, hydrocolloids have become more significant commercially and are extracted in huge amounts by hot water or alkaline extraction (Gomez et al. 2020).

Agar is the most common hydrocolloid that has been used from the ancient times. It was initially extracted in Japan using marine macroalgae (*Gracilaria corticata* and *Gelidium sp.*). Due to their excellent gelation property, agar consistently holds a higher market value compared to any other phycocolloids (Bixler and Porse 2011). The recorded annual sales are in the range of USD 200–280 million, with a volume of 11,000 tons. The structure of agar was initially believed to be a simple, sulfated poly-galactose polymer and the subunits of the structure consisting of agarose and agaropectin were demonstrated much later (Azevêdo et al. 2020). The number of repeating units of 1–3-P-D-galactopyranosyl-(1–4)-3,6-anhydro-a-L-galactopyranose forming the gel may depend on different factors such as environmental and seasonal conditions.

Alginate is another common polysaccharide produced by brown macroalgae which are colourless, odorless, and tasteless (Mohammed et al. 2020). They are located in the cell walls in a crystalline configuration parallel to the cellulose microfibrils. The alginate structure consists of 1–4-linked β-d-mannuronate and α-l-guluronate residues in different proportions and sequences. The structural composition of alginates depends on the species, geography, season, and location (Fertah et al. 2017). Alginate is mainly produced from the giant kelp *Macrocystis pyrifera, Sargassum wightii Turbinaria sp.* and *Ascophyllum nodosum*. The thallus of *Laminaria sp.* is mostly composed of 20–30% alginate with variation in composition based on seasonal changes. They are insoluble in water and have high viscosity. Currently, commercial alginates are also produced from macroalgae – *Laminaria digitata*, *Undaria sp.*, *Turbo cornutus, Spinula solidissima*, and *Dolabella auricula* through organic solvent extraction. Alginates are used as food additives, edible films and coatings on pastries, thickeners and emulsifiers in bakery products and confectionaries (Fertah et al. 2017). They are also used in the production of sauces, ice cream, milk products, and dressings at concentrations of 0.1–0.5% (Glicksman 2020). They are used as thickener in jellies and are also employed to remove proteins from beer and to retain the structure, taste, and stability of frozen products. In pharmaceutical sector, application, they help in faster wound healing as they support fluid collection and fosters clotting process enhancing accelerated epidermal regeneration.

Carrageenan is the hydrocolloid unique to most of the red macroalgae species which include *Kappaphycus alvarezii*, *Eucheuma sp.* and *Gigartina skottsbergii*. Carrageenan was first extracted from *Chondrus crispus* and widely used as a thickener (Therkelsen 1993). Philippines and Indonesia developed the pioneer cultivation technologies for macroalgae while developed countries such as China were cultivating (e.g. *Betaphycus gelatinum*) in the wild marine environments (FAO 2012). Cultivation of carragenophytes (carrageenan producing macroalgae) is challenging due to variations in cultivation parameters. *Kappaphycus sp.*, grows well in water exposed to slow to moderate tides and rocks, sand and coral substrates with higher light intensities whereas the *Eucheuma sp.* and Betaphycus sp. grows better in stronger tides in dark adapted conditions (Du Preez et al. 2020; Yermak et al. 2020). *Chondrus sp.* was found to grow in higher temperature zones with moderate light. In food industry, carrageenans are used as a stabilizer in dairy products, toothpastes and water emulsions. Both refined carrageenan (*E*407) and as semi-refined carrageenan (E407a) are approved by the European Union and are available as edible products for the food sector (Du Preez et al. 2020). The air freshener gels are formulated with carrageenan to hold the fragrance longer avoiding evaporation. They also help increase the shelf life of the milk product and to retain tenderness of meat, and poultry products. Carrageenan extracted from *Kappaphycus alvarezii* has shown anti-microbial, anti-viral and anti-cancer activity (Deepika 2017b). From animal models, it was demonstrated that carrageenan enhances the activity of killer cells and increase macrophage phagocytosis. Similarly, carrageenan from *Hypnea musciformis* showed anti-bacterial activity against *Staphylococcus aureus* and anti-fungal effect against *Candida albicans* (Yermak et al. 2020).

The global market of carrageenan was estimated to be USD 931.6 million in 2020 and is forecast to reach USD 1.2 billion by 2025 (5.6% CAGR) and to reach 3879 million tonnes production by 2027 (Forecast 2020). Macroalgae-derived carrageenan holds 2.3% of the total hydrocolloid market. Asia-Pacific manufactures have consistently performed as the top producers of carrageenan producing about 9900 tonnes annually. CP Kelco US Inc., Geltech Hayco Inc., Myeong Shin Chemical Ind. Co. Ltd., Gumindo Perkasa Industries, Quest International Philippines Corp., FMC Corporation and Soriano SA are some of the leading companies in carrageenan production.

2.7 Bio-Fertilizer

Synthetic agrochemicals (fungicides, pesticides, and herbicides) give immediate results; however, their continuous use in agriculture has an adverse impact on the quality of the soil, plant growth and the microbial communities (Uthirapandi et al. 2018; Dewi et al. 2020).

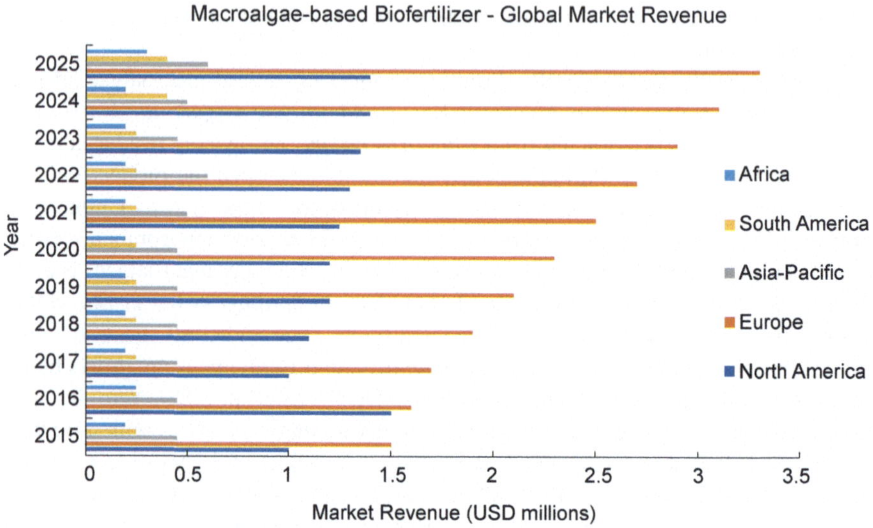

Fig. 2.3 Global macroalgae fertilizer market revenue (HexaResearch 2020)

To circumvent this, Macroalgae are effectively used as bio-fertilizers because they include high levels of organic matter, which leads to soil nutrient enrichment enhancing plant growth and productivity. Macroalgal extracts with a range of bioactivities serve as a potent nutrient source and accelerates germination of seeds, increasing the crop yield and resistance. Recent studies have reported wide applications of these macroalgae as sustainable organic fertilizers in agriculture and horticulture sector (HexaResearch 2020; Soares et al. 2020). The application of these extracts as a soil drench was found to be more effective on plant vigour than the foliar spray application as a supply of micro and macro elements, vitamins, fatty acids and growth regulators. The global macroalgal biofertilizer market was estimated to be USD 9.6 million in 2017 and is expected to grow at a CAGR of 7.5% from 2015 to 2025 (Fig. 2.3) (HexaResearch 2020). The European Union (EU) (Council Regulation (EEC) No. 2092/91) has permitted the use of macroalgae extracts as biofertilizer. Commercially, several macroalgae powder and liquid fertilizer concentrates are prepared from e.g. *Ascophyllum nodosum*, *Fucus* sp., *Sargassum* sp., *Laminaria* sp., *Ecklonia* and *Durvillea* (Amin et al. 2020). Some of the top companies in the macroalgae fertilizer market include Fox Farm, Espoma, Hydrofarm, Technaflora Plant Products and Maxi crop.

2.8 Animal and Aquaculture Feed

Macroalgae are a rich source of important macro and micronutrients, making them suitable as feed supplements. Indeed, copper, iodine, iron, potassium, and zinc are abundantly found in macroalgae (Fiedor and Burda 2014). A range of vitamins are also found in macroalgae, such as thiamine, niacin, nicotinate, cyanocobalamin, biotin, ascorbic acid, pantothenic acid and riboflavin (Campbell et al. 2020). As discussed above, they are also an excellent source of carotenoids such as β-carotene, lutein and astaxanthin. The nutritive value of macroalgae as animal feed or aquaculture feed depends on the requirements of the particular species. Many macroalgae species such as *Ulva lactuca* and *Gracilaria cliftonii* are approved by the FDA (Food and Drug Administration, US), EU (European Union) and CFIA (Canadian Food Inspection Agency) as safe feed ingredients (Campbell et al. 2020).

The application of macroalgae in animal feed has long been studied, dating as far back as the 1950's. Currently, animals and fishes are an important part of diet consumed by 90% of the world population (Morais et al. 2020). Commercially, macroalgae-based feed would highly help in increased quality and quantity of meat produced, promote or accelerate growth, improve reproduction ability. The use of dry macroalgae biomass as feed has proven to benefit the physiology by improving immune response, disease resistance, anti-viral and anti-bacterial action, gut function and probiotic colonization stimulation (Campbell et al. 2020). Macroalgae are also a relevant source of minerals and polyunsaturated fatty acids (PUFAs), especially eicosapentaenoic acid (EPA) and docosahexaenoic acid (DHA) which are known to impact positively on meat quality of lambs, pigs and broilers (Cabrita et al. 2016). In poultry industry, use of macroalgae as feed ingredients has increased popularity due to increased egg production and disease resistance (Michalak and Mahrose 2020).

2.9 Conclusion and Future Perspectives

Macroalgae are a huge untapped resource positioned at the nexus of global challenges. They are rich source of a range of bio-products that are highly valuable with vast industrial potential. The macroalgae derived carbohydrates are used for the fermentation process for bioethanol production; lipids used for bio-diesel production; proteins, fatty acids, pigments are used for nutraceutical and pharmaceutical applications. The multi-product bio-refinery approach would be ideal to reduce the production cost of seaweed products as they enable sustainable co-production of a range of products such as proteins, pigments, hydrocolloids, carbohydrates, biofuels and biogas production in a CO_2 neutral approach (Zhong et al. 2020).

Acknowledgments ARR acknowledges support and resources from the Improvement of Science & Technology Infrastructure in Higher Educational Institutions (FIST Project no: LSI-576/2013) from the Department of Science and Technology, Government of India and Centre of Excellence, Department of Biotechnology, Vignan's Foundation for Science, Technology and Research.

References

Ambati RR, Phang S-M, Ravi S, Aswathanarayana RG (2014) Astaxanthin: sources, extraction, stability, biological activities and its commercial applications—a review. Mar Drugs 12(1):128–152

Ambati RR, Gogisetty D, Aswathanarayana RG, Ravi S, Bikkina PN, Bo L, Yuepeng S (2019) Industrial potential of carotenoid pigments from microalgae: current trends and future prospects. Crit Rev Food Sci Nutr 59(12):1880–1902

Amin MS, Elshinawy MZ, Abdallah MM, El-Gawad A (2020) Effect of seaweed extract and biofertilizer on organic production of common bean seeds (Phaseolus vulgaris L.). Arab Universities Journal of Agricultural Sciences, 28(1), 265–273

Aryee AN, Agyei D, Akanbi TO (2018) Recovery and utilization of seaweed pigments in food processing. Curr Opin Food Sci 19:113–119

Azevêdo HV, Raimundo RA, Ferreira LS, Silva MM, Morales MA, Macedo DA, Gomes UU, Cavalcante DG (2020) Green synthesis of CoWO4 powders using agar-agar from red seaweed (Rhodophyta): structure, magnetic properties and battery-like behavior. Mater Chem Phys 242:122544

Baghel RS, Suthar P, Gajaria TK, Bhattacharya S, Anil A, Reddy C (2020) Seaweed biorefinery: a sustainable process for valorising the biomass of brown seaweed. J Clean Prod 121359:1–9

Balu M, Lingadurai K, Shanmugam P, Raja K, Teja NB, Vijayan V (2020) Biodiesel production from *Caulerpa racemosa* (macroalgae) oil. Indian J Geo Marine Sci 49(04):616–621

Bedoux G, Hardouin K, Burlot AS, Bourgougnon N (2014) Bioactive components from seaweeds: cosmetic applications and future development. Adv Bot Res Elsevier 71:345–378

Begum H, Yusoff FM, Banerjee S, Khatoon H, Shariff M (2016) Availability and utilization of pigments from microalgae. Crit Rev Food Sci Nutr 56(13):2209–2222

Bikker P, van Krimpen MM, van Wikselaar P, Houweling-Tan B, Scaccia N, van Hal JW, Huijgen WJ, Cone JW, López-Contreras AM (2016) Biorefinery of the green seaweed Ulva lactuca to produce animal feed, chemicals and biofuels. J Appl Phycol 28(6):3511–3525

Bixler HJ, Porse H (2011) A decade of change in the seaweed hydrocolloids industry. J Appl Phycol 23(3):321–335

Brennan L, Owende P (2010) Biofuels from microalgae—a review of technologies for production, processing, and extractions of biofuels and co-products. Renew Sust Energ Rev 14(2):557–577

Brown ES, Allsopp PJ, Magee PJ, Gill CI, Nitecki S, Strain CR, McSorley EM (2014) Seaweed and human health. Nutr Rev 72:205–216

Cabrita AR, Maia MR, Oliveira HM, Sousa-Pinto I, Almeida AA, Pinto E, Fonseca AJ (2016) Tracing seaweeds as mineral sources for farm-animals. J Appl Phycol 28(5):3135–3150

Campbell M, Ortuño J, Ford L, Davies DR, Koidis A, Walsh PJ, Theodoridou K (2020) The effect of ensiling on the nutritional composition and fermentation characteristics of Brown seaweeds as a ruminant feed ingredient. Animals 10(6):1019

Caro C, Pozo D (2015) Polysaccharide colloids as smart vehicles in cancer therapy. Curr Pharm Des 21(33):4822–4836

Chakraborty P, Kumar S, Dutta D, Gupta V (2009) Role of antioxidants in common health diseases. Res J Pharm Technol 2(2):238–244

Chen YC, Cheng CY, Liu CT, Sue YM, Chen TH, Hsu YH, Hwang PA, Chen H (2018) Alleviative effect of fucoxanthin-containing extract from brown seaweed Laminaria japonica on renal tubular cell apoptosis through upregulating Na+/H+ exchanger NHE1 in chronic kidney disease mice. J Ethnopharmacol 224:391–399

Dahiya S, Katakojwala R, Ramakrishna S, Mohan SV (2020) Biobased products and life cycle assessment in the context of circular economy and sustainability. Mater Circul Econ 2(1):1–28

Danesi EDG, Rangel-Yagui CDO, Carvalho JCM, Sato S (2002) An investigation of effect of replacing nitrate by urea in the growth and production of chlorophyll by *Spirulina platensis*. Biomass Bioenergy 23(4):261–269

de Amarante MCA, Braga ARC, Sala L, Kalil SJ (2020) Colour stability and antioxidant activity of C-phycocyanin-added ice creams after in vitro digestion. Food Res Int 137:109602
Deepika, C. (2016). Antidiabetic, anticoagulant, anticholinesterase inhibition, anti-tyrosinase and antiinflammatory activity of three selected seaweeds after eradication of phycocolloids from vedalai. Int J Pharma Biosci 0975-6299
Deepika C (2017a) Fucose-containing sulfated polysaccharides from *Sargassum wightii*: extraction technology and anticancer activity assessment. Int J Pharm Chem Biol Sci 7(3):272–277
Deepika C (2017b) Phytochemical characterization of *Sargassum wightii*, *Kappaphycus alvarezii* and *Gracilaria corticata* after phycocolloid extraction. Int J Pharm Chem Biol Sci 7(3):223–232
Deepika C, Bhaskar TCJ, Madhusudhanan J (2016) Antioxidant activities of a few common seaweeds from Gulf of Mannar and the effect of drying as the method of preservation. Indian J Eng Sci Technol 10(1):37
Déléris P, Nazih H, Bard JM (2016) Chapter 10 – Seaweeds in human health. In: *Seaweed in* health and disease prevention. Academic Press, Amsterdam, pp 319–367
Dewi OE, Andriyono S, Alamsjah MA (2020) Effect of commercial biofertilizer product of seaweed (*Euchema Cottonii*) towards the growth of Pakcoy mustard (Brassica Rapa L) and Dumbo catfish (*Clarias Gariepinus*) in Aquaponic system. J Marine Coast Sci 9(2):39–45
Du Preez R, Paul N, Mouatt P, Majzoub ME, Thomas T, Panchal SK, Brown L (2020) Carrageenans from the red seaweed *Sarconema filiforme* attenuate symptoms of diet-induced metabolic syndrome in rats. Mar Drugs 18(2):97
Energy, U.D.o. *MARINE ALGAE*. 2019
FAO (2012) FIGIS Global Production Statistics 1950–2012. from http://www.fao.org/fishery/statistics/en
Fernand F, Israel A, Skjermo J, Wichard T, Timmermans KR, Golberg A (2017) Offshore macroalgae biomass for bioenergy production: environmental aspects, technological achievements and challenges. Renew Sust Energ Rev 75:35–45
Fernandes TM, Gomes BB, Lanfer-Marquez UM (2007) Apparent absorption of chlorophyll from spinach in an assay with dogs. Innov Food Sci Emerg Technol 8(3):426–432
Fernandez-García C, Riosmena-Rodriguez R, Wysor B, Tejada OL, Cortes J (2011) Checklist of the Pacific marine macroalgae of Central America. Bot Mar 54(1):53–73
Fertah M, Belfkira A, Taourirte M, Brouillette F (2017) Extraction and characterization of sodium alginate from Moroccan *Laminaria digitata* brown seaweed. Arab J Chem 10:S3707–S3714
Fiedor J, Burda K (2014) Potential role of carotenoids as antioxidants in human health and disease. Nutrients 6(2):466–488
Frost A, Cullen M (1997) Marine debris on northern New South Wales beaches (Australia): sources and the role of beach usage. Mar Pollut Bull 34(5):348–352
Forecast MD (2020) Global Carrageenan Market - Market Data Forecast. Retrieved 04/11/2020, 2020, from https://www.marketdataforecast.com/market-reports/carrageenan-market
Ganesan AR, Tiwari U, Rajauria G (2019) Seaweed nutraceuticals and their therapeutic role in disease prevention. Food Sci Human Wellness 8(3):252–263
García-Poza S, Leandro A, Cotas C, Cotas J, Marques JC, Pereira L, Gonçalves AM (2020) The evolution road of seaweed aquaculture: cultivation technologies and the industry 4.0. Int J Environ Res Public Health 17(18):6528
Glazer AN (1988) [31] Phycobiliproteins. Methods in enzymology, vol 167. Elsevier, San Diego, pp 291–303
Glicksman M (2020) Food hydrocolloids. CRC Press, Boca Raton
Gomez L, Alvarez C, Zhao M, Tiwari U, Curtin J, Garcia-Vaquero M, Tiwari B (2020) Innovative processing strategies and technologies to obtain hydrocolloids from macroalgae for food applications. Carbohydr Polym 248:116784
Grossman AR, Bhaya D, Apt KE, Kehoe DM (1995) Light-harvesting complexes in oxygenic photosynthesis: diversity, control, and evolution. Annu Rev Genet 29(1):231–288
Guedes AC, Amaro HM, Malcata FX (2011) Microalgae as sources of carotenoids. Mar Drugs 9(4):625–644

Henríquez V, Escobar C, Galarza J, Gimpel J (2016) Carotenoids in microalgae. Carotenoids in nature. Springer, Cham, pp 219–237

Hessami MJ, Cheng SF, Ambati RR, Yin YH, Phang SM (2019) Bioethanol production from agarophyte red seaweed, Gelidium elegans, using a novel sample preparation method for analysing bioethanol content by gas chromatography. 3 Biotech 9(1):25

HexaResearch (2020) Global Seaweed Fertilizer Market Size And Forecast, by Form (Powder, Liquid), by Application (Farm, Garden), by Region, and Trend Analysis, 2015–2025. Retrieved 04/11/2020, 2020, from https://www.hexaresearch.com/research-report/seaweed-fertilizer-market

Hu IC (2019) Production of potential coproducts from microalgae. In Biofuels from Algae, Elsevier, pp. 345–358

Hughes A (2014) Scottish seaweed aquaculture; research, development and commercialisation. Proceedings of the seaweed for biofuel: towards a sustainable seaweed supply chain, Oban, UK, 21

Jacob-Lopes E, Maroneze MM, Deprá MC, Sartori RB, Dias RR, Zepka LQ (2019) Bioactive food compounds from microalgae: An innovative framework on industrial biorefineries. Current Opinion in Food Science, 25, 1–7

Ji S-Q, Wang B, Lu M, Li F-L (2016) Direct bioconversion of brown algae into ethanol by thermophilic bacterium *Defluviitalea phaphyphila*. Biotechnol Biofuels 9(1):81

Kannaujiya VK, Singh PR, Kumar D, Sinha RP (2020) Phycobiliproteins in microalgae: occurrence, distribution, and biosynthesis. Pigments from microalgae handbook. Springer, Cham, pp 43–68

Khan S, Siddique R, Sajjad W, Nabi G, Hayat KM, Duan P, Yao L (2017) Biodiesel production from algae to overcome the energy crisis. HAYATI J Biosci 24(4):163–167

Koca N, Karadeniz F, Burdurlu HS (2007) Effect of pH on chlorophyll degradation and colour loss in blanched green peas. Food Chem 100(2):609–615

Koizumi J, Takatani N, Kobayashi N, Mikami K, Miyashita K, Yamano Y, Wada A, Maoka T, Hosokawa M (2018) Carotenoid profiling of a red seaweed *Pyropia yezoensis*: insights into biosynthetic pathways in the order Bangiales. Mar Drugs 16(11):426

Krinsky NI, Landrum JT, Bone RA (2003) Biologic mechanisms of the protective role of lutein and zeaxanthin in the eye. Annu Rev Nutr 23(1):171–201

Kwak J-Y (2014) Fucoidan as a marine anticancer agent in preclinical development. Mar Drugs 12(2):851–870

Laurens LM, Nelson RS (2020) Sustainable technologies for seaweed conversion to biofuels and bioproducts. Sustainable seaweed technologies. Elsevier, pp 643–661

Leandro A, Pereira L, Gonçalves AM (2020) Diverse applications of marine macroalgae. Mar Drugs 18(1):17

Lee J, Seo E, Kweon D-H, Park K, Jin Y-S (2009) Fermentation of rice bran and defatted rice bran for butanol production using *Clostridium beijerinckii* NCIMB 8052. J Microbiol Biotechnol 19(5):482–490

Leong HY, Show PL, Lim MH, Ooi CW, Ling TC (2018) Natural red pigments from plants and their health benefits: a review. Food Rev Int 34(5):463–482

Li G, Chang Y, Chen L, Liu F, Ma S, Wang F, Zhang Y (2020) Process design and economic assessment of butanol production from lignocellulosic biomass via chemical looping gasification. Bioresour Technol 316:123906

Liu Z, Wang K, Chen Y, Tan T, Nielsen J (2020a) Third-generation biorefineries as the means to produce fuels and chemicals from CO2. Nat Catal 3(3):274–288

Liu J, Luthuli S, Yang Y, Cheng Y, Zhang Y, Wu M, Choi JI, Tong H (2020b) Therapeutic and nutraceutical potentials of a brown seaweed *Sargassum fusiforme*. Food Sci Nutr 8(10):5195–5205

Lourenço-Lopes C, Fraga-Corral M, Jimenez-Lopez C, Pereira AG, Garcia-Oliveira P, Carpena M, Prieto MA, Simal-Gandara J (2020) Metabolites from macroalgae and its applications in the cosmetic industry: a circular economy approach. Resources 9(9):101

MacArtain P, Gill CI, Brooks M, Campbell R, Rowland IR (2007) Nutritional value o edible seaweeds. Nutr Rev 65(12 Pt 1):535–543
Machado M, Machado S, Pimentel FB, Freitas V, Alves RC, Oliveira M (2020) Amino acid profile and protein quality assessment of macroalgae produced in an integrated multi-trophic aquaculture system. Foods 9(10):1382
Manirafasha E, Ndikubwimana T, Zeng X, Lu Y, Jing K (2016) Phycobiliprotein: potential microalgae derived pharmaceutical and biological reagent. Biochem Eng J 109:282–296
Manivasagan P, Bharathiraja S, Santha Moorthy M, Mondal S, Seo H, Dae Lee K, Oh J (2018) Marine natural pigments as potential sources for therapeutic applications. Crit Rev Biotechnol 38(5):745–761
Michalak I, Mahrose K (2020) Seaweeds, intact and processed, as a valuable component of poultry feeds. J Marine Sci Eng 8(8):620
Miyashita K, Beppu F, Hosokawa M, Liu X, Wang S (2020) Nutraceutical characteristics of the brown seaweed carotenoid fucoxanthin. Arch Biochem Biophys 686:108364
Mohammad JH, Ranga Rao A, Ravishankar GA (2019) Opportunities and challenges in seaweeds as feed stock for biofuel production. In: Ravishankar GA, Ranga Rao A (eds) Handbook of algal technologies and phytochemicals: volume-II: phycoremediation, biofuels and global biomass production, vol II. CRC Press, London, pp 39–50
Mohammed A, Rivers A, Stuckey DC, Ward K (2020) Alginate extraction from Sargassum seaweed in the Caribbean region: optimization using response surface methodology. Carbohydr Polym 245:116419
Morais T, Inácio A, Coutinho T, Ministro M, Cotas J, Pereira L, Bahcevandziev K (2020) Seaweed potential in the animal feed: a review. J Marine Sci Eng 8(8):559
Namvar F, Baharara J, Mahdi AA (2014) Antioxidant and anticancer activities of selected persian gulf algae. Indian J Clin Biochem 29:13–20
Newswire IG (2020) Hydrocolloids Market Worth $10.2 billion by 2027. Retrieved 04/11/2020, 2020, from https://www.globenewswire.com/news-release/2020/07/22/2065777/0/en/Hydrocolloids-Market-Worth-10-2-billion-by-2027-Exclusive-Report-Covering-Pre-and-Post-COVID-19-Market-Analysis-by-Meticulous-Research.html
Nguyen PKT, Das G, Kim J, Yoon HH (2020) Hydrogen production from macroalgae by simultaneous dark fermentation and microbial electrolysis cell. Bioresour Technol 315:123795
Ortega A, Geraldi NR, Alam I, Kamau AA, Acinas SG, Logares R, Gasol JM, Massana R, Krause-Jensen D, Duarte CM (2019) Important contribution of macroalgae to oceanic carbon sequestration. Nat Geosci 12(9):748–754
Outlook O-FA (2010) © OECD/FAO 2010
Pangestuti R, Siahaan EA (2018) Chapter 5 – Seaweed-derived carotenoids. In: Qin Y (ed) Bioactive seaweeds for food applications. Academic Press, London, pp 95–107
Patel NB, Tailor V, Rabadi M, Jain A, Kalasariya H (2020) Role of marine macroalgae in skin hydration and photoprotection benefits: a review. Int J Bot Stud 5(3):201–206
Pimentel FB, Alves RC, Rodrigues F, Oliveira MBPP (2018) Macroalgae-derived ingredients for cosmetic industry-an update. Cosmetics 5(1):2
Pulz O, Gross W (2004) Valuable products from biotechnology of microalgae. Appl Microbiol Biotechnol 65(6):635–648
Puspita M, Setyawidati NAR, Pangestuti R (2020) Hydrocolloid production of Indonesian seaweeds. Encyclop Marine Biotechnol 1:407–416
Ramachandra T, Hebbale D (2020) Bioethanol from macroalgae: prospects and challenges. Renew Sust Energ Rev 117:109479
Rammuni MN, Ariyadasa TU, Nimarshana PHV, Attalage RA (2019) Comparative assessment on the extraction of carotenoids from microalgal sources: Astaxanthin from *H. pluvialis* and β-carotene from *D. salina*. Food Chem 277:128–134
Richa RR, Kumari S, Singh K, Kannaujiya V, Singh G, Kesheri M, Sinha R (2011) Biotechnological potential of mycosporine-like amino acids and phycobiliproteins of cyanobacterial origin. Biotechnol Bioinformatics Bioeng 1:159–171

Sáez CA, Ronc F, Arati A, Moenne A, Moody AJ, Brown MT (2015) Copper-induced intra-specific oxidative damage and antioxidant responses in strains of the brown alga *Ectocarpus siliculosus* with different pollution histories. Aquat Toxicol 159:81–89

Sahayaraj K, Rajesh S, Asha A, Rathi J, Raja P (2014) Distribution and diversity assessment of the marine macroalgae at four southern districts of Tamil Nadu, India. Indian J Geo-Marine Sci 43(4):607–617

Sanchez-Machado DI, Lopez-Hernandez J, Paseiro-Losada P, Lopez-Cervantes J (2020) Fatty acids, total lipid, protein and ash contents of processed edible seaweeds. Food Chem 85:439–444

Schoefs BT (2003) Chlorophyll and carotenoid analysis in food products. A practical case-by-case view. TrAC Trends Anal Chem 22(6):335–339

Shama A, Joyce SG, Mari FD, Ranga Rao A, Ravishankar GA, Hudaa N (2019) Macroalgae and microalgae: novel sources of functional food and feed. In: Ravishankar GA, Rao AR (eds) Handbook of algal technologies and phytochemicals: volume-i: food, health and nutraceutical applications. CRC Press, Boca Raton, pp 207–219

Singh U, Ahluwalia A (2013) Microalgae: a promising tool for carbon sequestration. Mitigation and adaptation strategies for global change, vol 18(1). Springer, p 73–95

Soares C, Švarc-Gajić J, Oliva-Teles MT, Pinto E, Nastić N, Savić S, Almeida A, Delerue-Matos C (2020) Mineral composition of subcritical water extracts of Saccorhiza Polyschides, a brown seaweed used as fertilizer in the north of Portugal. J Marine Sci Eng 8(4):244

Spalding HL, Amado-Filho GM, Bahia RG, Ballantine DL, Fredericq S, Leichter JJ, Nelson WA, Slattery M, Tsuda RT (2019) Macroalgae. Mesophotic coral ecosystems. Springer, Cham, pp 507–536

Steffen W, Richardson K, Rockström J, Cornell SE, Fetzer I, Bennett EM, Biggs R, Carpenter SR, De Vries W, De Wit CA (2015) Planetary boundaries: guiding human development on a changing planet. Science 347(6223)

Suutari M, Leskinen E, Fagerstedt K, Kuparinen J, Kuuppo P, Blomster J (2015) Macroalgae in biofuel production. Phycol Res 63(1):1–18

Therkelsen GH (1993) Carrageenan. Elsevier, Industrial gums, pp 145–180

Torres MD, Flórez-Fernández N, Domínguez H (2019) Integral utilization of red seaweed for bioactive production. Mar Drugs 17(6):314

Uthirapandi V, Suriya S, Boomibalagan P, Eswaran S, Ramya SS, Vijayanand N, Kathiresan D (2018) Bio-fertilizer potential of seaweed liquid extracts of marine macro algae on growth and biochemical parameters of *Ocimum sanctum*. J Pharm Phytochem 7:3528–3532

Vega J, Bonomi-Barufi J, Gómez-Pinchetti JL, Figueroa FL (2020) Cyanobacteria and red macroalgae as potential sources of antioxidants and UV radiation-absorbing compounds for cosmeceutical applications. Mar Drugs 18(12):659

Whiteman G, Walker B, Perego P (2013) Planetary boundaries: ecological foundations for corporate sustainability. J Manag Stud 50(2):307–336

Yan X, Chuda Y, Suzuki M, Nagata T (1999) Fucoxanthin as the major antioxidant in Hijikia fusiformis, a common edible seaweed. Biosci Biotechnol Biochem 63(3):605–607

Yeremenko N (2004) Functional flexibility of photosystem-I in cyanobacteria. Universiteit van Amsterdam – IBED, Amsterdam

Yermak IM, Davydova VN, Kravchenko AO, Chistyulin DA, Pimenova EA, Glazunov VP (2020) Mucoadhesive properties of sulphated polysaccharides carrageenans from red seaweed families Gigartinaceae and Tichocarpaceae. Int J Biol Macromol 142:634–642

Young A, Britton G (2012) Carotenoids in photosynthesis. Springer Science & Business Media, Dordrecht, pp 1–489

Zhong J, Hu X, Yüksel S, Dincer H, Ubay GG (2020) Analyzing the investments strategies for renewable energies based on multi-criteria decision model. IEEE Access 8:118818–118840

Chapter 3
Palmaria Species: From Ecology and Cultivation to Its Use in Food and Health Benefits

Justine Dumay, Bruno Cognie, Joël Fleurence, Michèle Morançais, Vincent Turpin, Marta Castilla Gavilan, Yoran Le Strat, and Priscilla Decottignies

3.1 Introduction

Marine macroalgae have a long history of use as foods and colorants all around the world. Brown and red seaweeds are essentially harvested and cultivated for use in the food, agricultural or medicine sectors (Dawczynski et al. 2007; Shama et al. 2019). *Palmaria palmata* is one the most exploited seaweed. *Palmaria palmata* belongs to the phylum of Rhodophyta (commonly determined as Red seaweed), to the Class of Florideophyceae and the Palmariales Order. According to the World Register of Marine Species (WoRMS), *P. palmata* has been firstly described by Weber and Mohr in 1805. Although *P. palmata* has been traditionally harvested and consumed in north Europe and America (Mouritsen et al. 2013), its demand has increased over the last decades due to its successful use in the health and food industry (Fleurence 1999) as well as a feed in shellfish aquaculture (Castilla-Gavilán et al. 2019; García-Bueno et al. 2014). The seasonal availability and variation in its chemical composition (Fleurence 1999; Lüning 1993; Rødde et al. 2004) and the increasing pressure on its wild populations have led to several researches developing different cultivation techniques for this species (Edwards and Dring 2011; Le Gall et al. 2004; Pang and Lüning 2006).

While interest in *P. palmata* compounds is constantly evolving, their accessibility remains a problem and it is necessary to find new techniques for extracting these compounds. A detailed knowledge of the composition of the cell wall has made

J. Dumay (✉) · B. Cognie · J. Fleurence · M. Morançais · V. Turpin · M. C. Gavilan · Y. Le Strat · P. Decottignies
Nantes Université, Institut des Substances et Organismes de la Mer, ISOMER, Nantes, France
e-mail: Justine.dumay@univ-nantes.fr

A. Ranga Rao, G. A. Ravishankar (eds.), *Sustainable Global Resources of Seaweeds Volume 1*, https://doi.org/10.1007/978-3-030-91955-9_3

possible to orient the extraction processes towards specific techniques (enzymes, ultrasound, etc.) to increase the extraction yields of the target molecules.

3.2 Habitat, Ecological and Environmental Consideration

Palmaria palmata is a predominant species of rocky shores in North Atlantic Ocean. Growing in cold to temperate waters, *P. palmata* can be found from Arctic to New-Jersey in the USA and to Portugal and Spain in Europe. Records of *P. palmata* in North Pacific Ocean were recently validated on genetic basis; these are clearly distinguished from the genus *Devaleraea*, as described earlier considering anatomy and morphology (Kumagai et al. 2019; Skriptsova and Kalita 2020; van der Meer 1981; van der Meer and Bird 1985).

Palmaria palmata has a foliose thallus presenting a dichotomous or palmately shape and a skin or leather like texture (Braune and Guiry 2011). The fronds are about 25–50 cm in length (up to 1 m) and commonly 5 cm in width (Faes and Viejo 2003; Irvine and Guiry 2011; Jorde 1966). This species develops a very short stipe (<5 mm in length) which is attached to the substrates by a small disk-like holdfast (Fig. 3.1).

This general morphology presents a high degree of plasticity that may result in misidentification (Irvine and Guiry 2011). However, the second major morphotype with narrower fronds and commonly found in North-East Atlantic Ocean should be considered as an ecotype inhabiting sheltered and silty environments (Kraan and Guiry 2006; van der Meer 1987).

Fig. 3.1 *Palmaria plamata* picture in its French coastal habitat (Batz Sur Mer, France)

Palmaria palmata, as red algae, is adapted to low light habitats. However, the tolerance of this species to a large range of illumination levels leads to variation in the color of fronds, from red purple to pinky-red (Bird and van der Meer 1993; Mouritsen et al. 2013). In less illuminated habitats, fronds are thinner and darker.

Populations are found in lower intertidal and shallow subtidal zones, rarely up to a depth of 20 meters (Faes and Viejo 2003; Irvine and Guiry 2011; Jorde 1966; South et al. 1988). *Palmaria palmata* does well in flowing waters consistent with a main distribution in sheltered to moderately exposed areas.

Palmaria palmata is epilithic on rocks or shells (Bird and van der Meer 1993; Kraan and Guiry 2006; South et al. 1988) or epilithic on stipes of large brown seaweeds such as *Alaria esculenta*, *Cystoseira baccata*, *Devaleraea ramentacea*, *Fucus serratus*, *Laminaria digitata*, *L. hyberborea* or *Mastocarpus stellatus* (Faes and Viejo 2003; Garbary et al. 2012; Jorde 1966).

3.3 Cultivation

In the present section, methods for cultivation of sporelings or fronds in tanks or at sea and uses for bioremediation will be summarized.

3.3.1 Traditional Cultivation Techniques

3.3.1.1 Life Cycle Considerations

Palmaria palmata has a particular life cycle that can make it difficult to grow in facilities. It presents two main characteristics: (i) an asexual tetrasporophyte phase and a sexual gametophyte phase, and (ii) a strong sexual dimorphism: female gametophytes are microscopic encrusted thalli (3–4 cells of 100 μm diameter), while male gametophytes and tetrasporophytes are indistinguishable and can reach 30–50 cm (van der Meer and Todd 1980) (Fig. 3.2).

3.3.1.2 Spore Isolation, Germination and Plantlet Cultivation in Tanks

Tetrasporophytes are usually collected at low tide from wild populations. Cultivated fronds can also be used and manipulate to induce year-round tetrasporangia by changing photoperiod to short days and reducing temperature (Grote 2019). Wild or not, cleaned tetrasporophytes fragments should be processed at low temperature (1–10 °C) and irradiance (5–40 μmol m^{-2} s^{-1}) for spore releasing. Le Gall et al. (2004) desiccated thalli overnight in the darkness at 4 °C, while other authors subjected thalli to a photoperiod with 12–16 h of darkness at 10 °C (Edwards and Dring 2011). Pang and Lüning (2006) tested the same photoperiod and a lower

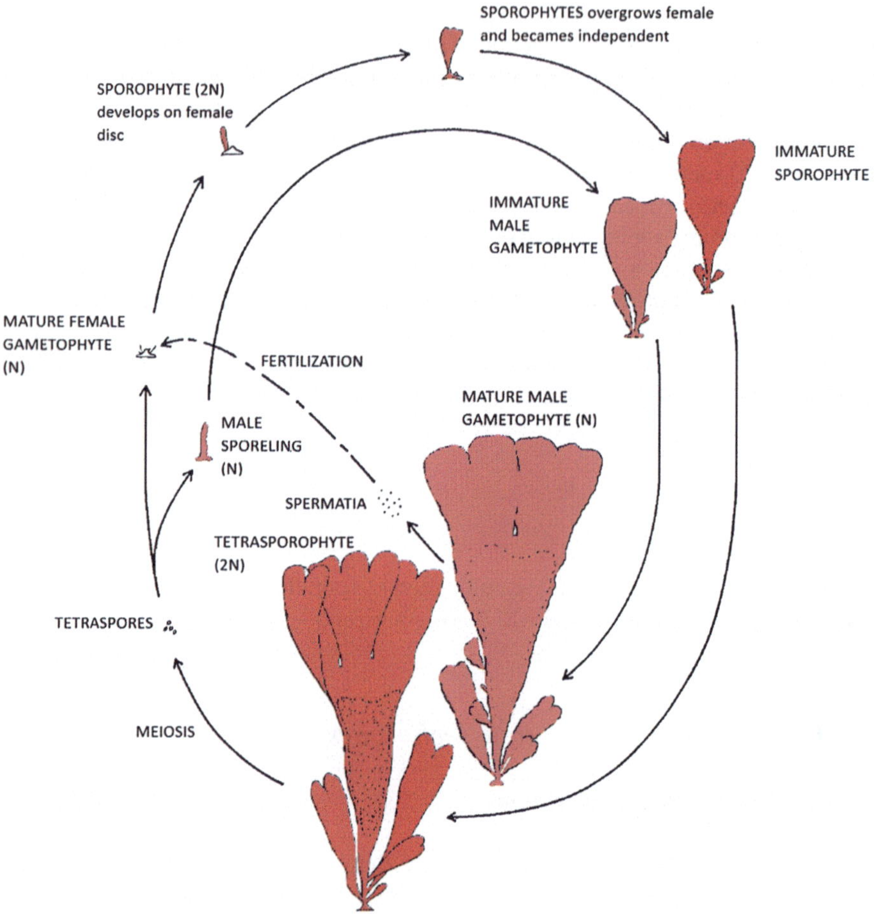

Fig. 3.2 The life cycle of *Palmaria palmata* . (Modified from van der Meer and Todd 1980)

temperature (1 °C) for spore release and reported that 80% of the spores were released during the dark periods. Thalli in UV-filtered aerated seawater can be maintained floating in suspension or placed above a substrate provided for the spore settlement: strings attached to collectors (Werner and Dring 2011) or plates (Edwards and Dring 2011; Le Gall et al. 2004).

For a successfully growth of germlings, 2–3 days after settlement, water at 10 °C should be aerated (improving CO_2 and nutrient uptake) and the volume can be progressively increased. Light is placed at 0.5 m above the water surface with a photoperiod of 12:12 h light/dark and an irradiance of 50 μmol m^{-2} s^{-1}. Every 7–15 days, depending on the volume, a water renewal and a throughout cleaning of the substrate (when a substrate is used) have to be carried out. The water needs to be enriched (nutrient media and vitamins) to ensure high survival and growth rates (Grote 2019; Le Gall et al. 2004; Pang and Lüning 2006). Germlings are then maintained at 10 °C with a minimum photoperiod of 12 h of light per day and a maximum irradiance of 60 μmol m^{-2} s^{-1} (Edwards and Dring 2011; Le Gall et al. 2004; Pang and Lüning 2006; Werner and Dring 2011).

When germlings are strong, irradiance can be increased to >200 μmol m^{-2} s^{-1} at a photoperiod of 16:8 h light/dark. As highlighted by Pang and Lüning (2006), in tumble culture, stocking density, circulation time and tank depth are important parameters that should be taken into account. They observed high growth rates with a stocking density of >4 Kg FW m^{-2} when high surface irradiances and short circulation times were supplied. At this stocking density, it can be accepted that *P. palmata* can double its biomass in 4 weeks (Grote 2019).

P. palmata can also be cultivated by vegetative growth of fragments of harvested fronds from wild populations at a moderate temperature (10 °C) and light intensity (Morgan et al. 1980a; Morgan and Simpson 1981a, 1981b). Fragments of thallus or meristematic tissues removed from the apical side of the fronds should be put in tumble culture in UV-filtered enriched seawater under aeration and an irradiance of 120 μmol m^{-2} s^{-1} (Sanderson 2015).

3.3.1.3 On-Growth at Sea

In open sea, system design and site selection play an important role (Grote 2019). Culture of *P. palmata* in open sea is carried out in seeding strings/nets or attaching sporelings and fragments. Edwards and Dring (2011) explain that before transfer to sea, a minimal size (2 mm length) of fronds must be reached, with a better survival rate for sporelings of 2–5 mm.

The site selection for successful cultivation and optimal growth of *P. palmata* is essential. For nutrient and CO_2 exchange, a location with a water current of 5–10 cm s^{-1} is necessary (Werner and Dring 2011) and inhibits epiphytic growth on fronds. Wave exposure should be moderate to avoid loss of sporelings and thalli. Therefore, longlines should be placed at about 1–10 m depth, to avoid at low tide wave action and high irradiances damaging the fronds.

Water temperature should be between 6 and 17 °C and culture sites should have low turbidity to avoid coverage of seaweed fronds which would reduce photosynthesis, nutrient uptake and growth (Grote 2019).

Furthermore, the control of early life stages, large-scale seedling production, cultivation site selection criteria and improvement of deployment and harvesting technologies for commercial-scale culture are important research topics to provide a basis for a growing European industry in order to achieve a predictable biomass production and an economically stable business (Grote 2019).

3.4 Bioremediation

Palmaria palmata has been demonstrated to have a potential for bioremediation of animal aquaculture effluents (Corey 2012; Manríquez-Hernández et al. 2016) on recirculating aquaculture systems or integrated multi-trophic aquaculture systems (Matos et al. 2006; Sanderson 2006). It has been observed that the cultivation of *P. palmata* as a biofilter result in an increased protein content of the seaweed due to the higher contents of NH_4^+ in the seawater from the animals ejections (Grote 2016). As an example, one hectare of *P. palmata* culture at sea can absorb 30% of the N from 500 t of salmon (Sanderson 2006). Therefore, its use for bioremediation could be a tool for the sustainable development of the aquaculture and for reducing production costs of this seaweed.

3.5 Biochemical Composition

Seaweeds are known to be rich sources in essential nutrients, such as minerals, vitamins, amino acids and proteins or poly-unsaturated fatty acids as example. Their specific polysaccharides, the fibers could also present great interests in food and feed targets. Biochemical composition of *P. palmata* interest scientist for more than 70 years (Morgan et al. 1980b) where values reported were quite similar than those obtained nowadays. Main variations have been reported to be due to seasonal and nutritional conditions. This part of the chapter is devoted to draw an overview of the composition of *P. palmata* commonly found in literature.

3.5.1 Proteins and Amino Acids

Proteins could be reported as varying from 8% to 35% of the dry biomass (Beacham et al. 2019; Rioux et al. 2017; Sánchez-Machado et al. 2004). But it seems very important to discuss about those results. Indeed, total proteins are mainly calculated by multiplying the total nitrogen by a N-Prot conversion factor. Historically, this N prot factor was evaluated as 6.25 and came from studies on dairy products (Jones 1931). Recently, several studies have demonstrated the inaccuracy of using this factor for seaweeds. If Lourenço et al. have highlighted the variation among species and phylums (4.59 for red seaweeds) (Lourenço et al. 2002), studies of Jard et al. and Bjarnadottir et al. determined respectively specific N prot factor for *P. palmata* of 4.92 and 4,7 (Bjarnadóttir et al. 2018; Jard et al. 2013). Seasons, climate, environmental conditions could have a huge influence on the protein content and amino acid composition. Less is often more, it becomes relevant to use a standard nitrogen conversion factor for all seaweeds. More and more, studies used the new seaweed universal N prot conversion factor of 5, as recently identified (Angell et al. 2016).

For a few moments, it could be useful to use both 6.25 and 5 conversion factors in order to compare recent results with older.

Amino acids of *P. palmata* composition is also widely described in many studies and they represent 19.2% of the dry biomass (Bjarnadóttir et al. 2018). To estimate the nutritional potential of *P. palmata*, essential amino acids represent on average 40% of total amino acids (from 36% to 45% according to studies) (Dumay and Morançais 2016) and Asp-Asn and Glu-Gln amino acids, responsible of the specific fifth taste “Umami” are the most important amino acids found in *P. palmata* with respectively average of 10.33% and 13% (Bjarnadóttir et al. 2018; Galland-Irmouli et al. 1999; Mišurcová et al. 2014).

3.5.2 *Pigments*

Palmaria palmata contains different pigments: chlorophyll a, α and β carotenes, carotenoids (lutein, zeaxanthin, violaxanthin and fucoxanthin) and phycoerythrin. The content of chlorophyll a ranges from 0.213 to 8.56 mg/g, α and β carotenes from 0.01 to 0.3 mg/g, carotenoids from 0.085 to 1.1 mg/g and phycoerythrin from 0.15 to 10.88 mg/g of dry biomass (Beacham et al. 2019; Dumay and Morançais 2016; Gallagher et al. 2020; Lalegerie et al. 2020; Schmid et al. 2017). The difference in contents are due to seasons, environmental conditions and also to the methods used for extraction and quantification. Chlorophylls and carotenoids are extracted with methanol, DMSO, acetone and quantified by spectrophotometry method with equations or by HPLC method. Phycoerythrin is extracted with phosphate buffer pH 6.8 and quantified either by the Beer and Eshel’s equations or Sampath-Wiley’s equations (Beer and Eshel 1985; Sampath-Wiley and Neefus 2007).

3.5.3 *Carbohydrates and Fibers*

Carbohydrates content could be ranged from 38% to 74% of the dry biomass (Beacham et al. 2019; García-Bueno et al. 2016; Lopes et al. 2020; Morgan et al. 1980b; Razi Parjikolaei et al. 2013). Variations are mainly linked to seasons, nutrients availability, and environmental conditions, as reported previously on proteins. Total fibers of *P. palmata* are reported to be around 5% of the dry biomass, where 60% of the total fibers are soluble (Cherry et al. 2019; MacArtain et al. 2007). Sugars constituting those carbohydrates are mainly xylose and galactose, main constituents of xylanes and galactanes (Bikker et al. 2020; Deniaud et al. 2003b). Regarding storage polysaccharides, fluoridoside and floridean starch have been found, up to 25% of the dry biomass (Bikker et al. 2020; MacArtain et al. 2007; Morgan et al. 1980b).

Table 3.1 Fatty Acids repartition into *Palmaria palmata*

	Saturated fatty acids	Mono-unsaturated fatty acids	Poly-unsaturated fatty acids	ω3 fatty acid	ω6 fatty acids
Lowest value	43.80%	4.90%	20.40%	25.52%	2.14%
Highest value	63.50%	16.10%	52.80%	51.90%	7.30%

Values are given in % of total Fatty Acids. Values adapted according to studies of Schmid et al. (2017), Lopes et al. (2020), Foseid et al. (2020), Bikker et al. (2020)

3.5.4 *Lipids and Vitamins*

As previously seen, lipid content and lipid classes distribution in *P. palmata* are varying according to seasons and environmental conditions (Schmid et al. 2017). Total content of lipid could be ranged from 0.2% to 3.8% of the dry biomass (Morgan et al. 1980b; Rioux et al. 2017) but most of the studies reported approximately 2% of the dry biomass (Beacham et al. 2019; Foseid et al. 2020; Sánchez-Machado et al. 2004). The global composition of lipids is reported in the Table 3.1.

Palmaria palmata is also reported to be a good source of demosterol and SQDG, well known for their interest in health (Morgan et al. 1980b; Schmid et al. 2017). *Palmaria palmata* is a good source of vitamins such as B2 (4.27–19.1 ppm), B3 (10–83 ppm), C (1.5–1080 ppm), E (22–162 ppm) and A (36.5 ppm) (Kraan 2013).

3.5.5 *Ash and Minerals*

Ash represent from 12 to 37% of the dry weight (Morgan et al. 1980b) and *P. palmata* could be considered as a good source of iron, magnesium, calcium and iodine (in comparison with terrestrial edible vegetables and fruits) (Morgan et al. 1980b). Jard et al. reported 26.2% dry weight of mineral content. Main minerals found are K (7.8–11.69%), Cl (10.4%), Mg (0.2–9.7%) (Jard et al. 2013; Kraan 2013; MacArtain et al. 2007).

3.6 Cell Wall Complexity: Hydrolysis and Processing

3.6.1 Palmaria Palmata *Cell Wall Composition*

As one of the most valuable seaweed gender, composition of the cell wall of *P. Palmata species* has been widely studied. It has been established that cell walls were composed of complex system of polysaccharides, proteins and polyphenols

(Deniaud et al. 2003a). Matricial polysaccharides were mainly composed of β (1→3) xylans, mix-linked β (1→3) β (1→4) xylans, galactans and cellulose (Lahaye et al. 2003). The presence of acidic groups, such as phosphates and sulfates ones, induces ionic interactions between polysaccharides and proteins (Matsuhiro and Urzúa 1996). Hydrogen bonds are also highly responsible for the strength of the building. Hence, studies reported "an acidic glycoprotein complex composed of xylose, galactose, proteins, sulfates and uronic acids" (Deniaud et al. 2003a). This acidity found in the xylan remains from the presence of linkages of xylans to sulfated and/or phosphorylated xylogactoproteins complexes (Deniaud et al. 2003b).

3.6.2 Cell Wall Disruption: A Way for Nutrients and Bioactive Accessibility

The knowledge and the understanding of the composition of the cell wall of *P. Palmata species* are a great improvement to induce its disruption in order to enhance the access to nutrients and bioactives. The use of enzymes is one of the process that could be implemented. Obviously, according to the cell wall composition, many studies report the use of xylanase to hydrolyse the cell wall (Deniaud et al. 2003b; Fleurence et al. 1995; Lahaye and Vigouroux 1992). More recently, researches were devoted to improve this disruption by the use of mathematical optimization (Dumay et al. 2013). Other enzymes could also be used, such as commercial mixes (Naseri et al. 2020; Wang et al. 2010). Most of the enzymes used in those studies came from microorganisms, especially fungi. Marrion et al. studied directly the effect of fermentation on the accessibility to *P. palmata*'s proteins (Marrion et al. 2003). Following this tendency, interest is nowadays growing to study ecological interactions between holobiont (fungi and or bacteria communities) and seaweeds (Burgunder-Delamare et al. 2020, Tourneroche et al. 2019) or the use of new physical processes.

3.7 Uses in Food and Health

3.7.1 Traditional and Present Day Uses

The red seaweed *P. palmata*, well known under the name of Dulse, is used as sea vegetable or food ingredient. This species was consumed by the Norwich populations since the Viking period (Delaney et al. 2016). According to Dr Prannic Rhatigan, Dulse has been used in the diet of local irish population for over

5000 years (Mouritsen et al. 2013). In this country Dulse is also called "Dillisk" or "Duileach" (Gaelic name). In France, Dulse is also named "Tellesk" in Brittany region (Mouritsen et al. 2013). The use of Dulse in popular cuisine persists in several European regions or countries (Brittany, Wales, Scotland, Iceland, Norway, Denmark, Ireland) (Mouritsen et al. 2018). In Northern Ireland, it was mainly added for tasting several cooking preparations such as mashed potatoes, soups or fish stews. In North Brittany (France), Dulse is also used as ingredient for the preparation of "seaweed bread". Others algae such as sea lettuce and laver are also integrated in the production of this bread. It is mainly available during the summer season and it is sold under the name of "baguette aux algues" by the French bakers (Fleurence 2016). In Wales, a similar bread prepared from another red seaweed (laver) is also proposed to the consumers. Dulse and Kombu can also be included as ingredient for the preparation of butter in Brittany. This product called in French "beurre aux algues" is mainly used during the fish cooking or spread on bread when tasting seafood. More anecdotal, the use of the seaweed *Rhodymenia palmata (*misnaming of *Palmaria palmata*) for the preparation of a drink is cited in 1868 by the French author Jules Verne in his book 20,000 leagues under the seas. The use of Dulse in a liquor preparation has been also reported. A seaweed-flavored aquavitae with unami and floral notes is obtained with the integration of *P. palmata* in the final preparation (Kraan 2016). So traditional applications are often close to current ones.

On the other hand, new uses have been also developed. For example, fresh Dulse, especially when it is very young, can be eat as a salad. Dulse can be also consumed as sea vegetable after the use of soft process such as ripening. In this process, the blades are tenderized by the action of the endogen enzymes. The product obtained is softer and more flavorful than the raw material.

The type of storage (dry or semi-dry storage) can also modify the flavor and the texture of Dulse (Stévant et al. 2020). The compounds affected by these storage processes are free amino acids, volatile molecules and main macronutriments (proteins, polysaccharides, lipids, minerals).

An alternative way to use Dulse as food has found into the concept of New Nordic Cuisine (Mouritsen et al. 2013). In this new application, aqueous extracts of Dulse are used to prepare the broth at the base of Japanese cuisine well known as Dashi. This last type of application preserves in particular the subtle floral aroma (e.g. violet) of the Dulse.

In France, the use of Dulse as human food is submitted to a specific regulation. It is also the case of other algal species frequently consumed (Fleurence 2016; Mabeau and Fleurence 1993). This regulation includes a positive list giving the Latin and trading names of seaweed authorized for human consumption (Table 3.2). Toxicological and bacterial criteria are also associated with this regulation (Tables 3.3 and 3.4.)

Table 3.2 Red seaweed authorized in France for an use as sea vegetables or ingredients (Full regulation also includes seaweed belonging to brown and green algae)

Red algae (Rhodophyta)	Trading name
Palmaria palmata	Dulse
Porphyra umbilicalis, P. tenera, P. yezoensis, P. dioica, P. purpurea, P. laciniata, P. leucosticta	Nori
Gracilaria verrucosa	Ogo-nori
Lithothamnium calcareum	Maërl

Table 3.3 Maximum level of toxic minerals for edible algae in France

Toxic mineral	Level (in mg/Kg of dry matter)
Inorganic arsenic	≤ 3
Lead	≤ 5
Cadmium	≤ 0.5
Tin	≤ 5
Mercury	≤ 0.1
Iodine	≤ 500

Table 3.4 Bacteriology criteria applied for edible algae in France (only on the dried seaweed)

Bacteria	cfu/g
Aerobes	$\leq 100,000$
Anaerobes	≤ 100
Fecal coliforms	≤ 10
Clostridium perfringens	≤ 1

cfu colony forming units

3.7.2 Health Aspects

Dulse has been mainly studied for its interest in human nutrition in regard to its protein content. However, the effects of the consumption of *P. palmata* on the health has also been looked. The impact of Dulse consuming on lipid metabolism and glycemia in 104 participants with hypercholesterolemia has been recently studied (Takase et al. 2020). This study concludes that taking *P. palmata* in the diet had not significant effect on serum LDL-Cholesterol or glycemic control. However, the administration of Dulse in the women diet significantly decreases the triglyceride level (−10%). On the other hand, the presence in *P. palmata* of peptides showing an inhibitory action of renin angiotensin system (blood high pressure activator system) has been described (Fitzgerald et al. 2012). The main active peptide (FR-25) inhibits 59% of renin activity at the concentration of 1 mg/ml. Experiences to include *P. palmata* protein hydrolysate in the bread has been also performed in regard to develop a functional food (Fitzgerald et al. 2014).

Except these aspects, Dulse extracts have been also studied for their antioxidant activity (Yuan et al. 2005b) . However, these extracts showed a weak reducing activity to compare the vitamin C (100-fold weaker). This result could be explained by the loss of ascorbic acid activity during the drying and the storage of the seaweed. Indeed, this observation is contradictory with the high vitamin C contents described in fresh seaweed (200–500 mg/kg of dry matter) (Morgan et al. 1980b). Despite this, the extracts studied prove to be very effective in trapping superoxide ions, thus revealing a significant antioxidant power (Yuan et al. 2005a). Although the nutritional value of *P. palmata* is undeniable, its impact on human health is still the subject of numerous studies, as is the case for most edible species.

3.8 Conclusion and Future Perspectives

This chapter maps out the multiple interests of P. *palmata*, whether in terms of ecological life, cultivation techniques, bioremediation, biochemical composition, nutritional interest and concerning the health sector. It appears that this species is widely found on the world northcoasts. In addition, to withstand supply pressures, this species benefits from ease of cultivation, whether in vegetative or reproductive multiplication. The fact that it is one of the four species of red algae authorized in Europe has made it possible to deepen the knowledge of this species, concerning its ecophysiology and biochemical composition, as well as the ways of use in food and health. Protein content, biological activity, nutritional aspect in food and feed, organoleptic consideration, bioremediation… are the virtues of *P. palmata*. One of the most important objectives for research in this field is now the accreditation of other species of red algae with qualities similar to *P. palmata*, such as for example the red alga *Grateloupia turuturu Yamada*, which has a similar biochemical composition. To this day still considered unauthorized for human consumption in Europe, it possesses a wide range of utilities for the future.

References

Angell AR, Mata L, de Nys R, Paul NA (2016) The protein content of seaweeds: a universal nitrogen-to-protein conversion factor of five. J Appl Phycol 28:511–524. https://doi.org/10.1007/s10811-015-0650-1

Beacham TA, Cole IS, DeDross LS, Raikova S, Chuck CJ, Macdonald J, Herrera L, Ali T, Airs RL, Landels A, Allen MJ (2019) Analysis of seaweeds from South West England as a biorefinery feedstock. Appl Sci 9:4456. https://doi.org/10.3390/app9204456

Beer S, Eshel A (1985) Determining phycoerythrin and phycocyanin concentrations in aqueous crude extracts of red algae. Aust J Mar Freshw Res 36:785–793

Bikker P, Stokvis L, van Krimpen MM, van Wikselaar PG, Cone JW (2020) Evaluation of seaweeds from marine waters in northwestern Europe for application in animal nutrition. Anim Feed Sci Technol 263:114460. https://doi.org/10.1016/j.anifeedsci.2020.114460

Bird CJ, van der Meer JP (1993) Systematics of economically important marine algae: a Canadian perspective. Can J Bot 71:361–369. https://doi.org/10.1139/b93-040

Bjarnadóttir M, Aðalbjörnsson BV, Nilsson A, Slizyte R, Roleda MY, Hreggviðsson GÓ, Friðjónsson ÓH, Jónsdóttir R (2018) *Palmaria palmata* as an alternative protein source: enzymatic protein extraction, amino acid composition, and nitrogen-to-protein conversion factor. J Appl Phycol 30:2061–2070. https://doi.org/10.1007/s10811-017-1351-8

Braune W, Guiry MD (2011) Seaweeds: a colour guide to common benthic green, brown and red algae of the world's oceans. Koeltz Botanical Books, Königstein

Burgunter-Delamare B, KleinJan H, Frioux C, Fremy E, Wagner M, Corre E, Le Salver A, Leroux C, Leblanc C, Boyen C, Siegel A, Dittami SM (2020) Metabolic complementarity between a brown alga and associated cultivable bacteria provide indications of beneficial interactions. Front Mar Sci 7:85. https://doi.org/10.3389/fmars.2020.00085

Castilla-Gavilán M, Cognie B, Ragueneau E, Turpin V, Decottignies P (2019) Evaluation of dried macrophytes as an alternative diet for the rearing of the sea urchin *Paracentrotus lividus* (Lamarck, 1816). Aquac Res 50:1762–1769. https://doi.org/10.1111/are.14045

Cherry P, O'Hara C, Magee PJ, McSorley EM, Allsopp PJ (2019) Risks and benefits of consuming edible seaweeds. Nutr Rev 77:307–329. https://doi.org/10.1093/nutrit/nuy066

Corey PE (2012) Nutrient removal by *Palmaria palmata* and *Chondrus crispus* in bioremediation of aquaculture effluent. Dalhousie University, Halifax

Dawczynski C, Schubert R, Jahreis G (2007) Amino acids, fatty acids, and dietary fibre in edible seaweed products. Food Chem 103:891–899

Delaney A, Frangoudes K, Ii S-A (2016) Chapter 2 – Society and seaweed: understanding the past and present. In: Seaweed in health and disease prevention. Academic, San Diego, pp 7–40

Deniaud E, Fleurence J, Lahaye M (2003a) Interaction of the max-linked b-(1,3) / b-(1,4)-D-xylans in the cell walls of *Palmaria plamata* (Rhodophyta). J Phycol 39:74–82

Deniaud E, Quemener B, Fleurence J, Lahaye M (2003b) Structural studies of the mix-linked β-(1→3)/β-(1→4)-d-xylans from the cell wall of *Palmaria palmata* (Rhodophyta). Int J Biol Macromol 33:9–18. https://doi.org/10.1016/s0141-8130(03)00058-8

Dumay J, Clément N, Morançais M, Fleurence J (2013) Optimization of hydrolysis conditions of *Palmaria palmata* to enhance R-phycoerythrin extraction. Bioresour Technol 131:21–27. https://doi.org/10.1016/j.biortech.2012.12.146

Dumay J, Morançais M (2016) Chapter 9 – Proteins and pigments. In: Seaweed in health and disease prevention. Academic, San Diego, pp 275–318

Edwards MD, Dring MJ (2011) Open-sea cultivation trial of the red alga, *Palmaria palmata* from seeded tetraspores in Strangford Lough, Northern Ireland. Aquaculture 317:203–209

Faes VA, Viejo RM (2003) Structure and dynamics of a population of *Palmaria palmata* (Rhodophyta) in Northern Spain 1. J Phycol 39:1038–1049

Fitzgerald C, Gallagher E, Doran L, Auty M, Prieto J, Hayes M (2014) Increasing the health benefits of bread: assessment of the physical and sensory qualities of bread formulated using a renin inhibitory *Palmaria palmata* protein hydrolysate. LWT – Food Sci Technol 56:398–405. https://doi.org/10.1016/j.lwt.2013.11.031

Fitzgerald C, Mora-Soler L, Gallagher E, O'Connor P, Prieto J, Soler-Vila A, Hayes M (2012) Isolation and characterization of bioactive pro-peptides with in vitro renin inhibitory activities from the macroalga *Palmaria palmata*. J Agric Food Chem 60:7421–7427. https://doi.org/10.1021/jf301361c

Fleurence J (1999) Seaweed proteins: biochemical, nutritional aspects and potential uses. Trends Food Sci Technol 10:25–28

Fleurence J (2016) Chapter 5 – Seaweeds as food. In: Seaweed in health and disease prevention. Academic, San Diego, pp 149–167

Fleurence J, Massiani L, Guyader O, Mabeau S (1995) Use of enzymatic cell wall degradation for improvement of protein extraction from *Chondrus crispus*, *Gracilaria verrucosa* and *Palmaria palmata*. J Appl Phycol 7:393–397

Foseid L, Natvik I, Devle H, Ekeberg D (2020) Identification of fatty acids in fractionated lipid extracts from *Palmaria palmata*, *Alaria esculenta* and *Saccharina latissima* by off-line SPE GC-MS. J Appl Phycol 32:4251–4262. https://doi.org/10.1007/s10811-020-02193-2

Gallagher JA, Adams JMM, Turner LB, Kirby ME, Toop TA, Mirza MW, Theodorou MK (2020) Bio-processing of macroalgae *Palmaria palmata*: metabolite fractionation from pressed fresh material and ensiling considerations for long-term storage. J Appl Phycol. https://doi.org/10.1007/s10811-020-02295-x

Galland-Irmouli AV, Fleurence J, Lamghari R, Luçon M, Rouxel C, Barbaroux O, Bronowicki JP, Villaume C, Guéant J-L (1999) Nutritional value of proteins from edible seaweed *Palmaria palmata* (Dulse). J Nutr Biochem 10:353–359

Garbary DJ, Beveridge LF, Flynn AD, White KL (2012) Population ecology of *Palmaria palmata* (Palmariales, Rhodophyta) from harvested and non-harvested shores on Digby Neck, Nova Scotia, Canada. Algae 27:33

García-Bueno N, Decottignies P, Turpin V, Dumay J, Paillard C, Stiger-Pouvreau V, Kervarec N, Pouchus Y-F, Marín-Atucha AA, Fleurence J (2014) Seasonal antibacterial activity of two red seaweeds, *Palmaria palmata* and *Grateloupia turuturu*, on European abalone pathogen *Vibrio harveyi*. Aquat Living Resour 27:83–89

García-Bueno N, Turpin V, Cognie B, Dumay J, Morançais M, Amat M, Pédron J-M, Atucha AM, Fleurence J, Decottignies P (2016) Can the European abalone Haliotis tuberculata survive on an invasive algae? A comparison of the nutritional value of the introduced *Grateloupia turuturu* and the native *Palmaria palmata*, for the commercial European abalone industry. J Appl Phycol 28:2427–2433. https://doi.org/10.1007/s10811-015-0741-z

Grote B (2016) Bioremediation of aquaculture wastewater: evaluating the prospects of the red alga *Palmaria palmata* (Rhodophyta) for nitrogen uptake. J Appl Phycol 28:3075–3082. https://doi.org/10.1007/s10811-016-0848-x

Grote B (2019) Recent developments in aquaculture of *Palmaria palmata* (Linnaeus) (Weber & Mohr 1805): cultivation and uses. Rev Aquac 11:25–41. https://doi.org/10.1111/raq.12224

Irvine LM, Guiry MD (2011) Palmariales. In: Seaweeds of the British Isles, volume 1 Rhodophyta, part 2A Cryptonemiales (Sensu Stricto), Palmariales, Rhodymeniales. Pelagic Publishing, Exeter, pp 65–74

Jard G, Marfaing H, Carrère H, Delgenes JP, Steyer JP, Dumas C (2013) French Brittany macroalgae screening: composition and methane potential for potential alternative sources of energy and products. Bioresour Technol 144:492–498. https://doi.org/10.1016/j.biortech.2013.06.114

Jones DB (1931) Factor for converting percentages of nitrogen in foods and feeds into percentages of proteins. USDA Circ, Washington, D.C., pp 1–22

Jorde I (1966) Algal associations of a coastal area south of Bergen, Norway. Sarsia 23:1–52

Kraan S (2013) Six - pigments and minor compounds in algae. In: Domínguez H (ed) Functional ingredients from algae for foods and nutraceuticals. Woodhead Publishing, Oxford, pp 205–251

Kraan S (2016) Chapter 6 – Seaweed and alcohol: biofuel or booze? In: Seaweed in health and disease prevention. Academic, San Diego, pp 169–184

Kraan S, Guiry MD (2006) A genetic investigation of two morphotypes of *Palmaria palmata* (Palmariales, Rhodophyceae) using rubisco spacer and ITS 1 and ITS 2 sequences. Cryptogam-Algol 27:17–30

Kumagai Y, Tsubouchi R, Miyabe Y, Takeda T, Adachi K, Yasui H, Kishimura H (2019) Complete sequence of mitochondrial DNA of red alga dulse *Palmaria palmata* (Linnaeus) Weber & Mohr in Japan. Mitochond DNA Part B 4:3177–3178

Lahaye M, Rondeau-Mouro C, Deniaud E, Buléon A (2003) Solid-state 13C NMR spectroscopy studies of xylans in the cell wall of *Palmaria palmata* (L. Kuntze, Rhodophyta). Carbohydr Res 338:1559–1569. https://doi.org/10.1016/s0008-6215(03)00241-6
Lahaye M, Vigouroux J (1992) Liquefaction of dulse (*Palmaria palmata* (L.) Kuntze) by a commercial enzyme preparation and a purified endo-b-1,4-D-xylanase. J Appl Phycol 4:329–337
Lalegerie F, Stiger-Pouvreau V, Connan S (2020) Temporal variation in pigment and mycosporine-like amino acid composition of the red macroalga *Palmaria palmata* from Brittany (France): hypothesis on the MAA biosynthesis pathway under high irradiance. J Appl Phycol 32:2641–2656. https://doi.org/10.1007/s10811-020-02075-7
Le Gall L, Pien S, Rusig A-M (2004) Cultivation of *Palmaria palmata* (Palmariales, Rhodophyta) from isolated spores in semi-controlled conditions. Aquaculture 229:181–191. https://doi.org/10.1016/S0044-8486(03)00390-9
Lopes D, Melo T, Rey F, Meneses J, Monteiro FL, Helguero LA, Abreu MH, Lillebø AI, Calado R, Domingues MR (2020) Valuing bioactive lipids from green, red and brown macroalgae from aquaculture, to foster functionality and biotechnological applications. Molecules 25:3883. https://doi.org/10.3390/molecules25173883
Lourenço SO, Barbarino E, De-Paula JC, da SPereira LO, Marquez UML (2002) Amino acid composition, protein content and calculation of nitrogen-to-protein conversion factors for 19 tropical seaweeds. Phycol Res 50:233–241. https://doi.org/10.1046/j.1440-1835.2002.00278.x
Lüning K (1993) Environmental and internal control of seasonal growth in seaweeds. Hydrobiologia 260–261:1–14
Mabeau S, Fleurence J (1993) Seaweed in food products: biochemical and nutritional aspects. Trends Food Sci Technol 4:103–107. https://doi.org/10.1016/0924-2244(93)90091-N
MacArtain P, Gill CIR, Brooks M, Campbell R, Rowland IR (2007) Nutritional value of edible seaweeds. Nutr Rev 65:535–543. https://doi.org/10.1111/j.1753-4887.2007.tb00278.x
Manríquez-Hernández JA, Duston J, Garbary DJ (2016) Effect of irradiance on bioremediation capacity of the red algae *Chondrus crispus* and *Palmaria palmata*. Aquac Int 24:39–55. https://doi.org/10.1007/s10499-015-9907-6
Marrion O, Schwertz A, Fleurence J, Guéant JL, Villaume C (2003) Improvement of the digestibility of the proteins of the red alga *Palmaria palmata* by physical processes and fermentation. Food Nahr 47:339–344. https://doi.org/10.1002/food.200390078
Matos J, Costa S, Rodrigues A, Pereira R, Sousa Pinto I (2006) Experimental integrated aquaculture of fish and red seaweeds in Northern Portugal. Aquaculture Seaweed Based Integr Maricul 252:31–42. https://doi.org/10.1016/j.aquaculture.2005.11.047
Matsuhiro B, Urzúa CC (1996) The acidic polysaccharide from *Palmaria decipiens* (Palmariales, Rhodophyta). Hydrobiologia 326:491–495. https://doi.org/10.1007/BF00047850
van der Meer JP, Todd ER (1980) The life history of *Palmaria palmata* in culture. A new type for the Rhodophyta. Can J Bot 58:1250–1256. https://doi.org/10.1139/b80-155
Mišurcová L, Buňka F, Vávra Ambrožová J, Machů L, Samek D, Kráčmar S (2014) Amino acid composition of algal products and its contribution to RDI. Food Chem 151:120–125. https://doi.org/10.1016/j.foodchem.2013.11.040
Morgan KC, Shacklock PF, Simpson FJ (1980a) Some aspects of the culture of *Palmaria palmata* in greenhouse tanks. Bot Mar Ger Fed Repub 23:12
Morgan KC, Simpson FJ (1981a) Cultivation of *Palmaria (Rhodymenia) palmata*: effect of high concentrations of nitrate and ammonium on growth and nitrogen uptake. Aquat Bot 11:167–171. https://doi.org/10.1016/0304-3770(81)90057-7
Morgan KC, Simpson FJ (1981b) The cultivation of *Palmaria palmata*. Effect of light intensity and temperature on growth and chemical composition. Bot Mar 24:547–552. https://doi.org/10.1515/botm.1981.24.10.547
Morgan KC, Wright JLC, Simpson FJ (1980b) Review of chemical constituents of the red alga *Palmaria palmata* (Dulse). Econ Bot 34:27–50

Mouritsen OG, Dawczynski C, Duelund L, Jahreis G, Vetter W, Schröder M (2013) On the human consumption of the red seaweed dulse (*Palmaria palmata* (L.) Weber & Mohr). J Appl Phycol 25:1777–1791. https://doi.org/10.1007/s10811-013-0014-7

Mouritsen OG, Rhatigan P, Pérez-Lloréns JL (2018) World cuisine of seaweeds: science meets gastronomy. Int J Gastron Food Sci 14:55–65. https://doi.org/10.1016/j.ijgfs.2018.09.002

Naseri A, Marinho GS, Holdt SL, Bartela JM, Jacobsen C (2020) Enzyme-assisted extraction and characterization of protein from red seaweed *Palmaria palmata*. Algal Res 47:101849. https://doi.org/10.1016/j.algal.2020.101849

Pang SJ, Lüning K (2006) Tank cultivation of the red alga *Palmaria palmata*: year-round induction of tetrasporangia, tetraspore release in darkness and mass cultivation of vegetative thalli. Aquaculture 252:20–30. https://doi.org/10.1016/j.aquaculture.2005.11.046

Razi Parjikolaei B, Kloster L, Bruhn A, Rasmussen MB, Frette X, Christensen KV (2013) Effect of light quality and nitrogen availability on the biomass production and pigment content of *Palmaria palmata* (Rhodophyta). Chem Eng Trans 32:967–972. https://doi.org/10.3303/CET1332162

Rioux L-E, Beaulieu L, Turgeon SL (2017) Seaweeds: a traditional ingredients for new gastronomic sensation. 30th Anniv. Spec. Issue. Food Hydrocolloids 68:255–265. https://doi.org/10.1016/j.foodhyd.2017.02.005

Rødde RSH, Vårum KM, Larsen BA, Myklestad SM (2004) Seasonal and geographical variation in the chemical composition of the red alga *Palmaria palmata* (L.) Kuntze. Bot Mar 47:125–133

Sampath-Wiley P, Neefus CD (2007) An improved method for estimating R-phycoerythrin and R-phycocyanin contents from crude aqueous extracts of *Porphyr*a (Bangiales, Rhodophyta). J Appl Phycol 19:123–129. https://doi.org/10.1007/s10811-006-9118-7

Sánchez-Machado DI, López-Cervantes J, López-Hernández J, Paseiro-Losada P (2004) Fatty acids, total lipid, protein and ash contents of processed edible seaweeds. Food Chem 85:439–444. https://doi.org/10.1016/j.foodchem.2003.08.001

Sanderson JC (2006) Reducing the environmental impact of seacage fish farming through cultivation of seaweed. PhD thesis report. SAMS University. 339p

Sanderson JC (2015) Maximising production of *Palmaria palmata* (Linnaeus) Weber & Mohr, 1805 (Final Report to MASTS). Tasmania

Schmid M, Guihéneuf F, Stengel DB (2017) Plasticity and remodelling of lipids support acclimation potential in two species of low-intertidal macroalgae, Fucus serratus (Phaeophyceae) and Palmaria palmata (Rhodophyta). Algal Res 26:104–114. https://doi.org/10.1016/j.algal.2017.07.004

Shama A, Joyce SG, Mari FD, Ranga Rao A, Ravishankar GA, Hudaa N (2019) Macroalgae and microalgae: novel sources of functional food and feed. In: Ravishankar GA, Ranga Rao A (eds) Handbook of algal technologies and phytochemicals: volume-I: food, health and nutraceutical applications. CRC Press, Boca Raton, pp 207–219

Skriptsova AV, Kalita TL (2020) A re-evaluation of *Palmaria* (Palmariaceae, Rhodophyta) in the North-West Pacific. Eur J Phycol 55:266–274

South GR, Tittley I, Farnham WF, Keats DW (1988) A survey of the benthic marine algae of southwestern New Brunswick. Canada Rhodora 90:419–451

Stévant P, Ólafsdóttir A, Déléris P, Dumay J, Fleurence J, Ingadóttir B, Jónsdóttir R, Ragueneau É, Rebours C, Rustad T (2020) Semi-dry storage as a maturation process for improving the sensory characteristics of the edible red seaweed dulse (*Palmaria palmata*). Algal Res 51:102048. https://doi.org/10.1016/j.algal.2020.102048

Takase T, Nakamura A, Miyoshi H, Koga M, Toyomaki A, Kusumi I, Kino R, Konishi Y, Kiso Y, Atsumi T (2020) Effects of *Palmaria palmata* on lipid metabolism and glycemic control in participants with hypercholesterolemia in a randomized double-blind placebo-controlled trial. Phytother Res 34:2303–2312. https://doi.org/10.1002/ptr.6680

Tourneroche A, Lami R, Hubas C, Blanchet E, Vallet M, Escoubeyrou K, Paris A, Prado S (2019) Bacterial–fungal interactions in the kelp endomicrobiota drive autoinducer-2 quorum sensing. Front Microbiol 10:1693. https://doi.org/10.3389/fmicb.2019.01693

van der Meer JP (1981) The life history of *Halosaccion ramentaceum*. Can J Bot 59:433–436
van der Meer JP (1987) Experimental hybridization of *Palmaria palmata* (Rhodophyta) from the northeast and Northwest Atlantic Ocean. Can J Bot 65:1451–1458
van der Meer JP, Bird CJ (1985) Palmaria mollis stat. Nov.: a newly recognized species of *Palmaria* (Rhodophyceae) from the Northeast Pacific Ocean. Can J Bot 63:398–403
Wang T, Jónsdóttir R, Kristinsson HG, Hreggvidsson GO, Jónsson JÓ, Thorkelsson G, Ólafsdóttir G (2010) Enzyme-enhanced extraction of antioxidant ingredients from red algae *Palmaria palmata*. LWT – Food Sci Technol 43:1387–1393. https://doi.org/10.1016/j.lwt.2010.05.010
Werner A, Dring M (2011) Cultivating *palmaria palmata*. Irish Sea Fisheries Board, Dublin
Yuan YV, Bone DE, Carrington MF (2005a) Antioxidant activity of dulse (*Palmaria palmata*) extract evaluated in vitro. Food Chem 91:485–494. https://doi.org/10.1016/j.foodchem.2004.04.039
Yuan YV, Carrington MF, Walsh NA (2005b) Extracts from dulse (*Palmaria palmata*) are effective antioxidants and inhibitors of cell proliferation in vitro. Food Chem Toxicol 43:1073–1081. https://doi.org/10.1016/j.fct.2005.02.012

Chapter 4
A Road to the Sustainable Seaweed Aquaculture

Glacio S. Araújo, Tiago Morais, João Cotas, Sara García-Poza, José W. A. Silva, Ana M. M. Gonçalves, and Leonel Pereira

4.1 Introduction

The seaweeds are common food supplement source from ancient times in the coastal areas mainly in the Oriental Asia. In the most recent times the seaweed are being exploited not only for the direct food consumption, but also, for other industries with a wide range of applications being the most prominent and commercial explored based in the unique seaweed polysaccharides (García-Poza et al. 2020). However, seaweeds synthesize many structural molecules, such as proteins, lipids and carbohydrates (primary metabolites) essential as food source. Furthermore they produce a wide range of bioacitive molecules that can be used in many industries (food, feed, agriculture, cosmetics, pharmaceutical and biotechnological) (Leandro et al. 2020b; Shama et al. 2019). This scenario provides a wide range of opportunites to mass produce the biomass and extract the valuable phycochemicals to meet

G. S. Araújo · J. W. A. Silva
Federal Institute of Education, Science and Technology of Ceará – IFCE, Aracati, Ceará, Brazil

T. Morais
Lusalgae, Lda, Incubadora de Empresas da Figueira da Foz, Figueira da Foz, Portugal

University of Coimbra, MARE – Marine and Environmental Sciences Centre, Department of Life Sciences, Coimbra, Portugal

J. Cotas · S. García-Poza · L. Pereira (✉)
University of Coimbra, MARE – Marine and Environmental Sciences Centre, Department of Life Sciences, Coimbra, Portugal
e-mail: leonel.pereira@uc.pt

A. M. M. Gonçalves
University of Coimbra, MARE – Marine and Environmental Sciences Centre, Department of Life Sciences, Coimbra, Portugal

Department of Biology and CESAM, University of Aveiro, Aveiro, Portugal

A. Ranga Rao, G. A. Ravishankar (eds.), *Sustainable Global Resources of Seaweeds Volume 1*, https://doi.org/10.1007/978-3-030-91955-9_4

the global demands (García-Poza et al. 2020). In addition, photosynthetic carbon accumulation Seawed cultivation provides additional advantage of carbon sequestration and address problems of climate change beside producing valuable feedstock to produce biofuels and raw materials (Hessami et al. 2019) for other industries Viz., food, feed, pharmaceuticals and fertilizers (Duarte et al. 2017).

Seaweeds are the basis of the ecological structure coastal area, supporting directly and indirectly a high number of aquatic species (Reisewitz et al. 2006).

Being autotrophic oranisms they offer support to a number of ecological resources, such as habitats, food and shelter to various trophic levels (herbivores, omnivorous and carnivorous, from invertebrates to vertebrates species) (Almanza and Buschmann 2013; Vásquez et al. 2014) and therefore, seaweeds are vital to support aquatic ecosystem.

Globally, the increasing demand for seaweeds and their sub-products have enhanced the interest in their production, rather than restricting only to wild collections. Harvesting of seaweeds from natural habitats is very dangerous from an ecological point of view, as uncontrolled wild collection can promote a negative impact in the ecosystems (Jung et al. 2013). This scenario supports industrial production of seaweeds hence attractive to many stakeholders to invest more in the production of different macroalgal species that can meet the needs of the economic sector (Ashkenazi et al. 2019). Moreover, it is extremely necessary to maximise the food production for feeding the rising human population on a planet through tapping of alternate renewable sources such as seaweed farming to supplement the limited production possibilities of food production through agricultural crops which demand fertile soil and water resources which are becoming scanty (Charrier et al. 2017).

In this chapter, the sustainable seaweed aquaculture is analyzed from seaweed economic point of view including the new cultivation technologies, sustainability and safety to meet the global industrial demands to support blue growth.

4.2 Seaweeds Economic Importance

Seaweeds are an exceptional source of raw material for industries such as food, feed, energy, biomolecules and livelihood for humans (Charrier et al. 2017; Mohammad et al. 2019). The use of seaweed biomass allows these industries to have alternative sources of raw material, with multiple advantages, such as lowering the production costs or increasing the value of the product. Seaweed can also be promoted as a specialty product for food uses. Today, it is estimated that more than 80% of the worldwide seaweed production and harvesting is destined for Human consumption, directly or as hydrocolloids (thickeners, gelling agents, etc.) (Rebours et al. 2014; Charrier et al. 2017). Seaweed are a rich source of natural compounds with multiple biological activities, namely as antioxidant and antibacterial. They are rich in vitamins (namely, A, B_1, B_2, B_{12}, C and E) and minerals, such as iodine. They also produce secondary metabolites of high commercial value and beneficial to human health. These characteristics turn seaweed in alternative food sources, since they can meet human nutritional needs and still be low on fat (Pereira 2010; Cardoso et al. 2014; Carvalho and Pereira 2014; Azevedo Fonseca 2016; Leandro et al. 2020a).

In recent years, we started to search for healthier ways of feeding ourselves. In addition, health care and life expectancy has increased. These facts boosted the studies on seaweed bioactive compounds (Plaza et al. 2008; 2009; Azevedo Fonseca 2016). Thus, seaweed represents a natural food with a high nutritional value, low in calories, and provide abundant biological activities supporting human health. All these characteristics are advantageous to the food industries to produce valuable products in a sustainable manner (Pereira 2010; Azevedo Fonseca 2016).

Seaweeds are also used both as fertilizers and as animal feed, hence an important alternative raw material for agricultural and livestock industries (Makkar et al. 2016; Charrier et al. 2017). Seaweed are being increasingly considered as beneficial for world agriculture, since they represent an organic, healthy, and nutritious raw material. They can be used by direct application of the biomass or in extracts. There is evidence that using extracts of brown seaweeds (*Ascophyllum nodosum* and *Sargassum muticum*, for instance) in lower concentrations (diluted extracts) improve agricultural crops (Silva 2001). In that way, it is advantageous for the enhancement of agricultural productivities, making the crops more efficient, healthier and reduce dependence on the conventional fertilizers and also supporting organic farming (Sousa et al. 2020).

There are research and published bibliography advocating for the use of seaweed in as livestock feeds. Research shows that seaweed should be used as feed additive and not be applied as a substitute of the typical feed, since the beneficial effects are registered when is usually used under 10% of the total concentration. Seaweed represents a strong alternative animal feed due to the present search for alternatives to the typical feed supplements and antibiotics, which are being highly regulated at a world level. Macroalgae are rich in protein, dietary fibers and phytochemicals which can be used not only to enhance the nutritional quality of animal feed, but to act as a substitute of antibiotics (Morais et al. 2020). The actual demand for renewable and sustainable energy sources that will not compromise on food and land resources can also be fulfilled by seaweeds. This is possible as seaweeds are fast growing, show high biomass yielding with elevated and free of charge productivity, when compared to other conventional biomass feedstock, as corn or soybean for example. However, there are questions related with the biosafety of the use of such biomass could not have a quality guarantee due to variations of nutritional values and risks of heavy metals accumulation (Morais et al. 2020).

In addition, the industrial sector uses seaweed biomass for nutraceuticals, cosmetics, biotechnological and pharmaceutical applications, thus propelling the growth of seaweed biotechnology (Mazarrasa et al. 2013; Morais et al. 2021). Seaweed's extracts and purified compounds demonstrate high potential to be incorporated in cosmetic formulae and pharmaceutical products, with various functions fortifying with natural ingredients as a substitute for the synthetic ones (Carvalho and Pereira 2014; L. Pereira 2018; Morais et al. 2021). There are diverse companies that already use seaweed extracts and compounds in their formulas. However, the monitoring of seaweed biochemical profile is a problem that seaweed-based cosmetic products must overcome. The development of seaweed cultivation and green extraction methods are the major questions for this subject, with the latest research showing promising results (Morais et al. 2021).

Currently, around 32.4 million tons of seaweeds per year (wet weight) are produced worldwide and, as a signal for the growth of the biotechnology market of seaweed products, seaweed-related patent applications have increased at 11%/year since 1990 (Mazarrasa et al. 2013; Charrier et al. 2017; FAO 2020). The production capacity value presents itself as the triple of almost 20 years ago, back to 2000. Asia leads the way in cultivation of seaweed, despite harvesting natural stock (as is current practice in non-Asian countries), with 99% of its production coming from aquaculture (Charrier et al. 2017; FAO 2020). As an example, in the European seaweed market, aquaculture only accounts for 32% of the production (Araújo et al. 2021).

There is still a problem of sustainable production technologies and profitability associated with seaweed economics which needs to be addressed (Steneck et al. 2002; Charrier et al. 2017; Araújo et al. 2021).

The major advantage in promoting seaweed cultivation is its ability to sequester carbndioxide at far exceeding levels compared to coastal vegetation and land plants. CO_2 Sequestration abilities of seaweed are higher up to 1.5 times that of seagrass meadows, salt marshes and mangroves (Krause-Jensen and Duarte 2016; Charrier et al. 2017) Seaweed also help in the removal of dissolved nutrients from coastal waters and coastal protection from erosion (Arkema et al. 2013; Charrier et al. 2017; Araújo et al. 2021). These bioremediation actions have an economic value, furthermore in the Green Deal Era. De Groot et al. (2012) estimated the value of coastal Ecosystem services provided by macroalgae to be over 28,000 $/ha/year, thereby providing opportunities in the global market (de Groot et al. 2012; Charrier et al. 2017).

4.3 Seaweed Aquaculture: The Development of a Millennial Technique into the Industry

Seaweed aquaculture technologic development have been growing dramatically in Asia over the past century and, more recently, have also developed a strong presence in the Americas and Europe (Kim et al. 2017; Ferdouse et al. 2018; García-Poza et al. 2020). Historical records demonstrates large-scale seaweed production have been operating in Asia for decades, mainly in nearshore cultivation techniques (Cheng 1969). Most of the seaweed production occurs in China, Indonesia, and other Asian countries (47.9%, 38.7%, and 12.8%, respectively, in 2016), mostly for human food and food additives (Goecke et al. 2020).

Still, the increasing global effort to develop these farms differs from country to country in terms of seaweed production and its marketing strategies. There is a higher demand for edible seaweed as a direct food product in the East, which generates farmer incomes higher than the resources obtained from the application of seaweed in the polysaccharide industry in Western countries (Hafting et al. 2015).

The nearshore cultivation technique is used by near-coastal population and it is the most common cultivation system to cultivate seaweeds, being a low cost and low productivity cultivation system. Although, this cultivation system is used by Asiatic population by centuries (Soto and Wurmann 2019). Considering the seaweed cultivation, it only need nutrients, seawater and light to initiate the seaweed cultivation (García-Poza et al. 2020).

However, it has been developed more cultivation methods, for example:

- Offshore, this cultivation is used along the times (similar to the nearshore cultivation) which seaweed can be cultivated on the sea floor (attached to hard substrate) or on long-lines (anchored lines or nets that are either seeded or have individuals tied to them for grow-out). The advantage is the installation and maintenance costs are very low, however, there is need of a land laboratory to do the long lines preparation (García-Poza et al. 2020).
- Inshore, this type of seaweed cultivation started in the 1970s–1980s. The advantage is the scope of observing and rapidly modify based on the cultivation conditions. However, a disadvantage of this type of cultivation are the high cost of construction and maintenance (García-Poza et al. 2020).

This seaweed cultivation can be also grouped with other aquatic species cultivation (for example, oysters, fish, shrimp) forming a multitrophic aquaculture (IMTA), which try to mimic the natural ecosystem to produce more than one species, and lowering the production costs ratio when compared to single species aquaculture (Granada et al. 2016; Knowler et al. 2020).

The global need to produce large quantities of seaweed will rise in the coming years, however, there is still a continuous optimization of the cultivation method to satisfy this increasing need, a sustainable and safe output of seaweed and its compounds until today (Buschmann et al. 2017; Camus et al. 2018). This thematics demands an integrative collaboration between academia and the aquaculture industry through research and development (R&D) centers, which has contributed to the joint development of research projects to enhance the profitability and sustainability of the seaweed cultivation industry (Hafting et al. 2012). Presently in most places the new onshore and offshore cultivation processes are not yet total environmentally viable and are economically unsustainable, due to the influence of abiotic and biotic factors (Peteiro et al. 2016; Buschmann et al. 2017). Inorder to address this issue there is need to enhance innovation in production for the benefit of farmers and industrial partners to develop globally acceptable quality biomass for international trade(García-Poza et al. 2020). This is the basis to the Aquaculture 4.0, which relates the seaweed cultivation systems coupled to a multidisciplinary engineering including computer aided automation thus leading to increased competitiveness and performance of aquaculture, minimizing total costs (García-Poza et al. 2020). Moreover, this new technological advancement reduces the costs and pollution in the aquaculture systems, with the advantage that permit to obtain higher yields of biomass with a known quality (Behroozi and Couturier 2019). These innovation in the aquaculture is a long step for the sustainability of the aquaculture system and biomass production.

4.4 Aquaculture Sustainability and Safety

Nowadays, with the blue and circular economy mindset, there is a need to guarantee the safety and sustainability of the seaweed aquaculture from an ecological and economic point of view.

4.4.1 Aquaculture Sustainability

For sustainable aquaculture, there is a need of a study from the beginning of the process to understand the economic and ecological impact in the ecosystem of the seaweed cultivation up to the seaweed biomass production phase, reducing the wastes and ecological impact.

At the beginning, there is a need an authorization of licensing by the local country authorities and also study the aspects of social and economic impacts and risk managements (García-Poza et al. 2020).

Wood et al. (2017) studied the installation of a seaweed aquaculture farm in the United Kingdom, according to licensing and environmental conditions. Based in this study, there is a need of lease the seabed and obtain a Maritime License from the national regulator. There are no impacts related to existing populations in the cultivation environments. It is unlikely that a small farm alone will have a major effect on the marine environment, in contrast very large farms or several small farms next to each other can have a more noticeable effect.

Pereira et al. (2021) evaluated the sustainability of the cultivation of the red macroalga *Hypnea pseudomusciformis* and its use in human food with the Association of Algae Producers of Flecheiras and Guajiru, in Ceará, Brazil. The authors found that the environmental indicators showed an efficient use of energy, nitrogen, and phosphorus, which impacted the algae biomass production by 383, 894 and 1860%, respectively. In addition, algae absorb carbon, do not pollute, and present a low risk to local biodiversity because they are native. The social indicators revealed that 51% of the investment is in the local community and the distribution of income is equal among workers. The farm has a high demand for labor, which is socially inclusive. Finally, the farm was highly profitable, with an Internal Rate of Return (IRR) of 119%, capital recovery in 1.2 years and positive externalities, generating an additional income of US$ 262.00 t^{-1}. This assay demonstrates that this project was excellent to develop into a commercial seaweed aquaculture, very well supported by the sustainability data.

However, when talking in the production of biomass, there is a need to guarantee that the biomass is fully exploited. The best opportunities to fully exploit the seaweed produced is as food industry (as food supplement) or in the biorefinery processes. This Biorefinery concept aims to fully use the biomass to produce several products and/or compounds together.

The step-by-step extraction of pigments, mannitol, phlorotannins, carbohydrates and residues of four species of brown algae *Ecklonia radiata* and *Undaria pinnatifida* was investigated by Zhang et al. (2020). The authors obtained yields, calculated based on the dry weight of each product in each species of seaweed, of 3.4–9.8% (pigments), 22.2–30.7% (mannitol), 0.1–5.1% (phlorotannins), 5.2–15.5% (alginates), 12.2–18.5% (other carbohydrates) and 13.5–19.5% (residual algae). The results indicated that brown seaweeds are potential candidates to be used in biorefineries in order to produce biomaterials, adding value and bioenergy.

Furthermore, Rudke et al. (2020) cite that in the future, it is expected that the extraction of carrageenan from the red macroalgae *Kappaphycus alvarezii*, in the biorefinery process, should be expanded to obtain ethanol, fertilizers, pigments, protein concentrates, among other products that can be obtained from the biomass of this species, which has a large volume of cultivation around the world.

4.4.2 Aquaculture Safety

The aquaculture has an important factor that is very dynamic, and most importantly the monitoring of quality of seawater and ensuring the safety levels needed for the biomass production free from contaminants(Lin et al. 2019; Ngajilo and Jeebhay 2019; Nuwansi et al. 2019 (Kumararaja et al. 2019). The anthropological actvities have enhanced the pollution levels in the marine ecosystem whieh is a matter of serious concern throughout the world (Shah and Shah 2020). They also add potential organic an inorganic components.

(Mawi et al. 2020). Moreover, the seaweeds can absorb heavy metals, incorporating them into the cell wall (López Losada et al. 2020) and some toxic organic compounds such as organo-halogenated (Leri et al. 2019), which are very dangerous to the exploitation of the biomass. There should be sufficient supportive regulations regarding the quality monitoring of the seawater to ensure the safety for seaweed cultivation. Only then that a reasonably safe product can be obtained by the farmers and producers.

4.5 Conclusions and Future Perspectives

Seaweeds have been gaining a high interest from diverse industries to be exploited as natural source of food, fuel, fertilizer and biochemicals.

The monitoring of the aquaculture system is essential to guarantee the biomass quality and safety for further exploitation.

Acknowledgments This work was financed by the Live Food Production Laboratory (LABPAV) and the Tropical Aquaculture Study Group (GEAQUI) of the Federal Institute of Education, Science and Technology of Ceará – IFCE, Campus Aracati, Ceará, Brazil. This work is financed by national funds through FCT - Foundation for Science and Technology, I.P., within the scope of the projects UIDB/04292/2020 – MARE - Marine and Environmental Sciences Centre and UIDP/50017/2020+UIDB/50017/2020 (by FCT/MTCES) granted to CESAM - Centre for Environmental and Marine Studies. João Cotas thanks to the European Regional Development Fund through the Interreg Atlantic Area Program, under the project NASPA (EAPA_451/2016). Sara García-Poza thanks to the project MENU - Marine Macroalgae: Alternative recipes for a daily nutritional diet (FA_05_2017_011) which co-financed this research, funded by the Blue Fund under Public Notice No. 5 - Blue Biotechnology. Ana M. M. Gonçalves acknowledges University of Coimbra for the contract IT057-18-7253.

References

Almanza V, Buschmann AH (2013) The ecological importance of *Macrocystis pyrifera* (Phaeophyta) forests towards a sustainable management and exploitation of Chilean coastal benthic co-management areas. Int J Environ Sustain Dev 12:341. https://doi.org/10.1504/IJESD.2013.056331

Araújo R, Calderón FV, López JS, Azevedo IC, Bruhn A, Fluch S, Tasende MG et al (2021) Current status of the algae production industry in Europe: an emerging sector of the blue bioeconomy. Front Mar Sci 7. https://doi.org/10.3389/fmars.2020.626389

Arkema KK, Guannel G, Verutes G, Wood SA, Guerry A, Ruckelshaus M, Kareiva P, Lacayo M, Silver JM (2013) Coastal habitats shield people and property from sea-level rise and storms. Nat Clim Chang 3:913–918. https://doi.org/10.1038/nclimate1944

Ashkenazi DY, Israel A, Abelson A (2019) A novel two-stage seaweed integrated multi-trophic aquaculture. Rev Aquac 11:246–262. https://doi.org/10.1111/raq.12238

Azevedo Fonseca J (2016) Aplicação de Algas na Indústria Alimentar e Farmacêutica. Universidade Fernando pessoa Faculdade de Ciências da Saúde Porto 75

Behroozi L, Couturier MF (2019) Prediction of water velocities in circular aquaculture tanks using an axisymmetric CFD model. Aquac Eng 85:114–128. https://doi.org/10.1016/j.aquaeng.2019.03.005

Buschmann AH, Camus C, Infante J, Neori A, Israel Á, Hernández-González MC, Pereda SV et al (2017) Seaweed production: overview of the global state of exploitation, farming and emerging research activity. Eur J Phycol 52. Taylor & Francis:391–406. https://doi.org/10.1080/09670262.2017.1365175

Camus C, Infante J, Buschmann AH (2018) Overview of 3 year precommercial seafarming of *Macrocystis pyrifera* along the Chilean coast. Rev Aquac 10:543–559. https://doi.org/10.1111/raq.12185

Cardoso S, Carvalho L, Silva P, Rodrigues M, Pereira O, Pereira L (2014) Bioproducts from seaweeds: a review with special focus on the Iberian Peninsula. Curr Org Chem 18:896–917. https://doi.org/10.2174/138527281807140515154116

Carvalho LG, Pereira L (2014) Review of marine algae as source of bioactive metabolites: a marine biotechnology approach. In: Marine algae: biodiversity, taxonomy, environmental assessment, and biotechnology. CRC Press, Boca Raton, pp 195–227. https://doi.org/10.1201/b17540

Charrier B, Abreu MH, Araujo R, Bruhn A, Coates JC, De Clerck O, Katsaros C, Robaina RR, Wichard T (2017) Furthering knowledge of seaweed growth and development to facilitate sustainable aquaculture. New Phytol 216:967–975. https://doi.org/10.1111/nph.14728

Cheng T-H (1969) Production of kelp - a major aspect of China's exploitation of the sea. Econ Bot 23:215–236. https://doi.org/10.1007/BF02860454

de Groot R, Brander L, van der Ploeg S, Costanza R, Bernard F, Braat L, Christie M et al (2012) Global estimates of the value of ecosystems and their services in monetary units. Ecosyst Serv 1:50–61. https://doi.org/10.1016/j.ecoser.2012.07.005

Duarte CM, Wu J, Xiao X, Bruhn A, Krause-Jensen D (2017) Can seaweed farming play a role in climate change mitigation and adaptation? Front Mar Sci 4. https://doi.org/10.3389/fmars.2017.00100

FAO (2020) The state of world fisheries and aquaculture 2020. FAO, Rome. https://doi.org/10.4060/ca9229en

Ferdouse F, Holdt SL, Smith R, Murúa P, Yang Z (2018) The global status of seaweed production, trade and utilization. FAO Globefish Research Programme, vol 124. Food and Agriculture Organization of the United Nations, Rome, p 120

García-Poza S, Leandro A, Cotas C, Cotas J, Marques JC, Pereira L, Gonçalves AMM (2020) The evolution road of seaweed aquaculture: cultivation technologies and the industry 4.0. Int J Environ Res Public Health 17:6528. https://doi.org/10.3390/ijerph17186528

Goecke F, Klemetsdal G, Ergon Å (2020) Cultivar development of kelps for commercial cultivation—past lessons and future prospects. Front Mar Sci 8. https://doi.org/10.3389/fmars.2020.00110

Granada L, Sousa N, Lopes S, Lemos MFL (2016) Is integrated multitrophic aquaculture the solution to the sectors' major challenges? – a review. Rev Aquac 8:283–300. https://doi.org/10.1111/raq.12093

Hafting JT, Critchley AT, Cornish ML, Hubley SA, Archibald AF (2012) On-land cultivation of functional seaweed products for human usage. J Appl Phycol 24:385–392. https://doi.org/10.1007/s10811-011-9720-1

Hafting JT, Craigie JS, Stengel DB, Loureiro RR, Buschmann AH, Yarish C, Edwards MD, Critchley AT (2015) Prospects and challenges for industrial production of seaweed bioactives. Edited by M. Graham. J Phycol 51:821–837. https://doi.org/10.1111/jpy.12326

Hessami MJ, Cheng SF, Ranga Rao A, Yin YH, Phang SM (2019) Bioethanol production from agarophyte red seaweed, *Gelidium elegans* using a novel sample preparation method for analysing bioethanol content by gas chromatography. 3 Biotech J 9(1):25

Jung KA, Lim S-R, Kim Y, Park JM (2013) Potentials of macroalgae as feedstocks for biorefinery. Bioresour Technol 135:182–190. https://doi.org/10.1016/j.biortech.2012.10.025

Kim JK, Yarish C, Hwang EK, Park M, Kim Y (2017) Seaweed aquaculture: cultivation technologies, challenges and its ecosystem services. Algae 32:1–13. https://doi.org/10.4490/algae.2017.32.3.3

Knowler D, Chopin T, Martínez-Espiñeira R, Neori A, Nobre A, Noce A, Reid G (2020) The economics of integrated multi-trophic aquaculture: where are we now and where do we need to go? Rev Aquac:raq.12399. https://doi.org/10.1111/raq.12399

Krause-Jensen D, Duarte CM (2016) Substantial role of macroalgae in marine carbon sequestration. Nat Geosci 9:737–742. https://doi.org/10.1038/ngeo2790

Kumararaja P, Suvana S, Saraswathy R, Lalitha N, Muralidhar M (2019) Mitigation of eutrophication through phosphate removal by aluminium pillared bentonite from aquaculture discharge water. Ocean Coast Manag 182:104951. https://doi.org/10.1016/j.ocecoaman.2019.104951

Leandro A, Diana P, Cotas J, Marques JC, Pereira L, Gonçalves AMM (2020a) Seaweed's bioactive candidate compounds to food industry and global food security. Life 10:140. https://doi.org/10.3390/life10080140

Leandro A, Pereira L, Gonçalves AMM (2020b) Diverse applications of marine macroalgae. Mar Drugs 18:17. https://doi.org/10.3390/md18010017

Leri AC, Dunigan MR, Wenrich RL, Ravel B (2019) Particulate organohalogens in edible brown seaweeds. Food Chem 272:126–132. https://doi.org/10.1016/j.foodchem.2018.08.050

Lin H, Chen Z, Hu J, Cucco A, Sun Z, Chen X, Huang L (2019) Impact of cage aquaculture on water exchange in Sansha Bay. Cont Shelf Res 188:103963. https://doi.org/10.1016/j.csr.2019.103963

López Losada R, Owsianiak M, Ögmundarson Ó, Fantke P (2020) Metal residues in macroalgae feedstock and implications for microbial fermentation. Biomass Bioenergy 142:105812. https://doi.org/10.1016/j.biombioe.2020.105812

Makkar HPS, Tran G, Heuzé V, Giger-Reverdin S, Lessire M, Lebas F, Ankers P (2016) Seaweeds for livestock diets: a review. Anim Feed Sci Technol 212:1–17. https://doi.org/10.1016/j.anifeedsci.2015.09.018

Mawi S, Krishnan S, Din MFMD, Arumugam N, Chelliapan S (2020) Bioremediation potential of macroalgae *Gracilaria edulis* and *Gracilaria changii* co-cultured with shrimp wastewater in an outdoor water recirculation system. Environ Technol Innov 17:100571. https://doi.org/10.1016/j.eti.2019.100571

Mazarrasa I, Olsen YS, Mayol E, Marbà N, Duarte CM (2013) Rapid growth of seaweed biotechnology provides opportunities for developing nations. Nat Biotechnol 31:591–592. https://doi.org/10.1038/nbt.2636

Mohammad JH, Ranga Rao A, Ravishankar GA (2019) Opportunities and challenges in seaweeds as feed stock for biofuel production. In: Ravishankar GA, Rao AR (eds) Handbook of algal technologies and phytochemicals: volume-II Phycoremediation, biofuels and global biomass production, vol II. CRC Press, Boca Raton, pp 39–50

Morais T, Inácio A, Coutinho T, Ministro M, Cotas J, Pereira L, Bahcevandziev K (2020) Seaweed potential in the animal feed: a review. J Mar Sci Eng 8:559. https://doi.org/10.3390/jmse8080559

Morais T, Cotas J, Pacheco D, Pereira L (2021) Seaweeds compounds: an ecosustainable source of cosmetic ingredients? Cosmetics 8:8. https://doi.org/10.3390/cosmetics8010008

Ngajilo D, Jeebhay MF (2019) Occupational injuries and diseases in aquaculture – a review of literature. Aquaculture 507:40–55. https://doi.org/10.1016/j.aquaculture.2019.03.053

Nuwansi KKT, Verma AK, Rathore G, Prakash C, Chandrakant MH, Prabhath GPWA (2019) Utilization of phytoremediated aquaculture wastewater for production of koi carp (*Cyprinus carpio* var. *koi*) and gotukola (*Centella asiatica*) in an aquaponics. Aquaculture 507:361–369. https://doi.org/10.1016/j.aquaculture.2019.04.053

Pereira L (2010) As algas marinhas e respectivas utilidades, Coimbra

Pereira L (2018) Seaweeds as source of bioactive substances and skin care therapy-cosmeceuticals, algotheraphy, and thalassotherapy. Cosmetics 5. https://doi.org/10.3390/cosmetics5040068

Pereira SA, Kimpara JM, Valenti WC (2021) Sustainability of the seaweed *Hypnea pseudomusciformis* farming in the tropical Southwestern Atlantic. Ecol Indic 121:107101. https://doi.org/10.1016/j.ecolind.2020.107101

Peteiro C, Sánchez N, Martínez B (2016) Mariculture of the Asian kelp Undaria pinnatifida and the native kelp *Saccharina latissima* along the Atlantic coast of Southern Europe: an overview. Algal Res 15:9–23. https://doi.org/10.1016/j.algal.2016.01.012

Plaza M, Cifuentes A, Ibáñez E (2008) In the search of new functional food ingredients from algae. Trends Food Sci Technol 19:31–39. https://doi.org/10.1016/j.tifs.2007.07.012

Plaza M, Herrero M, Cifuentes A, Ibáñez E (2009) Innovative natural functional ingredients from microalgae. J Agric Food Chem 57. American Chemical Society:7159–7170. https://doi.org/10.1021/jf901070g

Rebours C, Marinho-Soriano E, Zertuche-González JA, Hayashi L, Vásquez JA, Kradolfer P, Soriano G et al (2014) Seaweeds: an opportunity for wealth and sustainable livelihood for coastal communities. J Appl Phycol 26:1939–1951. https://doi.org/10.1007/s10811-014-0304-8

Reisewitz SE, Estes JA, Simenstad CA (2006) Indirect food web interactions: sea otters and kelp forest fishes in the Aleutian archipelago. Oecologia 146:623–631. https://doi.org/10.1007/s00442-005-0230-1

Rudke AR, de Andrade CJ, Ferreira SRS (2020) *Kappaphycus alvarezii* macroalgae: an unexplored and valuable biomass for green biorefinery conversion. Trends Food Sci Technol 103:214–224. https://doi.org/10.1016/j.tifs.2020.07.018

Shah A, Shah M (2020) Characterisation and bioremediation of wastewater: a review exploring bioremediation as a sustainable technique for pharmaceutical wastewater. Groundw Sustain Dev 11:100383. https://doi.org/10.1016/j.gsd.2020.100383

Shama A, Joyce SG, Mari FD, Ranga Rao A, Ravishankar GA, Hudaa N (2019) Macroalgae and microalgae: novel sources of functional food and feed. In: Ravishankar GA, Rao AR (eds) Handbook of algal technologies and phytochemicals: volume-I: food, health and nutraceutical applications. CRC Press, Boca Raton, pp 207–219

Silva JF (2001) Historical changes of bottom topography and tidal amplitude in the Ria de Aveiro, Portugal trends for future evolution. Clim Res 18:17–24

Soto D, Wurmann C (2019) Offshore aquaculture: a needed new frontier for farmed fish at sea. In: Boudreau PR, Brooks MR, Butler MJA, Charles A, Coffen-Smout S, Griffiths D, McAllister I et al (eds) The future of ocean governance and capacity development, pp 379–384. https://doi.org/10.1163/9789004380271_064

Sousa T, Cotas J, Bahcevandziev K, Pereira L (2020) Atualização científica da tradição: Revisitação científica aos métodos tradicionais de fertilização com macroalgas em Portugal. A Revista da Associação Portuguesa de Horticultura 136:38–41

Steneck RS, Graham MH, Bourque BJ, Corbett D, Erlandson JONM, Estes JA, Tegner MIAJ (2002) Kelp forest ecosystems: biodiversity, stability, resilience and future. Environ Conserv 29. Cambridge University Press:436–459

Vásquez JA, Zuñiga S, Tala F, Piaget N, Rodríguez DC, Alonso Vega JM (2014) Economic valuation of kelp forests in northern Chile: values of goods and services of the ecosystem. J Appl Phycol 26:1081–1088. https://doi.org/10.1007/s10811-013-0173-6

Wood D, Capuzzo E, Kirby D, Mooney-McAuley K, Kerrison P (2017) UK macroalgae aquaculture: what are the key environmental and licensing considerations? Mar Policy 83:29–39. https://doi.org/10.1016/j.marpol.2017.05.021

Zhang R, Yuen AKL, de Nys R, Masters AF, Maschmeyer T (2020) Step by step extraction of bio-actives from the brown seaweeds, *Carpophyllum flexuosum, Carpophyllum plumosum, Ecklonia radiata* and *Undaria pinnatifida*. Algal Res 52:102092. https://doi.org/10.1016/j.algal.2020.102092

Chapter 5
Seaweed Cultivation Technologies in Indonesia: Current Trends and Future Prospects

Ma'ruf Kasim, Abdul Muis Balubi, Wardha Jalil, Oce Astuti, Bahtiar, Wellem Muskita, Nurdiana, Onu La Ola, Abdul Rahman, Rahmad Sofyan Patadjai, Budiyanto, Utama Kurnia Pangerang, and Ruslaini

Abbreviations

HN Horizontal net
SGR Specific growth rate
VN Vertical net

5.1 Introduction

5.1.1 Seaweed Cultivation in Indonesia

Seaweed production in Indonesia is mostly from natural stocks. The carrying capacity of coastal waters is quite extensive. It is estimated that an area of 1.110.900 Ha is available for seaweed cultivation in Indonesia. Seaweed cultivation in Indonesian waters, are mainly restricted to provinces of Bali, Lampung, West Nusa Tenggara, East Nusa Tenggara, Riau, Sulawesi and Maluku. The cultivation of seaweed in Indonesia is still limited to several types of Eucheuma and Gracilaria. Seaweed cultivated in Indonesia has been *Eucheuma striatum* Tambolong strain, which was originally imported from the Philippines (Phang et al. 2010). Some of the most widely cultivated species are *Kappaphycus alvarezii*, *Eucheuma striatum*, and *Eucheuma denticulatum* (Sulu et al. 2003). Indonesia, until 2017, cultivated

M. Kasim (✉) · A. M. Balubi · O. Astuti · Bahtiar · W. Muskita · Nurdiana · O. La Ola
A. Rahman · R. S. Patadjai · Budiyanto · U. K. Pangerang · Ruslaini
Faculty of Fishery and Marine Sciences, Halu Oleo University, Kampus Bumi Tridarma UHO, Kendari, Southeast Sulawesi, Indonesia
e-mail: marufkasim@uho.ac.id

W. Jalil
Faculty of Fisheries, Dayanu Iksanuddin University, Baubau, Southeast Sulawesi, Indonesia

A. Ranga Rao, G. A. Ravishankar (eds.), *Sustainable Global Resources of Seaweeds Volume 1*, https://doi.org/10.1007/978-3-030-91955-9_5

seaweed production reached 17 ton/Ha. *Eucheuma* sp. exhibits a fairly fast growth (Azanza-Corrales et al. 1992), which has been adopted for largescale cultivation in Southeast Sulawesi, Indonesia. This has prompted research into of the species of *K. alvarezii*. The results of this study explained that the release of spores of seaweed *K. alvarezii* occurred in August and September and the natural spores were found attached to dead branch of corals. The development of seaweed spores as the embryo is fast, reaching a size of 10.64–13.38 cm in about 6 weeks (Kasim and Asnani 2012). *K. alvarezii* cultivation in Indonesia has brought new hope to farmers for economic development. They are adopting Maumere strain which is fast growing. Edison et al. (1999) explained that the selection of *K. alvarezii* seaweed strains for commercial cultivation. Munoz et al. (2004) revealed that the cultivation of *K. alvarezii* developed quite well and provided a fairly good growth in relation to various physical and chemical environmental factors, especially in the Yucatan area of Mexico. Although species of *K. alvarezii* grows well, the growth fluctuates seasonally, especially in the northwestern part of the Indian coast (Kumar et al. 2015). The production value of the *K. alvarezii* species also depends on the cultivation method used. This occurred in the south coastal waters of Rio de Jeneiro, Brazil, where the cultivation technique with tubular netting was better than the longline method (Goromel de Goes and Reis 2011). The results of the economic analysis of the cultivation method of *K. alvarezii* per hectare are very different depending on the method used, especially in the comparison of the use of the bottom line method and the raft monoline in the waters of the west coast of Visayas, Philippines (Samonte et al. 1993). In Indonesia, *K. alvarezii* cultivation, which initially used the longline method, has begun to switch to the cage method, in recent years.

5.1.2 Development of Cultivation Technology

Kappahycus alvarezii cultivated in Indonesia is being exported to various countries for until 2018, Indonesia was one of the largest seaweed producers, exporting 213.000 tons of dry seaweed. One of the parameters for the success of seaweed cultivation is growth. *K. alvarezii* growth is influenced by internal and external factors. Internal factors are genetics and plant physiology and external factors are environmental factors and cultivation methods. The success of seaweed cultivation depends on the environmental factors, quality of the seeds used, and methods of cultivation. Until now, in Indonesia, the method applied by the farmers is longline method. The method of planting with the longline is generally very simple. This method uses a nylon rope and uses used small plastic ball as a float. The advantage of this method is that it is a cheap investment and the seaweed receives enough sunlight. In addition, cultivation using this method works easier, costs less, and produces good quality seaweed.

The development of seaweed cultivation methods in Indonesia began in 1985 with bamboo rafts and the cultivation method has been developed into long-line method since 1992. Until now, the longline method has been used by farmers and it is estimated that it has been used in more than 80% coastal areas in Indonesia. Another method is the seaweed planting technique with an off-base system. The off-bottom system is carried out by directly sinking the seeds on the bottom and allowing them to grow naturally. The off-base system is by tying the seedlings with nylon rope stretched across the bamboo as a binding pole. Another method is cultivation with bamboo rafts. The raft method is a method that has been used for a long time by the community. This method is the first-generation method used by the seaweed cultivator community. When seaweed cultivation began to develop in Indonesia in the 70s and 80s, one of the methods commonly used was the raft method. Bamboo is used as a float and a pole is used for tying the rope. Bamboo is arranged into a rectangular shape in such a way as to maintain the shape. The width of each pole varies depending on the length of the existing bamboo (Kasim et al. 2019). However, all cultivation methods in use today are open access methods. This method provides open access for some herbivorous fish and turtles posing serious problem to biomass-yield. The activity of herbivorous fish and turtles eating seaweed can reduce total seaweed production by up to 60% (Kasim and Asnani 2012). The phenomenon of herbivorous pest attack is very prevalent in almost all commercial seaweed cultivation. In India, especially in Krusadai Island, the production of cultivation of seaweed, such as *Euchuema cottoni* and *E. denticulatum*, has decreased around 10% of its growth due to the attack of predatory pests, such as *Siganus javus* (Rabbit fish), *Acanthurus* sp. (Surgeon fish), *Cetoscarus* sp. (Parrot fish), and sea urchin *Tripneustes* sp. (Ganesan et al. 2006). Hurtado-Ponce (1992) recommended keeping the seaweed *Kappaphycus alverazii*, or known as the trade name *E. Cottoni*, into confinement which can prevent herbivore attacks. Cultivation with this protective method can be carried out with modifications according to the topography of the area and this has been done in several cultivation areas in the Philippines. Currently, the development of the floating cage method has begun in several locations in Indonesia. The floating cage that is used is generally made of PVC or bamboo pipes and nets. PVC or bamboo pipe is the main frame that also functions as a float. Meanwhile, nets are containers for cultivating seaweed. This method is quite simple and effective in protecting seaweed from herbivore attacks. However, this method requires a more comprehensive analysis to assess the business feasibility and future development prospects.

5.2 Data Collection Methods

5.2.1 Study Sites

This study was conducted in one of the largest seaweed cultivation areas in Lakeba coastal area, (50 48′78.2″ N, 122 056″26.3″ E), Southeast Sulawesi, Indonesia (Fig. 5.1). This field experiment performs during August–November 2019. The area

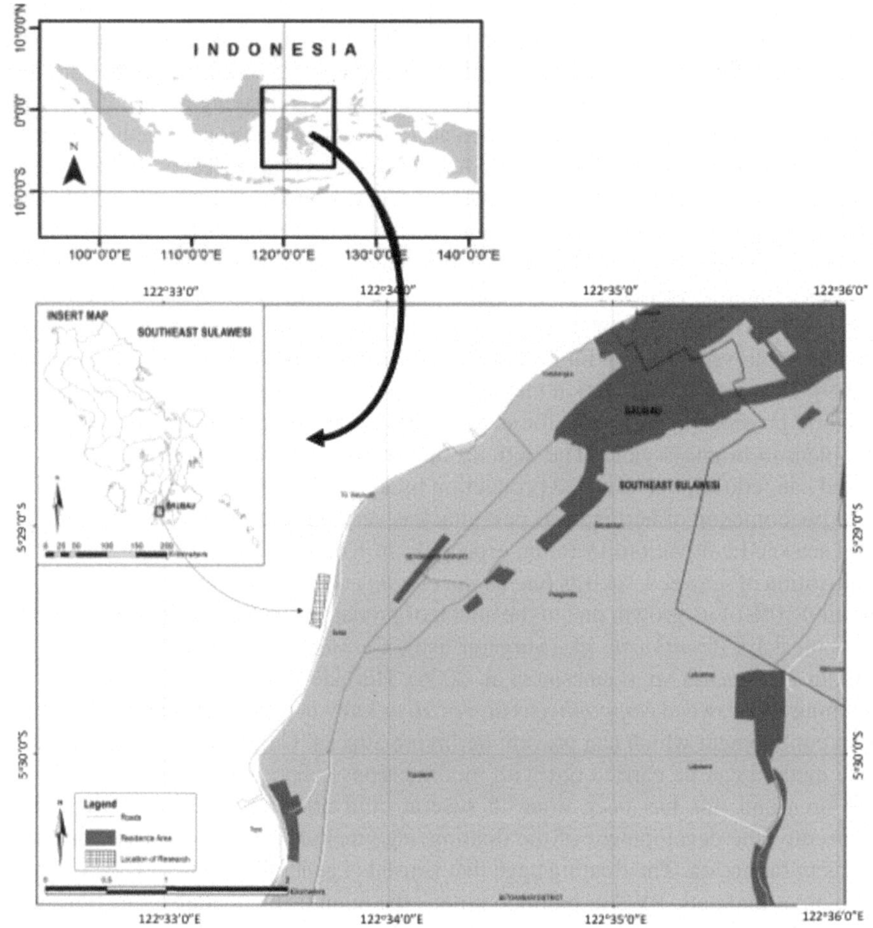

Fig. 5.1 Map of study sites

covered by sand and several rubble corals with little seagrass and seaweeds. The depth during low tides is 6 m and high tide is 8 m (Fig. 5.2).

5.2.2 Growth Rate Measurement

The measurements of total growth and daily growth rate were performed by periodic measurement of the weight of the thallus of *Kappahycus alvarezii* every 7 days. Weight measurement was carried out every 7 days with 6 observations so that the total maintenance time was 42 days. The tools used in this research are horizontal net (HN), vertical net (VN) and longline. Those tools are in the form of an

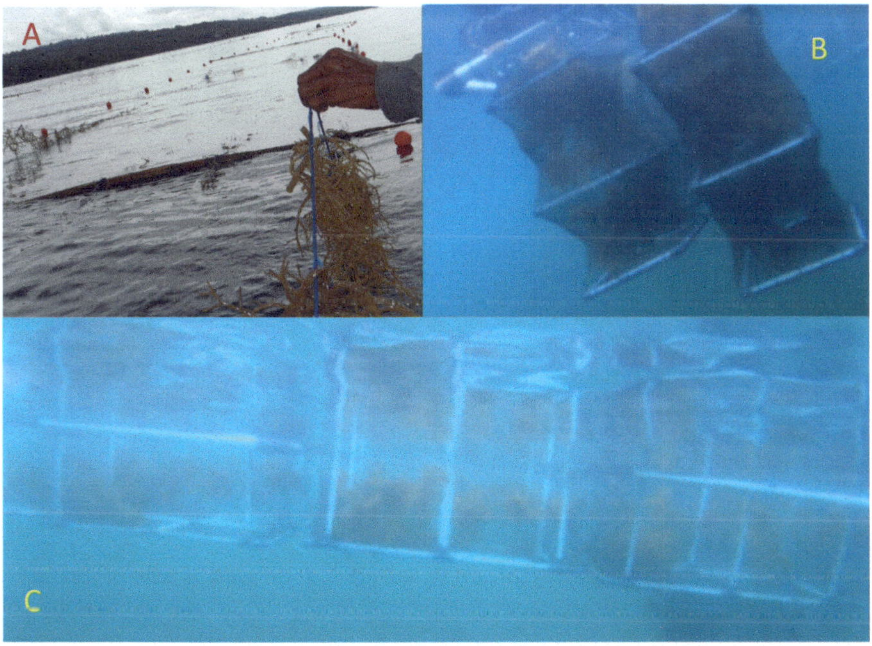

Fig. 5.2 An overview of seaweed cultivating with (**a**) longline, (**b**) vertical net and (**c**) horizontal net in study site

equilateral square or rectangular pocket of appropriate size and is installed lengthwise with one side floating and the other submerged under the water. The HN and VN consists of two main components, namely, the main frame of the PVC pipe and the protective net. The main frame is assembled to form an isosceles rectangle. The protective net is a net wall that encloses the two main frames. The size of the wall mesh can range from 1 cm to 2 cm and it is made of fiber or small nylon rope, monofilament rope, or multifilament rope. The HN size is 200 × 80 × 60 m VN size is 60 × 60 × 300 cm (Fig. 5.3). As a comparison, we conducted a study using the longline method (traditional methods). The longline tool used was 10 m long with 1 m distance between ropes (Fig. 5.4). Each HN, VN and long line were placed at five stations to represent the cultivation area. The seeds used were the seeds of *K. alvarezii* and *E. denticulatum* that were obtained from seaweed farmers. Each thallus of *K. alvarezii* and *E. denticulatum* weighed 20 g. The number of seeds needed for each HN, VN and Longline was 30 thalli with 3 replications. The total seed thallus used in this study was 180 thalli.

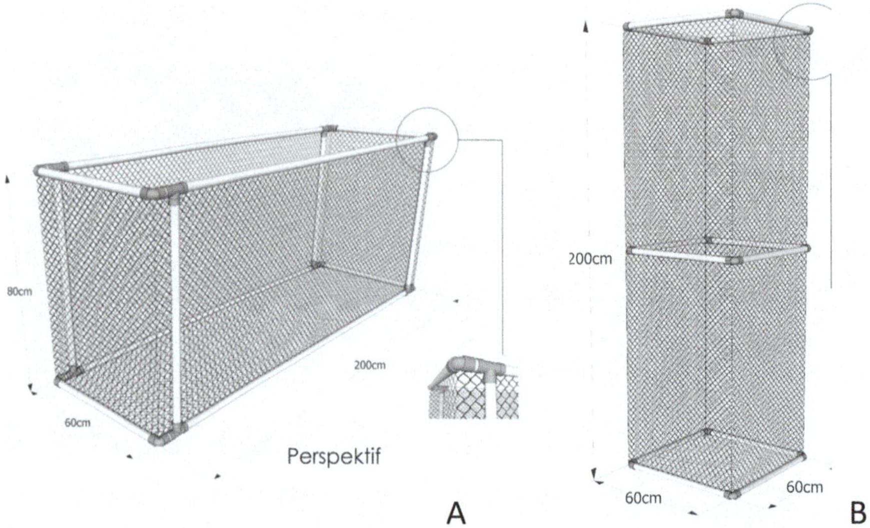

Fig. 5.3 Sketch of horizontal net (HN) (**a**) and vertical net (VN) (**b**) which is being developed by seaweed farmers in Indonesia

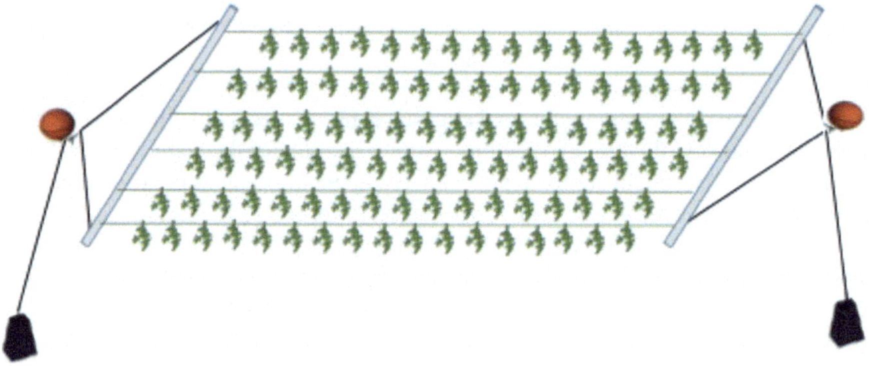

Fig. 5.4 Sketch of the longline method used by the Seaweed farmers

5.2.3 Data Analysis

The data taken were analyzed to determine absolute growth and specific growth rate (SGR). SGR can be calculated using the formula proposed by Luhan and Sollesta (2010).

$$\mathrm{SGR} = \left(\ln W_t / \ln W_o\right) / t \times 100\%$$

Where:

SGR = Specific Growth Rate (%/day),
W_t = weight after t days
W_o = initial weight
t = time in days

To analyze the comparison between HN, VN and longline, an analysis was carried out using SPSS ver. 24.

5.3 Analysis of Growth and Biomass Production

5.3.1 Seaweed Growth Using Horizontal Nets (HN) and Vertical Net (VN)

The use of the HN and VN as two of the cultivation innovations gives different results from the use of the longline. The results of the research that had been done showed that the total weights of *K. alvarezii* thallus in the HN from an average initial weight of about 20 g after being cultivated for 42 days showed an average value of 92 g, 98 g, 93 g, 98 g, 95 g, 134 g, and 149 g at stations 1, 2, 3, 4 and 5, respectively (Fig. 5.5).

While VN showed an average of 42 days 80 g, 81 g, 76 g, 74 g, 79 g, at stations 1, 2, 3, 4 and 5, respectively (Fig. 5.6).

The total weights of *E. denticulatum* thallus in the HN showed an average of 158 g, 132 g, 131 g, 128 g, 115 g, at stations 1, 2, 3, 4 and 5, respectively (Fig. 5.7). While corresponding set up for VN, resulted in average final yield of 105 g, 3 g, 84 g, 79 g, 84 g, at stations 1, 2, 3, 4 and 5, respectively in 42 days (Fig. 5.8). Analysis of the specific growth rate (SGR) of *E. denticulatum* thallus on HN, VN

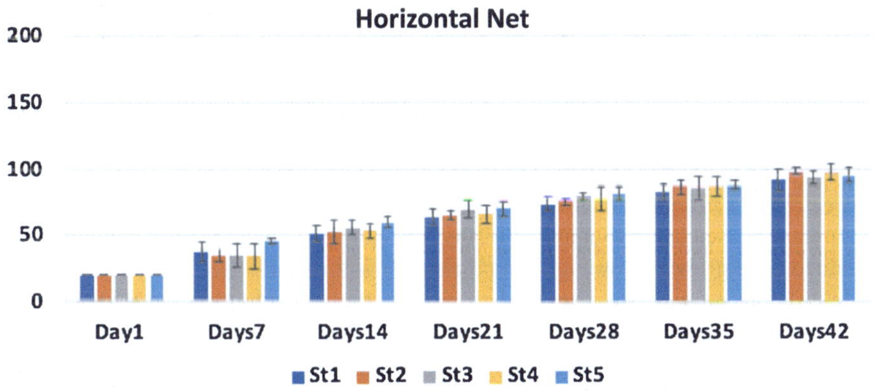

Fig. 5.5 Total growth of *K. alvarezii* cultivated with horizontal net (HN) (st = station)

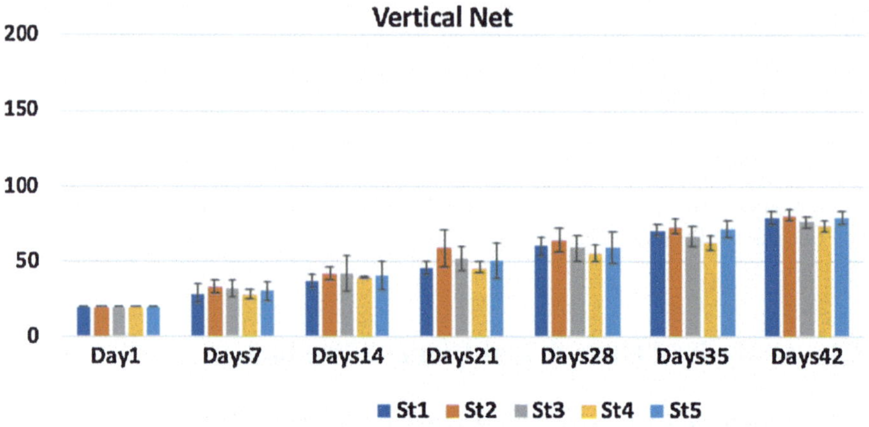

Fig. 5.6 Total growth of *K. alverazii* cultivated with vertical net (VN) (st = station)

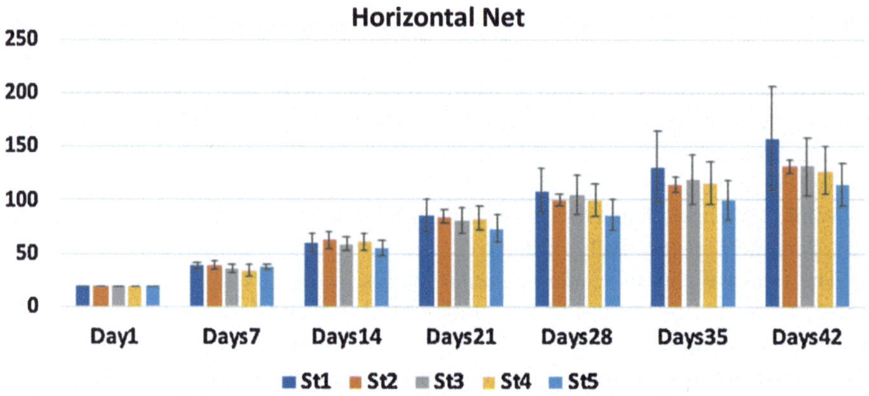

Fig. 5.7 Total growth of *E. denticulatum* cultivated with horizontal Net (VN) (st = station)

and Longline showed averages of 5.4, 5.0 and 3.0%/day, respectively (Fig. 5.9). While *K. alvarezii* thallus on HN, VN and longline was average 5.1, 4.9, 3.0%/day, respectively (Fig. 5.10). In the same location, the highest total growth of seaweed was obtained by using the Floating nets method (Fig. 5.11), from 20 g to an average increase of 143 g. Meanwhile, the total growth of seaweed that was cultivated using the long line method was 89 g (Ardila 2017). The data showed that HN and VN are more effective in seaweed cultivation compared to the long line method.

In general, seaweeds that are cultivated using these two cultivation methods still produce good growth. However, if a pest attack occurs, these two methods will provide different growth rate. During the research, the intensity of the pest attack was low so that the difference between the two methods was not significant. In the peak season for seaweed infestation, the difference in SGR can be very significantly different (Kasim et al. 2019). Seaweed pest attacks occur every year with fluctuating

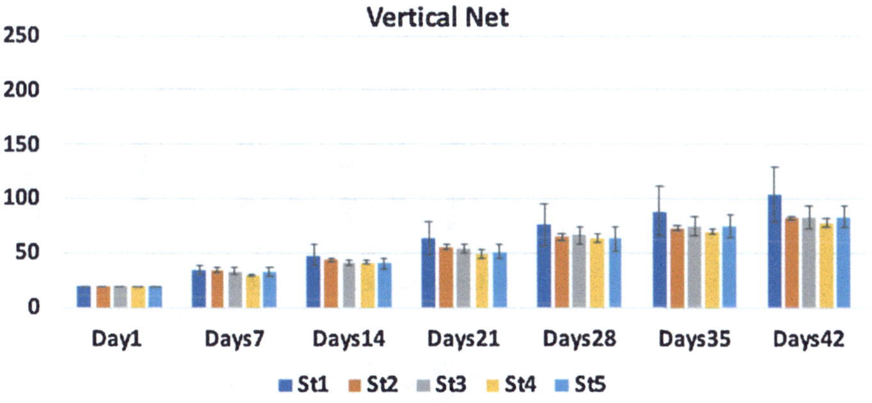

Fig. 5.8 Total growth of *E. denticulatum* cultivated with vertical net (VN) (st = station)

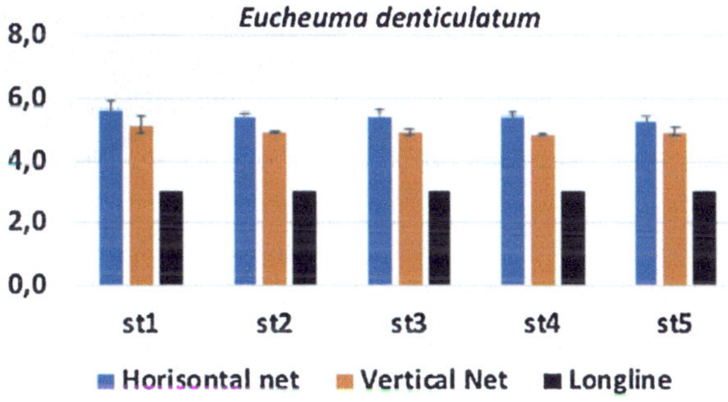

Fig. 5.9 Comparison of SGR of *Eucheuma denticultaum* cultivated by Horizontal net, vertical net and longline (st = station)

seasonal variations. March–June period is the low attack season because the *Siganus* sp. are not yet in the reproductive phase. Their reproductive season occurs from July–August each year in the waters in the study location. August – November period is the peak season for fish pest infestation, which makes the difference between the two methods very significant. This occurs because this period is a post-reproductive period that produces thousands of small fish that harbor near the coast to search for food. The period from December to February is the low season for pest attacks. This period is the big wave season, marked by an increase in the speed of currents and waves around the study area. Basically, seaweed cultivated on HN has more growth, than seaweed cultivated on longline, because HN can protect seaweed from herbivorous fish pests and marine debris. Hurtado-Ponce (1992) recommended the use of confinement cultivation methods in rearing seaweed to prevent the attack of herbivorous organisms. Kasim et al. (2016) explained that cultivation using floating confinement for *E. denticulatum* had an average production yield of 74–78 kg from the initial weight of 5 kg for each cage after cultivation for 40 days. Kasim et al. (2019) explained that the high growth of seaweed in the research location was maintained using a floating cage due to the construction of tools that protect seaweed from herbivorous fish. Besides that, facilitation of the availability of light in

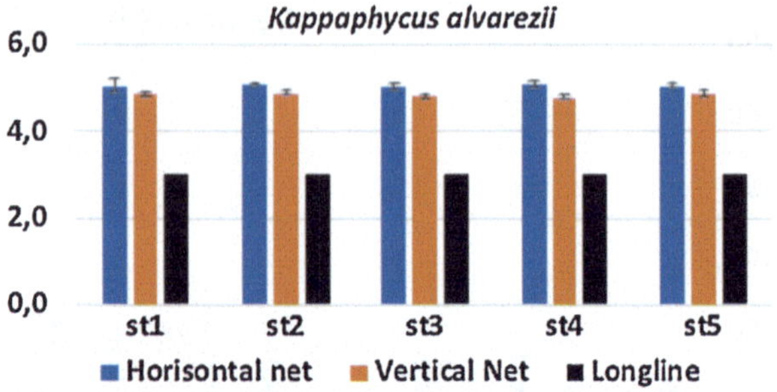

Fig. 5.10 Comparison of SGR of *Eucheuma denticultaum* cultivated by Horizontal net, vertical net and longline (st = station)

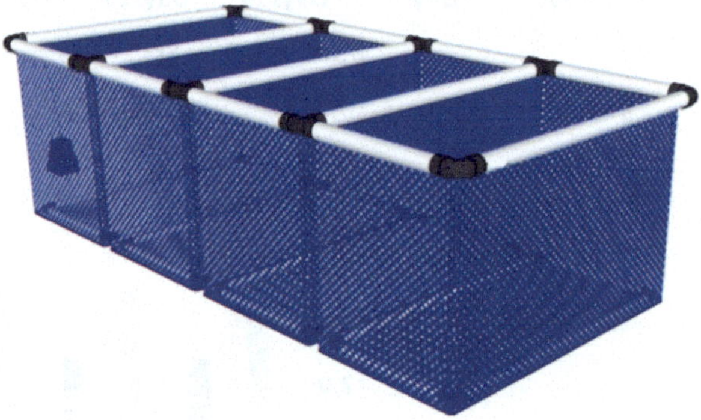

Fig. 5.11 Sketch of floating cage using in seaweed cultivation (Kasim and Mustafa 2017)

this method of cultivation also plays a very important role in enhancing the growth rate of seaweed (Budiyanto et al. 2019).

Apart from cultivation methods and seed quality, the growth of seaweed is also influenced by environmental factors. In general, the cause of the low growth of seaweed is a cultivation method that uses vegetative seeds. This can cause a decrease in quality, quantity, and susceptibility to disease (Indriani and Suminarsih 2003). Cultivation of *K. avarezii* and *Gracillaria verrucosa* in Indonesia, especially in South Sulawesi, has experienced very good development. This cultivation activity continues to strive to meet the growing demand for raw materials for carrageenan and agar. However, all obstacles and problems that occur at the cultivator level can be circumvented by improving the quality of the seaweed seeds. In general, the cultivation process can continue to use good seeds and by clone selection (Parenrengi et al. 2017; Hurtado and Cheney 2005). Kasim et al. (2017), stated that the growths of *K. alvarezii* cultivated in floating nets and longlines were different. The use of floating nets provided an average growth rate of *E. denticulatum* species in 45 days of cultivation and the weight changed from 5 kg of initial seed to 97, 73.3, and 47.5 g during August, September, and October, respectively. In the same month, the average growths of *K. alvarezii* were 61.8 g, 45.5 g, and 40 g respectively. Another factor that causes the low absolute growth of local seaweed seed is the repeated use of seeds (vegetative), which depletes the quality, quantity, and susceptibility to disease. Unstable environmental factors will lead to stress on the seaweed, resulting in a decrease in growth rate (Arisandi et al. 2011). Even weather and season conditions can affect growth and correlate with the process of disease occurrence in seaweed (Kuang and Xia 1996). One of the important environmental factors that also influences the growth rate of seaweed is the intensity of light that enters the surface of the water and is absorbed by the seaweed. This condition will affect the expanse of new cell walls that hardly change when there is expansion of growth inhibited by light. In cultivation activities, spacing also affects the growth of seaweed. The growth of *K. alvarezii* seaweed will be better at large space because of the good circulation which allows good nutrient absorption on each surface of the thallus (Prihaningrum et al. 2001). Another condition is that the increase in seaweed growth is driven by the presence of thallus biomass, which can still develop properly so that the nutrient absorption process by the thallus can run well (Yusuf 2004). The growth of *K. alvarezii* seaweed in North Gorontalo Coast was lower than 3% because in certain seasons, it had serious problems, including pest attack and disease (Pong-Masak et al. 2010).

Some researchers recommend that cultivated *K. alvarezii* should be kept in protective cages that prevent the seaweed from being attacked by herbivores. The decrease in daily growth rate is also related to thallus density in one cultured colony that affects growth (Hurtado et al. 2008). The highest SGR can occur with the availability of interactions between sunlight, temperature, and water movement that always occurs on the water surface (Santelices 1999). The specific growth rate of seaweed cultivated in the surface area has a better growth than at a certain depth. This condition is associated with surface water currents (Harrison and Hurd 2001; Santelices 1999). Methods that provide opportunities for good water movement will have a good impact on the growth of seaweed in addition to protecting seaweed from various harmful activities. According to Neish (2003), in shallower areas, water movement will experience turbulence, thereby increasing the availability of nutrients in water bodies (Neish 2003). This condition can cause the available nutrients in the deeper areas to be low enough so that it will have an impact on nutrient absorption in general. Glenn and Doty (1990) explained that nutrient input occurred rapidly due to the movement of water flowing between the thallus stacks. Seaweed will usually absorb as much ammonium as possible at lower depths and nitrate at higher depths (Bracken and Stachowicz 2006; Taylor et al. 1998). The highest SGR in cultivation in shallow areas is strongly influenced by solar radiation and water movement (Neish 2003). Growth and carrageenan content in the thallus of *K. alvarezii* in Philippines have a positive correlation with water flow, phosphate, and nitrate. The flow of water in the rainy season increases hydrodynamics, thereby increasing growth and carrageenan. In addition, phosphorus and nitrogen are nutrients that play an important role in the growth of seaweed and carrageenan (Orbita 2013).

5.3.2 New Innovation in Seaweed Cultivation

The use of confinement for cultivation of *Kappaphycus alvarezii* can be applied in areas that have high intensity of herbivorous fish attack. However, the durability of tools to support cultivation needs to be considered. The cultivation of *E. cottoni* along with oysters and snapper is one of the cultivation alternatives with a good confinement system (Biswajit et al. 2009). *Graciillaria gracilis* can be cultivated in closed containers to avoid various attacks by herbivores such as isopods (Smith et al. 2004). A floating cage is one of the technological designs used for seaweed cultivation so as to protect seaweed from pests while reducing the potential for ice-ice disease (Kasim et al. 2018; Kasim et al. 2016). Tools in various sizes and shapes were developed for seaweed cultivation. The basic materials of the floating cage are PVC pipe and multifilament netting. PVC pipe also functions as a float and the main frame to facilitate the placement of the floating cage in the desired area. The shape of the floating cage is designed in rectangular shape with various sizes. The surface of the raft is left open and the bottom and the entire side are covered by multifilament net (mesh size 1 cm) (Kasim et al. 2016). The process of placing seaweed in the floating cage is conducted by distributing seaweed directly in the cage without binding the thallus. During the harvesting process, the floating cage is directly pulled to the shallow coastal part so that the harvesting process is easier. In the development of the floating cage, this innovation has undergone several changes and is modified with its main design to facilitate seaweed cultivation. The size and materials used are also becoming better with cheaper investment. This development is based on the results of field tests at all appropriate topographic levels. Adjustments are also made based on direct suggestions from farmers according to the habits and effectiveness of tool use regarding assembly, setting, and planting and harvesting of seaweed. Some variants that can currently be developed include floating net rafts, floating bag nets, vertical net, horizontal net, basket net, long net, wall net, box net, and hanging net raft (Kasim et al. 2020). In general, all of these variants are cultivation tools that can protect seaweed from various pests and lower the potential for ice-ice disease. The principle of the tool is to cultivate seaweed in nets with a variety of designs so that the seaweed is protected from pests of herbivorous fish and turtles.

5.4 Conclusion and Recommendation

Seaweed cultivation methods play a very important role in increasing seaweed production. The increase in production is influenced by various internal and external factors. The overall synergy of these methods is very important to have an impact on increasing seaweed production. The Horizontal Net (HN) and Vertical Net (VN) method are two of the well-developed methods in Indonesia. Those methods are one of the cultivation innovations that can protect seaweed from various seaweed pests (*Siganidae* sp. and sea turtles). This indication can be seen from the different production values between HN, VN and longline. The growth rate of *K. alvarezii* cultivated by HN and VN was higher than that of the longline. Currently, longline is one of the most developed methods of cultivating seaweed in Indonesia. The relatively good production value in recent years has been disturbed by a decrease in production due to the attack of herbivorous fish and sea turtle. In current developments, aquaculture innovations have begun to use protection methods to protect seaweed from pests. This innovation is considered quite effective in increasing production value and cultivation effectiveness in an area in Indonesia.

The development of cultivation with protected methods has several different variations depending on the topography of the sea. However, HN and VN are a superior method that can be used in all marine topographies. The use of HN and VN are believed to not only increase production but also the welfare of coastal communities in Indonesia in the future.

Acknowledgments We are grateful to the Ministry of Research and Technology/BRIN of the Republic Indonesia, for the full research funding of the present research project. Gratitude for Fishery Laboratory, Faculty of Fishery and Marine Science Halu Oleo University for their assistance in sample analysis. Many thanks for LPPM UHO for facilitation of this research.

References

Ardila (2017) Comparison growth of seaweed (*Kappaphycus alvarezii*) from tissue culture cultivated by floating cage and long line, essay of Faculty of Fohsery and Marine Science. Halu Oleo University, Kendari, 71p

Arisandi A, Marsoedi, Nursyam H, Sartimbul A (2011) Rate and percentage of ice-ice in *Kappaphycus alvarezii* in the waters of Bluto Sumenep. Jurnal Ilmiah Perikanan dan Kelautan 3(1):47–51

Azanza-Corrales R, Mamauag SS, Alfiler E, Orolfo MJ (1992) Reproduction in *Eucheuma denticulatum* (Burman) Collins and *Kappaphycus alverazii* (Doty) Doty, in Danajon Reef Philippines. Aquaculture 103:29–34

Biswajit D, Suresh KM, Syda RG (2009) Integration of seaweed (*Kappaphycus alvarezii*) and pearl oyster (*Pinctada fucata*) along with Asian seabass (*Lates calcarifer*) in open sea floating cage off Andhra Pradesh coast, National Fisheries Development Board, http://eprints.cmfri.org.in/6070/1/10._Bis.pdf. Accessed 12 Mei 2020

Bracken MES, Stachowicz JJ (2006) Seaweed diversity enhances nitrogen uptake via complementary use of nitrate and ammonium. Ecology 87(9):2397–2403

Budiyanto, Kasim M, Abadi SY (2019) Growth and carrageenan content of local and tissue culture seed of *Kappaphycus alvarezii* cultivated in floating cage. AACL Bioflux 12(1):167–178

Edison JP, Ricardo TLP, Masao O (1999) Strain selection in *Kappaphycus alvarezii* var. alvarezii (Solieriaceae, Rhodophyta) using tetraspore progeny. J Appl Phycol 11(1):111–121. https://doi.org/10.1023/A:1008085614360

Ganesan M, Thiruppathi S, Nivedita Sahu, Rengarajan N, Veeragurunathan V, Jha B (2006) In situ observations on preferential grazing of seaweeds by some herbivores. Curr Sci 91:1256–1260

Glenn EP, Doty MS (1990) Growth of the seaweeds *Kappaphycus alvarezii*, *Kappaphycus striatum* and *Eucheuma denticulatum* as affected by environment in Hawaii. Aquaculture 84:245–255

Goromel de Goes H, Reis RP (2011) An initial comparison of tubular netting versus tie-tie methods of cultivation for *Kappaphycuss alvarezii* (Rhodophyta, Solieriaceae) on the South Coast of Rio de janeiro State, Brazil. J Appl Phycol 23 607–613

Harrison PJ, Hurd CI (2001) Nutrient physiology of seaweeds: application of concepts to aquaculture. Cah Biol Mar 42:71–82

Hurtado-Ponce AQ (1992) Cage culture of *Kappaphycus alvarezii* var tambalang (Gigartinales, Rhodophyceae). J Appl Phycol 4:311–313

Hurtado AQ, Cheney DP (2005) Propagule production of *Eucheuma Denticulatum* (Burman) Collins et Harvey by tissue culture. Bot Mar 46:338–341

Hurtado AQ, Critchley AT, Trespoey A, Bleicher-Lhonneur G (2008) Growth and carrageenan quality of *Kappaphycus striatum* var. sacol grown at different stocking densities, duration of culture and depth. J Appl Phycol 20:551–555

Indriani H, Suminarsih E (2003) Seaweed processing and marketing cultivation. Penebar Swadaya Press, Jakarta

Kasim M, Asnani (2012) Determination of the generative reproduction season and preference for spore placement, *Eucheuma cottoni*. Ilmu Kelautan 17(4):209–216

Kasim M, Mustafa A (2017) Comparison growth of *Kappaphycus alvarezii* (Rhodophyta, Solieriaceae) cultivation in floating cage and longline in Indonesia, Aquac Rep 6:49–55. http://dx.doi.org/10.1016/j.aqrep.2017.03.004.7

Kasim M, Mustafa A, Munier T (2016) The growth rate of seaweed (*Eucheuma denticulatum*) cultivated in longline and floating cage. AACL Bioflux 9(2):291–299

Kasim M, Jamil MR, Irawati (2017) Occurrence of macro-epiphyte on *Eucheuma spinosum* cultivated on floating cages. AACL Bioflux 10(3):633–639

Kasim M, Asjan, Effendy IJ, Wanurgayah, Ishak E (2018) Influence of initial weight of seeds in variation of growth and carrageenan content of *Eucheuma spinosum*. AACL Bioflux 11(4):1155–1163

Kasim M, Marlia A, Balubi AM, Djalil W (2019) Succession of epiphyte on thallus of *Kappaphycus alvarezii* (Rhodophyta) in horizontal net cage culture. AACL Bioflux 12(5):1710–1717

Kasim M, Balubi AM, Mustafa A, Nurdin R, Patadjai RS, Jalil W (2020) Floating cage: a new innovation of seaweed culture. In: Lu Q, Serajuddin M (eds) Emerging technologies, environment and research for sustainable aquaculture. IntechOpen. https://doi.org/10.5772/intechopen.90035

Kuang M, Xia BM (1996) Reproductive morphology of *Eucheuma gelatinae* (Esper) J.Agardh (Solieraceae, Giartinales Rhodophyta). Chin J Oceanol Limnol 14(1):31–36

Kumar KS, Ganesan K, Rao PVS, Thakur MC (2015) Seasonal studies on field cultivation of *Kappaphycus alvarezii* (Doty) Doty on the Northwest coast India. J Appl Phycol. https://doi.org/10.1007/s10811-015-0629y

Luhan MRJ, Sollesta H (2010) Growing the reproductive cells (carpospores) of the seaweeds, Lappaphycus striatum, in the laboratory until outplanting in the field and maturation to tetrasporophyte. J Appl Phycol 22:579–585

Munoz J, Freile-Pelegrin Y, Robledo D (2004) Mariculture of *Kappaphycus alvarezii* (Rhodophyta Solieriaceae) color strain in tropical water of Yucatan, Mexico. Aquaculutue 239:161–177

Neish IC (2003) The ABC of *Eucheuma* seaplant production. Monograph 1-0703. SuriaLink Infomedia, p 82

Orbita MLS (2013) Growth rate and carrageenan yield of *Kappaphycus alvarezii* (Rhodophyta, Gigartinales) cultivated in Kolambugan, Lanao del Norte, Mindanao, Philippines. AAB Bioflux 5(3):128–139

Parenrengi A, Fahrur M, Makmur, Redjeki HMS (2017) Selection of *Kappaphycus striatum* seaweed in efforts to increase seed growth rate for cultivation. J Riset Akuakultur 11(3):235–248

Phang SM, Yeong HY, Lim PE, Md Nor AR, Gan KT (2010) Commercial varieties of *Kappaphycus* and *Eucheuma* in Malaysia. Malays Jo Sci 29(3):214–224

Pong-Masak PR, Andi IJA, Hasnawi AMP, Mahatma L (2010) Land suitability analysis for seaweed cultivation development in Gusung Batua, Badi Island, Pangkep Regency, South Sulawesi. J Ris Akuakultur 5(2):299–316

Prihaningrum, Meiyana M, Evalawati (2001) Seaweed biology; seaweed cultivation technology (*Kappaphycus alvarezii*). Technical instructions. Ministry of Marine Affairs and Fisheries. Directorate General of Aquaculture. Marine Cultivation Center, Lampung, p 66

Samonte GPB, Hurtado-Ponce AQ, Caturao RD (1993) Economic analysis of bottom line and raft monoline culture of *Kappaphycus alvarezii* var. tambalang in Western Visayas, Philippines. Aquaculture 110:1–11

Santelices B (1999) A conceptual framework for marine agronomy. Hydrobiologia 398(399):15–23

Smith JE, Hunter CL, Conklin EJ, Most R, Sauvage T, Squair C, Smith CM (2004) Ecology of the invasive red alga *Gracilaria salicornia* (Rhodophyta) on O'ahu, Hawai'i. Pac Sci 58(2):325–343

Sulu R, Kumar L, Hay C, Pickering T (2003) *Kappaphycus* seaweed in the Pacifik: review of introduction and field testing proposed quarantine protocols. The Institute of Marine resources. The University of the South Pacific

Taylor RB, Peek JTA, Rees TAV (1998) Scaling of ammonium uptake by seaweeds to surface area: volume ratio: geographical variation and the role of uptake by passive diffusion. Mar Ecol Prog Ser 169:143–148

Yusuf MI (2004) Production, growth and carrageenan content of *Kappaphycus alvarezii* (doty) doty (1998) cultivated with different water media systems and seed thallus. (Dissertation) Postgraduate Program, Hasanudin University. Makassar. 59 p

Chapter 6
Biodiversity, Cultivation and Utilization of Seaweeds in Thailand: An Overview

Khanjanapaj Lewmanomont and Anong Chirapart

6.1 Introduction

Thailand is a tropical country in Southeast Asia surrounded by the countries Malaysia in the south, Myanmar in the north and west, Laos in the north and east, and Cambodia in the east. The long stretch of southern peninsular exposed to the Gulf of Thailand (Gulf of Siam in the map) and the South China Sea by the east and the Andaman Sea by the west (Fig. 6.1). The most conspicuous features of Thailand's terrain are high mountains, a central plain, and an upland plateau. Mountains cover much of northern Thailand and extend along the Myanmar border down through the Kra Isthmus and the Malay Peninsula. The central plain is a lowland area drained by the Chao Phraya River and its tributaries, the country's principal river system, which feeds into the delta at the head of the Bay of Bangkok. The Chao Phraya system drains about one-third of the nation's territory. In the northeastern part of the country the Khorat Plateau, a region of gently rolling low hills and shallow lakes, drains into the Mekong through the Mun River. The Mekong system empties into the South China Sea and includes a series of canals and dams. Together, the Chao Phraya and Mekong systems sustain Thailand's agricultural economy by supporting wet-rice cultivation and providing waterways for the transport of goods and people. In contrast, the distinguishing natural features of peninsular Thailand are long coastlines, offshore islands, and diminishing mangrove swamps (Anonymous 2018a).

The Gulf of Thailand is a semi-enclosed tropical sea located in the South China Sea (Pacific Ocean). It is separated from the South China Sea by two ridges that limit water exchanges with the open South China Sea. The Gulf may be divided into two portions, Upper Gulf and Lower Gulf. The Upper Gulf at the innermost area has an inverted U-shape. This part is the catchment basin of four large rivers on the

K. Lewmanomont (✉) · A. Chirapart
Faculty of Fisheries, Kasetsart University, Bangkok, Thailand
e-mail: khanlew@gmail.com

A. Ranga Rao, G. A. Ravishankar (eds.), *Sustainable Global Resources of Seaweeds Volume 1*, https://doi.org/10.1007/978-3-030-91955-9_6

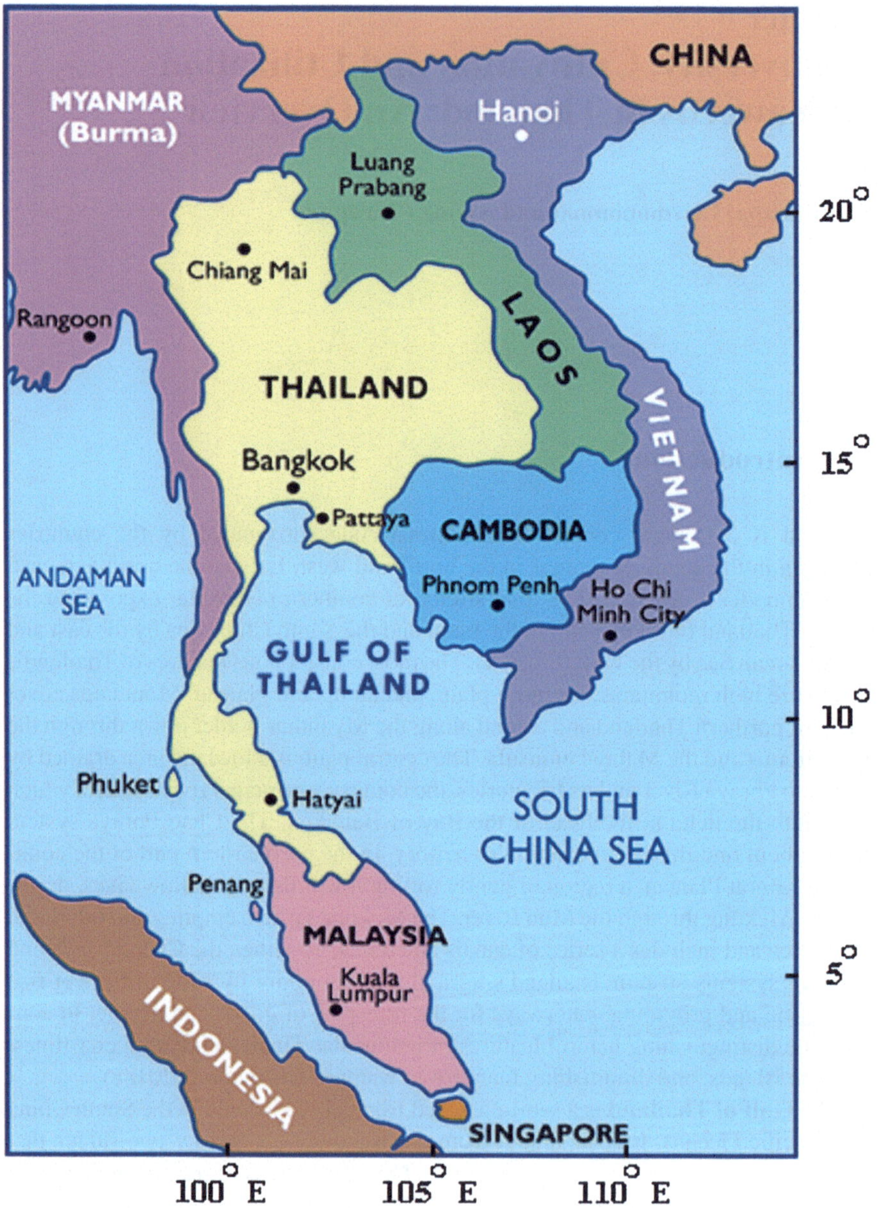

Fig. 6.1 Map of Thailand and adjacent countries

northern side and two on the western coast. Numerous rivers discharge freshwater and sediment into the Gulf. Among them, the Chao Phraya River has the biggest volume of transport next to the Mekong River. The depth of the Gulf of Thailand varies from 40–80 m. In the Upper Gulf, little mixing occurs between coastal and offshore waters. The innermost part of the Gulf is a large area of intertidal mudflats around the shores of a huge, shallow sea bay forming the estuary of the four major rivers. The area formerly supported extensive mangroves. While the largest areas have now been cleared for aquaculture and salt pans, much secondary mangrove still remains and is usually found as a narrow (10–100 m) fringe along the seaward margins (Wattayakorn 2006).

The Andaman Sea is located along the eastern side of the Indian Ocean between the Malay Peninsula and the Andaman and Nicobar Islands. The depth varies from 150–4198 m. In the coastal area, there are many beaches and little runoff along the northern stretch while the southern stretch is composed of a large area of mangrove forests and much runoff. The seafloor is covered with pebbles, gravel, and sand. The Andaman Sea is characterized by a seasonal reversing monsoon, a low surface salinity, and a strong internal wave. It is a part of the Asian monsoon system, which is dominated by prevailing southwest winds in summer and northeast winds in winter, and two transition periods appear in spring and autumn. Also, the Andaman Sea has been well documented as a productive sea because the upwelling phenomenon prevails the sea during northeast monsoon. The area is rich in three important components of marine ecosystems namely mangrove forests, seagrass beds and coral reefs (Limpsaichol n.d.)

6.2 Biodiversity of Marine Algae of Thailand

The coastline of Thailand is about 3148 km long, of which 1093 km is on the west (the Andaman Sea) coast, 2055 km on the Gulf of Thailand (Anonymous 2018b). Ranging from the sand beach, rocky shores to muddy sand beaches with dead coral fragments is suitable for algal attachment. Most intertidal marine algae experience desiccation and water abrasion. Many of them are uprooted by wave abrasion and cast ashore which makes the shore look dirty. Some are exposed during low tide and dried up, but can still survive after the new tide. These algae contain gelatinous substances to protect them from desiccation. Some groups that cannot tolerate desiccation will disappear during the dry season. Whereas, those that grow in the sublittoral zone are less likely to change because they are continuously covered by water. Generally, the seaweeds grow abundantly from September to April and are thereafter less abundant especially in the rainy season.

Based on the list of collections from 18 provinces along the coasts (Fig. 6.2), more than 500 taxa of marine algae from Thailand were recorded. Of which, 462 species have been identified. Many are still unidentified or reported as uncertain species. Among the identified species, 214 belong to the reds, 123 to the greens, 70 to the blue-greens (cyanobacteria), and 55 to the browns (unpublished report). The reds are the most common and abundant in species composition, while the browns are common and highest in their biomass. Those common genera found in almost all the provinces investigated are *Acanthophora, Bryopsis, Canistrocarpus, Caulerpa, Codium, Dictyota, Gelidiella, Gracilaria, Hypnea, Padina, Sargassum,*

Fig. 6.2 Map of Thailand showing some provinces along the coasts

Fig. 6.3 Some common species of Thai seaweeds. **(a)** *Acanthophora spicifera*. **(b)** *Bryopsis pennata*. **(c)** *Chondrophycus cartilagineus*. **(d)** *Gracilaria salicornia*. **(e)** *Hypnea pannosa*. **(f)** *Padina australis*. **(g)** *Sargassum oligocystum*. **(h)** *Ulva reticulata*

and *Ulva* (Fig. 6.3). Comparison of species composition along the Gulf of Thailand and the Andaman Sea indicates the higher in the Gulf over the Andaman due to the longer of coast lines along the east and the west coasts of the Gulf (Lewmanomont et al. 2007, 2013).

From the year 1998, a project on biodiversity of Thai Seas and Islands was initiated under Her Royal Highness Princess Maha Chakri Sirindhorn in cooperation with the Royal Thai Navy. Inventories of biological and physical diversities on the islands in the Gulf of Thailand as well as the Andaman Sea have been performed. That made an increase in numbers of biodiversity in Thailand including marine algal diversity. Many records of new and rare species have been reported such as, *Acrothamnion butlerae*, *Asparagopsis taxiformis*, *Champia salicornioides*, *Cottoniella filamentosa*, *Crouania attenuata*, *Gracilaria lantaensis*, *Gracilaria longirostris*, *Leptolyngbya crosbyana*, *Portieria hornemannii*, *Renouxia antillana*, *Rhodogorgon ramosissima* and *Trichosolen solomonensis* (Fig. 6.4) (Lewmanomont 2008; Lewmanomont and Chirapart 2004; Lewmanomont and Noiraksa 2010; Lewmanomont et al. 2007, 2013; Muangmai et al. 2014; Phang et al. 2016).

6.3 Cultivation and Utilization of Marine Algae in Thailand

In former time, the use of marine algae directly as food in Thailand is limited to particular areas, only those local people living along the seacoast. The majority of edible marine algae are those belonging to the genera *Gracilaria*, *Porphyra*, *Ulva*, and *Caulerpa*. These genera are exploited commercially, although they are only harvested from naturally occurring populations. After the year 2014 when sea grape farming, *Caulerpa lentillifera* and *Caulerpa corynephora* was initiated. They became popular as a healthy food and increased the extra profit of the farmers.

***Gracilaria*,** the most common red alga, is used as a source of agar. Twenty four species have been reported. The most common species used are: *Gracilaria fisheri* and *Gracilaria tenuistipitata* (current name *Agarophyton tenuistipitatum* (Gurgel et al. 2018)). These two species are common and previously recorded as abundant in southern part of Thailand, Songkhla and Pattani provinces. Before the year 2000 all of the production of *Gracilaria* was harvested from a natural occurrence in Pattani Bay and Songkhla Lagoon in a significant amount. At present, natural biomass has been decreasing. Cultivation in abandoned shrimp ponds and raceways becomes popular. Annual productivity, particularly of *G. fisheri*, has been reported at 1–5 tones for each farm, in which the selling price was \$US 0.3–0.5 kg^{-1} wet weight. Total production is around 100–200 tones dry weight per year. Except in 2017 when there was heavy flood making the production declined to only 30–35 tones. These are consumed by local people in the country and a certain amount is exported to Malaysia during the Ramadan (Ruangchuay et al. 2010).

Porphyra, an expensive marine red alga, is used as a foodstuff in soup and Chinese cuisine. The only species found is *Porphyra vietnamensis* (current name *Phycocalidia vietnamensis* (Santiañez and Wynne 2020)). It occurs only in the

Fig. 6.4 Some uncommon species of seaweeds recently reported from Thai waters. (**a**) *Asparagopsis taxiformis*. (**b**) *Cottoniella filamentosa*. (**c**) *Leptolyngbya crosbyana*. (**d**) *Portieria hornemannii*. (**e**) *Renouxia antillana*. (**f**) *Rhodogorgon ramosissima*

southern part of Thailand, Prachuab Khiri Khan, Songkhla, Pattani, and Narathiwas provinces. The annual commercial harvest is less than 100 kg dry weight from Songkhla province which is consumed within the country. Most of the products sold in Thailand are imported from Japan, China, and Korea which is increasing annually.

Caulerpa, is a green alga commonly used as salad vegetable. As it is rich in minerals, vitamins, trace elements, and bioactive substances, makes it popular as health food in recent years. The two most favorite species are *Caulerpa lentillifera* and *C. corynephora* due to their succulent texture (Fig. 6.5).

Fig. 6.5 Two popular species of *Caulerpa* as food and being used for farming. (**a**) *Caulerpa lentillifera*. (**b**) *Caulerpa corynephora*. (**c**) and (**d**) *Caulerpa* salad

Caulerpa lentillifera is known as "sea grape or green caviar". It is commonly found along the east coast of the Gulf of Thailand, especially in Trat province. This species was previously used as part of the recycling treatment of waste water from shrimp ponds. The ability of the plant to absorb large nutrient concentrations makes it suitable candidates for bio-filtration of aquaculture effluent. It was also found growing well in ponds of various habitats and has been successfully cultivated as human food recently. It is usually eaten fresh as a salad vegetable or dressed with special Thai hot sauce.

Caulerpa corynephora was found growing only in southern part of Thailand along the Andaman Sea coast, especially Krabi, Trang, and Satun provinces. This species has been consumed fresh by local people and is now cultured in ponds and cages.

6.3.1 Seaweed Farming

Previously, seaweed farming was not popular in Thailand. Only *Gracilaria* was accepted by local people in Songkhla and Pattani provinces. Until the year 2010, seaweeds especially sea lettuce, *Ulva rigida*, was successful for pond culture in Trat province, followed by sea grape, *Caulerpa lentillifera* in Petchaburi province in 2014. At present, seaweed farming is well known in many provinces in southern Thailand which provides an additional income to coastal fishermen.

6.3.1.1 *Caulerpa* Farming

Caulerpa farming is popular in many provinces in the south of Thailand especially in Petchaburi province where seaweed farming was initiated. More than 30 farms are active in this province. The farmers use those salt ponds that stopped producing salt and abandoned shrimp ponds for *Caulerpa* farming.

Two cultivation methods namely hanging net-frame and sowing or planting methods are commonly used.

Hanging Net-Frame Method

A 500 g of *Caulerpa* seedlings is spread on a 50 × 50 cm^2 net-frame weighting with bricks (Fig. 6.6a and b). Then cover with another sheath of the plastic net and tie with nylon string to prevent the movement of seedlings. Tie four corners of the

Fig. 6.6 Preparing frame for cultivation. (**a**) A net-frame of 50 × 50 cm^2 weighting with bricks. (**b**) A net-frame with 500 g of seedlings. (**c**) A net-frame being ready to hang in pond. (**d**) A net-frame being ready for harvesting

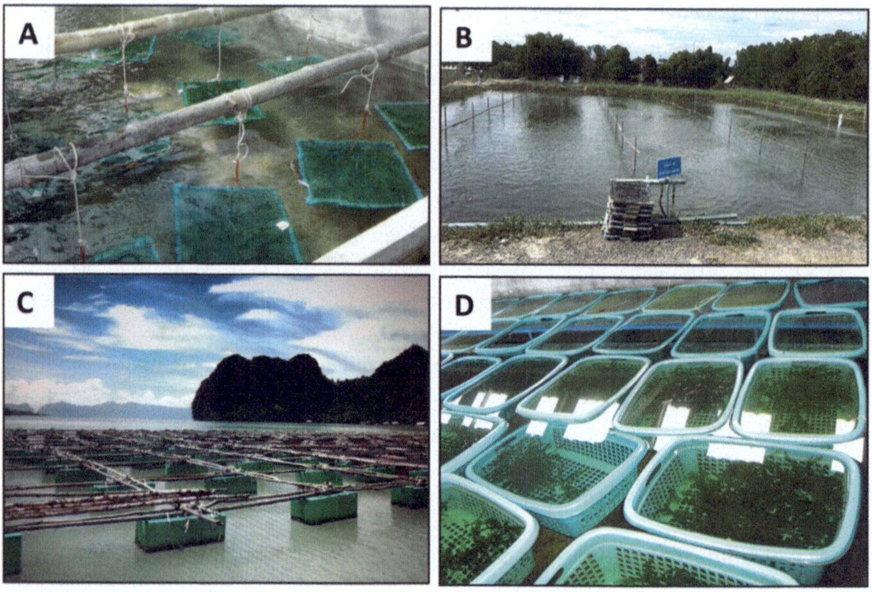

Fig. 6.7 Method of cultivation. (**a**) Hanging net-frames in concrete pond. (**b**) Hanging net-frames in earthen pond. (**c**) Net cage cultivation in bay. (**d**) Floating baskets in concrete pond

frame with nylon string for hanging in the pond (Fig. 6.6c). Prepare several frames based on the estimated production planned. Hang these *Caulerpa* net-frames in the concrete pond or earthen pond using a bamboo stick for holding at the interval of 2 feet apart (Fig. 6.7a and b). The frames can be adjustable for the depth from 50–100 cm from the water surface depending on water turbidity. It is necessary to shake the frames from time to time to remove a silt deposit.

After 6–8 weeks check the density of the plant by pulling the frame up to the surface of water to see whether the whole frame is covered densely enough for harvesting (Fig. 6.6d), then take the frame out of the water and cut the plants close to the net. The remaining plants on the frame can be cultivated again and harvesting can be done weekly afterward.

Sowing Method

The pond must be prepared using a sandy muddy bottom which frees from other aquatic weeds and animals. *Caulerpa* seedlings are sowed on the bottom of pond filled with seawater about 50–100 cm deep or uniformly planted at approximately 1 m intervals. The planting method is more efficient than sowing which will be carried away by the wind to one end of the pond. *C. lentillifera* is adaptable to a variety of environments making it suitable for cultivation in ponds. Regular seawater exchange is needed to maintain the necessary level of nutrients required for growth and maintain good water quality. After 6–8 weeks the plants can be harvested. *C. lentillifera* growing by sowing method has a significantly higher growth rate and higher biomass production than net-frame cultivation.

In the province of Satun, Trang, and Krabi, *C. lentillifera* is preferable to cultivate in cages in a calm bay by a hanging-frame method. Number of cages varies from 2–10 cages for small farms up to more than 100 cages for the big ones (Fig. 6.7c). While *C. corynephora* is cultivated in the basket floating on the surface of seawater in a cement tank (Fig. 6.7d).

C. lentillifera can yield up to 20 kg wet wt. m^{-2} $month^{-1}$. The average production is about one-ton wet wt. $month^{-1}farm^{-1}$. Total biomass production is more than 1000 tones wet wt. $year^{-1}$. The farmers can get a minimum profit not less than 30,000 baht $month^{-1}farm^{-1}$ (approximately $1000).

Seaweed culture has great advantages over other sea products. The plants can be harvested almost all year round. Only during the rainy season when salinity is lower than 25 ppt. the production will be dropped.

At present, many provinces have developed successfully seaweed farming which can provide an extra income to farmers up to 300 baht (about $10) per kg fresh. It can be restored at room temperature for about one week and be transported for sale in other areas. Therefore, seaweed farming will be a very promising future in Thailand not only increasing the profit of the community, but also improving environmental quality as well.

6.3.2 *Seaweed Biotechnology in Thailand*

Importance of seaweed in Thailand as a resource is based primarily on their economic uses. Its main use is as food products for human consumption and animal feed additives. Recently seaweed utilization in Thailand has been more attention; due to its contain high polysaccharides, dietary fibers, minerals, proteins, amino acids, vitamins, polyphenols, and carotenoid (Benjama and Masniyom 2011; Mahae et al. 2014; Yarnpakdee et al. 2015; Yangthong 2017). Some species are potential sources of novel antioxidants (Boonchum et al. 2011a).

6.3.2.1 Nutritional and Biochemical Uses

The green seaweed *Caulerpa* species is one of the interesting taxon for its nutritional composition. Nutritional compositions have been reported in several seaweed species (Ratana-arporn and Chirapart 2006; Benjama and Masniyom 2011, 2012; Mahae et al. 2014). Ratana-arporn and Chirapart (2006) evaluated the nutritional qualities of *Caulerpa lentillifera*, from culture ponds of coastal aquaculture stations in Amphor Ban Lam, Petchaburi province. Protein and ash contents were the two most abundant components in this seaweed with 12.49% and 24.21% (dry weight basis), respectively. *C. lentillifera* contained high amounts of minerals and essential amino acids; such as threoline, valine, lysine, leusine and isoleusine. Regarding the Dietary Reference Intake, this seaweed was notably rich in copper and iodine with 2200 and 1424 μg 100 g^{-1} dry weight, respectively. This alga was also potent in phosphorus, magnesium, manganese, potassium, and calcium. It was quite rich in vitamin A, vitamin E, and polyunsaturated fatty acids such as palmitoleic, linoleic, linolenic, and eicosanoic acids.

Benjama and Masniyom (2011) reported that *Ulva pertusa* and *U. intestinalis* from Pattani Bay in Southern Thailand contains high level of protein (14.6–19.5% DW), lipid (2.1–8.7% DW), ash (25.9–28.6% DW), soluble fiber (25.3–39.6% DW), insoluble fiber (21.8–33.5% DW) and total dietary fiber (51.3–62.2% DW). For mineral *U. pertusa* was rich in Mg, K and Ca, while *U. intestinalis* was Mg, K, Cl, Na, and Ca. The two *Ulva* species were rich in essential amino acids, leucine, valine, and arginine contents. Other red seaweeds *Gracilaria fisheri* and *G. tenuistipitata* contained lipid (1.7–3.6% DW), ash (7.9–22.9% DW), total dietary fiber (TDF) (57.5–64.0% DW), soluble dietary fiber (SDF) (15.6–18.8%DW) and insoluble dietary fiber (IDF) (38.9–45.2% DW); high levels of K and Cl with essential amino acid contents of arginine, leucine, and threonine (Benjama and Masniyom 2012). Mahae et al. (2014) reported the composition and content of sterol, unsaturated fatty acid, amino acid, and minerals in *U. rigida* and *U. intestinalis.* They stated that fucosterol (29,961 μg.100 g^{-1} DW) was found in *U. rigida,* while beta-sitosterol (2126 μg.100 g^{-1} DW) and desmosterol (923.50 μg.100 g^{-1} DW) were found in *U. intestinalis.* Eicosapentaenoic acid (EPA), omega-3 fatty acid, was detected only in *U. rigida.* Besides that, aspartic acid was dominant in *U. rigida,* whereas cyteine was dominant in *U. intestinalis.*

6.3.2.2 Antioxidative Substances

Currently, utilizations of antioxidative substance from seaweed are receiving most attention. Aqueous extracts of *Caulerpa racemosa* var. *macrophysa, Gracilaria tenuistipitata* var. *tenuistipitata, Sargassum* sp., and *Ulva lactuca* were evaluated for their antioxidant activities using DPPH (1, 1-diphenyl-2-picrylhydrazyl), hydroxyl radical ($OH^{\bullet}$) and superoxide anion ($O_2^{\bullet-}$) scavenging assays (Yangthong et al. 2009). These seaweeds contain high total phenolic contents (TPC) and showed the $O_2^{\bullet-}$ scavenging activity. Other extracts of *Sargassum binderi* (current name *Sargassum aquifolium*), *Amphiroa* sp., *Turbinaria conoides,* and *Halimeda macroloba* collected from the Gulf of Thailand showed antioxidant activity; and only *T. conoides* had a potential to antioxidative agent in nutraceutical products (Boonchum et al. 2011a). The extracts of *Laurencia mariannensis, Padina australis, Sargassum polycystym,* and *Caulerpa lentillifera* have been reported as a source of natural antioxidants; due to their high phenolic content and antioxidant activity (Praiboon and Chirapart 2014).

Recent years, several algal species have been reported to prevent oxidative damage by scavenging free radicals and active oxygen and hence able to prevent the occurrence of cancer cell formation. Therefore, algal species as alternative materials to extract natural antioxidative compounds have attracted much attention of biomedical scientists. There are some evidences that seaweed contain compounds with a relatively high antiproliferative activity. Moreover, there are much attention had been paid to the anticancer activity of seaweed sulfated polysaccharide. Praiboon et al. (2012) evaluated the anticancer potential of cold water extracted from *Gracilaria edulis* (current name *Hydropuntia edulis*); the cold-water extract exhibited cytotoxic activity against human cervical adenocarcinoma cell line (HeLa cells) with IC_{50} of 560 mg/ml but has low activity against human breast adenocarcinoma cell line MCF-7 cells.

6.3.2.3 Antimicrobial Substances

Several works reported on antimicrobial activity of seaweed extracts (Kantachumpoo and Chirapart 2010; Boonchum et al. 2011b; Srikong et al. 2015; Rattaya et al. 2015). The aqueous and acid extracts of *Sargassum polycystum* and *Colpomenia sinuosa* showed activity against *Candida albicans* (Kantachumpoo and Chirapart 2010). Other extracts of *Sargassum binderi*, *Amphiroa* sp., *Turbinaria conoides* and *Halimeda macroloba* showed an activity against *Staphylococcus aureus*, *S. epidermidis*, *Propionibacterium acnes*, *Proteus mirabilis* and *Candida albicans* (Boonchum et al. 2011b). Srikong et al. (2015) reported that the crude extracts of *Ulva intestinalis* and *Gracilaria fisheri* had the inhibitory activity against 13 microbial strains (*Vibrio alginolyticus* PSU VA 1, *V. parahaemolyticus* PSU 5124, *V. harveyi* PSU 4109, *Escherichia coli* ATCC 25922, *Pseudomonas aeruginosa* ATCC 27853, *Klebsiella pneumoniae*, *Salmonella typhi*, *Proteus mirabilis*, *Staphylococcus aureus* ATCC 29213, *Listeria monocytogenes* DMST 4553, *Bacillus cereus* TISTR 687, Methicillin resistant *S. aureus* NPRC 001R (MRSA 001R) and *Enterobacter faecalis* (ATCC 29212). Also, *Turbinaria ornata* and *Sargassum polycystum* extracts could inhibit *Staphylococcus aureus* with the extracts at 500 mg/L (Rattaya et al. 2015). Other brown seaweed extracts of *Padina minor*, *P. tetrastromatica* and *Lobophora australis* (current name *Lobophora nigrescens*) have ability to inhibit acne-inducing bacteria (*Propionibacterium acnes*) (Petchyothin et al. 2015).

Interestingly, there was a report on two new diterpenes, (1E, 2R*, 3R*, 4S*, 6E,18S*)-4,18-dihydroxydictyolactone (1) and 8α,11-dihydroxypachydictyol A (2), together with fucoxanthin (3) and a known diterpene 4α-hydroxycrenulatane (4), isolated from the methanol extract of *Dictyota* sp. collected from Bangsaen Beach, Chon Buri province (Jongaramruong and Kongkam 2007). The new diterpene compound 1 showed weak anti-tuberculosis activity, while the compound 2 displayed strong cytotoxicity against the NCI-H187 cell line (National Cancer Institute human small cell lung carcinoma) and potent anti-malarial activity. In addition, compound 3 was active against HSV-1 and malarial parasites.

Aqueous extract of *Turbinaria conoides* show anti-gastric ulcer and acute oral toxicity; the aqueous extract of this species could help protect the stomach from stress, and indomethacin, as well as HCl/ethanol, induced gastric ulcers (Boonchum et al. 2012). In addition, the sulfated galactans from *Gracilaria fisheri* can stimulate immune activity in shrimp (Wongprasert et al. 2014). However, the mechanism by which sulfated galactans stimulates immunity in shrimp has not been elucidated. Then, Rudtanatip et al. (2015) employed a shrimp haemocyte culture system to study the underlying immune stimulus mechanism of the sulfated galactans; and suggested that the sulfated galactans have a stimulatory effect on the immune activity in shrimp haemocytes. In addition, Boonsri et al. (2017) suggested that a protein extract from *Gracilaria fisheri* (GPE) can prevent acute hepatopancreatic necrosis disease (AHPND) infection in shrimp. The GPE has antibacterial activity against *Vibrio parahaemolyticus*, and that *G. fisheri* is a viable source of antibacterial substance that could be used as a feed supplement in shrimp culture to protect against or prevent AHPND.

6.3.2.4 Seaweed Used as Biosorption and Biofilter

Dried *Caulerpa lentillifera* has been used as biosorption of Cu^{2+}, Cd^{2+}, Pb^{2+}, and Zn^{2+} (Pavasant et al. 2006). This species was found to have adsorption capacity for a basic dye, Astrazon Blue FGRL (Marungrueng and Pavasant 2006). Fresh *C. lentillifera* has been used as a biofilter in a hatchery scale recirculating the aquaculture system for juvenile spotted babylons (*Babylonia areolata*). Under this system, *C. lentillifera* had a specific growth rate of 1.70–2.52% d^{-1} (Chaitanawisuti et al. 2011). In addition, Intawongse et al. (2018) reported that *C. corynephora*, collected from six local markets located near the Andaman coast of Krabi province, southern Thailand were good or excellent sources of Mn and Fe; however, none of the seaweeds were good or excellent sources of Cu and Zn. At present, *C. lentillifera* and *C. corynephora* are commercially farmed.

6.3.2.5 Seaweeds Used as Bioenergy Feedstock

In general, the main bioethanol feedstocks in Thailand are all landbased crops, with little work conducted to date on marine biomass sources. Because seaweeds are highly productive, its potential use is of interest. Recently, seaweed chemical composition and the potential for using some Thai seaweed species were evaluated for ethanol production (Chirapart et al. 2014; Nunraksa et al. 2015, 2018, 2019; Rattanasaensri et al. 2018). *Ulva intestinalis*, *Rhizoclonium riparium*, *Gracilaria salicornia* and *Gracilaria tenuistipitata* have been used as substrates for ethanol fermentation (Chirapart et al. 2014). The sugar components found in the seaweed species indicate the carbon source and ethanol production potential of the Thai seaweed. Nunraksa et al. (2015) proposed hydrochloric acid pretreatment to improve ethanol yield during fermentation using *G. tenuistipitata* as a substrate.

Because seaweeds contain a high content of degradable carbohydrates, making them a potential substrate to produce liquid fuels. However, during acid hydrolysis pretreatment, several variables affect the yield of sugar and by-products (5-hydroxymethylfurfural and levulinic acid), such as the concentration of substrate loading, which may impact yeast cell activities and ethanol production. Thus, it is important to decrease the by-products and increase the yield of fermentable sugar. Optimization of acid concentrations, reaction periods, and substrate concentration for hydrolysis of *G. fisheri* and *G. tenuistipitata* have been done to obtain a maximum of ethanol production and low production of by-products (Nunraksa et al. 2018, 2019). In general, fermentation of ethanol, sugars are converted to ethanol using yeasts or bacteria fermentation, for example, the yeast strain *Saccharomyces cerevisiae* TISTR 5339 (Chirapart et al. 2014; Nunraksa et al. 2015). Marine yeasts isolated from *G. fisheri* have been used for ethanol production (Rattanasaensri et al. 2018).

Thai seaweeds are known for their food value, having certain essential amino acids which are not found in land plants. Recently, some seaweeds have been shown to contain bioactive, antimicrobial, and antioxidative substances. Therefore, seaweed will be a new approach for Thai people as a health food, animal feed additive, cosmetics, pharmaceutical purposes, and biofilter as well.

References

Annonymous (2018a) Geography of Thailand. Wigipedia, the free encyclopedia. https://en.wigipedia.org/wiki/Geography_of_Thailand. Accessed 27 Sept 2020

Annonymous (2018b) Database of coastal erosion from 23 coastal provinces of Thailand. Department of Marine and Coastal Resources (in Thai). http://www.dmcr.go.th/detailLib/2394/. Accessed 27 Sept 2020

Benjama O, Masniyom P (2011) Nutritional composition and physicochemical properties of two green seaweeds (*Ulva pertusa* and *U. intestinalis*) from the Pattani Bay in Southern Thailand. Songklanakarin J Sci Technol 33(5):575–583

Benjama O, Masniyom P (2012) Biochemical composition and physicochemical properties of two red seaweeds (*Gracilaria fisheri* and *G. tenuistipitata*) from the Pattani Bay in Southern Thailand. Songklanakarin J Sci Technol 34(2):223–230

Boonchum W, Peerapornpisal Y, Vacharapiyasophon P, Pekkoh J, Pumas C, Jamjai U, Amornlerdpison D, Noiraksar T, Kanjanapothi D (2011a) Antioxidant activity of some seaweed from the Gulf of Thailand. Int J Agric Biol 13:95–99

Boonchum W, Peerapornpisal Y, Kanjanapothi D, Pekkoh J, Amornlerdpison D, Pumas C, Sangpaiboon P, Vacharapiyasophon P (2011b) Antimicrobial and anti-inflammatory properties of various seaweeds from the Gulf of Thailand. Int J Agric Biol 13:100–104

Boonchum W, Vacharapiyasophon P, Kanjanapothi D, Pekkoh J, Amornlerdpison D, Pumas C, Peerapornpisal Y (2012) Anti-gastric ulcer and acute oral toxicity of aqueous extract of *Turbinaria conoides* from the Gulf of Thailand. Chiang Mai J Sci 39(2):292–299

Boonsri N, Rudtanatip T, Withyachumnarnkul B, Wongprasert K (2017) Protein extract from red seaweed *Gracilaria fisheri* prevents acute hepatopancreatic necrosis disease (AHPND) infection in shrimp. J Appl Phycol 29:1597–1608

Chaitanawisuti N, Santhaweesuk W, Kritsanapuntu S (2011) Performance of the seaweeds *Gracilaria salicornia* and *Caulerpa lentillifera* as biofilters in a hatchery scale recirculating aquaculture system for juvenile spotted babylons (*Babylonia areolata*). Aquacult Int 19:1139–1150

Chirapart A, Praiboon J, Puangsombat P, Pattanapon C, Nunraksa N (2014) Chemical composition and ethanol production potential of Thai seaweed species. J Appl Phycol 26:979–986

Gurgel CFD, Norris JN, Schmidt WE, Le HN, Fredericq S (2018) Systematics of the Gracilariales (Rhodophyta) including new subfamilies, tribes, subgenera, and two new genera, *Agarophyton* gen. nov. and *Crassa* gen. nov. Phytotaxa 374(1):1–23

Intawongse M, Kongchouy N, Dean JR (2018) Bioaccessibility of heavy metals in the seaweed *Caulerpa racemosa* var. *corynephora*: human health risk from consumption. Instrum Sci Technol 46(6):628–644

Jongaramruong J, Kongkam N (2007) Novel diterpenes with cytotoxic, antimalarial and anti-tuberculosis activities from a brown alga *Dictyota* sp. J Asian Nat Prod Res 9(8):743–751

Kantachumpoo A, Chirapart A (2010) Components and antimicrobial activity of polysaccharides extracted from Thai brown seaweeds. Kasetsart J (Nat Sci) 44:220–233

Lewmanomont K (2008) *Caulerpa* and *Caulerpella* of Thailand. In: Phang SM, Lewmanomont K, Lim PE (eds) Taxonomy of southeast Asian seaweeds, Monograph series 2. Institute of Ocean and Earth Sciences, University of Malaysia, pp 1–14

Lewmanomont K, Chirapart A (2004) Additional records of *Gracilaria* from Thailand. In: Abbott IA, McDermid KJ (eds). Taxonomy of economic seaweeds with reference to the Pacific and other locations. A Publication of the California Sea Grant College System. vol. IX, p 201–210

Lewmanomont K, Noiraksa T. (2010). *Rhodogorgon and Renouxia* (Rhodogorgonales, Rhodophyta) the gorgonianlike marine red algae from Thailand. Algal Resources, 3: 73–80.

Lewmanomont K, Noiraksa T, Kaewsuralikit C (2007) Marine algae from Koh Kram and adjacent islands. Plant Genetic Conservation Project Under the Royal Initiation of Her Royal Highness Princess Maha Chakri Sirindhorn (RSPG). (in Thai)

Lewmanomont K, Noiraksa T, Praiboon J, Saengkaew J, Tokham S (2013) Marine algae from Similan Islands. Plant Genetic Conservation Project Under the Royal Initiation of Her Royal Highness Princess Maha Chakri Sirindhorn (RSPG). (in Thai)

Limpsaichol (n.d.) Oceanographic Features with Biological Indication in the Andaman Sea, Thailand. http://ivy3.epa.gov.tw/OMISAR/data/OMISAR/wksp.mtg/wom2.98/fullpaper/6E_PRAWEEN.htm. Accessed 27 Sept 2020

Mahae N, Chankaew W, Seechamnanturakit V (2014) Nutritional value of green seaweeds *Ulva rigida* and *Ulva intestinalis*. Kasetsart Uni Fish Res Bull 38(1):20–29

Marungrueng K, Pavasant P (2006) Removal of basic dye (Astrazon Blue FGRL) using macroalga *Caulerpa lentillifera*. J Environ Manag 78:268–274

Muangmai N, Zuccarello GC, Noiraksa T, Lewmanomont K (2014) A new flat *Gracilaria*: *Gracilaria lantaensis sp. nov.* (Gracilariales, Rhodophyta) from the Andaman coast of Thailand. Phycologia 53(2):137–145

Nunraksa N, Praiboon J, Puangsombat P, Chirapart A (2015) Effects of hydrochloric acid pretreatment on ethanol yield of the agarophyte, *Gracilaria tenuistipitata*. Kasetsart Univ Fish Res Bull 39(1):38–47

Nunraksa N, Rattanasaensri S, Praiboon J, Chirapart A (2018) Comparison of ethanol production from *Gracilaria fisheri* and *Gracilaria tenuistipitata* cultivated in aquaculture system in Thailand. J Appl Phycol 30:3319–3325

Nunraksa N, Rattanasaensri S, Praiboon J, Chirapart A (2019) Proximate composition and the production of fermentable sugars, levulinic acid, and HMF from *Gracilaria fisheri* and *Gracilaria tenuistipitata* cultivated in earthen ponds. J Appl Phycol 31:683–690

Pavasant P, Apiratikul R, Sungkhum V, Suthiparinyanont P, Wattanachira S, Marhaba TF (2006) Biosorption of Cu^{2+}, Cd^{2+}, Pb^{2+}, and Zn^{2+} using dried marine green macroalga *Caulerpa lentillifera*. Bioresour Technol 97:2321–2329

Petchyothin P, Praiboon J, Chirapart A (2015) Antibacterial activity of seaweed extracts against acne inducing bacteria (*Propionibacterium acnes*). KMUTT Res Dev J 38(3):273–282. (in Thai)

Phang SM, Yeong HY, Ganzon-Fortes ET, Lewmanomont K, Prathep A, Hau LN, Gerung GS, Tan KS (2016) Marine algae of the South China Sea bordered by Indonesia, Malaysia. Philippines, Singapore, Thailand and Vietnam. Raffles Bull Zool 34(Suppl Part I of II):13–59

Praiboon J, Chirapart A (2014) Phenolic content and antioxidative activity of solvent extracts from seaweeds. In: Proceedings of 52th Kasetsart University Annual Conference, Bangkok, p 181–191 (in Thai)

Praiboon J, Bhumibhamon O, Chirapart A, Akakabe Y (2012) Anticancer activities of cold water polysaccharide extracted from *Gracilaria edulis* (S.G. Gmelin) P.C. Silva. KKU Sci J 40(1):254–264

Ratana-arporn P, Chirapart A (2006) Nutritional evaluation of tropical green seaweeds *Caulerpa lentillifera* and *Ulva reticulata*. Kasetsart J (Nat Sci) 40(Suppl):75–83

Rattanasaensri S, Nunraksa N, Muangmai N, Praiboon J, Chirapart A (2018) Ethanol production from *Gracilaria fisheri* using three marine epiphytic yeast species. J Appl Phycol 30:3311–3317

Rattaya S, Benjakul S, Prodpran T (2015) Extraction, antioxidative, and antimicrobial activities of brown seaweed extracts, *Turbinaria ornata* and *Sargassum polycystum*, grown in Thailand. Int Aquat Res 7:1–16

Ruangchuay R, Lueangthuvapranit C, Nuchaikaew M (2010) Cultivation of *Gracilaria fisheri* (Xia & Abbott) Abbott, Zhang & Xia (Gracilariales, Rhodophyta) in abandoned shrimp ponds along the coast of Pattani Bay, southern Thailand. Algal Resour 3:185–192

Rudtanatip T, Withyachumnarnkul B, Wongprasert K (2015) Sulfated galactans from *Gracilaria fisheri* bind to shrimp haemocyte membrane proteins and stimulate the expression of immune genes. Fish Shellfish Immunol 47:231–238

Santiañez WJE, Wynne MJ (2020) Proposal of *Phycocalidia* Santiañez & M.J.Wynne nom. nov. to replace *Calidia* L.-E.Yang & J.Brodie nom. illeg. (Bangiales, Rhodophyta). Notulae Algarum 140:1–3,

Srikong W, Mittraparp-arthorn P, Rattanaporn O, Bovornreungroj N, Bovornreungroj P (2015) Antimicrobial activity of seaweed extracts from Pattani, Southeast coast of Thailand. Food Appl Biosci J 3(1):39–49

Wattayakorn G (2006) Environmental issues in the Gulf of Thailand. In: Wolanski E (ed) The environment in Asia Pacific harbours, chapter 16. Springer, pp 249–259

Wongprasert K, Rudtanatip T, Praiboon J (2014) Immunostimulatory activity of sulfated galactans isolated from the red seaweed *Gracilaria fisheri* and development of resistance against white spot syndrome virus (WSSV) in shrimp. Fish Shellfish Immunol 36(1):52–60

Yangthong M (2017) Chemical and mineral compositions of *Sargassum* spp. from Bo Mao beach, Chumphon province, Thailand. J Fish Environ 41(3):1–5

Yangthong M, Hutadilok-Towatana N, Phromkunthong W (2009) Antioxidant activities of four edible seaweeds from the southern coast of Thailand. Plant Foods Hum Nutr 64:218–223

Yarnpakdee S, Benjakul S, Kingwascharapong P (2015) Physico-chemical and gel properties of agar from *Gracilaria tenuistipitata* from the lake of Songkhla, Thailand. Food Hydrocoll 51:217–226

Chapter 7
Seaweeds of Vietnam: Opportunities for Commercial Production

Dang Diem Hong and Nguyen Cam Ha

Abbreviation

*cox*3	cytochrome a oxidase subunit III
cpSSRs	Chloroplast simple sequence repeats
DNA barcoding	deoxyribonucleic acid barcoding
GPSs	Global Positioning Systems
*psa*A	photosystem I P700 chlorophyll a appoprotein A1
SEASTAx	Southeast Asian Seaweed Taxonomy Consortium

7.1 Introduction

Marine macroalgae or seaweeds plant-like organisms that lived attached to rock or other hard substrata in coastal areas. Traditionally, it was taxonomically classified into three major groups, green algae (Chlorophyceae, about 1200 species), brown algae (Phaeophyceae - 1750 species), and red algae (Rhodophyceae – 6000 species) depending on their pigments, morphology, and biochemical composition (Farah et al. 2015).

D. D. Hong (✉)
Institute of Biotechnology, Vietnam Academy of Science and Technology (VAST), Hanoi, Vietnam

Graduate University of Science and Technology, VAST, Hanoi, Vietnam

ThuyLoi University, Hanoi, Vietnam
e-mail: ddhong@ibt.ac.vn; ddhong@tlu.edu.vn

N. C. Ha
Institute of Biotechnology, Vietnam Academy of Science and Technology (VAST), Hanoi, Vietnam

Graduate University of Science and Technology, VAST, Hanoi, Vietnam

A. Ranga Rao, G. A. Ravishankar (eds.), *Sustainable Global Resources of Seaweeds Volume 1*, https://doi.org/10.1007/978-3-030-91955-9_7

Seaweed is a group of "lower aquatic plants" living in the offshore and coastal areas; they play a very important role in marine ecology and human life. In addition to the environmental and ecological values, such as participating in the nutrient cycle of water bodies, being the habitat, shelter and foraging of many marine species, seaweed also has great value for human activities such as supplying raw materials for processing industries (raw materials to extract agar, alginate, carrageenan), as functional food for human and feed for animals, medicine, biofertilizer, biofuel (Dang et al. 2007a, b, 2011; Tan et al. 2013; Sze-Looi et al. 2014; Lee et al. 2014; Phaik et al. 2014; Le 2015a, b; Calvyn et al. 2017; Hoang et al. 2018; Ngo et al. 2018; Do et al. 2018; Dang et al. 2019; Shama et al. 2019; Mohammad et al. 2019; Hessami et al. 2019, Zhao et al. 2021). On the other hand, due to its large biomass, seaweed has created a large source of organic matter for the marine ecosystem. Seaweed not only provides primary products directly to the marine environment but also provides places for the biotic plankton, a very high biological productivity population. Therefore, there have been many studies conducted in Vietnam such as classification, ecology, farming, harvesting and processing of especially value added products from seaweed (Lee et al. 2014; Calvyn et al. 2017). The most important thing is to evaluate the potential of resources and the ability to exploit and cultivate economic seaweed species in the front-end islands for socio-economic development with the research mission by investigating, surveying and evaluating biodiversity and seaweed resources; exploitation potential, demand for using, cultivating and processing economic seaweed species in islands and coastal sea of Vietnam (Do et al. 2018).

Vietnam is a country with a tropical monsoon climate extending from 8°30′ to 24° 50′ N and has a coastline of approximately 3260 km. In addition, Vietnam also has many large and small islands. The Sea of Vietnam is located in the northwestern part of the South East Asia Sea. Vietnam's climate varies from subtropical in the northern part to tropical in the southern part of the country and is influenced by two seasonal wind patterns and two monsoon currents. Surface current flow, temperature, rainfall and salinity of coastal areas are very favorable for seaweed growth and ensure not only the diversity of seaweed species and composition as well as seaweed of Vietnam contains valuable high natural bioactive compounds (Dang et al. 2007a).

Vietnam is influenced by two seasonal wind types of monsoons and two monsoon currents. The first in the northeast monsoon usually appear from October to March every year and carries cold water currents. The second is the southwest monsoon, which usually occurs from May to September each year and is associated with warm water currents that influence from the south to the north (Huynh and Nguyen 1998). In the summer season, the temperature of coastal water is about 29–30 °C. On the other hand, the temperatures in the north and the south of Vietnam are around 23–25 °C and 27–30 °C, respectively, in winter. The salinity of coastal water in Vietnam is relatively low due to the inflows of the Son Coi and Mekong rivers into the northern and southern coasts of Vietnam, respectively. Coastal water in Vietnam varies from subtropical to tropical along the coastal zones (Dang et al. 2007a).

7.2 Seaweed Flora of Vietnam

Vietnam has a diverse seaweed flora. About 639 species of seaweeds (269 Rhodophyta, 143 Phaeophyta, 151 Chlorophyta, and 76 Cyanophyta) have been identified (Nguyen and Huynh 1993; Pham 1969). Among these, 310 species can be found along the coast of the Northern provinces and 484 in the southern provinces, while 156 are common to both areas (Huynh and Nguyen 1998).

The first checklist of marine algae of Vietnam was reported wherein a total of 204 species comprising 16 Cyanophyta, 48 Chlorophyta, 22 Ochrophyta, and 118 Rhodophyta were reported. Subsequently, several authors published several more species concerning with marine algae of Vietnam (Pham 1969, 1985; Nguyen 1997; Nguyen et al. 1993, 2000, 2013; Huynh and Nguyen 1998; Nguyen and Pham 2003; Le 2000, 2004; Abbott et al. 2002; Tsutsui et al. 2005; Le and Nguyen 2006; Le and Lin 2006; Dang and Hoang 2004; Dang et al. 2007a).

Up to now, many publications on new species of Vietnam or in the world continue to be published by Vietnamese scientists. According to paper of Nguyen and Le (2005) five species belonging to Rhodophyta and Pheophyta were identified and described. They are *Stenopeltis setchelliae* (Yamanoto) Itono & Yoshizaki, *Hementhocladia australis* Harvey, Liagora filiformis Fan & Li, *Wrangelia tagegana* Harvey, and *Hydroclathrus tenuis* Tseng & Lu which was collected from the coast of Quang Ngai (Ly Son Island), Khanh Hoa (Nha Trang Bay, Truong Sa Archipelago), Ninh Thuan (Ninh Hai) provinces. They are new records to the flora of Vietnam. Le et al. (2015) showed six taxa are new records for the Vietnam algal flora. They are the red algae *Titanophora pikeana* (Dickie) Feldmann from Cu Lao Xanh Island, *Laurencia natalensis* Kylin from Tho Chu Island, *Coelothrix irregularis* (Harvey) Børgesen from Con Dao Island, the green seaweed of *Caulerpa oligophylla* Montagne, *Caulerpa andamanensis* (W.R. Taylor) Draisma, Prudhomme et Sauvage from Phu Quy Island, and *Caulerpa falcifolia* Harvey & Bailey from Ly Son Island, Vietnam.

DNA barcoding has been a major advancement in the field of taxonomy, seeing much effort put into the barcoding of wide taxa of organisms, macro and microalgae including Rhodophyta, Chlorophyta, Ochrophyta seaweed (Tan et al. 2013); genetic diversity of *Kappaphycus* Doty and *Eucheuma* J. Agardh (Solieriaceae, Rhodophyta) in Southeast Asia including Vietnam was studied (Phaik et al. 2014). Development of chloroplast simple sequence repeats (cpSSRs) for the intraspecific study of *G. tenuistipitata* from different populations including Vietnam species was done (Sze-Looi et al. 2014). Still, the application of DNA barcoding has demonstrated our relatively poor taxonomic comprehension of these seaweeds, thus suggesting more in-depth efforts in taxonomic restructuring as well as establishment. The phylogenetic relationships of *Rosenvingea* (Scytosiphonaceae, Phaeophyceae) from Vietnam based on *cox*3 and *psa*A sequences were reported by Lee et al. (2014). On the other hand, the appearance of *Phycocalidia* (old name of *Porphyra*) *tanegashimensis* in Brazil was believed to be a recent introduction from the Indo-Pacific/

Table 7.1 Summary of number of families and taxa belonging to the different division of marine algae of Vietnam (Phang et al. 2016)

	Number of families estimated	Total number of taxa	Cyanophyta		Chlorophyta		Rhodophyta		Ochrophyta	
			Number of families	Total number of taxa	Number of families	Total number of taxa	Number of families	Total number of taxa	Number of families	Total number of taxa
Seaweed flora	78	805	10	65	21	182	36	409	11	149

North Pacific but without conclusive evidence. Thus, recently, haplotype networks of *Phycocalidia tanegashimensis* (Bangiales, Rhodophyta) including Vietnamese species indicate a probable invasion from the South China Sea to Brazil was published in 2021 by Zhao et al. (2021).

Phang et al. (2016) showed that the Vietnamese flora was also compared with that of Malaysia, Philippines, Singapore, Thailand and Vietnam. Until now, in Vietnam, a total of 805 species (with number of Families estimated 78) was reported (Table 7.1). Based on similarities in flora, about the diversity of marine algae in South East Asian sea bordered by Southeast Asian countries have indicated that in general, the recent efforts at compiling the checklist of Vietnam from all published records, has given Vietnam the highest number of taxa (805) followed by Philippines (631), and Malaysia (355).

Based on the composition, characteristics of species, distribution, natural conditions, the Vietnam coastline can be divided into two main geographical regions including the subtropical region in the North of Vietnam from Mong Cai to gorge Hai Van, the tropical region in South of Vietnam from gorge Hai Van to Ha Tien and with five subregions. According to Huynh and Nguyen (1998), the seaweed flora of Vietnam northern part can be divided into the two subsections as following:

1. The northern part of Tonkin Bay is from Quang Ninh province to Hai Phong city in which is the coldest part of the Vietnam Sea and is directly and strongly influenced by the cold water current. In this region, water temperature in winter ranges from 15 °C to 20 °C and increases from 25 °C to 30 °C in the summer season. Vietnam has the number of sunny hours from 1400–3000 hours/year. The average annual rainfall ranges from 1500 to 2000 mm. The air humidity is above 80%. Due to the influence of monsoons and the complexity of the topography, Vietnam often experiences disadvantages in weather such as storms, floods and droughts (Report to General Statistics Office 2020). There are 143 seaweed species of which 83 species are similar to the southern part of Tokin Bay (Huynh and Nguyen 1998). Sixty-eight species are similar to the seaweed flora of the southern part of Vietnam. In this region, seaweed is predominantly subtropical.
2. The southern part of Tokin Bay is from Hai Phong city to Ca Mau province in which in winter, the water temperature ranges from 18.5 °C to 21.7 °C and approximately 30 °C in the summer season. There are 185 seaweed species recorded in this region, of which 96 species appear in the southern part of Vietnam. Sixty species are common to both this region and the Northern part of Tokin Bay. The marine algae in this area predominantly belong to tropical subregions (Huynh and Nguyen 1998). Some of the main genera of seaweed commontly distributed along the coast of Vietnam showed in Table 7.2.

Table 7.3 shows the gaps in marine algal diversity information in South East Asia Sea including Vietnam and the tasks ahead. Due to pollution, climate change and

Table 7.2 Some of the main genera of seaweed commonly distributed along the coast of Vietnam, with the number of species corresponding to each genus in parentheses

Phyllum	Number of species corresponding to each genus
Rodophyta	*Acanthophora* (3), *Acrochaetium* (18), *Acrosorium* (1), *Actinotrichia* (1), *Ahnfeltia* (7), *Akalaphycus* (1), *Alsidium* (1), *Amphiroa* (6), *Anotrichium* (3), *Antithamnion* (1), *Antithamnionella* (3), *Antrocentrum* (1), *Asparagopsis* (1), *Bangia* (2), *Bangiopsis* (1), *Betaphycus* (1), *Bostrychia* (3), *Branchioglossum* (1), *Botryocladia* (2), *Caloglossa* (7), *Carpopeltis* (1), *Catenella* (3), *Centroceras* (2), *Ceratodictyon* (5), *Ceramium* (16), *Champia* (3), *Cheilosporum* (1), *Chondracanthus* (3), *Chondria* (7), *Chondrophycus* (3), *Chroodactylon* (1), *Claudea* (1), *Colacconema* (4), *Compsopogon* (1), *Corallina* (2), *Corallophila* (4), *Cottoniella* (1), *Crouania* (1), *Cryptonemia* (1), *Dasya* (4), *Demornema* (3), *Dichotomaria* (3), *Dictyurus* (1), *Diplothamnion* (1), *Erithrocladia* (4), *Eucheuma* (2, almost in Central part of Vietnam), *Exophyllum* (1), *Falkenbergia* (1), *Galaxaura* (4), *Ganonema* (3), *Gayliella* (3), *Gelidium* (11), *Gelidiella* (3), *Gigartine* (4), *Gibsmithia* (1), *Gloiopeltis* (2), *Gracilaria* (22), *Gracilariopsis* (8), *Grateloupia* (12), *Griffithsia* (3), *Gymnogongrus* (2), *Gymnothamnion* (1), *Halichrysis* (1), *Haloplegma* (1), *Halymenia* (4, mostly in southern part of Vietnam), *Helminthocladia* (1), *Herposiphonia* (7), *Hildenbrandia* (1), *Hydrolithon* (3), *Hydropuntia* (6), *Hypnea* (14), *Hypoglossum* (2), *Izziella* (1), *Jania* (12), *Kappaphycus* (3, almost in Central part of Vietnam), *Laurencia* (23, all almost in southern part of Vietnam), *Lamentaria* (1), *Leveillea* (1), *Lophosiphonia* (3), *Liagora* (4, almost in southern part of Vietnam), *Lithophyllum* (2), *Lithothamnion* (1), *Martensia* (3), *Mastophora* (2), *Melanamansia* (1), *Meristotheca* (1), *Mesophyllum* (2), *Metagoniolithon* (1), *Montemaria* (1), *Neogoniolithon* (2), *Neoizziella* (1), *Neomonospora* (1), *Neosiphonia* (8), *Neurymenia* (1), *Odonthalia* (1), *Palisada* (6), *Parviphycus* (2), *Peyssonnelia* (6), *Pleonosporium* (1), *Pneophyllum* (1), *Polyopes* (1), *Polysiphonia* (11), *Porphyra* (5), *Portieria* (2), *Prionitis* (1), *Pterocladia* (6, almost in southern part of Vietnam), *Reinboldiella* (1), *Rhodogorgon* (1), *Rhodymenia* (3), *Rodriguezella* (1), *Sahlingia* (1), *Scinaia* (1), *Solieria* (1), *Sonderopella* (1), *Spongoclonium* (1), *Spiriodia* (2), *Stylonema* (1), *Symphyocladia* (1), *Symphyocladia* (1), *Rhodymenia* (3), *Taenioma* (1), *Tayloriella* (1), *Titanophycus* (1), *Titanophora* (1), *Tolypiocladia* (2), *Tricleocarpa* (2), *Yamadaella* (1), *Yonagunia* (1), *Wrangelia* (3), *Wurdemannia* (1).
Ochrophyta	*Acrothrix* (1), *Asteronema* (1), *Canistrocarpus* (2), *Chilionema* (1), *Chnoospora* (2), *Chnoospora* (2), *Colpomenia* (2), *Dictyota* (12), *Dictiopteris* (4), *Distromium* (1), *Ectocarpus* (2), *Feldmania* (5), *Hermophysa* (1, almost in southern part of Vietnam), *Hormophysa* (1), *Hydroclathrus* (1), *Kuetzingiella* (1), *Lobophora* (1), *Mesospora* (1), *Myrionema* (1), *Nemacystus* (1), *Neoralfsia* (1), *Padina* (6), *Petalonia* (1), *Petroderma* (1), *Pylaiella* (1), *Ralfsia* (2), *Rosenvingea* (4), *Sargassum* (73), *Scytosiphon* (1), *Spatoglossum* (2), *Sphacelaria* (6), *Stypopodium* (1), *Turbinaria* (6, mostly in southern part of Vietnam),

(continued)

Table 7.2 (continued)

Phyllum	Number of species corresponding to each genus
Chlorophyta	*Acetabularia* (2), *Anadyomene* (3), *Avrainvillea* (4), *Boergesenia* (1), *Boodlea* (3), *Bornetella* (3), *Bryopsis* (6), *Caulerpa* (34), *Chaetomorpha* (10), *Chlorodesmis* (1), *Cladophoropsis* (4), *Cladophora* (27), *Codium* (10), *Derbesia* (2), *Dictyosphaeria* (3), *Enteromorpha* (11), *Gayralia* (1), *Geppella* (1), *Gomontia* (1), *Halimeda* (13, mostly in southern part of Vietnam), *Halicystis* (1), *Microdictyon* (4), *Monostroma* (1), *Neomeris* (3), *Ostreobium* (1), *Parvocaulis* (3), *Phyllodictyon* (1), *Pseudobryopsis* (1), *Pseudochlorodesmis* (1), *Rhipidosiphon* (1), *Rhipiliopsis* (1), *Rhizoclonium* (4), *Tydemannia* (1), *Trichosolen* (2), *Udotea* (4), *Ulothrix* (2), *Ulva* (14), *Ulvella* (2), *Valonia* (6), *Vaucheria* (1).
Cyanophyta	*Blennothrix* (2), *Brachytrichia* (2), *Coleofasciculus* (1), *Calothrix* (8), *Gloeotrichia* (1), *Heteroleibleinia* (1), *Hydrocoryne* (2), *Leibleinia* (2), *Leptolynbya* (2), *Lyngbya* (9), *Mastigocoleus* (1), *Microchaete* (2), *Nostoc* (1), *Oscillatoria* (7), *Phormidium* (9), *Planktolyngbya* (1), *Pseudanabaena* (1), *Richelia* (1), *Rivularia* (3), *Scytonema* (1), *Scytonematopsis* (2), *Symploca* (1), *Spirulina* (4), *Trichocoleus* (1)

Sources: Huynh and Nguyen (1998), Le and Dang (2005), Dang et al. (2007a), Nguyen et al. (2013), Phang et al. (2016), Dang et al. (2019), Zhao et al. (2021)

Table 7.3 Some gaps and tasks for enhancing marine algal diversity research in South East Asia countries (Phang et al. 2016)

Gap	Task
Checklists for each country (sub-regions) including location, Global Positioning Systems - GPSs data, ecological data	Expeditions Taxonomy workshop (e.g. Southeast Asian Seaweed Taxonomy Consortium - SEASTAx) Monographs
Historical biodiversity data	Compile archives/historical reports Cleaning house for publications Search herbaria worldwide
Methods (collection, processing, identification, systematics, phylogenetics)	Workshops for method standardization Distribution of research tasks amongst regional laboratories Close cooperation amongst herbaria; molecular systematics & phylogenetics
List of threatened, endangered, extinct species	Identify methods to determine threatened species Compile list of threatened species using periodic checklists
Regional marine protected areas	Identify habitats, regions for protection of the species

even overexploitation of transboundary natural seaweed populations, the management of natural seaweed resources needs to be solved together by all countries in the South East Asia Sea (Phang et al. 2016). The marine algae serve as an important source of revenue for Vietnam, especially for the poorer coastal and maritime communities, where *Kappaphycus*/*Eucheuma*, *Gracilaria* and *Caulerpa* farming brings in additional income.

7.3 Diversity of Coastal Seaweeds in the Geographical Regions of Vietnam

In terms of coastal geographical areas of Vietnam alone, 805 seaweed species have been found. Of these, cyanobacteria have 65 species (accounting for 7.64% of the total species), Rhodophyta: 409 species (48.1%), Ochrophyta: 149 species (17.53%), and Chlorophyta: 182 species (21.41%) (Phang et al. 2016; Dam 2021).

7.3.1 Diversity of Seaweed Species in the Islands of Vietnam

In the coastal and offshore islands of Vietnam, 883 different seaweed species have been found. Of which, there are 63 species of cyanobacteria, Rhodophyta: 430 species; Ochrophyta: 173 species and Chlorophyta: 217 species (Dam 2021).

7.3.2 Seaweed Resources in Vietnam

According to Nguyen (1993), the coastal area of Vietnam has a reserves of the genus of *Sargassum* nearly 35,000 tons and *Gracilaria* about 93,000 tons. In addition to these two basic groups, genus *Ulva* in Ly Son has reserves of about 700 tons of fresh (Dam 2021), *Caulerpa* in Phu Quy about 740 tons of fresh (Do et al. 2018, 2019), Con Co about 90 tons of fresh (Dam 2021) and *Halimeda* in Phu Quy about 630 tons of fresh (Do et al. 2018, 2019).

7.3.3 Economic Seaweed of Vietnam

For over 100 years, the traditional Vietnamese coastal people have harvested and utilized seaweeds. However, the use of seaweeds is still limited to people living in the coastal areas. Vietnam's seaweeds can be divided as follows: 31.7% of the total species are subtropical, while only about 40% of the total species are tropical. Sixty percent of the seaweeds of southern Vietnam have tropical affinities, while only 14.3% have subtropical affinities (Huynh and Nguyen 1998). The major economically important seaweed groups can be used for food and feed (humans and animals), materials for industry, traditional medicine, biofuels and biofertilizers. The species of *Sargassum*, *Gracilaria*, *Kappaphycus* and *Eucheuma*, *Ulva* genera are economically important seaweeds which occur in significant quantities in Vietnam (Dang et al. 2007a).

The investment in developing seaweed farming in recent years in Vietnam has been much improved however, the seaweed processing industry of Vietnam still has many limitations (Dam 2021).

7.4 Cultivation and Exploitation of Potential Seaweed Species in Vietnam

At the present, there are seven species cultivated and many other species harvested from the wild. For each species, there is an increase of biomass and reproductive seasons for harvesting. Depending on the Secondary metabolites content of each species, cultivation and harvesting techniques have been established as follows: (1) *Sargassum* harvesting can take place from April to September annually, four times a year and at four different depths such as 0.5–1 m, 1–1.5 m, 1.6–5 m, 6–20 m; (2) Red seaweed can be harvested from January to May, once a year for each species after releasing spores; (3) Green seaweed need to be harvested from March to June, each harvest have to be separated by every 35 days (Le 2015a).

7.4.1 *Cultivated* Gracilaria

The income from *Gracilaria* resources is mainly due to farming. Total production of Vietnam *Gracilaria* spp. is currently around 47,700 wet tons (equal 7000 tons dry per year) with 9830 hectares of cultivated area (Dang et al. 2019). Currently, two species of *Gracilaria* genus are cultivated at a commercial scale including *G. tenuistipitata* (about 42,700 wet tons (accounting for 89% of the total production of *Gracilaria* in the whole country), *G. firma* (about 4000 wet tons/year, accounting for 8.3% of the total *Gracilaria* production in the whole country) (Le and Nguyen 2010). This value was 11% lower than that from the past (Nguyen 1993; Dang et al. 2019). *Gracilaria* is cultured mainly in brackish water, semi closed lagoons, ponds, and river mouths. Cultivation typically still involves semi-intensive, extensive, or modified extensive farming (Huynh and Nguyen 1998; Le 2015a).

In Vietnam, there has been a reliable process of growing *Gracilaria asiatica* Chang et Xia (the former scientific name is *G. verrcusa* (Huds.) Papenf.) seaweed to yield 2 tons of dry seaweed/ha/year - as reported by Vietnam Ministry of Fisheries (https://www.vanbanphapluat.co/28tcn155-2000-quy-trinh-ky-thuat-trong-rong-cau; Nguyen 2000; Le 2015a).

However, cultivating *G. tenuistipitata* has the disadvantage of being harvested many 4–5 times per a year, the economic efficiency is low because of the low cost (about 4500 VND for 1 kg of dry) and the area is limited in the pond because this seaweed grows well at salinity from 5% to 25‰ and poor resistance to waves but it is creating jobs and increasing economic income for coastal people (Nguyen 1993; Le 2015a; Dang et al. 2019).

7.4.2 *Cultivated* Sargassum

The standing crop of *Sargassum* at coastal province Khanh Hoa, Vietnam reached about 7302 dried tons per year including 22 dominant species (Le et al. 2015) while in Vietnam, until now there are 73 species of *Sargassum* genus reported by Phang et al. (2016).

Many countries such as Japan, Korea, China have a method of propagating *Sargassum* seedlings by spores in the laboratory and then grown out in the nature field to restore and cultivate (Hwang et al. 2006, 2007; Li et al. 2009; Shao et al. 2009). *Sargassum* is a seaweed resource with a very large natural yield, their yield according to statistics is about 20,000 dry tons, its value is about 120 billion Viet Nam dong (VND). This seaweed has an important role in balancing coastal ecosystems such as absorption of nutrients in water to minimize nutrient pollution in coastal water environment, as shelter and breeding grounds for valuable seafood species such as shrimp, crab, squid, sea cucumber, and sea urchin (Phan 2018).

Nguyen (1997) has studied the classification system, morphological characteristics, structure and reproduction of the genus *Sargassum*. However, the information on the environmental effects on the development of *Sargassum* zygote which is necessary for the development of this seaweed culture in Vietnam is still limited. Many reports show that *Sargassum* eggs release into the environment in April–May and develop into seedlings while in the first stage from August to December, they grow very slowly (Nguyen 1997; Le et al. 2009; Le and Bui 2011; Le et al. 2015).

It is estimated that Vietnam has a coastal water area of several million hectares that is capable of cultivating seaweed (Huynh 2008a). If *Sargassum* is grown with an average yield of 5 dry tons/ha, the total yield obtained is about 5,000,000 tons dry/year for about one million ha, equivalent to 2,500,000 tons of hydrocarbons/year, or 2,000,000 tons of ethanol/year is expected (assuming about 80% of hydrogen carbon is converted to ethanol) (Le 2015a). Studies on the model type, seed density, water level depth, and seeding time were conducted with the following results: The yield of 1902.65 kg fresh is equivalent to 237.83 kg of dry according to the national standard TCVN 10371: 2014 with maximum humidity of 35%, impurities maximum 3%. The model and farming techniques of *Sargassum* seaweed are published in Vietnam (Nguyen 1997; Le 2015a, b).

7.4.3 *Cultivated* Kappaphycus/Eucheuma

In Vietnam, there is a process of cultivating *Kappaphycus alvareii* and *K. striatum/Eucheuma denticulatum* seaweed in the shrimp ponds and seawater surface of coastal areas with a yield of 13–15 dry tons/ha/year (Huynh 2008b).

There are some species belonging to *Kappaphycus/Eucheuma* genera which can be cultivated in the open sea such as *K. alvarezii*, *K. striatum*, *Eucheuma denticulatum*. The main method for culturing the above species is the method of floating rafts not only in Vietnam but also in Asian countries (Dang et al. 2008). Results have shown that the stocking density of 100–120 g/propagation or 1–1.3 kg/m^2, a duration of 60 days and grown in 40–80 cm below the water surface are suitable for their cultivation. While *K. alvarezii* can be grown in the cold season, *K. striatum* prefers the hot season. Carrageenophytes yield from species of the *Kappaphycus* genus reached approximately 12–17 tons dry/ha/year (Dang et al. 2010; Le 2015a, b).

7.4.4 *Cultivated* Ulva

The species of the genus *Ulva* is perfectly suitable for farming in ponds and lakes in Vietnam. They can grow well in areas where salinity changes from 5–35‰ at the temperature from 20–45 °C. The species of green seaweed can completely meet the demand of raw materials for industrial products such as bioethanol and fertilizers in Vietnam on a large and industrial scale. Studies and trials on different scales showed the model of farming for different species had approximately 25–32 tons dry/ha/year and a fully guaranteed 2000 VND/kg dry corresponding to 1 cent/1 kg (Le 2015a).

7.5 Selection of Suitable Farming Areas and Technology

7.5.1 Gracilaria *Seaweed*

The potential is based on areas of *Gracilaria* ponds and ponds in which *Gracilaria* could be cocultivated with other animal species including shrimp and fish. At present in Vietnam, *Gracilaria* yields are not high and labor is expensive. Assuming a yield of 2 tons dry/ha/year, the production yield of *Gracilaria* farming could potentially rearch 9.577.4 tons dry (Table 7.4, Fig. 7.1) (Le 2015a).

Table 7.4 The area and production yield of *Gracilaria* and *Kappaphycus* cultivation in 2015 in Vietnam (Le 2015a)

Provine	*Gracilaria* Cultivated area (ha)	*Gracilaria* Production yield (tons dry)	*Kappaphycus* Cultivated area (ha)	*Kappaphycus* Production yield (tons dry)
Quang Ninh	250	500	445	2672
Hai Phong	390	780	211	1264
Thai Binh	340	680	–	–
Nam Dinh	395	790	–	–
Ninh Binh	215	430	–	–
Thanh Hoa	61.9	123.8	66	393
Nghe An	191.3	382.6	140	842
Ha Tinh	60.4	120.8	49	295
Thua - Thien Hue	160.9	321.7	–	–
Quang Tri	16.4	32.7	86	518
Quang Binh	60.0	120.0	75	448
Da Nang	10.0	20.0	–	–
Quang Nam	125.0	250.0	–	–
Quang Ngai	8.0	16.0	–	–
Binh Dinh	150.0	300.0	1050	6220
Phu Yen	210.0	420.0	635	3970
Khanh Hoa	120.0	240.0	1120	6660
Ninh Thuan	71.7	143.4	974	6300
Binh Thuan	61.6	123.2	–	–
Vung Tau	42.0	84.0	1420	8740
Ca Mau	1849.6	3699.2	–	–
Kien Giang			1420	8740
Total	**4788.8**	**9577.4**	**7691**	**47,062**

7.5.2 Kappaphycus *Seaweed*

The actual coastal aquaculture areas in Vietnam are enormous. However, seaweed (*K. alvarezii*) is still new in Vietnam and is only commonly grown in some central provinces, with current production of about 2150 tons of dry/year on an area of 1049 ha with a yield of 2–4 dry tons/ha/year, of which the majority is for export. The total sea surface area can be increased on a large scale in many provinces such as Quang Ninh, Hai Phong, Thanh Hoa, Nghe An, Ha Tinh, Quang Binh, Quang Tri, Binh Dinh, Phu Yen, Khanh Hoa, Ninh Thuan, Kien Giang and can reach 7691 ha with estimated potential yield 47,062 tons dry biomass in Table 7.4, Fig. 7.1 (Huynh 2008a, b; Le 2015a).

Fig. 7.1 Illustration for the cultivation of *Kappaphycus alvarezii* seaweed in Phu Quy island district (Binh Thuan Province) (top row); Preliminary commercial processing of *Gracilaria* and *Hydropuntia* seaweed in Ly Son island district (Quang Ngai province) (second row); Harvesting *Ulva* sp. from nature in Ly Son island district (Quang Ngai province) (third row) and natural distribution of *Caulerpa racemose* in Phu Quy island district (Binh Thuan province) (fourth row) (Phung et al. 2019; Do et al. 2018, 2019)

7.5.3 Green Seaweed

For the development of brackish water shrimp farming in Vietnam 2015 and 2025 it has been shown that post-shrimp wastewater must be treated before reuse or discharge to reduce the effects of nitrate and phosphate pollution on the region of coastal sea. The area of the water treatment ponds in shrimp culture accounts for 10% of the total area of shrimp culture at any given culture period (Fig. 7.1). The use of green seaweed for rotational or alternating culture with shrimp will be very beneficial for shrimp because it both reduces excess nutrient pollution in shrimp ponds while increasing income from using algae biomass for exploitation of bioactive substances and other uses (Le 2015a).

Table 7.5 Expected seaweed area and production of Vietnam in 2015 (Le 2015a)

Seaweed	Natural		Cultivation		Total	
	Area (ha)	Production (tons dry)	Area (ha)	Production (tons dry)	Area (ha)	Production (tons dry)
Green seaweed	2810	949	536,702	2,481,632	539,512	2,482,581
Red seaweed	1593	1265	12,480	56,639	14,073	57,904
Brown seaweed	2229	9488	0	0	2229	9488
Total	6632	11,702	549,182	2,538,272	555,814	2,549,973

7.6 Potential Areas for Seaweed Development for Natural Seaweed Harvesting

The potential area for exploitation of the natural seaweed beds was calculated based on the field survey data from biomass of $0.25m^2$ combined with measurements using Google map images. The calculated results showed that the estimated area was about 6632 ha in 2010, and the exploited seaweed biomass was up to 11,702 tons dry as shown in Table 7.5 (Le 2015a).

7.7 Use of Seaweed in Vietnam

Agar production Vietnam started to produce agar in the 1960s of the last century in Hai Phong. In 1976, agar production was adopted in many different localities such as Thua Thien Hue, Nha Trang, Ho Chi Minh City. Materials for agar production are mainly species such as *Gracilaria, Gracilariopsis*, *Gelidiella* and focus on some species such as *Gracilaria verucosa*, *G. tenuistipitata*, *Gracilariopsis baillinae/Gracilaria heteroclada*, *Gelidiella acerosa.* Although the agar extraction technology in Vietnam has been much improved, the quality of the agar produced is still low in coagulation, and the akali content is still high, so it is rarely used in high technology industries. In particular, Vietnam does not have many agarose products to serve the actual domestic needs. This product continues to be imported (Huynh and Nguyen 1998; Dang et al. 2007b; Dam 2021).

Biomass of some *Gracilaria* species such as *G. tenuistipitata* and *G. firma* is important material for the production of agar. In recent years, the quality of agar has decreased over time (in years) in both its content and condensation (Le and Nguyen 2010; Dang et al. 2019). This may be due to many reasons, either due to post-harvest management or to the increased production demand for juvenile *Gracilaria* biomass.

Alginate production The main source of raw materials used for alginate extraction in Vietnam only focuses on some species of the *Sargassum* genus such as *S. mcclurei*, *S. kjellmanianum* (Dam 2021).

Carrageenan production The raw material of Rhodophyta seaweed that contains a lot of carrageenan in Vietnam is not much. *Kappaphycus alvarezii* containing carrageenane cultivated since 1993 and currently is being strongly developed in Vietnam, serving for export and domestic carrageenan production (Dang et al. 2010; Le 2015a).

7.8 Processing of Health Foods from Seaweeds

Seaweeds is used as traditional food in Vietnam for long time. Many species of seaweeds are of economic importance as food for humans and feed for animals such as cattle, poultry etc. Just like in some other Asian countries as Japan, China, Korea, in Vietnam, *Porphyra* is also commonly used as an ingredient in soups, it can be eaten as a raw vegetable or pickled or prepared for jelly, and is served in large and small restaurants in whole Vietnam. *Gracilaria* is used for raw food and used in cooking, and for the preparation of sour vegetables and jellies or soups. Meanwhile, *Kappaphycus* and *Betaphycus* are used to make jellies and cakes (Dang et al. 2007a).

Ulva, Caulerpa, Dermonema, Gloiopeltis spp. and *Sargassum* spp. (juvenile) is eaten raw as well as cooked vegetables. *Hypnea* is used to make jelly. Diets containing a high proportion of Nori (*Porphyra tenera*), Kombu (*Laminaria digitata*) can be used as a dietary supplement reported by Bocanegra et al. (2003, 2006). In the South and Central of Vietnam, *Gelidiella acerosa* is used to produce "Xu Xoa" which is used during hot summer in Vietnam (Tsutsui et al. 2005; Dang et al. 2007a). This species is also used as an ingredient of "Che", a kind of jelly. *Gracilaria eucheumatoides* is used as an ingredient in soft candy. A jelly made from industrially processed powder agar is called "Dong Xuong", is distinguished from "Xu Xoa" that is made with natural *Gelidiella acerosa* or *Gracilaria*. From the year 2000, in Vietnam, *Kappaphycus* began to be used regularly as an ingredient of "Che" in the Mekong Delta. *Gracilaria* collected from the sea has been used to make agar for domestic consumption or is called "Che" of Vietnam, similar to "Che" made from *Gelidiella acerosa*. *Ulva* sp. are gathered collectively along the central coast of Vietnam. It can not only be fried with pork or beef, but also used as an ingredient in soups. Vietnamese people love these dishes in their daily meals (Tsutsui et al. 2005; Dang et al. 2007a). It can be said that Vietnamese seaweed used as food has initially received attention with small and medium-sized processing facilities such as cooking jelly, jam, candy, seaweed tea and some large and big companies in Vietnam have successfully cultivated *Caulerpa lentilifera*, which is suitable for domestic consumption and export to Japan with high economic value (Dam 2021).

It was reported that *Sargassum* species produced secondary metabolites, for example, terpenoids, alginate, fucoidan, fucoxanthine, laminaran, polyphenols, plastoquinones, steroids, glycerids, etc., as well as possessed several therapeutic activities, such as antitumor, anticoagulant, anti-inflammatory and antioxidant (Dang et al. 2011, 2019).

7.9 Summary

We believe that in the coming time, along with the development of the scientific and technological revolution in Vietnam, Vietnam's seaweed industry will develop strongly and sustainably, well adapt to variable conditions of climate in Vietnam and make a great contribution to socio-economic development. In particular, Vietnam's economic seaweed will be grown on a large commercial scale with disease-free seeds, rich in nutrient with high nutritional values and highly resistant to climate change created using high technology of marine biotechnology, from genetic engineering and cellular techniques such as omics (genomics, metagenoimic, proteomics, algalomics etc.). The Vietnamese seaweed industry will fully exploit the great potential of using seaweed as food to protect human health and domestic animals, as raw materials for industry, as raw materials for use in medicine and pharmacy, as traditional medicine; exploitation of bioactive substances; as raw materials for the development of biofuels, making organic fertilizers and contributing to the reduction of climate change in Vietnam.

References

Abbott IA, Fisher J, McDermid KJ (2002) Newly reported and revised marine algae from the vicinity of Nha Trang, Vietnam. In: Abbott IA (ed) Taxonomy of economic seaweeds with reference to some Pacific species, vol 8. California Sea Grand College Program, La Jolla, pp 291–321

Bocanegra A, Nieto A, Blas B, Sanchez-Muniz FJ (2003) Diets containing a high percentage of Nori or Kombu algae are wellaccepted and efficiently utilized by growing rats but induce different degrees of histological changes in the liver and bowel. Food Chem Toxicol 41:1473–1489

Bocanegra A, Benedi J, Sanchez-Muniz FJ (2006) Differential effects of Kombu and Nori seaweed dietary supplementation on liver glutathione status in normo- and hypercholesterolaemic growing rats. Br J Nutr 95:696–702

Calvyn FAS, Put OA Jr, John B, Alecia B, Sung MB, Gerung GS, Hepburn CD, Dang DH, Zhengyu H, Hiroshi K, Phaik EL, Danilo L, Jin AL, Jaruwan M, Nelson WA, Jung HO, Siew MP, Dinabandhu S, Yuwadee P, Yufeng Y, IkKyo C (2017) Carbon dioxide mitigation potential of seaweed aqualculture beds (SABs). J Appl Phycol 29(5):2366–2373

Dam DT (2021) Biodiversity and seaweed resources in Vietnam. Vietnam J Sci Technol. ISSN 1859-4794. Number 4 year 2021 (745). 63rd year, 14–17 (in Vietnamese)

Dang DH, Hoang TMH (2004) Nutritional analysis of Vietnamese seaweeds for food and medicine. BioFactors Special issue: the proceeding of the 3rd international conference on food factors (ICoFF 03). IOS Press. ISSN 0951-6433 Coden BIFAEU. 22(1–4):323–325

Dang DH, Hoang MH, Pham NS (2007a) Seaweed from Vietnam used for functional food, medicine and biofertilizer. J Appl Phycol 19:817–826

Dang DH, Hien HM, Son PN (2007b) Effect of irradiation on the protein profile, protein content and quality of agar from *Gracilaria asiatica* Zhang et Xia (Rhodophyta). J Appl Phycol 19(6):809–815

Dang DH, Hoang MH, Ngo HT, Hoang SN, Huynh QN, Tran MD (2008) Analysis of the genetic variation of *Eucheuma* and *Kappaphycus* strains in Vietnam using RAPD markers. In: Phang SM, Lewmanomaon K, Phaik-Eem L (eds) Taxonomy of southeast Asian seaweeds, Monograph series 2. IOES, pp 69–82

Dang DH, Hoang MH, Ngo THT, Dinh TTH, Huynh QN (2010) Establish cultivation by mixing crop of different strains of *Eucheuma* and *Kappaphycus* species. J Marine Biosci Biotechnol 4(1):24–30

Dang DH, Hoang MH, Hoang TLA (2011) Studies on the analgesic and anti-inflammatory activities of *Sargassum swartzii* (turner) C. Agardh (Phaeophyta) and *Ulva reticulata* Forsskal (Chlorophyta) in experiment animal models. Afr J Biotechnol 10(12):2308–2313

Dang DH, Hoang TLA, Le TT, Nguyen CH, Dam DT, Do AD (2019) The genus *Gracilaria* in Vietnam. In: Phang, Song, Lim (eds) Institute of Ocean and Earth Sciences Monograph Series 17: taxonomy of southeast Asian Seaweed III, pp 29–46

Do AD, Do VK, Dam DT, Dang DH, Tran MD (2018) Potential resources, possibility to explore and grow economic seaweed in Islands of Vietnam. National Key Science and Technology Program 2016–2020. Science and Technology Research for marine and island management and marine economic development. Code KC09/16-20. Proceedings of the Report of the Mid-Term Conference Program KC09/16-20; pp.: 134–148 (in Vietnamese)

Do DA, Tran VH, Bui MT, Nguyen VH, Nguyen TM, Dang DH (2019) Species diversity of seaweed around Ly Sown Island, Quang Ngai province. J Hue Univ Nat Sci; ISSN 1859-1388; 128(1A): 51–72/hueuni-jns.v128i1A.5114 (in Vietnamese)

Farah DA, Abdullah A, Shahrul Hisham ZA, Chan KM (2015) Antioxidant activity of red algae *Kappaphycus alvarezii* and *Kappaphycus striatum*. Int Food Res J 22(5):1977–1984

Hessami MJ, Cheng SF, Ranga Rao A, Yeong HY, Siew MP (2019) Bioethanol production from agarophyte red seaweed, *Gelidium elegans* using a novel sample preparation method for analysing bioethanol content by gas chromatography. 3 Biotech J 9(1):25

Hoang MH, Nguyen CH, Ngo THT, Dang DH (2018) Hypolipidemic effect of *Sargassum swartzii* extract in macrophage and hepatocyte cells mediated by the regulation of peroxisome proliferator activated receptor α and γ. Res J Biotech 13(10):23–27

https://www.vanbanphapluat.co/28tcn155-2000-quy-trinh-ky-thuat-trong-rong-cau. Technical process for cultivation of *Gracilaria asitica* to productivity of 2 tones/ha/year issued by the Ministry of Fisheries

Huynh QN (2008a) Survey planning and propose solutions to develop sustainable cultivation of *Kappaphycus alvarezii*. The report of Ministry of Fisheries 2005–2007, p 240 (in Vietnamese)

Huynh QN (2008b) Investigate planning abd proposing solutions to develop sustainable seaweed culture. Report on project of Ministry fisheries 2005–2007; 280 pp. (in Vietnamese)

Huynh QN, Nguyen HD (1998) The seaweed resources of Vietnam. In: Critchley AT, Ohno M (eds) Seaweed resources of the world. Kanagawa International Fisheries Training Centre, Japan International Cooperation Agency, JICA, pp 60–69

Hwang EK, Park CS, Baek JM (2006) Artificial seed production and cultivation of the edible brown alga, *Sargassum fulvellum* (Turner) C. Agardh: developing a new species for seaweed cultivation in Korea. J Appl Phycol 18:251–257

Hwang EK, Baek JM, Park CS (2007) Assessment of optimal depth and photon irradiance for cultivation of the brown alga, *Sargassum fulvellum* (Turner) C. Agardh. J Appl Phycol 19:787–793

Le NH (2000) Notes on some new species of marine algae from Ninh Thuan province (South Vietnam). In: Collection of marine research works, vol 10. Vietnam Academy of Science and Technology, pp 141–148

Le NH (2004) Some new recorded species for the genus *Gracilaria* of Vietnam. In: Collection of marine research works, vol 14. Vietnam Academy of Science and Technology, pp 81–88

Le NH (2015a) Status and solutions to develop Vietnamese seaweed biomass as materials for bioethanol production. Dr. Hau Nhu Le's collection of published research works to celebrate the 38th anniversary (1978–2015) of the scientific career and the 60th birthday. Vietnam Academy of Science and Technology, Nature Science and Technology publish House, Nha Trang 28.11.2017: 269–279 (in Vietnamese)

Le NH (2015b) Artificial propagation and experimental cultivation of some *Sargassum* seaweed species in coastal areas, General Report; Direction of marine science and technology with subject code number VAST06.05 /13-14 (in Vietnamese)

Le NH, Bui ML (2011) Potential and solutions to develop seaweed resources in the Central provinces. International Workshop in Investigation and Research on Marine Resources and Environment, pg. 305–312 (in Vietnamese)

Le NH, Dang DH (2005) Rename of some species of *Gracilaria* cultivated at Dinh Vu and Cat Hai ponds (Hai Phong) based on the comparative morphology and ITS1 sequence analysis. J Biotechnol 3(2):245–256. (in Vietnamese)

Le NH, Lin SM (2006) *Gracilariopsis* nhatranggensis (Gracilariaceae, Rhodophyta) a new marine red alga from Nhatrang, southern Vietnam. Bot Stud 47:329–337

Le NH, Nguyen HD (2006) Contribution to the study of *Gracilaria* and related genera (*Gracilaria* - Rhodophyta) form Vietnam. Coast Mar Sci 30:214–221

Le NH, Nguyen HD (2010) Rong câu Việt Nam nguồn lợi và sử dụng (The Vietnamese *Gracilaria*, resources and utilization). Vietnamese Natural Science and Technology Publishing House, Hanoi, p 241. (in Vietnamese)

Le NH, Bui ML, Vo DT, Nguyen BK, Vo XM, Tran QT, Nguyen HD (2009) The *Sargassum* season of Khanh Hoa province. The third national scientific conference on ecology and biological resources. 958–962. Agricultural Publishing House (in Vietnamese)

Le NH, Bui ML, Tran VH, Vo TT (2015) New records of marine algae in Vietnam. Ocean Sci J 50(2):221–229

Lee KM, Dang DH, Sung MB (2014) Phylogenetic relationships of *Rosenvingea* (Scytosiphonaceae, Phaeophyceae) from Vietnam based on cox3 and psaA sequences. Research article. Algae 29(4):289–297

Li MZ, Ding G, Zhan DM, Yu B, Liu W, Wu HY (2009) A method for early production of large-size *Sargassum thunbergii* seedling in north China. Prog Fish Sci 30:75–82. (in Chinese with English abstract)

Mohammad JH, Ranga Rao A, Ravishankar GA (2019) Opportunities and challenges in seaweeds as feed stock for biofuel production. In: Ravishankar GA, Rao AR (eds) Handbook of algal technologies and phytochemicals: volume-II phycoremediation, biofuels and global biomass production, vol II. CRC Press, Boca Raton, pp 39–50

Ngo THT, Hoang TLA, Hoang TMH, Nguyen CH, Luu TT, Tran XK, Tran MD, Dang DH (2018) Preparation and evaluation of cream mask from Vietnamese seaweeds. J Cosmet Sci 69:1–16

Nguyen VT (1993) Seaweed resources in coastal areas of Vietnam. Conference of the study and the management of coastal zones of Vietnam. Hanoi, 48–53 (in Vietnamese)

Nguyen HD (1997) Sargassaceae in Vietnam resources and utility. Agriculture Publishing House, Hanoi, 199 pp. (in Vietnamese)

Nguyen XL (2000) Research on resources and technology for planting and exploiting some seaweed species. Ministry of Fisheries, 91 pp. (in Vietnamese)

Nguyen HD, Huynh QN (1993) Rong bien Vietnam (Marine algae of North Vietnam). Nha xuat ban KH-KT, Hanoi, p 364. (in Vietnamese)

Nguyen HD, Le NH (2005) Some new records of marine algae from Vietnam. J Mar Sci Technol 4(5):120–128

Nguyen HD, Pham HT (2003) Some new records of marine algae from Vietnam -2. Collect Mar Res Works 13:95–104
Nguyen HD, Huynh QN, Tran NB, Nguyen VT (1993) Marine algae of North Vietnam (Rong biển phía Bắc Việt Nam), vol 344. Science and Technical Publishing House, Hanoi. (in Vietnamese)
Nguyen HD, Pham HT, Nguyen XV (2000) New records of marine algae from Vietnam. Collect Mar Res Works 10:127–140
Nguyen TV, Le NH, Lin SM, Steen F, De Clerck O (2013) Checklist of the marine macroalgae of Vietnam. Bot Mar 56(3):207–227
Phaik EL, Ji T, Siew MP, Aluh N, Dang DH, Sunarpi H, Hurtado AQ (2014) Genetic diversity of *Kappaphycus* Doty and *Euchauma* J. Agardh (Solieriaceae, Rhodophyta) in Southeast Asia. J Appl Phycol 26:1253–1272
Pham HH (1969) Vietnamese seaweed (southern part). Saigon Learning Resource Center, 558 p (in Vietnamese)
Pham HH (1985) Plants of Phuquoc Island. Ho Chi Minh Publishing House, p 188. (in Vietnamese)
Phan BV (2018) Evaluation of *Sargassum* sp, as ingradients for cosmetic application. University of Science & Technology of Hanoi (USTH), Internship report for the degree of Master II, 32 pp
Phang SM, Yeong HY, Ganon-Fortes ET, Lewmanomont K, Prathep A, Hau LN, Gerung GS, Tan KS (2016) Marine algae of the South China Sea bordered by Indonesia, Malaysia, Philippines, Singapore, Thailand and Vietnam. Raffles Bull Zool Suppl 34:13–59
Phung VG, Do AD, Bui MT, Do VKH (2019) Development of a model of *Kappaphycus alvarezii* seaweed cultivation using the net cages in Phu Quy island, Binh Thuan province. Sci Technol Mar Fish 4:18–24
Report to General Statistics Office (2020). https://www.gso.gov.vn/en/homepage
Shama A, Joyce SG, Mari FD, Ranga Rao A, Ravishankar GA, Hudaa N (2019) Macroalgae and microalgae: novel sources of functional food and feed. In: Ravishankar GA, Rao AR (eds) Handbook of algal technologies and phytochemicals: volume-I: food, health and nutraceutical applications. CRC Press, Boca Raton, pp 207–219
Shao JP, Feng L, Ti FS (2009) Cultivation of the brown alga *Sagarssum horneri*: sexual reproduction and seeding production in tank culture under reduced solar irradiance in ambient temperature. J Appl Phycol 21:413–422
Sze-Looi S, Phaik-Eem L, Siew-Moi P, Weng-Wah L, Dang DH, Anchana P (2014) Developmet of chloroplast simple sequence repeats (cpSSRs) for the intraspecific study of *Gracilaria tenuistipitata* (Gracilariales, Rhodophyta) from different populations. BMC Res Notes 7:77
Tan J, Phaik-Eem L, Siew-Moi P, Dang DH, Sunarpi H, Hurtado AQ (2013) Assessment of four molecular markers as potential DNA barcodes for red algae *Kappaphycus Doty* and *Eucheuma J.* Agardh (Solieriaceae, Rhodophyta). PLoS One. (PONE-D-12-27271R1) 7(12):e52905
Tsutsui I, Huynh QN, Dinh Nguyen H, Arai S, Yoshida T (2005) The common marine plants of southern Vietnam. Japan Seaweed Association, Kochi, 250 pp
Zhao W, Dong L, Dang DH, Brodie J, Chen W-Z, Dam DT, Zhou W, Lu QQ, Zhang M-R, Yang LE (2021) Haplotype networks of *Phycocalidia tanegashimensis* (Bangiales, Rhodophyta) indicate a probable invasion from the South China Sea to Brazil. Mar Biodivers 51:33

Chapter 8
Seaweeds in Mauritius: Bioresources, Cultivation, Trade, and Multifarious Applications

Nadeem Nazurally, Sunita Facknath, Swaleha Hudaa Neetoo, Bhanooduth Lalljee, Ambati Ranga Rao, and Gokare A. Ravishankar

Abbreviations

EEZ	Exclusive Economic Zone
IMTA	Integrated Multitrophic Aquaculture
SIDS	Small Island Developing States

8.1 Introduction

8.1.1 Republic of Mauritius

The Republic of Mauritius is a tropical small Island Developing State (SIDS) of volcanic origin comprising numerous Islands located in the southwest Indian Ocean. Mauritius comprises a lagoonal area of around 300 km^2 within an exclusive

S. Facknath · S.H. Neetoo · B. Lalljee · N. Nazurally (✉)
Department of Agricultural and Food Science, Faculty of Agriculture, University of Mauritius, Reduit, Mauritius
e-mail: n.nazurally@uom.ac.mu

A. R. Rao
Department of Biotechnology, Vignan's Foundation of Science, Technology and Research (Deemed to be University), Vadlamudi, Guntur, Andhra Pradesh, India

G. A. Ravishankar
Dr. C. D. Sagar Centre for Life Sciences, Dayananda Sagar College of Engineering, Dayananda Sagar Institutions, Bangalore, Karnataka, India

A. Ranga Rao, G. A. Ravishankar (eds.), *Sustainable Global Resources of Seaweeds Volume 1*, https://doi.org/10.1007/978-3-030-91955-9_8

economic zone (EEZ) of 1.9 km^2 and an additional sea area of 396,000 km^2 jointly managed by Seychelles under the Joint Management Agreement (JMA). Mauritius has a population of roughly around 1.2 million residents, topped by a growing number of tourists who already surpassed the 1.3 million per year visiting Mauritius (Statistics Mauritius 2021) prior to the COVID-19 pandemic outbreaks in 2020. The ocean has been referred to as the pillar of the global economy (EDB 2021) and Mauritius has not been immune to developing this important sector.

8.1.2 Blue Economy of Mauritius

Since the independence in 1968, Mauritius have made considerable progress and is classified as a country with high income by the World Bank in 2019. Fisheries represent a very important sector in Mauritius and fisheries in Mauritius include artisanal fisheries around the main island Mauritius, Rodrigues, Agalega and St. Brandon islands, offshore semi-industrial fishing on the oceanic banks along the Mascarene Ridge stretching from St. Brandon to Saya de Malha and around the Chagos Archipelago, and foreign-flagged industrial purse seiners (Boistol et al. 2011). The goods and services of the Mauritian lagoon and seas are clearly critical to the social and economic welfare of the current development trends. The fact of having limited natural resources in Small Island Developing States (SIDS) has prompted new approaches in developing the Ocean Economy. Aquaculture is key among the various activities identified (EDB 2021). The Government of Mauritius is committed to ensuring responsible and sustainable development of the aquaculture industry in the island: capacity building and knowledge programs to ensure food security whereby self-sufficiency in fish products could be attained are underway. Thirty-one sites have been allocated for marine-based aquaculture, both for lagoonal and offshore exploitation, among which eight sites are already being exploited in the South-East of the Island, more precisely at Point-aux-Feuilles, one site over 16 years while four sites over for over 6 years exploitation while three sites are being planned with a potential of 5000 tons of yearly production. Mauritius is a member of the Indian Ocean Rim Association (IORA) which is mainly involved in Blue economy. The first conference was held in Mauritius during which much emphasis was laid on the development of blue economic sector of the big island. Development englobes job creation in the marine sector, to sustainably exploit oceans' resources and give a boost to the economy as a whole (IORA 2017). Mauritius draws 1.5% of Gross Domestic Progress (GDP) from the blue economy (SeaFood Source 2020). The blue economy of Mauritius includes (1) Fisheries; (2) Aquaculture; (3) Shipping; (4) Tourism; (5) Trade; (6) Marine biotechnology; (7) Trade among others. The Mauritian ocean territory holds enormous potential for economic growth of the country, provided that this resource is managed responsibly. Fisheries represent a very important sector in Mauritius and fisheries in Mauritius include artisanal fisheries around the main island Mauritius, Rodrigues, Agalega and St. Brandon islands, offshore semi-industrial fishing on the oceanic banks along the Mascarene Ridge stretching from St. Brandon to Saya de Malha and around the Chagos Archipelago, and foreign-flagged industrial purse seiners (Boistol et al. 2011). Meanwhile. seaweeds have long been an untapped resource that Mauritius approximately 435 species hold in terms of economic potential. The world seaweed industry has already reach US$8.8 billion in 2019 and Mauritius has the potential to develop this sector given the high diversity of seaweeds and well documented species. The recorded seaweed species are currently 435 species (59 brown algae, 108 green algae, and 268 red algae) (Jagtap 1993) and downloaded from the website (Guiry and Guiry 2011).

8.2 Seaweeds in Mauritius

Seaweeds or macroalgae refer to several species of macroscopic, multicellular, and marine algae that dwells and grow in the ocean (Shama et al. 2019). Some may be microscopic as well as those towering as large, massive underwater forests. They can be useful in the transportation of photosynthetic elements or simply act as support systems (The science of seaweeds, American Scientist, 2013). In Mauritius, seaweeds are primarily obtained from wild sources with little research attempted to cultivate selected species without any commercial scale production up to now. Seaweeds have been mainly used for different purposes; (1) Consumption; (2) Animal feed and; (3) Crude manure. Nowadays, the use of seaweeds has largely been diversified with the help of advanced technologies but yet to be exploited in the Island. It ranges from production of (1) biofuels, (2) cosmetics, (3) medicinal purposes, (4) refined edible products and (5) animal feeds. Due to the rising demands, aquaculture farms are new avenues for commercial exploitation of seaweeds. This also stands as an employment generator for the locals and aquaculture products as new source of exports.

8.2.1 Seaweed Diversity in Mauritius

The seaweed diversity in Mauritius is well documented with 435 species of seaweeds (Bolton et al. 2012). Namely, 59 brown algae (Phaeophyceae), 108 green algae (Chlorophyceae) and 268 red algae (Rhodophyceae). This is arguably an elevated figure, which designates the wide diversity of seaweeds in the lagoons of Mauritius. This recorded figure is approximately the same for the coastline of Tanzania (Bolton and Stegenga 2002; Bolton et al. 2004; Spalding et al. 2007). This shows an indication of extremely high species diversity on the island as well as in the marine environment of Mauritius. The work of Vaughan and Boergesen contributed largely to the figures (Bolton et al. 2012). Jagtap (1993) worked on 10 different sites around the island and came across 127 species of seaweeds. The abundance of seaweeds from his study shows that *Ulva, Cladophora, Boodlea, Sargassum,*

Turbinaria, Gracilaria, Hypnea, Digenea and Palisada (formerly *Laurencia)* are the dominant species, while in today's climate change era requires further and updated studies to confirm richness. This were reported in the sites in intertidal zones and lagoons around Mauritius. As compared to subtidal zones, *Asparagopsis, Halimeda, Turbinaria* and coralline red algae were in dominance in 1993.

8.3 Bioresources, Cultivation and Trade

As per the FAO (2020), aquaculture of seaweeds represented 97.1 percent by volume of the total of 32.4 million tons of wild-collected and cultivated combined and is bringing in US$7.4 billion each year in the net global revenue. The production of seaweeds has shown a growth of averagely 0.7 percent from 1970 to 2008 but global production of aquacultured macroalgae experienced relatively low growth with a decline of 0.7% in 2018 (FAO 2020). This was due to a decrease in tropical seaweeds and reduction in Southeast Asia while seaweeds farming in temperate and cold-water was yet on the rise (FAO 2020). Mauritius being a tropical Island can develop this sector into a flourishing market for international trade. As per FAO, 2020, the first 3 leading seaweed species being farmed are as follows (1. *Laminaria japonica*; 2. *Euchema* spp. and 3. *Gracilaria* spp.). Mauritius is known to host both *Euchema* spp. and *Gracillaria* spp. which could be cultured. However, the predation by herbivorous fishes remains a serious threat for this industry in lagoonal culture except in close land-based aquaculture system or enclosed coastal aquaculture. Seaweed cultivation using the rope techniques has already been attempted at research level in 2011 by the Mauritius Research and Innovation Council (MRIC). Seaweeds were attached to the rope and were laden on the seafloor of shallow water. This method is an easy one involving wooden poles tied with ropes (Fig. 8.1). Another method that was studied was the floating frames. In this way, frames (which were made up of bamboo) were hanged in the middle of the sea. The frames were anchored to the seafloor and ropes containing the seaweeds' seeds were attached to the frames. As such, Mauritius can be a potential site to introduce the seaweeds industry. Only one species (*Gracilaria salicornia*) was cultured by MRIC. The high diversity of seaweeds in Mauritius definitely opens new avenues for research and cultivation prospective in Mauritius.

8.3.1 Integrated Multitrophic Aquaculture (IMTA)

Aquaculture started back in 2002 with the setup of the fish farm at Pointe aux Feuilles namely Ferme Marine de Mahebourg Ltd. In 2014, the Government embarked on developing the blue economy with marine aquaculture being a priority for expansion. Marine aquaculture and Seafood Industry is a key among the various activities identified for future development contributing to national food security.

Fig. 8.1 Coral farm being cleaned by herbivorous juvenile fishes. While grazing is important to maintain coral nurseries in healthy state, seaweed farming in Mauritian waters needs to be protected to ensure no predation on cultured species

Marine aquaculture holds a tremendous potential, but it has also its disadvantages in terms of waste generations from excessive fish feeds and fish excretions (Nazurally 2018). Currently, Mauritius has only one operational floating cage fish farm with an output of around 3000 tons of fish per year and increasing to 5000 tons in the near future. Fish are exported to the United States, European Union and African countries while a sustainability approach to cultivate seaweeds within fin-fish farms could maintain and create new markets.

Fed marine aquacultures can be a threat to the coastal waters. There are input of both organic and inorganic matters in the systems. With excess of these substances, there can be a disbalance in the nutrient level of the systems as well as the surrounding waters. Seaweeds could play a vital role in sustainable nutrient cycling (Chopin et al. 2008) while ensuring both aquaculture of fin fish and seaweeds for commercial purposes. With eutrophication, Integrated Multitrophic Aquaculture (IMTA) can be a potential solution. This is the rearing of several organisms simultaneously on the same site. The IMTA systems are adopted in the African countries for abalone culture (Bolton et al. 2009) and fish culture in Tanzania with the use of the seaweed species *Ulva* spp. (Msuya et al. 2006). Furthermore, herbivorous fishes can be reared and directly fed with the seaweed cultured within the IMTA system (Bolton et al. 2009; Nobre et al. 2010). *Asparagopsis* sp., a local seaweed of Mauritius is

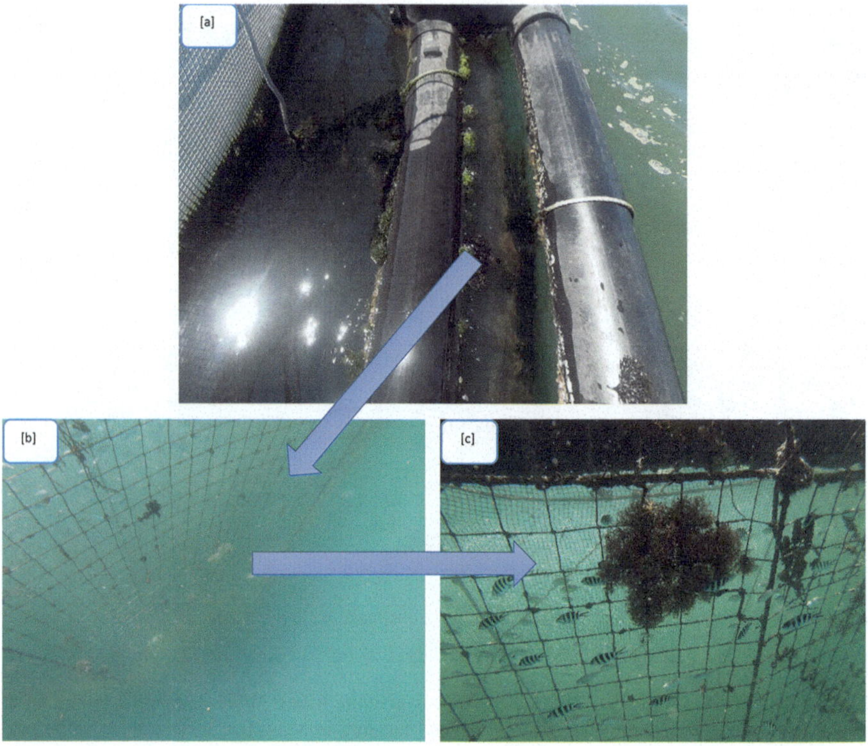

Fig. 8.2 Natural growths of various species of seaweeds on the outer net of the round shaped lagoonal floating cage aquaculture. (**a**) The structures supporting the fish nets with seaweeds growing; (**b**) outer net placed to protect the first net of floating cage fish farm; (**c**) seaweed attached to the outer net with high diversity of fish at the fish farm for a seaweed culture experiment

abundantly found around the island. Based on the various studies conducted worldwide, this species can be used in bioremediation of excess nutrient discharge from aquaculture farms (Schuenhof et al. 2006) and have many other useful uses (Kraan and Barrington 2005; Bolton et al. 2011).

Research conducted at the lagoonal floating cage aquaculture in Mauritius showed promising results where green seaweeds: *Ulva intestinalis.*, *Caulerpa zeyheri.*, brown seaweeds: *Padina boryana*, *Chnoospora minima* and red seaweeds: *Gracilaria corticata*, *Jania* sp. completely covered the outer net (10 cm mesh size) of the 2 nets of the fish farm in just 3 months. Being vertical very few echinoderms were observed to graze on the seaweeds with no sea urchins rather some starfish only. Surprisingly, the nets at the fish farm were not subject to high predation by herbivorous fishes which promotes the IMTA system as a preferred solution for seaweeds culture in Mauritius. The cylindrical net size varies from 5 m, 15 and 20 m radius on top and 10–15 m deep. Seaweeds have been observed to grow to a depth of 8–10 m only. The dried biomass production was an average 27 kg of seaweeds per m^2 on the net. All the ropes attached to each fish cages (maximum depth 10 m) were also observed to support natural seaweed growths (Fig. 8.2).

8.4 Trade and Potential of Multifarious Application of Seaweeds for Mauritius

Currently, seaweeds have no economic trade in Mauritius, rather few kgs are sold in local restaurants by fisher people specially the *Gracilaria* spp. and used exclusively by tourists and very few Mauritians. The potential to develop this important industry is enormous given the high diversity and density of species Mauritius host within its water. There are different ways that seaweeds can be prepared and marketed. It can be sold raw, dried or conceptualize into a product like in Mauritius, it was dried and made into pickles and still sold in some artisanal places. In Zanzibar, the seaweeds are dried and grinded into a powder. Prior to this, Vaseline and soaps are made. These products are sold locally. These 2 products are two potential seaweed value-added products that are made in Zanzibar. With the proper marketing strategies, they can reach the international markets.

8.4.1 Research, Development and Conservation Using Seaweeds

8.4.1.1 Sand Erosion

Seaweeds not only represents a valuable commodity for human use but are also crucial to the marine ecosystems such as feeds for herbivorous commercial fishes, supporting life at sea and breeding grounds of many fishes. While seaweeds have long been studied on their impacts and proliferation in the lagoonal waters of Mauritius, research are being undertaken by the University of Mauritius on the importance of seaweeds to defer the effect of sand erosion through dissipation of wave energy in the lagoon prior to reaching the shoreline.

8.4.1.2 Coral Farming

Coral reefs are often referred to rainforests in terms of their huge biodiversity (Rinkevich 2006). Similarly, to rainforests, coral reef in the ocean inhabits a wide variety of marine organisms. Right at the beginning of the Ordovician period (roughly 500 million years ago), it has been shown that corals have been able to overcome natural threats (Heeger et al. 2000), Yet threats like overgrowths of seaweeds (Nazurally 2018) and declining herbivorous fishes are hindering natural recruits and coral farming attempts (Fig. 8.3).

Fig. 8.3 Shallow table coral nursery with corals and naturally occurring seaweeds left uncleaned for 3 years and maintained naturally by herbivorous fishes who in turns attracts carnivores and help in keeping the coral predators away. The inter-relationship of seaweeds, corals and surrounding biodiversity helps keep a healthy environment

8.4.1.3 Biofuels

With the ever-growing demand for fuel and energy, new techniques must be employed to compensate for the non-renewable sources of energy. Global energy consumption was 400 quadrillion BTU in 2000 and is predicted to be almost doubled by 2030 (World Energy Outlook 2008). Therefore, much emphasis is laid on finding new ways of producing energy without harming the environment and with lesser carbon footprint (Sitompul et al. 2012a, b). Biomass conversion for the production of energy is a step of utmost importance. Seaweeds can be a sustainable source or marine biomass that can be used in the industry (Mohammad et al. 2019; Hessami et al. 2019). It happens that seaweeds have low content of lignin, but high level of carbohydrates and water. As such, seaweeds have good anaerobic degradability (Bruton et al. 2009; Bruhn et al. 2011; Chang et al. 2010). Both green and brown seaweeds can be good sources of biomass as they yield high levels of methane (Bruhn et al. 2011; Bruton et al. 2009; Kelly and Dworjanyn 2008). *Ulva Lactuca* can be a good candidate to produce biofuels. As per the study of McKendry (2002), it is more likely to have higher rate of growth as compared to terrestrial crops that can be used as feedstock. This marine plant is a potential biomass for high yields (Bruhn et al. 2011) as compared to banana stem wastes.

Mauritius is fast developing in marine aquaculture industry with a potential of 15% organic waste from total production. While in some places in Mauritius, excessive nutrient leaching into the lagoon by agricultural activities has sparked overgrowths of seaweeds causing nuisance upon decomposition and accumulation along the coastal zones (Leatherman 1997). A paper published in Mauritius (Nazurally 2018) investigated the biogas potential from fish waste and a mixture of seaweeds/seagrasses with promising results for sustainable energy production. Substrates were prepared for total solids of 10–12% and were analysed in Biochemical Methane Potential (BMP) assays. The inoculum used in all BMPs was acclimated sludge in a ratio of 1:3 for inoculum to substrate on a mass basis. There were in total four vessel reactors (VRs) with different compositions: VR1 100% fish wastes, VR2 60% fish wastes and 40% seagrass/macroalgae, VR3 40% fish wastes and 60% seagrass/macroalgae, and VR4 100% seagrass/macroalgae. VR5 acted as control with 100% inoculum. The maximum cumulative biogas productions (CBPs) reached 8288 ml for VR1, 8410 ml for VR2, 4236 ml for VR3 and 2746 ml for VR4 with a concentration of methane gas of 61.1, 65.07, 68.07, and 53.28%, respectively (Nazurally 2018). This potential for biogas production represents a clean source of cheap fuel for sustainable development in Small Island Developing States having aquaculture and seafood industries along coastal regions (Nazurally 2018).

8.4.1.4 Compost

Solid wastes and landfills are a major problem in Small Island Developing States with limited land space for disposal. Emphasis is laid upon the reduction of organic waste to landfill, which otherwise could be an important transformation into compost for agricultural and other sustainable use. The treatment and recycling are the eco-friendly bioremediations technologies (Artola et al. 2009). A study conducted by Mohee et al. 2013 compared the batch composting process of mixed *Ulva reticulata*, bagasse and broiler litter in two composting experiments with one mix containing double distilled water washed seaweeds supported that aerobic batch compositing of this species is feasible compared to other composting experiments in terms of the evolution of several parameters. While seawater can be a problem for seed germination and ways to remove the excess salt needs to be done in a sustainable manner for commercial scale production (Mazé et al. 1993; Taylor et al. 2001; Han et al. 2014). The quality of compost produced by the washed sample was superior to that produced from the unwashed seaweeds (Mohee et al. 2013). Composting of seaweeds, rather than disposal can be an excellent initiative to produce high-quality fertilizers (Mohee et al. 2013). The N:P:K content is quite elevated which is a good aspect for the growth and development of crops (Mohee et al. 2013). Seaweed can well be used in the fertilization of terrestrial cultivation. They are advertised as being sustainable, containing growth hormones and trace elements. to be the staple food for a major proportion of the global population.

8.4.2 Consumption of Seaweeds

Seaweeds are increasingly gaining interest in many countries due to their opinion as being an environmentally friendly food (FAO 2020) rich in iodine, iron and vitamin A (Tanna and Mishra 2019). Since the beginning of civilizations, seaweeds have been consumed by coastal communities (Yang et al. 2010; Teas et al. 2004). In the habitual diet of most Asian populations, seaweed is a major component (Jiménez et Sánchez 2000). This have gain popularity in the Western cultures also, Hawaii,

Table 8.1 Edible algae from Mauritius

Type of algae	Scientific name	Order/Genus
Brown	*Phaeophyceae*	*Saccharina* *Sargassum* *Dictyotales*
Red	*Rhodophyceae*	*Rhodymenia* *Gracilaria* *Asparagopsis*
Green	*Chlorophyceae*	*Ulva*

Brazil and California, whereby seaweeds are involved in several dishes. It is interesting that several food value seaweeds are found naturally in the Mauritius sea environment (Table 8.1). The are gaining popularity for food uses due to (1) the preference of Japanese who have influence the local diet; (2) invasion of Asian cuisine and (3) for the many heal benefits associated with its consumption (Cardoso et al. 1997; Yamori et al. 2001). Epidemiological evidence has shown that with the consumption of seaweeds, several diseases may be inhibited. Seaweeds as a food component have also shown that it increases fullness (satiety) and help reduce absorption of lipids and fats in the human body. As such, it can be a major component to help fight obesity. (MacArtain et al. 2007; Paiva et al. 2017).

In the years 1994 and 1995, over 2 million ton (dry matter) of seaweeds were produced and harvested (Zemke and Ohno 1999). A major part of the harvest was meant for consumption while the remaining were used for industrial purposes.

The addition of seaweeds in canned foods have been demonstrated as a preserving agent in response to Gram-Negative bacteria. This would be a way to reduce the addition of salt to the food given that Mauritius is already on the high list of high salt consumption and diabetic population. Studies have shown that the antimicrobial effects of seaweeds are linked with the antioxidant level present in them (Devi et al. 2008). Seaweeds contain high levels of dietary fibers (Table 8.2). However, this differs for each type of seaweed and research is crucial to understand the dietary fibers of the species present in Mauritius.

8.4.2.1 Livestock Feeds

Seaweeds have a long record of use as livestock feed worldwide and even in few coastal villages of Mauritius. They are known to have variable composition which depends on species and environmental conditions they are thriving in water (Makkar et al. 2016). Most essential amino acids are lacking in seaweeds except the sulphur containing amino acids (Makkar et al. 2016). The most abundant seaweeds found in Mauritius is the brown algae namely from the *Sargassum* genera (Jagtap 1993).

Asparagopsis taxiformis is also known as the climate change solution seaweeds due to its capacity to lower the methane emissions from livestock by just adding one to two percent of the seaweed in their diet (Machado et al. 2014). are abundant in

Table 8.2 Corresponding type of algae and dietary fibers

Type of algae	Type of dietary fibers
Brown	Alginates Fucans Laminarans
Green	Soluble ulvans Cellulose
Red	Galactans Agar Carageenans

both Mauritius and Rodrigues Island. While Rodrigues Island is very dependent on livestock and farming, *Asparagopsis taxiformis* could be added to feeds to reduce the carbon footprints of agricultural activities. While feeding this seaweed has also been subject to disadvantage given the high dosage of bromoform which could be carcinogenic to cattle (Daalder 2020).

8.5 Conclusions and Future Perspectives

Mauritius has a large diversity and density of important seaweeds for the potential commercialization and development of such industries. Seaweeds have long been an untapped important marine resource that Mauritius has yet to commercialized in various forms and products. Preliminary research on several aspects has demonstrated the benefits and potential sustainable solutions for the Small Island Developing States. Nevertheless, in-depth research needs to be undertaken with an emphasis on commercial production for both local and international markets. Mauritius has been twice subjected to Covid-19 lockdown (March 2020 and March 2021). The economic activity has been seriously jeopardized further advocating the advancement of the blue economy to tap such resources to ensure long-term exploitation and economic growths. Future perspectives lie in establishing solid local resource exploitation for commercialization as Covid-19 persists even in 2021. The propagation of marine finfish lagoonal aquaculture could be vital in establishing an IMTA with seaweed culture in mass quantity for large scale production.

Acknowledgement NN is thankful to the Mauritius Research and Innovation Council for partly financing his MPhil/PhD. The authors are thankful to the private firm La Ferme Marine de Mahebourg Ltd for continuous support in conducting research at their facilities. ARR acknowledges support and resources from the Improvement of Science & Technology Infrastructure in Higher Educational Institutions (FIST Project no: LSI-576/2013) by the Department of Science and Technology, Government of India and Centre of Excellence, Department of Biotechnology, Vignan's Foundation for Science, Technology and Research.

References

Artola A, Barrena R, Font X, Gabriel D, Gea T, Mudhoo A, Sánchez, A. (2009) Composting from a sustainable point of view: respirometric indices as a key parameter. In: Martín-Gil J (ed) Compost II, dynamic soil dynamic plant, vol 3, pp 1–16

Boistol L, Harper S, Booth S, Zeller D (2011) Reconstruction of marine fisheries catches for Mauritius and its outer. Fish Centre Res Rep 19(4):39

Bolton JJ, Stegenga H (2002) Seaweed biodiversity in South Africa. S Afr J Mar Sci 24:9–18. https://doi.org/10.2989/025776102784528402

Bolton JJ, Leliaert F, De Clerck O, Anderson RJ, Stegenga H, Engledow HE, Coppejans E (2004) Where is the western limit of the tropical Indian Ocean seaweed flora? An analysis of intertidal seaweed biogeography on the east coast of South Africa. Mar Biol 144:51–60. https://doi.org/10.1007/s00227-003-1182-9

Bolton JJ, Robertson-Andersson DV, Shuuluka D, Kandjengo L (2009) Growing *ulva* (chlorophyta) in integrated systems as a commercial crop for abalone feed in South Africa: a swot analysis. J Appl Phycol 21:575–583. https://doi.org/10.1007/s10811-008-9385-6

Bolton JJ, Andreakis N, Anderson RJ (2011) Molecular evidence for three separate cryptic introductions of the red seaweed Asparagopsis (Bonnemaisoniales, Rhodophyta) in South Africa. Afr J Mar Sci 33:263–271. https://doi.org/10.2989/1814232X.2011.600339

Bolton JJ, Bhagooli R, Lydiane M (2012) The Mauritian seaweed flora: diversity and potential for sustainable utilisation. Univ Maurit Res J:V18a-2012

Bruhn A, Dahl A, Nielsen HB, Nikolaisen L, Markager S, Olesen B, Arias C, Jensen D (2011) Bioenergy potential of Ulva lactuca, biomass yield, methane production and combustion. Bioresour Technol 102:2595–2604. https://doi.org/10.1016/j.biortech.2010.10.010

Bruton T, Lyons H, Lerat Y, Stanley M, Rasmussen MB (2009) A review of the potential of marine algae as a source of biofuel in Ireland. Sustainable Energy Ireland Report

Cardoso MA, Hamada GS, De Souza JMP, Tsugane S, Tokudome S (1997) Dietary patterns in Japanese migrants to southeastern Brazil and their descendants. J Epidemiol 7:198. https://doi.org/10.2188/jea.7.198

Chang HN, Kim N, Kang J, Jeong CM (2010) Biomass-derived volatile fatty acid platform for fuels and chemicals. Biotechnol Bioprocess Eng 15:1–10. https://doi.org/10.1007/s12257-009-3070-8

Chopin T, Robinson SMC, Troell M, Neori A, Buschmann AH, Fang J (2008) Multitrophic Integration for Sustainable Marine Aquaculture. In Jørgensen, SE, Fath BD (Editor-in-Chief), Ecological Engineering. Vol. [3] of Encyclopedia of Ecology 5, 2463–2475, Elsevier, Oxford. https://doi.org/10.1016/B978-008045405-4.00065-3

Daalder M (2020, February 19) How this seaweed could slash dairy emissions. Newsroom. Retrieved from https://www.newsroom.co.nz/2020/02/19/1035951/stewart-island-seaweed-could-cut-dairy-emissions

Devi KP, Suganthy N, Kesika P, Pandian SK (2008) Bioprotective properties of seaweeds: in vitro evaluation of antioxidant activity and antimicrobial activity against food borne bacteria in relation to polyphenolic content. BMC Complement Altern Med 8:38. https://doi.org/10.1186/1472-6882-8-38

Economic Development Board (2021). Available from: https://www.edbmauritius.org/node/19. Accessed 24 Feb 2021

Food and Agriculture Organisation of the United Nations (2020) The state of the world fisheries and aquaculture. Sustainability in Action. 224 pp. https://doi.org/10.4060/ca9229en

Guiry MD, Guiry GM (2011) AlgaeBase. World-wide electronic Publication. National University of Ireland, Galway. http://www.algaebase.org

Han W, Clarke W, Pratt S (2014) Composting of waste algae: a review. Waste Manag 34:1148–1155. https://doi.org/10.1016/j.wasman.2014.01.019

Heeger T, Sotto F, Gatus JJ. Laron C (2000) Community based coral farming for reef rehabilitation, biodiversity conservation and as a livelihood option for fisher folk. Proceedings ADSEA, SEAFDEC Philippines

Hessami MJ, Cheng SF, Ranga Rao A, Yin YH, Phang SM (2019) Bioethanol production from agarophyte red seaweed, *Gelidium elegans* using a novel sample preparation method for analysing bioethanol content by gas chromatography. 3-Biotech J 9(1):25

Indian Ocean Rim Association (IORA) (2017) Blue economy. Available from: https://www.iora.int/en

Jagtap TG (1993) Studies on littoral and sublittoral macrophytes around the Mauritius coast. Atoll Res Bull 382:23

Jiménez-Escrig A, Sánchez-Muniz FJ (2000) Dietary fibre from edible seaweeds: chemical structure, physicochemical properties and effects on cholesterol metabolism. Nutr Res 20:585. https://doi.org/10.1016/S0271-5317(00)00149-4

Kelly MS, Dworjanyn S (2008) The potential of marine biomass for anaerobic biogas production. Marine estate research report, the crown estate, Scotland. The Crown Estate, 103 pages, ISBN: 978-1-906410-05-6

Kraan S, Barrington KA (2005) Commercial farming of *Asparagopsis armata* (Bonnemaisoniales, Rhodophyta) in Ireland, maintenance of an introduced species? J Appl Phycol 17:103–110. https://doi.org/10.1007/s10811-005-2799-5

Leatherman SP (1997) Beach rating: a methodological approach. J Coast Res 13:253–258

MacArtain P, Gill CIR, Brooks M, Campbell R, Rowland IR (2007) Nutritional value of edible seaweeds. Nutr Rev 65:535. https://doi.org/10.1301/nr.2007.dec.535-543

Machado L, Magnusson M, Paul NA, Nys RD, Tomkins N (2014) Effects of marine and freshwater macroalgae on in vitro total gas and methane production. PLoS One 9(1):e85289. https://doi.org/10.1371/journal.pone.0085289

Makkar PS, Harinder TG, Heuze V, Giger-Reverdin S, Lessire M, Lebas F, Ankers P (2016) Seaweeds for livestock diets: a review. Anim Feed Sci Technol 212:1–17. https://doi.org/10.1016/j.anifeedsci.2015.09.0180377-8401

Mazé J, Morand P, Potoky P (1993) Stabilisation of 'Green tides' Ulva by a method of composting with a view to pollution limitation. J Appl Phycol 5:183–190. https://doi.org/10.1007/BF00004015

McKendry P (2002) Energy production from biomass (part 1): overview of biomass. Bioresour Technol 83:37–46. https://doi.org/10.1016/S0960-8524(01)00118-3

Mohammad JH, Ranga Rao A, Ravishankar GA (2019) Opportunities and challenges in seaweeds as feed stock for biofuel production. In: Ravishankar GA, Rao AR (eds) Handbook of algal technologies and phytochemicals: volume II Phycoremediation, biofuels and global biomass production. CRC Press, Boca Raton, pp 39–50

Mohee R, Mudhoo A, Peryagh C, Somaroo GD (2013) Composting of washed Ulva reticulata seaweed: process parameter variations and compost quality. Int J Environ Waste Manag 12(1):2013

Msuya FE, Kyewalyanga MS, Salum D (2006) The performance of the seaweed Ulva reticulata as a biofilter in a low-tech, low-cost, gravity generated water regime in Zanzibar, Tanzania. Aquaculture 254:284–292. https://doi.org/10.1016/j.aquaculture.2005.10.044

Nazurally N (2018) Anaerobic digestion of fish waste and seagrass/macroalgae: potential sustainable waste management for tropical Small Island Developing States. J Mater Cycles Waste Manage 20(3):1724–1735. https://doi.org/10.1007/s10163-018-0738-1

Nobre AM, Robertson-Andersson DV, Neori A, Sankar K (2010) Ecological-economic assessment of aquaculture options: comparison between abalone monoculture and integrated multi-trophic aquaculture of abalone and seaweeds. Aquaculture 306:116–126. https://doi.org/10.1016/J.AQUACULTURE.2010.06.002

Paiva L, Neto AI, Lima E, Baptista JJ (2017) Nutritional and functional bioactivity value of selected azorean macroalgae: *Ulva compressa*, *Ulva rigida*, *Gelidium microdon*, and *Pterocladiella capillacea*. J Food Sci 82(7):1757–1764. https://doi.org/10.1111/1750-3841.13778

Rinkevich B (2006) The coral gardening concept and the use of underwater nurseries: lessons learned from silvics and silviculture. CRC Press Taylor & Francis Group, pp 291–299

Schuenhoff A, Mata L, Santos R (2006) The tetrasporophyte of *Asparagopsis armataas* a novel seaweed biofilter. Aquaculture 252:3–11. https://doi.org/10.1016/j.aquaculture.2005.11.044

SeaFoodSource (2020) Mauritius charts growth path for fisheries and aquaculture sectors. By Shem Oirere. Available from: https://www.seafoodsource.com/news/supply-trade/mauritius-charts-fisheries-and-aquaculture-growth-path

Shama A, Joyce SG, Mari FD, Ranga Rao A, Ravishankar GA, Hudaa N (2019) Macroalgae and microalgae: novel sources of functional food and feed. In: Ravishankar GA, Rao AR (eds) Handbook of algal technologies and phytochemicals: volume-I: food, health and nutraceutical applications. CRC Press, Boca Raton, pp 207–219

Sitompul J, Bayu A, Soerawidjaja TH, Lee HW (2012a) Studies of biogas production from green seaweeds. Int J Environ Bioenergy 3(3):132–144

Sitompul J, Widayat P, Soerawidjaja TH (2012b) Evaluation and modification of processes for bioethanol separation and production. Int J Renew Energy Develop 1:15–22. https://doi.org/10.14710/ijred.1.1.15-22

Spalding MD, Fox HE, Allen GR, Davidson N, Ferdanñ ZA, Finlayson M, Halpern BS, Jorge MA, Lombana A, Lourie SA, Martin KD, Mcmanus E, Molnar J, Recchia CA, Robertson J (2007) Marine ecoregions of the world: a bioregionalization of coastal and shelf areas. Bioscience 57:573–583

Statistics Mauritius (2021). Available from: http://statsmauritius.govmu.org/English/Pages/default.aspx. Accessed 24 Feb 2021

Tanna B, Mishra A (2019) Nutraceutical potential of seaweed polysaccharides: structure, bioactivity, safety, and toxicity. Compr Rev Food Sci Food Saf 18(3):817–831. https://doi.org/10.1111/1541-4337.12441

Taylor R, Fletcher RL, Raven JA (2001) Preliminary studies on the growth of selected 'Green Tide' algae in laboratory culture: effects of irradiance, temperature, salinity and nutrients on growth rate. Bot Mar 44:327–336. https://doi.org/10.1515/BOT.2001.042

Teas J, Hebert JR, Fitton JH, Zimba PV (2004) Algae -- a poor man's HAART? Med Hypotheses 62:507. https://doi.org/10.1016/j.mehy.2003.10.005

World Energy Outlook (2008) Available from, www.worldenergyoutlook.org

Yamori Y, Miura A, Taira K (2001) Implications from and for food cultures for cardiovascular diseases: Japanese food, particularly Okinawan diets. Asia Pac J Clin Nutr 10:144. https://doi.org/10.1111/j.1440-6047.2001.00227.x

Yang YJ, Nam S-J, Kong G, Kim MK (2010) A case-control study on seaweed consumption and the risk of breast cancer. Br J Nutr 103:1345. https://doi.org/10.1017/S0007114509993242

Zemke-White W, Ohno M (1999) World seaweed utilisation: an end-of-century summary. J Appl Phycol 11:369–376. https://doi.org/10.1023/A:1008197610793

Chapter 9
Seaweed Resources and Their Cultivation in Iran

Jelveh Sohrabipour and Reza Rabiei

9.1 Introduction

Iran with 1560 m^3 renewable water per capita is under water stress based on the Falkenmark water stress index and during the period of 1994–2014 the average surface run-off in Iran was 52 billion m^3 which is 42% lesser than the long-term average. The cumulative groundwater shortage has increased from 65×10^9 m^3 to $109 \times 10^9 m^3$ (Moridi 2018). The agriculture sector uses 90% of the total allocated water and total area of agricultural lands has significantly increased in this 10-year period while this sector is not economically efficient. Many large cities in Iran with more than 37 million people suffer from water stress. These evidences show that Iran's water resources have a critical condition based on the UN water scarcity index (Moridi 2018). Use of fresh water in agricultural activities has increased from $44 \times 10^9 m^3$ in 1961 to $80 \times 10^9 m^3$ in 2001 and 86.5×10^9 m^3 in 2011. Population growth and increasing the large cities' population due to the migration of rural communities, as well as more than two decades of drought periods and also the climate change has put pressure on the drinking water resources of Iran's metropolises. Therefore, finding alternative water resources that will support the long-term agricultural activities sustainably is a very important issue in the country. Desalination plans are one of the options which can be noticed and a big project is underway in this area but desalination, especially in large scale, has many negative side effects on the environment. Seaweeds as marine plants can be considered as suitable bioresources in marine agronomy or mariculture to provide some of the human food needs (Craigic 2011; Cornish et al. 2017). Seaweeds have many uses for humans,

J. Sohrabipour (✉) · R. Rabiei
Department of Natural Resources Researches, Agriculture and Natural Resources Research and Education Center, Agricultural Research, Education and Extension Organization (AREEO), Bandar Abbas, Iran
e-mail: sohrabipour2@gmail.com

A. Ranga Rao, G. A. Ravishankar (eds.), *Sustainable Global Resources of Seaweeds Volume 1*, https://doi.org/10.1007/978-3-030-91955-9_9

particularly for food, feed, phycocolloids, fertilizer, and so on. These marine plants had been used in some countries especially in Asia for centuries and nowadays extensively are being investigated for many propose such as producing bio-fuel, nutraceutical, medicinal, personal care products and food applications (Kim et al. 2017; Shama et al. 2019; Mohammad et al. 2019). Based on the latest FAO reports about 51.3% of the total production of marine and coastal aquaculture belong to seaweeds and 99.5% of seaweed mariculture is concentrated in Asia (FAO 2018). Only eight genera of cultivated seaweeds provide 96.8% of global seaweed production. Asian countries such as China, Indonesia, Philipines, South Korea, Malaysia and Vietnam has produced % 99.5 of the total world production of seaweeds (Chopin and Tacon 2020; FAO 2020). In India, wild resources of seaweeds are used for agar and alginate production, whereas cultivated red seaweeds *Kappaphycus alvarezii* (Doty)Doty ex P.C. Silva is used for carrageenan production (Ganesan et al. 2019).

The country of Iran in south west of Asia has around 3000 km of marine coastline in its south and north boundaries. Nearly, 750 km of coastline is located in northern part adjacent to the Caspian Sea area, while the southern coastlines are around 2250 km in length along the Persian Gulf and Gulf of Oman (Pak and

Fig. 9.1 Location of the Persian Gulf and gulf of Oman

Farajzadeh 2007).The country in the south is connected to the Arabian Sea and the Indian Ocean via Strait of Hormuz in the Persian Gulf (Fig. 9.1). Around one third of the southern coast of Iran is adjacent to the sea waters of the Oman Gulf and the remained is adjacent to the waters of the Persian Gulf. The Persian Gulf, with an area of about 235,000 square kilometers (about 90,000 square miles) is located among the countries of Iran, Iraq, Kuwait, Saudi Arabia, Bahrain, Qatar and the United Arab Emirates. The present chapter provides an introduction to the Iranian seaweed resources and current efforts to create and develop cultivation and industry of seaweed in this country. Challenges and future prospects for establishment of seaweeds industry in Iran are also discussed.

9.2 Climatic Conditions of the Persian Gulf and the Gulf of Oman

High temperatures and low annual rainfall are the climatic features of the southern coasts of Iran. The average annual temperature in the area is higher than all parts of Iran.The absolute maximum temperature is around 52 °C, which is only 4 °C lower than the absolute maximum temperature of the world. The lowest air temperature in the coldest month of year (January) range between (16–19 °C) with the absolute minimum recorded (2–15 °C). The temperature of coastal waters is also affected by these conditions and varies between 10 and 39 °C. The waters of Persian Gulf are extremely stressful environment for marine organisms because of very high temperatures as well as high salinities.

9.3 Seaweed Resources of the Persian Gulf and Gulf of Oman

9.3.1 The Arabian Shores of the Persian Gulf

Various studies have been carried out on algal species of Saudi Arabia (Newton 1955a; Basson, 1979a, b, 1992; Basson et al., 1989; DeClerck and Coppejans 1996; Abdel-Kareem 2009a, b). Many species of algae also have been reported from the shores of Kuwait (Newton, 1955a, b; Al-Hasan and Jones, 1989; Jones 1986). From 1570 species reported in the catalog of Indian Ocean marine algae, 214 species belonged to the Persian Gulf region (Silva et al. 1996). John and Al-Thani (2014) published a revised checklist on benthic marine algae of the Persian Gulf, regard to all literature that had already been published from all over the Persian Gulf including the above-mentioned literature and also the reported species from Qatar and Abu Dabi (Dorgham 1990; Mshigeni & Dorgham 1987; Heiba et al. 1990; Dipper 1991; Dipper & Woodward 1993; Kureishy et al. 1995). In their report, most of the studies on the Iranian coastlines of the Persian Gulf (Sohrabipour and Rabiei 1996, 1999a, b; Sohrabipour and Rabiei 2007; Sohrabipour et al. 2004; Rabiei et al. 2005) also have been considered.

9.3.2 The Iranian Shores of the Persian Gulf

The history of seaweed study in Iran dates back to 1845, when Endlicher and Diesing (1845) reported six species of brown and red algae from Khark and Kish Islands in the Persian Gulf.The second list of Iranian seaweeds consist of species was reported by Børgesen (1939) which was based on Danish fisheries investigation in the Persian Gulf in 1937.Børgesen report included 14 new species and varieties from the Persian Gulf. Nizamuddin and Gessner (1970) reported 38 species from the Kish Island in the Persian Gulf which was based on the marine algae collection in the "Meteor" expedition on northern parts of the Arabian Sea.

The most knowledge on the marine algae species of the Iranian coastlines has been achieved after Islamic revolution when in 1996 the first report of marine algae species from the seashores of Hormozgan Province in south of Iran was published by Sohrabipour and Rabiei (1996), which included 16 new records of marine algae. Three years later in 1999 a list of 153 species (including macro and microalgae) were reported from the coast lines of Hormozgan Province (Sohrabipour and Rabiei 1999a). A bilingual book (Persian and English languages) about marine algae species of Iran was published by institute of Iranian fisheries science (2001), contained 155 species from all over Iranian coast line in Gulf of Oman and the Persian Gulf. Sohrabipour et al. (2004) reported 119 species of marine algae from intertidal regions of Bandar Lengeh in western part of Hormozgan Province with emphasis on their monthly distribution in the area. The first checklist of Iranian green marine algae (62 species) and their local distribution was published by Sohrabipour and Rabiei (2007). A checklist including 74 red algae from the Iranian coastlines of the Oman Gulf in Chabahar province was published by Sohrabipour and Rabiei (2008). In 2008 one economic new species of red algae, *Gracilariopsis persica*, belong to Gracilariaceae family were introduced from sea shores of Hormozgan Province based on molecular and morphological studies (Bellorin et al. 2008), the species already had been recorded as *Gracilariopsis longissima* from southern coastlines of Iran (Sohrabipour and Rabiei, 2005). In a checklist from Iran 309 taxa of macroalgae at species and infraspecific level was published and revealed that *Sargassum*, with 25 taxa was the most diverse genus of brown algae and the Rhodomelaceae with 36 taxa was the most diverse family of red algae in the area (Kokabi and Yousefzadi 2015). Totally all studies in recent two decades resulted in a number to 326 eukaryote species of marine algae (Table 9.1, Sohrabipour and Rabiei 2017) including 79 green algae (Table 9.2), 80 brown algae (Table 9.3) and 167 red algae (Table 9.4). More than 240 numbers of the identified species were the new reports from the southern coastlines of Iran in compare to the Silva checklists from the Persian Gulf (Silva et al. 1996).

Table 9.1 Number of Families, genera and species of algal phylum identified from coastlines of the Persian Gulf and Gulf of Oman in south of Iran

Algae phylum	Number of families	Number of genus	Number of Species
Chlorophyta (green algae)	16	22	79
Ochrophyta (brown algae)	9	27	80
Rhodophyta (red algae)	32	74	167
Total	57	123	326

Table 9.2 Number of Families, genera, and species of green algae identified from coastlines of the Persian Gulf and Gulf of Oman in the south of Iran

No.	Families of Cholorophyta	Genus	Species
1.	Boodleaceae	1	2
2.	Bryopsidaceae	1	5
3.	Caulerpaceae	1	13
4.	Cladophoraceae	3	23
5.	Codiaceae	1	9
6.	Derbesiaceae	1	1
7.	Dichotomosiphonaceae	1	2
8.	Halimedaceae	1	2
9.	Phaeophilaceae	1	1
10.	Polybelphararidaceae	1	1
11.	Polyphysaceae	2	3
12.	Siphonocladaceae	1	1
13.	Ulotrichaceae	1	1
14.	Ulvaceae	2	12
15.	Ulvellaceae	1	1
16.	Valoniaceae	3	3
Total	16	22	79

Table 9.3 Number of Families, genera and species of brown algae identified from coastlines of the Persian Gulf and Gulf of Oman in south of Iran

No.	Family of Ochrophyta	Genus	Species
1.	Acinetosporaceae	1	3
2.	Bachelotiaceae	1	1
3.	Chordariaceae	2	2
4.	Cystoseiraceae	4	5
5.	Dictyotaceae	9	27
6.	Ectocarpaceae	1	1
7.	Sargassaceae	3	30
8.	Scytosiphonaceae	6	8
9.	Sphacelariaceae	1	3
Total	9	27	80

Table 9.4 Number of Families, genera and species of red algae identified from coastlines of the Persian Gulf and Gulf of Oman in south of Iran

No.	Families of Rhodophyta	Genus	Species
1.	Acrochaetiaceae	1	2
2.	Ahnfeltiaceae	1	1
3.	Bonnemaisoniaceae	1	1
4.	Callithamniaceae	3	4
5.	Ceramiaceae	4	13
6.	Champiaceae	1	7
7.	Corallinaceae	4	7
8.	Corynomorphaceae	1	1
9.	Cystocloniaceae	1	10
10.	Dasyaceae	2	6
11.	Delesseriaceae	5	5
12.	Erythrotrichiaceae	3	3
13.	Furcellariaceae	1	1
14.	Galaxauraceae	3	3
15.	Gelidiacea	1	5
16.	Gelidiellaceae	1	3
17.	Gigartinaceae	2	2
18.	Gracilariaceae	2	17
19.	Halymeniaceae	2	5
20.	Hymenocladiaceae	1	1
21.	Liagoraceae	3	5
22.	Lomentariaceae	2	4
23.	Phylophoraceae	1	1
24.	Rhodomelaceae	16	37
25.	Rhodymenyaceae	2	3
26.	Sarcomeniaceae	2	2
27.	Scinaiaceae	1	6
28.	Sebdeniaceae	1	1
29.	Solieriaceae	3	7
30.	Stylonemataceae	2	2
31.	Spyridiacceae	1	1
32.	Wrangeliaceae	2	2
Total	32	74	167

By identifying the potential sources of seaweed and understanding their environmental importance and economic value, basic strategies can be achieved for the creation and sustainable development of the algae industry, which will improve the socio-economic conditions in local coastal communities.

9.4 Preliminary Basic Studies for Seaweed Cultivation in Iran

In an ecological study, production of 13 species of red algae in an area of about 18 hectares of intertidal regions of Oman gulf (Chabahar province) was estimated by 865.2 tons/year (Qaranjik et al. 2011). The highest amount was 828 g.m^{-2}in February and the lowest was 90.2 g. m^{-2} in May. The highest production belonged to *Gracilaria corticata* (33.1%) and the lowest production rate (1%) was for *Hypnea pannosa*. The highest amount of algae growth in the studied area was recorded in winter (February and March) and the lowest amount was in the middle spring(May) (Qaranjik et al. 2011).

In a survey of three stations along the coasts of the Bandar Lengeh in Hormozgan Province, production and ecological characteristics of the dominant brown species including *Polycladia myrica* (= Cystoseira myrica), *Sargassum boveanum, Padina australis, and Colpomenia sinuosa* were studied and their average total biomass was estimated (Sohrabipour et al. 2003). *Padina australis* and *Colpomenia sinuosa* showed the higher growth in autumn and winter while *S. boveanum* showed the higher growth (12.24 kg dw/m^2.year) during June to August. *Polycladia myrica* with annual production of 10.77 (kg dw/m^2.year) was the permanent brown algae with widespread distribution in the studied area.

Another ecological study was on *Gracilaria salicornia* (red algae). The species formed small communities in northeast of the Qeshm island in Persian Gulf. In this study, seasonal changes in biomass and two communities of the species were studied since October 2001 for one year. The Monthly production and percentage cover of the species were measured in different depths of the intertidal regions of the two communities. The Results revealed that there are significant differences in percentage cover and production (dry weight) of *Gracilaria salicornia* with respect to depth ranges (shore elevation) ($P < 0.01$). The highest dry production (598.9 ± 67.2 g/m^2) and percentage cover (% 69.12 ± 3.7) of *Gracilaria salicornia* were determined between 2.7–3.5 m. depth of sea water (Rabiei and Sohrabipour 2007).

9.5 Ecological and Economic Importance of Seaweeds

Seaweed as a completely halophyte plant in addition to the role of production in the food chain of aquatic ecosystems, with their colloidal substances and various minerals and vitamins and metabolites, as well as growth hormones have many applications in food, pharmaceutical, forage and fertilizer industries. Seaweeds contain valuable polysaccharides such as agar, carrageenan and alginate, which are widely used in food, pharmaceutical, paper, textile, leather and cosmetic industries.

Algae in addition to their role as food suppliers in marine and ocean ecosystems and positive environmental effects, are also of great importance in terms of food, medicine, industry and agriculture. In terms of nutritional value, algae have high levels of vitamins and minerals needed by humans, as well as compounds with diverse medicinal properties. Reducing blood pressure and harmful fats, weight loss, preventing myocardial infarction, combating osteoporosis, and meeting the needs of essential trace elements needed by humans cause the medicinal properties of algae.

Because of the mentioned properties, 90% of the seaweed production in Asian countries is allocated to direct human food consumption. This is very common in China, Japan, the Philippines, Indonesia, Thailand, and most Southeast Asian countries (McHugh 2003). Consumption of algae and their extracts as algal fertilizers is also common in many countries. Seaweed fertilizers provide the necessary rare minerals, increase the quantity and quality of products, increase the root defense against worms, nematodes, and other pests and also increase the tolerance of plants to environmental stresses such as low and high temperatures and salt stresses (EL-Boukhari et al. 2020; McHugh 2003). Uses of algae as animal feed shows their beneficial effects in terms of nutritional value. In countries such as England, Scotland, Ireland, and France, algae had long been used to raise livestock such as horses, cattle, and pigs (McHugh 2003).

Industrial use of algae is also due to the stabilizing and thickening properties of extracted biopolymers such as agar, carrageenan, alginate, and some other biopolymers. Due to their biological origin, algae are used with more confidence in various products. These algal compounds are exploited to make a variety of health and beauty products such as pills, capsules, medicated syrups, dental impressions and radiological imaging materials. These biopolymers also are used in the food industry to make canned food, dairy products, confectionery, marmalades, jellies, juices and to process meat products such as sausages, hot dogs, and hams. Agarose, which is a purified form of agar, is used in the preparation of culture media in biological research, pathological diagnostics, and preparing of electrophorese media (McHugh 2003).

Algae have a great effect on maintaining the health of coastal ecosystems by absorbing heavy metal and consuming nutrients in the effluents of shrimp, fish, and agricultural farms. Some harmful microalgae, such as red tide-producing species, also cause significant damage to the aquatic sector due to mismanagement of effluents and sewage entering into the sea. This phenomenon has caused severe problems and damage in the southern coasts of Iran in recent years.

9.6 Socio-Economic Importance of Seaweed

Despite considerable allocations for the development of the southern coastal region of Iran during the past years, the situation of rural communities and even the people living on the outskirts of cities in this areas has not yet been improved. Unemployment widely exists in both educated and uneducated groups of youth. In many developed countries especially in the countries such as China, Japan, Korea and most developing southeastern Asian countries traditional cultivation of algae have enabled them to create a niche for survival, revival, development and marketing of algal products and supporting employment for their rural communities(Dogma et al., 1990). The presence of several economically valuable seaweed in the southern coastal waters of Iran is great opportunity for improving the socioeconomic situations of rural communities.

9.7 Commercial Seaweeds Resources in Iran

In recent years, to evaluate the possibility of seaweed farming and estimation of crop production in economically view, some of the economic species such as *Gracilariopsis persica* and *Gracilaria corticata* as agarophytic algae, *Sargassum boveanum* and *S. ilicifolium* as alginophyte species, *Hypnae flagelliformis* and *Hypnea musciformis* as carrageenophyte species have been experimentally cultivated in the intertidal waters or in drainage canals of shrimp farms effluents in south of Iran (Sohrabipour et al. 2003, Rabiei et al., 2005; Rabiei and Sohrabipour 2007, 2008; Farahpour et al. 2010).

9.8 Seaweed Cultivation in Iran

The idea of seaweed Cultivation in Iran was developed by Rabiei and Sohrabipour, who were working in the Agriculture and Natural Resources Research Center of Hormozgan in the south of Iran. They were experts in marine algae taxonomy and ecology. After a few years of collecting and identifying the marine algae resources in southern coastlines of Iran and doing some studies on ecological aspects of some economic seaweeds of the studied area, they focused on the experimental cultivation of *Sargassum spp* (brown algae) and *Gracilaria corticata* (red algae). After finding a members of *Gracilariopsis* genus from the seashores of Banar Abbas, the capital of Hormozgan provinces, they named it as *Gracilariopsis longissima* and focused on growth phenology of the species and experimentally grew it in the intertiadal pools of the area (Fig. 9.2) and the growth rate, the effective environmental factors, monthly fluctuation of physicochemical factors and annual yield of the species were evaluated (Sohrabipour and Rabiei 2008). Later during a collaboration with E.C. Oliveira from the University of São Paulo and A. M. Bellorin from the University of Cape Town, South Africa, who were studying Indian Ocean economic seaweeds using molecular methods, it revealed that the samples which had been identified as *Gracilariopsis longissima* are belong to a new species which was named *Gracilariopsis persica* Bellorin, Sohrabipour et E.C. Oliviera and introduced as new species for the Iranian seashores of the Persian Gulf (Bellorin et al. 2008). In 2006 to 2007 they did trial cultivation of this species along with *Hypnea flagelliformis*, *Gracilaria corticata* and *Sargassum boveanum* in the ponds and drainage canals of shrimp farms in the Tiab at eastern part of Hormozgan Province. The

Fig. 9.2 *Gracilariopsis persica* cultivation in tiadal pools of Bandar Abbas in Hormozgan province (Persian Gulf)

Fig. 9.3 Cultivation of *G. persica* in drainage canals of shrimp farm effluents (Tiab, Hormozgan province)

results showed that only *Gracilraia corticata* has a negligible growth in the pond condition and *S. boveanum* did not able to grow neither in pond nor in drainage canals. The two species *Gracilariopsis persica* and *Hypnea flagelliformis* both able to grow in drainage small canals between ponds. *G.persica* was able grow on monoline plastic ropes (Fig. 9.3) but *H. flagelliformis* showed good growth in the plastic baskets.

In another study on cultivation of *H. flagelliformis* and *Gracilariopsis persica* in shrimp farms in Kolahi and Tiab ports, the maximum crop growth rates for *G. persica* and *H. flageliformis* in drainage canals were measured 83 and 35 ($g.m^{-2}.d^{-1}$) respectively (Rabiei et al. 2014).

During the years 2008 to 2010, a pilot plan project to promote the cultivation of economic algae in Hormozgan province was carried out on the intertidal regions of Bandar Abbas. The red alga *G. persica* with 106(tonnes. fw. $hect^{-1}.year^{-1}$) showed the highest annual yieldand followed by *S. boveanum* and *H. flagelliformis* with 48

Fig. 9.4 Cultivation and harvesting of *G. persica* in tidal pools of Bandar Abbas (Hormozgan) (2009–2010)

Fig. 9.5 Cultivation of *S. boveanum* in tidal area of Bushehr province (2009–2010)

and 20(tons.hect $^{-1}$.year $^{-1}$) respectively (Fig. 9.4). In same time a project was carried out in Bushehr province in southwestern part of Iran both in tidal area and in shrimp farms of this province that was focused on *Sargassum boveanum*. and *Gracilaria corticata*. Result of this project showed Sargassum boveanum grows well in autumn and winter and within 50 days the weight of each algae seedling increased from 15 g to 500 g and if no unexpected factors occur, more than 80 tons of fresh algae can be harvested per hectare (Fig. 9.5), but shrimp farm was not suitable for Sargassum cultivation and seedlings could not settle and were destroyed in the drainage canals and shrimp pond due to high salinity, mud cover and lack of waves.

One of the other common species of the red algae in southern coasts of Iran is *Hypnea musciformis*, which was cultivated experimentally in 2009 on the shores of Tis, Chabahar (Southestern of Iran), its production increased from 500 g of initial weight to 4600 g in 60 days (Fig. 9.6), which showed a 9-fold increase in biomass (Farahpour et al. 2010), showed that *Hypnea flagelliformis* can be used to treat the effluent of shrimp farms by reducing of the nitrogen and phosphorus content in sea water.

In another pilot research during the years 2015–2017, the possibility of producing *Gracilariopsis persica* in tidal farms and also storing its seedlings in the

Fig. 9.6 Cultivation of *Hypnea musciformis* in tidal area of Tis, Chabahar, southeast of Iran (2009–2010)

Fig. 9.7 Creating greenhouse with winter season condition to keep seedlings of *G. persica* in hot season (2016)

unfavorable season and hot months of the year was investigated. This project, which was financially supported by one of the subsidiary companies of the Ministry of Energy of Iran, was implemented by a knowledge-based company in the field of algae (Zist Darya Parvaran Anahita). This study showed that in hot and unfavorable months of the year, by providing greenhouse condition with the temperature and light intensity similar to winter season (Fig. 9.7), algal seedlings can be survive and maintained until about the next planting season. Another way to achieve the purpose of this study was to transfer some ropes with algae seedlings to offshores area, where the water temperature is favorable in hot season. For this purpose, these ropes were installed in a cage near the fish breeding cages (Fig. 9.8), which were about

Fig. 9.8 Cultivation of the *G. persica* at offshore region near the fish cages in Qeshm island (2016)

3 km away from the tidal zone, and the results showed that the time to reach maximum growth was reduced from 45 days to 25 days. Production was also higher and the survival and growth period of seaweed expanded from mid-May in the tidal area to mid-September in the offshore area continued. Unfortunately, this research ended with the influx of turtles and grazing of the seaweeds, and fish farmers did not no longer allow to continue the work.

9.9 Economic Values of Iranian Seaweed Resources

As formerly mentioned the red alga *Gracilariopsis persica* which has been introduced as a new species from Iranian shores of the Persian Gulf (Bellorin et al. 2008) is one of the native agarophytic algae from the southern coast of Iran and can be a good candidate for commercial cultivation and agar production. Based on the studies on agar properties of the species (Salehi et al. 2011) this species has high quality of agar and agarose biopolymers. In addition some studies showed that the species can be used as feed for livestock, shrimp, fish and poultry. In parallel with the above mentioned activities on seaweeds cultivations, several research studies have been conducted on their nutritional effects on livestock, poultry, fishes and shrimp. Studies on poultries showed that *Gracilariopsis persica* has satisfactory results on laying hens, especially egg quality, and cause a significant increase in iodine and egg yolk, beneficial fats (HDL) and shell resistance. in the other hand dcreased the undesirable factors such as triglycerides and harmful fats such as LDL and had no adverse effect on growth and weight of chickens (Vosough-Sharifi et al. 2011; Abbaspour et al. 2015; Kazemi et al. 2018; Safavi et al. 2019). No difference was observed in the consumption of algae-containing diets with the control diet, indicates that up to 5% of *Gracilariopsis persica* can be used in the diet of laying hens without any negative effect on yield. A study on nutritional effects of *Sargassum ilicifolium* on sheep showed that different levels of the species has significantly positive effects on triglyceride level, which shows that this species can be used as an unconventional feed up to 30% in sheep forage (Valikamal et al. 2010). The positive effect of algae on shrimp has also been determined (Hafezieh et al. 2017; Tamadoni Jahromi et al. 2021). A study by Moezzi et al. (2017) showed that several species of macroscopic algae common in the southern coasts of Iran have a significant inhibitory effect on the dinoflagellate alga *Cochlodinium polykrikoides*, which was the main cause of red tide on the coast of Iran during the years 2008–2010, and by killing many of the marine organisms causd irreparable damage to the country's fisheries economy. During an study on bio-methane production from *Gracilariopsis persica* it concluded that agar wastes of *G. persica* showed more bio-gas production than the dried raw materials of the species (Hessami et al. 2019). Five steroid compounds were including 22-dehydrocholesterol, cholesterol, stigmasterol, β-sitosterol, and fucosterol separated and identified from *G. persica* (Saeidnia et al. 2012).

In some studies antiviral, antimicrobial, antioxidant, antidiabetic and anticancer effects of some were seaweeds in Iranian coasts have been investigated which shows the pharmaceutical importance of these marine bieresources (Ghanadi et al. 2007; Zandi 2011, 2013). Phytochemical composition of brown, red and green marine algae of the Persian Gulf in Iranian coastlines also has revealed the nutraceutical values of these marine plants (Pirian et al. 2017a, 2017b, 2017c, 2018a, b, 2020; Zarei et al. 2021). Content of fatty acids and amino acids and protein of *Ulva* spp. (Pirian et al. 2018b; Sohrabipour et al. 2020), antidiabetic and antioxidant activities of some brown and red macroalgae (Pirian et al. 2016), Nutritional and phytochemical properties of some common seaweeds of the Persian Gulf have been studied (Pirian et al. 2017a, b, c, Zarei Jeliani et al. 2017). Inhibitory effects of some species of seaweeds on α-amylase enzyme activity also has revealed the antidiabetic effects of the some marine algae of the Persian Gulf (Pirian et al. 2018a, 2017b).

9.10 Problems and Prospects

Unfortunately, in recent years due to the construction of a coastal boulevard along the city of Bandar Abbas, habitat of the two species *Gracilariopsis persica* and *Hypnea flageliiformis* has severely damaged and the density and aboundant of these two species in natural habitats has decreased significantly compared to the past. So, it is not possible to achieve the necessary and sufficient amount of seedling to grow these valuable seaweeds in the scale of large farms and this is one of the main and important problems in the development of the algae industry in the country. One of the suggested ways to solve this problem is vegetative propagation to produce enough amount of seedling that be able to expand the seaweed farms. It mean that a desirable amount of healthy thallus of the grown seaweeds in the favorable months of the year collected and kept in the storage hall (Fig. 9.9), which is equipped with facilities to control temperature, light and other physicochemical conditions to save them in unfavorable month of the year, as seedlings in next cycle of seaweed farming.

Seaweed cultivation in offshore area where the seawater temperature flactuation is not considerable is another way to grow economic seaweed for longer time and more crop production. But in this case we need high budget for anchores, cages, rops boat and etc. By integrating the fish cage culture techniques with seaweed cultivation in offshore areas can maje more cost-efective the aquculture activities.

Fig. 9.9 Job opportunities for women, men, fishermen, educated youth in seaweed farming activities

Micropropagation methods such as spore culture and tissue culture of the economic seaweed are the other techniques to produce seedlings but these technologies are expensive.

If we can do these work successfully, it could be an effective step in expanding the seaweed cultivation industry in the vast coastal waters of the country and will be able to create job opportunities and income generation in the field of planting, holding, harvesting and processing of various products from these marine algae in the coastal provinces of the south and north of the Iran (Fig. 9.9).

By preventing the entry of urban and rural domestic wastewater as well as the effluent of factories located in coastal areas, in addition to preserving the marine ecosystem, better and more algal growth can be guaranteed.

Acknowledgement We would like to express our sincere gratitude to the directors of the Research Institute of Forests and Rangelands of Iran (RIFRI) and Agriculture and Natural Resources Research Center of Hormozgan Province for providing the necessary facilities for research on Iranian seaweed during two decades. The achievements of these researches on seaweed of the Persian Gulf and the gulf of Oman, provided basic information for the development of the phycology science in Iran and a relative knowledge on these marine resources and also the possible of cultivating these seaweeds.

Authors special thanks to Prof. Gokare A. Ravishankar and Assoc. Prof. Ranga Rao Ambati, Editors of Sustainable Global Resources of Seaweeds: Industrial Perspectives, Volume-I: Bioresources, cultivation, trade and multifarious applications, Springer-Nature publishers for giving us the opportunity and this chapter is focused to Iranian marine algae resources and their economic values.

References

Abbaspour B, Sharifi SD, Mohammadi-Sangcheshmeh A (2015) Dietary supplementation of *Gracilariopsis persica* is associated with some quality related sera and egg yolk parameters in laying quails. J Sci Food Agr 95:643–648

Abdel-Kareem MSM (2009a) New algal records from the Arabian gulf coast of Saudi Arabia. Bot Res Int 2:268–276

Abdel-Kareem MSM (2009b) Phenetic studies and new records of *Sargassum* species (Fucales, Phaeophyceae) from the Arabian Gulf coast of Saudi Arabia. Acad J Plant Sci 2:173–181

AL-Hasan RH, Jones WE (1989) Marine algal flora and seagrasses of the coast of Kuwait. J Univ Kuwait(Sc.) 16:289–340

Basson PW (1979a) Marine algae of the Arabian Gulf coast of Saudi Arabia (first half). Bot Mar 22:47–64

Basson PW (1979b) Marine algae of the Arabian Gulf coast of Saudi Arabia (second half). Bot Mar 22:65–82

Basson PW (1992) Checklist of marine algae of the Arabian Gulf. J Univ Kuwait (Sci) 19:217–230

Basson PW, Mohamed SA, Arora DK (1989) A survey of the benthic algae of Bahrain. Bot Mar 32:27–40

Bellorin AM, Buriyo A, Sohrabipour J, Oliveira MC, Oliveira EC (2008) *Gracilariopsis mclachlanii* sp. nov.and *Gracilariopsis persica* sp. nov.of the Gracilariaceae (Gracilariales, Rhodophyceae) from the Indian Ocean. J Phycol 44:1022–1032

Børgesen F (1939) Marine algae from the Iranian Gulf especially from the innermost part near Bushire and the Island Kharg. In: Jesse K, Spark R (eds) Danish scientific investigations in Iran part 1. Munksgaard, Copenhagen, pp 42–141

Chopin T, Tacon A (2020) Importance of seaweeds and extractive species in global aquaculture production. Rev Fish Sci Aquacult:1–10

Cornish M, Critchley AT, Mouritsen O (2017) Consumption of seaweeds and the human brain. J Appl Phycol 29:2377–2398

Craigie JSJ (2011) Seaweed extract stimuli in plant science and agriculture. J Appl Phycol 23:371–393

DeClerck O, Coppejans, E (1996) Marine algae of the Jubail Marine Wildlife Sanctuary, Saudi Arabia. – In: Krupp, F., A.H. Abuzinada & I.A. Nadar: A marine wildlife sanctuary for the Arabian Gulf. Environmental research and conservation following the 1991 Gulf War Oil Spill, pp. 199–289. – NCWCD, Riyadh and Senckenberg Res. Inst., Frankfurt

Dipper, F (1991) Colonisation and natural changes in a newly established 'artificial reef' in Gulf waters. – In: Elliott, M. & J.-P. Ducrotoy: Estuaries and coasts: spatial and temporal intercomparisons, pp. 259–264. Olsen & Olsen, Fredensborg

Dipper F, Woodward T (1993) The living seas marine life of the southern gulf, 3rd edn. Motivate Publ, Dubai, p 95

Dogma Jr IJ, Trono Jr GC, & Tabbada RA (Eds.) (1990) Culture and use of algae in Southeast Asia. Proceedings of the Symposium on Culture and Utilization of Algae in Southeast Asia, 8-11 December1981, Tigbauan, Iloilo, Philippines [Book]. Aquaculture Department, Southeast Asian Fisheries Development Center. https://hdl.handle.net/10862/1817

Dorgham MM (1990) Distribution of benthic algae in the Qatari Arabian Gulf. Bull Nat Inst Oceanogr Fish 16:49–63

El-Boukhari MEM, Barakate M, Bouhia Y, Lyamlouli K (2020) Trends in seaweed extract based biostimulants: manufacturing process and beneficial effect on soil-plant systems. Plant 9(359):1–23

Endlicher SL, Diesing CM (1845) Enumeratioalgarum, quasadoraminsulaeKarek, sinus Persici, legit TheodorusKotschy. Bot Zeit 3:268–269

FAO (2018) The state of world fisheries and aquaculture: meeting the sustainable development goals. FAO, Rome, pp 1–210

FAO (2020) The state of world fisheries and aquaculture: sustainability in action. FAO, Rome, pp 1–206

Farahpour M, Abkenar A, Notash G (2010) Alga cultural survey on three species *Sargassum ilicifolim, Hypnea musciformis* and *Cystoseira indica* in Chabahar seashores, Iran. Iranian J Med Aromat Plants 3(49):296–304

Ganesan M, Trivedi N, Gupta V, Madhav SV, Reddy CRK, Levine IA (2019) Seaweed resources in India – current status of diversity and cultivation: prospects and challenges. Bot Mar 62(5):463–482

Hafezieh M, Ajdehakosh A, Hosseini SH (2017) Dietary effects of seaweed *Sargassum ilicifolium* on reducing chlosterol level of white leg shrimp (*Liptopenaeus vannamei*). Iranian J Fish Sci 16(4):1248–1256

Heiba HI, Dorgham MM, Al-Nagdy SA, Rizk AM (1990) Phytochemical studies on the marine algae of Qatar, Arabian Gulf. Qatar Univ Sci Bull 10:99–113

Hessami MJ, Phang SM, Sohrabipour J, Fazeli Zafar F, Aslanzadeh S (2019) The bio-methane potential of whole plant and solid residues of two species of red seaweeds: *Gracilaria manilaensis* and *Gracilariopsis persica*. Algal Res 42:1–7

John DM, AL-Thani RF (2014) Benthic marine algae of the Arabian Gulf: A critical review and analysis of distribution and diversity patterns. Nova Hedwigia 98(3–4):341–392

Jones DA (1986) A field guide to the sea shores of Kuwait and the Arabian Gulf. Univ. Kuwait/ Blandford Press, Kuwait & Dorset, p 160

Kazemi M, Abediankenari A, Rabiei R (2018) Effect of marine macroalgae on growth performance and immune response in rainbow trout fingerlings. J Fish Sci Technol (Jfst) 7(1):9–16. http://jfst.modares.ac.ir/article-6-12442-en.html

Kim JK, Yarish C, Hwang EK, Park M, Kim Y (2017) Seaweed aquaculture: cultivation technologies, challenges and its ecosystem services. Algae 32:1–13

Kokabi M, Yousefzadi M (2015) Checklist of the marine macroalgae of Iran. Botanica Marina 58(4):307–320

Kureishy, TW, Abdel-Moati, MAR, Al-Moftah, AR (1995) Marine algae as bioindicators of pollution level in the Arabian Gulf. Qatar University Sci J 15:215–221

McHugh DJ (2003) A guide to the seaweed industry. FAO Fisheries Technical Paper 441, Rome 105

Moezzi M, Abdolalian E, Rohani GK, Fourooghifard H, Mortazavi MS, Zahedi MR (2017) Study algicidial activity of 6 species seaweed from Persian Gulf on growth *Cochlodinium polykrikoides*. Iran Sci Fish J 25(5):87–97

Mohammad JH, Ranga Rao A, Ravishankar GA (2019) Opportunities and challenges in seaweeds as feed stock for biofuel production. In: Ravishankar GA, Rao AR (eds) Handbook of algal technologies and phytochemicals. volume II phycoremediation, biofuels and global biomass production. CRC Press, Boca Raton, pp 39–50

Moridi A (2018) State of water resources in Iran. Int J Hydrol 1(4):111–114

Mshigeni KE, Dorgham M (1987) Benthic marine algae of Qatar. A preliminary survey. UNESCO Reg. Office, Doha, Qatar, pp 1–78

Newton L (1955a) The marine algae of Kuwait. In: Dickinson V (ed) The wild flowers of Kuwait and Bahrain. George Allen &Unwin, London, pp 100–102

Newton L (1955b) The marine algae of Bahrain. In: Dickinson V (ed) The wild flowers of Kuwait and Bahrain. George Allen &Unwin, London, pp 141–147

Nizamuddin M, Gessner F (1970) The marine algae of northern part of the Arabian Sea and of the Persian Gulf. – "Meteor" Forschungs-ErgebnisseReihe D. Biologie 6:1–42

Pak A, Farajzadeh M (2007) Iran's integrated coastal management plan: Persian Gulf, Oman Sea, and southern Caspian Sea coastlines. Ocean Coast Manag 50(9):754–773

Pirian K, Zarei Jeliani Z, Sohrabipour J, Arman M, Faghihi MM, Yousefzadi M (2017a) Nutritional and bioactivity evaluation of common seaweed species from the Persian Gulf. Iran J Sci Technol Transact A Sci. https://doi.org/10.1007/s40995-017-0383-x

Pirian K, Moein S, Sohrabipour J, Rabiei R, Blomster J (2017b) Antidiabetic and antioxidant activities of brown and red macroalgae from the Persian Gulf. J Appl Phycol 29(6):3151–3159. https://doi.org/10.1007/s10811-017-1152-0

Pirian K, Piri K, Sohrabipour J, Tamadoni Jahromi S, Rabiei R (2017c) Evaluation of chemical components and physicochemical properties of two green macroalgae species *Ulva intestinalis* and *Ulva linza* from Persian Gulf. Iran J Med Aromatic Plants 33(1):62–72

Pirian K, Piri K, Sohrabipour J (2018a) Evaluation of antioxidant and a-amylase inhibitory activity of two species *Sargassum angustifolium* and *Palisada perforata*. J Mol Cell Res (Iran J Biol) 31(2):158–171

Pirian K, Piri K, Sohrabipour J, Blomster J (2018b) Three species of *Ulva* (Ulvophyceae) from the Persian Gulf as potential sources of protein, essential amino acids and fatty acids. Phycol Res 66(2):149–154. https://doi.org/10.1111/pre.12212

Qaranjik BM, Wynne M, Xia B, Khajeh S, Keyanmehr H, Hosseini MR (2011) The biomass of the medicinal red algae (Rhodophyta) in the intertidal zone of the Chabaharcoasts. J Iran Fish Sci (in Persian language) 20(3):103–112

Rabiei R, Sohrabipour J (2007) The change of biomass and cover percentage of *Gracilaria salicornia (C. Agardh)* Dawson and its dispersion in seashore of Persian Gulf. Iran J Med Aromatic Plants Res 23(1):23–38

Rabiei R, Sohrabipour J, Nejadsattari T, Assadi M, Majd A (2005) The study of species diversity in *Gracilaria salicornia* association in northeast of Qeshm island. Pajouhesh-va-Sazandegi 66:85–93

Rabiei R, Sohrabipour J, Assadi M, Nejad T, Majd A (2008) Morphological and anatomical study of *Gracilaria salicornia* (C. Agardh) Dawson (Gracilariaceae-Rhodophyta) in the Persian Gulf seashores (Qeshm island). Pajouhesh-va-Sazandegi 75:47–53

Rabiei R, Phang SM,Yeong HY, Lim PE, Ajdari D, Zarshenas GA, Sohrabipour J (2014). Bioremediation efficiency andbiochemical composition of *Ulva reticulata* Forsskål (Chlorophyta) cultivated in shrimp (*Penaeus monodon*) hatchery effluent. J Iran Fish Sci 13 (3):621–639

Saeidnia S, Permeh P, Gohari AR, Mashinchian-Moradi A (2012) *Gracilariopsis persica* from Persian Gulf contains bioactive sterols. Iran J Pharm Res 11(3):845–849

Safavi SV, Abediankenari A, Tabarsa M, Esmaeili M (2019) Effect of sulfated polysaccharides extracted from marine macroalgae (*Ulva intestinalis and Gracilariopsis persica*) on growth performance, fatty acid profile, and immune response of rainbow trout (*Oncorhynchus mykiss*). J Appl Phycol 31:4021–4035

Salehi P, Dashti Y, Mozaraati Tajabadi F, Sefidkon F, Rabiei R (2011) Structural and compositional characteristics of a sulfated galactan from the red alga *Gracilariopsis persica*. Carbohydr Polym 83(4):1570–1574

Shama A, Joyce SG, Mari FD, Ranga Rao A, Ravishankar GA, Hudaa N (2019) Macroalgae and microalgae: novel sources of functional food and feed. In: Ravishankar GA, Rao AR (eds) Handbook of algal technologies and phytochemicals: volume-I: food, health and nutraceutical applications. CRC Press, Boca Raton, pp 207–219

Silva PC, Basson PW, Moe RL (1996) Catalogue of the benthic marine algae of the Indian Ocean. Univ California Publ Bot 79:1–1259

Sohrabipour J, Rabiei R (1996) New records of algae from Persian Gulf and flora of Iran. Iranian J Bot 7:95–115

Sohrabipour J, Rabiei R (1999a) A list of the algae of sea shores of Persian Gulf and Oman Sea in the Hormozgan Province. Iranian J Bot 8:131–162

Sohrabipour J, Rabiei R (1999b) A list of marine algae from seashores of Iran (Hormozgan Province). Qatar Univ Sci J 19:312–337

Sohrabipour J, Rabiei R (2005) Morphology and anatomy of *Gracilariopsis longissima* in the Persian Gulf (Iranian coast lines). Pajouhesh-va-Sazandegi 77:2–7

Sohrabipour J, Rabiei R (2007) The checklist of green algae of the Iranian coastal lines of the Persian Gulf and Gulf of Oman. Iranian J Bot 13:146–149

Sohrabipour J, Rabiei R (2008) Rhodophyta of Oman Gulf (South East of Iran). Iranian J Bot 14:70–74

Sohrabipour J, Rabiei R (2017) Algal vegetation in southern coastline of Iran. Iran Nature 2(1):62–68

Sohrabipour J, Nezhadsatari T, Ghahraman A, Asadi M (2003) An investigation on identification and primary production of brown algae and effect of ecological factors on the vegetation in seashores of Bandar Lengeh. Pajouhesh-va-Sazandegi 16(2):44–58

Sohrabipour J, Nejadsatari T, Assadi M, Rabiei R (2004) The marine algae of the southern coast of Iran, Persian Gulf, Lengeh Area. Iran J Bot 10:83–93

Tamadoni Jahromi S, Pourmozaffar S, Jahanbakhshi A, Rameshi H, Gozari M, Khodadadi M, Sohrabipour J et al (2021) Effect of different levels of dietary *Sargassum cristaefolium* on growth performance, hematological parameters, histological structure of hepatopancreas and intestinal microbiota of *Litopenaeus vannamei*. Aquaculture 533:736130. https://doi.org/10.1016/j.aquaculture.2020.736130

Valikamal AR, Gourchi T, Farahpour M, Gharehbash AM, Samii R (2010) The effect of replacing different level of *Sargassum ilicifolium* algae against silage on the performance and blood and ruminal factors in the lambs of zel. Iran J Anim Sci 279–285

Vosough-Sharifi O, Yaghoubfar A, Sharifi SD, Mirzadeh G, Askari F (2011) study on the possibility of *Gracilariopsis Persica* utilization in layer diet. J anim Prod 14:1–10

Zaei Jeliani Z, Yousefzadi M, Sohrabipour J, Toiserkani H (2017) Growth, phytochemicals, and optimal timing of planting *Gracilariopsis persica*: an economic red seaweed. J Appl Phycol:1–9. https://doi.org/10.1007/s10811-017-1217-0

Zarei Jeliani Z, Sohrabipour J, Soltani M, Rabiei R, Yousefzadi MJ (2021) Seasonal variations in growth and phytochemical compounds of cultivated red alga, *Hypnea flagelliformis*, in southern coastlines of Iran. J Appl Phycol (accepted)

Chapter 10
Seaweeds in Ireland: Main Components, Applications, and Industrial Prospects

X. Zhu, A. B. Soro, B. K. Tiwari, and M. Garcia-Vaquero

10.1 Introduction

Seaweeds have been commercially harvested for thousands of years in more than 50 countries, in waters ranging from cold to tropical. The global market of seaweeds, with a current annual growth rate of 8.9%, it is estimated to reach a value of around USD 22.13 billion by 2024 (Zhu et al. 2021a). In 2016, all the seaweeds harvested in Ireland summed up to 29,500 tons of biomass, accounting for 11% of the overall European seaweed market, making Ireland the third most productive country in Europe, right behind Norway and France (Table 10.1) (FAO 2018). Seaweed species can be classified within 3 phyla namelyRhodophyta (red algae), Phaeophyta (brown algae), and Chlorophyta (green algae). Amongst all the seaweed species available, the Irish industry identified 9 of them (*Ascophyllum nodosum*, *Laminaria digitata*, *Laminaria hyperborea*, *Palmaria palmata*, *Chondrus crispus*, *Himanthalia elongata*, *Fucus serratus*, *Lithothamnion corallioides*, and *Ulva lactuca*) as priorities regarding further exploitation opportunities of the biomass as a source of high-value molecules with added health benefits or bioactive compounds (Troy et al. 2017).

Seaweeds are traditional low caloric foods and their composition includes mainly polysaccharides and dietary fibres, proteins, polyunsaturated fatty acids (PUFA),

---

X. Zhu
TEAGASC Food Research Centre, Ashtown, Dublin, Ireland

School of Biosystems and Food Engineering, University College Dublin, Belfield, Dublin, Ireland

A. B. Soro · B. K. Tiwari
TEAGASC Food Research Centre, Ashtown, Dublin, Ireland

M. Garcia-Vaquero (✉)
Section Food and Nutrition, School of Agriculture and Food Science, University College Dublin, Belfield, Dublin, Ireland
e-mail: marco.garciavaquero@ucd.ie

A. Ranga Rao, G. A. Ravishankar (eds.), *Sustainable Global Resources of Seaweeds Volume 1*, https://doi.org/10.1007/978-3-030-91955-9_10

Table 10.1 Top European countries producing wild seaweed biomass as well as main species produced

Country	Main seaweed species	Tonnage of biomass (wet weight)
Norway	Aquatic plants, brown seaweeds, rockweed.	169,407
France	Brown seaweeds, north European kelp, tangle.	55,041
Ireland	North Atlantic rockweed, north European kelp, red seaweeds.	29,500
Iceland	Rockweed, north European kelp, tangle.	17,985
Russian Federation	Aquatic plants, brown seaweeds, north European kelp, red seaweeds.	14,022
Spain	Brown seaweeds, *Gelidium* seaweeds, green seaweeds, ribbonednori, wakame.	3493
Portugal	Red seaweeds.	2328
Italy	Green seaweeds, red seaweeds.	1200
Estonia	Red seaweeds.	348

Adapted from Monagail and Morrison (2020) and reproduced with permission from Springer

and other compounds including pigments, vitamins, minerals and phenolic compounds. The industrial use of this biomass has shifted over the years. At all stages, the 'potential' of the algal industry has been viewed as being larger than its actual scale (Hafting et al. 2015). Moreover, the huge amount of bioactive compounds discovered in seaweeds anticipates promising prospects for the seaweed industry in Ireland and globally in the near future. These algal compounds can be used as functional products, such as value-added ingredients for food and feed (Fleurence 2016; Shama et al. 2019), cosmaceuticals (Wijesinghe and Jeon 2011), nutraceuticals (Shannon and Abu-Ghannam 2019), pharmaceuticals (Moualee and Pradhanang 2019; Rajauria et al. 2016), as well as low value applications, such as bioenergy,ensuring the full utilization of the biomass (Fernand et al. 2017). Currently, seaweed biorefinery has been proposed and multiple exploitation models are currently being developed aiming to minimize waste and environmental impacts of the seaweed industry while utilizing fully these natural resources. To be able to develop these exploitation models, a deep knowledge of seaweed composition, potential health benefits of seaweed compounds, as well as extraction technologies available for the recovery of these molecules, will be relevant in determining the future application of the biomass and derivedcompounds as summarized below.

10.2 Seaweeds as a Source of Major Nutritional Compounds

Seaweeds or macroalgae are a large and diverse group of organisms with over 10,000 species identified worldwide, being only 5% of them currently exploited for food or animal feed applications (Chojnacka et al. 2012). The consumption of seaweeds has a long tradition in many Asian countries, with a documented use of whole or processed algae over millennia (Fleurence et al. 2012). In Western countries, such as coastal areas of Ireland, Brittany, and Iceland, there was a discontinued tradition for eating seaweeds or incorporating them as ingredients or additives to improve the sensory attributes of foods, such as bread (Holdt and Kraan 2011; Mouritsen et al. 2013). The recent change of consumers' perception of seaweeds as healthy, nutritious, and tasty food commodities has increased the demand of food products and dietary supplements containing algae in recent decades (Mouritsen et al. 2013; Roleda et al. 2019; Lafarga et al. 2021).

In terms of their composition, overall, the scientific literature describes the macroalgal biomass is rich in carbohydrates (up to 60%), with medium or high amounts of proteins (10–47%), low in lipids (1–3%), and variable contents of minerals (7–38%) and pigments (Dominguez and Loret 2019; Kraan 2013). Moreover, seaweeds can adapt to the rapid changes of the marine environmental conditions, i.e. changes in temperature, solar radiation, by producing unique secondary metabolites including polysaccharides, proteins, lipids, and phenolic compounds (Collins et al. 2016; García-Vaquero et al. 2017). Prolonged exposure of macroalgae to environmental stressors, such as fluctuations in water level, solar radiation and temperature, can lead to the formation of reactive oxygen species and other free radicals in the biomass. As a defense mechanism, the stressed macroalgal biomass produces high amounts of antioxidant compounds, such as phenolic compounds and sulphated polysaccharides amongst others, trying to maintain the integrity of the cellular structures (Roleda et al. 2019). Thus, this macroalgal biomass enriched in antioxidant compounds could be incorporated as food ingredients or supplements, providing additional health benefits to those of basic nutrition namely nutraceuticals or functional foods (Garcia-Vaquero and Hayes 2016) that could help in the prevention of chronic diseases, such as cancer, cardiovascular diseases, obesity and diabetes (Déléris et al. 2016; Lordan et al. 2011).

Understanding the composition of the major macronutrients of the macroalgal biomass (carbohydrates, proteins, lipids), as well as other minor compounds and micronutrients produced by macroalgae, will help to understand and elucidate the future utilization of this biomass and their derived compounds and associated health benefit for food and nutraceutical applications.

10.2.1 Carbohydrates

Carbohydrates are one of the major components of macroalgae, with overall contents ranging from 4–76% of the dry weight (DW) of seaweeds (Holdt and Kraan 2011). Significant variations in the overall amount of carbohydrates in seaweeds have been described depending on the species, the season of collection, and even parts of the biomass sampled (Kim 2012; Men'shova et al. 2012; Skriptsova et al. 2012). The main type of carbohydrates described in each seaweed class (red, brown or green macroalgae) together with their biological functions in the biomass and the future uses or applications of these compounds in the food industry are summarized in Table 10.2.

Table 10.2 Main seaweed carbohydrates, biological functions in the biomass and current industrial applications/research gaps

Macroalgae	Carbohydrates	Function in algae[a]	Industrial applications	References
Rhodophyta (red macroalgae)	*Floridean starch*	Storage	Further research is needed to develop applications	Hlima et al. (2019)
	Cellulose	Structural component cell-wall	Emulsifier	Lakshmi et al. (2017)
			Thickener	
			Protective colloid in food industries	
			Film	
	Mannan	Structural component cell-wall and storage	Further research is needed to develop applications	de Jesus Raposo et al. (2015)
	Xylan	Structural component cell-wall	Packaging materials	Sedlmeyer (2011)
			Ingredients for improving water holding capacities	
			Encapsulation of active components or as a slow-release agent	
	Sulphated galactans		κ-Carrageenan (to avoid whey segregation)	Kraan (2012); Kumar et al. (2008)
			λ-Carrageenan (firmness and texture)	
			Gelling carrageenans (gelatin or pectin alternatives in desserts)	
			Agar, as medium for cell culture	
			Galactans are used in the composition of drugs; drug release process and in tissue engineering	

Ochrophyta (brown macroalgae)	*Laminarin*	Energy reserve present in reserve vacuoles	Enhanced quality and shelf life of pork meat when used as feedstock	Øverland et al. (2019)
	Mannitol	Translocation and storage	Food coating	Qin (2018)
	Alginic acid	Structural component cell-wall	Beer foam stabilizers	Goyanes and D'Accorso (2017); Hernandez-Carmona et al. (2013)
			Welding rods	
			Pill disintegrators	
			Protein imprinting	
	Fucoidans		Further research is needed to develop applications	Lorbeer et al. (2015)
Chlorophyta (green macroalgae)	*Starch*	Storage	Animal feed	Wolf (2010)
			Pulp for paper making	
			Film and packaging material	
	Cellulose, mannan and xylan	Structural component cell-wall	Same applications as above	de Jesus Raposo et al. (2015); Lakshmi et al. (2017); Sedlmeyer (2011)
	Sulphated xyloarabinogalactan		Prebiotics for foodstuffs	Delattre et al. (2005)
	Sulphated glucuronoxylorhamnan		Possible application for cholesterol reduction	Husni (2018)
	Sulphated glucuronoxylorhamnogalactan		Further research is needed to develop applications	Kamble et al. (2018)
	Ulvan		Can be used for developing edible films and hydrogels	Chiellini and Morelli (2011); Ganesan et al. (2018)

This table is reproduced from Gomez et al. (2020) with permission from Elsevier

[a]Functions if these compounds in macroalgae adapted from García-Vaquero et al. (2017), Kumar et al. (2008), Ojima (2013)

In general, and based on their functions in seaweeds, polysaccharides can be classified as (1) structural carbohydrates, involved in the maintenance of the structural integrity of the macroalgal cell walls; or (2) energy reserve polysaccharides, intracellular polysaccharides present inside reserve vacuoles and functions related to energy storage within the algal cells (García-Vaquero et al. 2017; Kumar et al. 2008; Ojima 2013). Once these compounds are extracted from their original matrices these carbohydrates can (1) exert biological benefits, and thus be used as bioactive compounds; (2) possess relevant physical properties that allow their use as stabilizers, thickeners, and emulsifiers; or (3) both, being valued compounds for both their physical and biological attributes when included in food (Lafarga et al. 2020).In fact, most macroalgal polysaccharides currently used as hydrocolloids, mainly alginate, agar, and carrageenan, have shown also promising biological properties (Gomez et al. 2020) expanding their range of applications to multiple industries including food, pharmaceutical, and cosmaceuticals. Other polysaccharides, such as fucoidans and laminarins, have recently gained momentum in the scientific literature due to their wide range and promising biological activities including anti-inflammatory, anticoagulant, antioxidant, antiviral, antitumor, antiapoptosis, antiproliferative, and immunostimulatory when assayed *in vitro* and/or *in vivo* models (García-Vaquero et al. 2017).

The industrial application of these carbohydrates in the food industry relies on the extraction of these high-value compounds from the seaweed biomass.These extraction processes should ideally be selective, achieve high yields of compounds, minimize the generation of waste and reduce the use of solvents and energy during the extraction process. Thus, increased research activity and industry interest has focused in recent years on using innovative technologies, including enzyme-assisted extraction, ultrasound-assisted extraction, microwave-assisted extraction, supercritical fluid extraction, ultra-high pressure extraction, and pressurized fluid extraction (Gomez et al. 2020). Moreover, recent technological trends in the sector include the use of a combination of technologies, applying multiple extraction forces in the biomass either sequentially or simultaneously aiming to increase the yields of polysaccharides extracted from the biomass. Garcia-Vaquero et al. (2020) explored ultrasounds, microwaves, and the simultaneous application of both extraction forces to extract fucoidan from *A. nodosum*. The authors demonstrated that both innovative technologies alone had a significant influence on the extraction of fucoidan, however, the yields of fucoidan extracted from this brown macroalgae using a simultaneous application of both ultrasounds and microwaves achieved yields of fucoidan of approximately 10 and three fold compared to the use of ultrasounds and microwaves alone, respectively (Garcia-Vaquero et al. 2020).

10.2.2 Proteins

Seaweeds have been increasingly studied as a source of proteins. Proteins are essential nutrients in both animal and human nutrition, as well as one of the most expensive ingredients in the diet. Thus, increased research interest has focused on proteins

Table 10.3 Total protein concentrations (% dry weight) and proposed extraction methods in selected brown, red, and green seaweed species

Macroalgae (sp.)	Protein content (% DW)	Protein extraction methods (main results)	References
Phaeophyta (brown macroalgae)			
Ascophyllum nodosum	3–15%	Ultrasound and a combination of acid and alkali solvents (59% of protein recovery)	Fleurence (1999); Kadam et al. (2017)
Laminaria digitata	3.8%	Enzyme hydrolysis and fermentation (97% of protein recovery)	Hou et al. (2015)
Saccharina latissima	4–16.1%	pH shift (protein yield of 25.1 ± 0.9%)	Mols-Mortensen et al. (2017); Harrysson et al. (2018)
Rhodophyta (red macroalgae)			
Palmaria palmata	11.9–21.9%	Combination of enzyme and alkaline (>90% of recovery)	Naseri et al. (2020), Galland-Irmouli et al. (1999)
Porphyra umbilicalis	13.5–20.6%	pH shift (protein yield of 22.6 ± 7.3%)	Biancarosa et al. (2017), Harrysson et al. (2018)
Gracilaria sp.	9.2–12.6%	Ultrasound combined with an alkali solvent (protein yield of 86%)	Chan and Matanjun (2017); Kazir et al. (2019)
Chlorophyta (green macroalgae)			
Ulva sp	11.2–15%	Pulsed electric fields (seven-fold increase in total protein extracted compared to osmotic shock samples)	Biancarosa et al. (2017); Robin et al. (2018)
Ulva lactuca	11.2–15%	Ultrasound using water as a solvent (protein yield of 19.6 ± 0.8%)	Biancarosa et al. (2017); Harrysson et al. (2018)

from seaweeds as novel and cheap alternative sources of protein (Nunes et al. 2014; Fleurence et al. 2018). However, the exploitation of seaweeds as a source of protein still offers challenges, mainly related to the wide ranges of protein content described in different seaweed species, ranging between 3 and 47% (Harnedy and Fitzgerald 2013); as well as the influences of external factors, such as season of harvest and place of collection (Zhu et al. 2021a), on the protein contents of the biomass.

Moreover, the methodology used to determine the levels of protein in seaweeds also represents a huge variation in the protein levels reported in the scientific literature. Commonly, the protein contents of bseaweeds have been assessed by measuring the nitrogen content of the biomass and then multiplying it by specific conversion factors. In terms of seaweeds, these conversion factors range from 3.75 to 5.72 (Lourenço et al. 2002). However, as some of the seaweeds are regarded as edible green food, many studies are using 6.25, the general nitrogen-to-protein factor, in their studies. Macroalgae may have non-protein nitrogen, including free amino acids, chlorophyll, nitrates, ammonium ions, and nucleic acids, and thus, the protein content calculation with 6.25 often results in an over estimation of algal protein (Makkar et al. 2016). Table 10.3 summarizes total protein concentrations (% DW) in selected brown, red, and green seaweed species. Generally, the protein content decreases in the order of seaweed group: red>green>brown (Černá 2011).

The most promising seaweed protein families,from a biotechnological point of view,are lectins and phycobilli-proteins. Lectins are carbohydrate binding proteins contributing in cell communication, including recognition of foreign or cancerous cells (Ziółkowska and Wlodawer 2006). Some red algal lectins have painkilling effects, anti-inflammatory, anti-cancer properties, and the potential to develop antiviral agents with activity against HIV and coronavirus (Singh and Walia 2018).

Other compounds derived from algal proteins, such as amino acids (AA) and bioactive peptides, are being studied in recent years due to their promising biological benefits when included in foods, but also as cosmetic, pharmaceutical, and other applications in the biotechnologicalindustry (Harnedy and Fitzgerald 2011). Macroalgae contain all AA (Matanjun et al. 2009) and are regarded as an excellent source of essential amino acids (EAA),especially red and green species (Organization 1991; Wong and Cheung 2000). Contradictory results have been reported on the levels of particular AA between seaweed species. Methionine and cysteine were described to be at higher levels in red seaweeds compared to green and brown seaweeds (Qasim 1991), but an opposite behavior was reported by Gressler et al. (2011). Gaillard et al. (2018) studied the amino acid content in 9 seaweed species and determined that glutamine was the most abundant AA followed by Asp and Ala. In general, the seaweeds were rich in threonine, serine, glycine, valine, leucine, lysine, and arginine (on average 4.58 g/16 g N, ranging from 4.06 to 4.99 g/16 g N), and low in cysteine, histidine, and methionine (on average 1.70 g/16 g N, ranging-from 1.24 to 2.28 g/16 g N).The AA concentrations of crude protein varied with the seasons, species, and methods of AA analysis used.

Moreover, seaweeds as well as other protein-rich products can be ideal candidates for the generation of bioactive peptides or cryptides (Garcia-Vaquero et al. 2019). Bioactive peptides are sequences of 2 to 30 amino acids in length that have hormone-like beneficial properties once these compounds are released from their original or parent proteins through several hydrolytic processes (Garcia-Vaquero et al. 2019). In Japan, a number of functional foods have been already commercialized as permitted by the Japanese Ministry of Health and Welfare (Arai 2000). Although there is a growing trend in isolating proteinand bioactive peptides from marine algae, the number of biologically active compounds generated from seaweed is still limited. Most seaweed bioactive compounds, including protein, are contained inside the algal cells by a highly rigid and structurally complex cell wall and thus, this has been described as one of the major obstaclesfor an efficient utilizationand digestibility of algal proteins (Fleurence et al. 2012; Harnedy and Fitzgerald 2011). There is a need to extract these compounds from seaweed, optimizing these processes in terms of resources (i.e. chemicals, solvents, energy…) consumed during the extraction, while maintaining or increasing the yields of protein achieved. Some protein extraction processes extracted from the recent scientific literature are summarized in Table 10.3. Some of the potential technologies with promising applications improving the extraction of protein from seaweeds include the use of ultrasound (Kadam et al. 2017), microwaves (Chemat 2012), supercritical fluid extraction (Liang and Fan 2013) and ultrahigh-pressure extraction (Xi et al. 2013) amongst others.

10.2.3 Lipids

Lipids are essential macronutrients in the diet as a source of energy and other antioxidants, such as tocopherols and carotenoids (Bialek et al. 2017). Lipids in seaweeds are present at relatively low levels (1–5% of DW), and thus, fresh seaweeds can be considered as low-energy foods. The health benefits of any source of dietary fat are linked to the fatty acid composition of the lipids, especially the content of PUFA. PUFA, such as eicosapentaenoic acid and arachidonic acid, are proved to be beneficial in controlling blood pressure and blood clotting and reducing the risk of cardiovascular diseases, osteoporosis, and diabetes (Maeda et al. 2008). The optimum dietary ratio of PUFAs (omega-6 to omega-3) of 5:1 is recommended by the European Nutritional Society, however, most of the European diets failed to meet these recommendations (Holdt and Kraan 2011). Fish oils have been utilized as the source of omega-3 or n-3 PUFAs (Rubio-Rodríguez et al. 2010). However, due to the environmental, economic, sustainable and food safety concerns, novel sources of PUFAs are currently required (Vannuccini et al. 2019).

As shown in Table 10.4, total PUFAs in selected seaweeds range from 16.64 to 23.8% of the total amount of fatty acids.*Undaria pinnatifida* has eicosapentaenoic acid at concentrations of up to 0.14 g per 100 g DW of seaweeds' thallus (Khan et al. 2007). Gosch et al. (2012) reported that the green seaweed *Derbesia tenuissima* had high levels of fatty acids (39.58 mg/g DW) with a high proportion of PUFA (n-3) (31% of total fatty acid), which are suitable as nutraceuticals or fish oil replacers.Regarding biofuel production, *Spatoglossum macrodontum* had the highest fatty acid contents (37.40 mg/g DW) with a high content of C18:1, which is suitable as a biofuel feedstock (Gosch et al. 2012).

Generally, macro- and micronutrients, the lipid content and fatty acid composition in seaweeds are influenced by their intrinsic traits andalso climatological conditions (light intensity, seawater salinity, and temperature). Generally, red and brown seaweeds are rich in eicosapentaenoic and arachidonic acids, and green seaweeds like *Ulva pertusa* predominantly contain hexadecatetraenoic, oleic, and palmitic acids (Ortiz et al. 2006). Regarding the environmental conditions affecting the biomass, high light intensity and low salinity results in a decreased level of total fatty acids within the same species (Floreto and Teshima 1998). Furthermore, Ortiz et al. (2006) reported a strong influence of the temperature of the water (particularly low temperatures) on the fatty acid levels produced in different seaweed species.

There is an increased interest in the extraction of lipids from seaweed biomass. Traditionally the recovery of lipophilic compounds has been achieved by using a Soxhlet apparatus or soaking the biomass in organic solvents (Dickson et al. 2020). Recent developments in lipid extraction include the use of supercritical fluid extraction,as well as the application of ultrasounds and microwaves (Cravotto et al. 2008).

Table 10.4 Summary of fatty acid composition (expressed as % of total fatty acids) and method of extraction from selected seaweed species

	Saccharina japonica	*Palisada flagellifera*	*Ulva fasciata*
Extraction methods	Supercritical carbon dioxide	Chloroform/methanol	Chloroform/methanol
Fatty acids (% total fatty acid)			
(C14:0) Myristic acid	11.36 ± 0.29	6.4 ± 0.0	0.7 ± 0.0
(C16:0) palmitic acid	20.60 ± 0.86	37.6 ± 0.4	45.0 ± 3.4
(C17:0) Heptadecanoic acid	4.13 ± 0.12	0.3 ± 0.0	0.4 ± 0.0
(C18:0) stearic acid	2.68 ± 0.15	1.1 ± 0.1	0.7 ± 0.1
(C20:0) Arachidic acid	1.02 ± 0.11	0.2 ± 0.1	0.2 ± 0.1
(C22:0) Behenic acid	n.d.	n.d.	2.0 ± 0.3
(C23:0) Tricosanoic acid	10.31 ± 0.22	n.d.	n.d.
(C24:0) Lignoceric acid	n.d.	n.d.	0.3 ± 0.0
Total SFA	50.66 ± 1.08	45.6 ± 0.6	49 ± 3.9
(C14:1) Mystoleic acid	n.d.	0.7 ± 0.1	0.6 ± 0.1
(C16:1) Palmitoleic acid	2.82 ± 0.18	1.4 ± 0.1	2.4 ± 0.1
(C17:1) cis-10-Heptadecenoic acid	0.16 ± 0.01	0.4 ± 0.0	0.8 ± 0.2
(C20:1) Gondoic acid	n.d.	n.d.	0.2 ± 0.1
(C18:1n9c) Elaidic acid	29.65 ± 0.35	7.7 ± 0.2	2.4 ± 0.1
Total MUFA	32.71 ± 0.70	11.5 ± 0.4	17.2 ± 0.4
(C18:2n6c) linoleic acid	7.30 ± 0.21	1.7 ± 0.1	5.9 ± 0.7
(C18:3n6) r-Linolenic acid	2.04 ± 0.13	1.5 ± 0.2	5.4 ± 1.4
(C18:3n3) Linolenic acid	2.71 ± 0.15	2.6 ± 0.1	7.8 ± 1.5
(C20:3n6) cis-8,11,14-Eicosatrienoic acid	0.44 ± 0.07	0.6 ± 0.0	0.2 ± 0.0
(C20:4n6) arachidonic acid methyl ester	n.d.	16.2 ± 0.3	0.5 ± 0.2
(C20:5n3) EPA	4.14 ± 0.18	0.6 ± 0.7	2.1 ± 0.2
C22:6n3 DHA	n.d.	0.6 ± 0.2	0.4 ± 0.4
Total PUFA	16.64 ± 0.71	23.8 ± 1.6	22.3 ± 4.4
Nutritional index			
Σ n-3	6.85 ± 0.33	21.4 ± 6.7	16.6 ± 3.1
Σ n-6	9.78 ± 0.41	18.9 ± 0.1	7.5 ± 0.3

Table 10.4 (continued)

	Saccharina japonica	*Palisada flagellifera*	*Ulva fasciata*
Ratio n-6/n-3	1.4	0.9	0.5
References	Getachew et al. (2018)	Santos et al. (2019)	Santos et al. (2019)

"n.d." stands for not detected

10.3 Macroalgae as a Source of Other Minor Compounds

Apart from the aforementioned dominant macronutrients, minerals, vitamins, and other pigments are also highly valued compounds present in seaweeds. Seaweeds contain significant amounts of minerals (I, Ca, Fe, Zn, Mn, and Cu), which perform many important biological functions, including assisting in cell transportation and other metabolic processes (Mišurcová et al. 2011). Seaweeds can absorb inorganic substances from the environment due to the interactions of these compounds with cell surface polysaccharides (Yoshioka et al. 2007). Similar to the levels of macronutrients, the mineral profiles of seaweeds are also really variable depending on the type of seaweeds, season and place of collection (Teas et al. 2004). High contents of minerals have been reported in the literature in certain seaweed species. I.e. in *Ulva clathrata* biomass collected from Mexico the mineral contents of this seaweed represented approximately 49.6% of the dry matter of the biomass (Peña-Rodríguez et al. 2011). Mišurcová et al. (2011) summarized the mineral contents in selected marine algae and claimed that many of the seaweed species can be excellent contributors of these compounds based on the reference daily intakes (RDIs) in EU countries, United States, Australia, and Asia, especially when seaweeds are used as sources of I and Fe (Mišurcová et al. 2011).

Other micronutrients of nutritional relevance produced by seaweeds are vitamins. Vitamins are essential catalysts in human body that must be obtained from the diet, as they can only be synthesized to a limited extent. Certain vitamins from seaweeds have antioxidant activity, while other vitamins may contribute to maintaining and promoting health by providing additional health benefits, such as decreasing blood pressure, prevention of cardiovascular diseases, or reducing the risk of cancer (Škrovánková 2011). Ortiz et al. (2006) claimed that 100 g of seaweed provides more than the recommended daily requirements of vitamin A, B_2, B_{12}, and two-thirds of the recommended vitamin C levels (Ortiz et al. 2006). Many vitamins have been detected and measured in seaweeds, including L-ascorbic acid (vitamin C), thiamine (vitamin B_1), riboflavin (vitamin B_2), cobalamin (vitamin B_{12}), folic acid and its derivatives, tocopherols (vitamin E), and carotenoids (Mišurcová 2012). Generally, red and green seaweeds are rich in vitamin C ranging from 500 mg/kg to 3000 mg/kg DW. Levels of up to 2000 mg/kg DW have been reported from the red seaweed *Eucheuma denticulatum* and 3000 mg/kg DW in the green seaweed *Enteromorpha flexuosa* (Mcdermid and Stuercke 2003). Other red seaweeds, such as *Palmaria* and *Porphyra* spp. contain a large amount of vitamins of the B group

(Mabeau and Fleurence 1993). Although lipid content in seaweed is low, seaweed fats contain high levels of vitamin E. Generally, brown seaweeds contain more α-tocopherol (also β- and γ-tocopherols) than red and green algae which contain only α-tocopherol. Ortiz et al. (2009) reported a high amount of vitamin E detected in kelp *Macrocystis pyrifera* with 132.77 mg/100 g fat of α-tocopherol from total tocol content of 145.72 mg/100 g fat. Regarding the vitamin content in seaweeds, it also varies depending on the seaweed species and the aforementioned environmental factors. I.e., the highest levels of vitamin E in *Eisenia arborea*collected in Mexico were appreciated during the month of September (Hernández-Carmona et al. 2009). Furthermore, other factors affecting the biomass including the concentration of certain compounds in the sea, the depth of growth of the biomass or the sunlight intensity at particular sea depths in which the biomass is growing will affect the vitamin content of the biomass even within the same species (Smith et al. 2007).

Other minor compounds from seaweeds include pigments such as chlorophylls, carotenoids and xanthophylls with promising potential applications in the food, cosmetics and pharmaceutical industries. Algal pigments can be applied in foods to impart the desired sensorial properties in food products, as well as for their health benefits including antioxidant, antidiabetic, immune-modulatory, and anti-angiogenic (Manivasagan et al. 2018). In order to meet the industry demands of algal pigments, many novel extraction techniques have been developed in recent years. Supercritical fluids were employed to obtain carotenoids, fucoxanthin, and phlorotannins from *Saccharina japonica* (Saravana et al. 2017). Microwave, ultrasound, and pulsed electric fields were reported to increase the extraction yields and reduce processing time to get carotenoids from *Undaria pinnatifida* (Zhu et al. 2017).

Other minor constituents from seaweeds with health benefits, such as anti-diabetic, anti-inflammatory and anti-microbial, are phenolic compounds (Generalić Mekinić et al. 2019; Nwosu et al. 2011; Lopes et al. 2012). Phenolic compounds encompass a widely diverse chemical group of compounds including phlorotannins, bromophenols, flavonoids, and terpenoids produced by seaweeds. Phlorotannins are the main phenolic compounds present in brown seaweeds. The highest total phenolic content of 47.16 mg galic acid equivalent/g dry *Ascophyllum nodosum* after post-harvest treatments was reported by Zhu et al. (2021b). Other large proportions of phenolic compounds detected in green and red seaweeds are bromophenols, flavonoids, phenolics acids, and terpenoids (Wells et al. 2017). As an example of the chemical diversity of these compounds, the chemical structures of the main phlorotannins produced by brown macroalgae are represented in Fig. 10.1.

10.4 Prospects of the Seaweed Industry

Public awareness towards sustainability hasrecently increased in response to current global challenges, such as climate change, intensive production, poor waste management, and overpopulation, among others. To reduce the negative impact of these challenges in the environment, population, and economy, new alternatives and

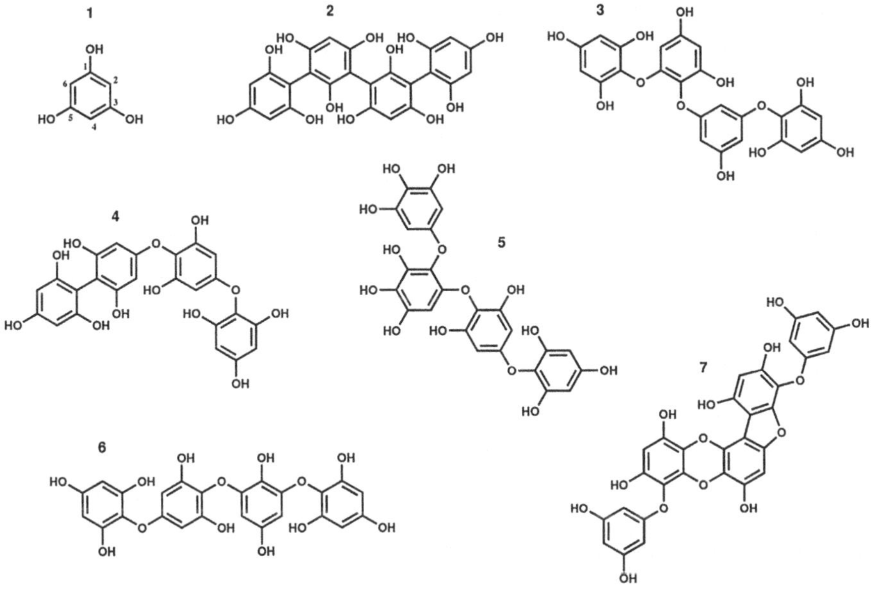

Fig. 10.1 Chemical structures of phlorotannins as described by Lopes et al. (2012) originally published by PLOS ONE. Numbers inside the image represent: phloroglucinol (1), tetrafucol A (2), tetraphlorethol B (3), fucodiphlorethol A (4), tetrafuhalol A (5), tetraisofuhalol (6), and phlorofucofuroeckol (7)

production systems are currently being considered in replacement of conventional processes and products. In the past decades, seaweed production and cultivation have increased the interest of the scientific and industrial communities as a more sustainable system compared to traditional agricultural practices. Moreover, seaweedshave been demonstrated as excellent source of nutrients and have a multipurpose nature (e.g. animal feed, biofuel, biomass) (Tiwari and Troy 2015; Mohammad et al. 2019; Hessami et al. 2019). Additionally, the global market of seaweed aquaculture and wild capture has been increasing with an annual production of 31.1 million tonnes at the global level. This rising trend is explained by the increased consumer acceptance of seaweeds as food and the broad applicability of these raw materials and theirby-products in biotechnological, pharmaceutical, and chemical industries (Kraan 2020).

The cultivation of seaweeds is now facing the challenge of variability and availability of biomass. In general, aquaculture of seaweeds usually enhances the content of protein and also decreases the pressure of harvesting large quantities of biomass from wild populations. However, in Europe, more than 97% of seaweed production is harvested from the wild in Norway, France, and Ireland (Grote 2019). This current scenario represents a challenge in utilizing seaweed protein, but can also be an opportunity to develop the seaweed industry further. Eco-intensification and sustainability of seaweed cultivation can be improved through seaweed permaculture by recreating seaweed forest habitat and other ecosystems water environments. For example, *Rhodymenia pseudopalmata* showed an increase of total protein content after 3 days of cultivation (18.7% DW) in fishpond effluents rich in ammonium (NH_4+) than it harvested from the wild (9.4% DW) (Pliego-Cortés et al. 2019). Seaweed permaculture allows seaweeds to absorb the NH_4+ that is produced by seafood production system, and meanwhile, reduces the negative impact on the ecosystems due to the capacity of seaweeds to absorb inorganic nutrients (Grote 2019). The high protein content of seaweed biomass can be recycled as feed. *U. lactuca* (37.4% DW) and *Gracilaria conferta* (29.4% DW) produced by seaweed permaculture were used to feed sea urchins as a sole feed. Compared with traditional fish diet, *U. lactuca* enhanced somatic and gonad growth, and improved the protein assimilation efficiency at levels of 81% compared to urchins receiving formulated diets (59%) (Shpigel et al. 2018).

Seaweed cultivation is commonly carried out in open water and requires a prior optimisation to assure the productivity of the process and the continuous monitoring of different cultivation factors (e.g. seawater temperature, available nutrients, water movement and quality) to maintain its commercial quality. The physiological processes of the seaweed can affect the biomass production in the seaweed aquaculture and therefore, a deeper understanding of the physiological and metabolic mechanisms of seaweed could improve its productivity (Hayashi et al. 2020). Additionally, controlling environmental and process parameters of the seaweed cultivation in open water can be difficult due to climate conditions and negative actions of wild animals and marine microbiota. In this way, cultivation in open water could be replaced by an intensive system in tanks or ponds which facilitates the monitoring of the biomass and prevents unnecessary depreciation of the production as a result of natural causes. Nevertheless, the maintenance of seaweed on tanks is expensive and can lead to the stress of cultivated seaweeds and thereby, a previous selection of the seaweed species is required (Hayashi et al. 2020; Kim et al. 2017). Another strategy that could be applied in seaweed cultivation is the implementation of an integrated multi-trophic aquaculture (IMTA) in a closed environment like ponds, tanks, bays, and coastal lagoons. IMTA consist of farming fish or seafoodin a shared environment with other living individuals (e.g. macroalgae, mussels, sea-cucumber) to increase the productivity of the cultivation system, enhancing the sustainability of the process and product, and generate enriched products and/or species with high commercial value. In open-water systems, IMTA cultivation has been also contemplated; however, the impact of this type of cultivation in the environment and the final product has been difficult to measure due to the unpredictability of nutrients and high dilution (Ramli et al. 2020).

The harvesting of wild seaweeds is also being currently carried out.Small countries with a long harvesting tradition like Ireland are the focus of international and large-scale companies. Ireland's shores meet suitable conditions for the proliferation of a large range of seaweed species. In addition, the cultivation of seaweeds is limited as a result of the important tradition of harvesting wild species and the pristine state of the Irish's shores. A growth rate in seaweed cultivation together with an increase in automation and scaling up processes are expected in many of these countries which would lead to an increase in intensive production (Monagail and Morrison 2020).

In general, the cultivation of seaweeds has been demonstrated to be a sustainable system beneficial for the ecosystems.Some of these positive effects of seaweed proliferation in the environment relate to water purification,production of nutrients and oxygen, and the reduction of greenhouse gas emissions and the eutrophication phenomenon (Buschmann et al. 2017). Although seaweed cultivation has been considered as a potential strategy to reduce anthropogenic CO_2, a portion of this materialized gas into the organic matter has been observed to be easily decomposed back to CO_2 gas and therefore, further research is necessary to understand the underlying mechanisms of CO_2 assimilation by seaweeds. Other concerns have been raised in the last few years with regard to seaweed aquaculture. Global environmental changes,such as increased atmospheric CO_2 and temperature, and intensification of ultraviolet light levels and ocean acidification are likely to affect the productivity of seaweed cultivation worldwide (Chung et al. 2017). Furthermore, the intensive cultivation of seaweed could also entailsevere repercussions like the hybridism between wild and domesticated seaweed strains and the proliferation of parasites or pathogens that could lead to disease outbreaks within the seaweed industry (Buschmann et al. 2017; Hayashi et al. 2020). The implementation of IMTA could reduce some effects of global climate change and the selection of seaweed species could prevent the spread of infectious diseases (Buschmann and Camus 2019). Thus, there is a need for strategies or alternatives to be applied in the seaweed cultivation that could mitigate the already mentioned negative impacts.

10.5 Conclusions and Future Perspectives

Seaweeds, either from wild or aquaculture, are valuable resources. During their cultivation, seaweeds can efficiently absorb CO_2 from the seawater and thus, have the potential to neutralize greenhouse gas emissions. Moreover, seaweeds produce a wide range of bioactives mainly carbohydrates, proteins and lipids, as well as other minor compounds that have expanded the interest of this biomass for multiple applications as pharmaceuticals, cosmetics, functional foods, fertilizers, animal feeds and biofuels. Currently Ireland has a long coastline, a wide variety of seaweed species and over 97% of Irish seaweeds are wildly harvested, thus, the development of seaweed farms or permaculture systems are promising businesses in the near future. However, there are still many challenges remaining with respect to the use of seaweeds for the production of high value compounds. These challenges refer to (1) the raw biomass, including aspects of biomass production, availability and chemical or compositional stability of the raw biomass for future industrial use; (2) industrial utilization of these natural resources by developing biorefinery exploitation models, including the use of pre-treatments and novel or disruptive technologies aiming to utilise the full biomass. Moreover, these new seaweed exploitation models should be evaluated from economic, social, and environmental perspectives to contribute to the circular economy of Ireland, addressing current challenges, such as the recycling of nutrients,while alleviating current and upcoming environmental challenges related to greenhouse emissions.

Acknowledgements The authors would like to acknowledge UCD-CSC Scholarship Scheme supported by University College Dublin (UCD) and China Scholarship Council (CSC) for this study. The researchers also acknowledge funding from the transnational ERA-NET Co-fund on Blue Bioeconomy (BlueBio) and the Department of Agriculture Food and the Marine (DAFM) as part of the project BIOCARB-4-FOOD.

References

Arai S (2000) Functional food science in Japan: state of the art. Biofactors 12:13–16

Bialek A, Bialek M, Jelinska M, Tokarz A (2017) Fatty acid composition and oxidative characteristics of novel edible oils in Poland. CyTA-J Food 15:1–8

Biancarosa I, Espe M, Bruckner CG, Heesch S, Liland N, Waagbø R, Torstensen B, Lock EJ (2017) Amino acid composition, protein content, and nitrogen-to-protein conversion factors of 21 seaweed species from Norwegian waters. J Appl Phycol 29:1001–1009

Buschmann AH, Camus C (2019) An introduction to farming and biomass utilisation of marine macroalgae. Phycologia 58:443–445

Buschmann AH, Camus C, Infante J, Neori A, Israel Á, Hernández-González MC, Pereda SV, Gomez-Pinchetti JL, Golberg A, Tadmor-Shalev N, Critchley AT (2017) Seaweed production: overview of the global state of exploitation, farming and emerging research activity. Eur J Phycol 52:391–406

Černá M (2011) Seaweed proteins and amino acids as nutraceuticals. Adv Food Nutr Res 64:297–312

Chan PT, Matanjun P (2017) Chemical composition and physicochemical properties of tropical red seaweed, *Gracilaria changii*. Food Chem 221:302–310

Chemat F (2012) Microwave-assisted extraction for bioactive compounds. Springer

Chiellini F, Morelli A (2011) *Ulvan*: a versatile platform of biomaterials from renewable resources. Biomater Phys Chem:75–98

Chojnacka K, Saeid A, Witowska Z, Tuhy L (2012) Biologically active compounds in seaweed extracts-the prospects for the application. The Open Conference Proceedings J

Chung IK, Sondak CFA, Beardall J (2017) The future of seaweed aquaculture in a rapidly changing world. Eur J Phycol 52:495–505

Collins KG, Fitzgerald GF, Stanton C, Ross RP (2016) Looking beyond the terrestrial: the potential of seaweed derived bioactives to treat non-communicable diseases. Mar Drugs 14:60

Cravotto G, Boffa L, Mantegna S, Perego P, Avogadro M, Cintas P (2008) Improved extraction of vegetable oils under high-intensity ultrasound and/or microwaves. Ultrason Sonochem 15:898–902

de Jesus Raposo MF, De Morais AMB, De Morais RMSC (2015) Marine polysaccharides from algae with potential biomedical applications. Mar Drugs 13:2967–3028

Delattre C, Michaud P, Courtois B, Courtois J (2005) Oligosaccharides engineering from plants and algae: applications in biotechnology and therapeutics. Minerva Biotecnologica 17:107

Deleris P, Nazih H, Bard JM (2016) Seaweeds in human health; In: Fleurence J and Levine I (Eds.) Seaweed in health and disease prevention. PP: 319-367, Academic Press, USA.

Dickson R, Brigljevic B, Lim H, Liu J (2020) Maximizing the sustainability of a macroalgae biorefinery: a superstructure optimization of a volatile fatty acid platform. Green Chem 22:4174–4186

Dominguez H, Loret EP (2019) *Ulva lactuca*, a source of troubles and potential riches. Mar Drugs 17:357

FAO (2018) FAO yearbook: Fishery and aquaculture statistics. 2016

Fernand F, Israel A, Skjermo J, Wichard T, Timmermans KR, Golberg A (2017) Offshore macroalgae biomass for bioenergy production: environmental aspects, technological achievements and challenges. Renew Sust Energ Rev 75:35–45

Fleurence J (1999) Seaweed proteins: biochemical, nutritional aspects and potential uses. Trends Food Sci Technol 10:25–28

Fleurence J (2016) Seaweeds as food. In: Seaweed in health and disease prevention, pp 149–167

Fleurence J, Morançais M, Dumay J (2018) Seaweed proteins. Elsevier, Proteins in food processing

Fleurence J, Morançais M, Dumay J, Decottignies P, Turpin V, Munier M, Garcia-Bueno N, Jaouen P (2012) What are the prospects for using seaweed in human nutrition and for marine animals raised through aquaculture? Trends Food Sci Technol 27:57–61

Floreto E, Teshima S (1998) The fatty acid composition of seaweeds exposed to different levels of light intensity and salinity. Bot Mar 41:467–482

Gaillard C, Bhatti HS, Novoa-Garrido M, Lind V, Roleda MY, Weisbjerg MR (2018) Amino acid profiles of nine seaweed species and their in situ degradability in dairy cows. Anim Feed Sci Technol 241:210–222

Galland-Irmouli AV, Fleurence J, Lamghari R, Luçon M, Rouxel C, Barbaroux O, Bronowicki JP, Villaume C, Guéant JL (1999) Nutritional value of proteins from edible seaweed *Palmaria palmata* (Dulse). J Nutr Biochem 10:353–359

Ganesan AR, Shanmugam M, Bhat R (2018) Producing novel edible films from semi refined carrageenan (SRC) and *ulvan* polysaccharides for potential food applications. Int J Biol Macromol 112:1164–1170

Garcia-Vaquero M, Hayes M (2016) Red and green macroalgae for fish and animal feed and human functional food development. Food Rev Intl 32:15–45

Garcia-Vaquero M, Mora L, Hayes M (2019) In vitro and in silico approaches to generating and identifying angiotensin-converting enzyme I inhibitory peptides from green macroalga *Ulva lactuca*. Mar Drugs 17:204

García-Vaquero M, Rajauria G, O'Doherty JV, Sweeney T (2017) Polysaccharides from macroalgae: recent advances, innovative technologies and challenges in extraction and purification. Food Res Int 99:1011–1020

Garcia-Vaquero M, Ummat V, Tiwari B, Rajauria G (2020) Exploring ultrasound, microwave and ultrasound–microwave assisted extraction technologies to increase the extraction of bioactive compounds and antioxidants from brown macroalgae. Mar Drugs 18:172

Generalić Mekinić I, Skroza D, Šimat V, Hamed I, Čagalj M, Popović Perković Z (2019) Phenolic content of brown algae (Pheophyceae) species: Extraction, identification, and quantification. Biomolecules 9:244

Getachew AT, Saravana PS, Cho YJ, Woo HC, Chun BS (2018) Concurrent extraction of oil from roasted coffee (*Coffea arabica*) and fucoxanthin from brown seaweed (*Saccharina japonica*) using supercritical carbon dioxide. J CO2 Utilizat 25:137–146

Gomez LP, Alvarez C, Zhao M, Tiwari U, Curtin J, Garcia-Vaquero M, Tiwari BK (2020) Innovative processing strategies and technologies to obtain hydrocolloids from macroalgae for food applications. Carbohydr Polym 248:116784

Gosch BJ, Magnusson M, Paul NA, de Nys R (2012) Total lipid and fatty acid composition of seaweeds for the selection of species for oil-based biofuel and bioproducts. Gcb Bioenergy 4:919–930

Goyanes NS, D` Accorso NB (2017) Industrial applications of renewable biomass products- Past, Present and Future. Springer International Publishing AG, Springer, Cham. PP: 332.

Gressler V, Fujii MT, Martins AP, Colepicolo P, Mancini-Filho J, Pinto E (2011) Biochemical composition of two red seaweed species grown on the Brazilian coast. J Sci Food Agric 91:1687–1692

Grote B (2019) Recent developments in aquaculture of *Palmaria palmata* (Linnaeus)(Weber & Mohr 1805): cultivation and uses. Rev Aquac 11:25–41

Hafting JT, Craigie JS, Stengel DB, Loureiro RR, Buschmann AH, Yarish C, Edwards MD, Critchley AT (2015) Prospects and challenges for industrial production of seaweed bioactives. J Phycol 51:821–837

Harnedy PA, Fitzgerald RJ (2011) Bioactive proteins, peptides, and amino acids from macroalgae 1. J Phycol 47:218–232

Harnedy PA, Fitzgerald RJ (2013) Extraction of protein from the macroalga *Palmaria palmata*. LWT-Food Sci Technol 51:375–382

Harrysson H, Hayes M, Eimer F, Carlsson NG, Toth GB, Undeland I (2018) Production of protein extracts from Swedish red, green, and brown seaweeds, *Porphyra umbilicalis* Kützing, *Ulva lactuca* Linnaeus, and *Saccharina latissima* (Linnaeus) JV Lamouroux using three different methods. J Appl Phycol 30:3565–3580

Hayashi L, Cantarino SDJ, Critchley AT (2020) Challenges to the future domestication of seaweeds as cultivated species: understanding their physiological processes for large-scale production. Seaweeds Around the World, State of Art and Perspectives

Hernández-Carmona G, Carrillo-Domínguez S, Arvizu-Higuera DL, Rodríguez-Montesinos YE, Murillo-Álvarez JI, Muñoz-Ochoa M, Castillo-Domínguez RM (2009) Monthly variation in the chemical composition of *Eisenia arborea* JE Areschoug. J Appl Phycol 21:607–616

Hernandez-Carmona G, Freile-Pelegrín Y, Hernández-Garibay E (2013) Conventional and alternative technologies for the extraction of algal polysaccharides. Elsevier, Functional ingredients from algae for foods and nutraceuticals

Hessami MJ, Cheng SF, Ranga Rao A, Yin YH, Phang SM (2019) Bioethanol production from agarophyte red seaweed, *Gelidium elegans* using a novel sample preparation method for analysing bioethanol content by gas chromatography. 3-Biotech J 9(1):25

Hlima HB, Dammak M, Karkouch N, Hentati F, Laroche C, Michaud P, Fendri I, Abdelkafi S (2019) Optimal cultivation towards enhanced biomass and floridean starch production by *Porphyridium marinum*. Int J Biol Macromol 129:152–161

Holdt S, Kraan S (2011) Bioactive compounds in seaweed: functional food applications and legislation. J Appl Phycol 23:543–597

Hou X, Hansen JH, Bjerre A-B (2015) Integrated bioethanol and protein production from brown seaweed *Laminaria digitata*. Bioresour Technol 197:310–317

Husni A (2018) Therapeutic potential of seaweed polysaccharides for diabetes mellitus. Seaweed Biomaterials. IntechOpen, Riejeka, pp 27–45

Kadam SU, Álvarez C, Tiwari BK, O'Donnell CP (2017) Extraction and characterization of protein from Irish brown seaweed *Ascophyllum nodosum*. Food Res Int 99:1021–1027

Kamble P, Cheriyamundath S, Lopus M, Sirisha VL (2018) Chemical characteristics, antioxidant and anticancer potential of sulfated polysaccharides from *Chlamydomonas reinhardtii*. J Appl Phycol 30:1641–1653

Kazir M, Abuhassira Y, Robin A, Nahor O, Luo J, Israel A, Golberg A, Livney YD (2019) Extraction of proteins from two marine macroalgae, *Ulva* sp. and *Gracilaria* sp., for food application, and evaluating digestibility, amino acid composition and antioxidant properties of the protein concentrates. Food Hydrocoll 87:194–203

Khan MNA, Cho JY, Lee MC, Kang JY, Park NG, Fujii H, Hong YK (2007) Isolation of two anti-inflammatory and one pro-inflammatory polyunsaturated fatty acids from the brown seaweed *Undaria pinnatifida*. J Agric Food Chem 55:6984–6988

Kim JK, Yarish C, Hwang EK, Park M, Kim Y (2017) Seaweed aquaculture: cultivation technologies, challenges and its ecosystem services. Algae 32:1–13

Kim KT (2012) Seasonal variation of seaweed components and novel biological function of fucoidan extracted from brown algae in Quebec Ph.D thesis, Universite Laval, Quebec City, Canada

Kraan S (2012) Algal polysaccharides, novel applications and outlook. Carbohydrates-comprehensive studies on glycobiology and glycotechnology, IntechOpen

Kraan S (2013) Pigments and minor compounds in algae. Elsevier, Functional ingredients from algae for foods and nutraceuticals

Kraan S (2020) Seaweed resources, collection, and cultivation with respect to sustainability. Sustainable Seaweed Technologies

Kumar CS, Ganesan P, Suresh PV, Bhaskar N (2008) Seaweeds as a source of nutritionally beneficial compounds-a review. J Food Sci Technol 45:1

Lafarga T, Acién-Fernández FG, Garcia-Vaquero M (2020) Bioactive peptides and carbohydrates from seaweed for food applications: natural occurrence, isolation, purification, and identification. Algal Res 48:101909

Lafarga T, Rodríguez-Bermúdez R, Morillas-España A, Villaró S, García-Vaquero M, Morán L, Sánchez-Zurano A, González-López CV, Acién-Fernández FG (2021) Consumer knowledge and attitudes towards microalgae as food: the case of Spain. Algal Res 54:102174

Lakshmi DS, Trivedi N, Reddy C (2017) Synthesis and characterization of seaweed cellulose derived carboxymethyl cellulose. Carbohydr Polym 157:1604–1610

Liang X, Fan Q (2013) Application of sub-critical water extraction in pharmaceutical industry. J Mat Sci Chem Eng 1:1

Lopes G, Sousa C, Silva LR, Pinto E, Andrade PB, Bernardo J, Mouga T, Valentão P (2012) Can phlorotannins purified extracts constitute a novel pharmacological alternative for microbial infections with associated inflammatory conditions? PLoS One 7:e31145

Lorbeer AJ, Lahnstein J, Fincher GB, Su P, Zhang W (2015) Kinetics of conventional and microwave-assisted fucoidan extractions from the brown alga, *Ecklonia radiata*. J Appl Phycol 27:2079–2087

Lordan S, Ross RP, Stanton C (2011) Marine bioactives as functional food ingredients: potential to reduce the incidence of chronic diseases. Mar Drugs 9:1056–1100

Lourenço SO, Barbarino E, De-Paula JC, Pereira LODS, Marquez UML (2002) Amino acid composition, protein content and calculation of nitrogen-to-protein conversion factors for 19 tropical seaweeds. Phycol Res 50:233–241

Mabeau S, Fleurence J (1993) Seaweed in food products: biochemical and nutritional aspects. Trends Food Sci Technol 4:103–107

Maeda H, Tsukui T, Sashima T, Hosokawa M, Miyashita K (2008) Seaweed carotenoid, fucoxanthin, as a multi-functional nutrient. Asia Pacific J Clin Nutrit 17

Makkar HP, Tran G, Heuzé V, Giger-Reverdin S, Lessire M, Lebas F, Ankers P (2016) Seaweeds for livestock diets: a review. Anim Feed Sci Technol 212:1–17

Manivasagan P, Bharathiraja S, Santha Moorthy M, Mondal S, Seo H, Dae Lee K, Oh J (2018) Marine natural pigments as potential sources for therapeutic applications. Crit Rev Biotechnol 38:745–761

Matanjun P, Mohamed S, Mustapha NM, Muhammad K (2009) Nutrient content of tropical edible seaweeds, *Eucheuma cottonii, Caulerpa lentillifera* and *Sargassum polycystum*. J Appl Phycol 21:75–80

Mcdermid KJ, Stuercke B (2003) Nutritional composition of edible Hawaiian seaweeds. J Appl Phycol 15:513–524

Men'shova RV, Ermakova SP, Rachidi SM, Al-Hajje AH, Zvyagintseva TN, Kanaan HM (2012) Seasonal variations of the composition, structural features, and antitumor properties of polysaccharides from *Padina pavonica* (Lebanon) as a function of composition. Chem Nat Compounds 47:870–875

Mišurcová L (2012) Chemical composition of seaweeds. Wiley Online Library

Mišurcová L, Machů L, Orsavová J (2011) Seaweed minerals as nutraceuticals. Adv Food Nutr Res 64:371–390

Mohammad JH, Ranga Rao A, Ravishankar GA (2019) Opportunities and challenges in seaweeds as feed stock for biofuel production. In: Ravishankar GA, Rao AR (eds) Handbook of algal technologies and phytochemicals: volume-II Phycoremediation, biofuels and global biomass production, volume II. CRC Press, USA, pp 39–50

Mols-Mortensen A, Jacobsen C, Holdt SL (2017) Variation in growth, yield and protein concentration in *Saccharina latissima* (Laminariales, Phaeophyceae) cultivated with different wave and current exposures in the Faroe Islands. J Appl Phycol 29:2277–2286

Monagail MM, Morrison L (2020) The seaweed resources of Ireland: a twenty-first century perspective. J Appl Phycol 32:1287–1300

Moualee, A. & Pradhanang, T. 2019. Seaweed consumption and its effect on breast cancer

Mouritsen OG, Dawczynski C, Duelund L, Jahreis G, Vetter W, Schröder M (2013) On the human consumption of the red seaweed dulse (*Palmaria palmata* (L.) Weber & Mohr). J Appl Phycol 25:1777–1791

Naseri A, Marinho GS, Holdt SL, Bartela JM, Jacobsen C (2020) Enzyme-assisted extraction and characterization of protein from red seaweed *Palmaria palmata*. Algal Res 47:101849

Nunes AJ, Sá MV, Browdy CL, Vazquez-Anon M (2014) Practical supplementation of shrimp and fish feeds with crystalline amino acids. Aquaculture 431:20–27

Nwosu F, Morris J, Lund VA, Stewart D, Ross HA, McDougall GJ (2011) Anti-proliferative and potential anti-diabetic effects of phenolic-rich extracts from edible marine algae. Food Chem 126:1006–1012

Ojima T (2013) Polysaccharide-degrading enzymes from herbivorous marine invertebrates. Elsevier, Marine enzymes for biocatalysis

Ortiz J, Romero N, Robert P, Araya J, Lopez-Hernández J, Bozzo C, Navarrete E, Osorio A, Rios A (2006) Dietary fiber, amino acid, fatty acid and tocopherol contents of the edible seaweeds Ulva lactuca and Durvillaea Antarctica. Food Chem 99:98–104

Ortiz J, Uquiche E, Robert P, Romero N, Quitral V, Llantén C (2009) Functional and nutritional value of the Chilean seaweeds *Codium fragile, Gracilaria chilensis* and *Macrocystis pyrifera*. Eur J Lipid Sci Technol 111:320–327

Øverland M, Mydland LT, Skrede A (2019) Marine macroalgae as sources of protein and bioactive compounds in feed for monogastric animals. J Sci Food Agric 99:13–24

Peña-Rodríguez A, Mawhinney TP, Ricque-Marie D, Cruz-Suárez LE (2011) Chemical composition of cultivated seaweed *Ulva clathrata* (Roth) C. Agardh. Food Chem 129:491–498

Pliego-Cortés H, Bedoux G, Boulho R, Taupin L, Freile-Pelegrín Y, Bourgougnon N, Robledo D (2019) Stress tolerance and photoadaptation to solar radiation in *Rhodymenia pseudopalmata* (Rhodophyta) through mycosporine-like amino acids, phenolic compounds, and pigments in an integrated multi-trophic aquaculture system. Algal Res 41:101542

Qasim R (1991) Amino acid composition of some common seaweeds. Pak J Pharm Sci 4:49–54

Qin Y (2018) Applications of bioactive seaweed substances in functional food products. Bioactive seaweeds for food applications. Elsevier

Rajauria G, Foley B, Abu-Ghannam N (2016) Identification and characterization of phenolic antioxidant compounds from brown Irish seaweed *Himanthalia elongata* using LC-DAD–ESI-MS/MS. Innovative Food Sci Emerg Technol 37:261–268

Ramli NM, Verreth JAJ, Yusoff FM, Nurulhuda K, Nagao N, Verdegem MC (2020) Integration of algae to improve nitrogenous waste Management in Recirculating Aquaculture Systems: a review. Front Bioeng Biotechnol 8:1004

Robin A, Kazir M, Sack M, Israel A, Frey W, Mueller G, Livney YD, Golberg A (2018) Functional protein concentrates extracted from the green marine macroalga *Ulva* sp., by high voltage pulsed electric fields and mechanical press. ACS Sustain Chem Eng 6:13696–13705

Roleda MY, Marfaing H, Desnica N, Jónsdóttir R, Skjermo J, Rebours C, Nitschke U (2019) Variations in polyphenol and heavy metal contents of wild-harvested and cultivated seaweed bulk biomass: health risk assessment and implication for food applications. Food Control 95:121–134

Rubio-Rodríguez N, Beltrán S, Jaime I, Sara M, Sanz MT, Carballido JR (2010) Production of omega-3 polyunsaturated fatty acid concentrates: a review. Innovative Food Sci Emerg Technol 11:1–12

Santos JP, Guihéneuf F, Fleming G, Chow F, Stengel DB (2019) Temporal stability in lipid classes and fatty acid profiles of three seaweed species from the north-eastern coast of Brazil. Algal Res 41:101572

Saravana PS, Getachew AT, Cho YJ, Choi JH, Park YB, Woo HC, Chun BS (2017) Influence of co-solvents on fucoxanthin and phlorotannin recovery from brown seaweed using supercritical CO2. J Supercrit Fluids 120:295–303

Sedlmeyer F (2011) Xylan as by-product of biorefineries: characteristics and potential use for food applications. Food Hydrocoll 25:1891–1898

Shama A, Joyce SG, Mari FD, Ranga Rao A, Ravishankar GA, Hudaa N (2019) Macroalgae and microalgae: novel sources of functional food and feed. In: Ravishankar GA, Rao AR (eds) Handbook of algal technologies and phytochemicals: Volume-I: food, health and nutraceutical applications. CRC Press, USA, pp 207–219

Shannon E, Abu-Ghannam N (2019) Seaweeds as nutraceuticals for health and nutrition. Phycologia 58:563–577

Shpigel M, Shauli L, Odintsov V, Ashkenazi N, Ben-Ezra D (2018) *Ulva lactuca* biofilter from a land-based integrated multi trophic aquaculture (IMTA) system as a sole food source for the tropical sea urchin *Tripneustes gratilla elatensis*. Aquaculture 496:221–231

Singh RS, Walia AK (2018) Lectins from red algae and their biomedical potential. J Appl Phycol 30:1833–1858

Skriptsova AV, Shevchenko NM, Tarbeeva DV, Zvyagintseva TN (2012) Comparative study of polysaccharides from reproductive and sterile tissues of five brown seaweeds. Mar Biotechnol 14:304–311

Škrovánková S (2011) Seaweed vitamins as nutraceuticals. Adv Food Nutr Res 64:357–369

Smith AG, Croft MT, Moulin M, Webb ME (2007) Plants need their vitamins too. Curr Opin Plant Biol 10:266–275

Teas J, Pino S, Critchley A, Braverman LE (2004) Variability of iodine content in common commercially available edible seaweeds. Thyroid 14:836–841

Tiwari BK, Troy DJ (2015) Seaweed sustainability–food and nonfood applications. In: Seaweed sustainability. Academic press, pp 1–6

Troy DJ, Tiwari BK, Hayes M, Ross RP, Stanton C, Johnson M, Stengel D, O'Doherty JV, Fitzgerald RJ, McSorley E, Kerry JP (2017) Marine functional foods research initiative (NutraMara)

Vannuccini S, Kavallari A, Bellù LG, Müller M, Wisser D (2019) Understanding the impacts of climate change for fisheries and aquaculture: global and regional supply and demand trends and prospects. Impacts Climate Change on Fisheries and Aquaculture 41

Wells ML, Potin P, Craigie JS, Raven JA, Merchant SS, Helliwell KE, Smith AG, Camire ME, Brawley SH (2017) Algae as nutritional and functional food sources: revisiting our understanding. J Appl Phycol 29:949–982

Wijesinghe W, Jeon YJ (2011) Biological activities and potential cosmeceutical applications of bioactive components from brown seaweeds: a review. Phytochem Rev 10:431–443

Wolf B (2010) Polysaccharide functionality through extrusion processing. Curr Opin Colloid Interface Sci 15:50–54

Wong KH, Cheung PC (2000) Nutritional evaluation of some subtropical red and green seaweeds: part I—proximate composition, amino acid profiles and some physico-chemical properties. Food Chem 71:475–482

World Health Organization (1991) Protein Quality Evaluation: Report of the Joint FAO/WHO Expert Consultation, Bethesda, Md., USA 4–8 December 1989. Food & Agriculture Org

Xi J, Xue Y, Xu Y, Shen Y (2013) Artificial neural network modeling and optimization of ultrahigh pressure extraction of green tea polyphenols. Food Chem 141:320–326

Yoshioka Y, Satoh H, Mitani M (2007) Theoretical study on electronic structures of FeOO, FeOOH, FeO (H2O), and FeO in hemes: as intermediate models of dioxygen reduction in cytochrome c oxidase. J Inorg Biochem 101:1410–1427

Zhu X, Healy L, Zhang Z, Maguire J, Sun DW, Tiwari BK (2021a) Novel post-harvest processing strategies for value-added applications of marine algae. J Sci Food Agric

Zhu X, Zhang Z, Hinds LM, Sun DW, Tiwari BK (2021b) Applications of ultrasound to enhance fluidized bed drying of *Ascophyllum Nodosum*: drying kinetics and product quality assessment. Ultrason Sonochem 70:105298

Zhu Z, Wu Q, Di X, Li S, Barba FJ, Koubaa M, Roohinejad S, Xiong X, He J (2017) Multistage recovery process of seaweed pigments: investigation of ultrasound assisted extraction and ultra-filtration performances. Food Bioprod Process 104:40–47

Ziółkowska NE, Wlodawer A (2006) Structural studies of algal lectins with anti-HIV activity. Acta Biochim Pol 53:617–626

Chapter 11
Kappaphycus Seaweed Farming in Semporna, Sabah, Malaysia: A Review of Farming System, Economic Analysis and Risk Management

M. I. Khoo, M. Mahmud, W. Wee, R. A. Saufi, and S. W. Lee

11.1 Introduction

Seaweed has gained international attention as a valuable and multi-purpose marine commodity. Initially, the macroalgae are harvested from the wild and farmed as a food source. Now, Rodophytes are the most valuable seaweed species that are cultivated for their high carrageenan (*Kappaphycus* and *Eucheuma*) and agar content (*Gracilaria*) (Diego et al. 2015). Carrageenan is a good binding and gelling agent widely used in food processing, pharmaceutical and cosmetic industries (Bixler and Porse 2011). In addition, seaweed are used as food additives, pet food, livestock feed, fertilizer and biofuel (McHugh 2003; Shama et al. 2019; Mohammad and Ranga Rao 2019; Hessami et al. 2019). In Malaysia, commercial seaweed farming are carried out in several locations such as Pulau Sayak, Kedah (*Gracilaria*) and Eastern Sabah mainly Semporna (*Kappaphycus* and *Eucheuma*). *Gracilaria* sp. is an ingredient in local delicacies such as the *Kerabu Sareh*. It is also used as a biofilter in many aquaculture farms and shelter for shrimp and crab seedlings. However, *Gracilaria* production is lower compared to *Kappaphycus* and *Eucheuma* in Malaysia (Nor et al. 2020).

M. I. Khoo (✉) · M. Mahmud · S. W. Lee
Faculty of Agro Based Industry, Universiti Malaysia Kelantan (Jeli Campus), Jeli, Kelantan, Malaysia
e-mail: mohd@umk.edu.my

W. Wee
Department of Basic Knowledge and Entrepreneurship, Center of Fundamental and Continuing Education, Universiti Malaysia Terengganu, Kuala Nerus, Terengganu, Malaysia

R. A. Saufi
Faculty of Entrepreneurship and Business, Universiti Malaysia Kelantan (City Campus), Pengkalan Chepa, Kelantan, Malaysia

A. Ranga Rao, G. A. Ravishankar (eds.), *Sustainable Global Resources of Seaweeds Volume 1*, https://doi.org/10.1007/978-3-030-91955-9_11

Fig. 11.1 Fresh and dried seaweed sold in Semporna, Sabah

In Semporna, dried *Kappaphycus* and *Eucheuma* are packed in small plastics (500 g to 1 kg) and sold to tourists as souveniers and food supplement (Fig. 11.1). Besides, the locals use *Kappaphycus* and *Eucheuma* to prepare seaweed beverage for their daily consumption. Generally, these seaweed species are sold in dried form to the local carrageenan processing plant or exported to the Philippines and Indonesia.

The demand for carrageenan saw a spike after World War II due to the drop in the supply of Irish moss (*Chondrus crispus*) from Canada, Ireland, Portugal, Spain and France (Trono 1993; Diego et al. 2015). At the same time, there was a lack of natural supply of *Gigartina* or *Iridaea* from South America and Southern Europe (Trono 1993; Diego et al. 2015). Therefore, suppliers were forced to look for other potential seaweeds as an alternative to the dwindling wild stock of Irish moss and *Gigartina*. Subsequently, they discovered *Kappaphycus* and *Eucheuma* as potential replacements that are naturally found in abundance and widely cultivated in the southern Philippines. Thereafter, *Kappaphycus* and *Eucheuma* became the main commodity and aquaculture species in the country. Gradually, the Philippines replaced Canada as the top supplier of carrageenophytes in the world. The success story of *Kappaphycus* and *Eucheuma* farming in the Philippines begin to spread and were adopted by other nearby countries like Malaysia and Indonesia, followed by Tanzania, Vietnam and over 20 countries within 35 years (Bindu and Levine 2011). Consequently, the world annual seaweed production amounted to 183,000 tons dry weight by 2009 (Bixler and Porse 2011). Since carrageenophytes have became a world commodity, the challenges revolving around seaweed farming have also increased including disease outbreaks (ice-ice disease), predation by herbivorous fish, and tropical storms, resulting in the volatility of seaweed prices (Diego et al. 2015). The inconsistency in seaweed price has become a burden to small scale farmers to sustain their farm operation. In 2008, a seaweed price bubble was reported which saw a triple increase from USD 0.60 to USD 1.80 per kg of dried seaweed (Barta 2008). The manipulation of seaweed price ended with a price crash that heavily impacted the small-scale farmers (Barta 2008). At present, dry seaweed farm gate price in Semporna is around USD 0.60 per kg.

The history of seaweed farming particularly the *Kappaphycus* species in East Malaysia started in the 1970s with the migration of Filipinos to Sabah due to the civil war that broke out in the southern Philippines (Sade et al. 2006; Nor et al. 2016). They introduced the traditional methods in seaweed farming to support their livelihood. Slowly, seaweed farming was adopted by the local Sabahan to earn a living. However, the lack of farming knowledge and experience had a negative impact on the local seaweed industry. Moreover, the seaweed farming in Semporna is seasonal. The best months are February to September where the water exchange in the sites are rapid. Seaweed farming is avoided during the rainy season because the macroalgae will be infected by ice-ice disease due to constant change in salinity. Low seaweed price is another factor that becomes a major constraint in the seaweed farming industry. Middlemen are responsible for the low income of seaweed farmers. However, the establishment of government policies in improving the living standards of seaweed farmers through the Sabah Fisheries Department seemed promising. One of the systems that was introduced into the Semporna seaweed farming community is the Seaweed Cluster Project (SCP) (Nor et al. 2016). The SCP was able to improve the seaweed production, increase farmers' income by utilizing environmentally friendly methods to improve seaweed quality (Nor et al. 2016). Therefore, this paper will discuss the farming systems, economic analysis and risk management of *Kappaphycus* seaweed farming in Semporna, Sabah, Malaysia.

11.2 *Kappaphycus* Farming Systems in Semporna

11.2.1 *Kappaphycus* Life Cyle

Kappaphycus sp. has a triphasic life cycle (Fig. 11.2) consisting of several stages such as gametophyte (n), carposporophyte (2n) and tetrasporophyte (2n) (Bast 2014; Lim et al. 2014a, b). According to Bast (2014), female and male gametophytes produce carpogonium and spermatia in their thallus. Fertilization occurs in the trichogyne (female thalli) that forms a cystocarp containing carpospores at the carposporophyte stage. The carpospores undergo mitotic division leading to the formation of tetrasporophyte (Sahoo and Yarish 2005; Bast 2014). Meiosis takes place in tetrasprophytes generating tetraspores (Sahoo and Yarish 2005; Bast 2014). When tetraspores are released, they develop into gametophytes to complete the life cycle

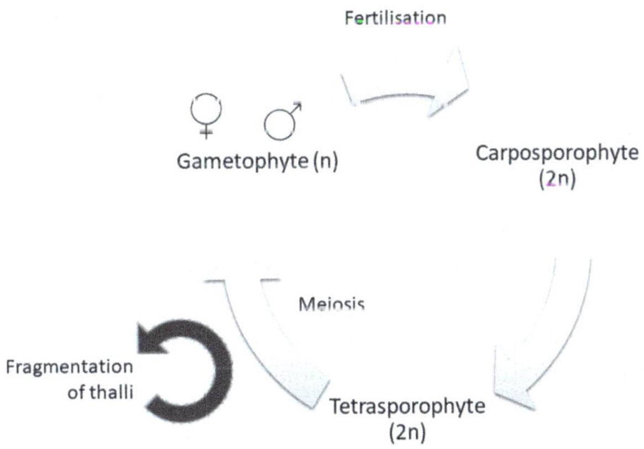

Fig. 11.2 Summary of *Kappaphycus* life cycle (modified from Bast 2014)

(Bast 2014; Lim et al. 2014a, b). However, it is not necessary for *Kappaphycus* to undergo sexual life cycle for propagation purposes. It is a clonal sea plant that can propagate via fragmentation, a less laborious farming method. Fragmentation is a process where a small part of the seaweed thalli can develop into a mature plant (Bast 2014).

There are generally four steps involved in *Kappaphycus* sp. farming which include site selection, selection of cultivation method, farm maintenance followed by harvesting and post-harvest handling (Sahoo and Yarish 2005; Hurd et al. 2014; Lim et al. 2014a, b; Rashilah et al. 2015).

11.2.2 Kappaphycus Sp. Site Selection

A typical *Kappaphycus* sp. farm should possess several characteristics. Firstly, the cultivation site should be located in an area protected from strong winds and currents (Sahoo and Yarish 2005; Rashilah et al. 2015). Farms that are located near islands and reefs are beneficial for farms since these areas act as buffer zones especially during the monsoon seasons, hence reducing the damage caused by strong waves (Lim et al. 2014a, b; Rashilah et al. 2015). *Kappaphycus* sp. farms should also be located far away from fresh water sources such as rivers, creeks estuaries and other sources of pollution (Sahoo and Yarish 2005; Rashilah et al. 2015). Clear water is essential for *Kappaphycus* sp. propagation to ease sunlight penetration into the sea for the macroalgae to carry out photosynthesis (Hurd et al. 2014). Good water motion (20–40 m/min) is another important characteristic for a seaweed farm site, considering that water turnover replenishes nutrients and facilitate their uptake while removing bio-waste from the seaweed which is believed to promote thalli robustness, therefore, increasing their carrageenan yield (Sahoo and Yarish 2005; Lim et al. 2014a, b). Good water movement also prevents extreme fluctuation in water parameters such as temperature, salinity, pH and dissolved gases (Sahoo and Yarish 2005). Water temperature should remain within 25–30 °C and salinity more than 30 ppt (Sahoo and Yarish 2005; Hurd et al. 2014). Water depth that is suitable for *Kappaphycus* sp. cultivation is between 0.5 m to 10 m where the macroalgae should remain underwater even at the lowest tides (Sahoo and Yarish 2005; Rashilah et al. 2015). *Kappaphycus* sp. grow best in coarse sandy to corally sea bottom (Sahoo and Yarish 2005; Hurd et al. 2014). Coarse-sand bottom is preferable because it is convenient for farmers to install mangrove stakes for seaweed plot preparation and prevent thalli from being covered with silt (Hurd et al. 2014). Substrates such as seagrasses, foreign seaweed, large stones, corals and grazers such as sea urchins are sometimes removed by farmers to ease farm operation activities (Hurd et al. 2014). The absence of seagrass will minimize competition and herbivory (Lim et al. 2014a, b). A seaweed farm should also obtain approval from local authorities to avoid problems in the future. For example, seaweed should be cultivated in the Aquaculture Industry Zone (ZIA) (Rashilah et al. 2015). Infrastructures such as the seaweed platform are also necessary at the seaweed farm to act as lodging for the farmers and seaweed farm management (Rashilah et al. 2015). Accessibility to essential infrastructure, resources and manpower are also important in selecting the site for *Kappaphycus sp* cultivation (Lim et al. 2014a, b).

11.2.3 Kappaphycus *Sp. Cultivation Methods*

There are several steps involved in the cultivation of *Kappaphycus* sp. The first part (Fig 11.3a) is the selection of seedlings (Rashilah et al. 2015; Radulovich et al. 2015). Seedlings are usually 20–30 days of age with apparent budding that are healthy, free of pollutants and diseases (Rashilah et al. 2015). The seedlings are either purchased from other seaweed farms nearby or obtained from the nursery developed by farmers themselves. Once the seedlings have been selected, they are cut into smaller sizes (20–150 g) (Rashilah et al. 2015; Radulovich et al. 2015). The seedlings are tied to cultivation lines (Fig 11.3b) by using the tie-tie system, where two seedlings are tied to each point on a rope spaced at 0.2–0.35 cm (Lim et al. 2014a, b; Rashilah et al. 2015; Radulovich et al. 2015). The next step is supplying the seedlings with fertilizer (Rashilah et al. 2015). However, it is optional due to the mixed reviews regarding the effectiveness of fertilizer in improving seaweed growth especially due to the extra cost incurred. The cultivation lines are then transferred to the prepared seaweed plots (Rashilah et al. 2015; Radulovich et al. 2015).

The second part in the cultivation of *Kappaphycus* sp. is selecting a suitable cultivation method. The most common methods used by seaweed farmers in Malaysia include the floating line, fixed off-bottom line and raft methods (Hurtado et al. 2014; Lim et al. 2014a, b; Rashilah et al. 2015).

Fig. 11.3 (**a**) Seedling selection. (**b**) Seedling preparation and tying them on cultivation lines

Fig. 11.4 Floating line cultivation method

The most popular and highly recommended method, particularly among seaweed farmers in Semporna, Sabah is the floating long line method (Fig. 11.4) (Lim et al. 2014a, b; Rashilah et al. 2015). This method is suitable in deeper waters (2–5 m) as an alternative to the fixed, off-bottom method when space is limited (Sahoo and Yarish 2005; Rashilah et al. 2015). There are several variations to this method. However, they generally involve stretching long nylon lines carrying seedlings (10 m – 1 km) parallel to each other (Sahoo and Yarish 2005; Lim et al. 2014a, b; Rashilah et al. 2015; Radulovich et al. 2015). The ends of these lines are held in place by wooden stakes, mangrove spikes, concrete anchors or bamboo with a length of 2 m that are driven deep into the sea bottom (Sahoo and Yarish 2005; Lim et al. 2014a, b). Some farmers also opt to attach their cultivation ropes to a much thicker and more durable motherlines that are stretched between two stakes and anchored with heavy rocks to keep the monolines in place (Rashilah et al. 2015). The space between the cultivation lines can range from 0.5 m to 5 m (Sahoo and Yarish 2005; Lim et al. 2014a, b; Radulovich et al. 2015). The four corners of the seaweed plot are anchored to wooden stakes (Sahoo and Yarish 2005). Foam or plastic floats are used to ensure the buoyancy of the cultivation lines with the number of floats varying depending on the length of the rope (Lim et al. 2014a, b; Rashilah et al. 2015). Previously, *Polyethylene terephthalate* or used mineral water bottles were used as floats while plastic raffia strings acted as the tie. The Department of Fisheries of Sabah attempted to mitigate this conventional method by using plastic floats and single nylon string as tie tie as they are more durable and environmentally friendly (Rashilah et al. 2015). This method is advantageous as it can be adopted in deeper waters and in areas with the uneven seabed. Moreover, it improves seaweed growth rates to minimize loss from grazing by benthic animals. (Sahoo and Yarish 2005; Lim et al. 2014a, b). Nonetheless, there are a few downsides to this method mainly due to the high cost of environmentally friendly materials such as nylon ropes and plastic floats as well as the construction of platforms for harvesting and processing of seaweeds. Furthermore, the floating lines are easily damaged by passing boats especially at night (Lim et al. 2014a, b).

The fixed, off-bottom cultivation method is also a popular and convenient option for seaweed farmers (Sahoo and Yarish 2005; Hurtado et al. 2014; Lim et al. 2014a, b; Rashilah et al. 2015; Radulovich et al. 2015). As its name suggests, seaweed cultivated using this method will be positioned near the bottom of the sea instead of the water surface like the floating line method (Fig. 11.5). It is suitable in shallow

Fig. 11.5 Fixed, off-bottom cultivation method

Fig. 11.6 Floating raft method

waters with a depth of 0.3–0.5 m (Rashilah et al. 2015; Radulovich et al. 2015). The cultivation lines attached with seedlings (10 m–1 km) are arranged in a similar manner as the floating line method (Sahoo and Yarish 2005; Lim et al. 2014a, b; Rashilah et al. 2015; Radulovich et al. 2015). The ends of the ropes are tied to bamboo or mangrove stakes that are driven deep into the substrata with a distance of 0.5 – 1 m between the rows (Sahoo and Yarish 2005; Lim et al. 2014a, b; Radulovich et al. 2015). Seaweed farmers often opt for this method because it is relatively of low cost because they can access the farm at low tides allowing for easy installation of the plot as well as convenient farm management and maintenance (Sahoo and Yarish 2005; Lim et al. 2014a, b). However, this method can only be used in shallow areas where seaweed are exposed during low tides (Lim et al. 2014a, b; Radulovich et al. 2015).

The floating raft or basket method (Fig. 11.6) is characterized by the presence of a bamboo frame that holds the cultivation lines instead of stakes or motherlines. It is commonly used in deep waters (3 – 10 m) where the raft (3 × 3 m @ 4 × 4 m) can be positioned at the water surface or submerged underwater (>50 cm) (Sahoo and Yarish 2005; Lim et al. 2014a, b; Rashilah et al. 2015; Radulovich et al. 2015). Experiments have been carried out by setting the raft vertically or at an angle to optimise the sunlight absorption with depth (Radulovich et al. 2015). The raft is held in position by anchoring stakes or ropes attached to the sea bottom by heavy rocks (Sahoo and Yarish 2005; Lim et al. 2014a; b). Floats are installed to improve the buoyancy of the raft as seaweed mature and increase in weight (Lim et al. 2014a, b).

The benefits of this method are similar to the floating line method since the raft is used in deep waters and areas with uneven seabed. Seaweeds are exposed to wave-induced water movement that promotes nutrient absorption and herbivory by benthic organisms are reduced (Sahoo and Yarish 2005). Furthermore, fishing nets can also be installed at the bottom of the raft to prevent predation (Lim et al. 2014a, b). The mobility of the raft enables seedlings to be planted onshore and the rafts can also be moved to other locations in bad weather (Lim et al. 2014a, b). However, the development of bamboo frames is labour intensive while the raw materials are not always readily available (Lim et al. 2014a, b). Therefore, this method is uncommon among seaweed farmers in Malaysia.

In recent years, a new seaweed cultivation method has been developed in Malaysia known as the floating basket method. This method is still in the experimental stage and expected to reduce herbivory by predators such as *ikan belawis* and turtles. However, the development of the baskets is costly (LKIM 2009). In addition, efforts have also been done to make *Kappaphycus spp.* cultivation possible onshore. The raceway culture system (Zuldin 2015) and tank culture system (Wahidatul and Rossita 2015) have been developed for this purpose. Despite its advantages, these systems have not been widely practiced because they require further research to make them commercially viable for seaweed farmers (Zuldin 2015; Wahidatul and Rossita 2015).

11.2.4 **Kappaphycus** ***Sp. Farm Maintenance***

Farm maintenance is crucial in ensuring the success of a *Kappaphycus sp* farm. Seaweed farmers must be aware of any aspects that could have a negative impact on their produce. For example, epiphytes are algae species that compete with *Kappaphycus sp* for essential requirements such as nutrients light and space (Sahoo and Yarish 2005). It is necessary to remove them from the surface of the seaweed by gently shaking the seaweed at the water surface so the epiphytes will wash away (Lim et al. 2014a, b; Rashilah et al. 2015). Unhealthy plants that have been infected by ice ice disease also needs to be removed and replaced with healthy seedlings (Lim et al. 2014a, b). Besides, benthic grazers such as sea urchin, starfish, siganid fish and turtles are also a nuisance to seaweed farm operators because they have been found to consume large quantities of seaweed leading to significant losses of biomass (Sahoo and Yarish 2005). Therefore, farmers must find a solution to get rid of them from the seaweed farm. Furthermore, domestic waste that floats into the farm from nearby villages should be removed since they are potential shelters for pests such as rabbitfish and pufferfish (Lim et al. 2014a, b; Rashilah et al. 2015). On top of that, the pristine water condition must always be maintained to ensure that seaweed is not affected by the drastic change in temperature and receive optimal sunlight for its growth (Rashilah et al. 2015). Last but not least, repairs on damaged cultivation lines, floats or stakes need to be carried out immediately by the seaweed farmers (Sahoo and Yarish 2005; Lim et al. 2014a, b; Rashilah et al. 2015)

11.2.5 Kappaphycus *Sp. Harvesting and Post-Harvest Handling*

A cultivation cycle for *Kappaphycus sp* usually ranges from 30 to 60 days (Sahoo and Yarish 2005; Lim et al. 2014a, b; Rashilah et al. 2015; Radulovich et al. 2015). At the end of the cycle, *Kappaphycus sp* would have grown up to 1 kg per seedling (Sahoo and Yarish 2005). The total harvest is the common practice in *Kappaphycus* cultivation due to several reasons. Firstly, the seaweed which has attained its maximum growth and may experience losses of biomass as a result of seasonal changes. Secondly, mature *Kappaphycus* normally possess a higher carrageenan content compared to younger shoots (Radulovich et al. 2015). Harvesting of *Kappaphycus sp* are done by manually, where the cultivation lines are placed on a boat and transported to the platform for further processing (Lim et al. 2014a, b). This procedure often yields higher quality seaweed since farmers can wash away sea-borne contaminants such as fouling, opportunistic animals and epiphytes, tying strings as well as other sea debris from their produce (Lim et al. 2014a, b; Radulovich et al. 2015). Damaged and unwanted parts are removed at this stage while healthy plants are selected to be used as seedlings for the next cultivation cycle and the rest will be dried or sold fresh at the local market (Sahoo and Yarish 2005; Radulovich et al. 2015).

Traditionally, the drying of *Kappaphycus sp* is carried out on the platform (Fig. 11.7) where the seaweed is hung or spread evenly on the ground (Sahoo and Yarish 2005; Lim et al. 2014a, b). *Kappaphycus sp* is turned over regularly to promote desiccation which generally takes two to three days and are protected from rains or storms by covering them with large canvas (Sahoo and Yarish 2005; Lim et al. 2014a, b, Radulovich et al. 2015). A recent method of drying of *Kappaphycus sp* is by using forced-wind and heat-assisted solar driers. (Majid et al. 2013; Radulovich et al. 2015). *Kappaphycus sp* is dried until it reaches 30–35% moisture content (Sahoo and Yarish 2005; Radulovich et al., 2015). The dried *Kappaphycus sp* are loaded into plastic sacks for storage, marketing or further processing for carrageenan recovery (Sahoo and Yarish 2005). In Malaysia, seaweed farmers are required to abide by the Good Aquaculture Practices (GAqP) to ensure the quality, the safety of seaweed for consumption while obtaining the best-selling price for their product (Rashilah et al. 2015). Dried *Kappaphycus sp* is processed into

Fig. 11.7 *Kappaphycus* drying

refined or semi-refined carrageenan, the latter being a more popular product due to the lower cost and less complicated procedure (Lim et al. 2014a, b). There were three semi-refined carrageenan factories that have been established in Malaysia namely Omnigel Sdn. Bhd, Tacara Sdn. Bhd and Lucky Frontier Sdn. Bhd. (Lim et al. 2014a, b). Seaweed farmers can opt to transport their produce directly to the factories which can be costly or sell them to middlemen who will then ferry the dried *Kappaphycus sp* to the semi-refined carrageenan factories (Lim et al. 2014a, b).

11.3 Economic Analysis of *Kappaphycus* Farming System in Semporna

Semporna is a famous tourist destination for hundreds to more than two hundred thousand visitors annually, as it is the gateway to many beautiful islands like Mabul Island, Bohey Dulang and Sipadan Island. Besides being a tourist spot, Semporna is located in the Coral Triangle, surrounded by Malaysia, Indonesia and Philippines which are the main world seaweed producers. Approximately 80% of the world seaweed production comes from this area (Lunkapis and Danny 2016). In order to increase the seaweed production in Malaysia, government agencies, non-government organizations, private companies, farmer associations and individuals are working together to improve the local seaweed farming system. This section will elaborate on the economic analysis of each seaweed farming system in Semporna, Sabah. Overall, there are four main seaweed farming systems at Semporna, namely cluster system through farmers association, cluster system through village committee, subsistence and traditional system and mini estate system via private companies (Fig. 11.8). All the systems adopted the long line method, off-bottom method or both in their respective farms. Other methods such as the basket method were recently introduced but the production is not very promising so far. A detail production cost of *Kappaphycus* seaweed is illustrated in Table 11.1. The initial cost for *Kappaphycus* seaweed farming is around USD 5710, whereas the total income per month is approximately USD 700. Meanwhile, the return of investment (ROI) of *Kappaphycus* seaweed farming is within eight to nine months.

11.4 Risk Management of *Kappaphycus* Seaweed Farming in Semporna

11.4.1 Risk in Malaysian Seaweed Supply Chain

The risks in Malaysian seaweed supply chain can be divided into two broad categories: macro and micro risks (Ho et al. 2015) or catastrophic and operational risks (Mohan, Sodhi and Tang 2012). Macro risks exist outside of the supply chain

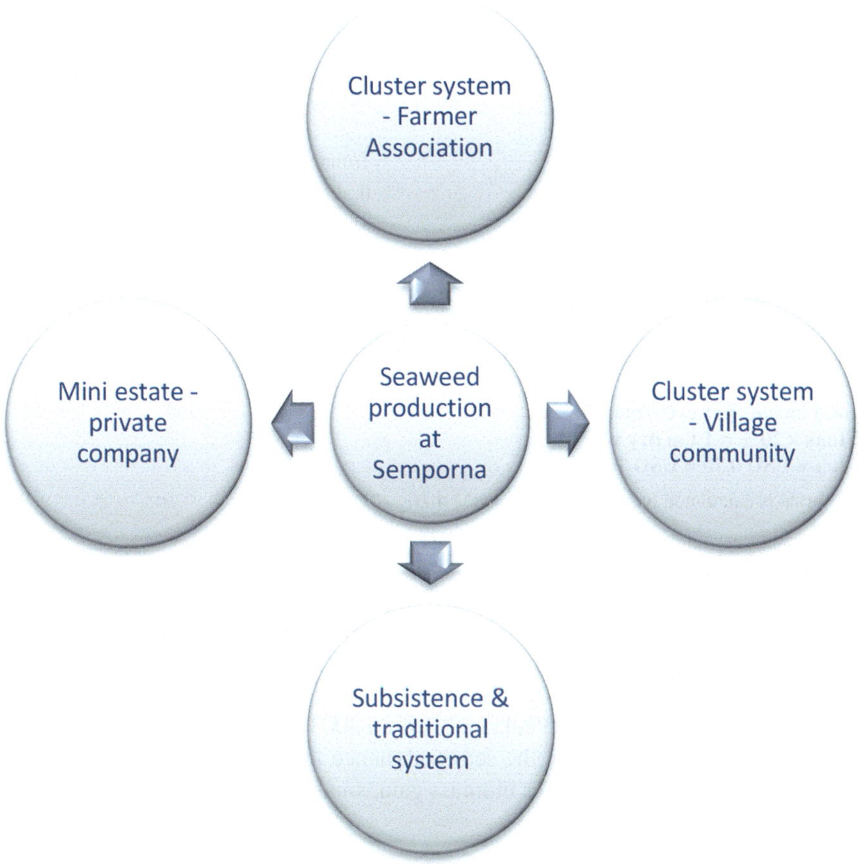

Fig. 11.8 Four types of seaweed production systems at Semporna, Sabah

but significantly impact the seaweed supply chain e.g. climate change, natural disaster, political instability and economic uncertainty. Meanwhile, micro factors refer to risks that might exist or caused by the activities performed directly by the actors in the said supply chain, which can be further divided into five subcategories: demand, process, supply, financial, logistics and infrastructure as illustrated in Fig. 11.9.

11.4.2 Macro Risks

There is a low probability for macro risks to take place but the impact can be severe upon occurrence (Ho et al. 2015). In Sabah, six main macro risks have been identified.

Table 11.1 Economic analysis of seaweed farming via long line method (100 m × 150 m = 50 lines) and fixed off-bottom method

No	Item	USD	(USD) Total
1.	PE braided rope	80/210 m	2000.00
2.	PE rope (tie-tie)		500.00
3.	Plastic floaters	1/unit	550.00
4.	Seedlings	0.3/kg	200.00
5.	Mangrove stakes	4/unit	10.00
6.	Fibre boat		300.00
Subtotal expenses			**3560.00**
7.	Engine (15 hp) (optional)		2000.00
8.	Fuel (optional)	20/month	150.00
Total expenses			**5710.00**
Total income per cycle/month: **5 tons × 20% = 1 ton dry weight** **1 ton × USD 0.70 = USD 700**			**700**

[a]The price is calculated based on USD 1 = RM 4.20

11.4.2.1 Rough Wind

Eastern Sabah is exposed to strong winds from October to February during the northeast monsoon season (Peng 2011). During this period, the wind influences the sea current, thus, affecting environmental parameters such as seawater temperature, salinity, and light penetration (Paula and Pereira 2003). The unfavourable underwater condition induces stress on the seaweed, hence impacting their growth performance exceeding 10% (i.e. lower biomass gain, smaller thalli) (Muñoz et al. 2004).

11.4.2.2 Strong Sunlight

Depending on the cultivation technique, the effects of strong sunlight towards seaweed might be different. The common practice in Semporna, Sabah is the long line method which exposes the seaweed to intense sunlight, especially during low tide. Generally, about 90% of water loss in the seaweed biomass were recorded as early as two hours upon exposure to sunlight and will continue until it resubmerged at high tide (Schagerl and Möstl 2011).

11.4.2.3 Heavy Rain

Monsoon season is not only accompanied by strong winds but also heavy rainfall, thus, reducing the salinity and pH of the seawater (Zuldin 2015, Radulovich et. al. 2015) Furthermore, photosynthetic rate of seaweed may drop up to 50% (Schagerl and Möstl 2011), resulting in lower productivity (Rajasegar 2003).

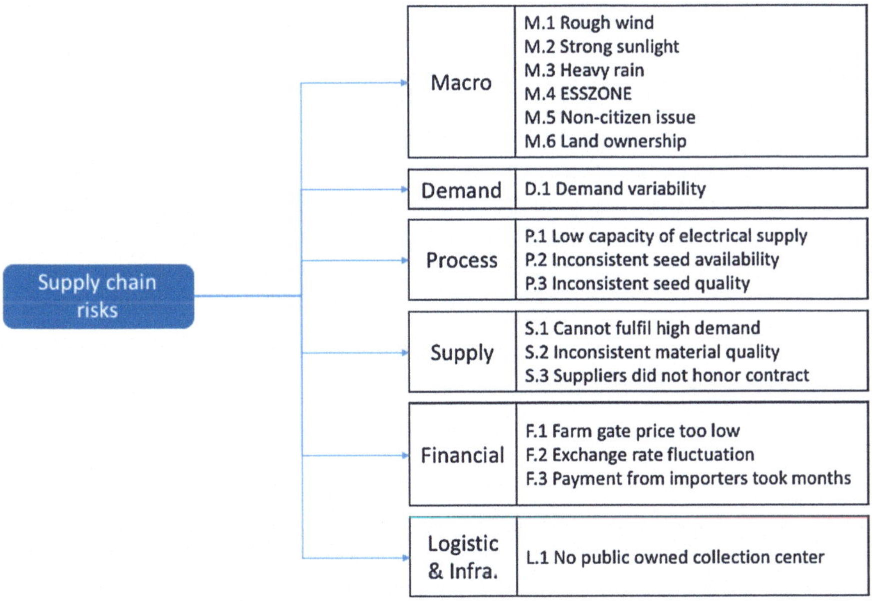

Fig. 11.9 Identified risks in Sabah seaweed industry

11.4.2.4 Eastern Sabah Security Zone (ESSZONE)

The Eastern Sabah Security Command (ESSCOM) and a special security zone, ESSZONE that covers 10 districts in the East Coast of Sabah including all major seaweed cultivation areas (Kudat, Lahad Datu, Kunak, Semporna and Tawau) were established to ensure security at Sabah waters especially from terrorist activities (Dollah et al. 2016). Paul (2008) and Kleindorfer and Saad (2005) include terrorism or political instability as one of the major causes of supply chain disruptions alongside equipment failures and natural hazards, which are applicable in the case of *Kappaphycus* production.

11.4.2.5 Non-citizen Issue

Sea-nomads or *Palauh*, have been residing in Eastern Sabah especially Kudat, Sandakan, Lahad Datu and Semporna for many generations (Datu Eranza et al. 2015) and its population has reached tens of thousands throughout the years. They are responsible for producing 90% of seaweed cultivated in Tun Sakaran Marine Park with the value of around RM5.573 million in 2015 (Dollah et al. 2016). The local seaweed industry benefited greatly from this specific demographic group and eliminating them in one way or another from the chain may bring catastrophic effect to the industry.

11.4.2.6 Land Ownership

Any individual or private entity who are interested to apply for a seaweed cultivation grant under the National Key Economic Area (NKEA) must present a valid land ownership certificate together with an application form to the central government in Putrajaya. However, in Sabah, all matters related to land is managed exclusively by the state goverment according to the Land Ordinance Cap. 68 enforced in 1930 (the State Attorney-General's Chambers of Sabah 2019). The central government requires a valid land grant before any financial assistance can be provided, which is also subjected to a long bureaucracy process and disagreement between policymakers.

11.4.3 Demand Risk

Sodhi (2005) discusses demand risk as last minute change in order quantity which prompts the company to minimize their inventory. Negative demand risk occurs when final quantity ordered by the end customer is lower than the inventory resulting in a surplus. In contrast, a positive demand is when the final quantity ordered is higher than the inventory owned by the company, indicating its expansion.

11.4.4 Process Risk

Seaweed harvested in Sabah are either dried or transformed into semi-refined carrageenan before being exported. Process risk for producing raw dried seaweed is minimal since it requires mainly sunlight and can be conducted on the platform. On the other hand, semi-refined carrageenan production is a complex chemical process, which require stable electricity and clean water supply (Mulyati 2015). As a result, these factories are often located far from the seaweed farms and collection sites, increasing the cost for stakeholders along the supply chain.

11.4.5 Supply Risk

Located at the beginning of the chain, supply risk is frequently divided into disruption risk and operational risk (Tang 2006). The former can be identified as external to the organization but internal to the supply chain (Craighead et al. 2007; Kleindorfer and Saad 2005), whereas the latter are internal to the organization (Ritchie and Brindley 2007). Due to its location on the chain, supply risk is interrelated with demand risk and process risk, thus, considered as a macro risk.

11.4.5.1 Inability to Fulfil Higher Demand

Even though the Sabah seaweed production in 2016 was 205,988.70 mT (Department of Fisheries 2017) and continues to increase annually, the numbers remain lower than the demand. Customers will then opt for competitors like Indonesia and the Philippines, which produce more than ten million mT seaweed annually and are capable of fulfilling higher global demand (Buschmann et al. 2017).

11.4.5.2 Inconsistent Material Quality

Semi-carrageenan factories require dried carragenophytes with a moisture content of 35% to 38%, and free of contaminant such as tie-tie made of raffia strings to produce high-quality output. However, the standard of dried seaweed sent to the factory are often inconsistent and hard to maintain, leading to extra cost and time for further processing, thus, risking not meeting their clients' demands in time.

11.4.5.3 Breach of Contract

As a mitigation strategy, processors usually draw a contract with seaweed agents. However, this approach seemed inefficient because agents do not take the agreement seriously by not adhering to the quality standard, price and amount of raw materials.

11.4.6 Financial Risk

Financial risks include exchange rate (Chopra and Sodhi 2004; Hahn and Kuhn 2012; Manuj and Mentzer 2008; Tummala and Schoenherr 2011), price fluctuation (Cucchiella and Gastaldi 2006) and financial inflow (Chopra and Sodhi 2004; Tsai 2008).

11.4.6.1 Deflated Farm Gate Price

Since there is no market regulation for seaweed, its price is determined by the free market. Ex-farm price was approximately RM1.60 in 2015 until 2017 and increased to RM3.75 in 2018 (Kassim 2018). Moreover, the existence of multilayer agents worsens the ex-factory price, which will be substantially reduced before reaching the farmers.

11.4.6.2 Exchange Rate Fluctuation

This particular risk is exclusive to processors and major middlemen who export seaweed to eastern European countries, Philippines, and China. Nevertheless, a contract is often signed prior to the transaction, thus, reducing losses due to exchange rate.

11.4.6.3 Delayed Payment

After the raw dried seaweed are delivered to the factory, agents do not receive their money immediately. Their payment will be delayed until the processors are paid by their clients from Holland, Austria, Russia, or Romania which can take up to six months.

11.4.7 Logistic and Infrastructure Risk

All raw dried seaweed agents are responsible for their own logistics, especially storage. Upon collecting the seaweed from the sea, the produce will be delivered to the warehouse for grading and further processing if necessary. However, grading and pricing are not standardized or regulated by any governing body, resulting in variability in quality and price from one agent to another.

11.5 Conclusion

Malaysia used to be the main producer of *Kappaphycus* sp. in the world. However, the position was taken over by Indonesia and the Philippines in recent years. The Malaysian government has identified the main issues of the local seaweed production and initiatives to improve the *Kappaphycus* industry such as introducing an effective farming system and providing incentives to farmers with the aim to make Malaysia one of the main global seaweed producers once again.

References

Barta P (2008) Indonesia got soaked when the seaweed bubble burst. Wall Street J. http://online.wsj.com/news/articles/SB122454073909251775

Bast F (2014) An illustrated review on cultivation and life history of agronomically important seaplants. In: Seaweed: mineral composition, nutritional and antioxidant benefits and agricultural uses. Nova Publishers, New York, pp 39–70

Bindu MS, Levine IA (2011) The commercial red seaweed *Kappaphycus alvarezii* – an overview on farming and environment. J Appl Phycol 23(4):789–796

Bixler HJ, Porse H (2011) A decade of change in the seaweed hydrocolloids industry. J Appl Phycol 23(3):321–325

Buschmann AH, Camus C, Infante J, Neori A, Israel A, Hernández-González, M. C. … Critchley, A. T. (2017) Seaweed production: an overview of the global state of exploitation, farming and emerging research activity. Eur J Phycol 52(4):391–406. https://doi.org/10.1080/09670262.2017.1365175

Chopra S, Sodhi MS (2004) Managing risk to avoid supply-chain breakdown. MIT Sloan Manag Rev 46:53–61. https://doi.org/10.1108/IJOPM-10-2012-0449

Craighead CW, Blackhurst J, Rungtusanatham MJ, Handfield RB (2007) The severity of supply chain disruptions: design characteristics and mitigation capabilities. Decis Sci 38(1):131–156. https://doi.org/10.1111/j.1540-5915.2007.00151.x

Cucchiella F, Gastaldi M (2006) Risk management in the supply chain: a real option approach. J Manuf Technol Manag 17(6):700–720. https://doi.org/10.1108/17410380610678756

Datu Eranza DR, Bahron A, Alin J (2015) Sustaining seaweed farming in Malaysia. Inter J Business Manag 3(3):201–205

Department of Fisheries (2017) Pengeluaran Rumpai Laut Mengikut Bulan Dan Negeri

Diego V, Cai J, Hishamunda N, Ridler N, Neish IC, Hurtado AQ, Msuya FE, Krishnan M, Narayanakumar R, Kronen M, Robledo D, Gasca-Leyva E, Fraga J (2015) The economics of *Kappaphycus* seaweed cultivation in developing countries: A comparative analysis of farming systems. Aquacult Econom Manag 19:251–277

Dollah R, Wan Hassan WS, Peters D, Othman Z (2016) Old threats, new approach and National Security in Malaysia: issues and challenges in dealing with cross-border crime in East Coast of Sabah. Mediterr J Soc Sci 7(3):178–186. https://doi.org/10.5901/mjss.2016.v7n3s1p178

Hahn GJ, Kuhn H (2012) Value-based performance and risk management in supply chains: A robust optimization approach. Int J Prod Econ 139(1):135–144. https://doi.org/10.1016/j.ijpe.2011.04.002

Hessami MJ, Cheng SF, Ranga Rao A, Yin YH, Phang SM (2019) Bioethanol production from agarophyte red seaweed, *Gelidium elegans* using a novel sample preparation method for analysing bioethanol content by gas chromatography. Biotech J 9(1):25

Ho W, Zheng T, Yildiz H, Talluri S (2015) Supply chain risk management: a literature review. Int J Prod Res 53(16):5031–5069. https://doi.org/10.1080/00207543.2015.1030467

Hurd CL, Harrison PJ, Bischof K, Lobban CS (2014) Seaweed ecology and physiology. Cambridge University Press

Hurtado AQ, Gerung GS, Yasir S, Critchley AT (2014) Cultivation of tropical red seaweeds in the BIMP-EAGA region. J Appl Phycol 26(2):707–718

Kassim A (2018) Status semasa pembangunan rumpai laut negara. In *Seminar Pembangunan Industri Rumpai Laut Malaysia*. Semporna, Sabah

Kleindorfer PR, Saad GH (2005) Managing Disruption Risks in Supply Chains- ProQuest 14(1):53–68. https://doi.org/10.1111/j.1937-5956.2005.tb00009.x

Lembaga Kemajuan Ikan Malaysia (LKIM) (2009) Program pembangunan rumpai laut menggunakan kaedah raga. In: Paper presented during the MIMA seminar on developing the seaweed aquaculture sector, 27 Oct 2009. Maritime Institute of Malaysia, Organiser

Lim PE, Tan I, Adibi R, Phang SM (2014a) A guide to *Kappaphycus* and *Eucheuma* seaweeds in Malaysia. University of Malaya Press

Lim PE, Tan J, Phang SM, Adibi R (2014b) A guide to *Kappaphycus* and *Eucheuma* seaweeds in Malaysia / Garis Panduan Mengenai Rumpai Laut *Kappaphycus* dan *Eucheuma* di Malaysia. University of Malaya

Lunkapis GJ, Danny HW (2016) Preliminary impact assessment of seaweed cultivation by the coastal communities in Sabah, Malaysia. Transaction on Science and Technology 3(2–2):413–420

Majid MKA, Rafiq S, Sakari M, Sulaiman J, Md S (2013) Seaweed Modeling for drying and the efficiency as heavy metal removal in *Kappaphycus* striatum variety Sacol using solar dryer. Int J Environ Bioenergy:2165–8958

Manuj I, Mentzer JT (2008) Global supply chain risk management. J Bus Logist 29(1):133–155

McHugh DJ (2003) A guide to the seaweed industry. FAO fisheries technical paper no. 441. Food and Agriculture Organization of the United Nations, Rome, Italy

Mohammad JH, Ranga Rao A, Ravishankar GA (2019) Opportunities and challenges in seaweeds as feed stock for biofuel production. In: Ravishankar GA, Rao AR (eds) Handbook of algal technologies and phytochemicals: volume-II Phycoremediation, biofuels and global biomass production, Volume II. CRC Press, USA, pp 39–50

Mulyati H (2015) Supply chain risk management study of the Indonesian seaweed industry

Muñoz J, Freile-Pelegrín Y, Robledo D (2004) Mariculture of *Kappaphycus alvarezii* (Rhodophyta, Solieriaceae) color strains in tropical waters of Yucatán, México. Aquaculture 239:161–177. https://doi.org/10.1016/j.aquaculture.2004.05.043

Nor AM, Gray TS, Caldwell GS, Stead SM (2016) Is a cooperative approach to seaweed farming effectual? An analysis of the seaweed cluster project (SCP), Malaysia. J Appl Phycol. 22nd International Seaweed Symposium, Copenhagen

Nor AM, Gray TS, Caldwell GS, Stead SM (2020) A value chain analysis of Malaysia's seaweed industry. J Appl Phycol 32(4):2161–2171

Paul J (2008) Supply chain disruption: risk and management. SCMS J Indian Manag 5(1):109–112

Paula EJ, Pereira RTL (2003) Factors affecting the growth rate of *Kappaphycus alvarezii* (Doty) Doty ex P. Silva (Rhodophyta, Solieriaceae) in sub-tropical waters of São Paulo State, Brazil. In: Proceedings of the 17th International Seaweed Symposium, Cape Town, South Africa, 28 January-2 February 2001, pp 381–388. Retrieved from https://www.cabdirect.org/cabdirect/abstract/20033183350

Peng H (2011) Seasonal monsoon effects on seaweed production in Semporna. Universiti Malaysia Sabah. Retrieved from http://eprints.ums.edu.my/13178/1/ae0000002825.pdf

Radulovich R, Neori A, Valderrama D, Reddy CRK, Cronin H, Forster J (2015) Farming of seaweeds. In Seaweed Sustainability (pp. 27–59). Academic Press

Rajasegar M (2003) Physico-chemical characteristics of the Vellar estuary in relation to shrimp farming. J Environm Biolo 24(1):95–101. Retrieved from http://www.ncbi.nlm.nih.gov/pubmed/12974418

Rashilah M, Azhar K, Nur Fazliana MN (2015) Industri Rumpai Laut Negara. Penerbit Institut Penyelidikan dan Kemajuan Pertanian Malaysia (MARDI)

Ritchie B, Brindley C (2007) Supply chain risk management and performance. Int J Oper Prod Manag 27(3):303–322. https://doi.org/10.1108/01443570710725563

Sade A, Ali I, Mohd Ariff MR (2006) The seaweed industry in Sabah East Malaysia. Jati 11:97–107

Sahoo D, Yarish C (2005) Mariculture of seaweeds. Phycological Methods: Algal Culturing Techniques. Academic Press, New York, pp 219–237

Schagerl M, Möstl M (2011) Drought stress, rain and recovery of the intertidal seaweed *Fucus spiralis*. Mar Biol 158:2471–2479. https://doi.org/10.1007/s00227-011-1748-x

Shama A, Joyce SG, Mari FD, Ranga Rao A, Ravishankar GA, Hudaa N (2019) Macroalgae and microalgae: novel sources of functional food and feed. In: Ravishankar GA, Rao AR (eds) Handbook of algal technologies and phytochemicals: Volume-I: food, health and nutraceutical applications. CRC Press, USA, pp 207–219

Sodhi M (2005) Managing demand risk in tactical supply chain planning for a global consumer electronics company. Prod Oper Manag 14(1):69–79. Retrieved from https://www.scopus.com/record/display.uri?eid=2-s2.0-19844381041&origin=inward&txGid=3898c731238ceb5c7d4cfce07cfb30b7

Sodhi MS, Tang CS (2012) Managing supply chain risk (Vol. 172). https://doi.org/10.1007/978-1-4614-3238-8

Tang CS (2006) Perspectives in supply chain risk management. Int J Prod Econ 103(2):451–488. https://doi.org/10.1016/j.ijpe.2005.12.006

The State Attorney-General's Chambers of Sabah (2019) Land Ordinance Cap. 68. Official Website of The State Attorney-General's Chambers. https://sagc.sabah.gov.my/?q=en/content/land-ordinance-cap-68

Trono GC (1993) *Eucheuma* and *Kappaphycus*: taxonomy and cultivation. In: Ono M, Critchley AT (eds) Seaweed cultivation and marine ranching. Kanagawa International Fisheries Training Center, Japan International Cooperative Agency, Yokosuka, Japan, pp 75–88

Tsai CY (2008) On supply chain cash flow risks. Decis Support Syst 44(4):1031–1042. https://doi.org/10.1016/j.dss.2007.12.006

Tummala R, Schoenherr T (2011) Assessing and managing risks using the supply chain risk management process (SCRMP). Supply Chain Manag Inter J 16(6):474–483. https://doi.org/10.1108/13598541111171165

Wahidatul HZ, Rossita S (2015) Performance of red seaweed (*Kappaphycus* sp.) cultivated using tank culture system. J Fisheries Aquatic Sci 10(1):1

Zuldin W (2015) The Cultivation of Red Seaweed (Rhodophytes, *Kappaphycus* spp.) in Raceway Culture System. In International Conference on Aquaculture & Fisheries (pp. 1–31)

Chapter 12
Seaweed Production Companies in Korea: An Overview

Chan Sun Park and Eun Kyoung Hwang

Abbreviations

FAO (Food and Agriculture Organization of the United Nations)

12.1 Introduction

Korea has a well-established seaweed industry ranked fourth among the world's seaweed-producing countries and is the world's leader in *Pyropia* exports (Hwang et al. 2019). The impetus for the seaweed industry's growth in Korea stems from a combination of a national preference for seaweed cultivation and significant investment in industry capabilities and technologies.

While seaweed production in Korea is primarily a source of income for fishermen, it is also an important food source for the abalone aquaculture industry. Cultivated seaweeds are also used to regenerate marine forests that aid the recovery of coastal marine ecosystems.

Harvested seaweeds have a wide range of uses, including bioenergy production, the production of various natural food additives, the development of health supplements, and beauty products. The use of seaweeds for the production of beauty and cosmetics has increased in recent years, and there is a clear trend of diversification of culture species to meet the demands of this industry.

C. S. Park
Department of Marine and Fisheries Resources, Mokpo National University, Muan, Republic of Korea

E. K. Hwang (✉)
Fisheries Seed and Breeding Research Institute, National Institute of Fisheries Science, Haenam, Jeollanamdo, Republic of Korea
e-mail: ekh215@korea.kr

A. Ranga Rao, G. A. Ravishankar (eds.), *Sustainable Global Resources of Seaweeds Volume 1*, https://doi.org/10.1007/978-3-030-91955-9_12

As the seaweed industry expands, it has become increasingly industrialized and has increasingly become segmented into specialized areas such as production, processing, distribution and export. This change in the industrial structure is expected to accelerate as the existing seaweed farming population ages, and fewer younger farmers become engaged in the seaweed industry.

Increasing world demand for natural foods such as seaweeds (sea vegetables), containing beneficial bioactive compounds and produced using environmentally friendly farming practices, is expected to contribute significantly to the expansion of the Korean seaweed industry in the future.

In this Chapter, the Korean seaweed cultivation industry is discussed in terms of existing cultivation production, processing and export technologies, and opportunities for future development.

12.1.1 History

The first Korean laver (*Pyropia* sp) (Gim) was recorded as a souvenir of the Taeindo in Gwangyang-gun, South Province in 1481 according to the Donggukyeojiseungnam (a book of Korean geography from the Chosun Dynasty). Cultivation of laver began between 1623 and 1649 (Bae 1991; Chung 1937; Sohn 1996; Hwang et al. 2019). In 1928, a horizontal net system was devised (Kang and Ko 1977) and the use of synthetic nets started in the 1960s (Yoo 1964). Seaweed farming in Korea expanded rapidly in the 1970s. This rapid and somewhat uncontrolled expansion resulted in low-quality seaweed products, genetic degradation, failure of environmental adaptability, and increased incidence of disease (Hwang et al. 2019). These issues emphasized the need for investment into research on seaweed biology and breeding to support industry growth. Further significant industry growth occurred in the 2000s, as abalone farming increased, creating a demand for farmed kelp as abalone feed. Around the same time, export demand for *Pyropia* increased significantly, leading to increased *Pyropia* production.

To maximize production opportunities, research on seaweed farming has not been limited to local species. Transplantation of Japanese varieties of *S. japonica* into Korea occurred in the 1968 (Chang and Geon 1970), and selective breeding to extend cultivation periods has recently been undertaken (Hwang et al. 2017). Newly introduced species include three green seaweeds, *Codium* (Hwang et al. 2008b), *Capsosiphon* (Sohn 1998; Hwang et al. 2003), and *Ulva* (Yoon et al. 2003; Park and Hwang 2011); five brown seaweeds, *Ecklonia* (Hwang et al. 2013), *Eisenia, Costaria, Undariopsis* (Hwang et al. 2011), and *Sargassum* (Sohn 1998; Hwang et al. 2007, b); and one red seaweed, *Gracilariopsis* (Sohn 2009).

12.1.2 Production Trends

In 2019 the Korean aquaculture industry produced 1,812,765 tons of seaweed, accounting for 76% of its total aquaculture production (Table 12.1). Seaweed production in 2019 was 214 times greater than the total seaweed harvest from wild fisheries (Table 12.2).

Seaweed farming in Korea is concentrated on the west side of the south coast (Sohn 1998) where almost 90% of total seaweed cultivation occurs (Fig. 12.1).

Pyropia is the most valuable of the cultivated species within the Korean seaweed industry and is the industry's largest export earner. The increase in export demand for seaweed exports has become a driving force in the seaweed industry. Increasing numbers of production, processing and export companies are now being established to service this demand.

The supply of fresh brown seaweed dominates the domestic market as abalone feeds. The low cost of producing seaweeds makes them an attractive feed option for abalone producers (Hwang et al. 2013). Hence, a large amount of kelp production (over 65% of total production) in 2019 was directed at meeting the expanding abalone industry's demands (Fig. 12.2).

In a world where consumers are becoming increasingly aware of environmental sustainability, the industry must also meet the challenges of developing and demonstrating more eco-friendly practices. A Korean seaweed company recently obtained ASC-MSC (Aquaculture Stewardship Council-Marine Stewardship Council) certification for their production techniques. More farms are expected to seek similar certification in the future as the Korean seaweed industry develops in line with consumer demands.

Table 12.1 Aquaculture production of Korea in 2019

Contents	Production (ton-wet wt.)	Value (1000 US$)
Seaweeds	1,812,765 (76.4%)	852,279 (30.7%)
Shellfish	435,163 (18.3%)	939,721 (33.8%)
Fish	85,217 (3.6%)	818,163 (29.4%)
Crustacean	7543 (0.3%)	105,470 (3.8%)
Other animals	31,312 (1.3%)	64,848 (2.3%)
Sum	**2,371,999**	**2,780,481**

Data from the Ministry of Ocean and Fisheries in 2020 available from https://www.mof.go.kr/statPortal/main/portalMain.do

Table 12.2 Production and value of Korean seaweed aquaculture and fisheries in 2019

	Aquaculture		Fisheries	
Species	Production (ton-wet wt.)	Value (1000 US$)	Production (ton-wet wt.)	Value (1000 US$)
Saccharina japonica	662,557	88,742	0	142
Pyropia spp.	606,873	561,407	73	16,890
Undaria pinnatifida	494,947	173,755	4952	11,162
Sargassum fusiforme	33,477	14,832	552	638
Ulva spp.	6321	5841	287	568
Capsosiphon	3386	4635	23	48
Codium fragile	3258	1801	249	712
Gracilariopsis	1769	875	0	177
Sargassum fulvellum	176	387	131	281
Other algae	1	6	2167	3337
Sum	1,812,765	852,278	8449	16,890

Data from the Ministry of Ocean and Fisheries in 2020 available from https://www.mof.go.kr/statPortal/main/portalMain.do

12.1.3 Characteristics

The Korean seaweed farming industry has historically been based on family-run businesses rather than large companies. Each household is responsible for a given cultivation site, and a local fisheries village's clerk coordinates the community production. Recently, an amalgamation of growers has occurred, resulting in a decrease in the number of farming households and an increase in farming companies (Fig. 12.3). However, the coastal area licensed for seaweed cultivation has maintained relatively constant through government control (Fig. 12.4).

Changes in farm operation have been particularly marked for *Pyropia*, where production and processing have become separated industries. Groups of farmers provide raw product to processors that can operate at a scale required to meet processing and export requirements.

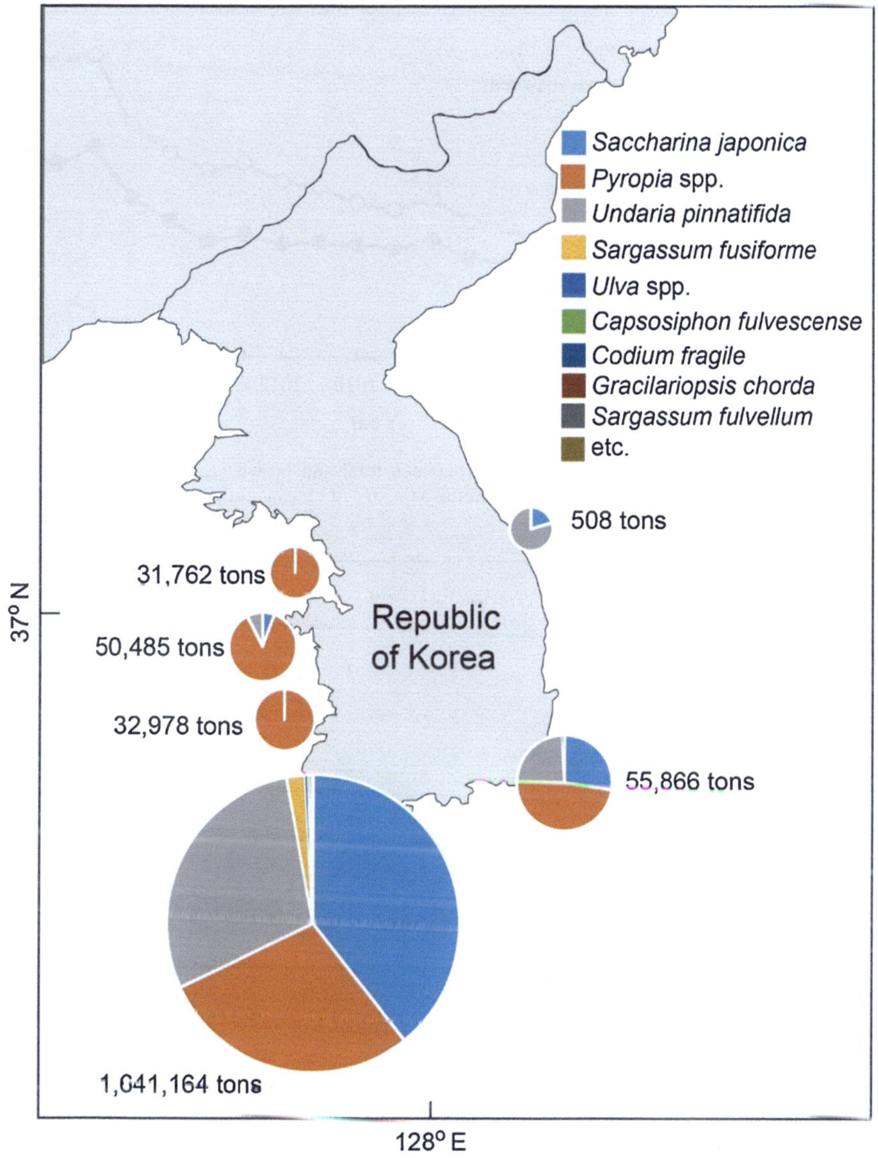

Fig. 12.1 A map showing the regional distribution of Korean seaweed production in 2019

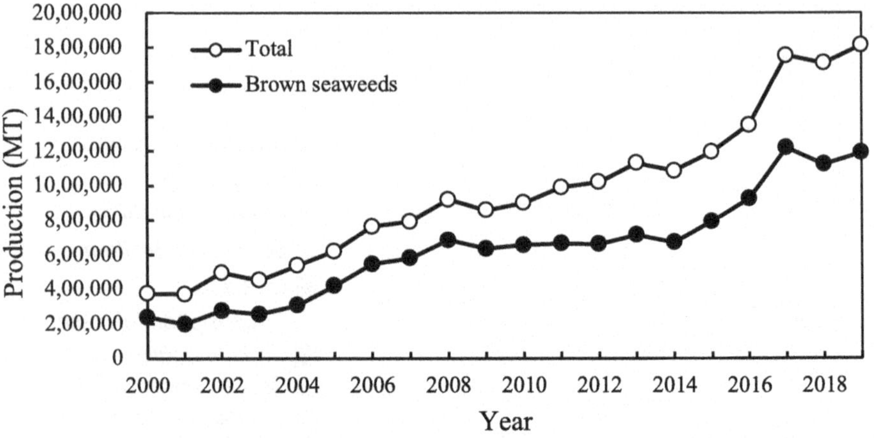

Fig. 12.2 Comparison of production trends between total and brown seaweeds farmed in the Korean waters from 2000 to 2019. (Data from the Ministry of Ocean and Fisheries 2000–2020)

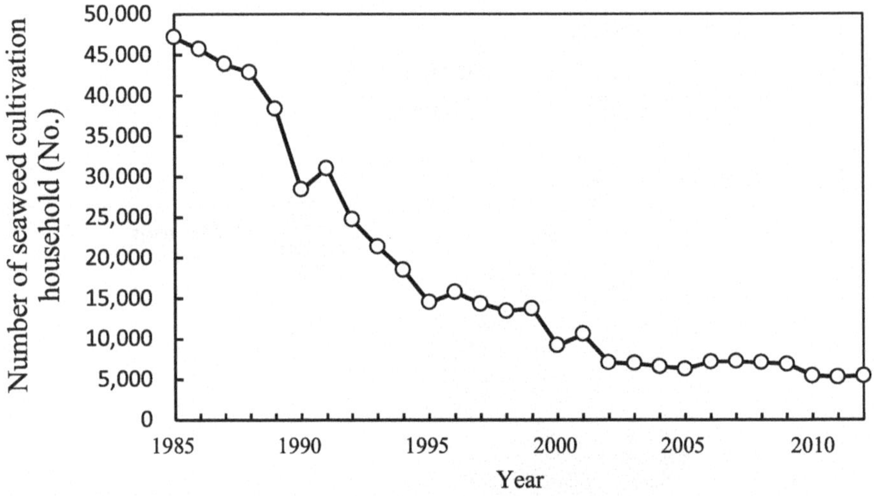

Fig. 12.3 Changes in the number of seaweed cultivation households from 1985 to 2012 in Korea

12.2 Seaweed Production Technologies

12.2.1 Seed Production

Seed production systems are well established in Korea. While some seaweed farmers produce their own seedlings for cultivation many purchase them from companies specialisng in seed production. The seaweed seed production industry is now well established in Korea and is worth US$24.7 million per year (Table 12.3).

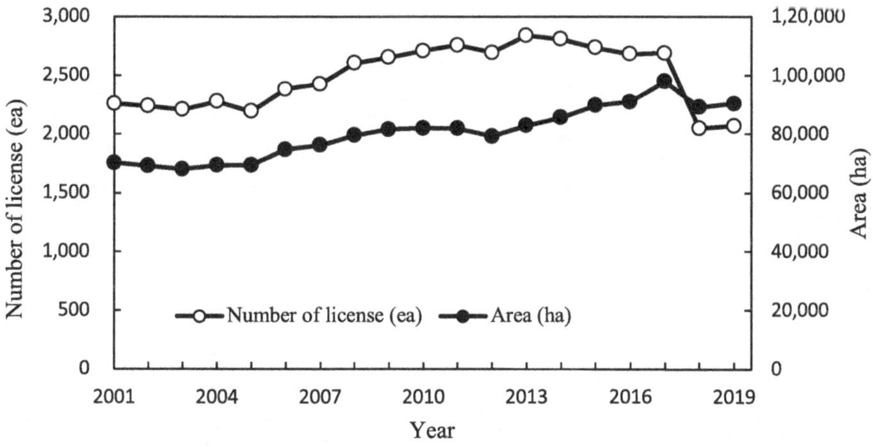

Fig. 12.4 Number of licenses and area of seaweed cultivation from 2001 to 2019 in Korea

Table 12.3 Number and value of seaweed seed production companies in Korea

Species	Amount of Seed Production (Box[a] or seed frames[b])	Value (1000 US$)	Number of company
Pyropia spp.	4,300,000[a]	12,900	161
Undaria pinnatifida	272,770[b]	6408	65
Saccharina japonica	304,000[b]	5239	
Costaria costata	2000[b]	40	10
Sargassum fulvellum	1000[b]	100	
Ecklonia cava	400[b]	20	
Ecklonia stolonifera	200[b]	10	
Sum	**4,880,370**	**24,718**	**236**

Data from the Korea Maritime Institute 2019
[a]A box includes 50 oyster shells bearing conchocelis
[b]A seed frame is 45 × 55 cm, 200 m of seed fibers included

12.2.2 *Pyropia* Spp.

Several species of *Pyropia* are cultivated in Korea. The key species include *P. yezoensis*, *P. seriata* and *P. dentata*.

Pyropia seed production companies can deliver seed in two formats; either as a free-living conchocelis or as a shell conchocelis. The majority of conchocelis companies supply shell conchocelis with just five companies supplying the free-living conchocelis. Most of the *Pyropia* seedling companies also grow *Pyropia* (Fig. 12.5).

Pyropia is cultured by either a fixed pole or a floating net system (Sohn 1993). After harvesting, *Pyropia* is processed into dried and seasoned laver and is exported

Fig. 12.5 Artificial seed production and cultivation process for *Pyropia*. (**a**) Free-living conchocelis culture. (**b**) Oyster shell bearing conchocelis filaments culture in horizontal culture tank. (**c**) Oyster shell bearing conchocelis filaments culture in vertical concrete tank. (**d**) Outdoor seeding showing the slender mosquito net bags (arrow, included oyster shell pieces to facilitate the release of the inner conchospores) with seeding nets on sea surface. (**e**) Seeding wheel with cultivation nets, immersed in a seawater tank containing oyster shell bearing conchospores. (**f**) Fixed pole system. (**g**) Floating system with cultivation nets exposed to the air. H. Harvest

in each state. There are ca. 320 companies producing dried laver and ca. 1000 producing seasoned laver.

In 2019 there were 29 Korean *Pyropia* export-related companies (Korean *Pyropia* Export Association), exporting US$ 487 million (Table 12.4) of laver. Korea larva production now accounts for 58.3% of the global laver market. Korea exported 48,268 tons of seaweed products in 2019 of which 27,245 was larva (Table 12.4).

A significant increase in *Pyropia* production occurred after the introduction of the UPOV (International Union for the Protection of New Variety of Plant) system in 2012. With the ability to protect plant varietals, the Korean government committed resources to develop new strains of *Pyropia*. That development has led to 13 new cultivars being established. These cultivars and the associated breeding technologies have been supplied to the aquaculture industry and have contributed directly to productivity improvement,

Table 12.4 Seaweed exports from Korea (unit: ton-wet wt., 1000US$)

	2000		2005		2010		2015		2019	
Content	Export	Value	Export	Value	Export	Value	Export	Value	Export	Value
Pyropia	5178	31,015	7580	54,243	9560	105,196	17,694	304,868	27,245	487,350
Undaria	12,950	29,748	21,454	29,868	11,967	19,038	14,107	29,036	16,276	33,606
Sargassum fusiforme	5521	30,353	4137	23,938	2668	29,622	1988	24,276	2115	17,189
Saccharina	1038	4611	1455	4888	1479	5987	638	4215	980	5626
Others	1455	8953	1932	12,591	2122	12,051	1481	9633	1652	11,056
Total	**26,142**	**104,680**	**36,558**	**125,528**	**27,796**	**171,894**	**35,908**	**372,028**	**48,268**	**554,827**

Data from the Korea Customs Service in 2020 available at https://www.customs.go.kr

Fig. 12.6 Artificial seed production and cultivation process for *Undaria* and *Saccharina*. (**a**) Sporophylls of *U. pinnatifida*. (**b**) Sorus of *S. japonica*. (**c**) Tank culture of kelp seedlings. (**d**) Nursery culture. (**e**) Young sporophytes of kelp after nursery culture. (**f**) Automatic sewing machine for insertion of seed fibres into the cultivation rope. (**g**,**h**) Harvesting of *U. pinnatifida* and *S. japonica*

12.2.3 Undaria and Saccharina

Production of *Undaria* in Korea has steadily increased since 1964 and approached 494,947 MT in 2019 (Table 12.2, Fig. 12.6).

There has been considerable research interest in strain development of *Undaria* to both improve product quality and respond to the challenges associated with climate change. NIFS (National Institute of Fisheries Science) have developed 5 new cultivars.

Saccharina is not native to Korea and was introduced for cultivation in 1970 and 1973 from Hokkaido, Japan. *Saccharina* is cultured on the eastern and southern coast of Korea and used for human foods and for abalone feeds. Production of *Saccharina* reached 662,557 tons-wet wt. in 2019 (Table 12.2, Fig. 12.6).

Since the 2000s, *Undaria* and *Saccharina* production has increased annually due to demand for abalone feeds.

A single company controls the production of the kelp seedlings (*Undaria* and *Saccharina*). In 2019, to prevent a decline in seedling value, the Kelp Seedlings Association agreed to freeze the kelp seed production to a level of 400,000 seed frames (a frame is 45x55 cm, 200 m of seed fiber included) per year.

12.2.4 *Sargassum* Spp.

Sargassum fusiforme (formally *Hizikia fusiformis*) cultivation, which was initiated in the early 1980s, mainly for export, and more than 90% of production was exported to Japan.

Normally artificial seeding is not used for *Sargassum* production; instead, young fronds of *S. fusiforme* are collected from the wild by seed collectors for on-growing on farms (Hwang and Park 2020). Multiple generations of plants can be grown using a method of holdfast regenerations developed by Hwang et al. (1999). This technique is now widely used. Production levels of *S. fusiforme* vary depending on the demand from the main export market, Japan.

Unlike *S. fusiforme*, *S. fulvellum* is mainly cultured for domestic use as cooked salad or a soup ingredient. Most aquaculture companies manage the own mature thalli for seeding, collect embryos and use them for cultivation (Hwang et al. 2006, 2007) However some purchase seedlings from commercial nurseries. The production of *S. fusiforme* and *S. fulvellum* was 33,477 and 176 tons respectively in 2019 (Table 12.2, Fig. 12.7).

12.2.5 Other Algae

In addition to the culture of *Pyropia* and the large brown seaweeds described above, cultivation techniques have been developed for a range of other seaweeds, including; *Capsosiphon fulvescense, Codium fragile, Ulva* spp. and *Ecklonia* spp. (Fig. 12.8).

With its unique flavour and soft texture, *C. fulvescens* is traditionally used in Korea in soup (Sohn 1998). Large-scale cultivation of this alga has succeeded both in laboratory and field experiments (Hwang et al. 2003). Currently, commercial cultivation occurs around Wando, on the southwestern coast of Korea. Hwang et al.

Fig. 12.7 Seed production and cultivation processes for *Sargassum* spp. (**a**) Insertion of young fronds of *S. fusiforme*. (**b**) After harvesting, holdfasts of *S. fusiforme* remain on the culture rope without any epiphytes for regeneration induction next cultivation season. (**c**) Zygotes of *S. fulvellum* attached on seed fibres in culture tanks. (**d**) Fully grown *S. fulvellum* on a marine farm

Fig. 12.8 Cultivation of *Ulva, Capsosiphon, Codium, Ecklonia* spp. and *Gracilariopsis*. (**a**) *Ulva* spp. (**b**) *Capsosiphon fulvescense*. (**c**) *Codium fragile*. (**d**) *Ecklonia stolonifera*. (**e**) *Ecklonia cava*. (**f**) *Gracilariopsis chorda*

(2008a) reported that the alga has the potential to be further exploited as human food and as an ingredient in formulated food.

C. fragile is often produced by the settlement of wild zygotes onto culture lines. However, Hwang et al. (2008b) reported that using this alga's regenerative capability could be a more reliable method for producing seed stock and have published methods for the artificial seed production and nursery culture for this species. The use of this medullary filament seeding methods increased production from less than 1 kg-fresh wt. to ca. 7 kg-fresh wt. per 1 m culture rope (Hwang and Park 2020).

The *Ulva* complex of species includes *U. prolifera*, *U. compressa*, *U. intestinalis*, *U. linza*, *U. clathrata. U. prolifera* and *U. linza* that are cultivated on the southwestern coast, and the southeastern coasts of Korea.

Ecklonia spp. are also cultivated in Korea. However, production levels are very low and consequently are not recorded in the national statistics. *E. stolonifera* is used as food, abalone feed, and for sea forest construction (Hwang and Park 2020), while *E. cava* is also used as for the restoration of sea forest. *Ecklonia* spp. are increasing interest to the abalone industry as they are perennial and continue to grow and produce biomass for abalone feed during the summer when other brown seaweed species are not abundant (Hwang et al. 2013).

Gracilariopsis chorda is cultured as human food and eaten in a steamed salad. Asexual propagation (regeneration) and spore seeding methods have been developed for the cultivation of this species (Sohn 2009), leading to commercial production of 1769 tons in 2019 (Table 12.2).

12.3 Seaweed Breeding and Cultivars

Seaweed farming is continually developing in Korea, and there is a constant demand for improved seaweed varieties and cultivation techniques that deliver increased productivity or higher quality seaweeds. Climate change is also playing a role in driving the development of new varieties, as the productivity of some existing varieties has reduced in some regions due to warming seawater. Seaweed production companies in Korea are therefore supportive of breeding studies and rapidly adopt new varieties for aquaculture production.

In 2002, Korea joined the International Union for the Protection of New Varieties of Plants (UPOV) (Park et al. 2016; Hwang et al. 2019, 2020) and has applied the varieties protection system to seaweed since 2012. To date, 20 new seaweed cultivars have been registered, including 14 *Pyropia*, 5 *Undaria*, and 1 *Saccharina* varieties (Table 12.5). These cultivars have contributed directly to industrial development and increased seaweed production significantly in Korea.

Pyropia spp. are the most important seaweeds in Korea, with 20 species and two subspecies recorded (Lee and Kang 1986; Hwang and Lee 2001; Kim and Kim 2011; Hwang et al. 2019). The goal of breeding *Pyropia* is to develop fast-growing, temperature-tolerant cultivars which are resistant to diseases and are also rich in highly desirable secondary metabolites (e.g., antioxidants, amino acids, vitamins,

Table 12.5 List of registration seaweed cultivars in Korea (2012–2020)

Genera	Species	No. Registration	Selected(S), Hybrid(H) or Mutation(M)	Name of Cultivar	Year
Pyropia	*P. yezoensis*	APVP-1	S	Pulmunoeul	2014
		APVP-6	S	Haepoong no. 1	2015
		APVP-7	S	Pulmuhaesim	2015
		APVP-8	S	Sugwawon no. 104	2015
		APVP-11	S	Sugwawon no. 105	2017
		APVP-13	M	Jeonsu no. 1	2017
		APVP-15	S	Sugwawon no. 108	2018
		APVP-16	S	Haepoong no. 2	2019
		APVP-17	S	Sugwawon no. 106	2019
		APVP-18	S	Sugwawon no. 109	2019
		APVP-19	S	Sugwawon no.110	2019
		APVP-20	S	Heungcho no. 3	2020
	P. seriata	APVP-12	S	Haemodol no. 1	2017
		APVP-14	S	Shinpoong no. 1	2018
Undaria	*U. pinnatifida*	APVP-2	S	Sugwawon haeorum	2014
		APVP-3	S	Sugwawon bibari	2014
		APVP-4	S	Sugwawon cheonghae	2014
		APVP-9	S	Sugwawon no. 201	2015
		APVP-10	H	Sugwawon no. 202	2016
Saccharina	*S. japonica*	APVP-5	S	Jeongwan no. 1	2014

Data from Aquatic Plant Variety Center, National Institute of Fisheries Science (2019) available at http://nifs.go.kr/apvc/index.ap

etc.). Three methods are used for *Pyropia* breeding; selection (Park and Hwang 2014), hybridisation (Kim 2001), and mutation through mutagen induction (Lee and Choi 2018) or radiation treatment (Lee et al. 2019).

Among 13 *Pyropia* cultivars registered for plant variety protection in Korea, 12 have been grown commercially. This includes a variety isolated by Park and Hwang (2015) (2014) (*P. yezoensis*-AP1), which was resistant to the pathogen *Pythium porphyrae* that causes red rot disease in cultivation. An additional 12 applications for new varieties are currently being examined for registration (Hwang et al. 2019). Completion of the mitochondrial genome sequencing in *P. yezoensis* and *P. retorta* is underway. It is hoped that this will provide novel molecular markers for systematic study and will help to differentiate cultivars (Hwang et al. 2014, 2018b; Kim et al. 2018).

Undaria pinnatifida occurs throughout eastern Asia (Kang 1966) and is widely cultivated in Korea. The expanding abalone industry in Korea is closely associated with the seaweed aquaculture industry (Hwang et al. 2009), and *Undaria* is in high demand as a fresh feed for abalone. However, this species' natural growth period is not long enough to cover the period for which fresh seaweed is required for feeding abalone. One goal of breeding *Undaria* in Korea is to develop high-temperature resistant varieties that can be grown as abalone feed during the summer to extend the abalone growing season. Some success has been achieved by rearing hybrids of *U. peterseniana* (Kjellman) Okamura (male), and *U. pinnatifida* (female) to extend the algal growing season (Hwang et al. 2011, 2012). In trials, this hybrid continued to grow after the growth of *U. pinnatifida* ceased in April (Hwang et al. 2014). The trials showed that the hybrid also provided greater biomass than *U. pinnatifida.* Difficulties in mass-producing artificial seedings of hybrids have delayed the commercial uptake of this variety. However, these issues are expected to be resolved in the near future. All five *Undaria* cultivars that have been registered for plant variety protection, were developed by selection or hybridisation (Table 12.5).

The increasing demand for kelp feed from abalone farmers has also led to the rapid expansion of *S. japonica* cultivation, with farming area increasing by 671% between 2001 to 2015, to 9147 ha (Hwang et al. 2019). Technological advancements that have supported the kelp farming industry's development include selective breeding, seedling-rearing and 'autumn sporeling-rearing' (Sohn 1998). Hwang et al. (2017, 2018a) demonstrated that cultivars displayed different morphological traits, temperature tolerance, and resistance to wave action when they grew in different environments. There is, therefore, the potential to increase productivity on seaweed farms by developing unique cultivars for specific environments. To date, only one *Saccharina* cultivar has been registered for variety protection in Korea (Table 12.5). A second cultivar, developed from consecutive selection for three generations (F_3) in *S. japonica,* has been shown to have an extended cultivation period (Hwang et al. 2017). This cultivar 'Sugwawon No. 301' is currently being registered, and once this process is completed, it will be distributed to seaweed growers.

12.4 Processing Companies

Traditionally Korean people used seaweeds in their raw or sun-dried forms for food (Hwang et al. 2020). Since the 1980s many different seaweed food products have been developed, e.g., machine-dried *Pyropia*, toasted *Pyropia* seasoning, *Saccharina* jam, salted or cut *Undaria*, sun-dried *Undaria* and *S. fusiforme*, etc. Several processed fast foods and various packaged goods (seaweed soup, seaweed powders for rice, seaweed noodles, etc.) have also been manufactured (Fig. 12.9). Almost all harvested *Pyropia* is machine processed to form dried sheets. With increased *Pyropia* production, the number of seaweed processing companies rapidly increased

Fig. 12.9 Seaweed processing and products. (**a**) Auction of *Pyropia* on harvest boats. (**b**) Processing of dried *Pyropia*. (**c**) Processing of roasted and salted *Pyropia*. (**d**) *Pyropia* products. (**e**) Kelp products. (**f**) Sea-polyphenol products for health care

during the 2010s, and the production of dried *Pyropia* reached ca. 200 million sheets (21 cm × 19 cm, 2.5 g-dry wt.) in 2017. The export of *Pyropia* is now the second most valuable marine product exported from Korea, with a value of approximately US$ 487 million in 2019 (Table 12.2).

Increased competition and changing market demands mean that Korean seaweed processors must continue to innovate to produce new products and more efficient processes. One *Pyropia* processor, Moosankim Co Ltd., has developed an acid-free dried laver in order to produce a more eco-friendly product. Although production is still small, the process has a similar cost to traditional processing techniques and is serving to demonstrate the potential for change in the industry.

Salted *Undaria* was one of the most important processed seaweed products in Korea between the 1970s and 1990s and was manufactured mainly on Korea's southwest coast. A decrease in the export of salted *Undaria* to Japan has led to a change in production from salted to dried *Undaria*. Dried *Undaria* has become a common product in Korea and is included in many processed foods, snacks, and well-being products that utilise the nutritional properties of *Undaria*.

Boiled and sun-dried *Sargassum fusiforme* is also a commercially important product for export to Japan. Sun-dried *S. fusiforme* is used as a raw material for secondary products such as seasoning. Some of the pigments of *S. fusiforme* are lost during processing, and the fronds are dyed using pigments extracted from *Ecklonia cava* to maintain their appearance (Sohn 1998). *S. japonica* is mainly sold as a chopped dried product. Auctions of dried *S. japonica* are held in June and July in Wando, on the southwestern coast of Korea.

Ulva prolifera is dried in the form of sheets similar to *Pyropia* and has recently been processed into salt and oil. Various processed fast foods, snacks, and instant salads also have been manufactured from *Ulva*.

Seanol (sea polyphenol) is extracted from *E. cava* and has been commercialised for use in cosmetics, medical food, and treatment to manage human health and lifestyle products.

The extraction of agar-agar from *Gelidium* has historically been an important industry in Korea and has consistently been a leading export item. However, the agar processing industry has declined significantly in recent years (Sohn 1998) because most of the processing plants have moved offshore. There are a few remaining agar processing plants, and agar-agar exports account for only ca. US$three million in exports annually.

In addition to food uses, the market for health supplements such as tablets, extracts, jelly, and powder using kelp is expanding. Sales of beauty products (soaps, shampoos, body washes, bathing agents, etc.) using various seaweeds are also increasing, and in some districts, local governments have created thalassotherapy using seaweed. Other seaweed uses in Korea include medicines, food additives, livestock feeds, organic fertilisers, biodegradable materials, pulp, eco-friendly building materials, bio-energy sources, and materials for seaforest restoration.

With the increasing consumption of seaweed products, there have been calls for risk-benefit analyses to evaluate any potential health risks (Barbier et al. 2019). Analysis including microbial and chemical risk assessment of seaweed products such as bacteria and heavy metals are now required to ensure that food safety requirements are met both for the raw product and during processing (Food Safety Act 17809).

12.5 Conclusion and Future Perspectives

Korean seaweed cultivation has continued to grow since 1970s. This growth is the result of the development of seaweed cultivation technology, which has focused on; increasing scale, reducing labour and the efficient use of technology. Technology advances include the development of automated harvesting and processing technologies and increasing productivity through developing more productive varieties. Looking to the future, seaweed cultivation will focus on deriving value from improved product quality and through more eco-friendly farming technologies rather than simply increasing the quantity of seaweed produced. The impacts of climate change on mariculture have reinforced the need to continue to develop varieties that can adapt to these new conditions. The continued development of new high-quality and disease-resistant varieties is helping the Korean seaweed industry meet these challenges. The Korean government has recognized the importance of seaweed breeding and has created the 'Golden Seed Project' supporting seaweed cultivars' development to increase seaweed export earnings.

In addition to technological advances that lead to increased productivity from marine farms, there is a growing interest in indoor seaweed culture systems. The development of indoor culture systems will help to secure the competitiveness of the industry and enable year-round production of seaweeds that compete with terrestrial vegetables. Locating such systems close to markets will meet consumer demand for fresh products while reducing the carbon emissions associated with transporting fresh materials from distant ports.

The seaweed industry must also adapt to meet changing consumer demands as the demand for functional foods and high-value seaweed extracts continue to rise. Research to support varieties that have a higher proportion of functional ingredients will increase the value of the industry without the need to increase production area or biomass.

In a world where consumers are increasingly aware of environmental sustainability as a key part of product quality, the industry must also meet the challenges of developing and demonstrating more eco-friendly practices. A seaweed company in Korea obtained the ASC-MSC (Aquaculture Stewardship Council-Marine Stewardship Council) certification in 2020 and more companies are expected to do so in the future as the Korean seaweed industry develops in line with the demands of eco-friendly and health-conscious consumers.

In summary, the key challenges facing the future development of the Korean seaweed industry are: Understanding the environmental capacity for production to better enable the management and maintenance of the seaweed farming environment.

1. Raising awareness of environmental management with seaweed farmers and developing and promoting more environmentally sustainable practices.
2. Developing seaweed varieties that will allow the industry to adapt to the challenge of climate change.
3. Developing value-added functional seaweed varieties that add value to the industry without the need to increase production area.
4. Developing new processed products and fresh convenience products to meet changes in consumer habits and facilitate the expansion of export markets.
5. Increasing the adoption of technological solutions to address the issues of an aging fishing population and rising labor costs.
6. Developing food safety protocols that meet international standards.
7. Implementing administrative and financial support at the national level to secure the seaweed cultivation industry's national and international status.

Acknowledgments This research was supported by a grant from the National Institute of Fisheries Science, Republic of Korea (R2021005). The authors would like to thank Dr. Philip Heath (Tisbe Ltd., New Zealand) for reviewing the English.

References

Bae SH (1991) The origin and development process of laver culture industry in Korea. Bull Korean Fish Soc 24:153–166

Barbier M, Charrier B, Araujo R et al (2019) Phycomorph European guidelines for a sustainable aquaculture of seaweeds. Eur Co Sci Tech, Nice, p 194

Chang JW, Geon SH (1970) Studies on the culture of *Laminaria*. (1) on the transplantation of tangle *Laminaria religiosa* Miyabe in temperate zone (the coast of Ilsan-dong, Ulsan city). Bull Fish Res Dev Agency 5:63–75

Chung MK (1937) Laver of Chosun. Fisheries of Chosun 144:1–4

FAO (2017) Yearbooks of fishery statistics summary tables. ftp://www.fao.org/fishery/publications/yearbooks/en down load 4 Jan 2021

Hwang EK, Amano H, Park CS (2008a) Assessment of the nutritional value of *Capsosiphon fulvescens* (Chlorophyta): developing a new species of marine macroalgae for cultivation in Korea. J Appl Phycol 20:147–151

Hwang EK, Baek JM, Park CS (2007) Assessment of optimal depth and photo irradiance for cultivation of the brown alga, *Sargassum fulvellum* (Turner) C Agardh. J Appl Phycol 19:787–793

Hwang EK, Baek JM, Park CS (2008b) Cultivation of the green alga, *Codium fragile* (Suringar) Hariot, by artificial seed production in Korea. J Appl Phycol 20:469–475

Hwang EK, Baek JM, Park CS (2009) The mass cultivation of *Ecklonia stolonifera* Okamura as a summer feed for the abalone industry in Korea. J Appl Phycol 21:585–590

Hwang EK, Cho YC, Sohn CH (1999) Reuse of holdfasts in *Hizikia* cultivation. J Korean Fish Soc 32:112–116

Hwang EK, Choi HG, Kim JK (2020) Seaweed resources of Korea. Bot Mar 63:395–405

Hwang EK, Gong YG, Hwang IK et al (2013) Cultivation of the two perennial brown algae *Ecklonia cava* and *E. stolonifera* for abalone feeds in Korea. J Appl Phycol 25:825–829

Hwang EK, Gong YG, Park CS (2011) Cultivation of a brown alga, *Undariopsis peterseniana* (Kjellman) Miyabe and Okamura, as a warm-temperature species by artificial seed production in Korea. J Appl Phycol 23:449–455

Hwang EK, Gong YG, Park CS (2012) Cultivation of a hybrid of free-living gametophytes between *Undariopsis peterseniana* and *Undaria pinnatifida*: morphological aspects and cultivation period. J Appl Phycol 24:401–408

Hwang EK, Ha DS, Park CS (2017) Strain selection and initiation timing influence the cultivation period of *Saccharina japonica* and their impact on the abalone feed industry in Korea. J Appl Phycol 29:2297–2305

Hwang EK, Hwang IK, Park EJ et al (2014) Development and cultivation of F_2 hybrid between *Undariopsis peterseniana* and *Undaria pinnatifida* for abalone feed and commercial mariculture in Korea. J Appl Phycol 26:747–752

Hwang EK, Liu F, Lee KH et al (2018a) Comparison of the cultivation performance between Korean (Sugwawon no. 301) and Chinese strains (Huangguan no. 1) of kelp *Saccharina japonica* in an aquaculture farm in Korea. Algae 33:101–108

Hwang EK, Park CS (2020) Seaweed cultivation and utilization of Korea. Algae 35:107–121

Hwang EK, Park CS, Baek JM (2006) Artificial seed production and cultivation of the edible brown alga, *Sargassum fulvellum* (Turner) C Agardh: Developing a new species for seaweed cultivation in Korea. J Appl Phycol 18:251–257

Hwang EK, Yi YH, Shin WJ et al (2003) Growth and maturation of a green alga, *Capsosiphone fulvescens*, as a new candidate for seaweed cultivation in Korea. In: Champman RO, Anderson RJ, Vreeland VJ, Davison IR (eds) Seventeenth Int seaweed Symp. Oxford University Press, Oxford, pp 59–64

Hwang EK, Yotsukura N, Pang SJ et al (2019) Seaweed breeding programs and progress in eastern Asian countries. Phycologia 58:484–495

Hwang IK, Kim SO, Hwang MS et al (2018b) Intraspecific variation of gene structure in the mitochondrial large subunit ribosomal RNA and cytochrome c oxidase subunit 1 of *Pyropia yezoensis* (Bangiales, Rhodophyta). Algae 33:49–54

Hwang MS, Lee IK (2001) Taxonomy of the genus *Porphyra* (Bangiales, Rhodophyta) from Korea. Algae 16:233–273

Kang JW (1966) On the geographical distribution of marine algae in Korea. Bull Pusan Fish Coll 7:1–136

Kang JW, Ko NP (1977) Seaweed cultivation. Taehwa Publish, Busan

Kim HS, Kim SM (2011) Algal flora of Korea. Vol. 4, No. 1. Primitive red algae. National Institute of Biological Resources, Incheon, p 138

Kim NG (2001) Physiological study on the hybrid by interspecific crossing between *Porphyra pseudolinearis* and *P. dentata* (Bangiales, Rhodophyta), two dioecious species in culture. J Aquac 13:353–357

Kim SM, Choi HG, Hwang MS et al (2018) Biogeographic pattern of four endemic *Pyropia* from the east coast of Korea, including a new species, *Pyropia retorta* (Bangiaceae, Rhodophyta). Algae 33:55–68

Lee HJ, Choi JI (2018) Isolation and characterization of a high-growth-rate strain in *Pyropia yezoensis* induced by ethyl methane sulfonate. J Appl Phycol 30:2513–2522

Lee HJ, Park EJ, Choi JI (2019) Isolation, morphological characteristics and proteomic profile analysis of thermo-tolerant *Pyropia yezoensis* mutant in response to high-temperature stress. Ocean Sci J 54:65–78

Lee IK, Kang JW (1986) A check list of marine algae in Korea. Korean J Phycol 1:311–325

Park CS, Hwang EK (2011) An investigation of the relationship between sediment particles size and the development of green algal mats (*Ulva prolifera*) on the intertidal flats of Muan, Korea. J Appl Phycol 23:515–522

Park CS, Hwang EK (2014) Isolation and evaluation of a strain of *Pyropia yezoensis* (Bangiales, Rhodophyta) resistant to red rot disease. J Appl Phycol 26:811–817

Park CS, Hwang EK (2015) Biochemical scharacterization of *Pyropia yezoensis*-AP1 strain accompanies the resistance reaction to the red rot disease pathogen, *Pythium porphyrae*. J Appl Phycol 27:2149–2156

Park CW, Choi KJ, Soh EH et al (2016) Study on the future development direction of plant variety protection system in Korea. Korean J Breed Sci 48:11–21

Sohn CH (1993) *Porphyra*, *Undaria* and *Hizikia* cultivation in Korea. Korean J Phycol 8:207–216

Sohn CH (1996) Historical review on seaweed cultivation of Korea. Algae 11:357–364

Sohn CH (1998) The seaweed resources of Korea. In: Critchley AT, Ohno M (eds) . Seaweed resources of the world, Japan Int Coop Age, pp 15–33

Sohn CH (2009) Seaweed cultivation. Dain Communications, Busan, p 283

Yoo JG (1964) Experiment of PVA net in *Porphyra* cultivation. Tech Rep Natl Fish Res Dev Age:36

Yoon JT, Cho YC, Gong YG (2003) A study on the cultivation of *Enteromorpha prolifera* (Müller) J. Agardh, Chlorophyta in Korea. J Aquac 16:44–50

Chapter 13
Prospects and Challenges in Commercialization of Seaweeds in Bangladesh

Nowrin Akter Shaika, Jinnath Rehana Ritu, Saleha Khan, and Ambati Ranga Rao

Abbreviations

AIG	Alternative Income Generating
AIS	Agricultural Information Service
BARC	Bangladesh Agricultural Research Council
BARI	Bangladesh Agricultural Research Institute
BFRI	Bangladesh Fisheries Research Institute
BoB	Bay of Bengal
BOD	Biochemical Oxygen Demand
Ca	Calcium
Cd	Cadmium
Cl	Chlorine
COD	Chemical Oxygen Demand
Cu	Copper
FAO	The Food and Agriculture Organization
K	Potassium
MFTS	Marine Fisheries and Technology Station
Mg	Magnesium
Na	Sodium
NACA	The Network of Aquaculture Centres in Asia-Pacific
NGO	Non-governmental organizations
Ni	Nickel

N. A. Shaika · J. R. Ritu · S. Khan (✉)
Department of Fisheries Management, Bangladesh Agricultural University, Mymensingh, Bangladesh
e-mail: salehakhan@bau.edu.bd

A. R. Rao (✉)
Department of Biotechnology, Vignan's Foundation of Science, Technology and Research (Deemed to be University), Vadlamudi, Guntur, Andhra Pradesh, India
e-mail: arangarao99@gmail.com

A. Ranga Rao, G. A. Ravishankar (eds.), *Sustainable Global Resources of Seaweeds Volume 1*, https://doi.org/10.1007/978-3-030-91955-9_13

P	Phosphorus
S	Sulfur
SDGs	Sustainable Development Goals
UNDP	United Nations Development Programme
Zn	Zinc

13.1 Introduction

Marine resources are significant for the economy just opening up new vistas of opportunity for nations with advanced technologies. In the twenty-first century, several nations are investing in the proper utilization and exploitation of marine resources for sustainable economic activities, in this context, Bangladesh has huge potential to gain significantly. The Bay of Bengal in Bangladesh is blessed with extensive marine resources including various finfish, shellfish, and aquatic plants which provide food and livelihood for the millions of coastal people and offer auspicious opportunities for the economic development of the country. At present, seaweed is an important promising marine resource in Bangladesh that is emerging possibilities for the evolvement of the blue economy.

Seaweeds, or marine benthic macroalgae, are crucial primary producers in the marine aquatic food chain. Typically, seaweeds grow in the intertidal and sub-tidal regions of the sea and live attached to rock or other substrata in coastal areas. Seaweeds possess a high concentration of essential vitamins, proteins, lipids, minerals, polysaccharides, enzymes, and trace elements of food and nutritional value (Shama et al. 2019). The seaweeds are natural partners in supporting the ecology and environment in the ocean ecosystem. They harness solar energy into biomass and in turn support biotic and abiotic components of the planet earth.

Bangladesh offers enough substrates and habitats for the culture of various kinds of seaweeds in a rich environment comprising of sandy and muddy beaches, mangrove swamps, and estuarian regions covering nearly 710 km long coastline and a 25,000 sq. km coastal area. Seaweeds grow naturally in the littoral and sub-littoral zones in the south-eastern part of the mainland, the Cox's Bazar coast, and in the rocky sub-strata of the Saint Martin, an offshore island of the country. Sundarban coast and Kuakata areas of Patuakhali district are also recognized as congruent places for growing seaweed. The seaweed flora in Bangladesh is highly diversified. Earlier, around 200 species of seaweeds belonging to 77 genera have been recorded in Cox's Bazar coast of the Bay of Bengal, which is comprised of 94 species of red seaweeds (Rhodophyta), 59 brown seaweeds (Phaeophyta), and 47 green seaweeds (Chlorophyta) (Islam et al. 2020).

These autotrophic marine macroalgae produce versatile and multipurpose biomass apt for multiple applications. Meanwhile, seaweeds are globally considered as a multibillion-dollar industry, providing nutritional supplements, food, fertilizers, therapeutic protection, and extraction of valued bio-chemicals such as agar, agarose, alginate, and carrageenan. However, despite its great potential and commercial importance, the production and use of seaweeds have not yet been established in Bangladesh. Local tribal people partly used to consume them as food. Traditionally in Bangladesh, seaweeds are known to the ethnic communities (e.g., Mog or Rakhyine) and people of St. Martin Island as 'Hejla' (Siddiqui et al. 2019). In recent years, few restaurants in Bangladesh engaged in the preparation of Chinese foods have been using seaweeds as fresh salads, cooked vegetables, and in curry dishes.

Seaweed, if systematically cultured and explored, could bloom as a profitable economic activity in Bangladesh. The Government of Bangladesh has already destined seaweed as a major component of mariculture in the national development plan including Sustainable Development Goals (SDGs). Also, extensive surveys and research need to be conducted through improving harvesting techniques, seeding of suitable coastal areas, and creation of artificial habitats to identify suitable sites for large-scale seaweed culture. This book chapter is an update of the present scenario of seaweed status in Bangladesh that describes prospects and challenges for the advancement of a seaweed aquaculture industry (Shama et al. 2019).

13.2 Distribution of Seaweeds in Bangladesh Coast

Bangladesh is abundantly rich and resourceful part of the Indo-Burma biodiversity hotspot with illustrious marine resources. This region supports luxuriant growth of red, brown, and green seaweeds, being opulent with vast seaweed biomass predominantly in the south-eastern and south-western coasts (Fig. 13.1). The natural growth of seaweeds along the whole Bangladesh coast, commonly in the St. Martin Island, Cox's Bazar and in the Sundarbans Mangrove Forest is well explored.

Particularly, Saint Martin Island is a staggering place for naturally occurring seaweeds.The Island is considered as the hotspot of seaweed due to its rocky bottom structure, stable hydrology, better transparency, greater light penetration which result in increased photosynthesis and high growth of macroalgae there (Fig. 13.2).

Contrariwise, the least diversity of green, brown and red macroalgae was seen in Bakkhali, Sonadia, Teknaf, Kuakata, Inani and Sundarbans (Fig. 13.3). Sundarbans provide a favourable condition by meeting the criteria of required salinity (2–34 ppt), p^H (7.5–8.5), and temperature (20–30 ° C) (Satpati et al. 2012; COAST Trust 2013). Hence, a new natural seaweed bed from Nuniarchhara to Nazirartek areas of the Bakkhali river and Moheshkhali Channel estuary of Cox's Bazar is discovered by the Seaweed Research Team of the Marine Fisheries and Technology Station (MFTS), Bangladesh Fisheries Research Institute, Cox's Bazar and very recently. Further, they traced another seaweed bed in Gongamoti, the Kuakata area (Islam 2020).

In Bangladesh, seaweeds are available in the summer, winter and spring seasons. And based on the monthly distribution pattern, seaweeds are mostly available from October to April when the conditions are favourable for its growth and spore settlement. But the highest abundance was found to occur from January to March as the ideal environmental conditions prevail during this time of the year (Fig. 13.4). A similar finding of the highest abundance of seaweed was also reported by FAO/NACA 1996 for other Asian countries.

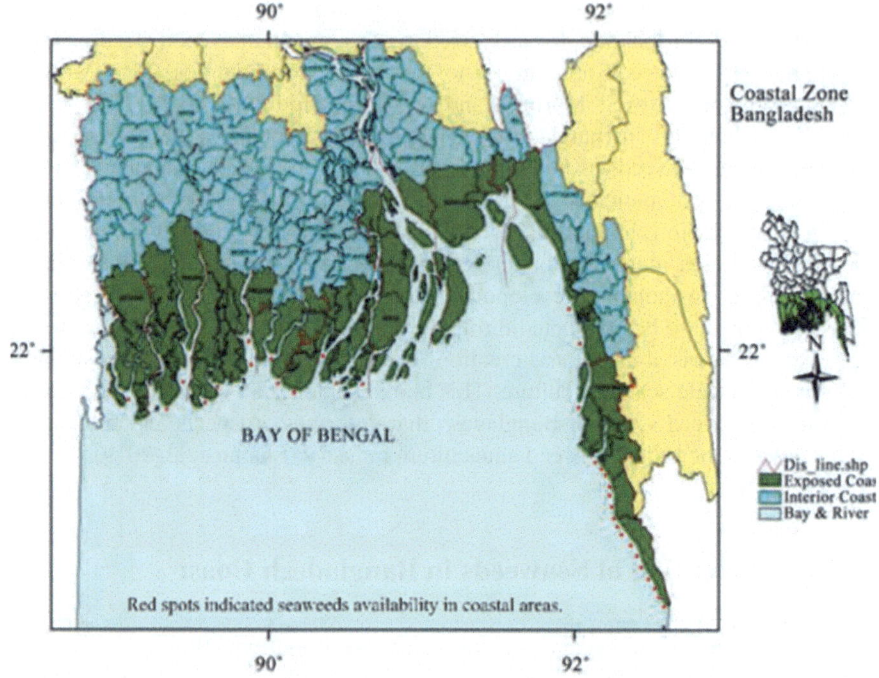

Fig. 13.1 Map indicating seaweeds availability in coastal zone of Bangladesh (Islam et al. 2020)

Fig. 13.2 St. Martin Island; an exemplary site for naturally growing seaweeds (Sarkar et al. 2016; The Financial Express, November 24 2018)

13.3 Seaweed Species Diversity in Bangladesh

Diversity, growth and propagation of seaweeds depend on specific water quality requirements like water transparency, water temperature, salinity, dissolved oxygen, pH, nutrients, etc. The diversity of marine algae in the south-eastern and south-western coasts of Bangladesh showed that the members of Rhodophyta were dominant contributing 69.52% followed by Phaeophyta (35.27%) and Chlorophyta (28.21%) (Fig. 13.5).

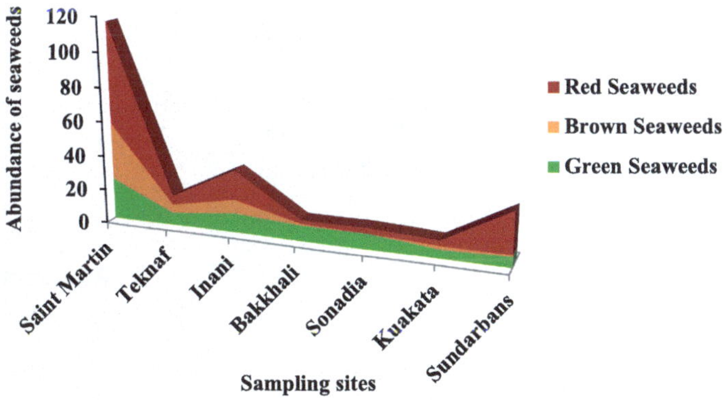

Fig. 13.3 Distribution of seaweeds at different coastal sites of Bangladesh (Islam et al. 2020)

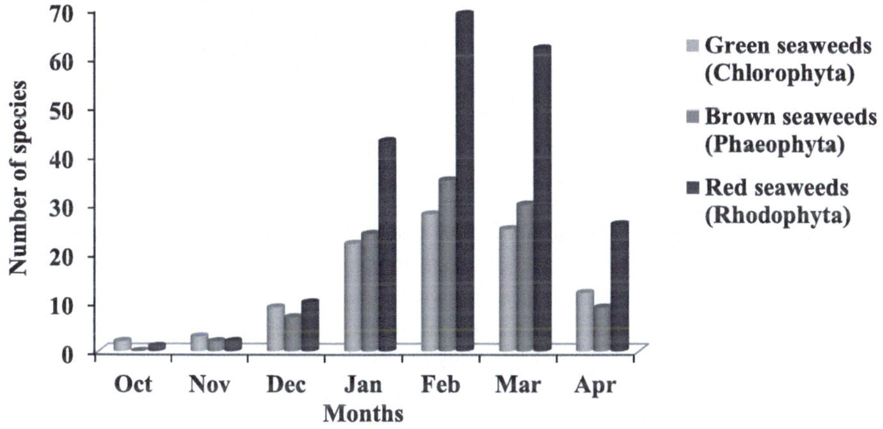

Fig. 13.4 Monthly distribution pattern of green, brown and red seaweeds in Bangladesh (Islam et al. 2020)

It is reported that about 5000 metric tons of seaweed biomass are annually traceable within Bangladesh territory (Sarkar et al. 2016). Approximately 155 seaweed species are available in Cox's Bazar and about 60 seaweed species are traced from Sundarbans (Siddiqui et al. 2019). Of more importance, Bangladesh's Saint Martin Island is a substantive source of naturally growing seaweeds with about 140 species. Seaweeds are more plethoric on the west coast than on the eastern coast of the St. Martin Island. *Sargassum, Hypnea, Halymenia, Caulerpa,* and *Cladophora* are the common species found in the St. Martin Island. The diversity and abundance of seaweed species in Saint Martin Island are embodied in Table 13.1.

Among the available seaweed species in Bangladesh, 19 species of 14 genera are devised as commercially emergent (Table 13.2). The seasonal abundance of such commercially important species in Bangladesh are figured out graphically and the picture of commercially relevant species is also shown (Figs. 13.6 and 13.7).

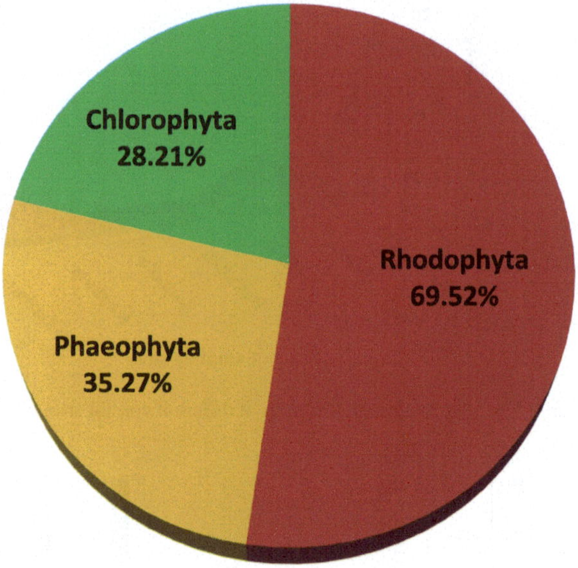

Fig. 13.5 Percentage of seaweed biodiversity in Bangladesh (Islam et al. 2020)

13.4 Seaweed Aquaculture Progress in Bangladesh

Seaweed aquaculture technologies have prospered dramatically over the past 70 years mostly in Asia. Over 29.9 million metric tons production is emerged by several methods in Asia, standing as the world leader in seaweed cultivation (Kim et al. 2017). History says, seaweed cultivation started primarily in Tokyo, Japan in 1670 and its commercial farming began in 1940. Thereafter, many other Asian countries including the Philippines, Thailand, Vietnam also initiated the cultivation.

However, seaweed culture methods are yet to be introduced in Bangladesh. As a diversification activity in mariculture, seaweed cultivation has huge prospects all along the coasts of Bangladesh. In Zafar 2005, Mr. Zafar first started seaweed culture in Bangladesh (Islam 2020) whereas the green and red seaweeds could be congruous for culture in the country, according to the seaweed experts.

Seaweed is being cultivated in Nuniarchhara and Moheshkhali Channel, Cox's Bazar. Seaweeds such as *Hypnea* sp., *Enteromorpha* sp., *Kappaphycus* sp. and *Gracilaria* sp. can be successfully cultivated by vegetative propagation methods in long-line ropes and nets (Uddin et al. 2021). In three locations of Cox's Bazar coast, Saint Martin Island, Inani and Bakkhali, the culture method of the red seaweed *Hypnea* sp. was evaluated by Islam et al. (2017) with a net method of 4 × 4 m coir rope net. *Hypnea* is the most procurable species, rich in iodine. This red seaweed is also being cultured in the tidal coastline of Nuniarchhara beach (Islam 2020). However, the commonly practiced seaweed culture methods are evident in St. Martin Island and Cox's Bazar (Fig. 13.8).

Table 13.1 Species diversity and abundance of seaweeds in St. Martin Island, Chittagong (Billah et al. 2018)

Class	Order	Family	Genus	Species	Abundance
Chlorophyceae	Cladophorales	Cladophoraceae	*Cladophora*	*Cladophora prolifera*	+++++
	Codiales	Codiaceae	*Codium*	*Codium geppei*	++++
	Siphonales	Codiaceae	*Avrainvillea*	*Avrainvillea amadelpha*	+++
	Caulerpales	Caulerpaceae	*Caulerpa*	*Caulerpa cactoides*	++++
				Caulerpa chemnitzschi	++++
				Caulerpa chemnitzschi var. *Occidentalis*	++++
				Caulerpa racemosa var. *Clavifera*	+++++
				Caulerpa sertularioides	++++
				Caulerpa taxifolia	++++
	Caulerpales	Udotiaceae	*Halimeda*	*Halimeda discoidea*	+++
				Halimeda opuntia	+++++
Phaeophyceae	Dictyotales	Dictyotaceae	*Dictyota*	*Dictyota atomaria*	++++
				Dictyopteris australis	+++
			Padina	*Padina tenuis*	++
				Padina pavonica	+++
				Padina gymnospora	++++
			Pocockiella	*Pocockiella variegata*	+++++
			Spatoglossum	*Spatoglossum variabile*	++++
				Spatoglossum asperum	++++
	Dictyosiphonales	Chnoosporaceae	*Chnoospora*	*Chnoospora implexa*	+++++
	Fucales	Sargassaceae	*Sargassum*	*Sargassum coriifolium*	+++++
				Sargassum olygocystum	+++++
				Sargassum swartzii	+++++
				Sargassum tenerrimum	+++++
				Sargassum pallidum	+++++

(continued)

Table 13.1 (continued)

Class	Order	Family	Genus	Species	Abundance
Rhodophyceae	Nemalionales	Chaetangiaceae	*Galaxura*	*Galaxura fastigiata*	++++
	Cryptonemiales	Coralliniaceae	*Jania*	*Jania ungulata*	+++
	Gigartinales	Hypneaceae	*Hypnea*	*Hypnea musciformis*	+++++
	Bonnemaisoniales	Bonnemaisoniaceae	*Asparagopsis*	*Asparagopsis taxiformis*	++++
	Ceramiales	Dasyaceae	*Dasya*	*Dasya corymbifera*	++++
				Dasya pedicellata	++++
		Rhodomelaceae	*Gracilaria*	*Gracilaria spinuligera*	+++
			Neurymenia	*Neurymenia fraxinifolia*	++++
		Delesseriaceae	*Vanvoorstia*	*Vanvoorstia coccinea*	+++++
	Cryptonemiales	Grateloupiaceae	*Halymenia*	*Halymenia floresia*	++++
				Halymenia floridana	+++
		Squamariaceae	*Peyssonnelia*	*Peyssonnelia polymorpha*	+++++

N.B.: Ascending of (+) sign remarks more availability of species

Table 13.2 Commercially emergent Seaweeds of Bangladesh

Sl. No.	Genus	Species	Type
1	*Gelidiella*	*Gelidiella tenuissima*	Red seaweed
2	*Gelidium*	*Gelidium pusillum*	Red seaweed
3	*Hypnea*	*Hypnea pannosa*	Red seaweed
4	*Halymenia*	*Halymenia discoidea*	Red seaweed
5	*Catenella*	*Catenella spp.*	Red seaweed
6	*Porphyra*	*Porphyra spp.*	Red seaweed
7	*Gelidium*	*Gelidium amansii*	Red seaweed
8	*Caulerpa*	*Caulerpa racemosa*	Green seaweed
9	*Enteromorpha*	*Enteromorpha moniligera*	Green seaweed
10	*Codium*	*Codium fragile*	Green seaweed
11	*Enteromorpha*	*Enteromorpha intestinalis*	Green seaweed
12	*Sargassum*	*Sargassum spp.*	Brown seaweed
13	*Hydroclathrus*	*Hydroclathrus clathratus*	Brown seaweed
14	*Padina*	*Padina tetrastromatica*	Brown seaweed

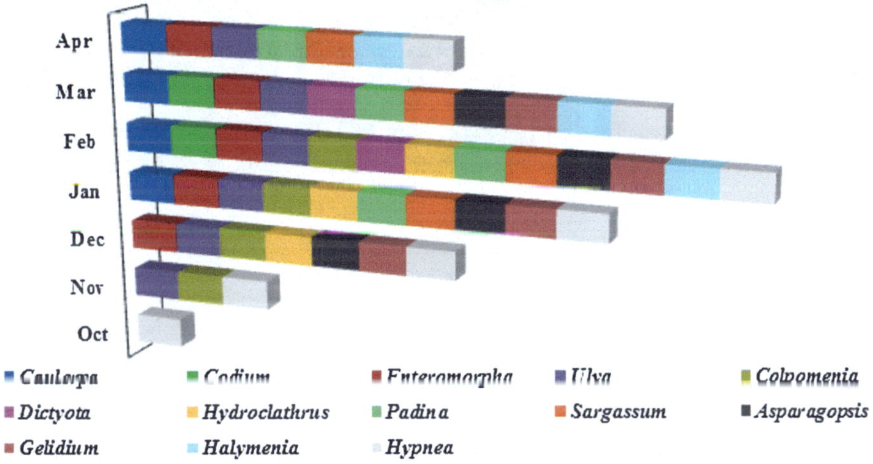

Fig. 13.6 Seasonal abundance of commercially important seaweeds in Bangladesh (Islam et al. 2020)

13.5 Prospects of Seaweed in Bangladesh

Sea is an enriched repository of diversified resources which exerts immense potentialities to make contribution in the national and global economy. Among them, seaweed is considered as an excellent benison of the sea. It has been identified as valuable cash crop in the economy of China, Indonesia, Japan, Korea, Malaysia and the Philippines of Asia where a majority of the biomass is obtained from the seaweed cultivation industries (Ferdouse et al. 2018). Bangladesh is far behind in the case of seaweed cultivation, which is a matter of concern and also provides future

Fig. 13.7 Different types of commercially important seaweeds in Bangladesh (Islam et al. 2020); (**a**). *Sargassum*, (**b**). *Ulva*, (**c**). *Caulerpa*, (**d**). *Hydroclathrus*, (**e**). *Padina*, (**f**). *Porphyra,* (**g**). *Hypnea*, (**h**). *Gelidium*, (**i**). *Enteromorpha*, (**j**). *Colpomenia*, (**k**). *Halymenia*, and (**l**). *Codium*

opportunities. Cognizant of this fact, the Government of Bangladesh has already taken some initiatives and made people aware of spreading and modernization of culture technique without harming environmental sustainability and marine biodiversity which is going to be a turning point for achieving 8 sustainable development Goals (SDGs), namely, (i) no poverty, (ii) zero hunger, (iii) gender equality, (iv) decent work and economic growth, (v) sustainable consumption and production, (vi) climate action, (vii) life below water, and (viii) partnerships for the goals (Hossain et al. 2021).

Fig. 13.8 Various seaweed culture methods practiced in Saint Martin Island and Cox's Bazar, Bangladesh (Siddiqui et al. 2019); (**a**). Attached net system; (**b**). Net culture system long; (**c**). Line rope system, and (**d**). Long line floating system of *Caulerpa* sp.

13.5.1 Favorable Tropical Maritime Climate and Natural Resources for Seaweed Culture in Bangladesh

Bangladesh is one of the largest deltaic countries with a unique geographical location as well as a vast reservoir of natural resources and which has become an attraction of the world as an abode of beauty amalgamated with diversified ecosystem (Fig. 13.9). It is bestowed with 710 km extended coastline, a resourceful exclusive economic zone (200 nautical miles) along with a 24,800 square nautical miles continental shelf (Ahmad 2019). In addition, Bangladesh has regained its sovereignty of 118,813 km^2 of the Bay of Bengal (BoB) from Myanmar and India respectively after a prolonged debate over maritime boundaries which will tend to favour profuse growth of seaweed in the near future (Hussain et al. 2019).

Fig. 13.9 Seaweed, an untapped treasure of the economy of Bangladesh (Photo (**a**) is taken from (Ghose 2020) and photo (**b**) is taken from (Hossain et al. 2021))

13.5.2 Export Potentiality of Bangladesh Seaweed in International Market

The demand for seaweed is an increasing trend in the international arena. Seaweed has boundless beneficent medical values and immense food values that's why it can be exported as seafood. Approximately 390 tons of biomass in wet weight is produced from seaweed farming in the coastal waters of Bangladesh which is equivalent to 97.5 tons in dry weight (Hossain et al. 2021). Moreover, Sarker 1992 had reported an annual harvest of about 1500 tons of seaweeds from the natural habitat of the St. Martin Island. Although, the government of Bangladesh has prohibited the collection of seaweed from St. Martin Island, still, the poor inhabitants around St. Martin Island collect seaweed in very small quantities from April to May, a small portion of which are exported to China and Myanmar but there is no well-established statistical data on it (Ahmed and Taparhudee 2010). It is worth mentioning that some Asian countries like Japan, Thailand, Malaysia, and some of the European countries, the United States of America, etc. are importing seaweed every year (Ahmad 2019). So, Bangladesh could have a golden opportunity to explore the seaweed market through triggering its cultivation system as well as strengthening its research hub. In consequence, Bangladesh could focus on seaweed cultivation because it acts as a source of delicious food items for human consumption as well as a significant source of foreign exchange earnings.

13.5.3 Cost-Effectiveness of Seaweed Cultivation in Bangladesh

One more advantage of increasing seaweed cultivation is to support the economic empowerment of poor people of coastal areas. Although fisherfolk of Bangladesh is cultivating seaweed at a very lower scale as their alternative livelihood strategy during the banning period of fishing, it could have an opportunity to flourish as an exclusive sector as well. Furthermore, in aquaculture, the biggest hurdle is fish feed and its availability whereas seaweed is a simpler, economically viable technique and environment friendly in this regard. Most interestingly, it can be grown by using indigenous materials like bamboo and rope (Siddiqui et al. 2019) that ultimately. Moreover, it is an environmentally friendly activity of sustainable production of biomass free from any use of chemicals without harming the environment especially fisheries resources. Considering the above points, seaweed production in Bangladesh may be conducive to bring in a radical change in economic activity.

13.5.4 Potentiality of Women Empowerment through Seaweed Farming in Bangladesh

In the current era of globalization, the participation of women in every developmental sector is commendable and a matter of pride. They are spearheading shoulder to shoulder with men in the workplace as well as contributing to the family which imparts them an exalted and more esteemed image in the society. Spontaneous participation of women in seaweed production may be another blooming example. In most developing countries, women's participation in seaweed farming is higher than that of men which may be a role model for Bangladesh. Women are more devoted, enduring, and resilient as well as more interested to gain managerial knowledge of seaweed farming to sustain resources together with the environment (Msuya and Hurtado 2017). They are involved in different activities of seaweed farming from cultivation to processing and producing diversified value-added products. In addition, in general, seaweed beds are located in the shallow part of the waters which makes it convenient to women to manage that consummately triggers the potentiality of women involvement in seaweed farming (Ferdouse et al. 2018). Some research studies highlighted the involvement of women in seaweed farming in different countries like in India, Africa (Tanzania, Kenya), South East Asian countries (The Philippines, Indonesia, and Malaysia), etc. (Msuya and Hurtado 2017). Therefore, Bangladesh has great potential to start seaweed cultivation commercially by involving and providing assistance associated with basic training to women. Different international, national, and non-governmental organizations are taking responsibility to raise awareness among women and to make them interested in starting seaweed farming for financial development which has life-changing contribution. Such an inspiring woman is Mrs. Mariam Begum who achieved success and solvency in life by adopting seaweed cultivation training by the combined support of Feed the Future's partnership with the Department of Fisheries, Bangladesh, and Falcon International Ltd. (Feed the Future 2020). In this way, women are turned into skilled manpower of seaweed farming, contributing to the regional and global economy by their diligence, intellectuality, and strong willpower (Fig. 13.10).

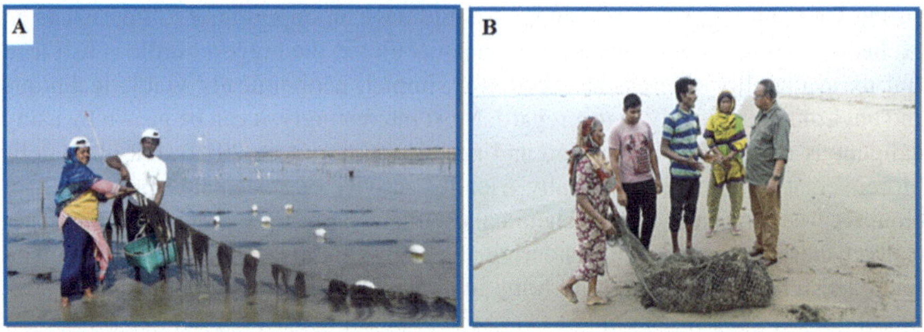

Fig. 13.10 Women involvement in Seaweed cultivation in Bangladesh (Photo (**a**) is taken by World Fish https://www.feedthefuture.gov/article/success-through-seaweed-in-bangladesh/ and photo (**b**) is taken from (Seraj 2018)

13.5.5 Socio-Economic Benefits of Seaweed for Coastal Communities in Bangladesh

UNDP 2000 indicated that 40% of the world population are projected to live within 100 km of a coastline and will be increased more in the coming half-century. In Bangladesh, it is estimated that approximately 36 million people live in the coastal area, considered as a highly populated and most flourished expanse of the land (Ahmad 2019). Thus, stretching of seaweed cultivation in the coastal area will play an efficacious role in changing the fortunes of the coastal people as well as in the economy (Fig. 13.11) Despite these advantages very few people of south-eastern and south-western coasts of Bangladesh are engaged in seaweed cultivation (Siddiqui et al. 2019).

It may be a good option for utilization and alternative income-generating (AIG) source of coastal communities (Fig. 13.12). Another research group of Bangladesh, Hossain et al. (2021) showed the potentiality of seaweed in ameliorating ecosystem health and ensuring food security of the coastal people of Bangladesh..

In summary, by embracing the seaweed culture technique, people will be aware of the sustainable use of natural resources which help them to transform themselves into skilled manpower and they even be able to keep contributing to ensure nutritional security for themselves, their family and the world as a whole. Thus, this unfolded resource needs exploration to open a new gateway for coastal people by enriching human and indigenous resources that will create milestones in the global arena in the future.

Fig. 13.11 Seaweed, a blessing for the unheeded people of the coastal region, Bangladesh (Algae World News 2015)

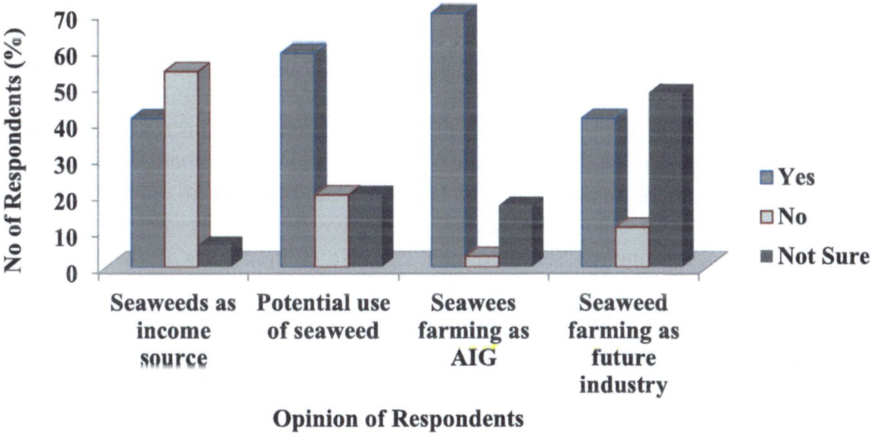

Fig. 13.12 Bar graph shows responses of Coastal communities in Bangladesh toward potentiality of seaweed farming (Islam and Haroon 2017)

13.6 Uses of Seaweeds and its Application in Bangladesh

Seaweeds are the wealth of the ocean and are valued greatly for their diverse applications worldwide as significant marine resources. They produce useful biomass for multiple applications in a broad variety of formats for direct human consumption or processed into food additives and nutraceuticals, biofuels, feeds, fertilizers, cosmetics, and medicines, amongst others (Mohammad et al. 2019; Hessami et al. 2019). Hence, global demand for seaweeds has been thriving together with the rise in usage beyond prior traditional applications.

13.6.1 Seaweeds as Human Food

The use of seaweed as human food has dynamic roots around the world for centuries, possibly millennia. In many countries, seaweed has been considered a staple food for a very long time (Shama et al. 2019). *Caulerpa, Codium, Enteromorpha, Ulva,* and many other types of green seaweeds are exclusively utilized as a food source and often taken as fresh salads or cooked as vegetables. Fresh, dried, and processed seaweeds are also used for the preparation of seaweed masala, seaweed salad, seaweed pickle, seaweed wafer, seaweed jelly, etc. In Japan, approximately 100,000 tons of seaweeds are consumed annually alone in the name of several products like Nori, Kombu (konbu), and Wakame. The Japanese, the Chinese, the Hawaiians, and the Filipinos have been using seaweed in their diets for centuries and consider it as a food of great delicacy. Multiple seaweeds have also been an essential segment of Asian cuisine.

13.6.2 Seaweeds as Supplementary Feed and Fertilizer

Results of different studies with seaweed supplementary feed indicate the higher and effective growth performance of fish than the use of commercial feed. In an efficient aquaculture, the production of seaweed plays a mysterious role being influential to the cost-effective supplementary fish feed and also to the fish health and immunity. Again, the seaweed meals can be used as supplements to the daily rations of poultry, cattle, and other farm animals as seaweeds are accessible sources of minerals and trace elements. Seaweed meal can also be blended with fish meal and used as poultry feed.

Seaweeds are appropriate in organic agriculture to be used as fertilizer. Hence, the application of seaweeds as manure is a usual practice in coastal areas throughout the world.

13.6.3 Medicinal and Pharmacological Properties of Seaweeds

The uses of seaweed polymer extract play a milestone in the field of pharmacy, medicine, and biochemistry. Seaweeds have curative powers against various biological diseases like anti-microbial, anti-fungal, anti-viral, anti-coagulant, anti-allergic, anti-fouling, anti-cancer, and anti-oxidant activities. Their known medicinal effects have been taken for the treatment of arthritis, colds, influenza, and tuberculosis. Some seaweed compounds restrain cholesterol levels, high blood pressure, and impede strokes. Also, seaweeds have a rich level of iodine, justifying the low incidence of hypothyroidism and goiter in the people of Asiatic coasts (Islam 2020).

13.6.4 Seaweeds in Wastewater Treatment

In modern times, seaweeds are used in the treatment of sewage and agricultural wastes by removing most of the nutrients efficiently from the wastewater. Seaweeds can remove nutrients like nitrogen and phosphorus and check eutrophication in freshwater, brackish water, and the marine environment under a standard treatment process. It is being used to avert the effluents from shrimp culture also. For instance, *Gracilaria verrucosa* has the potency to remove BOD and COD levels whereas *Ulva fasciata* has more potency for dismissal of ammonia. The other application of seaweeds is entrusted with the removal of toxic metals such as Cd, Zn, Ni, and Cu from industrial wastewater.

13.6.5 Seaweeds in Cosmetic Applications

Currently, seaweeds have become a key arsenal in cosmetic products such as creams, powders, soaps, shampoos, and sprays. Many companies are producing seaweed powder containing seaweed extracts for beauty and body care. Extracts of brown seaweed are widely used in massage therapy, thalassotherapy, which banishes impurities from the body and simultaneously balances the pH of the skin. Many seaweed extracted compounds are supposed to be valuable in various cosmetic applications and some are now becoming commercially important.

13.6.6 Seaweeds as Renewable Energy Supplier

The unexploited seaweed biomass is applied for the generation of biogas through anaerobic digestion to methane and it has been practiced in most of the developed countries. For various end purposes (i.e. heating, cooking, or electricity generation) biogas can be reasonably used as a clean-burning fuel.

However, the continuity of seaweed application in various economically valuable ways is quite antithetic in Bangladesh except for utilization by a few ethnic groups (e.g., Mog or Rakhyine tribal communities) living in the coastal areas. Mog people consume seaweeds as salad, soup, sauce, and vegetables. Besides, seaweeds have been utilized by the people of St. Martin Island. Macroalgae are most significantly harvested and processed on the Island for marketing to Myanmar and are sometimes used as a medicinal food for post-pregnant females and young ladies. Beyond this type of application, boiled seaweeds are traditionally eaten by the adult female for sound health and rotten seaweeds are applied as plant manure for vegetable cultivation. Therefore, the potential application of naturally occurring seaweeds on the coast of Bangladesh is shown in Table 13.3.

Table 13.3 Potential application of seaweeds naturally grown on the coasts of Bangladesh (Islam and Haroon 2016)

Applications	Seaweeds
Edible	*Cladophora prolifera, Caulerpa sp., Codium geppei, Dictyota atomaria, Dictyopteris australis, Gracilaria sp., Hypnea musciformis, Hydroclathrus sp., Halymenia sp., Padina sp., Ulva lactuca*
Medicinal/pharmaceutical (anti-bacterial, anti-fungal, anti-tumor properties)	*Caulerpa taxifolia, Codium geppei, Dictyota atomaria, Hydroclathrus sp., Halimeda sp., Gracilaria sp., Padina sp., Sargassum sp., Ulva lactuca*
Industrial (agar, alginate)	*Gracilaria spinuligera, Sargassum sp.*
Agriculture (animal feed and fertilizer)	*Cladophora sp., Codium geppei, Dictyota atomaria, Gracilaria sp., Hydroclathrus sp., Hypnea sp., Halimeda sp., Halymenia sp., Padina spp, Sargassum sp., Ulva lactuca*

Alongside, Hossain et al. (2021) observed seaweeds role as the basis of marine food webs and application as a true food source for crustaceans, mollusks, fishes, sea cucumbers and sea urchins and further added its operation as the spawning ground for different types of fishes and invertebrates.

The incipience in seaweed application through product and process evolvement could help in fulfilling the food and nutritional surety of the Bangladeshi people as well as enhancing the value of whole fisheries export. Nowadays, different types of functional food, value-added food, and personal care products have been prepared by government institutions, Non-Governmental Organizations (NGOs), and in the private sector in Bangladesh. Seaweed cake, pickle, jelly, soup, salad, etc. are manufactured by MFTS (DoF 2014). COAST Trust, a local NGO, also prepares various functional and value-added food products. Effective and environmentally sustainable utilization of seaweed can promote diversified economic sectors with potential application in biofuel, food, feed, pharmaceutical, manure, and cosmetic sectors that ultimately leverage the development of blue economy in Bangladesh (Fig. 13.13).

13.7 Challenges of Seaweed Culture Development in Bangladesh

The advancement of seaweed culture largely relies on skilled manpower with excellent farm management capabilities. Bangladesh has enormous potential in the availability of manpower, but their lack of skill is hampering the augmentation of seaweed farming. Hereafter, this is the veritable time to scale up the production by boosting the skills of coastal people through training programs and demonstrations.

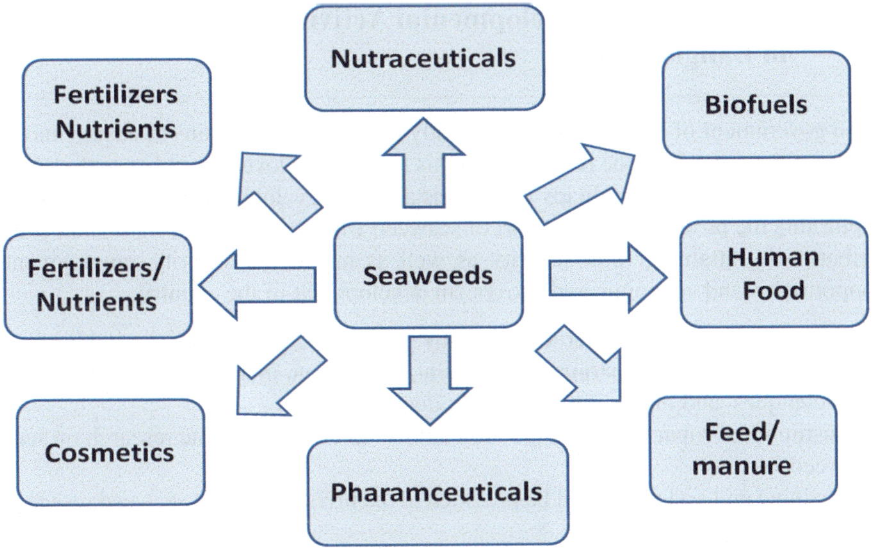

Fig. 13.13 Proper and sustainable application of seaweed promote the blue economy of Bangladesh

Furthermore, the lack of extensive receptive and supportive social and economic condition is also a prime constraint in the way of developing seaweed cultivation where demand, market condition, resource availability, and property rights are the most important issues for the economic side and in the context of the social side, consumer trends, tastes, mechanisms of resource allocation and distribution are considered important (Ahmed and Taparhudee 2010). Considering all of these issues amalgamated with biological, technical, and environmental factors can bring success in the large-scale cultivation of seaweed.

In addition, the upliftment of seaweed mass culture largely depends on the marketing system of a country. But still, there is no formal marketing system in Bangladesh that is the main driver of economic growth. Furthermore, the low price of seaweed is another major problem in the coastal areas. In the global market, the international price of seaweed is about $16/kg (BDT 1280/kg) according to the Agricultural Information Service (AIS) (Ghose 2020). Ghose (2020) also added that Falcon International Ltd. has earned $16,000 by exporting 1000 kg dry seaweed from Bangladesh in 2015, but farmers got a very little amount of that only Tk 56–80 per kg of dry seaweed. For this reason, a proper market system is needed to get fair prices of the seaweed. In addition, lack of awareness as well as collaboration with foreign agencies is another gap in the path of seaweed development in Bangladesh. In summary, to keep a positive image in the global seaweed market it is necessary to address these challenges which can be expected to further enrich our indigenous resources of seaweed.

13.8 Research and Developmental Activities on Seaweed in Bangladesh

The government of Bangladesh; has already adopted a lot of plans to ensure maximum utilization of marine resources. In this regard, the government has undertaken timely and appropriate policies for the sustainable development of the country by upturning the production and export of seaweed which will make a significant contribution by attaining food security as well as nutritional security, employment opportunity, and economic and ecosystem development of the country.

- The government has undertaken initiatives for the identification of suitable areas for seaweed farming, strain development, cultivation, innovation of post-harvest techniques, and increment of mass production.
- Institutional capacity-building action will be taken to invigorate research on seaweed farming.
- Training and assistance will be provided to the private sector in seaweed research and advancement.
- Measures will be taken to strengthen local and export market connectivity of seaweed products and provide incentives and loan assistance to private and individual entrepreneurs (National Agricultural Policy 2018).

Meanwhile, Bangladesh Fisheries Research Institute (BFRI) commenced research on seaweed from 2013 through its Marine Fisheries & Technology Station (MFTS), Cox's Bazar (Islam 2020). In addition, to their attempted development of seaweed culture techniques and evaluation of nutritional value and formulation of value-added product, BFRI implemented a project on "Seaweed Culture & Seaweed Product Development in Bangladesh Coast" from 2018 (Islam 2020). Over and above, some educational institutions of Bangladesh have also made their contribution through their valuable research activities on seaweeds.

Another project has been running from 2016 to 2021, by the Bangladesh Agricultural Research Institute (BARI) on the cultivation of seaweed in the coastal areas of Cox's Bazar which is directed by Bangladesh Agricultural Research Council (BARC) and funded by the Agricultural Research Foundation (Aziz 2021) (Fig. 13.14). Thus, seaweed production initiatives are moving towards success through the prudent and far-sighted leadership of the government of Bangladesh with the participation of governmental and non-governmental organizations as well.

13.9 Conclusion and Future Perspectives

Seaweeds have the prospects to upraise as an exclusive export-oriented industry in Bangladesh. A huge seaweed industry with endless possibilities can fructify through proper exploitation of resources and expansion of cultivation practices. To date, as

Fig. 13.14 Seaweed cultivation in Cox's Bazar by Bangladesh Agricultural Research Institute (Aziz 2021)

there is no extensive assessment of seaweed diversity and distribution on the Bangladesh coast, more comprehensive research on commercially important seaweeds and their application has to be conducted to make the status of marine macroalgae feasible and to conserve their diversity from being depleted. In this context, increasing seaweed aquaculture for diverse uses asserts proper knowledge respecting the feasible prospects and challenges engaged in setting up a market chain and production unit. Together with the government, if industrial entrepreneurs of the concerned arena of seaweeds come forward, they can certainly uncover a plausible door of a modern world in the blue economy which will prosper the national economy of Bangladesh. Seaweed culture eventually can be a sector of the durable economy along with confirming food security, enhancing employment opportunity, conservation of marine biodiversity, and resilience of the dependent coastal fishing communities and thus can play a pivotal role in the existent efforts to catalyze sustainable prosperity in the country.

Acknowledgments NAS, JRR and SK are grateful to the Bangladesh Fisheries Research Institute (BFRI) for providing information regarding seaweed resources of Bangladesh. ARR acknowledges Centre of Excellence, Department of Biotechnology, Vignan's Foundation for Science, Technology, and Research, and the FIST Project no: LSI 576/2013, Department of Science and Technology, Government of India for providing the facilities to prepare this manuscript.

References

Ahmad H (2019) Bangladesh coastal zone management status and future. J Coast Zone Manag 22(1):1–7

Ahmed N, Taparhudee W (2010) Seaweed cultivation in Bangladesh: problems and potentials. Kasetsart University Fisheries Research Bulletin, Thailand 28:13–21

Algae World News (2015) Seaweeds exploration vital for economy. https://news.algaeworld.org. Accessed 10 Oct 2015

Aziz (2021) Seaweed farming is the new trend in Cox's Bazar. https://www.dhakatribune.com. Accessed 10th Mar 2021

Billah MM, Kader MA, Mahmud SS et al (2018) Diversity and distribution of seaweeds in Saint Martin Island, Bangladesh. Inter J Fish and Aqua Stud 6(6):166–169

Coast Trust (2013) Potentials of seaweeds. COAST Trust, Shyamoli, Dhaka

DoF (2014) National Fish Week Compendium (in Bengali). Department of Fisheries, Ministry of Fisheries and Livestock, Bangladesh pp 144

FAO/NACA (1996) Regional study and workshop on the taxonomy, ecology and processing of economically important red seaweeds (NACA environment and aquaculture development series No. 3). Network of Aquaculture Centre in Asia-Pacific, Bangkok, Thailand

Feed the Future (2020) Success through seaweed in Bangladesh. https://www.feedthefuture.gov. Accessed 25 June 2020

Ferdouse F, Holdt SL, Smith R et al (2018) The global status of seaweed production, trade and utilization. Food and Agriculture Organization of the United Nations, Rome 124:1–120

Ghose TK (2020) Seaweed farming: a latent opportunity with immense potential. https://www.tbsnews.net. Accessed 5 Aug 2020

Hessami MJ, Cheng SF, Ranga Rao A, Yin YH, Phang SM (2019) Bioethanol production from agarophyte red seaweed, *Gelidium elegans* using a novel sample preparation method for analysing bioethanol content by gas chromatography. Biotech J 9(1):25

Hossain MS, Sharifuzzaman SM, Nobi MN et al (2021) Seaweeds farming for sustainable development goals and blue economy in Bangladesh. Elsevier. Mar Pol J 128:1–9

Hussain MG, Uddin SA, Failler P (2019) Potentials for the development of blue economy: prospects and challenges of mariculture in Bangladesh. J Bangla Agri 9(1):7–21

Islam MM (2020) Potential uses and prospects of seaweed culture in coastal areas of Bangladesh. https://dailyasianage.com. Accessed 22 Nov 2020

Islam S, Haroon TA (2016) Seaweed as an alternative income generation activity for coastal communities of Bangladesh. In: Conference: 2nd Global Summit on Aquaculture & Fisheries, Kuala Lumpur, Malaysia, 11–13 July 2016

Islam S, Haroon TA (2017) Seaweed aquaculture: an alternative income generation option to improve the livelihood of the southeast coastal communities of Bangladesh. Oceanogr Fish Open Access J 4(5):555648

Islam MM, Hoq ME, Haque MA et al (2020) Seaweeds of Bangladesh coast. Cox's bazar, Bangladesh, Bangladesh Fisheries Research Institute, p 136

Islam MM, Khan MSK, Hasan J (2017) Seaweed *Hypnea* sp. culture in Cox's bazar coast. Bangladesh. J Zool 45(1):37–46

Kim JK, Yarish C, Hwang EK (2017) Seaweed aquaculture: cultivation technologies, challenges and its ecosystem services. Algae J 32(1):1–13

Mohammad JH, Ranga Rao A, Ravishankar GA (2019) Opportunities and challenges in seaweeds as feed stock for biofuel production. In: Ravishankar GA, Rao AR (eds) Handbook of algal technologies and phytochemicals: Phycoremediation, biofuels and global biomass production, Volume-II. CRC Press, USA, pp 39–50

Msuya FE, Hurtado AQ (2017) The role of women in seaweed aquaculture in the Western Indian Ocean and South-East Asia. Euro J Phyco 52(4):482–494

National Agricultural Policy (2018) Ministry of agriculture, government of the people's republic of Bangladesh. https://moa.gov.bd/sites/default/files/files/moa.portal.gov.bd/policies

Sarkar MSI, Kamal M, Hasan MM et al (2016) Present status of naturally occurring seaweed flora and their utilization in Bangladesh. Res AgricLivest Fish 3(1):203–216
Sarker MN (1992) Studies on the red sea weeds in Bangladesh. In: Regional workshop on the taxonomy, ecology and processing of commercially important red sea weeds, held at Kasetsart University in Bangkok, Thailand. Organized by FAO/NACA and France Govt
Satpati GG, Barman N, Pal R (2012) Morpho-taxonomic account of some common seaweeds from Indian Sundarbans mangrove forest and inner island area. J Alga Biom Utln 3(4):45–51
Seraj S (2018) Seaweed door to blue economy. https://www.thedailystar.net. Accessed 02 Aug 2018
Shama A, Joyce SG, Mari FD, Ranga Rao A, Ravishankar GA, Hudaa N (2019) Macroalgae and microalgae: novel sources of functional food and feed. In: Ravishankar GA, Rao AR (eds) Handbook of algal technologies and phytochemicals, volume-I: Food, health and nutraceutical applications. CRC Press, USA, pp 207–219
Siddiqui AAM, Kashem MA, Islam MA et al (2019) Commercially important seaweed cultivation and its potentials for the coastal areas of Cox's bazar, Bangladesh. Inter J Fish and Aqua Stud 7(5):463–470
The Financial Express (2018) Seaweed farming should be brought under purview of fisheries ministry. https://thefinancialexpress.com.bd. Accessed 24 Nov 2018
Uddin SA, Hussain MG, Al MA et al (2021) On the potential and constraints of mariculture development in Bangladesh. Aquacul Int J. 29:575–593
UNDP, UNEP, World Bank, World Resources Institute (2000) World resources 2000–2001: people and ecosystems: the fraying web of life. World Resource Institute, Washington, DC
Zafar M (2005) Seaweed culture in Bangladesh holds promise. INFOFISH Int 1:18–10

Chapter 14
Seaweed Industries and Products in the UK: A Brief Review

Elisa Capuzzo

Abbreviations

Cefas	Centre for Environment, Fisheries and Aquaculture Science
EIA	Environmental Impact Assessment
MMO	Marine Management Organisation
MSC	Marine Stewardship Council
R&D	Research and Development
SAMS	Scottish Association for Marine Science
SSIA	Scottish Seaweed Industry Association
UK	United Kingdom
UKAS	United Kingdom Accreditation Service

14.1 Introduction

Macroalgae, or seaweeds, are marine macroscopic algae which live on seashores and in shallow seas (in the sub and inter tidal area) around the world. Based on their pigmentation, seaweed can be divided into three groups: red (Rhodophyta), brown (Phaeophyta) and green (Chlorophyta). There are over 10,000 species of seaweed in the world (approximately 7000 red, 2000 brown and 1700 green), and around 6% of these species can be found along the UK shores (Bunker et al. 2017). Brown seaweed represents the largest and most abundant seaweed type in British waters, with kelp forming large forests in subtidal areas and wracks covering large areas of the intertidal zone (Stanley et al. 2019). Seaweeds have high ecological and economic importance; they are ecosystem engineers and provide multiple ecosystem services including provision of habitat for other species, acting as carbon and nutrient sinks (bioremediation), and providing coastal protection by dampening

E. Capuzzo (✉)
Centre for the Environment, Fisheries and Aquaculture Science (Cefas), Weymouth, UK
e-mail: elisa.capuzzo@cefas.co.uk

A. Ranga Rao, G. A. Ravishankar (eds.), *Sustainable Global Resources of Seaweeds Volume 1*, https://doi.org/10.1007/978-3-030-91955-9_14

waves and currents (Duarte et al. 2017; Naylor et al. 2021). At the same time, macroalgae have been used by humans for millennia, with earliest records as far back as 14,000 years ago (Dillehay et al. 2008). Initially seaweeds were harvested from the wild, however, over the course of the last 60 years, the increased demand for seaweed has shifted the production towards aquaculture sources (Buchholz et al. 2012). Seaweeds are produced mainly for human consumption and for flavouring (80% of global production; West et al. 2016), although they can have multiple uses, including for extraction of phycocolloids, animal feed, fertilizers, water remediation (West et al. 2016; Shama et al. 2019).

14.2 History of Production and Trade in Seaweeds in the UK

In the UK, seaweeds have traditionally been used for centuries as a food source, animal feed, and soil enricher. Seaweeds, whole or extracts, can be found in traditional food such as the Welsh laverbread, made with *Porphyra umbilicalis*. Macroalgae have not only been providing nutrients and vitamins to human diets but also to animals, for example sheep, which were (and still are) allowed to graze on fresh or cast seaweed or have seaweed added to their feed (Adams 2016). Historically, addition of seaweed to poor soil helped to provide nutrients, organic carbon and alginic acids, retaining moisture in the soil (Adams 2016).

Seaweeds in the UK have played a relevant socio-economic role through the years, constituting an important element of coastal economies, particularly for island communities (Forsythe 2006). In the 17th and 18th centuries, brown seaweeds (e.g. *Laminaria* spp., *Fucus* spp. and *Ascophyllum* sp.) were burned for the production of soda ash, an alkaline product which was used in the glass industry, for soap production and for bleaching linen (Mesnildrey et al. 2012; Mouritsen 2013; Hothersall 2012). The practice was introduced in the south-west of England (Isle of Scilly) in the 1680s and spread to Scotland (Kintyre and Orkney) in the second half of the 1700s and became highly profitable for landlords (Hothersall 2012). Local families or tenants gathered, dried and burnt kelp in shallow stone-lined pits or kelp kilns (still visible today in Northern Ireland and Scotland) to produce a molten slag, which once cooled and broken into lumps, was shipped to industrial centres such as Bristol or Liverpool (Hothersall 2012; Forsythe 2006). During peak production, up to 25,000 t of kelp were harvested annually in summer months in Scotland, with an estimated 10,000 families involved in activities (Burrows et al. 2018; Hothersall 2012). At the start of the nineteenth century, new processes for making alkali (not relying on seaweed) were developed; however, macroalgae continued to be exploited for the extraction of iodine and the production of tincture of iodine, with 20 iodine factories in operation in Glasgow until the start of the twentieth century (Hothersall 2012; Mesnildrey et al. 2012). After World War II, seaweeds were mainly harvested for production of hydrocolloids, particularly alginic acid, a gelling and thickening agent in food and non-food applications (Mesnildrey et al. 2012). First extracted by a Scottish chemist in 1883, alginates were produced at an industrial scale in Scotland from the 1930s, and production continued until 2009, when the last plant at Girvan was sold (Burrows et al. 2018). At the peak of production (in the early 1970s) the workforce involved in alginate production at the two main Scottish sites numbered approximately 900 people (James 2010).

Throughout the centuries, demand for seaweed has been met by harvesting of wild resources; however, the recent resurgence of interest in seaweed and potential future increase in demand of seaweed biomass is likely going to be met by farming rather than the harvest of wild resources (Capuzzo and McKie 2016). In fact, the English Aquaculture Strategy (Seafood 2040) suggests a growth aspiration of 13,000 mt (wet weight) of seaweed produced per year by 2040 for England alone.

14.3 Seaweed Production from Harvesting of Wild Resources

There are no records of seaweed production for the UK although seaweeds are harvested in various parts of the UK (Capuzzo and McKie 2016); however, it is estimated that around 15,000 t (wet weight) of seaweed are harvested from the wild in Scotland annually, mainly in the Outer Hebrides (Seaweed Review Steering Group for Scotland[1]). The main species commercially harvested from wild resources in the UK are listed in Table 14.1.

Table 14.1 Seaweed species commercially harvested from the wild in the UK

Brown seaweed	Red seaweed	Green seaweed
	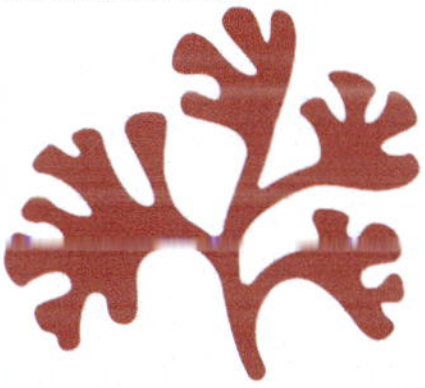	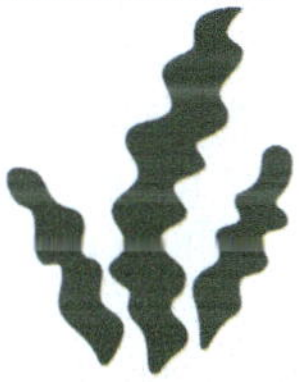
Saccharina latissima	*Palmaria palmata*	*Ulva* spp.
Laminaria digitata	*Chondrus crispus*	
Laminaria hyperborea	*Porphyra* spp.	
Himanthalia elongata	*Osmundea pinnatifida*	
Fucus vesiculosus	*Mastocarpus stellatus*	
Fucus serratus	*Lithothamnium* sp.	
Fucus spiralis		
Ascophyllum nodosum		
Alaria esculenta		
Pelvetia canaliculata		

[1] https://www.gov.scot/publications/seaweed-review-steering-group-minutes-september-2019/

Harvesting of seaweed from wild resources in the UK is carried out mainly at a small-artisanal scale (by hand-cutting or gathering of beach-cast), except for *Ascophyllum nodosum* (rockweed) which is harvested at a medium to large scale in Scotland in the Outer Hebrides (Burrows et al. 2018). Rockweed is collected either by hand or by using small, specialised cutting boats. For example, The Hebridean Seaweed Company manually harvest rockweed when the tide is out by encircling the chosen cutting area with a rope or net and then cutting within this area; with the return of the tide the seaweed floats and form a large circular bail which is towed to a sheltered area by a small boat for loading into a lorry.[2] For mechanical harvesting, the same company uses vessels which cuts the seaweed as the fronds float above the seabed; the harvested seaweed is then filled into sacks and towed by a small boat to a loading area. Another company (Uist Asco[3]) harvests rockweed by hand with a serrated sickle during low tide, or with boat and a customized rake with an extended handle, during low or high tide. Dried and milled rockweed biomass is used in animal feeds, soil enhancement, alginate, cosmetic and nutraceutical industries (Burrows et al. 2018). It is estimated that the available standing stock of *A. nodosum* in the Outer Hebrides is 170,000 t with a maximum recommended annual landing of 15,000 t (Burrows et al. 2010).

Regulation for harvesting of wild seaweed in the UK is defined by each devolved administration (i.e. England, Scotland, Wales and Northern Ireland) and their statutory environmental bodies (see for example Marine Scotland 2016; Natural Resources Wales 2018). In general, commercial operations of harvesting of living or beach-cast seaweed in the wild require the permission of the relevant landowner. If the latter is The Crown Estate, a lease for wild harvesting operations is issued. Based on how the seaweed are removed (e.g. hand-picked, hand-cut, by mechanicals methods), and the circumstances and scale of the removal (e.g. species, location, quantity to remove), a marine license may also be required. During the licensing process, the regulator considers the application and its potential implications against the needs of protecting the environment, human health and the legitimate use of the sea. Foraging of seaweed in small quantities for personal use does not require a lease or a license. Of fundamental importance is that seaweed harvesting is carried out sustainably; however, determining the quantity and frequency of biomass which can be harvested sustainably (particularly for medium-large operations) is complex, due to the spatial and temporal variability of the resource (Burrow et al. 2018). For example, Burrow and co-authors (2018) have been adopting predictive models (based on the specific environmental requirements of a given seaweed) to estimate standing stocks of *A. nodosum*, *L. hyperborea*, *L. digitata*, *S. latissima* and *Saccorhiza polyschides* in the Hebrides and determine potential harvest quantities.

Best practice guidelines (or code of conduct) for harvesting of seaweed from wild sources in the UK include: adoption of sharp tools to cut the frond; collection

---

[2] http://www.hebrideanseaweed.co.uk/harvesting-theprocess.html

[3] https://www.uistasco.com/our-sustainability/sustainable-seaweed-harvesting-methods/

of less than 1/3 of an individual seaweed leaving holdfast attached; harvesting sparsely, rotating the harvesting area; harvesting during the active growth season (outside reproduction phase); avoiding collection of by-catch; limiting and/or avoiding harvesting in vulnerable areas (Marine Scotland 2016; Natural Resources Wales 2018; Bailey and Owen 2014). Depending on the scale of extraction, mitigation and enhancement measures may also be needed such as adoption of monitoring programmes with pre- and post-harvesting surveys.

14.4 Seaweed Production from Aquaculture Sources

Seaweed farming in the UK is still limited although there is a growing number of commercial farms in Northern Ireland (1), Scotland (3), Wales (1) and England (4), in addition to pilot farms for Research & Development in Scotland (2 sites of the Scottish Association for Marine Science, SAMS) and Northern Ireland (1 site of Queen's University, Belfast). Several other seaweed companies are also currently going through the marine licensing process to start commercial seaweed farming in the UK, in a shift from harvesting of wild resources to farming. Industry organisations, such as the Scottish Seaweed Industry Association (SSIA), have been developed to support this growing industry and associated value chain. The Seaweed Alliance was funded in 2019 as a platform for stakeholders to engage, drive and steer policy and raise public awareness around seaweed. In addition, the Algae-UK[4] network has been established to support the community of scientists and engineers working with or interested in industrial biotechnology applications for algae (although this is not limited to farmed seaweed but include all types of algae).

Initial experimental trials of seaweed aquaculture carried out in Scotland in 2004 aimed to test the ability of farmed *S. latissima* and *P. palmata* to bioremediate salmon farm sites (Stanley et al. 2019). Subsequently, research focused on the potential of using seaweed biomass for production of bioenergy (biogas and bioethanol), for example, under the SeaGas, MacroFuel, and MacroBioCrude projects. More recently, interest in seaweed aquaculture involves food and higher value products (Stanley et al. 2019).

Seaweed species farmed in the UK for both commercial and scientific purposes include *Saccharina latissima*, *Laminaria digitata*, *Laminaria hyperborea*, *Alaria esculenta*, *Palmaria palmata* and *Ulva* spp. Cultivation of kelp species has shown to be successful with *S. latissima* the focus of many research projects (Stanley et al. 2019). As an example, in 2017, Queen's University Belfast (in Northern Ireland) grew and harvested 20 t of *S. latissima*, the UK's largest harvested seaweed batch, as part of the SeaGas project. *A. esculenta* is cultivated in Scotland while the waters off the southern part of England are less suitable for cultivation of this species due to higher water temperature, which in summer exceed 18 °C (MMO 2019).

[4] https://www.algae-uk.org.uk

Cultivation of *P. palmata* is considered unreliable due to difficulties in controlling its lifecycle; however, cultivation issues are being investigated due to the high market demand for this species (Stanley et al. 2019). Cultivation of *Porphyra/Pyropia* spp. has been trialled in Scotland, but is still at the Research and Development stage due to its complex lifecycle and expensive hatchery process (Stanley et al. 2019). Research has also been carried out on the cultivation of *Osmundea pinnatifida*, suggesting that tank cultivation (rather than at sea) may be the best solution for cultivation of this species (Stanley et al. 2019).

In terms of cultivation techniques adopted in the UK, seeding stock to begin a farm is provided by fertile wild seaweeds which should be collected locally to the farm. In the hatchery the fertile material is made to release spores, which grow into male/female gametophytes. After fertilization, the culture is seeded on twine spools and reared for 6–8 weeks in the hatchery until juvenile seaweeds are up to 1 cm long (Stanley et al. 2019). The twine is then helically wrapped around a rope, and out-planted at a seaweed farm. Alternatively, the juvenile seaweeds are seeded directly on a rope with the aid of a binder (bio-glue); in this latter case the hatchery phase is reduced as the cultivation surface is immediately deployed at sea (Kerrison et al. 2018; Stanley et al. 2019). There is currently only one commercial seaweed hatchery in the UK, based in Scotland, which is also acting as biological bank (biobank). Other research institutions (e.g. Queen's University Belfast; the Centre for Sustainable Aquatic Research of Swansea University) have macroalgae hatchery facilities for research and development.

In the UK, seaweeds are out-planted at sea in the autumn through the winter (October to February) and are generally harvested between April and June, although exact timings depends on the end use of the biomass, its chemical composition, presence of biofouling and growth rate of the crop (Stanley et al. 2019). Commonly, seaweed farms are characterised by longline systems with moorings (made of concrete or eco-anchors) every 100 m; the line is approximately at 1.5 m below the surface and is loose (so it can be easily pulled for inspection; Stanley et al. 2019). Other systems used are grid-based system, currently adopted at the cultivation sites in Argyll Scotland (operated by SAMS), or adapted mussel longlines (Stanley et al. 2019). Although longlines are an effective system for cultivating seaweeds, new farming technologies are being investigated to improve efficiency, yields, and particularly to withstand offshore exposed conditions (Bak et al. 2020).

To set up a seaweed farm in the UK, two permissions must be obtained before development can be introduced to the marine environment: a lease from the relevant landowner (e.g. The Crown Estate), and a marine licence from the relevant regulatory body (which varies depending on whether the farm is in England, Scotland, Wales or Northern Ireland; Wood et al. 2017). Within the application process there may be additional licensing requirements (e.g. Habitat Regulations Assessment; Marine Protected Areas Assessment), or the need for an Environmental Impact Assessment (EIA), depending on the scale, nature and location of the proposed development, and the risk it may pose to the local environment (Wood et al. 2017).

Encouraging cultivation results (such as from the SeaGas project) could help the further development of the industry and in opening-up the supply chain for seaweed biomass. The UK has potential to expand the seaweed aquaculture industry thanks to suitable environmental conditions, extensive coastline, regulatory frameworks and R&D. For example, based on modelling of suitable environmental conditions for growth of seaweed, it was estimated that over 50% of English coastal waters offer suitable condition for growth and farming of *S. latissima* and *L. digitata*, although only 9% are suitable for growth of *A. esculenta* (MMO 2019). However, for this potential to realise there are several knowledge gaps that need addressing and which are discussed in Sect. 14.7.

14.5 Seaweed-Related Businesses in the UK

The seaweed industry in the UK has expanded in the last decade, with more products becoming available through main and smaller retailers and directly from producers (Stanley et al. 2019). The UK is the main importer of seaweeds in Europe, for human consumption,[5] particularly of nori, likely the result of a rise in popularity of sushi restaurants and food retail (Stanley et al. 2019). Positive press coverage around consumption of seaweed is also helping to corroborate consumers perception of seaweed as a healthy and nutritional food, not just for humans but also for animals.

Current seaweed businesses in the UK were identified in this study by a web search and through information available in reports and feasibility studies (e.g. Capuzzo and McKie 2016; Stanley et al. 2019). For the web search keywords adopted included "seaweed", "business", "product", "buy", "UK", "Northern Ireland", "Scotland", "England", "Wales", in different combinations (e.g. "seaweed product UK"). The search focused on: (1) businesses which harvest and/or farm seaweeds in the UK; and (2) UK-based businesses which process seaweed or provide seaweed-related services. The result showed that many businesses harvested/farmed as well as processed seaweeds, while some exclusively produced or exclusively processed seaweed biomass. Businesses which sold seaweed-products (without producing and/or processing seaweed) were not considered in this study.

The search identified 74 seaweed-related businesses in the UK of which 35% were based in England, 34% in Scotland, 18% in Wales, 8% in Northern Ireland and a remaining 5% for which it is was not possible to determine the exact location in the UK. A further 12 businesses were identified, however, it was not clear whether they were active (e.g. waiting for the approval of a marine license application), while 3 businesses previously identified by Capuzzo and McKie (2016) were dissolved. Furthermore, in the previous analysis of seaweed businesses in the UK, Capuzzo and McKie (2016) highlighted the presence of 27 UK-based companies; even accounting for businesses which may have been omitted in this and the

[5] https://www.cbi.eu/market-information/natural-ingredients-health-products/seaweed/market-potential

previous search (e.g. due to lack of website or not picked up by the search terms), it appears that in the last 5 years (2016–2021), the number of seaweed business in the UK has increased and likely doubled.

For each of the businesses identified the method of production of the seaweed biomass was investigated (e.g. wild harvest, aquaculture), the provenance of the biomass and the main species used. It was estimated that 69% of the businesses produced/used seaweed harvested from the wild (equivalent to 50 businesses), while only 10% (7 companies) produced/used seaweed from aquaculture sources only. In addition, 3% produced/used seaweed from both sources (wild harvest and aquaculture) while 18% of the businesses (equivalent to 13) did not provide information on the source of the seaweed used. These figures are in line with the status of the seaweed aquaculture in the UK which is still in its infancy (see Sect. 14.4). In terms of provenance, 29% of the businesses used seaweed harvested/farmed in Scotland (particularly in the Hebrides), 15% seaweed harvested/farmed in England, 15% in Wales, and 6% in Northern Ireland; 8% used seaweed harvested/farmed outside the UK (mainly Ireland, France, Spain) with 6% using seaweed biomass which was partially from the UK and partially from outside the UK. The remaining 21% (15 businesses) did not provide information on the provenance of the seaweed.

The seaweed species most frequently produced/used by seaweed-businesses in the UK is *A. nodosum* (Fig. 14.1), which as seen in Sect. 14.3, is harvested at a medium to large scale in Scotland in the Outer Hebrides and used in several products. Kelp and wrack species (e.g. *Fucus*, *Laminaria* and *Saccharina*) are other

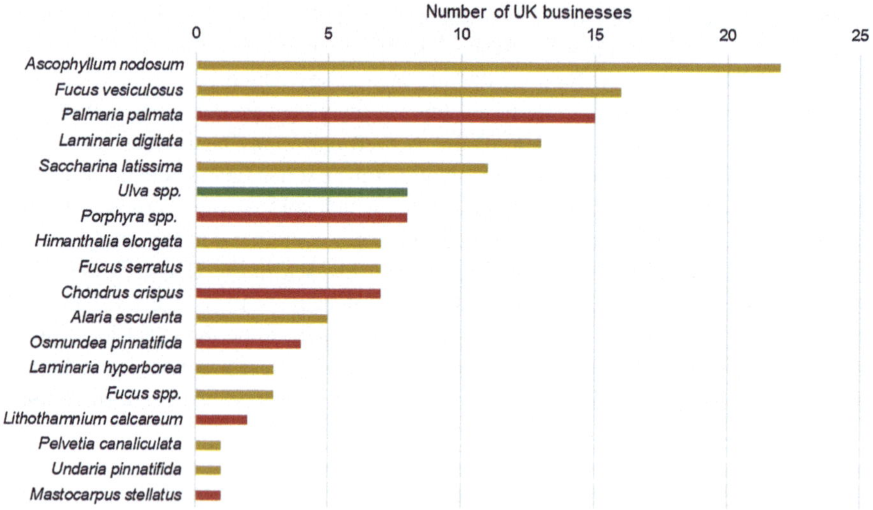

Fig. 14.1 . Number of businesses in the UK which produce/process each seaweed species as listed in Table 14.1 (note, that some businesses produce/process more than one species of seaweed). The colour of the bars refers to the algae groups (brown, red and green; see Sect. 14.1). It is possible that *Lithothamnium calcareum* indicated as ingredient by two businesses is *Phymatolithon calcareum* or *Lithothamnium corallioides*

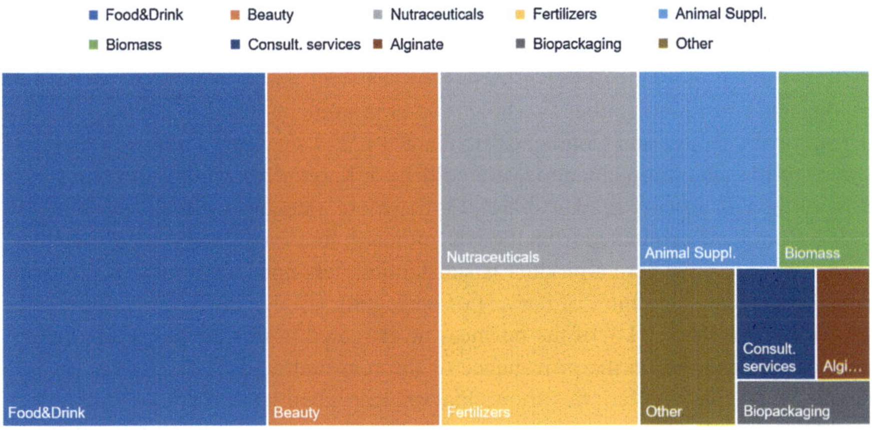

Fig. 14.2 . Relative proportion of seaweed-related businesses in the UK by product/service provided

important brown seaweeds which are produced/used in UK businesses. Within the red seaweeds, *P. palmata* is used by 15 businesses while *Porphyra* species by 8 (mainly located in Wales). *Undaria pinnatifida* (a non-native species in the UK) is used by one business, although imported from harvest of wild resources in Ireland. Some of the businesses investigated (24%) did not specify the species used or provided only common names (e.g. kelp) rather than scientific names.

The 74 seaweed-related businesses were then classified based on their product/s or services provided (Fig. 14.2). Approximately a third of the UK businesses (30%) targeted the food and drink sector (of which 23% targeting food and 7% drinks), with 19% producing products for the beauty industry, and 13% nutraceuticals (supplements). Fertilizers (soil enhancers) and products for animals (supplements and shampoos) were produced by 10 and 9% respectively of the businesses. Some producers of seaweed (6% of the total number of businesses) provided bulk seaweed biomass (processed or unprocessed), while others (5%) were offering consultation and support services for seaweed aquaculture (Fig. 14.2). A couple of businesses focused on alginate extraction and blending, with other two working on bio packaging. The 'other' category in Fig. 14.2 includes production of paper greeting cards with seaweed, seaweed candles and fish feed for aquarium. Of the 74 businesses, 16 produced products under multiple of these categories (e.g. food and nutraceuticals).

14.6 Seaweed Products and Processes in the UK

Seaweed is now a common food in the Western diet (in addition to use in traditional food in coastal areas), thanks to its high nutritional value and an expansion of the health-food industry (Bouga and Combet 2015). The distribution of the current seaweed-businesses in the UK (Fig. 14.2) supports this view highlighting a 'resurgence' of seaweed in the British diet (Bouga and Combet 2015). UK retailers (as well as online shopping) offer a variety of seaweed and edible algae to the UK market and consumers (either as whole food or ingredients). For example, in a product survey by Bouga and Combet (2015) in 2014, 224 seaweed-containing products (food, drinks and supplements) were sold by UK retailers online (covering over 82% of the UK grocery market share). Furthermore, the authors identified that 63% of these products originated from the UK, although they cautioned that 'origination' is the last country where a product is substantially changed and many of the products did not state a country of harvest or packaging (Bouga and Combet 2015). As shown in Sect. 14.5, 21% of the business investigated in this present study did not provide information on the provenance of the seaweed while 24% did not include details of the species used; this shows that unclear labelling remains an issue which could have implications particularly for food products when estimating iodine and other nutrients intake. In terms of seaweed species used in products, Bouga and Combet (2015) also observed that *A. nodosum* was the most common seaweed used.

14.6.1 Food and Drink

The UK businesses in the food category of this study provided seaweed mainly as dry product (as whole leaf, pieces, flakes, granules and powder of single or mixed species) and occasionally fresh or frozen products. Seaweeds were also frequently sold as seasoning, mixed with salt and/or herbs. Other food products containing seaweeds and produced by UK businesses included: cracker, biscuits, sauces, paste, pesto, tapenade, relish, snacks, chips, laverbread, flavoured butter, popcorn, nuts, infused oils, chocolate, and cheese. For the main part of these products, harvested seaweed received minimal processing, mainly consisting of quality inspection of the fresh product, followed by a gentle drying process (if required) to remove moisture, preserve flavour, vitamins and minerals, and packaging. Where applicable, dry seaweed was milled at different sizes. The final product was often tested for quality and nutrient composition, reported in the product label. Some products were marked as organic, for example approved by the UK Soil Association, and adhered to Marine Stewardship Council (MSC) environmental standards for sustainability.

Interestingly, 20% of the food products identified in the Bouga and Combet (2015) study were bread, cake, pizza base, and biscuits where seaweeds were included as an ingredient. This present study did not identify any bread, cake or pizza base products containing seaweed suggesting that the food category was under-represented in the online search carried out in Sect. 14.5. This could be due to the product/s not clearly stating presence of seaweed as ingredient and/or different keywords adopted in the search between the two studies.

In terms of seaweed drinks, the main products were spirits (gin, rum, whiskey) where seaweed was used as one of the herbals, beer, tea and powders to mix with water for drinks.

14.6.2 Beauty

The second most frequent category of products (19% of the businesses) targeted the beauty industry, providing a vast selection of products for body and hair containing seaweed. This included: soap (including exfoliating); lip balm and conditioner; body/skin care (balm, oil, scrub, wash, lotion, spray, moisturiser, exfoliant, toner); hand care (sanitiser, wash, lotion, cream); hair care (solid and liquid shampoo, conditioner); foot care (soak, cream); face care (cleansing, wash, toner, oil, mask, serum, eye balm, shave oil, beard oil and wax); dry seaweed (with or without salt, whole or powder) for bath and shower gel. These products contained seaweed in different forms such as dry seaweed (as whole leaf, fragments, powder) or as extract (purified and distilled from harvested seaweed) added to the other ingredients. Multiple products provided specifications of being free-from palm oil, parabens etc. as well as cruelty free (no animal testing).

14.6.3 Nutraceuticals

The main type of nutraceutical was represented by supplements (e.g. for iodine) in the form of capsules or granules. The process involved in the preparation of the seaweed biomass for production of supplements generally involved a gentle drying at low temperature (not altering the seaweed properties) and milling into a fine powder which was then encapsulated (in certain cases after a purification process to remove heavy metals). Some businesses specified in their website to have obtained organic certification (e.g. by the Soil Association), and that seaweed powder was checked by accredited (UKAS) laboratories in the UK for nutritional, contamination, pesticide, microbial, allergen content and other analysis.

14.6.4 Fertilizers

Products for improving plant growth (10% of the businesses identified) were available as liquid (extracts and concentrate) or solid form (e.g. seaweed 'meal') and included: plant growth enhancers, promoters, and stimulant; soil conditioners and growth enhancers. Some of these products were prepared by drying and cold-pressing seaweeds (particularly *A. nodosum*) while others by fermenting the seaweed in barrels for several months followed by cold-pressing and bottling of the resulting liquid. Some of these products had obtained an organic certification (e.g. Organic Farmers & Growers organic certification).

14.6.5 Products for Animals

Businesses targeting animals (9%) developed supplements for cat, dog, horse, poultry, beef and calf in the form of granules, meal, tablets, powder and tonic. Businesses were mainly based in England and Northern Ireland and used a variety of seaweed including kelps, wracks, and red seaweed (e.g. *P. palmata* and *C. crispus*). Two of the companies identified produced dog shampoos, containing seaweed.

14.6.6 Other Products/Services

Around 6% of the businesses produced seaweed (either by harvesting wild resources or aquaculture), subsequently sold as bulk biomass to be processed by other businesses. It was not always clear from their websites whether the biomass was available as wet or dry, except in some cases where it was specified that seaweeds were gently dried and milled to customer's specification.

Five out of the 74 businesses provided services and consultancy related to seaweed aquaculture, such as site selection surveys, support with obtaining a marine licence for seaweed aquaculture, farm design and installation including moorings inspection and removal, support with farming operations including equipment supply, provision of nursery services for seaweed, and seaweed industry and market development.

Two further businesses, both based in Scotland, focused on alginate extraction and blending, while another two businesses (one in Scotland and one in England) targeted development of bio-packaging and bio-degradable sachets (e.g. for sauce, water) with products currently in development. The business based in Scotland is also adopting a biorefinery process to extract bioactives for nutraceuticals and cosmeceuticals, as well as nutritional ingredients and supplements, including functional protein. Biorefining allows to optimise the processing of seaweed biomass, improving production economics while reducing waste produced; it is an essential step for extracting natural commodities of commercial value from seaweed, such as biochemicals, nutritional ingredients, polymers and minerals (Stanley et al. 2019).

14.7 Conclusions and Future Perspectives

The seaweed industry in the UK is still in its infancy, however, as suggested by this and previous studies (Burrows et al. 2018; Capuzzo et al. 2019; Huntington and Cappell 2020), it has the potential to thrive and support a range of businesses. Farming of seaweed would contribute towards different ecosystem services and resource production, helping address policy objectives (for example, in relation to Sustainable Development Goals, the UK 25 Year Environmental Plan, and national aquaculture strategies).

The interest in seaweed and the potential future increase in demand for seaweed biomass are likely to be met by farming rather than natural harvest (Capuzzo and McKie 2016). In fact, the UK is home to many seaweed species, has an extensive coastline with suitable environmental conditions for seaweed aquaculture and regulatory frameworks in place to ensure the sustainable development of the industry. However, for the industry to grow and realise this vision, several technical, market and supply chain challenges need to be addressed, which are summarised below (Capuzzo et al. 2019; Huntington and Cappell 2020; Stanley et al. 2019).

- **Technical**: poor knowledge of productivity, yield, reliability of farmed seaweed species, and of available suitable sites; limited available robust algae cultivars; lack of generally available technical knowledge for growing seaweed; presence of technical challenges particularly for offshore farming (e.g. infrastructure, logistic and servicing).
- **Financial**: requirement of access to business support and innovation funds, essential for small-medium enterprise to start a business; poor knowledge of farm running costs.
- **Disease, biosecurity and food safety**: poor knowledge of the risk of disease for seaweed farmed in the UK, as well as for biosecurity (invasive species) and genetic structure of wild seaweed populations; limited information on the presence of biotoxins and any potential microbial contamination in seaweed products; no standards required by law for seaweed products.
- **Regulatory and social**: complex and lengthy licensing process and the need for updating/reviewing regulatory requirements; need of effective spatial planning to address conflicts with other stakeholders (e.g. fishing industry, conservation areas) and local communities over the use of coastal areas.
- **Environmental**: knowledge gaps around environmental impacts of farming activities (which is affecting release of marine licences); lack of data on the environmental effects (positive and negative) of farming.
- **Market**: lack of knowledge of market demand and structure in the UK, and consumer preferences towards seaweed products.

Future research and investments should aim to address the above challenges which are not exclusive to the UK seaweed industry but are also shared by other countries (e.g. in Europe, see Barbier et al. 2019). Therefore, international collaborations and knowledge exchange should be encouraged to tackle these knowledge gaps.

Acknowledgments The author thanks Keith Jeffery (Cefas) for critical reading of the manuscript and Fiona Trappe (The Seaweed Alliance) for providing detailed and insightful input.

References

Adams J (2016) An overview on seaweed uses in the UK: past, present and future. Seaweed Resour 36:16–21

Bailey L, Owen K (2014) Seaweed harvesting. Natural England's Advice, July 2014, p 20

Bak UG, Gregersen Ó, Infante J (2020) Technical challenges for offshore cultivation of kelp species: lessons learned and future directions. Bot Mar 63(4):341–353. https://doi.org/10.1515/bot-2019-0005

Barbier M, Charrier B, Araujo R, Holdt SL, Jacquemin B, Rebours C (2019) PEGASUS – PHYCOMORPH European guidelines for a sustainable aquaculture of seaweeds, COST Action FA1406 In: Barbier M, Charrier B (eds) Roscoff, France, p 194. https://doi.org/10.21411/2c3w-yc73

Bouga M, Combet E (2015) Emergence of seaweed and seaweed-containing foods in the UK: focus on labeling, iodine content, toxicity and nutrition. Foods 4:240–253. https://doi.org/10.3390/foods4020240

Buchholz CM, Krause G, Buck BH (2012) Aquaculture production of seaweeds and its economic relevance. Chapter 22 Seaweed and man, 22.1. In Seaweed biology, Wiencke and Bischof eds. Ecol Stud 219:471–493. https://doi.org/10.1007/978-3-642-28451-9_22

Bunker F, Maggs CA, Brodie JA, Bunker AR (2017) Seaweeds of Britain and Ireland, 2nd edn. Wild Nature Press, Plymouth, p 312

Burrows MT, Macleod M, Orr K (2010) Mapping the intertidal seaweed resources of the Outer Hebrides. Scottish Association for Marine Science Internal Report No. 269. Report commissioned by Scottish Enterprise and Highlands and Islands Enterprise, p 45

Burrows MT, Fox CJ, Moore P, Smale D, Sotheran I, Benson A, Greenhill L, Martino S, Parker A, Thompson E, Allen CJ (2018) Wild seaweed harvesting as a diversification opportunity for fishermen. A report by SRSL for HIE, p 171

Capuzzo E, McKie T (2016) Seaweed in the UK and abroad – status, products, limitations, gaps and Cefas role. Cefas contract report FC002I. Report commissioned by Defra, p 66

Capuzzo E, Mengo E, Kennerley A (2019) Seaweed farming feasibility within the Dorset and East Devon FLAG area. Report 1 – potential species, farming methods, end uses and benefits. Cefas Contract C7747, European Maritime and Fisheries Fund (EMFF) grant number ENG2805, p 78

Dillehay T, Ramírez C, Pino M, Collins MB, Rossen J, Pino-Navarro JD (2008) Monte Verde: seaweed, food, medicine, and the peopling of South America. Science 320:784–786. https://doi.org/10.1126/science.1156533

Duarte CM, Wu J, Xiao X, Bruhn A, Krause-Jensen D (2017) Can seaweed farming play a role in climate change mitigation and adaptation? Front Mar Sci 4:100. https://doi.org/10.3389/fmars.2017.00100

Forsythe W (2006) The archaeology of the kelp industry in the Northern Islands of Ireland. Int J Naut Archaeol 35(2):218–229. https://doi.org/10.1111/j.1095-9270.2006.00104.x

Hothersall S (2012) The Scottish kelp industry and its archaeology. Hist Argyll 2012:32–36

Huntington T, Cappell R (2020) English aquaculture strategy. Final Report. Produced by Poseidon Aquatic Resources Management Ltd for the Seafish Industry Authority, p 88

James MA (2010) A review of initiatives and related R&D undertaken in the UK and internationally regarding the use of macroalgae as a basis for biofuel production and other non-food uses relevant to Scotland. Report commissioned by Marine Scotland, p 79

Kerrison PD, Stanley MS, Hughes AD (2018) Textile substrate seeding of *Saccharina latissima* sporophytes using a binder: an effective method for the aquaculture of kelp. Algal Res 33:352–357. https://doi.org/10.1016/j.algal.2018.06.005

Marine Scotland (2016) Wild seaweed harvesting. Strategic environmental assessment. Environmental Report. November 2016, p 261

Mesnildrey L, Jacob C, Frangoudes K, Reunavot M, Lesueur M (2012) Seaweed industry in France. Report. Interreg program Netalgae. Les publications du Pôle halieutique Agrocampus Ouest 9, p 34

MMO (2019) Identification of areas of aquaculture potential in English waters. A report produced for the Marine Management Organisation by the Centre for Environment Fisheries and Aquaculture Science (Cefas), MMO Project No: 1184, p 107

Mouritsen OG (2013) Seaweeds edible, available and sustainable. The University of Chicago Press, Chicago/London, p 287

Natural Resource Wales (2018) Detailed guidance for seaweed harvesting – hand gathering. Guidance note, p 33

Naylor RL, Hardy RW, Buschmann AH, Bush SR, Cao L, Klinger DH, Little DC, Lubchenco J, Shumway SE, Troell M (2021) A 20-year retrospective review of global aquaculture. Nature 591:551–563. https://doi.org/10.1038/s41586-021-03308-6

Shama A, Joyce SG, Mari FD, Ranga Rao A, Ravishankar GA, Hudaa N (2019) Macroalgae and microalgae: novel sources of functional food and feed. In: Ravishankar GA, Ranga Rao A (eds) Handbook of algal technologies and phytochemicals, Food, Health and Nutraceutical Applications, vol I. CRC Press, USA, pp 207–219

Stanley MS, Kerrison PK, Macleod AM, Rolin C, Farley I, Parker A, Billing S-L, Burrows M, Allen C (2019) Seaweed farming feasibility study for Argyll & Bute. A report by SRSL for Argyll & Bute Council, p 190

West J, Calumpong HP, Martin G (2016) Seaweeds. In: Chapter 14 in The first global integrated marine assessment. World ocean assessment I. United Nations. Available at: http://www.un.org/Depts/los/global_reporting/WOA_RegProcess.htm

Wood D, Capuzzo E, Kirby D, Mooney-McAuley K, Kerrison P (2017) UK macroalgae aquaculture: what are the key environmental and licensing considerations? Mar Policy 83:29–39. https://doi.org/10.1016/j.marpol.2017.05.021

Chapter 15
Utilisation of Seaweeds in the Australian Market – Commercialisation Strategies: Current Trends and Future Prospects

Charu Deepika, Juliane Wolf, Navid Moheimani, Ben Hankamer, Brian von Herzen, and Ambati Ranga Rao

Abbreviations

AABW	Antarctic Bottom Water
ABARES	Australian Bureau of Agricultural and Resource Economics
AIMS	Australian Institute of Marine Sciences
ASI	Australian Seaweed Institute
CAGR	Compound Annual Growth Rate
CFIA	Canadian Food Inspection Agency
CRC-P	Cooperative Research Centres Projects
CSIRO	Commonwealth Scientific and Industrial Research Organisation
FSANZ	Food Standards Australia New Zealand
GHG	Green-House Gas
GVP	Gross Value Production
IRR	Internal Rate of Return
ISFR	Implementation Subcommittee for Food Regulation
JCU	James Cook University
KIRDO	King Island Regional Development Organisation

C. Deepika · J. Wolf · B. Hankamer
Institute of Molecular Bioscience, The University of Queensland, Brisbane, QLD, Australia

N. Moheimani
Algae R&D Centre, Harry Butler Institute, Environmental and Conservation Sciences, Murdoch University, Murdoch, WA, Australia

B. von Herzen
Climate Foundation, Woodford, QLD, Australia

A. R. Rao (✉)
Department of Biotechnology, Vignan's Foundation of Science, Technology and Research (Deemed to be University), Vadlamudi, Guntur, Andhra Pradesh, India
e-mail: arangarao99@gmail.com

A. Ranga Rao, G. A. Ravishankar (eds.), *Sustainable Global Resources of Seaweeds Volume 1*, https://doi.org/10.1007/978-3-030-91955-9_15

NADW	North Atlantic Deep Water
QACs	Quaternary Ammonium Cations
RIRDC	Rural Industries Research and Development Corporation
TELCA	Techno-economic and Life Cycle Analysis
UNSW	University of New South Wales

15.1 Introduction

Australia is a global hotspot for seaweed biodiversity and has a rich and diverse history of using seaweeds dating back at least 65,000 years, including an extensive traditional knowledge of Australian natural resources. References (Western science/non-traditional owners) on the identification and collection of seaweeds in Australia date back to 1840 and have continued since then (Glicksman 1987; Norman et al. 1988; Fitzgerald 1991; Blanche 1992; Butler et al. 2020; Kinley et al. 2020; Layton et al. 2020; Vranken et al. 2021). The variations in local growth conditions due to climate change have had a huge impact on geographic distribution of species worldwide (Zemke-White and Ohno 1999; Yuan et al. 2020). The richness of the Australian marine flora is attributed to a combination of the convergence of three ocean waters, including the Indonesian through flow, the Indian Ocean and the southern ocean along with the diverse temperature ranges found around the Australian continent including the reefs (Beardall and Raven 2004; Perin and Lean 2004). The rise in temperature due to global heating has been reported to have a negative effect on the native seaweeds of chlorophyta in the Great Barrier Reef (Fig. 15.1d) and is a serious threat to coral bleaching and the health of reef systems (Sala and Knowlton 2006; Burke et al. 2011; van Ginneken 2019; Webster and Gorsuch 2020). For example, *Macrocystis pyrifera* (Giant Kelp) populations are found to be shifting south to maintain their temperature ranges in a warmer ocean due to nutrient deprivation caused by rising water temperatures (McIntyre et al. 2009; Yuan et al. 2020). However, in Eastern Tasmania is no shoreline further south that provides a habitat for *Macrocystis* kelp. As a result, this population has diminished by 95% over recent decades (Johnson et al. 2011; Mabin et al. 2019). Preservation of marine biodiversity endangered by environmental stress is a key issue which urgently requires the introduction of effective policies and conservation targets in line with the guidance of the Intergovernmental Panel on Climate change (Rilov et al. 2020).

The Great Southern Reef reportedly has the richest marine flora in the world, with a biodiversity estimated to cover more than 50,000 species (includes several endemic species) of which only ~16,000 species have been identified to date (Table 15.1, Fig. 15.1a), yet not many of the commercial products in the Australian market are produced from such native species resources (Wernberg 2015).

The sustainable production of seaweed-based bioproducts from domestic Australian species offers significant potential to drive the regeneration of macroalgae (seaweed) biodiverse abundance by promoting the demand for seaweed mariculture around the coastal stretch of Australia (Winberg et al. 2009; Cabrita et al. 2016). This increasing demand in turn would support expansion of the seaweed

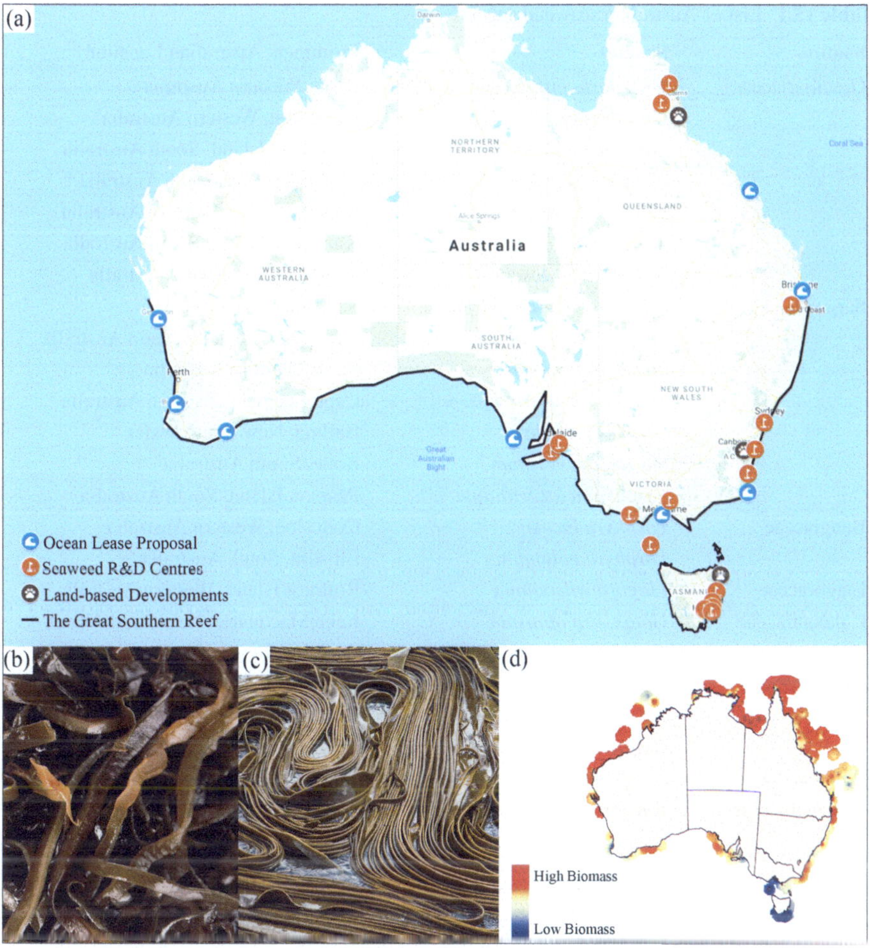

Fig. 15.1 Seaweed biodiversity in Australia. (**a**) The current state of the Australian seaweed industry (created in Google maps; indicating the location of ocean lease proposals, seaweed R&D centres/research hubs, land-based developments according to the Australian Seaweed Institute (ASI 2020) and The Great Southern Reef (thick black line; 8000 km) (Wernberg 2015). Examples of prominent Australian seaweeds in use – (**b**) *Sargassum fusiforme* and (**c**) *Durvillaea antarctica*. (**d**) Distribution of reef biodiversity in the reefs within the Australian coastline based on the survey done by Reef Life Survey (2010–2015) (SOE 2016)

sectors, R&D and product development from native Australian seaweed species to expand domestic and export markets and reduce reliance on imports (*Sargassum fusiforme* for biofertilizers and fucoidans, Fig. 15.1b). Currently, there is no specific central hub for seaweed collection or preservation in Australia, but seaweeds are mostly categorized as sub-projects in the Commonwealth Scientific and Industrial Research Organisation (CSIRO) and the Australian Institute of Marine Sciences (AIMS) (ASI 2020). Figure 15.1 summarises the current state of the seaweed

Table 15.1 List of Australian seaweed species

Family	Species	Prominent Australian Location
Graciliariaceae	*Gracilaria cliftonii*	Perth, Western Australia
	Gracilaria flagelliformis	Geraldton, Western Australia
	Gracilaria secundata	Kangaroo Island, South Australia
	Curdiea angustata	Encounter Bay, South Australia
	Curdiea irvineae	Cape Leeuwin, Western Australia
	Melanthalia obtusata	Kangaroo Island, South Australia
	Melanthalia abscissa	Wedge Island, South Australia
Sargassaceae	*Sargassum heteromorphum*	San Remo, Victoria
	Sargassum decipiens	Cape Naturaliste, Western Australia
	Sargassum varians	Perth, Western Australia
	Sargassum verruculosum	Cape Leeuwin, Western Australia
	Sargassum fallax	Ballina, New South Wales
	Sargassum vestitum	Robe, South Australia
	Sargassum lacerifolium	Pearson Island, South Australia
Bangiaceae	*Porphyra lucasii*	Cottesloe, Western Australia
	Porphyra columbina	Elliston, South Australia
Liagoraceae	*Liagora wilsoniana*	Rottnest Island, Western Australia
Galaxauraceae	*Galaxaura obtusata*	Lake Macquarie, New South Wales
	Galaxaura marginata	Freycinet Peninsula, Tasmania
Gelidiaceae	*Pterocladia lucida*	Coffs Harbour, New South Wales
	Pterocladia capillacea	Stradbroke Island, Queensland
	Gelidium australe	Perth, Western Australia
Polyidaceae	*Rhodopeltis australis*	Rottnest Island, Western Australia
Halymeniaceae	*Carpopeltis elata*	Archipelago, Western Australia
	Carpopeltis phyllophora	Phillip Island, Victoria
	Halymenia kraftii	Jervis Bay, New South Wales
	Codiophyllum flabelliforme	Albany, Western Australia
	Grateloupia turuturu	Bicheno, Tasmania
	Gelinaria ulvoidea	Houtman Abrolhos, Western Australia
	Polyopes constrictus	Sleaford Bay, South Australia
	Thamnoclonium dichotomum	Ballina, New South Wales
Kallymeniaceae	*Kallymenia cribrosa*	Flinders, Victoria
	Kallymenia tasmanica	Gulf St Vincent, South Australia
	Cirrulicarpus polycoelioides	Maria Island, Tasmania
	Polycoelia laciniata	Flinders, Victoria
	Callophyllis rangiferina	Champion Bay, Western Australia
	Callophyllis lambertii	Ceduna, South Australia
	Thamnophyllis lacerata	Gulf St Vincent, South Australia
Phyllophoraceae	*Stenogramme interrupta*	Nuyts Reef, South Australia
Nemastomataceae	*Tsengia feredayae*	Nuyts Reef, South Australia
Gigartinaceae	*Rhodoglossum gigartinoides*	Hamelin Bay, Western Australia
	Gigartina recurva	Recherche Bay, Tasmania
	Gigartina muelleriana	Port Phillip, Victoria

(continued)

Table 15.1 (continued)

Family	Species	Prominent Australian Location
Dicranemiaceae	*Dicranema revolutum*	Shark Bay, Western Australia
	Peltasta australis	Cape Woolamai, Victoria
Sarcodiaceae	*Sarcodia marginata*	Port Elliott, South Australia
Acrotylaceae	*Hennedya crispa*	Geraldton, Western Australia
Areschougiaceae	*Betaphycus speciosum*	Perth, Western Australia
	Callophycus dorsiferus	Cape Leeuwin, Western Australia
	Callophycus oppositifolius	Geraldton, Western Australia
	Callophycus harveyanus	Dongara, Western Australia
	Erythroclonium sonderi	Houtman Abrolhos, Western Australia
	Areschougia congesta	Hamelin Bay, Western Australia
Plocamiaceae	*Plocamium angustum*	Ceduna, South Australia
	Plocamium dilatatum	Victor Harbor, South Australia
	Plocamium patagiatum	Great Australian Bight, South Australia
	Plocamium mertensii	Nickol Bay, Western Australia
	Plocamium preissianum	Wilsons Promontory, Victoria
	Plocamium cartilagineum	Shark Bay, Western Australia
Phacelocarpaceae	*Phacelocarpus peperocarpus*	Esperance, Western Australia
Cystocloniaceae	*Craspedocarpus blepharicarpus*	Phillip Island, Victoria
	Craspedocarpus venosus	Fremantle, Western Australia
	Rhodophyllis multipartita	Gabo Island, Victoria
Mychodeaceae	*Mychodea aciculare*	Cape Riche, Western Australia
Hypnaeaceae	*Hypnea ramentacea*	Dongara, Western Australia
Bonnemaisoniaceae	*Asparogopsis armata*	Perth, Western Australia
	Asparogopsis taxiformis	Spencer Gulf in South Australia
	Delisea pulchra	Ballina, New South Wales
	Delisea plumosa	Port Davey to Bicheno, Tasmania
	Ptilonia australasica	Williamstown, Victoria
	Ptilonia subulifera	Walkerville, Victoria
Corallinaceae	*Amphiroa anceps*	Bicheno, Tasmania
	Amphiroa gracilis	Kalbarri, Western Australia
	Corallina officinalis	Tasmania
	Haliptilon roseum	Shark Bay, Western Australia
	Arthrocardia wardi	Norah Head, New South Wales
	Cheilosporum sagittatum	Perth, Western Australia
	Metagoniolithon stelliferum	Wilsons Promontory, Victoria
	Metagoniolithon radiatum	Dongara, Western Australia
	Spongites hyperellus	Western Port, Victoria
	Synarthrophyton patena	Rottnest Island, Western Australia
	Metamastophora flabellata	Kalbarri, Western Australia
	Mastophoropsis canaliculata	Encounter Bay, South Australia
	Phymatolithon masonianum	Cape Buffon, South Australia
	Sporolithon durum	Rottnest Island, Western Australia

market in Australia highlighting the regions in which R&D centres, land- and ocean-based development sites are located. The current research and development projects listed by the Australian Seaweed Institute (ASI) (ASI 2020) for example include (1) the development of *D. potatorum* farms (Bull Kelp) (in collaboration with Tassal Pty Ltd. and the University of Tasmania) (Fig. 15.1c); (2) the Australian Aquatic Plant Names Standard – FRDC project; (3) Harnessing seaweed genes to mitigate methane emissions from livestock (USC – ARC Discovery Project); (4) Operation Crayweed (UNSW restoration project, Sydney); (5) Seaweed Farming for SDGs Workshops; and (6) Anti-cancer properties and Cosmeceuticals from red algae (Griffith University, Queensland). Larger consortia grouped in Cooperative Research Centres (CRCs) that are currently working on native Australian seaweeds include the Blue Economy CRC, Future Food Systems CRC, Northern Australia CRC, Marine Bioproducts CRC and Coastal Communities CRC. The Climate Foundation has also conducted a multi-year study of temperate seaweed production rates in higher nutrient zones in collaboration with the University of Tasmania over the past 3 years and the ability to use deepwater irrigation of seaweed cultivation to rescue production presently of tropical red seaweeds and potentially with giant kelps in the future. This collaboration has resulted in a reservoir of warm-tolerant genotypes of *Macrocystis* at the University of Tasmania developed with the support of the Climate Foundation and other organisations. In the tropics, work has focused on restoring natural upwelling to ensure replete macrophyte production rates. This highlights that clearly more research and development is required in areas such as seaweed cultivation, product development, packaging and storage to progress technology development in the global race to supply the growing seaweed market. Two of the youngest companies in the Australian seaweed market include CH_4 Global Pty Ltd. in South Australia and Sea Forest Pty Ltd. in Tasmania. Kelp Industry Pty Ltd situated on King Island, Tasmania is considered to be a significant exporter of bull kelp (*D. potatorum*) for development of bio-fertilizers (KIRDO 2020). According to the Australian Bureau of Statistics (ABS) seaweed exports from Australia for use in fertilizers and cosmetics are valued at AUD $1.5 million.

15.2 Australian Seaweed Products

15.2.1 Food

There are several seaweed food products on the Australian market that have been popular for over a decade (Fig. 15.2). They are consumed as snacks, superfoods and garnishings infused in several authentic foods (Fleurence et al. 2018). This includes Golden Kelp Granules (100% Australian Sea Kelp Granules, Golden Kelp), Golden Kelp Powder and Smoked Golden Kelp Granules from SeaHealth Products Pty Ltd. Most of the seaweed food products available in Australia are imported. One of the first domestic manufacturing facility for food-grade seaweed products was launched in May 2020 on the New South Wales South Coast using edible seaweeds from the

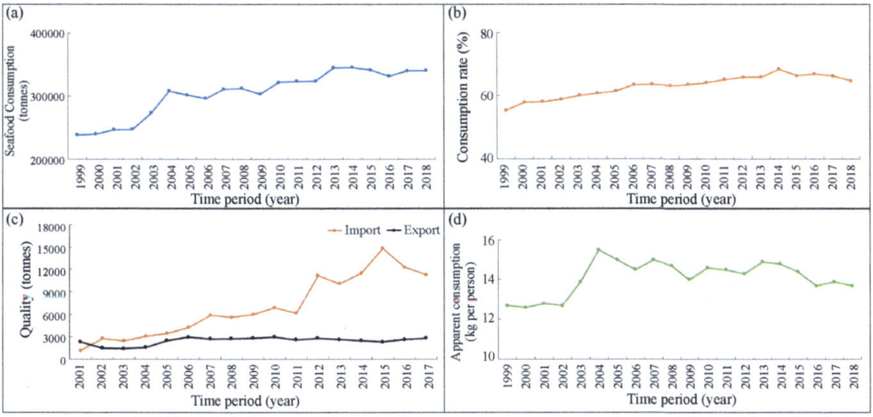

Fig. 15.2 Apparent consumption of seaweeds, Australia (1998–99 to 2017–18) – (**a**) Seafood consumption including seaweeds (tonnes) over time (Department of Agriculture 2019). (**b**) Imported seaweed consumption rate (%) over time (Department of Agriculture 2019). (**c**) Seaweed import and export – Australia from 2001–2017, data from Australian Bureau of Agricultural and Resource Economics (ABARES) (ABARES 2020). (**d**) Apparent consumption over time (Department of Agriculture 2019)

family of Chlorophyta e.g. *Undaria pinnatifida*, *Caulerpa lentillifera* (Glicksman 1987, Lee and Momdjian 1998).

Market analysis has shown that in the past 5 years, food-grade seaweeds have experienced an expanding market as 70% of the Australian population is reported to focus more on health-conscious diets; women were found to be more health-conscious than men (Birch et al. 2019b). Seaweeds are considered a functional food that is reported to offer a number of health benefits (Tiwari and Troy 2015). Traditionally, First Nations communities in Australia regularly consumed seaweeds which were reported to improve digestive track and bone health, as well as preventing chronic diseases, such as cardiovascular disease, cancer, diabetes, dysmenorrhea and metabolic syndrome (Hughes et al. 2018). Several Australian species were also considered highly nutritious (Tanna and Mishra 2019). Specifically, seaweed species from Western Australia have been found to be non-toxic, rich in antioxidants, proteins, carbohydrates, phenols, vitamins, minerals and contain beneficial micronutrients (Maeda et al. 2008; Marinho-Soriano et al. 2009; Fiedor and Burda 2014). Major considerations that make seaweeds attractive to the food-based seaweed industry are that macroalgae are a good source of protein (12–38.7%), essential amino acids and dietary fibres (25–75%), which are claimed to support the prevention of colon cancer, constipation, hypercholesterolemia and obesity (Fertah et al. 2017; Huisman 2000). Examples of such seaweed candidates include *Padina fraseri* (brown seaweed, Victoria), *Dilophus marginatus* (Noosa, Queensland), *Dictyopteris acrostichoides* (Warrnambool, Victoria), *Dictyopteris australis* (Dampier, Western Australia) and *Padina elegans* (Dongara, Western Australia and Pearson Island, South Australia) (Edgar 1997; Phillips et al. 1999; Huisman 2000).

Bioactives in seaweeds are also reported to exhibit a range of properties such as anti-hypertensive, anti-oxidant, anti-thrombotic, anti-microbial, and immunomodulatory traits (Huisman 2000). However, the validity of these claims will have to be verified to ensure that the products marketed on this basis meet appropriate food safety, nutraceutical, cosmeceutical and cosmetic standards.

In addition, their hydrocolloids such as alginates or carrageenan have been marketed as thickening and gelling agents with multiple applications in different food products such as ice creams, jellies and shakes. These have been reported to have additional benefits including anti-inflammatory, anti-coagulant, antioxidant, anti-proliferative, and immuno-stimulatory activities (Table 15.2) (Eom et al. 2011; Fiedor and Burda 2014; Deepika 2019). *Laminaria spp.* and *Laminaria japonica* from Tasmania are used as a source for laminarin and dietary fibre (Barbot et al. 2015; Fertah et al. 2017; Doh et al. 2020).

There are also some major reported concerns and risks associated with seaweed consumption including allergic reactions either caused by the seaweed themselves (kainoids) or by associated contaminants or pathogens like bacteria, fungi or viruses that are not destroyed during processing or heat treatment steps (i.e. in the case of fresh use in salad dressings (Afshin et al. 2019; Murray et al. 2020)). Food Standards Australia New Zealand (FSANZ) regularly reviews and examines product safety by assessing biochemical composition and is responsible for acting on any issues regarding regulatory compliance (Van der Spiegel et al. 2013). Barriers for the growth of food-grade seaweed products are taste and colour. Seaweeds have different sensory profiles and they respond differently when cooked or processed (Campbell et al. 2020; Arioli et al. 2021). The addition of seaweed extracts to traditional dishes does not negatively impact flavour and was also found to improve texture, appearance and the colour of foods. In contrast, the use of dried macroalgae in baked foods such as pies and biscuits, where the key drivers are the colour and smell, are less appealing to consumers due to their effect on colour, odour and taste.

Other parameters that affect the seaweed-based food market include relatively low product availability and affordability (Andrade et al. 2021). This negative impact was substantially lowered by the introduction of seaweed snacks (Figueroa et al. 2021). Seaweed snacks available in Australia include Abaskus Seaweed crisps, Ceres roasted seaweeds, Honest Sea salted nori chips, Coles roasted seaweeds and rice crackers, Doldori seaweed chips and others.

To ensure consumer safety moving forward, local regulatory bodies are engaged in the monitoring of health benefits of specific products from seaweeds that are to be approved for human consumption and use. There are variations in policies and protocols between countries and they are enforced by different entities, e.g. Food Standards Australia New Zealand (FSANZ), Implementation Subcommittee for Food Regulation (ISFR; subcommittee Australian and New Zealand), Food Safety and Standards Authority of India, Centre for Food Safety – Hong Kong, European Food Safety Authority and Canadian Food Inspection Agency (CFIA). Clearly, the nutritional profiles and health claims related to the food products will have to be scientifically substantiated prior to market release, safety is given supreme importance. However, classifications of these guidelines will likely remain country

Table 15.2 Examples of seaweeds used in the food industry

Seaweed	Common name in market	Dietary fiber (g/100 g wet weight)	Food applications	References
***Laminaria* sp.**(*e.g. Laminaria japonica*)	Aokombu, Cow's tail, Horsetail kelp, Kombu, Oarweed kelp, Sea girdles, sea rod/s, Sea tangle, Wild kelp, Split-blade kelp.	TF = 6.2 SF = 5.4 IF = 0.8	Salads, soups, fried snack, sauces, Sushi, beverages (tea).	Kolb et al. (2004), Yan et al. (2004), Shirosaki and Koyama (2011), and Chen et al. (2018)
***Undaria* sp.**(*e.g. Undaria pinnatifida*)	Apron-ribbon vegetable, Sea mustard, Wakame.	TF = 3.4 SF = 2.9 IF = 0.5	Miso soup, salads	Kolb et al. (2004)
***Sargassum* sp.**(*e.g. Sargassum fusiforme*)	Binder's Sargassum weed, Common kelp, Gulf-weed, Deer tail grass, Sheep-nest grass.	TF = 49.2[a] SF = 32.9[a] IF = 16.3[a]	Vegetable soup, stir fries	Chen et al. (2012), Yende et al. (2014), Deepika (2017), Husni et al. (2019), Ibrahim et al. (2020), and Liu et al. (2020b)
***Ulva* sp.**(*e.g. Ulva lactuca*)	Sea lettuce, Green Laver.	TF = 3.8 SF = 2.1 IF = 1.7	Stews, salad dressings, nutrient supplements	Nagaoka et al. (1999), Shibata et al. (2000), and Mise et al. (2011)
***Alaria* sp.**(*e.g. Alaria esculenta*)	Atlantic wakame, Bladderlocks, Drilly kelp, Irish wakame, Stringy kelp.	–	Salads, vegetable	Stevant et al. (2017), Blikra et al. (2019), and Afonso et al. (2021)
***Porphyra* sp.**(*e.g. Prophyra umbilicalis*)	Nori, Amanori, Gim.	TF = 3.8 SF = 3.0 IF = 1.0	Vegetable, garnish, salads, pickles	Eom et al. (2011), and Lehner et al. (2016)
***Ecklonia* sp.**(*e.g. Ecklonia cava*)	Common kelp, leather kelp, paddle weed, sea bamboo, Kajime, kamtae	TF = 43.1[a] SF = 8.4[a] IF = 34.7[a]	Vegetable, garnish, salads, pickles	Rengasamy et al. (2013), and Kulkarni et al. (2019)
***Fucus* sp.**(*e.g. Fucus vesiculosus*)	Arctic wrack, Red fucus, Black-tang, Bladder fucus, Bubble kelp, Lady wrack, Spiraling rockweed.	–	Food additives, flavourings, food supplements	Diaz-Rubio et al. (2009), Torres-Escribano et al. (2011), and Andre et al. (2020)
***Durvillaea* spp.**(*e.g. Durvillaea Antarctica*)	Southern bull-kelp, Cape kelp.	TF = 70.1[a] SF = 27.7[a] IF = 43.7[a]	Stews, salads	Spolaore, Joannis-Cassan et al. (2006), Kuda et al. (2007), and Uribe et al. (2018)
***Ascophyllum* spp.**(*e.g. Ascophyllum nodosum*)	Knobbed wrack, Pigweed, Rockweed, Sea whistle.	TF = 8.8 SF = 7.5 IF = 1.3	Alginate, seaweed meal	Gollety et al. (2010), Kadam et al. (2015), and Ronan et al. (2017)

TF total fiber, *SF* soluble fiber, *IF* insoluble fiber
[a]denotes values in % dry weight

specific. For example, the most common brown seaweed *Sargassum fusiforme*, generally called as *hijiki* which is used in sushi and salads, is banned in Australia due to high arsenic content, but has no restrictions in European countries. Apart from the heavy metals, there are specific species import and export restrictions in Australia (Fitzgerald 1991; Weigle et al. 2005).

15.2.2 Biofertilizers and Biostimulants

The application of seaweeds as commercial fertilizers by western settlers began around the twentieth century in Australia whereas the United Kingdom has been using them since at least the eighteenth century (Norman et al. 1988). Commonly used seaweed liquid fertilizers are derived from the aqueous extracts of *Ecklonia maxima, Durvillaea potatorum* and *Macrocystis pyrifera* (Ibrahim 2013; Amin et al. 2020). The approximate cost of a liquid seaweed fertilizer is between AUD \$10–30 L^{-1} and has a recommended application rate of 0.5–20 L ha^{-1} (refers to undiluted Seasol) depending on the crop and its development stage (Proffitt and Campbell-Clause 2012). The seaweed liquid extracts were previously extracted through a cold cell burst method after incubation of the bulk seaweed biomass at low ambient temperatures for ~3 days. More recently other extraction technologies to speed up the recovery of metabolites have been developed (e.g. acid/alkaline hydrolysis (Sultana et al. 2018; Soares et al. 2020)). The chemical composition of diverse Australian seaweeds has been increasingly studied over the last two decades. For example, the European Union (EU) since 2017 has categorized liquid seaweed extracts as biostimulants (with a stricter regulatory framework) which increased their product cost and market value. Biostimulants are defined as either liquids or soluble powders that alter biosynthesis pathways in plants to modulate nutrient uptake and use mechanisms, or enhance resistance, to abiotic stress (salt stress, temperature, moisture) which in turn increases crop yield and quality (Fig. 15.3) (Vessey 2003; Amin et al. 2020). Seaweed-derived biostimulants are reported to have naturally high levels of polyphenols, phytohormones (cytokinins and auxins), mannitol, amino acids and antioxidants (Uthirapandi et al. 2018). Specific primary and secondary metabolites in the seaweeds have been shown to provide resistance against biotic stress (pests and pathogens such as insects, nematodes, fungi, bacteria and viruses) which has mainly been attributed to the origin of seaweeds (sea/oceans; highly saline environment) (Sultana et al. 2018; Uthirapandi et al. 2018).

Tasbond Pty Ltd (Launceston, Tasmania) was an early company in Australia that launched organic fertilizers from seaweed including two popular products that have been on the market for over the past 27 years (Seasol International Pty Ltd.) – 'Seasol' and 'Powerfeed'. Seasol is derived from a mixture of brown kelp extracts from different parts of the world, and include – *Durvillaea potatorum* (Bull kelp, King Island, Tasmania), *Durvillaea Antarctica* (Bull Kelp, imported from Chile)

Fig. 15.3 Benefits of liquid seaweed extracts in tomato farming (Uthirapandi et al. 2018). These are reported to reduce the effect of heat stress (Plant care today 2021), promote flowering (Nick 2017), increase fruit quality (Smith 2021), protect from predators (Heather Rhoades 2020), provide disease resistance (Wiley 2021), enhance germination (Biospin 2018), improve root nodulation (Heather Rhoades 2021) and promote symbiotic relationships (XiaoZhi 2021)

and Knotted Kelp *Ascophyllum nodosum* (North Atlantic Ocean, Canada). The seaweed fertilizer company 'NATRAkelp Pty Ltd.' (https://www.natrakelp.com/) based in Queensland, was launched around the 1980s and manufactures a range of products using different kelp varieties available in north-west Queensland. NATRAkelp products use a fermentation technology to optimize the preservation of alginate in kelp and supplement the final product with trace metals to achieve higher crop yields. Some of the other fertilizer companies in Australia include Vitec Natural Australia, Vitec Organics Pty Ltd., Natrasol No Frills Fertiliser and Seaweed Enterprises Australia Pty Ltd.

In Queensland, Australia, Seasol™ is widely used in tomato production and claims to enhance symbionts (by providing essential nutrients to the soil microbiota), to revitalise nutrient composition of soil, to prevent nutrient leaching, to aid in faster root structure differentiation (alginate has high water holding capacity that retains moisture) and plant development (leaf and flowers), and to improve stress/disease resistance (Blanz et al. 2019). Furthermore, sulphated polyunsaturated fatty acids (rich in brown seaweeds), steroids and plant growth hormones are also secreted in seaweeds (Uthirapandi et al. 2018). For example, seaweed extracts from *Laminaria* are undergoing commercialisation for their quaternary ammonium cations (QACs) that provide buffering capacity for the roots, and act as osmoprotectants that help agriculture in marshy lands around Australia. Tropical red seaweeds are commonly used in the Philippines as foliar biostimulants, but such application in Australia has yet to be substantially developed. Despite the many red seaweeds that are native to Queensland, Northern Territories and Western Australia, this area seems to be relatively unexplored (Hurtado et al. 2021; Naughtin et al. 2021; Pascual et al. 2021).

Enhanced germination rates, increased soil aeration, faster root and shoot growth, and higher chlorophyll yields compared to other commercial fertilizers widely used within Australia, were recorded using seaweed fertilizers (Nkemka and Murto 2012). Thus, Australian seaweed extracts have gained increased market value and research attention for their use in the agriculture sector (HexaResearch 2020). The gaps in knowledge to be addressed include crop-specific dose-response studies, local seaweed strain-based effectiveness in different geographical locations, genetic studies on toxins and specific growth hormone associated genes (Nabti et al. 2017). Furthermore, it is critical to understand that the efficiency and shelf-life of liquid seaweed fertilizers are variable and depend on the source, extraction treatment and final product formulation. Timeframes for gaining approvals and product commercialisation of seaweed derived biostimulants are favourable due to established and rapid testing methods, and the fact that imported products (e.g. *Ascophyllum nodosum* extract as a biostimulant) are already being extensively used in Australian horticulture (Weigle et al. 2005; Nayar and Bott 2014; Blanz et al. 2019). The expansion of local seaweed-based fertilizers could favour both the agriculture sector and seaweed mariculture development if overall productivity and sustainability are improved (Blunden and Wildgoose 1977; Nabti et al. 2017; Herforth et al. 2020). Conducting integrated Techno-economic and Life-Cycle Analysis (TELCA) on such products is important (Rupawalla et al. 2021).

15.2.3 Cosmetics

Most of the brown seaweed (Phaeophyta) extracts are used in the development of cosmetics due to their richness in amino acids, lipids, polyphenols, terpenoids and pigments (carotenoids) which are reported to confer anti-oxidant, anti-inflammatory and anti-aging properties (Patel et al. 2020). More importantly, extensive toxicology studies have characterised the safety of specific production lines to ensure safety for cosmetics products and increase consumer acceptance (Yokoi et al. 2008; Saez et al. 2015).

The Australian beauty and personal care products market is forecast to reach AUD $8.64 billion by 2025. The current key players in the seaweed-based cosmetic industry are Lush Cosmetics, PhycoHealth, Beeseline, Delizioso Skincare, and Dr. Emerald Skincare. The seaweed-based bioactives are widely infused in different cosmetics for their range of reported benefits including anti-aging properties, skin whitening by inhibiting melanin secretion, and moisturising using hydrocolloids (Pimentel et al. 2018). The hydrocolloids derived from brown and red seaweeds such as agar, alginates and carrageenans are used in skin serum, shampoos, lotions and creams as gelling agents (Fertah et al. 2017; Rehm and Moradali 2018). Carrageenan extracts from *Chondrus crispus* are used as gelling agents (Bedoux

Fig. 15.4 Seaweed-based cosmetics. (**a**) Four commonly used seaweeds for development of cosmetics in Australia – *Ascophyllum nodosum, Halidrys siliquosa, Undaria pinnatifida* and *Chondrus crispus* (Edgar 1997; Phillips et al. 1999; Huisman 2000). (**b**) Examples of seaweed-based cosmetic products in the Australian market from different brands (included with the permission from respective brands)– PhycoHealth (PhycoHealth 2021), Ren Clean Skincare (Parsons 2017) and OceanWell (OceanWell 2021). (Image courtesy – PhycoHealth Pty Ltd., Ren Clean Skincare Pty Ltd. and OceanWell Pty Ltd.)

et al. 2014; Mourelle et al. 2021). More recently cosmetic products were developed and commercialized successfully with key ingredients being seaweed-based metabolites (e.g. Lush Cosmetics Pty Ltd). Australia is well-known for fresh handmade moisturizer for cosmetics markets (Skin Shangri La) and face mask (BB Seaweed) produced from dried kelp (*Laminaria sp.*) imported from the UK (Fertah et al. 2017; Chen et al. 2018). The presence of minerals, trace elements and essential vitamins in the brown seaweed in addition to their hydrocolloids was also reported to promote the texture of the skin and moisturise it (Fig. 15.4). The excellent water holding capacity of the hydrocolloids together with the presence of polyphenols is designed to help retain the skin elasticity to prevent the appearance of fine lines or wrinkles and help reduce aging-related skin texture and discoloration (Chakraborty et al. 2009; Deepika et al. 2016; Vega et al. 2020). Continuous exposure to sunlight increases melanin secretion due to the damage caused by the UV-radiation resulting in skin tanning (Sathasivam and Ki 2018). Seaweed-derived pigments such as fucoxanthin from *Laminaria japonica* are reported to act as tyrosinase-inhibitor and provide protection against the harmful UV rays (Chen et al. 2018).

PhycoHealth Pty Ltd. in New South Wales markets a range of different skincare products derived from Kelp which include SEAFIBER-3 Jar (Docosohexanoeic acid rich omega-3 capsules), PhycoDerm® NOURISH daily moisturiser, PhycoDerm® SOOTHE seaweed gel serum, PhycoDerm® SHIELD daytime barrier cream, Seaweed clay mask, PhycoDerm® CLEANSE. In 2018, MECCA Pty

Ltd. (MECCA 2018) launched seaweed-based cosmetics which include - Atlantic Kelp and Microalgae Bath Oil (developed from Kelp extract), Elemis Pro-Collagen Hydra-Gel Eye Masks (brown seaweed, *Padina pavonica* extract infused with hyaluronic acid) (Packer 2017), Hyaluronic Marine Dew It All Eye Gel (elasticity-building Japanese seaweed extract), La Mer's Crème de la Mer - Miracle Broth (developed using fermented kelp) and Skyn Iceland's Berry Lip Fix (developed from Hawaiian red macroalgae) (Bedoux et al. 2014; Pimentel et al. 2018; Lourenco-Lopes et al. 2020).

15.2.4 Wastewater Treatment

One of the main issues associated with agriculture (e.g. abattoir, piggeries and dairy) as well as the aquaculture industries is wastewater treatment (Moheimani et al. 2018). Inadequate wastewater treatment can cause significant environmental issues. Most of the currently existing conventional wastewater treatment processes are relatively expensive and energy intensive (Chuka-ogwude et al. 2020a). Algae have the ability to take up inorganic nutrients such as nitrogen and phosphorus with several studies describing the suitability of microalgal-based wastewater treatment approaches (Chuka-ogwude et al. 2020b; Matos et al. 2021; Vadiveloo et al. 2021). Freshwater macroalgal cultures have also been proposed as a way to treat domestic (Liu et al. 2020b) and agricultural (Nwoba et al. 2017) effluents. Seaweeds have been shown to provide a method to treat intensive marine aquaculture wastewater (Castine et al. 2013). Many intensive aquaculture industries generate and release high concentrations of nutrients, which if not treated, can cause serious eutrophication of coastal waters and negatively affect the ecosystem by generating toxic algal bloom and in due course, hypoxia. A series of characteristics are required when choosing algae strains for effluent treatment: (a) local isolation; (b) wide geographical distribution; (c) fast growth rate (Wolf et al. 2015; Sivakaminathan et al. 2018); (d) tolerance across a wide range of conditions (i.e. salinity, nutrients); and (e) reliable and stable year-round culture (Chuka-ogwude et al. 2020b). Previous studies have indicated that macroalgae can store high concentration of nitrogen (He et al. 2008). Freshwater and seawater macroalgal species tested for wastewater treatments include Rhodophyta (*Gracilaria* spp., *Chondrus* sp., *Palmaria* sp.) and Chlorophyta (*Ulva* spp., *Chladophora* spp. *Oodogonium* spp., *Rhizoclonium* spp.) (Nwoba et al. 2017; Arumugam et al. 2018). Ultimately a sustainable aquaculture industry requires an appropriate ecosystem management approach to maintain the natural interactions of species and allow the ecosystem to function sustainably (Bloomberg 2020). In aquaculture (e.g. fin-fish farms and prawn farms) the generated N- and P-rich wastewater can be treated by associated seaweed cultivation to support the development of sustainable mariculture. The seaweed produced as a by-product can generate additional revenue streams in the form of animal feeds or fertilizer (Michalak and Mahrose 2020). Australia's in-land aquaculture industry is also expanding rapidly. Providing suitable quality control mechanisms are in place, an integrated seaweed wastewater treatment approach combined with revenue generation from grown biomass offers a path towards more socially and economically feasible processes.

15.3 Australian Seaweed Market: Imports and Commercialisation Targets

Currently, Australia imports around 5000 tonnes of seaweed annually (value AUD $14 million) in both dried and frozen forms for use in the food manufacturing industries. The frozen forms of seaweed have a higher value of about AUD $8 wet-kg^{-1} compared to dried seaweeds (AUD $5.4) (Winberg et al. 2009; Bogard et al. 2019). Dried seaweed is mostly imported from Ireland (~4000 tonnes in 2007) while most of the frozen products were imported from Asian countries (~280 tonnes in 2007). With advancements in bioprocess technologies, Australian industries have focused on pharmaceuticals and nutraceuticals as key sectors (Table 15.3). Marinova Pty Ltd. based in Tasmania produces organic fucoidan and different custom seaweed extracts. Their key products are derived from *Undaria pinnatifida* and *Fucus vesiculosus*. Biorefinery approaches designed to produce multiple products from biomass also offer a significant opportunity (HexaResearch 2020) and could connect into food, feed, biochemical and biofertilizer markets. The aquaculture industry (fish farming) is one such model to assure high-value, high-tech economic opportunity (Kim 2018), which has the potential to improve environmental sustainability. However, significant work is required to optimise such biorefineries and validate them using techno-economic and life-cycle assessments to ensure the economically, socially and environmentally friendly production of food, cosmetics, aquaculture feeds, biofuels, bio-fertilizers and pharmaceuticals (Sharma and Sharma 2017; Carus and Dammer 2018; Mohammad et al. 2019; Roles et al. 2020; Rupawalla et al. 2021).

Table 15.3 Required time for commercialisation of seaweed-based products in various sectors in the Australia market (Nayar and Bott 2014; Birch et al. 2019a; Hessami et al. 2019; Forecast 2020; HexaResearch 2020; MordorIntelligence 2021)

Australian market sectors	Market value	Commercialisation time	
Food	Very high	2–5 years	Short
Bio-products/organics	High	3–5 years	Medium
Cosmetics	High	4–5 years	Medium
Animal or aquaculture feed	High	1–3 years	Short
Biofuels	Low	3–5 years	Medium
Bio-fertilizers	High	1–3 years	Medium
Pharmaceuticals	Very high	10–15 years	Long

The demand for healthier foods, especially foods that address nutritional deficiencies and demand for organic or environmentally sustainable production methods are the main reasons for the success of the seaweed-based food market in Australia (Birch et al. 2019b; Bogard et al. 2019). The finished or processed products such as seaweed salads with dressing and spices is currently a popular product in seafood retail outlets in both cities and regional areas of Australia (Garcia-Poza et al. 2020). The retail price of this frozen product imported from Japan is currently sold at approximately AUD $18 wet-kg^{-1}, which is comparable to the lower end value of fish products from an aquaculture system (Fig. 15.5). Consequently, there is significant potential to market native Australian seaweeds instead of relying on imports, but domestic manufacturing facilities are still limited due to strict quality control and food safety regulations governing cultivation and processing, as well as

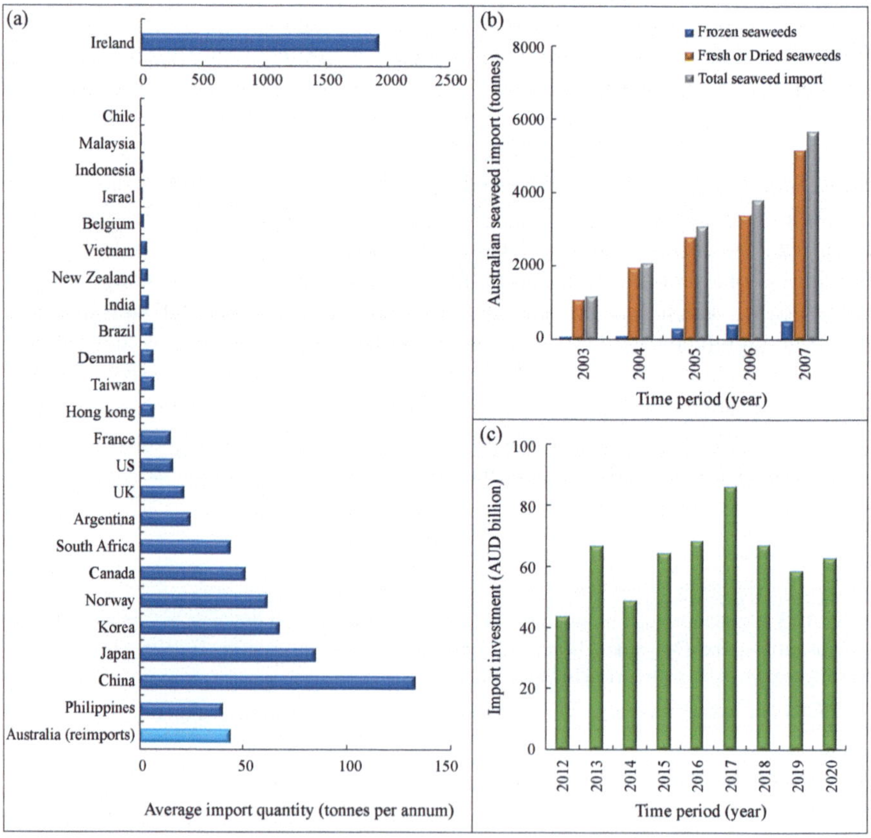

Fig. 15.5 Australian seaweed imports. (**a**) Australian seaweed imports according to country between 2003 and 2007 (Winberg et al. 2009). (**b**) Amount of seaweed imported in frozen and dried forms between 2002 and 2007 (Ferdouse et al. 2018). (**c**) Trend in seaweed import from Vietnam to Australia between 2010 and 2020. (ASI 2020) (Data adapted from – ABARES, Australia)

environmental standards, consumer acceptance and market competitiveness (Buck and Buchholz 2004; Gosavi et al. 2004; Daume 2006; Lopes et al. 2020). The key market determinants in the seaweed sector include price, quality, the products organic/sustainable production profiles for niche markets, carbon credits, application diversification and associated production risks and market volatility. The potentials for high end markets such as pharmaceuticals will open new opportunities and competitiveness in food production (coproduction for multiple product extraction) and help unlock local seaweed species potential (Pereira et al. 2020). Australia also has the opportunity to build on its strong national food and product brand. The Made in Australia label might also be used to ensure nationally defined quality standards especially considering the low dissolved metals contents of offshore good pristine Australian waters.

15.4 National Seaweed Industry Strategy in Australia

AgriFutures Australia formerly known as the Rural Industries Research and Development Corporation (RIRDC, launched in 1990) recently funded a blueprint for an Australia-based emerging seaweed industry. The primary aim of the National Seaweed Strategy is to create new coastal jobs, protect and regenerate waterways and contribute to greenhouse gas emissions reduction (Fig. 15.6). The current global market for seaweed products such as foods, cosmetics, nutraceuticals, animal feed and fertilizer are estimated at over USD $11 billion and it is expected to increase

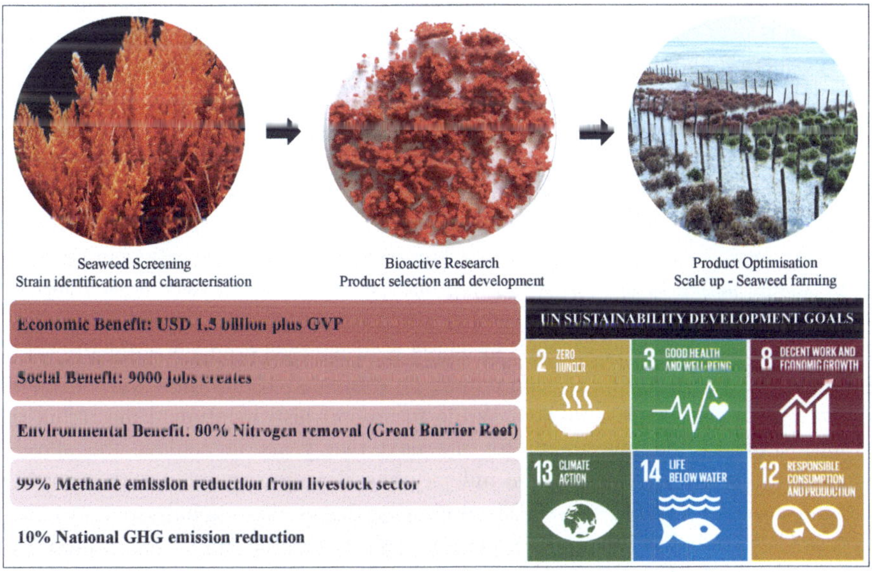

Fig. 15.6 Typical strategy to develop high value and sustainable macroalgae products. Examples of economic, social and environmental benefits (e.g. National seaweed strategy 2040 – Australian Seaweed Institute) (Kim 2018; Azzopardi 2019; Emont 2020; Roque et al. 2021)

two-fold by 2025 (Hepsy 2020). Australia has ideal growing conditions (a vast ocean estate and sunlight), a strong food and health brand, which supports export opportunities for high-value bio-products. Local researchers, entrepreneurs and the emerging sector would however benefit from a strategic industry development plan (AgriFutures 2020).

In response, the recent strategy developed by AgriFutures, Australia claims to grow the Australian seaweed industry to at least USD $100 million by 2025, and further to over USD $1 billion by 2040 (Kelly 2020). Their proposed action plan incorporates a series of steps targeting issues to be addressed on the way. These include an extensive literature review of native strains, identifying the gaps and barriers to market, prioritizing the seaweed industries research needs and product focused funding, mechanisms to improve nation-wide collaborations and engage stakeholders with a clear roadmap of the commercialisation pathway, industry leadership opportunities, production scale-up and innovation. Their Australian Seaweed Industry target for 2025 is AUD $100 million, 1200 direct jobs and a domestic greenhouse gas emissions reduction of 3% (Kelly 2020). The initiative targets reaching a USD $1.5 billion industry by 2040, creating 9000 jobs and to achieve a 10% emissions reduction target (Kelly 2020).

The National Seaweed Industry Strategy was published as a part of a report of the Australian seaweed industry blueprint, which was aimed to influence Federal and State Government aquaculture and environmental policy and regulatory departments. The report highlights the potential seaweed mariculture locations in Australia which includes South Australia, Tasmania, Southern Western Australia, Southern NSW and regional Queensland (Britton 2020). The report promotes the Australian seaweed, *Asparagopsis taxiformis* and *Asparagopsis armata* as an example, which was reported to reduce the methane emissions from livestock. As a greenhouse gas, methane is over 80 times more potent than carbon dioxide in trapping heat over a 20-year period and is responsible for nearly a quarter of global warming. In 2014, researchers at the Commonwealth Scientific and Industrial Research Organisation (CSIRO) and James Cook University (JCU) reported that feeding cattle with a seaweed-incorporated diet, reduced their methane enteric emissions by ~99% compared to current feeding practices. Among the twenty different edible seaweed studies, *Asparagopsis* spp. (red macroalgae) was found to have the most potency. However, others, such as the giant kelp *Macrocystis pyrifera*, actually can produce more of the active bromoforms, but retains less in its tissues than *Asparagopsis*. Translating this discovery into a global climate solution will require large-scale cultivation facilities for production of whole seaweed feed products. Globally, ~1.5 billion cattle and 1.2 billion sheep require an estimated 100 million wet tonnes of *Asparagopsis taxiformis* annually. While the report also mentioned the influx of around $8.1 million investment for the next two years to achieve the above-mentioned goals (Kelly 2020), the industry has already responded with a market capitalisation exceeding AU$100M combining the market capitalisations of Sea Forest, CH4 Global and Symbrosia along with the rest of the *Asparagopsis* industry, not to mention the FutureFeed business models.

15.5 Ocean Stratification in Australia

Ocean stratification globally has increased about 5.3–5.8% between the period 1960–2018 (0.9% per decade). This has resulted largely from climate warming (>90%), increasing temperature and affecting salinity, mostly due to variable rates of evaporation and precipitation (Li et al. 2020). The reduced mixing in ocean layers due to man-made climate change (CO_2 emissions), increases temperatures, accelerates ocean acidification and impacts the marine food chain. This warming and associated stratification in turn inhibits the macronutrient supply for kelp forests and seaweed mariculture and thus limits the export of fixed carbon from the mixed layer of the surface ocean to the deep ocean. The elevated temperature and low nutrient levels for most of the subtropical oceans (low N and P offshore) disturbs the habitat by decreasing specific communities offshore while increasing them inshore in eutrophic zones (Capotondi et al. 2012; Yamaguchi and Suga 2019). In particular, offshore algae production is limited by reduced natural upwelling, reducing primary productivity in subtropical seas. These disturbances shift natural species compositions, with more tolerant species surviving and new competitive species invading and out-competing original inhabitants, resulting in aquatic plant imbalances nearshore or in cyanobacteria blooms that can produce toxins especially under declining macronutrient conditions. Other species nearshore, experience oxygen-depleted environments that affect their growth, change behaviour and increase diseases and mortality of fish and other marine animals (e.g. shellfish).

Nutrients such as N and P from point sources (e.g. human waste, agriculture and industry effluents) fertilize nearshore harbours, bays, estuaries and inlets in a process called eutrophication. These nutrients stimulate photosynthesis, which increases the growth of algae and other photosynthetic organisms which result in more organic material sinking into deep water and to the sediment (Climate Council 2016; Froehlich et al. 2019). Animals and many microbes eating or decomposing this organic material respire, consuming oxygen (Yamaguchi and Suga 2019).

Lenton et al. 2015 developed a model to enable the investigation and quantification of changes in Australia's marine environment (0–200 m depth) in response to different emissions scenarios and stated that Australian oceans will be warmer and more acidic if the GHG emission trajectories are not controlled. Li et al. 2020 incorporated the spatial complexity of the ocean density in addition to other factors from previous studies and quantified ocean stratification up to 2000 m depth (10 times deeper). They used a mathematical model that measures the N^2 value (squared buoyancy frequency) and concluded that ocean stratification has already increased in recent decades and is set to continue increasing in the western Pacific Ocean. The New South Wales Government has been monitoring the changes along their coastlines through offshore mapping (high-resolution 3D digital elevation surfaces of the seabed) and started a 4-year project called SeaBed NSW (to collect the structure and composition of the seabed) in 2019. Several studies have provided evidence of green tides near shore consisting of marine green seaweeds (Chlorophya; e.g. *Ulva prolifera*) abundant in eutrophicated coastal waters (Ye et al. 2011) inshore. These seaweed blooms are recognizable by large blades of algae that may wash up onto the shoreline, significantly observed in Peel Inlet, Western Australia (Ye et al. 2011). Green tides can asphyxiate aquacultures or disrupt traditional artisanal fisheries (Cabre et al. 2015). Seaweed harvesting could therefore be an effective solution as a potential raw material for multiple product streams. Beyond bioproducts, seaweed mariculture helps absorb greenhouse gases (30–60 times higher than land plants) with the potential for carbon sinks from residual seaweed in in the middle and deep ocean (Antoine de Ramon et al. 2012, Froehlich et al. 2019). Regarding biogeochemical constraints, the Climate Foundation has estimated the sustainable deep water oxygen fluxes into the abyssal oceans could support up to 5 gigatons of biomass oxidation per year from the Antarctic Bottom Water (AABW) fluxes. An additional 5 gigatons per year of biomass oxidation is made possible by the oxygen flux through annual North Atlantic Deep Water (NADW) inflows and the associated Labrador Current. Several additional gigatons of marine biomass could be additionally sequestered in anoxic basins of the world each year, including the Black Sea, the Cariaco Basin, and possibly the Santa Barbara Channel. Offshore mariculture creates an opportunity to restore natural upwelling with deep water irrigation of seaweeds regenerating algae production and associated fish productivity while concurrently reducing the temperature in the mixed layer (Kapetsky et al. 2013). The regenerated seaweed biomass ensures robust and resilient reproduction throughout the year (Climate Foundation 2021). In the past few decades, nearly 20 different research groups have shown interest in investigating the restoration of natural upwelling in the oceans. In addition, accelerated stratification indicates a critical need for better marine management, which can be met by the development of products using native strains (mariculture products with Australia-owned IP) with support from state governments and impact funders.

15.6 Future Perspectives

The supply of seaweed to existing Australian markets is largely supported by imported products, as the domestic seaweed industry is still small and localized (Byrnes et al. 2013; Huisman and Millar 2013; Birch et al. 2019b). There are a large range of potential high-value bio-products and commodities that can be derived from seaweeds (Herforth et al. 2020). The development opportunities in this sector includes the food and feed production from integrated multi-trophic mariculture (OECD-FAO 2010; Nabti et al. 2017, Amin et al. 2020). Australia has only recently started to expand the window of opportunity (e.g. CRC projects) to compete as a world leader in sustainable marine technologies developing integrated production systems and seaweed mariculture (one of the highly-funded fields of research in many industrialized countries). For example, the 'SEAPURA' project (2001–2004) focussing on Seaweeds purifying effluents from fish farms received funding of ~ EUR 1.5 million by RIRDC Core Funds and the EU in 2001 (Luning 2001; Pang et al. 2003).

Some of the limitations of the present Australian seaweed industry include lack of protocols for classification and characterisation of seaweed species, lack of databases for strain-specific edibility analyses, lack of optimized cultivation conditions, cost-effective integration of diverse production systems, identification of processing and packaging opportunities, detailed scoping of current market opportunities and potential new markets (Morais et al. 2020; Palmieri and Forleo 2020; Pereira et al. 2020). With commitment from the Australian government, research organisations and regulatory bodies, the market value of seaweed-based products could easily expand over time. With an elevated demand for healthier and more environmentally sustainable food production and processing, seaweed mariculture is now featuring more prominently in global food production (IMARC 2021). Yet these new technologies need to be developed in close partnership with First Nations peoples and respective research organisations, complying with different jurisdictions. For the industry to be globally competitive, intellectual property needs to be secured within Australia that will enable capitalisation with increasing global investment in seaweed mariculture.

Acknowledgements CD, JW and BH thank the Australian Research Council (Linkage Grant LP180100269) and The University of Queensland, Australia (International Scholarship) for financial support. BVH thanks the Woka Foundation and Agrifutures for the support to cultivate offshore and experimentally measure carbon export from residual seaweed production, and IMAS at UTAS for their research collaboration. ARR acknowledges support and resources from the Improvement of Science & Technology Infrastructure in Higher Educational Institutions (FIST Project no: LSI-576/2013) from the Department of Science and Technology, Government of India and Centre of Excellence, Department of Biotechnology, Vignan's Foundation for Science, Technology and Research. Authors thank Prof. Gokare A. Ravishankar, Editor of Seaweed volumes for his critical reading, editing and revising the manuscript.

References

ABARES (Australian Bureau of Agricultural and Resource Economics) (2020) Australian seaweed production. Retrieved 07 July 2021, from https://www.agriculture.gov.au/abares/research-topics/fisheries/fisheries-and-aquaculture-statistics/australian-seaweed-production

Afonso C, Matos J, Guarda I, Gomes-Bispo A, Gomes R, Cardoso C, Gueifão S, Delgado I, Coelho I, Castanheira I (2021) Bioactive and nutritional potential of *Alaria esculenta* and *Saccharina latissima*. J Appl Phycol 33(1):501–513

Afshin A, Sur PJ, Fay KA, Cornaby L, Ferrara G, Salama JS, Mullany EC, Abate KH, Abbafati C, Abebe Z (2019) Health effects of dietary risks in 195 countries, 1990–2017: a systematic analysis for the Global Burden of Disease Study 2017. Lancet 393(10184):1958–1972

AgriFutures (2020) RIRDC annual report 2019–20. Retrieved 28 May 2021, from https://www.transparency.gov.au/annual-reports/rural-industries-research-and-development-corporation/reporting-year/2019-20-28

Amin MS, Metwally HG, Elshinawy MZ, Abdallah MF (2020) Effect of seaweed extract and biofertilizer on organic production of common bean seeds (*Phaseolus vulgaris L.*). Arab Univ J Agric Sci. AUJASCI 22(1):265–272, 2082

Andrade MA, Barbosa CH, Souza VG, Coelhoso IM, Reboleira J, Bernardino S, Ganhão R, Mendes S, Fernando AL, Vilarinho F (2021) Novel active food packaging films based on whey

protein incorporated with seaweed extract: development, characterization, and application in fresh poultry meat. Coatings 11(2):229
André R, Guedes L, Melo R, Ascensão L, Pacheco R, Vaz PD, Serralheiro ML (2020) Effect of food preparations on in vitro bioactivities and chemical components of *Fucus vesiculosus*. Foods 9(7):955
Antoine de Ramon NY, Chynoweth DP, Capron ME, Stewart JR, Hasan MA (2012) Negative carbon via ocean afforestation. Process Saf Environ Prot 90(6):467–474
Arioli T, Mattner SW, Hepworth G, McClintock D, McClinock R (2021) Effect of seaweed extract application on wine grape yield in Australia. J Appl Phycol 33(3):1–9
Arumugam N, Chelliapan S, Kamyab H, Thirugnana S, Othman N, Nasri NS (2018) Treatment of wastewater using seaweed: a review. Int J Environ Res Public Health 15(12):2851
ASI (2020) The current state of the Australian Seaweed Industry. Retrieved 09 June 2021, from https://www.australianseaweedinstitute.com.au/current-state-of-the-australian-seaweed-industry
Azzopardi B (2019) *Asparagopsis taxiformis* (Red Algae). Retrieved 07 July 2021, from https://www.atlantisgozo.com/asparagopsis-taxiformis-red-algae/
Barbot YN, Thomsen C, Thomsen L, Benz R (2015) Anaerobic digestion of *Laminaria japonica* waste from industrial production residues in laboratory-and pilot-scale. Mar Drugs 13(9):5947–5975
Beardall J, Raven JA (2004) The potential effects of global climate change on microalgal photosynthesis, growth and ecology. Phycologia 43(1):26–40
Bedoux G, Hardouin K, Burlot AS, Bourgougnon N (2014) Bioactive components from seaweeds: cosmetic applications and future development. Adv Bot Res, Elsevier 71:345–378
Biospin B (2018) NMR and MRI Technology Identifies Factors in Seed Germination. Retrieved 07 July 2021, from https://www.news-medical.net/whitepaper/20180828/NMR-and-MRI-Technology-Identifies-Factors-in-Seed-Germination.aspx
Birch D, Skallerud K, Paul NA (2019a) Who are the future seaweed consumers in a Western society? Insights from Australia. Br Food J 121(2):603–615
Birch D, Skallerud K, Paul N (2019b) Who eats seaweed? An Australian perspective. J Int Food Agribusiness Mark 31(4):329–351
Blanche K (1992) Preliminary observations on the distribution and abundance of seaweed flies (Diptera: Coelopidae) on beaches in the Gosford district of New South Wales, Australia. Aust J Ecol 17(1):27–34
Blanz M, Ascough P, Mainland I, Martin P, Taggart MA, Dieterich B, Wishart J, Sayle KL, Raab A, Feldmann J (2019) Seaweed fertilisation impacts the chemical and isotopic composition of barley: implications for analyses of archaeological skeletal remains. J Archaeol Sci 104:34–44
Blikra MJ, Løvdal T, Vaka MR, Roiha IS, Lunestad BT, Lindseth C, Skipnes D (2019) Assessment of food quality and microbial safety of brown macroalgae (*Alaria esculenta* and *Saccharina latissima*). J Sci Food Agric 99(3):1198–1206
Bloomberg (2020) Seaweed Cultivation Market worth $30.2 billion by 2025. Retrieved 06 Aug 2021, from https://www.bloomberg.com/press-releases/2020-10-15/seaweed-cultivation-market-worth-30-2-billion-by-2025-exclusive-report-by-marketsandmarkets
Blunden G, Wildgoose PB (1977) The effects of aqueous seaweed extract and kinetin on potato yields. J Sci Food Agric 28(2):121–125
Bogard JR, Farmery AK, Baird DL, Hendrie GA, Zhou S (2019) Linking production and consumption: the role for fish and seafood in a healthy and sustainable Australian diet. Nutrients 11(8):1766
Britton D (2020) Effects of ocean acidification and warming on shallow subtidal temperate seaweed assemblages in eastern Tasmania, Australia: implications for the blacklip abalone (*Haliotis rubra*). University of Tasmania, pp 40–84
Buck BH, Buchholz CM (2004) The offshore-ring: a new system design for the open ocean aquaculture of macroalgae. J Appl Phycol 16(5):355–368
Burke L, Reytar K, Spalding M, Perry A (2011) Reefs at risk revisited. World Resources Institute, Washington, DC, pp 1–112

Butler CL, Lucieer VL, Wotherspoon SJ, Johnson CR (2020) Multi-decadal decline in cover of giant kelp *Macrocystis pyrifera* at the southern limit of its Australian range. Mar Ecol Prog Ser 653:1–18

Byrnes L, Brown C, Foster J, Wagner LD (2013) Australian renewable energy policy: barriers and challenges. Renew Energy 60:711–721

Cabre A, Marinov I, Leung S (2015) Consistent global responses of marine ecosystems to future climate change across the IPCC AR5 earth system models. Clim Dyn 45(5):1253–1280

Cabrita AR, Maia MR, Oliveira HM, Sousa-Pinto I, Almeida AA, Pinto E, Fonseca AJ (2016) Tracing seaweeds as mineral sources for farm-animals. J Appl Phycol 28(5):3135–3150

Campbell M, Ortuño J, Ford L, Davies DR, Koidis A, Walsh PJ, Theodoridou K (2020) The effect of ensiling on the nutritional composition and fermentation characteristics of brown seaweeds as a ruminant feed ingredient. Animals 10(6):1019

Capotondi A, Alexander MA, Bond NA, Curchitser EN, Scott JD (2012) Enhanced upper ocean stratification with climate change in the CMIP3 models. J Geophys Res Oceans 117(C4)

Carus M, Dammer L (2018) The circular bioeconomy—concepts, opportunities, and limitations. Ind Biotechnol 14(2):83–91

Castine SA, McKinnon AD, Paul NA, Trott LA, de Nys R (2013) Wastewater treatment for land-based aquaculture: improvements and value-adding alternatives in model systems from Australia. Aquac Environ Interact 4(3):285–300

Chakraborty P, Kumar S, Dutta D, Gupta V (2009) Role of antioxidants in common health diseases. Res J Pharm Technol 2(2):238–244

Chen X, Nie W, Yu G, Li Y, Hu Y, Lu J, Jin L (2012) Antitumor and immunomodulatory activity of polysaccharides from *Sargassum fusiforme*. Food Chem Toxicol 50(3–4):695–700

Chen Y-C, Cheng C-Y, Liu C-T, Sue Y-M, Chen T-H, Hsu Y-H, Hwang P-A, Chen C-H (2018) Alleviative effect of fucoxanthin-containing extract from brown seaweed *Laminaria japonica* on renal tubular cell apoptosis through upregulating Na+/H+ exchanger NHE1 in chronic kidney disease mice. J Ethnopharmacol 224:391–399

Chuka-ogwude D, Ogbonna J, Moheimani NR (2020a) A review on microalgal culture to treat anaerobic digestate food waste effluent. Algal Res 47:101841

Chuka-ogwude D, Ogbonna J, Borowitzka MA, Moheimani NR (2020b) Screening, acclimation and ammonia tolerance of microalgae grown in food waste digestate. J Appl Phycol 32(6):3775–3785

Climate Council (2016) How seaweed can kelp us tackle climate change. Retrieved 06 June 2021, from https://www.climatecouncil.org.au/seaweed-climate-change/

Climate Foundation (2021) "What is Marine Permaculture?" Retrieved 07 Aug 2021, from https://www.climatefoundation.org/what-is-marine-permaculture.html

Daume S (2006) The roles of bacteria and micro and macro algae in abalone aquaculture: a review. J Shellfish Res 25(1):151–157

Deepika C (2017) Fucose-containing sulfated polysaccharides from *Sargassum wightii*: extraction technology and anticancer activity assessment. Int J Pharm Chem Biol Sci 7(3):272–277

Deepika C (2019) FTIR, SEM, EDS and GCMS metabolite profiling of macroalgae – *Sargassum wightii*. Int Res J Eng Technol (IRJET) 6(03):6791–6797

Deepika C, Bhaskar TCJ, Madhusudhanan J (2016) Antioxidant activities of a few common seaweeds from Gulf of Mannar and the effect of drying as the method of preservation. Indian J Eng Sci Technol 10(1):37

Department of Agriculture (2019) Australia's consumption of seafood. Retrieved 14 June 2021, from https://www.agriculture.gov.au/abares/research-topics/fisheries/fisheries-and-aquaculture-statistics/seafood-consumption-2018

Díaz-Rubio ME, Pérez-Jiménez J, Saura-Calixto F (2009) Dietary fiber and antioxidant capacity in *Fucus vesiculosus* products. Int J Food Sci Nutr 60(Suppl. 2):23–34

Doh H, Dunno KD, Whiteside WS (2020) Preparation of novel seaweed nanocomposite film from brown seaweeds *Laminaria japonica* and *Sargassum natans*. Food Hydrocoll 105:105744

Edgar GJ (1997) Australian marine life: the plants and animals of temperate waters. Reed Books, Melbourne, p 544

Emont JCC (2020) Cows make climate change worse. Could seaweed help? Retrieved 07 July 2021, from https://www.wsj.com/articles/cows-make-climate-change-worse-could-seaweed-help-11604152802

Eom S-H, Kang Y-M, Park J-H, Yu D-U, Jeong E-T, Lee M-S, Kim Y-M (2011) Enhancement of polyphenol content and antioxidant activity of brown alga *Eisenia bicyclis* extract by microbial fermentation. Fish Aquat Sci 14(3):192–197

Ferdouse F, Holdt SL, Smith R, Murúa P, Yang Z (2018) The global status of seaweed production, trade and utilization. Globefish Res Programme 124:I

Fertah M, Belfkira A, Taourirte M, Brouillette F (2017) Extraction and characterization of sodium alginate from Moroccan *Laminaria digitata* brown seaweed. Arab J Chem 10:S3707–S3714

Fiedor J, Burda K (2014) Potential role of carotenoids as antioxidants in human health and disease. Nutrients 6(2):466–488

Figueroa V, Farfán M, Aguilera J (2021) Seaweeds as novel foods and source of culinary flavors. Food Rev Int 1–26

Fitzgerald EA (1991) The seaweed rebellion: federal-state/provincial conflicts over offshore energy development in the United States, Canada, and Australia. Conn J Int Law 7:255

Fleurence J, Morançais M, Dumay J (2018) Seaweed proteins. Proteins in food processing. pp 245–262. Woodhead Publishing, France

Forecast MD (2020) Global carrageenan market. Retrieved 04 Nov 2020, from https://www.marketdataforecast.com/market-reports/carrageenan-market

Froehlich HE, Afflerbach JC, Frazier M, Halpern BS (2019) Blue growth potential to mitigate climate change through seaweed offsetting. Curr Biol 29(18):3087–3093

García-Poza S, Leandro A, Cotas C, Cotas J, Marques JC, Pereira L, Gonçalves AM (2020) The evolution road of seaweed aquaculture: cultivation technologies and the industry 4.0. Int J Environ Res Public Health 17(18):6528

Glicksman M (1987) Utilization of seaweed hydrocolloids in the food industry. Twelfth international seaweed symposium. Springer, pp 69. 563

Golléty C, Riera P, Davoult D (2010) Complexity of the food web structure of the *Ascophyllum nodosum* zone evidenced by a δ13C and δ15N study. J Sea Res 64(3):304–312

Gosavi K, Sammut J, Gifford S, Jankowski J (2004) Macroalgal biomonitors of trace metal contamination in acid sulfate soil aquaculture ponds. Sci Total Environ 324(1–3):25–39

He P, Xu S, Zhang H, Wen S, Dai Y, Lin S, Yarish C (2008) Bioremediation efficiency in the removal of dissolved inorganic nutrients by the red seaweed, *Porphyra yezoensis*, cultivated in the open sea. Water Res 42(4–5):1281–1289

Heather Rhoades L (2020) Tomato hornworm – organic control of hornworms. Retrieved 07 July 2021, from https://www.gardeningknowhow.com/edible/vegetables/tomato/tomato-hornworms-control.htm

Heather Rhoades ST (2021) Nitrogen nodules and nitrogen fixing plants. Retrieved 07 July 2021, from https://www.gardeningknowhow.com/garden-how-to/soil-fertilizers/nitrogen-nodules-and-nitrogen-fixing-plants.htm

Hepsy M (2020) $10,000,000 seaweed industry by 2025 – AgriFutures Australia and the Australian Seaweed Institute are developing a blueprint for Australia's seaweed industry. Retrieved 14 June 2021, from https://thefarmermagazine.com.au/10-million-dollar-seaweed-industry-by-2025/

Herforth A, Bai Y, Venkat A, Mahrt K, Ebel A, Masters W (2020) Cost and affordability of healthy diets across and within countries. Background paper for the State of Food Security and Nutrition in the World 2020. FAO, Food and Agricultural Organization, Rome

Hessami MJ, Cheng SF, Ranga Rao A, Yin YH, Phang SM (2019) Bioethanol production from agarophyte red seaweed, *Gelidium elegans* using a novel sample preparation method for analysing bioethanol content by gas chromatography. 3- Biotech J 9(1):25

HexaResearch (2020) Global seaweed fertilizer market size and forecast, by form (powder, liquid), by application (farm, garden), by region, and trend analysis, 2015–2025. Retrieved 04 Nov 2020, from https://www.hexaresearch.com/research-report/seaweed-fertilizer-market

Hughes LJ, Black LJ, Sherriff JL, Dunlop E, Strobel N, Lucas RM, Bornman JF (2018) Vitamin D content of Australian native food plants and Australian-grown edible seaweed. Nutrients 10(7):876

Huisman JM (2000) Marine plants of Australia, University of Western Australia Press in association with Australian Biological Resources Study. Book. 1–55. Western Australia, Australia

Huisman JM, Millar AJ (2013) Australian seaweed collections: use and misuse. Phycologia 52(1):2–5

Hurtado AQ, Neish IC, Ali MKM, Norrie J, Pereira L, Michalak I, Shukla PS, Critchley AT (2021) Extracts of seaweeds used as biostimulants on land and sea crops—an efficacious, phyconomic, circular blue economy: with special reference to *Ascophyllum* (brown) and *Kappaphycus* (red) seaweeds. In: Biostimulants for crops from seed germination to plant development. Elsevier: 263–288. Elsevier, USA

Husni A, Lailatussifa R, Isnansetyo A (2019) *Sargassum hystrix* as a source of functional food to improve blood biochemistry profiles of rats under stress. Prev Nutr Food Sci 24(2):150

Ibrahim ZR (2013) Effect of foliar spray of ascorbic acid, Zn, seaweed extracts (Sea) force and biofertilizers (EM-1) on vegetative growth and root growth of olive (*Olea europaea* L.) transplants cv. HojBlanca. Int J Pure Appl Sci Technol 17(2):79

Ibrahim RY, Hammad HB, Gaafar AA, Saber AA (2020) The possible role of the seaweed *Sargassum vulgare* as a promising functional food ingredient minimizing aspartame associated toxicity in rats. Int J Environ Health Res 1–20

IMARC (2021) Seaweed market: global industry trends, share, size, growth, opportunity and forecast 2021–2026. Retrieved 06 Aug 2021, from https://www.imarcgroup.com/seaweed-market

Johnson CR, Banks SC, Barrett NS, Cazassus F, Dunstan PK, Edgar GJ, Frusher SD, Gardner C, Haddon M, Helidoniotis F, Hill KL (2011) Climate change cascades: shifts in oceanography, species' ranges and subtidal marine community dynamics in eastern Tasmania. J Exp Mar Biol Ecol 400(1–2):17–32

Kadam SU, Pankaj S, Tiwari BK, Cullen P, O'Donnell CP (2015) Development of biopolymer-based gelatin and casein films incorporating brown seaweed *Ascophyllum nodosum* extract. Food Packag Shelf Life 6:68–74

Kapetsky JM, Aguilar-Manjarrez J, Jenness J (2013) A global assessment of offshore mariculture potential from a spatial perspective. FAO fisheries and aquaculture technical paper, (549), p I

Kelly J (2020) Australian seaweed industry blueprint. Retrieved 14 June 2021, from https://www.agrifutures.com.au/wp-content/uploads/2020/09/20-072.pdf

Kim RC (2018) Can creating shared value (CSV) and the United Nations Sustainable Development Goals (UN SDGs) collaborate for a better world? Insights from East Asia. Sustainability 10(11):4128

Kinley RD, Martinez-Fernandez G, Matthews MK, de Nys R, Magnusson M, Tomkins NW (2020) Mitigating the carbon footprint and improving productivity of ruminant livestock agriculture using a red seaweed. J Clean Prod 259:120836

KIRDO (2020) Bull Kelp *Durvillaea Potatorum* – King Island Kelp industry. Retrieved 14 June 2021, from https://kingisland.net.au/live/kelpindustry/

Koca N, Karadeniz F, Burdurlu HS (2007) Effect of pH on chlorophyll degradation and colour loss in blanched green peas. Food Chem 100(2):609–615

Kolb N, Vallorani L, Milanović N, Stocchi V (2004) Evaluation of marine algae wakame (*Undaria pinnatifida*) and kombu (*Laminaria digitata japonica*) as food supplements. Food Technol Biotechnol 42(1):57–61

Kulkarni MG, Rengasamy KR, Pendota SC, Gruz J, Plačková L, Novák O, Doležal K, Van Staden J (2019) Bioactive molecules derived from smoke and seaweed *Ecklonia maxima* showing phytohormone-like activity in *Spinacia oleracea* L. New Biotechnol 48:83–89

Layton C, Coleman MA, Marzinelli EM, Steinberg PD, Swearer SE, Vergés A, Wernberg T, Johnson CR (2020) Kelp forest restoration in Australia. Front Mar Sci 7:74

Lee BW, Momdjian K (1998) Potential opportunities for increased production of processed Asian foods in Australia. Rural Industries Research and Development Corporation. RIRDC Publication No 98/64; RIRDC Project No CON-6A. pp 1–47

Lehner M, Mont O, Heiskanen E (2016) Nudging–a promising tool for sustainable consumption behaviour? J Clean Prod 134:166–177

Lenton A, McInnes KL, O'Grady JG (2015) Marine projections of warming and ocean acidification in the Australasian region. Aust Meteorol Oceanogr J 65(1):1–28

Li G, Cheng L, Zhu J, Trenberth KE, Mann ME, Abraham JP (2020) Increasing ocean stratification over the past half-century. Nat Clim Chang 10(12):1116–1123

Liu J, Luthuli S, Yang Y, Cheng Y, Zhang Y, Wu M, Choi JI, Tong H (2020b) Therapeutic and nutraceutical potentials of a brown seaweed *Sargassum fusiforme*. Food Sci Nutr 8:5195–5205

Lopes D, Melo T, Rey F, Meneses J, Monteiro FL, Helguero LA, Abreu MH, Lillebø AI, Calado R, Domingues MR (2020) Valuing bioactive lipids from green, red and brown macroalgae from aquaculture, to foster functionality and biotechnological applications. Molecules 25(17):3883

Lourenço-Lopes C, Fraga-Corral M, Jimenez-Lopez C, Pereira AG, Garcia-Oliveira P, Carpena M, Prieto MA, Simal-Gandara J (2020) Metabolites from macroalgae and its applications in the cosmetic industry: a circular economy approach. Resources 9(9):101

Luning K (2001) SEAPURA: seaweeds purifying effluents from fish farms, an EU project coordinated by the Wattenmeerstation Sylt. Wadden Sea Newsl 2001:20–21

Mabin CJ, Johnson CR, Wright JT (2019) Physiological response to temperature, light, and nitrates in the giant kelp Macrocystis pyrifera from Tasmania, Australia. Mar Ecol Prog Ser 614:1–19

Maeda H, Tsukui T, Sashima T, Hosokawa M, Miyashita K (2008) Seaweed carotenoid, fucoxanthin, as a multi-functional nutrient. Asia Pac J Clin Nutr 17:7701–7706

Marinho-Soriano E, Nunes S, Carneiro M, Pereira D (2009) Nutrients removal from aquaculture wastewater using the macroalgae *Gracilaria birdiae*. Biomass Bioenergy 33(2):327–331

Matos ÂP, Vadiveloo A, Bahri PA, Moheimani NR (2021) Anaerobic digestate abattoir effluent (ADAE), a suitable source of nutrients for *Arthrospira platensis* cultivation. Algal Res 54:102216

McIntyre RE, Adams MA, Grierson PF (2009) Nitrogen mineralization potential in rewetted soils from a semi-arid stream landscape, north-west Australia. J Arid Environ 73(1):48–54

MECCA (2018) The in-tray kelp is on the way. Retrieved 11 June 2021, from https://www.mecca.com.au/the-mecca-memo/the-in-tray/kelp-seaweed-trend.html

Michalak I, Mahrose K (2020) Seaweeds, intact and processed, as a valuable component of poultry feeds. J Mar Sci Eng 8(8):620

Mise T, Ueda M, Yasumoto T (2011) Production of fucoxanthin-rich powder from *Cladosiphon okamuranus*. Adv J Food Sci Technol 3(1):73–76

Mohammad JH, Ranga Rao A, Ravishankar GA (2019) Opportunities and challenges in seaweeds as feed stock for biofuel production. In: Ravishankar GA, Ranga Rao A (eds) Handbook of algal technologies and phytochemicals, vol. II. Phycoremediation, biofuels and global biomass production, vol. II. CRC Press, USA, pp 39–50

Moheimani NR, Vadiveloo A, Ayre JM, Pluske JR (2018) Nutritional profile and in vitro digestibility of microalgae grown in anaerobically digested piggery effluent. Algal Res 35:362–369

Morais T, Inácio A, Coutinho T, Ministro M, Cotas J, Pereira L, Bahcevandziev K (2020) Seaweed potential in the animal feed: a review. J Mar Sci Eng 8(8):559

MordorIntelligence (2021) Australia beauty and personal care products market – growth, trends, covid-19 impact, and forecasts (2021–2026). Retrieved 11 June 2021, from https://www.mordorintelligence.com/industry-reports/australia-beauty-and-personal-care-products-market-industry

Mourelle ML, Gómez CP, Legido JL (2021) Role of algal derived compounds in pharmaceuticals and cosmetics. Recent advances in micro and macroalgal processing: food and health perspectives. Wiley, USA, pp 537–603

Murray CJ, Aravkin AY, Zheng P, Abbafati C, Abbas KM, Abbasi-Kangevari M, Abd-Allah F, Abdelalim A, Abdollahi M, Abdollahpour I (2020) Global burden of 87 risk factors in 204 countries and territories, 1990–2019: a systematic analysis for the Global Burden of Disease Study 2019. Lancet 396(10258):1223–1249

Nabti E, Jha B, Hartmann A (2017) Impact of seaweeds on agricultural crop production as biofertilizer. Int J Environ Sci Technol 14(5):1119–1134

Nagaoka M, Shibata H, Kimura-Takagi I, Hashimoto S, Kimura K, Makino T, Aiyama R, Ueyama S, Yokokura T (1999) Structural study of fucoidan from *Cladosiphon okamuranus* Tokida. Glycoconj J 16(1):19–26

Naughtin C, Moyle CL, Pandey V, Renando C, Poruschi L, Torres de Oliveira R, Doan N, Schleiger E (2021) A new chapter: opportunities to seed new industries for Queensland over the coming decade. CSIRO and Queensland University of Technology, Brisbane, pp 1–85

Nayar S, Bott K (2014) Current status of global cultivated seaweed production and markets. World Aquaculture 45(2):32–37

Nick (2017) When to remove the growing tip – feeding for upper truss growth. Retrieved 07 July 2021, from https://www.tomatogrowing.co.uk/newsletter/when-to-remove-the-growing-tip/

Nkemka V, Murto M (2012) Exploring strategies for seaweed hydrolysis: effect on methane potential and heavy metal mobilisation. Process Biochem 47(12):2523–2526

Norman J, Pickford C, Sanders T, Waller M (1988) Human intake of arsenic and iodine from seaweed-based food supplements and health foods available in the UK. Food Addit Contam 5(1):103–109

Nwoba EG, Moheimani NR, Ubi BE, Ogbonna JC, Vadiveloo A, Pluske JR, Huisman JM (2017) Macroalgae culture to treat anaerobic digestion piggery effluent (ADPE). Bioresour Technol 227:15–23

OceanWell (2021) OceanWell. Care Now! Retrieved 16 June 2021, from https://www.oceanwell.de/en/

OECD-FAO. Organisation for Economic Co-operation Development and the Food and Agriculture Organization (2010) OECD-FAO Agricultural Outlook. p 17

Packer A (2017) Seaweed beauty products from Elemis, John Frieda and more. Retrieved 16 June 2021, from https://www.express.co.uk/life-style/style/832739/Seaweed-beauty-products-Elemis-John-Frieda

Palmieri N, Forleo MB (2020) The potential of edible seaweed within the western diet. A segmentation of Italian consumers. Int J Gastron Food Sci 20:100202

Pang S, Luening K, Garcia Reina G, Pinchetti J, Santos R, Sousa Pinto I, Lòpez-Figueroa F, Dion P, Lognoné V, Dring M (2003) Species diversification and improvement of aquatic production in seaweeds purifying effluents from integrated-fish farms (EU project SEAPURA)

Parsons S (2017) Seaweed: unlocking the business secrets of the deep. Retrieved 16 June 2021. From https://cosmeticsbusiness.com/news/article_page/Seaweed_Unlocking_the_business_secrets_of_the_deep/130137

Pascual, PRL, Carabio DE, Abello NFH, Remedios EA, Pascual VU (2021) Enhanced assimilation rate due to seaweed biostimulants improves growth and yield of rice beans (*Vigna umbellata*). Agron Res 19

Patel NB, Tailor V, Rabadi M, Jain A, Kalasariya H (2020) Role of marine macroalgae in skin hydration and photoprotection benefits: a review. Int J Bot Stud 5(03):201–206

Pereira SA, Kimpara JM, Valenti WC (2020) A bioeconomic analysis of the potential of seaweed *Hypnea pseudomusciformis* farming to different targeted markets. Aquac Econ Manag 24(4).1–19

Perin S, Lean DR (2004) The effects of ultraviolet-B radiation on freshwater ecosystems of the Arctic: influence from stratospheric ozone depletion and climate change. Environ Rev 12(1):1–70

Phillips J, Conacher C, Horrocks J (1999) Marine macroalgae from the Gulf of Carpentaria, tropical northern Australia. Aust Syst Bot 12(3):449–478

PhycoHealth (2021) Phycoderm skin care made from Australian seaweed. Retrieved 07 July 2021, From https://www.phycohealth.com/collections/seaweed-skincare-by-phycoderm

Pimentel FB, Alves RC, Rodrigues F, Beatriz M, Oliveira PP (2018) Macroalgae-derived ingredients for cosmetic industry—an update. Cosmetics 5(1):2

Plant Care Today (2021) Tomato leaf curl: virus | causes | remedies for tomato leaves curling. From https://plantcaretoday.com/tomato-leaf-curl.html

Proffitt T, Campbell-Clause J (2012) Managing grapevine nutrition and vineyard soil health. Wines of Western Australia, Claremont, pp 1–32

Rehm BH, Moradali MF (2018) Alginates and their biomedical applications. 1-268, Springer Nature Singapore Pte Ltd, Springer, Singapore

Rengasamy KR, Aderogba MA, Amoo SO, Stirk WA, Van Staden J (2013) Potential antiradical and alpha-glucosidase inhibitors from *Ecklonia maxima* (Osbeck) Papenfuss. Food Chem 141(2):1412–1415

Rilov G, Fraschetti S, Gissi E, Pipitone C, Badalamenti F, Tamburello L, Menini E, Goriup P, Mazaris AD, Garrabou J, Benedetti-Cecchi L (2020) A fast-moving target: achieving marine conservation goals under shifting climate and policies. Ecol Appl 30(1):e02009

Roles J, Yarnold J, Wolf J, Stephens E, Hussey K, Hankamer B (2020) Charting a development path to deliver cost competitive microalgae-based fuels. Algal Res 45:101721

Ronan JM, Stengel DB, Raab A, Feldmann J, O'Hea L, Bralatei E, McGovern E (2017) High proportions of inorganic arsenic in *Laminaria digitata* but not in *Ascophyllum nodosum* samples from Ireland. Chemosphere 186:17–23

Roque BM, Venegas M, Kinley RD, de Nys R, Duarte TL, Yang X, Kebreab E (2021) Red seaweed (*Asparagopsis taxiformis*) supplementation reduces enteric methane by over 80 percent in beef steers. PLoS One 16(3):e0247820

Rupawalla Z, Robinson N, Schmidt S, Li S, Carruthers S, Buisset E, Roles J, Hankamer B, Wolf J (2021) Algae biofertilisers promote sustainable food production and a circular nutrient economy–An integrated empirical-modelling study. Sci Total Environ 148913

Saez CA, Roncarati F, Moenne A, Moody AJ, Brown MT (2015) Copper-induced intra-specific oxidative damage and antioxidant responses in strains of the brown alga *Ectocarpus siliculosus* with different pollution histories. Aquat Toxicol 159:81–89

Sala E, Knowlton N (2006) Global marine biodiversity trends. Annu Rev Environ Resour 31:93–122

Sathasivam R, Ki J-S (2018) A review of the biological activities of microalgal carotenoids and their potential use in healthcare and cosmetic industries. Mar Drugs 16(1):26

Sharma P, Sharma N (2017) Industrial and biotechnological applications of algae: a review. J Adv Plant Biol 1(1):01

Shibata H, Kimura-Takagi I, Nagaoka M, Hashimoto S, Aiyama R, Iha M, Ueyama S, Yokokura T (2000) Properties of fucoidan from *Cladosiphon okamuranus* tokida in gastric mucosal protection. Biofactors 11(4):235–245

Shirosaki M, Koyama T (2011) *Laminaria japonica* as a food for the prevention of obesity and diabetes. Adv Food Nutr Res 64:199–212

Sivakaminathan S, Hankamer B, Wolf J, Yarnold J (2018) High-throughput optimisation of light-driven microalgae biotechnologies. Sci Rep 8(1):1–13

Smith K (2021) How far apart should tomatoes be planted? Retrieved 07 July 2021, from https://www.hgtv.com/outdoors/flowers-and-plants/vegetables/how-far-apart-should-tomatoes-planted

Soares C, Švarc-Gajić J, Oliva-Teles MT, Pinto E, Nastić N, Savić S, Almeida A, Delerue-Matos C (2020) Mineral composition of subcritical water extracts of *Saccorhiza Polyschides*, a brown seaweed used as fertilizer in the North of Portugal. Journal of Marine Science and Engineering 8(4):244

SOE. State of the Environment (2016) Australia State of Environment 2016. Retrieved 09 June 2021, from https://soe.environment.gov.au/theme/marine-environment/framework/state-and-trends

Spolaore P, Joannis-Cassan C, Duran E, Isambert A (2006) Commercial applications of microalgae. J Biosci Bioeng 101(2):87–96

Stévant P, Marfaing H, Rustad T, Sandbakken I, Fleurence J, Chapman A (2017) Nutritional value of the kelps *Alaria esculenta* and *Saccharina latissima* and effects of short-term storage on biomass quality. J Appl Phycol 29(5):2417–2426

Sultana V, Tariq S, Hira K, Tariq A, Ara J, Tariq R, Ehteshamul-Haque S (2018) Seaweed bio-fertilizer for the management of root rotting fungi and root knot nematodes affecting cotton crop. Pak J Bot 50(6):2409–2412

Tanna B, Mishra A (2019) Nutraceutical potential of seaweed polysaccharides: structure, bioactivity, safety, and toxicity. Compr Rev Food Sci Food Saf 18(3):817–831

Tiwari BK, Troy DJ (2015) Seaweed sustainability–food and non-food applications. Seaweed sustainability. Elsevier, Academic, Elsevier, USA, pp 1–6

Torres-Escribano S, Denis S, Blanquet-Diot S, Calatayud M, Barrios L, Vélez D, Alric M, Montoro R (2011) Comparison of a static and a dynamic in vitro model to estimate the bioaccessibility of As, Cd, Pb and Hg from food reference materials *Fucus sp.*(IAEA-140/TM) and Lobster hepatopancreas (TORT-2). Sci Total Environ 409(3):604–611

Uribe E, Vega-Gálvez A, Vargas N, Pasten A, Rodríguez K, Ah-Hen KS (2018) Phytochemical components and amino acid profile of brown seaweed *Durvillaea antarctica* as affected by air drying temperature. J Food Sci Technol 55(12):4792–4801

Uthirapandi V, Suriya S, Boomibalagan P, Eswaran S, Ramya SS, Vijayanand N, Kathiresan D (2018) Bio-fertilizer potential of seaweed liquid extracts of marine macro algae on growth and biochemical parameters of *Ocimum sanctum*. J Pharmacol Phytochem 7:3528–3532

Vadiveloo A, Foster L, Kwambai C, Bahri PA, Moheimani NR (2021) Microalgae cultivation for the treatment of anaerobically digested municipal centrate (ADMC) and anaerobically digested abattoir effluent (ADAE). Sci Total Environ 775:145853

Van der Spiegel M, Noordam M, Van der Fels-Klerx H (2013) Safety of novel protein sources (insects, microalgae, seaweed, duckweed, and rapeseed) and legislative aspects for their application in food and feed production. Compr Rev Food Sci Food Saf 12(6):662–678

van Ginneken V (2019) The application of the seaweeds in neutralizing the ocean acidification as a long-term multifaceted challenge. J Geosci Environ Prot 7(12):126

Vega J, Bonomi-Barufi J, Gómez-Pinchetti JL, Figueroa FL (2020) Cyanobacteria and red macroalgae as potential sources of antioxidants and UV radiation-absorbing compounds for cosmeceutical applications. Mar Drugs 18(12):659

Vessey JK (2003) Plant growth promoting rhizobacteria as biofertilizers. Plant Soil 255(2):571–586

Vranken S, Wernberg T, Scheben A, Severn-Ellis A, Batley J, Philipp BE, Edwards D, Wheeler D, Coleman MA (2021) Genotype-environment mismatch of kelp forests under climate change. Mol Ecol 00:1–17

Webster N, Gorsuch H (2020) Monitoring additional values within the Reef 2050 Integrated Monitoring and Reporting Program: final report of the microbes expert group

Weigle SM, Smith LD, Carlton JT, Pederson J (2005) Assessing the risk of introducing exotic species via the live marine species trade. Conserv Biol 19(1):213–223

WernbergT(2015)TheGreatSouthernReef.Retrieved09June2021,fromhttps://theconversation.com/australias-other-reef-is-worth-more-than-10-billion-a-year-but-have-you-heard-of-it-45600

Wiley D (2021) 10 Common tomato plant diseases to watch out for this summer. From https://www.bhg.com/gardening/vegetable/vegetables/tomato-plant-diseases/

Winberg PC, Ghosh D, Tapsell L (2009) Seaweed culture in integrated multi-trophic aquaculture-Nutritional benefits and systems for Australia. RIRDC Publication No. 09/005, pp 1–31

Wolf J, Ross IL, Radzun KA, Jakob G, Stephens E, Hankamer B (2015) High-throughput screen for high performance microalgae strain selection and integrated media design. Algal Res 11:313–325

XiaoZhi LH (2021) Scientists report first step away from fertilizers. Retrieved 7 July 2021, from http://staceylab.missouri.edu/scientists-report-first-step-away-from-fertilizers/

Yamaguchi R, Suga T (2019) Trend and variability in global upper-ocean stratification since the 1960s. J Geophys Res Oceans 124(12):8933–8948

Yan X, Zheng L, Chen H, Lin W, Zhang W (2004) Enriched accumulation and biotransformation of selenium in the edible seaweed *Laminaria japonica*. J Agric Food Chem 52(21):6460–6464

Ye NH, Zhang XW, Mao YZ, Liang CW, Xu D, Zou J, Zhuang ZM, Wang QY (2011) 'Green tides' are overwhelming the coastline of our blue planet: taking the world's largest example. Ecol Res 26(3):477–485

Yende SR, Harle UN, Chaugule BB (2014) Therapeutic potential and health benefits of *Sargassum* species. Pharmacogn Rev 8(15):1

Yokoi K, Shizuka H, Konomi A (2008) Toxic effects of consumption of 'hijiki' seaweed in rats, Wiley Online Library. FASEB J 22(01):705

Yuan Z, Liu D, Masqué P, Zhao M, Song X, Keesing JK (2020) Phytoplankton responses to climate-induced warming and interdecadal oscillation in North-Western Australia. Paleoceanogr Paleoclimatol 35(3):1–12

Zemke-White WL, Ohno M (1999) World seaweed utilization: an end-of-century summary. J Appl Phycol 11(4):369–376

Chapter 16
Seaweeds: The Ecological Roles, the Economic Benefits and the Threats for Changing the Carbon Cycle

Gour Gopal Satpati, Sarban Sengupta, and Ruma Pal

Abbreviations

ACE-I	Angiotensin-I-converting enzyme
BCA	Blue carbon assessment
$CaCO_3$	Calcium carbonate
CCA	Crustose coralline algae
CO_2	Carbon-di-oxide
COVID	Corona virus diseases
DENV	Dengue virus
DHA	Docosahexaenoic acid
DNA	Deoxyribonucleic acid
DPPH	2,2-diphenyl-1-picrylhydrazyl
EGE	Enhanced greenhouse effect
EPA	Eicosapentaenoic acid
FAO	Food and agricultural organization
HCMV	Human cytomegalovirus
HIV	Human immunodeficiency virus
HPV	Human papilloma virus
HRV	Human rhinovirus
HSV	Herpes simplex virus
PUFA	polyunsaturated fatty acids
RGW	Runaway global warming

G. G. Satpati
Department of Botany, Bangabasi Evening College, Kolkata, West Bengal, India

S. Sengupta · R. Pal (✉)
Phycology Laboratory, Department of Botany, Centre of Advanced Study (CAS), University of Calcutta, Kolkata, West Bengal, India
e-mail: rpalcu@rediffmail.com

A. Ranga Rao, G. A. Ravishankar (eds.), *Sustainable Global Resources of Seaweeds Volume 1*, https://doi.org/10.1007/978-3-030-91955-9_16

RNA	Ribonucleic acid
SPs	Sulfated polysaccharides
WHO	World Health Organization

16.1 Introduction

Seaweeds or marine macroalgae are the primary producers in marine ecosystem. They commonly occur in the intertidal, deep and shallow waters and found at the depths of 150 m (Rameshkumar and Rajaram 2019). Sometimes they are predominantly found around brackish water estuaries and ponds (Satpati et al. 2012). Seaweeds are broadly categorized as red, brown and green belong to the division Rhodophyta, Phaeophyta and Chlorophyta (Fig. 16.1).

Collecting and harvesting of seaweeds and their usefulness throughout the world as food and commercial products have become an important issue for sustainable livelihood of the coastal communities. In coastal ecosystem, seaweeds play an important role in primary biomass production having an essential ecological function as substrata or habitat for marine organisms like fishes, invertebrates, birds and mammals (Graham et al. 2007). Seaweeds also help to prevent coastline erosion and influences marine biodiversity leading to economic, ecological and social changes at domestic and global scales (Rebours et al. 2014). On the other hand, deforestation demolishes natural forests and other crucial carbon sinks and causes global warming. Hence, seaweed farming or aquaculture helps to counter climate change through fixing atmospheric carbon-di-oxide (CO_2) in a sustainable way. Seaweed also amends deoxygenation, acidification to protect marine biodiversity as source of foods and livelihoods for coastal communities.

The book chapter summarizes socio-economic and environmental benefits of seaweed farming or aquaculture as source of bioactives for nutraceuticals and pharmaceuticals. Furthermore, the chapter deals with the role of seaweeds for changing global carbon cycle and fight against climate change.

16.2 Seaweed Farming and Socio-economic Benefits

The economic and ecological importance of seaweeds are broad spectrum, which includes food, fodder, fuel, fertilizer, pharmaceuticals, nutraceuticals, cosmetics, stabilizers etc. aimed the socio-economic benefits of rural and urban people. Therefore, commercialization of seaweeds for producing value added products depends on seaweed farming in coastal areas. Most of the coastal communities are dependent on fishing, which is not economically sufficient and may need additional income. Hence, seaweed cultivation can be an alternative source of income for better livelihood of the coastal communities. Naylor (1976) showed that seaweed farming in a large area is more profitable than agricultural farming. Later, Indonesia

Fig. 16.1 Plate showing various seaweeds: (**a**, **b**) *Sargassum cristaefolium*, (**c**) *Hypnea musciformis*, (**d**) *Padina dubla*, (**e**) *P. gymnospora*, (**f**) *Tricleocarpa fragilis*

started seaweed farming with the help of international agencies to overcome the socio-economic problems of marginalized coastal populations (Trono et al. 1980). For the socio-economic development of the coastal villages, several government and private organizations have initiated seaweed farming along the coastal areas of low and middle income countries including Indonesia, Philippines, Vietnam,

Tanzania, Kiribati and India (Valderrama 2012). The economic benefits and challenges associated with seaweed farming in the coastal areas of many countries are monitored by food and agricultural organization (FAO). Seaweed farming in Indonesia and Philippines started with *Kappaphycus alvarezii* and *Eucheuma denticulatum* and it was observed that for *K. alvarezii* profit was three times higher than expected (Valderrama 2012). In 2007–2009, the seaweed farmers showed net profit of $5000 per year, which was 33% higher than the average income suggested economic benefits of seaweed farming (Waters et al. 2019). Seaweed farming in Solomon Islands, Tanzania and Philippines is empowered by women and considered to be an important source of income for them. China and Indonesia contributed 87% of the total seaweed farming among the Asian countries. Seaweed harvesting by boat and diving are found to be better for fishermen or fisherwomen than harvesting by hand in low tide (Waters et al. 2019). It has been found that seaweed farming is a successful earning occupation for the fisherwomen of the coastal communities of Ramanathapuram district, Gulf of Mannar and Palk Bay coast of Tamilnadu, in India. (Rameshkumar and Rajaram 2019). Moreover, seaweed farming helps to improve the standard of living of fishers' households. During cultivation and processing, women become engaged in tying seedlings, preparation of planting materials, drying, cleaning and biomass harvesting, which makes them economically independent. Women and their children mostly carry out seaweed harvesting with a net bag during low tide. After harvesting, biomass is left on the seashore for sun drying for 2–3 days. The dried biomass then sold to an intermediate collector for industrial application (Rebours et al. 2014). Thus, seaweed has significant role in supporting the livelihoods of many people in and around the coastal communities.

16.3 Environmental Benefits of Seaweed Aquaculture

Seaweeds are designated as the 'promising plants of the millennium' due to their ability to grow without the application of fertilizer and use of fresh water resources which are fast depleting. In addition, seaweed cultivation provides many underwater ecosystem services such as creating habitats for fish and crayfish species (Hasselström et al. 2020). The successful mitigation of coastal eutrophication by sequestering nutrients can be achieved by seaweed cultivation. In another way, large-scale cultivation of seaweeds can serve as a potential carbon sink and may sequester carbon to mitigate ocean acidification (Krause-Jensen et al. 2015). Naturally grown macroalgal mats of *Chaetomorpha, Ulva, Gracilaria, Polysiphonia* and *Catenella* were found to be high in carbon sequestration in the brackish water habitats of Indian Sundarbans (Gorain et al. 2018). In China, large-scale cultivation of seaweed promoted phosphorus uptake, which leads to mitigate coastal eutrophication (Xiao et al. 2017). For the European bioeconomy, seaweed aquaculture has recommended as blue growth by providing low carbon and renewable products (European Commission 2012). In Europe and specially in Sweden, seaweed cultivation has gained popularity due to its positive environmental impacts such as carbon mitigation, coastal eutrophication, nitrogen and phosphorus recovery, and formation of fossil-based raw substitute from biomass (Hasselström et al. 2020). Thus, seaweed can compete with other mitigating conventional crops in terms of cost-effectiveness.

16.4 Ecology of Kelp Forests and the Threats Faced by Them Especially Their Extinctions

Seaweeds particularly those belonging to the Phaeophyta play an important role in providing food and shelter to myriad organisms especially in cold-water marine benthic habitats where they form some of the largest biogenic structures (Dayton 1985). Kelp forests (mainly brown algae belonging to the order Laminariales of Phaeophyta), support a diverse associated biota such as other algae and epibiota, crustaceans, molluscs, echinoderms, fishes and even marine mammals which collectively make this one of the most ecologically diverse and productive ecosystems of the world. Overall, three basic groups or guilds of kelp have been identified depending upon the height of canopy fronds representing in Table 16.1 (Dayton 1985; Steneck et al. 2002).

Table 16.1 Table depicting the various kelp species, their characteristics and locations where they form canopies

Serial No.	Place	Genera	Type of kelp	Length of fronds
1.	Western North Atlantic Coast	*Laminaria* sp.	Prostate	Upto 10 m
2.	Aleutian Islands and Alaska	*Alaria* sp., *Laminaria* sp., *Macrocystis* sp., *Pterygophora* sp., *Thalassiophyllum* sp., *Laminaria* sp., *Eisenia* sp. and *Agarum* sp.	Canopy, stipitate, prostrate	Upto 10 m long, >45 m long
3.	Southern California	*Macrocystis* sp., *Pterygophora* sp., *Laminaria* sp., *Eisenia* sp., *Pelagophycus* sp., *Egregia* sp. and *Agarum* sp.	Canopy, stipitate, prostrate	Upto 10 m long, >45 m long
4.	Pacific Coast of Asia	*Alaria fistulosa* and *Laminaria* sp	Stipitate	5–10 m long
5.	South Africa	*Ecklonia maxima*, *Laminaria* sp. and *Macrocystis* sp.	Stipitate, Prostrate, Canopy	10–45 m long
6	Australia and NewZeaLand	*Ecklonia* sp. and *Macrocystis* sp.	Stipitate, Canopy	10–45 m long
7.	Western Coast of South America	*Macrocystis* sp. and *Lessonia* sp.	Canopy, Stipitate	10–45 m long

Submerged canopies have been known to reduce turbulence and flow (Ackerman and Okubo 1993; Koch et al. 2009), increase the bottom shear stress and dampen wave energy (Mendez and Losada 2004) thereby providing refuge to a variety of fauna. About 16 genera and 20 species of kelps are available along the California coast of North America (Abbott and Hollenberg 1976).

Large quantities of kelp biomass provide a significant source of nutrition to the inhabitants of coastal marine ecosystems where they materialize into kelp forests. But this transfer of energy and biomass occurs mainly via food webs based on macroalgal detritus (Duggins et al. 1989) because herbivores rarely consume more than 10% of the living biomass (Mann 1973).

In the Southern ocean, a significant portion of the carbon fixed by the benthic brown algae and their epiphytic diatoms is transferred to the food chain especially and even to the pelagic food web (Dunton 2001). In the study, there is a clear dependence on carbon fixed by benthic algae in Antarctic waters that even extends to the pelagic webs. Thus the benthic macroalgae of the Antarctic waters not only serve as a refuge for various species (Richardson 1979), but also together its epiphytic diatoms, provide a valuable carbon source that is readily assimilated by the local fauna. The results are consistent with similar previous studies conducted in the Arctic that examined the role of kelp there (Dunton and Schell 1987) and in the surrounding Sub Arctic (Duggins et al. 1989).

Seaweeds especially kelp forest communities around the world are facing a variety of threats especially from anthropogenic disturbances. Mussels compete with them for anchoring space (Robles et al. 1995) while sea urchins (Echinoidea) have been known to negatively impact standing crops of macroalgae by grazing on them (Lawrence 1975). Their predators include sea otters (Lensink 1962) and spiny lobsters- *Panulirus interruptus* (Tegner and Levin 1983). Anthropogenic extermination of the urchin's predators- the sea otter because of hunting for the fur trade (Lensink 1962) and the spiny lobster because of the fishing industry (Ling et al. 2009) have led to their population explosion resulting in the overgrazing and deforestation of kelp beds globally (Fig. 16.2). Efforts to artificially reintroduce the sea otters to places where they have become locally extinct have in some cases proved futile because of the homing instinct of sea otters. However, it is worthwhile to note that the sea otters have naturally re-established themselves along much of their original range in the Americas following a blanket ban on their hunting. And with them, gradually the kelp forests have also returned.

In the low latitudes, other maroalgae such as *Cystoseira* sp., *Sargassum* sp., *Fucus* sp., (Phaeophyta), *Ulva* sp. (Chlorophyta) etc. are reported as dominant genera (Steneck et al. 2002). Mention may be made of the "Sargasso Sea" a region of the Atlantic Ocean (with its characteristic floating masses of *Sargassum* sp.) bounded by four ocean currents on four sides forming a natural gyre. Unlike other seas it does not have land boundaries and is demarcated by its characteristic brown *Sargassum* seaweed and clear water (Stow 2004). The Sargasso Sea is also characterized by high species diversity (Hulburt et al. 1960; Dewey 1971) thereby indicating that species-prey relations are also very complicated (Sheldon et al. 1973).

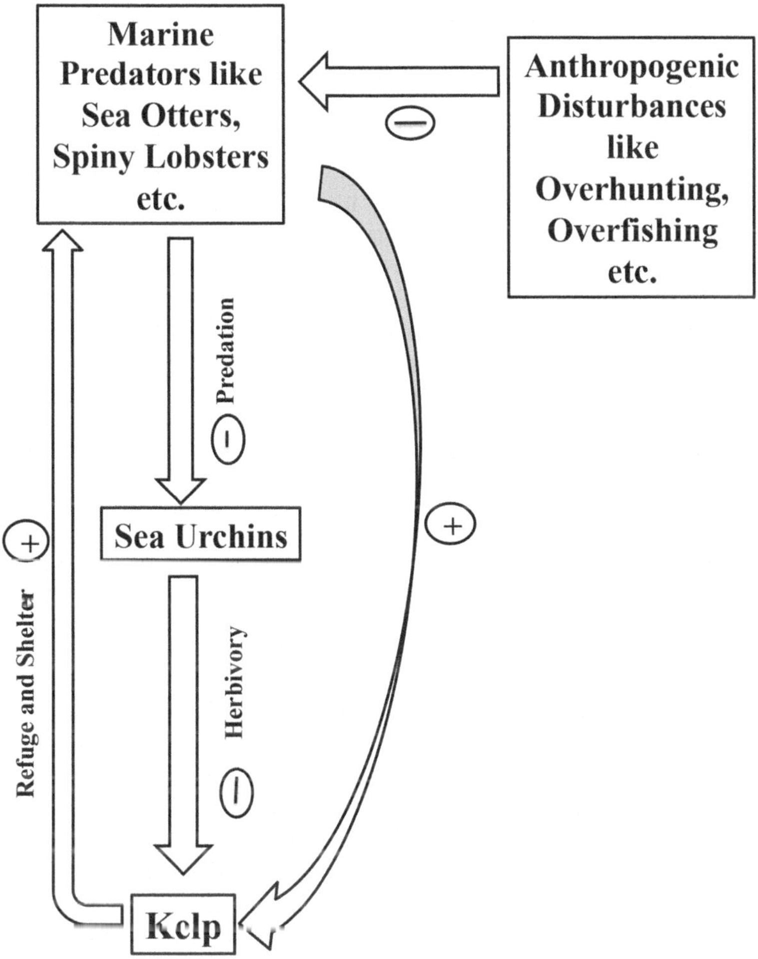

Fig. 16.2 Flow chart depicting the mechanism of loss of kelp forest ecosystems and the mechanisms behind it

16.5 Importance of Seaweed Bioactives as Nutraceuticals and Pharmaceuticals for Nutrition and Health

Seaweeds have enormous importance in nutraceuticals and pharmaceuticals industry as they contain important bioactives like pigments, vitamins, polysaccharides, phytochemicals, lipid and fatty acids, protein, peptide and amino acids. In the past few decades, seaweeds performed their role as nutraceuticals and functional food with high dietary benefits. Furthermore, seaweeds have high therapeutic importance as they contain metabolites of biological activity. Survey of World Health

Organization (WHO) has indicated that people in developed countries like United States, Australia and various European nations suffer from vitamin and mineral deficiency which leads to obesity, type-2 diabetes, cancer and metabolic syndromes. At the same time, countries like Japan, China, India and Indonesia where seaweed is consumed regularly, have comparatively less metabolic diseases and obesity (Nanri et al. 2017). Additionally, seaweeds are used to prevent goitre, viral infections and non-communicable diseases like inflammation and hypertension (Rosenfeld 2000). Also, routine consumption of seaweed extract obtained from *Laminaria japonica*, *Fucus vesiculosus* and *Macrocystis pyrifera* in combination with vitamin B6, zinc and manganese reduced the severe effects of osteoarthritis in a mixed population (Myers et al. 2016). It has been found that regular consumption of *Undaria* as dietary supplement can lower the risk of breast cancer in women (Teas et al. 2013). Other than these, bioactive compounds obtained from seaweeds have antibacterial, antiviral, antioxidant, anticoagulant and antitumor properties too (Ganesan et al. 2019).

16.5.1 Protein and Amino Acids

Dried seaweed biomass constitutes 5% to 50% protein, of which red seaweeds contain the highest percentage (20–35%) than green (10–20%) and brown (5–10%) seaweeds. Therefore, the protein content in seaweed is comparatively higher than other plant proteins like pulses and soybean (Dhargalkar 2015). It has also been found that seaweed protein can compete with animal proteins like egg and meat. For example, *Porphyra*, *Laminaria* and *Undaria* have high amino acid pool compared to egg and meat (Shannon and Abu-Ghannam 2019). The essential and non-essential amino acids in seaweeds recorded were threonine, arginine, alanine, aspartic acid and glutamic acid with rare occurrence of methionine and phenylalanine. Red seaweeds contain high ratio of essential and non-essential amino acids (0.98–1.02) followed by green (0.72–0.97) and brown seaweed (0.73) (Abirami and Kowsalya 2012). It has been reported that protein-derived peptides from seaweeds have several health benefits including antihypertensive, antioxidant, and antidiabetic properties (Admassu et al. 2018). Many bioactive compounds obtained from seaweeds like taurine, carnosine, glutathione and microsporine showed antiapoptotic and antioxidant activities in the rat brain (Harnedy and Fitz Gerald 2011; Aydin et al. 2016). Moreover, biopeptides obtained from seaweeds showed anticancer, antiatherosclerotic and immunomodulatory activities (Fan et al. 2014). Phycobiliproteins like phycocyanin, phycoerythrin and allophycocyanin obtained from *Porphyra yezoensis* showed host-pathogen interaction against virus, bacteria and simultaneously induced apoptosis through immunomodulation reaction, which prevents cancers (Fontenelle et al. 2018).

16.5.2 Lipids and Fatty Acids

Seaweeds are well known as low energy foods. Moreover they contain large amount of long chain fatty acids especially polyunsaturated fatty acids (PUFA), which is an important component of cell membrane and precursor of eicosanoids that can prevent the risk factor of many diseases like diabetes, cancer, osteoporosis and cardiovascular diseases (Misurcova, et al. 2011). Among the PUFA, omega-3-fatty acids (ω-3) and omega-6-fatty acids (ω-6) are of great interest due to their potential activities against cardiovascular diseases and obesity related problems (Ganesan et al. 2019). Seaweeds, especially *Ulva*, *Acanthophora*, and *Gracilaria* which are rich in ω-3- fatty acids showed anti-inflammatory, antihypertensive, anti-hyperlipidemic activities and inhibition of angiotensin-I-converting enzyme (ACE-I) (Fontenelle et al. 2018). The precursor of α-linolenic acid, eicosapentaenoic acid (EPA) and docosahexaenoic acid (DHA) obtained from *Ulva* species showed significant antioxidant and anti-inflammatory activities (McCauley et al. 2018). The lipid composition and fatty acid profile of *Ulva rigida*, *U. lactuca*, *U. prolifera*, *U. intestinalis*, *Chaetomorpha linum* and *Rhizoclonium africanum* showed significant amount of saturated and unsaturated fatty acids, which has the potential to act against many diseases (Satpati and Pal 2011; Satpati et al. 2015; Cardoso et al. 2017).

16.5.3 Sulfated Polysaccharides

Sulfated polysaccharides (SPs) are a major group of long chain polysaccharides, abundant in green, brown and red seaweeds including *Ulva*, *Enteromorpha*, *Laminaria*, *Monostroma*, *Caulerpa*, *Codium*, *Gracilaria* etc. Among different SPs, ulvan, fucoidan, fucan sulphate, carrageenan and laminaran have wider applications in pharmaceutical and nutraceuticals for their biological values. The existence of ulvan in green algae, laminaran, fucoidan or fucan sulfate in brown algae and carrageenan in red algae exhibit antioxidant, anti-inflammatory, anticancer, antidiabetic, anticoagulant, antibacterial and antiviral properties (Satpati 2020a). It has been reported that carrageenan can fight against respiratory diseases including Corona virus diseases (COVID-19) and also reduces the risk of other diseases in children and adults by boosting immunity (Satpati 2020b). Carrageenan obtained from *Hypnea*, *Eucheuma*, *Gigartina*, *Chondrus* and *Gracilaria* have potent immunomodulatory activity against human papilloma virus (HPV), human rhinovirus (HRV) and herpes simplex virus (HSV) (Satpati 2020a; Satpati 2020b). Similarly, laminaran, another group of SPs widely found in *Fucus* and *Ascophyllum* inhibits reverse transcriptase activity of human immunodeficiency virus (HIV) (Moran-Santibanez et al. 2016). The brown algae, rich in fucans or fucoidan sulphate showed bioactivity against many deoxyribonucleic acid (DNA) and ribonucleic acid (RNA) viruses including HIV, HSV-1, HSV-2, dengue virus (DENV), human cytomegalovirus (HCMV) and measles virus by inhibiting viral-induced syncytium formation

(Moran-Santibanez et al. 2016). These SPs obtained from seaweed may also reduce the risk of novel corona virus disease (COVID-19), by boosting immunity in children as well as in adults (Satpati 2020a).

16.5.4 Vitamins and Minerals

Seaweeds are rich in vitamins, including fat and water-soluble vitamins like vitamin A and E, B_1 (thiamine), B_2 (riboflavin), B_{12} (cobalamin) and C (ascorbic acid). Moreover, seaweeds are good source of essential minerals including trace and tracer elements like calcium, magnesium, iron, phosphorus, potassium, zinc, copper, manganese, selenium, and fluoride (Škrovánková 2011; Shannon and Abu-Ghannam 2019). Investigations have shown that vitamins are essential for biochemical and antioxidant activity. In addition, vitamin C can decrease blood pressure, beta-carotene (with vitamin A activity) lower the risk of cardiovascular diseases and vitamin C, E, A have anticancer activity (Škrovánková 2011). It has been reported that *Porphyra*, *Himanthalia* and *Gracilaria* are rich in vitamin C than vegetables whereas *Codium* and *Gracilaria* contains high carotenoids or vitamin A than carrots (Ganesan et al. 2019). Seaweeds like *Palmaria palmata*, *Ulva lactuca* contains relatively high iron and calcium than a portion of lean beef or cheddar cheese (Shannon and Abu-Ghannam 2019). The essential vitamins and minerals obtained from various seaweeds have powerful antioxidant, anticancer, antihypercholesterolemic activities and may cure breast cancerc and neurodegenerative diseases (Teas et al. 2013).

16.5.5 Secondary Metabolites

The secondary metabolites produced in macroalgal tissues are grouped into phenolic compounds, halogenated compounds, sterols, terpenes; and small peptides (Rosa et al. 2020). Seaweeds contain a wide range of polyphenol compounds, which includes phlorotannins, phloroglucinol, gallic acid, carotenoids and quercetin. It has been reported that brown seaweeds like *Hypnea*, *Eucheuma*, *Sargassum* and *Ecklonia* possesses high amount of phloroglucinol and showed potential antioxidant, antidiabetic, anti-inflammatory, intiproliferative, anti-HIV and anti-Alzheimer's activities including other health benefits (Nagarajan and Mathaiyan 2015; Rajauria et al. 2013). Phlorotannins are another group of phytochemicals obtained from *Ecklonia cava*, which are found to be active against superoxide, free radicals, 2,2-diphenyl-1-picrylhydrazyl (DPPH) and inhibits lipid peroxidation (Li et al. 2009). In another study, phlorotannins showed great pharmaceutical potential against dyslipidemia than lovastatin in an *in vivo* model (Rosa et al. 2020). A rare phlorotannin, Octaphlorethol A obtained from *Ishige foliacea* significantly inhibited melanin synthesis and tyrosinase activity and showed potential application in skin-whitening formulations (Kim et al. 2015). Halogenated terpenes and bromophenols, fucoxanthin, fucosterol, kahalalide F, phycoerythrin, Griffithsin, a group of secondary metabolites abundant in green, brown and red seaweeds, showed many biological activities including anti-inflammatory, antidiabetic, antitumor, antimicrobial, antiviral, antihypertensive, antiobesity and neuroprotective activities (Rosa et al. 2020).

16.6 The Position of Seaweeds in the Changing Global Carbon Cycle and Their Role as Allies in the Battle against Climate Change

The global carbon cycle takes place between five interconnected pools, the atmospheric, the biotic, the pedologic, the geologic and by far the largest the oceanic pool (Lal 2008; Sengupta et al. 2017). However human activities have upset this delicate balance of nature. Between the period of 1800 and 1994 the terrestrial biosphere was a source of 39 ± 0.4 PgC (Sabine et al. 2004). Continuously increasing CO_2 levels are responsible for the phenomenon known as enhanced greenhouse effect (EGE) and runaway global warming (RGW). Remediation especially the capture and sequestration of carbon in stable reservoirs is the need of the hour. The ocean is by far the largest sink of CO_2 accounting for ~48% of the total fossil fuel and cement-manufacturing related emissions for the same period of 1800–1994.

In the biosphere, vegetated coastal habitats are characterized by the presence of macrophytes both submerged (like sea-grasses and macro-algae) and emergent/emerged vegetation (mangroves and salt marshes) and occupy a narrow fringe (from the intertidal zone to ~40 m deep) along the shores of all the continents. Globally they cover an area of 2.3–7.0 million square kilometers with macro-algae being the largest contributors especially in carbon fixation and primary productivity (Duarte et al. 2013).

However, since macro-algal communities mostly occupy rocky coastal habitats they are unable to account for significant carbon burial. Instead the carbon they fix (~0.19–0.64 PgC/y) is at least partly exported to deeper waters (Dierssen et al. 2009; Kennedy et al. 2010; Duarte et al. 2013). Primarily due to the large uncertainties in their size or area covered (~10 fold) and large variation in carbon fluxes, there is a ten fold variation around the estimates of vegetated coastal habitats to carbon sequestration in deep sea sediments ranging from 73 to 866 TgC/y that subsequently represents between 3% and 33.33% of oceanic CO_2 uptake (Duarte 2017). The problem is particularly acute for macroalgal beds, which constitute the majority of these vegetated coastal habitats. Consequently, their importance in "blue carbon assessments (BCA)" has largely been overlooked.

But according to Smith (1981), in the ocean, areas dominated by macrophyte ecosystems are more important carbon sinks as compared to phytoplankton based ecosystems. This is because of the fact that despite accounting for only 5% of oceanic primary production and occupying only 2×10^6 km^2 out of the 3.6×10^8 km^2 (Whittaker and Likens 1973) they account for two thirds of the oceanic biomass (Smith 1981). Thus the biomass per unit area occupied by macrophytes is 400 times that of plankton. Phytoplankton may have high turnover rates and may dominate the spatial extent of oceanic ecosystems and thus be biologically important, but because ecosystems with rapid turnover of stored carbon are relatively ineffective carbon sinks, therefore carbon sequestration from macroalgal beds may be a significant yet overlooked part of the oceanic carbon uptake.

Increasing CO_2 concentrations are also responsible for reducing the pH of the water column by a phenomenon known as ocean acidification (Doney et al. 2009). Ocean acidification poses a special threat to calcifying organisms like corals, echinoderms, molluscs and calcyfing algae because it makes it more difficult for them to deposit calcium carbonate ($CaCO_3$). Several roles have been suggested for calcification in macroalgae but the most putative include

(i) Provision of mechanical resistance to herbivores and minimizing of tissue damage to tissues (Littler and Littler 1980; Padilla 1993)
(ii) Increasing the capacity for nutrient and bicarbonate assimilation via the generation of protons (McConnaughey and Whelan 1997)
(iii) Providing protection against excessive irradiance (Burger and Schagerl 2010)
(iv) Improving photosynthetic performance (McConnaughey 1998)

Thus calcification is very important for the physiological and ecological fitness of calcifying macroalge which form an important constitutent of benthic subtidal and intertidal habitats. Obligate calcifyers like the crustose coralline algae (CCA) are a major calcifying component of marine benthic habitats at all depths within the photic zone in almost every habitat type (Adey and Macintyre 1973; Littler et al. 1985; Steneck 1986) and are especially vulnerable to ccean acidification (Johnson et al. 2012). This is because they deposit $CaCO_3$ in form of high magnesium calcite needles. But as the surface saturation state of calcite is less than that of aragonite, those that precipitate the former (coralline algae) are more likely to face problems in depositing $CaCO_3$ skeletons (Kleypas et al. 1999). The response of calcifying brown macroalgae which are not obligate calcifyers but still important $CaCO_3$ producers is a bit complicated. Increasing CO_2 concentrations have been known to reduce calcification (Russell et al. 2009) while increasing the productivity (Reiskind et al. 1988). *Padina* sp. is one of the two genera of phaeophytes that calcify. *Padina* sp. produce aragonite needles in concentric bands known as rings on the surface of fan shaped thalli (Okazaki et al. 1986) and are important producers of both organic matter and $CaCO_3$. Johnson et al. (2012) evaluated the effects of increasing CO_2 concentrations on *Padina* sp. by studying the effects of volcanic CO_2 gradients on the abundance of *Padina* sp. and the most common grazers the sea urchins (Echinoidea). The authors observed a dramatic ecological shift with the total loss of sea urchins and coralloid algae on increasing CO_2 concentrations (moving near CO_2 vents) concomitant with an increasing abundance of *Padina* sp. The results are in accordance with a previous study that documented a similar loss of coralline algae and sea urchins along with a simultaneous rise in the abundance of Phaeophytes like *Padina* sp. along volcanic CO_2 vents in the Mediterranean (Hall-Spencer et al. 2008).

Ocean macroalgae afforestation aims to reduce atmospheric CO_2 concentrations through expanding natural populations of macroalgae. The macroalgae absorb CO_2 through photosynthesis and fix it into biomass which is then harvested to produce biomass and biomethane (N'Yeurt et al. 2012).

16.7 Conclusion and Future Perspectives

In conclusion, seaweed farming has generated a positive outcome for socio-economic developments of coastal communities for food and livelihoods. Low capital, short grow out cycles and low technological requirements helps seaweed farming most popular among millions of coastal peoples. As foodstuff, seaweed has paid great attention around the globe for its nutritional benefits but still it is consumed by a number of people in certain areas of the world today. As 'promising plants of the millennium' seaweed produces several bioactive compounds including polysaccharides, fatty acids, amino acids etc. having biotechnological, nutraceutical and pharmaceutical importance for health benefits. In this regard, seaweed farming helps to enhance the standard of living of poor people in and around coastal areas. Significantly seaweed in marine ecosystem counters the climate change as the higher plant communities continuously diminished due to deforestation. Subsequently, it protects the marine biodiversity and lowers the risk of acidification, deoxygenation and other environmental changes to maintain the ecological balance of nature. For a better future, seaweed farming needs to be more vigorous in coastal areas for both economic and ecological benefits of coastal communities and the global environment.

References

Abbott IA, Hollenberg GJ (1976) Marine algae of California. Stanford University Press, Stanford

Abirami RG, Kowsalya S (2012) Phytochemical screening, microbial load and antimicrobial activity of underexploited seaweeds. Int Res J Microbiol 3:328–332

Ackerman JD, Okubo A (1993) Reduced mixing in a marine macrophyte canopy. Funct Ecol 7:305–309. https://doi.org/10.2307/2390209

Adey WH, Macintyre IG (1973) *Crustose coralline* algae: a re-evaluation in the geological sciences. GSA Bull 84:883–904. https://doi.org/10.1130/0016-7606

Admassu H, Zhao W, Yang R, Gasmalla MAA, Alsir E (2018) Development of functional foods: seaweeds (algae) untouched potential and alternative resource- a review. Int J Sci Technol Res 4:108–115

Aydin AF, Çoban J, Doğan-Ekici I, Betül-Kalaz E, Doğru-Abbasoğlu S, Uysal M (2016) Carnosine and taurine treatments diminished brain oxidative stress and apoptosis in D-galactose aging model. Metab Brain Dis 31:337–345. https://doi.org/10.1007/s11011-015-9755-0

Burger K, Schagerl M (2010) Morphological studies of the brown alga *Padina pavonica* (L.) Thivy. Unpublished master thesis. Department of Marine Biology, University of Vienna, Vienna

Cardoso C, Ripol A, Afonso C, Freire M, Varela J, Quental-Ferreira H, Pousao-Ferreira P, Bandarra N (2017) Fatty acid profiles of the main lipid classes of green seaweeds from fish pond aquaculture. Food Sci Nutr 5:1186–1194. https://doi.org/10.1002/fsn3.511

Dayton PK (1985) Ecology of kelp communities. Annu Rev Ecol Syst 16:215–245. https://doi.org/10.1146/annurev.es.16.110185.001243

Dewey GB (1971) The annual cycle in quantity and composition of the zooplankton of the Sargasso Sea off Bermuda. I. The upper 500M. Limnol Oceanogr 16:219–240. https://doi.org/10.4319/lo.1971.16.2.0219

Dhargalkar V (2015) Uses of seaweeds in the Indian diet for sustenance and well-being. Sci Cult 80:192–202

Dierssen HM, Zimmerman RC, Drake LA, Burdige DJ (2009) Potential export of unattached benthic macroalgae to the deep sea through wind-driven Langmuir circulation. Geophys Res Lett 36:L04602. https://doi.org/10.1029/2008GL036188

Doney SC, Fabry VJ, Feely RA, Kleypas JA (2009) Ocean acidification: the other CO_2 problem. Annu Rev Mar Sci 1:169–192. https://doi.org/10.1146/annurev.marine.010908.163834

Duarte CM (2017) Reviews and syntheses: hidden forests, the role of vegetated coastal habitats on the ocean carbon budget. Biogeosciences 14:301–310. https://doi.org/10.5194/bg-14-301-2017

Duarte CM, Losada IJ, Hendriks IE, Mazarrasa I, Marbà N (2013) The role of coastal plant communities for climate change mitigation and adaptation. Nat Clim Chang 3:961–968. https://doi.org/10.1038/nclimate1970

Duggins DO, Simenstad CA, Estes JA (1989) Magnification of secondary production by kelp detritus in coastal marine ecosystems. Science 245:170–173. https://doi.org/10.1126/science.245.4914.170

Dunton KH (2001) $\delta^{15}N$ and $\delta^{13}C$ measurements of Antarctic Peninsula fauna: trophic relationships and assimilation of benthic seaweeds. Am Zool 41:99–112. https://doi.org/10.1093/icb/41.1.99

Dunton KH, Schell DM (1987) Dependence of consumers on macroalgal (*Laminaria solidungula*) carbon in an Arctic kelp community: $\delta^{13}C$ evidence. Mar Biol 93:615–625. https://doi.org/10.1007/BF00392799

European Commission (2012) Innovating for sustainable growth: a bioeconomy for Europe – communication from the Commission to the European Parliament, the Council, the European Economic and Social Committee and the Committee of the Regions (Brussels 13.2.2012)

Fan X, Bai L, Zhu L, Yang L, Zhang X (2014) Marine algae-derived bioactive peptides for human nutrition and health. J Agric Food Chem 62(38):9211–9222. https://doi.org/10.1021/jf502420h

Fontenelle TPC, Lima GC, Mesquita JX, de Souza Lopes JL, de Brito TV, Júnior, Fd. CV, Sales AB et al (2018) Lectin obtained from the red seaweed *Bryothamnion triquetrum*: secondary structure and anti-inflammatory activity in mice. Int J Biol Macromol 112:1122–1130. https://doi.org/10.1016/j.ijbiomac.2018.02.058

Ganesan AR, Tiwari U, Rajauria G (2019) Seaweed nutraceuticals and their therapeutic role in disease prevention. Food Sci Human Wellness 8:252–263. https://doi.org/10.1016/j.fshw.2019.08.001

Gorain PC, Sengupta S, Satpati GG, Paul S, Tripathi S, Pal R (2018) Carbon sequestration in macroalgal mats of brackish-water habitats in Indian Sunderbans: potential as renewable organic resource. Sci Total Environ 626:689–702. https://doi.org/10.1016/j.scitotenv.2018.01.106

Graham MH, Vásquez JA, Buschmann AH (2007) Global ecology of the giant kelp Macrocystis: from ecotypes to ecosystems. Oceanogr Mar Biol 45:39–88. https://doi.org/10.1201/9781420050943-4

Hall-Spencer JM, Rodolfo-Metalpa R, Martin S, Ransome E, Fine M, Turner SM, Rowley SJ et al (2008) Volcanic carbon dioxide vents show ecosystem effects of ocean acidification. Nature 454:96–99. https://doi.org/10.1038/nature07051

Harnedy PA, Fitz Gerald RJ (2011) Bioactive proteins, peptides, and amino acids from macroalgae. J Phycol 47:218–232. https://doi.org/10.1111/j.1529-8817.2011.00969.x

Hasselström L, Thomas JB, Nordström J, Cervin G, Nylund GM, Pavia H, Gröndahl F (2020) Socioeconomic prospects of a seaweed bioeconomy in Sweden. Sci Rep 10:1610. https://doi.org/10.1038/s41598-020-58389-6

Hulburt EM, Ryteier JH, Guillard RRL (1960) The phytoplankton of the Sargasso Sea off Bermuda. ICES J Mar Sci 25:115–128. https://doi.org/10.1093/icesjms/25.2.115

Johnson VR, Russell BD, Fabricius KE, Brownlee C, Hall-Spencer JM (2012) Temperate and tropical brown macroalgae thrive, despite decalcification, along natural CO_2 gradients. Glob Chang Biol 18:2792–2803. https://doi.org/10.1111/j.1365-2486.2012.02716.x

Kennedy H, Beggins J, Duarte CM, Fourqurean JW, Holmer M, Marba N, Middleburg JJ (2010) Seagrass sediments as a global carbon sink: isotopic constraints. Glob Biogeochem Cycles 24:GB4026. https://doi.org/10.1029/2010GB003848

Kim KN, Yang HM, Kang SM, Ahn G, Roh SW, Lee W, Kim D, Jeon YJ (2015) Whitening effect of octaphlorethol A isolated from Ishige foliacea in an *in vivo* zebra fish model. J Microbiol Biotechnol 25:448–451. https://doi.org/10.4014/jmb.1409.09033

Kleypas JA, Buddemeier RW, Archer D, Gattuso J-P, Langdon C, Opdyke BN (1999) Geochemical consequences of increased atmospheric carbon dioxide on coral reefs. Science 284:118–120. https://doi.org/10.1126/science.284.5411.118

Koch EW, Barbier EB, Silliman BR, Reed DJ, Perillo GME, Hacker SD, Granek EF et al (2009) Non-linearity in ecosystem services: temporal and spatial variability in coastal protection. Front Ecol Environ 7:29–37. https://doi.org/10.1890/080126

Krause-Jensen D, Duarte CM, Hendriks IE, Meire L, Blicher ME, Marbà N, Sejr MK (2015) Macroalgae contribute to nested mosaics of pH variability in a subarctic fjord. Biogeosciences 12:4895–4911. https://doi.org/10.5194/bg-12-4895-2015

Lal R (2008) Sequestration of atmospheric CO_2 in global carbon pools. Energy Environ Sci 1:86–100. https://doi.org/10.1039/B809492F

Lawrence JM (1975) On the relationships between marine plants and sea urchins. In: Volume 13 of Oceanography and marine biology. G. Allen and Unwin publishers, Australia, p 74

Lensink CJ (1962) The history and status of sea otters in Alaska. Doctoral thesis. Purdue University, Lafayette

Li Y, Qian ZJ, Ryu B, Lee SH, Kim MM, Kim SK (2009) Chemical components and its antioxidant properties *in vitro*: an edible marine brown alga, *Ecklonia cava*. Bioorg Med Chem 17:1963–1973. https://doi.org/10.1016/j.bmc.2009.01.031

Ling SD, Johnson CR, Frusher SD, Ridgway KR (2009) Overfishing reduces resilience of kelp beds to climate-driven catastrophic phase shift. Proc Natl Acad Sci U S A 106:22341–22345. https://doi.org/10.1073/pnas.0907529106

Littler MM, Littler DS (1980) The evolution of thallus form and survival strategies in benthic marine macroalgae: field and laboratory tests of a functional form model. Am Nat 116:25–44

Littler MM, Littler DS, Blair SM, Norris JN (1985) Deepest known plant life discovered on an uncharted seamount. Science 227:57–59. https://doi.org/10.1126/science.227.4682.57

Mann KH (1973) Seaweeds: their productivity and strategy for growth. Science 182:975–981. https://doi.org/10.1126/science.182.4116.975

McCauley JI, Winberg PC, Meyer BJ, Skropeta D (2018) Effects of nutrients and processing on the nutritionally important metabolites of *Ulva* sp. (Chlorophyta). Algal Res 35:586–594. https://doi.org/j.algal.2018.09.016

McConnaughey T (1998) Acid secretion, calcification, and photosynthetic carbon concentrating mechanisms. Can J Bot 76:1119–1126. https://doi.org/10.1139/b98-066

McConnaughey AT, Whelan JF (1997) Calcification generates protons for nutrient and bicarbonate uptake. Earth Sci Rev 42:95–117. https://doi.org/10.1016/S0012-8252(96)00036-0

Mendez FJ, Losada IJ (2004) An empirical model to estimate the propagation of random breaking and nonbreaking waves over vegetation fields. Coast Eng 51(2):103–118. https://doi.org/10.1016/j.coastaleng.2003.11.003

Misurcova L, Ambrozova J, Samek D (2011) Seaweed lipids as nutraceuticals. Adv Food Nutr Res 64:339–355. https://doi.org/10.1016/B978-0-12-387669-0.00027-2

Moran-Santibanez K, Cruz-Suarez LE, Ricque-Marie D, Robledo D, Freile-Pelegrin Y, Pena-Hernandez MA, Rodriguez-Padilla C et al (2016) Synergistic effects of sulfated polysaccharides from Mexican seaweeds against measles virus. BioMed Res Int 1–11. https://doi.org/10.1155/2016/8502123

Myers SP, Mulder AM, Baker DG, Robinson SR, Rolfe MI, Brooks L, Fitton JH (2016) Effects of fucoidan from Fucus vesiculosus in reducing symptoms of osteoarthritis: a randomized placebo-controlled trial. Biol: Targets Ther 10:81–88. https://doi.org/10.2147/BTT.S95165

N'Yeurt ADR, Chynoweth DP, Capron ME, Stewart JR, Hasan MA (2012) Negative carbon via ocean afforestation. Process Saf Environ Prot 90:467–474. https://doi.org/10.1016/j.psep.2012.10.008

Nagarajan S, Mathaiyan M (2015) Emerging novel anti HIV biomolecules from marine algae: an overview. J Appl Pharm Sci 5:153–158. https://doi.org/10.7324/JAPS.2015.50928

Nanri A, Mizoue T, Shimazu T, Ishihara J, Takachi R, Noda M, Iso H et al (2017) Dietary patterns and all-cause, cancer, and cardiovascular disease mortality in Japanese men and women: the Japan public health center-based prospective study. PLoS One 12:e0174848. https://doi.org/10.1371/journal.pone.0174848

Naylor J (1976) Production, trade and utilization of seaweeds and seaweed products. FAO fisheries technical paper no. 159. Food and Agriculture Organization of the United Nations, Rome, p 73

Okazaki M, Pentecost A, Tanaka Y, Miyata M (1986) A study of calcium carbonate deposition in the genus *Padina* (Phaeophyceae, Dictyotales). Br Phycol J 21:217–224. https://doi.org/10.1080/00071618600650251

Padilla DK (1993) Rip stop in marine algae: minimizing the consequences of herbivore damage. Evol Ecol 7:634–644. https://doi.org/10.1007/BF01237826

Rajauria G, Jaiswal AK, Abu-Gannam N, Gupta S (2013) Antimicrobial, antioxidant and free radical-scavenging capacity of brown seaweed *Himanthalia elongata* from western coast of Ireland. J Food Biochem 37:322–335. https://doi.org/10.1111/j.1745-4514.2012.00663.x

Rameshkumar S, Rajaram R (2019) Chapter 22-Impact of seaweed farming on socioeconomic development of a fishing community in Palk Bay, Southeast coast of India. In: Ramkumar M, Arthur James R, Menier D, Kumaraswamy K (eds) Coastal zone management: global perspectives, regional processes, local issues. Elsevier, pp 501–513

Rebours C, Marinho-Soriano E, Zertuche-Gonzalez JA, Hayashi L, Vasquez JA, Kradolfer P, Soriano G et al (2014) Seaweeds: an opportunity for wealth and sustainable livelihood for coastal communities. J Appl Phycol 26:1939–1951. https://doi.org/10.1007/s10811-014-0304-8

Reiskind JB, Seamon PT, Bowes G (1988) Alternative methods of photosynthetic carbon assimilation in marine macroalgae. Plant Physiol 87:686–692. https://doi.org/10.1104/pp.87.3.686

Richardson MG (1979) The distribution of Antarctic marine macro-algae related to depth and substrate. Br Antarct Survey Bull 49:1–13

Robles CD, Sherwood-Stephens R, Alvarado M (1995) Responses of a key intertidal predator to varying recruitment of its prey. Ecology 76:565–579. https://doi.org/10.2307/1941214

Rosa GP, Tavares WR, Sousa PMC, Pages AK, Seca AML, Pinto DCGA (2020) Seaweed secondary metabolites with beneficial health effects: an overview of successes in *in vivo* studies and clinical trials. Mar Drugs 18(1):8. https://doi.org/10.3390/md18010008

Rosenfeld L (2000) Discovery and early uses of iodine. J Chem Educ 77:984. https://doi.org/10.1021/ed077p984

Russell BD, Thompson Jo-Anne I, Falkenberg L, Connell SD (2009) Synergistic effects of climate change and local stressors: CO_2 and nutrient-driven change in subtidal rocky habitats. Glob Chang Biol 15:2153–2162. https://doi.org/10.1111/j.1365-2486.2009.01886.x

Sabine CL, Feely RA, Gruber N, Key RM, Lee K, Bullister JL, Wanninkhof R et al (2004) The oceanic sink for anthropogenic CO_2. Science 305:367–371. https://doi.org/10.1126/science.1097403

Satpati GG (2020a) Algal sulfated polysaccharides: potent immunomodulators against COVID-19 in pandemic 2020. Biosci Biotechnol Res Asia 17(3):601–605. https://doi.org/10.13005/bbra/2863

Satpati GG (2020b) A preliminary report on plant based immunity against SARS-CoV-2 (COVID-19) in pandemic 2020. Res J Biotechnol 15(10):174–176

Satpati GG, Pal R (2011) Biochemical composition and lipid characterization of marine green alga Ulva rigida- a nutritional approach. J Algal Biomass Util 2:10–13

Satpati GG, Barman N, Pal R (2012) Morphotaxonomic account of some common seaweed from Indian Sundarbans mangrove forest and inner island area. J Algal Biomass Util 3:45–51

Satpati GG, Kanjilal S, Prasad RBN, Pal R (2015) Rapid accumulation of total lipid in Rhizoclonium africanum Kutzing as a biodiesel feedstock under nutrient limitations and the associated changes at cellular level. Int J Microbiol 1–13. https://doi.org/10.1155/2015/275035

Sengupta S, Gorain PC, Pal R (2017) Aspects and prospects of algal carbon capture and sequestration in ecosystems: a review. Chem Ecol 33:695–707. https://doi.org/10.1080/02757540.2017.1359262

Shannon E, Abu-Ghannam N (2019) Seaweeds as nutraceuticals for health and nutrition. Phycologia 58:563–577. https://doi.org/10.1080/00318884.2019.1640533

Sheldon RW, Sutcliffe WH Jr, Prakash A (1973) The production of particles in the surface waters of the ocean with particular reference to the Sargasso Sea. Limnol Oceanogr 18(5):719–733. https://doi.org/10.4319/lo.1973.18.5.0719

Škrovánková S (2011) Seaweed vitamins as nutraceuticals. Adv Food Nutr Res 64:357–369. https://doi.org/10.1016/B978-0-12-387669-0.00028-4

Smith SV (1981) Marine macrophytes as a global carbon sink. Science 211:838–840. https://doi.org/10.1126/science.211.4484.838

Steneck RS (1986) The ecology of coralline algal crusts: convergent patterns and adaptive strategies. Annu Rev Ecol Syst 17:273–303. https://doi.org/10.1146/annurev.es.17.110186.001421

Steneck R, Graham MH, Bourque BJ, Corbett D, Erlandson JM, Estes JA, Tegner MJ (2002) Kelp forest ecosystems: biodiversity, stability, resilience and future. Environ Conserv 29(4):436–459. https://doi.org/10.1017/S0376892902000322

Stow D (2004) Encyclopedia of the oceans. Oxford University Press, Oxford

Teas J, Vena S, Cone DL, Irhimeh M (2013) The consumption of seaweed as a protective factor in the etiology of breast cancer: proof of principle. J Appl Phycol 25:771–779. https://doi.org/10.1007/s10811-012-9931-0

Tegner MJ, Levin LA (1983) Spiny lobsters and sea urchins: analysis of a predator-prey interaction. J Exp Mar Biol Ecol 73:125–150. https://doi.org/10.1016/0022-0981(83)90079-5

Trono GC, Rabanal HR, I Santika (1980) Seaweed farming. FAO/UNDP South China Sea Fisheries Development and Coordinating Programme, SCS/80/WP/91. Manila, p 56

Valderrama D (2012) Social and economic dimensions of seaweed farming: a global review. IIFET 2012 Tanzania Proceedings, pp 1–11

Waters TJ, Lionata H, Prasetyo Wibowo T, Jones R, Theuerkauf S, Usman S, Amin I, Ilman M (2019) Coastal conservation and sustainable livelihoods through seaweed aquaculture in Indonesia: a guide for buyers, conservation practitioners, and farmers, Version 1. The Nature Conservancy, Arlington/Jakarta, pp 1–45

Whittaker RH, Likens GE (1973) Carbon in the biota. Brookhaven Symp Biol 30:281–302

Xiao X, Agusti S, Lin F, Li K, Pan Y, Yu Y, Zheng Y et al (2017) Nutrient removal from Chinese coastal waters by large-scale seaweed aquaculture. Sci Rep 7:46613. https://doi.org/10.1038/srep46613

Chapter 17
Seaweed Farming-Global Scenario: Socio-economic Aspects

Muhammad Sajjad Iqbal

Abbreviations

EEZ	Exclusive Economic Zone
FAO	Food and Agricultural Organization
IMTA	Integrated multi-trophic aquaculture
SLF	Seaweed liquid fertilizer

17.1 Introduction to Commercial Applications of Seaweeds

Seaweed, which is a renewable marine resource, has been underutilized essentially due to the lack of realization of its economic potential. The vast diversity of habitats, including estuarine mangrove vegetation, sandy beaches, rocky shores, deep tide pools, cliffs and caves, coral substrates, and artificial offshore structures, provide ample support for diverse seaweed groups to thrive in intertidal and subtidal waters. Seaweeds can be differentiated mainly through three taxonomic groups due to the presence of pigment composition viz., Chlorophyta (green algae), Rhodophyta (red algae), and Ochrophyta (brown algae). Each phylum is composed of thousands of species. Food, folk remedies, dyes, and fertilizers traditionally use seaweed in their confection. In the nutraceutical, pharmaceutical, and biotechnological industries, there are some applications to hydrocolloids, for instance, alginate, carrageenan, and agar are used due to their gelling properties (Rhein-Knudsen et al. 2015). During the past three decades, several applications of seaweed as functional foods &, nutraceuticals are gaining popularity (Shama et al. 2019). Furthermore, to produce therapeutic products, seaweeds have been targeted for obtaining metabolites with biological activity (Davis and Vasanthi 2011). The seaweed polysaccharides are a prominent source of dietary fibers, (Holdt and Kraan 2011) and, minerals

M. S. Iqbal (✉)
Department of Botany, University of Gujrat, Gujrat, Pakistan
e-mail: drsajjad.iqbal@uog.edu.pk

A. Ranga Rao, G. A. Ravishankar (eds.), *Sustainable Global Resources of Seaweeds Volume 1*, https://doi.org/10.1007/978-3-030-91955-9_17

(Bouga and Combet 2015). They are also sources of natural pigments, cosmetics and therapeutically active products.

The value addition through improved cultivar has opened up renewed employment opportunities in the seaweed-based industries. Moreover, micro-propagation utilizing protoplast and tissue culture of seaweed, for multiplication of elite germplasm, is being adopted in the seaweed aquaculture. Additionally, these techniques will be helpful for conservation, locally extinct species, ruthlessly harvested and exploited, due to high demand. Likewise, fast-growing tetrasporophytes have been used to develop the method of mass cultivation from vegetative fragments as well as spores with high yield (Mantri et al. 2009). Seaweed cultivation neither requires any irrigation of water nor any fertilizer, rather it helps to oxygenate the seawater and functions as a CO_2 sink.

17.2 Seaweed Farming

17.2.1 Seaweed Aquaculture: Global Overview

The global annual production of seaweeds is increasing, reaching, 31.2 million tons (fresh weight) in the year 2016 (Ferdouse et al. 2018). Of this, just 3.5% was harvested from natural populations, and 96.5% was produced in aquaculture, representing 27% of the worlds' total aquaculture production (Goecke et al. 2020). The majority of this production happened in China, Indonesia, and other Asian countries (47.9%, 38.7% and 12.8% of the worldwide production in 2016, respectively), mainly for human food and food additives. A year wise trend is showing in Fig. 17.1 while percent contribution by various countries for seaweed production is depicted in Fig. 17.2 (FAO 2017).

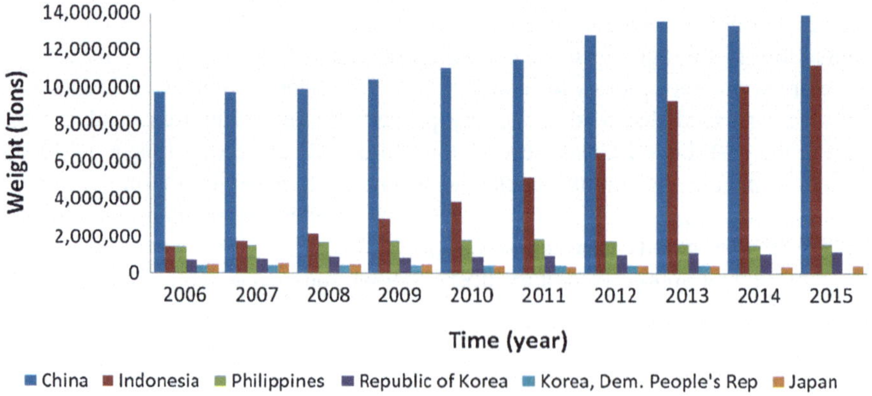

Fig. 17.1 Global main country producers, in tons. (Adapted from FAO—The global status of 2018 (García-Poza et al. 2020))

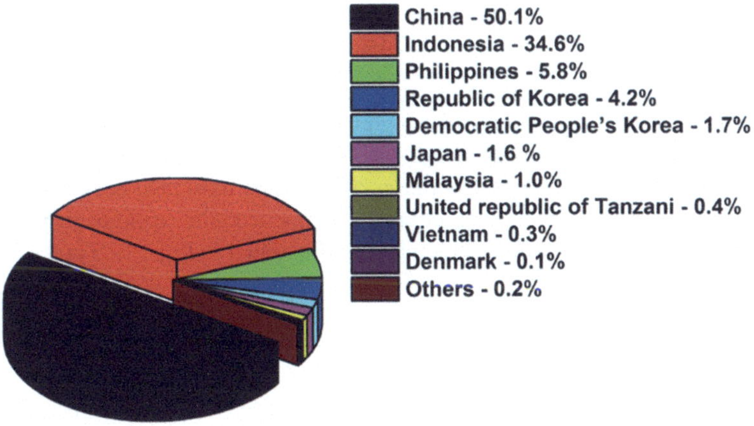

Fig. 17.2 Seaweed production countries (FAO 2017). (Adapted from FAO, Fisheries and aquaculture software. FishStatJ-software for fishery statistical time series)

The total aquaculture production of seaweeds exceeded more than double in the last 20 years and the total potential has been suggested to be 1000–100,000 million tons. But the main practice outside Asia is still to harvest natural stocks. Besides the developments in seaweed aquaculture in countries such as China, Japan, Korea, Indonesia, and the Philippines, there are also pilot-scale and pre-commercial farming projects for selected brown and red algae in Europe, Latin America, for instance in Chile and Brazil, the USA, and parts of Africa.

17.2.2 Environmental Requirements for Seaweed Aquaculture

The main environmental requirement for seaweed cultivation is seawater with quality assessment without contamination. The seaweed needs to be native to the location of aquaculture. Seaweed growth is always influenced by environmental conditions such as temperature, solar radiation, salinity, pH, and nutrient availability (Guo et al. 2015). Seaweed production site should have enough nutrients and light, as well as optimal salinity and temperatures. However, distinct species of seaweeds need different environmental conditions. Moreover, as life cycles are often complex, it is crucial to know the optimum or tolerable conditions to maximize seaweed production.

17.2.3 Farming Practices and Methods

Prominently, the following methods are employed for the large scale cultivation of seaweeds; Onshore Cultivation, (Hafting et al. 2012), Offshore Cultivation (Buck et al. 2018), Near shore Cultivation (Soto and Wurmann 2019), and IMTA Cultivation (Chávez-Crooker and Obreque-Contreras 2010). They have also been discussed in detail in this volume. These are predominantly practiced the world over. Some representative examples of farming practices are illustrated in Figs. 17.3 and 17.4.

Fig. 17.3 Large-scale seaweed farming in a multi-trophic aquaculture region of the coast of China. Sanggou Bay, a 130 km2bay in northern China annually produces 100 tonnes (fresh weight) of fed fish, 130,000 tonnes of bivalves, 2000 tonnes of abalone and 800,000 tonnes of kelp, for a total production of ~7000 tonnes km^{-2} $year^{-1}$. (Photo courtesy of Max Troell)

Fig. 17.4 A draft of a proposed land-based Integrated Multi-Trophic Aquaculture (IMTA) farm gravity-fed by Atlantic Ocean water (Green Sahara) a vision of the late Guillermo García-Blairsy Reina. (Courtesy of Bioagramar Foundation)

17.3 Saline Aquaculture

Saline aquaculture is land-based aquaculture utilizing saline groundwater. Sources include ephemeral and permanent saline lakes, saline water obtained with coal seam gas, and saline groundwater extracted from aquifers. Earthen or plastic-lined ponds, raceways, and tanks can be utilized (Allan et al. 2009). This technique is also helpful to promote marine algae culture in inland saline water (ISW) and may become an extra source of income and raw material with nominal investment.

17.4 Social Advantages of Seaweed Farming

Economic development is mostly concerned with job creation or employment, working conditions, investment, the standard of living, wealth, and quality of life improvement. Production of various seaweed species is increasing year by year which also needs comprehensive planning to yield maximum potential for local communities (Fig. 17.5). It is mainly concerned with policy formulation, planning design, strategy setting, and resource allocation. Seaweed farming is now a well-established industry that brings revenue and provides coastal people especially women, an opportunity to earn an income for themselves and their families. Food, pharmaceutical, cosmetic, and textile industries, among others, use seaweed extracts such as carrageenan, agar, and alginates as gelling substances, stabilizers, and emulsifiers. They are also used in industrial products like perfumes, shampoos, toothpaste, medicines, ice cream, milk shakes, and yoghurt.

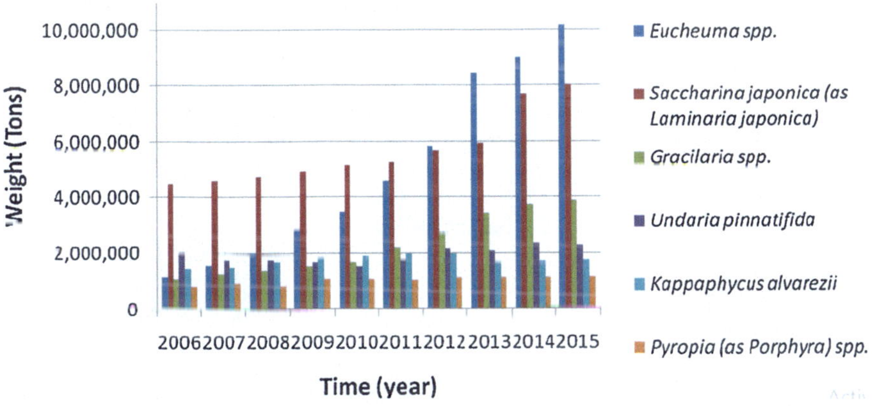

Fig. 17.5 Year wise global production of some prominent seaweed species (Ferdouse et al. 2018; García-Poza et al. 2020)

17.5 Benefits of Seaweed Farming Observed in Coastal Communities

Since its start, seaweed farming has become an important economic activity of the costal people, empowered farmers to improve their standard of living. Seaweed farming has also given recognition to women as they are involved in cultivation, while men usually assist in harvesting, carrying wet seaweed from the farms to the drying places, and carrying dry seaweed to the points of sale. This makes seaweed farming business a women centric activity. Although monthly income varies from country to country, the farmers in India, Philippines, Indonesia, Tanzania etc., continue seaweed farming business as a source of economic activity which provides additional income. As far as sustainability is concerned, many factors are beyond the ability of farmers to control e.g. demand in international markets, liquidity, and capacity of local marketing organizations as well as physical conditions that determine the supply of seaweeds (Hedberga et al. 2018). A sketch of the marketing channel is provided in Fig.17.6 which is working successfully in some states in India that need to be replicated in other countries for realizing the maximum potential of seaweed farming.

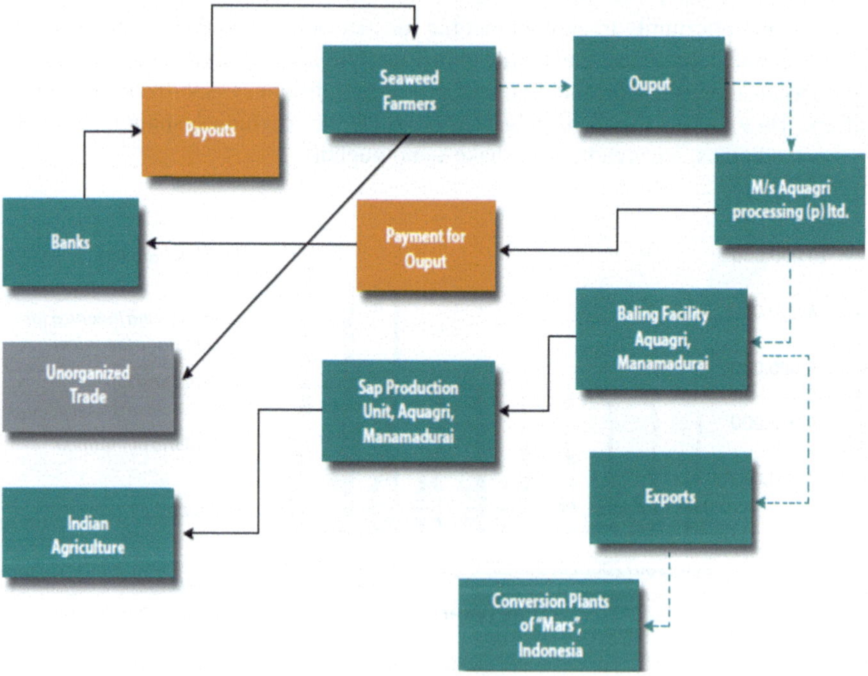

Fig. 17.6 An ideal marketing channel of seaweed farming in India

17.6 Possible Social and Economic Industrial Advantages of Seaweed Farming

Several socio-economic benefits are associated with seaweed farming in the form of value-added products as the most relevant livelihood strategy for everyone across the world (Figs. 17.7, and 17.8). Prominent value added products and their sources are shown below as representative examples;

Fig. 17.7 Cultivation of seaweeds. (**a**) Cultivation of *Ulva* on net. (Photo credit: R.M. Oza) (**b**) Cultivation of *Hydropuntia edulis* by long line rope method. (Photo credit: M. Ganesan) (**c**) Cultivation of *Hydropuntia edulis* by raft method. (Photo credit: M. Ganesan) (**d**) Cultivation of *Hypnea* by long line rope method. (Photo credit: M. Ganesan)

Fig. 17.8 Representing samples of (**a**) *Kappaphycus alvarezii,* (**b**) *Eucheuma isiformi and* (**c**) *Gelidiella acerosa*

17.6.1 Carbohydrates

Phycocolloids- Agar(*Gelidium* and *Gracilaria*)**, Alginate** *Ascophyllum, Laminaria* and *Macrocystis*,**Carragenan** Kappaphycus and *Eucheuma, Acanthophora spicifera, Grateloupia indica, Halymenia porphyroides, Halymenia venusta,* and *Hypnea musciformis*, etc.

17.6.2 Source of Proteins

Few green and red seaweeds such as *Ulva fasciata, U. rigida, Porphyra vietnamensis,* and *Centrocer asclavulatum, Acanthophora* contain rich protein 16–30% with essential amino acids.

17.6.3 Source of Minerals, Vitamins Etc.

Various types of seaweeds such as *Caulerpa, Codium, Hydroclathrus, Sargassum, Porphyra, Gracilaria, Acanthophora* and *Laurancia* are utilized as food in Japan, Indonesia, China, Philippines, and other countries of Indo-pacific regions. Nutritionally, they are rich in minerals (iodine, calcium), dietary fibre, vitamin B12, fucozanthin, fucosterol, and phlorotannin. Seaweeds are eaten in the form of salads, curry, soup or vegetables.

17.6.4 Uses as Fodder

Seaweeds such as *Ulva, Sargassum, Padina, Dictyota, Gracilaria*, and *Hypnea* could be utilized as fodder meals for cattle, poultry, and other farm animals.

17.6.5 Seaweeds as Biofertilizers

Coastal communities utilized the seaweeds as biofertilizers, due to their content of potassium, nitrogen, growth-promoting hormones, micronutrients, etc. Seaweeds are also used as a liquid fertilizer (SLF) which is biodegradable, non-toxic, non-polluting, and non-hazardous to humans, animals, and birds. *Fucus, Laminaria, Ascophyllum,* and *Sargassum* are used as raw materials in biofertilizer products like Maxicrop (UK), Kelpak 66 (South Africa), seagrow (New Zealand), Agifert (Norway), Plantozyme Shaktizyme (India), in the world markets.

17.6.6 Seaweeds as a Source of Medicine

Seaweeds are also well known for treating goiter, glandular diseases, wounds, burns, scurvy, and rashes. Laminaria and Sargassum are recognized for curing against cancer in China, Undaria to inhibit herpes simplex virus, breast cancer, and HIV (Khan and Satam 2003). *Corallina* species have been used in bone replacement therapy (Stein and Borden 1984). Goiter can be cured by *Asparagopsis taxiformis* and *Sarconema* species, while cardiovascular surgery can be performed by heparin, a seaweed extract; Carrageenan in surgical gellies, demulcents, and anti-acid tablets for treating hepatitis and ulcer (Anonymous 2021). *Digenea simplex* (red alga) releases kainic acid which is a central nervous system stimulant in neuroexcitotoxic and epileptogenic conditions. Codium species extraction has also possessed anticoagulant activity (Deacon-Smith et al. 1985).

17.7 Seaweeds for Waste Water Treatment

Seaweeds can absorb heavy metal ions such as zinc and cadmium from wastewater (Gujarat Ecology Commission 2012).

17.8 Socio-economic Impacts on Wellbeing of Local Communities

As seaweed farming generates employment by providing small-scale, family operations over the corporate, and plantation-style farms, it is considered as an alternative to fisheries. The FAO report (Valderrama 2012) indicated that seaweed farming along with aquaculture substantially increases income and reduces poverty. Some of the challenges are extremely important as low prices in some parts of the world like in Tanzania and the Solomon Islands. In Indonesian, Philippine and Mexican markets prices are in the range of US$ 0.60 through $ 1.40/kg while in Tanzania, India and the Solomon Islands prices never exceeded $ 0.38. Therefore, revenue generation is extremely important for sustainable seaweed farming. Timely availability and access of farming material, reduction in diseases like *ice-ice* disease of *K. alvarezii*, and new cultivars, capacity building of the farmers on the farm management are prominent one. Deeper-water farming promotion and enhanced farm management skills could also play key role in order to reduce production costs and improve price-negotiating capabilities.

17.9 Recent Global Trade of Seaweeds

To maintain current food consumption trends we will need 50–70% increase in world food production. It will also utilize ocean resources to produce large quantity of seaweeds for human food, animal feed and biofuels (Shama et al. 2019; Mohammad et al. 2019; Hessami et al. 2019). It is also expected that 14% per year would generate 500 million tons dry weight by 2050 as compared to 9% at present, adding about 10% to the world's present supply of food, generating revenues and improving environmental quality (WBG 2016).

World leaders in seaweed trade and utilization are including Asia (China, Thailand, the Republic of Korea, Malaysia, Indonesia, Singapore, and the Philippines); South America (Chile); Europe (the European Union, Denmark); and Africa (South Africa, Morocco and Tanzania). While its cultivation takes place in about 50 countries including Argentina, Canada, France, Japan, Mexico, Portugal, North Korea, Spain, Russia and the USA. In hydrocolloids production China is leading followed by Europe and North America as major share of the global alginate market. Markets of Asia-Pacific are increasingly more attractive for consumption of processed foods and medicines. Processed food in three different forms are available i.e., (a) edible seaweed products, (b) dried seaweed as raw materal and (c) the hydrocolloids, agar/alginates and carrageenan and agar agar (FAO 2018).

Today the global seaweed industry is worth more than USD 6 billion per annum (approximately 12 million tonnes per annum in volume) of which some 85% comprises food products for human consumption. Seaweed derived extracts (carrageenan, agar and alginates) make up almost 40% of the world's hydrocolloid market in terms of foods, the rest comes from certain animals, microbes and land plants. After china, Japanese farms are able to provide Nori, Kombu and Wakame. Nori is used for making Suchi. Japan is also a leading importer of seaweed from the republic of Korea. Indonesia reported an output from 1.2 million tonnes in 2006 to 11.3 million tonnes in 2015 and the sector in this country has expanded at 8 percent per annum with output more than doubling in this period. In Thailand and Viet Nam considerable industry is established with export and import with China, Japan, the republic of Korea and Taiwan.

According to UNComtrade the latest trade data of Seaweeds and other algae exports reflected that US $2.652 Million were involved, in which seaweed were contributed US $909 and seaweed based hydrocolloids as US $1743. Seaweeds and other algae was the world's 4528th most traded product. Prominently China, Indonesia, Republic of Korea, Philippines, Chili, Spain, France, USA, Germany and UK and rest of the world (US $432 Million) contributed 16%. Likewise the share in global trade has changes country to country (Tables 17.1 and 17.2) (Cai 2021).

Imports in 2019 showed the top ten countries which contributed about 62% while remaining supply is from rest of the world (37.54%). China is again in the top of the list with 15.34% of worth US $ 445 million followed by Japan and USA as major importers of seaweeds based hydrocolloids. In case of seaweeds China and Japan

Table 17.1 Contribution of seaweeds and seaweed based hydrocolloids exports in world trade from leading courtiers of the globe

Seaweeds and seaweed-based hydrocolloids			Seaweeds[1]			Seaweed-based hydrocolloids[2]		
Exporter	Million USD	Share of world (%)	Exporter	Million USD	Share of world (%)	Exporter	Million USD	Share of world (%)
1. China	578	21.79	1. Rep. of Korea	278	30.55	1. China	523	30.00
2. Indonesia	329	12.39	2. Indonesia	218	24.01	2. Philippines	214	12.28
3. Rep. of Korea	320	12.08	3. Chile	86	9.43	3. Spain	138	7.91
4. Philippines	252	9.52	4. China	55	6.03	4. Chile	123	7.06
5. Chile	209	7.87	5. Philippines	38	4.23	5. France	114	6.53
6. Spain	145	5.48	6. Ireland	33	3.60	6. Indonesia	110	6.34
7. France	124	4.68	7. Peru	22	2.43	7. USA	84	4.82
8. USA	102	3.85	8. Japan	21	2.33	8. Germany	76	4.39
9. Germany	82	3.11	9. USA[3]	18	1.98	9. UK	65	3.75
10. UK	78	2.93	10. Canada	18	1.97	10. South Korea	43	2.45
Rest of the world	*432*	*16.30*	*Rest of the world*	*36*	*3.93*	*Rest of the world*	*252*	*14.47*
World	**2 652**	**100.00**	**World**	**909**	**100.00**	**World**	**1 743**	**100.00**

Source: UNComtrade Access June 12, 2021 (Cai 2021)

[1] Seaweeds include cultivated and wild collected commodities under HS120220, HS120221 and HS120229.

[2] Seaweed-based hydrocolloids include HS130231 (agar), HS130239 (primarily carrageenan) and HS391310 (alginate).

are again the main importers with US $ 342 (29.47%) and 241 (20.80%), respectively. Rest of the world is also contributed US $ 236 (20.34%). The top most importers for seaweed hydrocolloids was United States of America US $ 225 million and 12.96%. As far as tariffs are concerned Seaweeds and other algae was 6.05%, making it the 3563rd lowest tariff using the HS6 product classification. Ranking: Seaweeds and other algae ranks 2690th in the Product Complexity Index (PCI).

Table 17.2 Contribution of seaweeds and seaweed based hydrocolloids imports in world trade from leading courtiers of the globe

Seaweeds and seaweed-based hydrocolloids			Seaweeds[1]			Seaweed-based hydrocolloids[2]		
Importer	Million USD	Share of world (%)	Importer	Million USD	Share of world (%)	Importer	Million USD	Share of world (%)
1. China	445	15.34	1. China	342	29.47	1. United States of America	225	12.96
2. Japan	341	11.76	2. Japan	241	20.80	2. Germany	112	6.44
3. United States of America	320	11.04	3. United States of America	95	8.17	3. China	103	5.93
4. Germany	124	4.27	4. Thailand	55	4.74	4. Spain	101	5.80
5. Spain	120	4.15	5. Taiwan Province of China	48	4.15	5. Japan	100	5.75
6. Russian Federation	116	3.99	6. France	35	3.02	6. Russian Federation	87	5.00
7. Thailand	112	3.86	7. Australia	30	2.57	7. United Kingdom	59	3.37
8. France	86	2.97	8. Russian Federation	29	2.47	8. Thailand	57	3.27
9. United Kingdom	80	2.76	9. Republic of Korea	29	2.46	9. Denmark	54	3.11
10. Denmark	67	2.32	10. United Kingdom	21	1.83	10. France	51	2.93
Rest of the world	*1 088*	*37.54*	*Rest of the world*	*236*	20.34	*Rest of the world*	*791*	45.45
World	**2 899**	**100.00**	**World**	**1 159**	**100.00**	**World**	**1 740**	**100.00**

Source: UNComtrade Access June 12, 2021 (Cai 2021)

[1] Seaweeds include cultivated and wild collected commodities under HS120220, HS120221 and HS120229.

[2] Seaweed-based hydrocolloids include HS130231 (agar), HS130239 (primarily carrageenan) and HS391310 (alginate).

17.10 Global Challenges

To meet new standards of farming, seaweed cultivation should be coupled with advance knowledge of engineering and computer science along online servers and/or workstations with the most appropriate software (Garcia-Poza et al., 2020). Aquaculture 4.0 technologies are a sustainable alternative that needs to be promoted. There is a need to develop a mechanism that can provide an efficient working system to monitor the relevant environmental factors (e.g., temperature, nutrient concentration, photoperiod, carbon dioxide, light intensity, and water flow rate) and also involvement during harvest and biomass processing.

There are some challenges with other activities such as, wild capture fisheries, hazards to protected species, renewable energy development (wind farms), and non-availability of suitable areas for offshore aquaculture. Usage of GIS tools is becoming very fruitful in resolving potential conflicts. Large-scale seaweed farming infrastructure by loose ropes and moorings also posed a serious threat to protected species like various marine mammals and sea turtles. The need is to promote entrepreneurial seaweed farming with clear knowledge of microbial (bacterial, viral, and fungal) pathogens, which would affect the productivities. Control measures are required for the epiphytes e.g., small crustaceans, hydroids, or other filamentous algae that can colonize the surface of farmed seaweed fronds, reducing seaweed production rates and degrading the quality of the product.

17.11 Research and Development

Seaweed research activities provide a wide window to explore new taxa which need taxonomical characterization not only morphologically but also through microscopic and molecular analysis. There are several habitats along the sea, in islands and subtidal water zones that remain to be explored to open up new avenues for socio-economic development through scientific knowledge and industrial application. Conservation efforts are also required for several endemic and endangered taxa. Further, the innovations in seaweed cultivation should adopt multidisciplinary approach such as (i) computational fluid dynamics (CFD); (ii) mechanical and chemical engineering; (iii) informatics and electrotechnical engineering; and (iv) biological sciences and engineering. This new industry 4.0 based aquaculture system would produce quality- biomass with high yields in a cost effective manner (Garcia-Poza et al. 2020).

17.12 Conclusions and Future Prospects

This neglected renewable marine resource requires efficient seaweed biodiversity mapping, taxonomical identification of various taxa distribute globally, To facilitate this activity the adoption of latest techniques of remote sensing and geographic information system are suggested.

The future impact of seaweed farming is directly proportional to the prices of the seaweed harvest and value addition. To counteract the effect of low prices, farmers should engage in deep water methods and the production of value-added products from biomass. Availability of disease free strains and high yield cultivars through research-intensive efforts is the need of the hour. The microfinance schemes that have proved to be so successful in places such as Bangladesh and India may expand to other parts of the world. Despite low prices, seaweed farming is a profitable venture for coastal communities, who are otherwise deprived of income opportunities.

They need to be trained as entrepreneurs and imparted needed skills for a profitable employment. In conclusion, there is need is to provide elite germplasm, impart cultivation skills backed up with innovative technologies for a gainful farming activities. Linking the biomass production with the production of valuable products will be a profitable proposition.

References

Allan GL, Fielder DS, Fitzsimmons KM, Applebaum SL, Raizada S (2009) Inland saline aquaculture. Woodhead Publishing Limited, Cambridge

Anonymous (2021) Agribusiness and marketing assistance services. Department of Agriculture. Elliptical Road, Diliman, Quezon City, Metro Manila, Philippines

Bouga M, Combet E (2015) Emergence of seaweed and seaweed-containing foods in the UK: focus on labeling, iodine content, toxicity and nutrition. Foods 4:240–253

Buck BH, Troell MF, Krause G, Angel DL, Grote B, Chopin T (2018) State of the art and challenges for offshore integrated multi-trophic aquaculture (IMTA). Front Mar Sci 5:1–21

Cai J (2021) (FAO). Global status of seaweed production, trade and utilization. Seaweed innovation forum Belize 28 May, 2021

Chávez-Crooker P, Obreque-Contreras J (2010) Bioremediation of aquaculture wastes. Curr Opin Biotechnol 21:313–317

Davis GDJ, Vasanthi AHR (2011) Seaweed metabolite database (SWMD): a database of natural compounds from marine algae. Bioinformation 5:361–364

Deacon-Smith RA, Lee-Potter JP, Rogers DJ (1985) Anticoagulant activity in extracts of Britishmarine algae. Bot Mar 28:228–333

FAO (2017) From Food and Agriculture Organization of the United Nations, Fisheries and Aquaculture software. FishStatJ-software for fishery statistical time series. In: FAO Fisheries and Aquaculture Department (online). Rome. Updated 21 July 2016. http://www.fao.org/fishery/

FAO (2018) The global status of seaweed production, trade and utilization. Globefish Res Programme Rome 124:120

Ferdouse F, Holdt SL, Smith R, Murúa P, Yang Z (2018) The global status of seaweed production, trade and utilization. FAO Globefish Res Programme 124:120

García-Poza S, Leandro A, Cotas C, Cotas J, Marques JC, Pereira L, Gonçalves AMM (2020) The evolution road of seaweed aquaculture: cultivation technologies and the industry 4.0. Int J Environ Res Pub Health 17:6528. https://doi.org/10.3390/ijerph17186528

Goecke F, Klemetsdal G, Ergon A (2020) Cultivar development of kelps for commercial cultivation-Past lessons and future prospects. Front Mar Sci 8:1–17

Gujarat Ecology Commission (2012) Study of seaweed diversity along the islands of Gulf of Kachchh. Government of Gujarat, Gujarat

Guo H, Yao J, Sun Z, Duan D (2015) Effect of temperature, irradiance on the growth of the green alga *Caulerpa lentillifera* (Bryopsidophyceae, Chlorophyta). J Appl Phycol 27:879–885

Hafting JT, Critchley AT, Cornish ML, Hubley SA, Archibald AF (2012) Onland cultivation of functional seaweed products for human usage. J Appl Phycol 24:385–392

Hedberga N, von Schreeba K, Charisiadoua S, Jiddawib NS, Tedengrena M, Nordlunda LM (2018) Habitat preference for seaweed farming-A case study from Zanzibar. Tanzania Ocean Coast Manag 154:186–195. https://doi.org/10.1016/j.ocecoaman.2018.01.016

Hessami MJ, Cheng SF, Ranga Rao A, Yin YH, Phang SM (2019) Bioethanol production from agarophyte red seaweed, *Gelidium elegans* using a novel sample preparation method for analysing bioethanol content by gas chromatography. 3- Biotech Journal 9(1):25

Holdt SL, Kraan S (2011) Bioactive compounds in seaweed: functional food applications and legislation. J Appl Phycol 23:543–597

Khan SI, Satam SB (2003) Seaweed mariculture: scope and potential in India. Aquac Asia VIII:4

Mantri VA, Thakur MC, Gupta MK, Reddy CRK, Jha B (2009) The carpospore culture of industrially important red alga *Gracilaria dura* (Gracilariales, Rhodophyta). Aquaculture 297:85–90

Mohammad JH, Ranga Rao A, Ravishankar GA (2019) Opportunities and challenges in seaweeds as feed stock for biofuel production. In: Handbook of algal technologies and phytochemicals, vol II. Phycoremediation, biofuels and global biomass production. CRC Press, USA, pp 39–50

Rhein-Knudsen N, Ale MT, Meyer AS (2015) Seaweed hydrocolloid production: an update on enzyme assisted extraction and modification technologies. Mar Drugs 13:3340–3359

Shama A, Joyce SG, Mari FD, Ranga Rao A, Ravishankar GA, Hudaa N (2019) Macroalgae and microalgae: novel sources of functional food and feed. In: Ravishankar GA, Ranga Rao A (eds) Handbook of algal technologies and phytochemicals, vol I. Food, Health and Nutraceutical Applications. CRC Press, USA, pp 207–219

Soto D, Wurmann C (2019) Offshore aquaculture: a needed new frontier for farmed fish at sea. In: The future of ocean governance and capacity development. Brill|Nijho: Leiden, pp 379–384

Stein JR, Borden CA (1984) Causative and beneficial algae in human disease conditions. A review. Phycologia 23:485–501

Valderrama D (2012) Social and economic dimensions of seaweed aquaculture: a global review. FAO Fisheries Technical Paper. Food and Agriculture Organization of the United Nations, Rome

WBG (2016) World Bank Group Environment and Natural Resoures. Seaweed aquaculture for food security, income generation and environmental health in tropical developing countries, pp 1–16

Part II
Seaweeds and Their Products for Animals and Aquaculture Industry

Chapter 18
Seaweeds and Their Products for the Health of Livestock

Stefan Kraan

Abbreviations

ALT	Alanine aminotransferase
AST	Asparate aminotrasferase
BWG	Body weight gain
COX-2	Cyclooxygenase-2
CP	Crude protein
DCs	Dendritic cells
DHA	Docosahexaenoic acid
DM	Dry matter
EE	Energy efficiency
EFSA	European food safety authority
EPA	Eicosapentaenoic acid
FCR	Feed conversion ratio
FI	Feed intake
GALT	Gut-associated lymphoid tissue
Hb	Hemoglobin
IFN	Interferon-γ
IFN-γ	Interferon-gamma
Ig	Immunoglobulin
IL	Interleukin
IL-1	Interleukin-1
IL-1β	Interleukin-1β
iNOS	Inducible nitric oxide synthase
KDa	Kilodalton
LC-n-3-PUFAs	n-3 long chain polyunsaturated fatty acids
LDH	Lactate dehydrogenase

S. Kraan (✉)
The Seaweed Company Blue Turtle Ltd., Mausrevagh, Kilcoona, Headford, Ireland
e-mail: stefan.kraan@theseaweedcompany.com

A. Ranga Rao, G. A. Ravishankar (eds.), *Sustainable Global Resources of Seaweeds Volume 1*, https://doi.org/10.1007/978-3-030-91955-9_18

LPS	Lipopolysaccharides
mRNA	Messenger ribonucleic acid
NDV	Newcastle disease virus
NFE	Nitrogen-free extract
NF-κB	Nuclear translocation, followed by
NO	Nitric oxide
OM	Organic matter
PGE2	Prostaglandin E2
PUFAs	Polyunsaturated fatty acids
RBC	Red blood cells
SCFAs	Short-chain fatty acids
SOD	Superoxide dismutase
TG	Triglycerides
TLR	Toll-like receptors
TNF-α	Tumor necrosis factor alpha
WBC	White blood cells

18.1 Introduction

Seaweeds of macroalgae have been used to improve the health and wellbeing of cattle for over a millennium, mainly in Asia. The first mention in Europe concerning the use of seaweed as cattle feed dates back to the Greeks in 45 BC (Indergaard and Minsaas 1991). For several centuries drift brown seaweed has traditionally been used as feed for farm animals, with benefits derived from its content in minerals, trace elements, and vitamins which are essential nutrients for animal growth. Use seaweeds requires few simple preparation processes, consisting of washing, milling, and drying before storage (Kraan and Guiry 2006). Unfortunately, these practices vanished in western Europe in the early twentieth Century with factors like industrial progress and lack of local knowledge of natural recourses playing a major role. Nevertheless, it seems seaweed usage for animal feed, health and wellbeing is going through some sort of renaissance in the twenty-first Century with increases in use and applications (Roque et al. 2019; Torres et al. 2020). Several studies have evaluated the utilization of seaweed and seaweed by-products for feed purposes (Indergaard and Minsaas 1991; Holdt and Kraan 2011; Shama et al. 2019). Trials that incorporated seaweed meal into different animal diets showed different benefits including an increase in the iodine content of the eggs when incorporated into poultry diets; increases in milk production in dairy cows; and the enhancement of sheep weight during the winter season, increasing lambs' birth weights and wool production (Holdt and Kraan 2011). In this chapter, we will cover the use of extracts and ingredients from seaweed in promoting health conditions in the livestock with far-reaching economic importance.

18.2 Poultry

Poultry being an important source of food, the benefits of enhancing productivity can not be underestimated. In this context, the use of seaweeds as poultry feed is a highly beneficial economic activity. Asar (1972) demonstrated that a 4.0% seaweed inclusion in the poultry feed enhanced the body weight gain significantly. Gu et al. (1988) concluded that 2.0% of marine algae meal improved broiler performance and dressing percentage, while Ventura et al. (1994) studied the effect of inclusion of *Ulva rigida* at 0.0%, 10.0%, 20.0%, and 30.0% on chicken performance. It was reported that *U. rigida* decreased feed intake (FI) and body weight gain (BWG) and they concluded that it is harmful to be included in the diet at a level higher than 10.0%. Abudabos et al. (2013) evaluated the effect of substituting 1.0% and 3.0% of corn with seaweed (*U. lactuca*) on performance, etc, and concluded that cornmeal can be replaced with Ulva and would improve weight gain and performance. Michalak et al. (2011) used *Ulva* species, *Enteromorpha prolifera* and *Cladophora* for the enrichment of feed. This resulted in heavier birds, heavier eggs, and improved nutrient value in the egg corresponding to the micronutrient tested and improved yolk colour suggesting benefits for layer production can be gained. Both Aguilar-Briseño et al. (2015) and Ibraheem et al. (2012) reported that *Ulva* (*U. clathrata, U. lactuca*) and *Caulerpa racemosa* exhibited an antiviral effect on Newcastle disease virus (NDV). In addition, *Ulva* had significant inhibitory effects on bacteria and fungi in both studies, highlighting an opportunity for use of this seaweed in poultry for anti-viral and gut health applications.

Yan et al. (2011) reported improvements in body weight and lower mortality of broilers infected with *Salmonella enteritidis* when birds were fed marine brown seaweed sodium alginate oligosaccharides (prepared with purified alginate lyase in the form of the sodium salt) at the rate of 0.04% and 0.2% of the diet. The positive results from this work included a decrease in caecal *Salmonella* spp. and significant increases in antibodies specific to *Salmonella* spp., 10 days after treatment.

Several feeding trials confirmed the beneficial effect of seaweed by-products as a novel animal feed supplement. When *Undaria pinnatifida* and *Hizikia fusiformis* waste products fermented with *Bacillus subtilis* and *Aspergillus oryzae* were supplemented to broilers, the body weight gain, gain:feed and immune response of seaweed by-product fed animals were higher, concomitantly mortality rate was lower compared to control group (Choi et al. 2014). Abbaspour et al. (2015) showed increased egg production and quality with using red algae in the diet (increased follicle production, reduced cholesterol and increased triglycerides). Changes were attributed to hypertriglyceridemia and hypocholesterolemia during biochemical transformation in laying quail. The feed supplementation at 12.5–15%, with *Gracilaria* wastes on carcass characteristics and production efficiency of ducks was confirmed with observation of lowering the fat content and raising the meat antioxidant status. Furthermore, it lowered the feed costs and increased farmers' income compared with the basal diet (Santoso et al. 2016). A reduction was reported in *E. coli* in caecal microflora samples, which was associated with increased lactic acid bacteria when brown algae was fed to broilers (Zhu et al. 2015). Antibacterial properties in laying hens, using *Chondrus crispus* and *Sarcodiotheca gaudichaudii* to reduce *Bifidobacterium* spp. and *Clostridium perfringens*, while supporting the growth of gram positive bacteria *Bifidobacterium longum* and *Streptococcus salivariu* have been demonstrated by Kulshreshtha et al. (2014). Moreover, it resulted in improved yolk weight and feed conversion ratio (FCR).

18.3 Dairy and Beef and Other Ruminants

Diet supplementation of breeding Hanwoo cows at 10% with fermented *U. pinnatifida* by-product for 2 months before parturition until weaning of their calves resulted in greater weaning weight and average daily gain, increased suckling calves weaning weight, average daily gain, serum and colostrum immunoglobulin G level in the first parity and elevation of moisture, crude fat and crude protein content of colostrum in both parities in breeding cows (Islam et al. 2016). Hong et al. (2015) studied the effect of supplementing with brown seaweed by-products up to 4% of the basal diet in Holstein dairy cows and evaluated in *in vitro* batch culture rumen fermentation. The pH tended to be higher for the higher level of supplementation, whereas the concentration of ammonia nitrogen was lower compared with the control and the volatile fatty acid concentration was hardly affected. Dry matter intake, daily gain, and feed efficiency during the transition were not affected. The concentration of plasma progesterone levels was similar, the concentration in 4% the seaweed by-product treatment increased to 158% compared with the initial level of the study. Triiodothyronine and thyroxine levels were also higher and seaweed supplementation did not affect milk yield and composition (Hong et al. 2015). The effect of 2% *U. pinnatifida* by-product supplemented diet significantly improved the average daily gain and gain:feed ratio as well as serum immunoglobulin G concentration in Hanwoo steers. Chemical composition, quality of meat, and carcass yield were unaffected and meat cholesterol, the myristic acid, and palmitoleic acid concentration were reduced, whereas the concentration of stearic acid and linolenic acid increased (Hwang et al. 2014). Bendary et al. (2013) reported that seaweed treatment (Kelp and fucus mix) showed significantly better digestibility coefficients of DM, OM, CP, EE, and NFE and subsequent nutritive values followed by premix treatment, while the control treatment revealed the lowest digestibility.

Sargassum spp. can be used up to 30% in the diets of growing sheep and goats without affecting intake, growth performance, and diet digestibility (Casas-Valdez et al. 2006; Marín et al. 2003, 2009). Feeding Sargassum increased water consumption, it also tended to decrease the concentration of volatile fatty acids (Marín et al. 2009). In dairy cows, a by-product of agar production from *Phyllophora* added (100 g) to a diet deficient in copper, zinc, and cobalt could increase milk yield by 4.4% and milk fat content by 0.24% units (Tolokonnikov et al. 1992). Arieli et al. (1993) demonstrated that *U. lactuca* can be fed to male lambs at up to 20% without adversely affecting the palatability of the diet. It had low protein degradability (40%) and also a moderate energy digestibility (60%). *Ulva* was comparable to a medium to low-quality forage and suitable for use with feeds that have a high energy/low protein content such as cereal grains. Also, *Cladophora linum* can be included at 20% in the diet of growing lambs (partially replacing barley) and had a slightly depressing effect on growth and feed conversion ratio (Ktita et al. 2010). Ramadan et al. 2020 showed in Barki sheep (*Ovis aries*) that feeding whole seaweed *Sargassum latifolium* at 0–4% in the diet that after 40 days heat stress (solar experiment 8–17) vs. mild temperature without solar exposure improved that leukocytosis, ESR, proinflammatory cytokines, HSP70 body weight gain, kidney function, blood anti-oxidant function went up.

18.4 Swine

Dierick et al. (2009) reported an improvement in pig gut health and an increase of iodine in meat using seaweed while O'Sullivan et al. (2010) demonstrated antibacterial effect and prebiotic effect using brown seaweed. Positive effects against scouring and diarrhea and reduced ammonia output were shown by Williams et al. (2001) and Reilly et al. (2008). Strong anti-helminth working against parasitic and intestinal worms was shown by Higa & Kuniyoshi (2000). In another study, the effects were evaluated of sow feed supplemented with 30 g/day of an extract of *A. nodosum* and *Fucus sp.* from the 85th day of gestation until weaning on the liver and lymphoid organs of piglets. The relative population of CD4 + CD8 + T cells was higher in piglets from treated sows in the thymus, spleen, mesenteric node, liver, and in peripheral blood, thus suggesting an important effect of maternal diet on the immune status of 40-day-old piglets (Azizi et al. 2018). In growing pigs (29 kg LW), 0.8% seaweed enhanced the immune function. Pigs were sensitized with the subcutaneous inoculum of sheep red blood cells at days 42 and 49. Seaweed increased the saliva IgA production five times more than the control after 56 days (Katayama et al., 2011). Several studies have reported that brown seaweeds have a positive influence on gut morphology. Supplementation with *Ecklonia cava* (0.05% and 0.15% of dietary inclusion), linearly improved villi height in the ileum (Choi et al. 2017). The effects of seaweed dietary supplementation on the improvement in antioxidant status and the decrease in inflammatory conditions may contribute to reducing energy and amino acid expenditure (Corino et al. 2019). In general, laminarin and fucoidan present in the brown seaweeds stimulated the growth of *Lactobacilli* and reduced the enterobacteria population or *E. coli*. Brown seaweed supplements supported the growth of *Bifidobacteria* species in the ileum in piglets. Gut health is modulated by laminarin and/or fucoidans, with the microbial production of short-chain fatty acids (SCFAs), butyrate (Corino et al. 2019).

18.5 Shrimp and Fish

Studies on the inclusion of the seaweeds *Porphyra yezoensis*, *A. nodosum,* and *U. pertusa* in the diets of fingerlings of red sea bream, *Pagrus major* showed an increase in body weight, feed utilization, and muscle protein deposition (Mustafa et al. 1995). Similar results were found by Valente et al. (2006) with two *Gracilaria spp.*, and *U. rigida* inclusion in the diets of juveniles of European sea bass, *Dicentrarchus labrax*. Experiments using the red alga *P. dioica* as a fish feed ingredient for rainbow trout *Oncorhynchus mykiss* showed that inclusion levels of 10% were feasible without any significant effects on growth or composition (Soler et al. 2009). The addition of the brown alga *Macrocystis pyrifera* in rainbow trout diets does not enhance the quantity of protein and lipid contents at muscle level but an addition of 3–6% contributes to increase the level of PUFAs, especially EPA, DHA, and LIN. Thus, the use of macroalgae meals might help to increase lipid quality content in the fish (Dantagnan et al. 2009). *Ascophyllum nodosum* meal used as binders for their alginates have immune-stimulatory effects as seen in a significantly higher level of lysozymes in Salmon after feeding a wet feed with *Ascophyllum* (Gabrielsen and Austreng 1998).

Seaweed supplementation can increase the concentration of long chain omega-3 PUFA (LC n-3 PUFAs) in various animal species, though this effect has never been assessed in salmon. Wilke et al. (2015) demonstrated that by adding seaweed blends to diets of farmed salmon total fatty acid concentrations were significantly higher (+30%, $p < 0.05$) in the flesh of the fish that were fed seaweed. Flesh concentrations of eicosapentaenoic acid (EPA) and DHA were 30% and 62% higher ($p < 0.05$). This study reveals a possible way to substantially increase EPA and DHA concentrations in farmed salmon using seaweed addition to the diet.

Zeynali et al. (2020) reported that with the inclusion of seaweed (*Sargassum ilicifolium*) meal 3–6–9% in the feed of the fish Asian sea bass (Lates calcarifer) after 6 weeks there was an improved growth and increase in; pancreatic enzyme activities serum Ig; (alternative) complement pathway components, lysozyme Ig in skin mucus, liver SOD, IL-1 and mRNA. Yeganeh and Adel (2019) reported after feeding whole seaweed (*Sargassum ilicifolium*) at 10% inclusion in the feed of the fish, great sturgeon (Huso huso), that after 8 weeks there was an improved growth and increase in serum protein, lysozyme, IgM, respiratory burst, complement Hb, RBC, WBC TG, LDH, AST, ALT, blood cholesterol and survival upon Yersinia ruckeri infection (14d. infection). Rivera et al. (2002) observed bactericidal activity of kelp preparations against *Vibrio sp.* at concentrations as low as 0.1% in shrimp. Takahasi et al. (1998) has observed the ability of fucoidan-containing meals and extracts to control WSSV in the Japanese kuruma shrimp. Cruz-Suarez et al. (2000) noted the same effects using Mexican kelp (*Macrocystis pyrifera*).

18.6 Effects on Physiological Parameters Using Whole Seaweed

18.6.1 Antioxidant Status

Using 2% addition of the brown seaweed *A. nodosum* powder in the diets of goat by Kannan et al. (2007) demonstrated an increased antioxidant status. Similarly, a 4-week treatment of rats with the brown seaweed *F. vesiculosis* stimulated increased serum paraoxonase and superoxide dismutase activities, thus leading to an increased anti-oxidant status (Zaragozá et al. 2008).

18.6.2 Blood Cell Counts and Immune Status

Increased counts were observed when chicken and fish were fed the kelp *S. japonica* or *S. oligocystum*, respectively (Bai et al. 2019). Such addition to animal feed is not only associated with increased growth and feed conversion ratios in chicken and fish but also enhanced status of innate and adaptive immune defenses and immune responsiveness and survival after infectious challenge (Zeynali et al. 2020; Yeganeh and Adel 2019; Zeraatpisheh et al. 2018). See also Sect. 18.7.1 effects of fucoidan.

18.6.3 Inflammation

Adding *S. latifolium* to sheep feed caused a reduced inflammatory response to the LPS challenge and increased blood antioxidant defense capacity (Ramadan et al. 2020). This treatment also mediated a reduced inflammatory response to heat stress challenges in these sheep. Reduction of heat stress-related effects on leukocyte oxidative and phagocytic function using *A. nodosum* was shown by Saker et al. (2004). A study with the shrimp *M. japonicas*, indicated that oral supplementation of the kelp *S. japonica* significantly increased survival upon White Spot Syndrome virus infection (Imai and Takahashi 2020). including superoxide production and antioxidative phenoloxidase activity upon appropriate stimulation.

18.7 Brown Seaweed Extracts

Although whole seaweed is often used as feed additive due to the simple way of incorporating it in animal feed diets a lot of research has been undertaken on the specific polysaccharides of brown seaweeds such as fucoidan and laminarin. Most attention goes to fucoidan as its bioactivity is most interesting and studied. Fucoidan attained GRAS-status and was EFSA approved up to 250 mg/day (Cherry et al. 2019). Laminarin, a β-glucan polysaccharide and alginate, a linear acidic soluble polysaccharide are also covered but less studies are published.

18.7.1 Main Health Attributes of Fucoidan in Respect of Animal Health

Fucoidans are a group of polysaccharides (fucans) primarily composed of sulphated L-fucose with less than 10% of other monosaccharides. They are widely found in the cell walls of brown seaweed, but not in other algae or higher plants (Berteau and Mulloy 2003). The prime function of fucoidan in cell walls is reinforcing and protecting against desiccation when the seaweed is exposed at low tide. Most brown temperate seaweeds contain between 8% and 20% of the dry weight in fucoidan of which the brown wrack *F. vesiculosus* contains the highest concentration of fucoidans of 20% on a dry weight basis (Holdt and Kraan 2011; van Weelden et al. 2019). Fucoidans are known to possess numerous biological properties with potential health applications (Berteau and Mulloy 2003). The list of biomedical applications of fucoidan for health is long (see for review Laurienzo 2010; Holdt and Kraan 2011). The biological activity (e.g. antioxidant and anti-coagulant) of sulphated polysaccharides is not only related to molecular weight and sulphated ester content (role in the charge of the molecule) but also glucuronic acid and fucose content, together with the position of the sulphate groups on the sugar residues (Berteau and Mulloy 2003; Li et al. 2005; Zhao et al. 2008). The molecular weight of fucoidan molecules range from 50–80 KDa in *U. pinnatifida* and *F. vesiculosus* respectively to 1920 kDa in *Cladosiphon* species (Fitton et al. 2015).

The most interesting aspect of fucoidan from an animal health perspective is gut health and the immune system modulation, and the numerous important biological effects of fucoidans that are related to their ability to modify cell surface properties (Usov et al. 2001). The ingestion of fucoidan has also been shown to increase the anti-pathogenic activity of granulocytes and macrophages in healthy people (Myers et al. 2011). Increased immune activity after fucoidan ingestion can even eliminate the tropical protozoan parasite *Leishmania*, as seen in a mouse model (Kar et al. 2011). Fucoidans can also promote the maturation of dendritic cells, activate NK cells, and promote cytotoxic activity (Zhang et al. 2015; Vetvicka and Vetvickova 2017). Following oral dosing of mice with fucoidan, splenocytes showed significant increases in interleukin-2 expression and phagocytic activity similar to effects by equivalent doses of yeast and mushroom derived (1,3/1,6)-β-D-glucans. Adaptive immunity was also affected by fucoidan as evidenced by significant increases in antibody titers in mice vaccinated with ovalbumin (Halling et al. 2015). Fucoidan has been found to restore the immune functions of immune-suppressed mice, act as an immunomodulator directly on the macrophage, T lymphocyte, B cell, natural killer cells (Wang et al. 1994), promote the recovery of immunologic function in irradiated rats (Wu et al. 2003), induce the production of interleukin (IL-1) and interferon-γ (IFN-γ) *in vitro*, and promote the primary antibody response in sheep red blood cells *in vivo* (Yang et al. 1995). Fucoidan has been reported to increase levels of the chemokine SDF1 in the blood and elicit the release of stem cells into the peripheral circulation when used intravenously in animal models (Sweeney et al. 2002).

The mechanism by which fucoidan protects cells is through the activation of the host immune responses and the benefits observed are attributed to its ability to modulate cellular immune function. Through binding to receptors such as Toll-like receptors (TLRs) on dendritic cells (DCs), macrophages, and other monocytes, and then activates them to release pro-inflammatory factors, cytokines, and chemokines, which can help the host to form a strong immune response and achieve multi-channel and multi-level regulation of the immune system (Akira et al. 2006). Wang (2019) has reviewed the biological function of fucoidan mode of action and therapeutic effects. Some anti-viral properties of sulphated fucans have also been characterised, for example inhibition of bovine viral diarrhoea virus (Iqbal et al. 2000). Many studies suggest that fucoidan has potential for use as an anti-inflammatory agent. A study showed that fucoidan treatment led to less severe symptoms in the early stages of *Staphylococcus aureus*-triggered arthritis in mice, but delayed phagocyte recruitment and decreased clearance of the bacterium (Verdrengh et al. 2000).

In addition, fucoidan is an excellent natural antioxidant and presents significant antioxidant activity in experiments *in vitro* from seaweed sources such as *F. vesiculosus* (Ruperez et al. 2002), *S. japonica* (Zhao et al. 2004), and *E. kurome* (Xu et al. 2001). Furthermore, fucoidan shows a good anti-inflammatory working as shown *in vivo* in rats (Cumashi et al. 2007) and as prophylactic against enteral prion infections (Doh-Ura et al. 2007). Polysaccharides, such as fucoidan, have been successfully used in drug delivery as oral antibiotics to inhibit the growth of *S. aureus* and *E. coli* and to prevent the adhesion of *Helicobacter pylori* biofilms in the gastric mucosa (Besednova et al. 2015). Oral administration of fucoidan has its effects via gut microbiome modulation, via interaction with gut epithelial wall and/or modulation of gut-associated lymphoid tissue (GALT), as well to intestinal fucoidan absorption and distribution to different tissues. Modulation of the gut microbiome has been proven in agricultural production animals, mouse models, and human clinical trials. The absorption of fucoidan through the small intestine was demonstrated both in mouse models and clinical trials (Nagamine et al. 2014). Inherently fucoidan is a competitive binding agent for envelope viruses, preventing cellular entrance (Hayashi et al. 2007). In its capacity to enhance the immune system, fucoidan has been shown to have improved outcome in viral load (Araya et al. 2011) and serum antibody levels and survival (Hayashi et al. 2007). Mechanistically, fucoidan treatment is thought to induce NF-κB nuclear translocation, followed by iNOS and COX-2 transcription in macrophages. Fucoidan potentiates this effect via scavenger receptor A and TLR4. In unstimulated macrophages, fucoidan induces the secretion of pro-inflammatory cytokines IFN-γ, TNF-α, and IL-1β and inflammatory mediators NO and PGE2 (Lee et al. 2020; Takeda et al. 2012). When lymphocytes and macrophages are pre-treated with fucoidan and subsequently stimulated with lipopolysaccharides (LPS) or other pro-inflammatory stimuli, fucoidan blunts the pro-inflammatory reaction or otherwise renders an anti-inflammatory effect and results in inhibition of NF-κB translocation and lower levels of pro-inflammatory cytokines and mediators (Sanjeewa et al. 2019; Jayawardena et al. 2019; Ni et al. 2020).

18.7.2 Feed Applications of Fucoidan and Other Polysaccharides

Since the 1950s brown seaweeds such as *A. nodosum* and *Fucus* sp. have been used as animal feed supplement (Mac Monagail and Morrison 2020) although the benefits and improved production have not directly been linked to fucoidan. In the search for new applications to reduce and prevent the use of antibiotics, stimulation of the immune system has been proven as an effective route (Corino et al. 2019). In this antibiotics-replacement quest, brown seaweed polysaccharides, and especially fucoidan as a dietary supplement was found to improve the immune response of farmed shrimp, fish, poultry, pigs/piglets and even ruminants (Table 18.1). Fucoidan reduces susceptibility to infection and improves growth performance.

18.8 Conclusion and Future Prospects

Overall, the evidence examining the effect of seaweed extracts and in particular fucoidan on growth performance in farm animals is equivocal, either as a direct physiological effect or as a probiotic throughout health (Corino et al. 2019). Reasons for the differences in responses observed between studies may be due to factors such as, differences in dietary inclusion levels, variations in the types and purity of seaweed extracts evaluated and differences in the species, age and health status of animals used, etc. Additional studies examining the effects of purified polysaccharides from seaweeds indicate the positive effects of fucoidan on gut health and immune system and indirect on health and well-being as well as production in farmed animals. An enhanced antiviral effect of fucoidan is seen with the oral application of fucoidan. These studies, therefore, lay a foundation for the development of fucoidan as a new generation of polysaccharide immunomodulator for use in animal husbandry to improve the production and health status of the farmed animals. It is envisaged that seaweed extracts or purified brown seaweed polysaccharides such as fucoidan will form an inherent part of the feeding strategy in farmed animals in the future.

Table 18.1 Overview of seaweed polysaccharide extracts and pure polysaccharide compounds and their effects after supplementation in animal diets on growth, survival, and physiology

Compound/format	Effective concentrations	Seaweed	Experimental model	Significant findings seaweed supplementation	References
Poultry					
Sodium alginate oligosaccharides	0.04% and 0.2% of diet	*Brown seaweeds, Ascophyllum, Laminaria Macrocystis*	Arbor Acres broiler chickens	Decrease in caecal Salmonella, Increase in Salmonella antibodies in unchallenged animals: Increase in IFN-γ, IL-10, IL-1β	Yan et al. (2011)
Sodium alginate oligosaccharides	0.2%	*Brown seaweeds, Ascophyllum, Laminaria Macrocystis*	Arbor Acres broiler chickens	Increase in feed intake and growth Increase in plasma glutathione (GSH) levels, serum IgM levels Increase in malondialdehyde (MDA) content Increase in the number of beneficial lactic acid bacteria in the cecal Decrease in number of harmful *E. coli*, Increase in the content of acetic acid, lactic acid and total volatile fatty acids in the cecum	Zhu et al. (2015)

(continued)

Table 18.1 (continued)

Compound/format	Effective concentrations	Seaweed	Experimental model	Significant findings seaweed supplementation	References
Commercial extract	0.5% and 3% in feed; 41 week	*Ascophyllum nodosum* *Chondrus crispus*	Lohmann LSL-Lite and Lohmann Brown-Lite hens, heat stress	Short-term decrease in feed intake; Increase in feed/egg efficiency Strain-dependent production, Increase in feed efficiency, Increase in heat stress resistance, improved ALP, ALT, GGT liver parameters	Borzouie et al. (2020)
Seaweed powder + anti-bacterial peptides (cecropin)	1–5% of basal diet	*Laminaria japonica*	Arbor Acres broiler chicks	Synergistic effect of seaweed + cecropin: Increase in Abs, increased lymphocytes, microbiota: Increase in Lactobacillus, Decrease *E. coli* Increase in feed conversion ratio	Bai et al. (2019)
By products and fermented seaweed by products	0.5%	*Undaria pinnatifida* *Hizikia fusiformis*	Ross male broilers chicks 1 day old	Increase in body weight gain, FCR, immunoglobulin; Decreased mortality	Choi et al. (2014)
Fucoidan	0.25 and 16 μg	*Cladosiphon okamurans*	11 day old chicken embryo's	90% infection rate of New Castle disease virus	Trejo-Avilla et al. (2016)

Compound/format	Effective concentrations	Seaweed	Experimental model	Significant findings seaweed supplementation	References
Crude fucoidan	500 ppm	*Sargassum tenerrimum*	One-day-old Ross 308 male broiler chickens	Increased body weight gain, FCR, immunoglobulin None of the treatments had significant effect on FI and mortality. Decreased concentration of triglycerides and very low density lipoprotein cholesterol (VLDL-c)	Shoikayan et al. (2019)
Bovine/ruminants					
Laminarin	1 g/day Pre-weaning (0–62 days) & post-weaning (63–93 days)	*Laminaria*	Holstein Friesian bull calves	Decreased growth Increase in serum haptoglobin Decrease in lymphocyte levels Decrease in stimulated IFN-γ (in vitro challenges)	McDonnell et al. (2019)
Commercial extract	1% feed, 27 days prior to and during exp. (10 days)	*Ascophyllum nodosum*	Crossbred wether lamb (*Ovis aries*); heat stress	Decrease in heat stress-induced reduction of phagocyte oxidative burst Increased SOD Decrease in heat stress-induced changes in GSH-peroxidase activity Decrease in lipid peroxidation Increase in leukocyte phagocytosis	Saker et al. (2004)

(continued)

Table 18.1 (continued)

Compound/format	Effective concentrations	Seaweed	Experimental model	Significant findings seaweed supplementation	References
Commercial extract	2% in feed, 3 week	*Ascophyllum nodosum*	Spanish and Boer x Spanish goat; stress by transport and feed withholding	No difference in cortisol, WBC and subset counts, but decrease in Eo No difference in phagocytosis Decreased lipid peroxidation; decrease SOD (strain diff.) Increased glutathione peroxidase (10–15%)	Kannan et al. (2007)
Fermented Undaria by product	10% of diet	*Undaria pinnatifida*	Hanwoo cows 2 months before parturition until weaning of calves	Increased weaning weight, daily gain, serum and colostrum immunoglobulin G, moisture, crude fat and crude protein in colostrum	Islam et al. (2016)
Undaria by products	2% of diet	*Undaria pinnatifida*	Hanwoo steers	Increase in daily gain feed: gain ratio serum immunoglobulin G, stearic and linolenic acid in meat no difference in chemical composition, quality of meat and carcass yield Decrease in cholesterol, myristic acid, palmitoleic acid	Hwang et al. (2014)
Swine					
Laminarin, fucoidan and ash	Lam 1 g, Fuc 0.8 g, d 107 of gestation until weaning (d 26)	*Laminaria digitata*	Pregnant + lactating sows	Increase in colostrum IgA Increased piglets serum IgG	Leonard et al. (2012)
Laminarin, fucoidan and ash	Lam 1 g, Fuc 0.8 g, d 83 of gestation until weaning (d 28)	*Laminaria digitata*	Pregnant + lactating sows	Increased piglets villus height in the jejunum and ileum	Heim et al. (2015)

Compound/format	Effective concentrations	Seaweed	Experimental model	Significant findings seaweed supplementation	References
Extract	D 83 of gestation until weaning (d 28)	*Ascophyllum nodosum* and *Fucus vesiculosis*	Pregnant + lactating sows	Increased piglets CD^{4+} and CD^{8+} T cells	Azizi et al. (2018)
Extract	LAM (0.15–0.30 g/kg) FUC (0.24 g/kg) LAM + FUC (0.15 + 0.24 and 0.30 + 0.24 g/kg)	*Laminaria digitata Fucus vesiculosis*	Weaned piglets (24 d age)	Increased digestibility, feed efficiency, weight gain, lactobacilli Decrease in *E. coli*	O'Dorehty et al. (2010), Reilly et al. (2008), and McDonnel et al. (2010)
Alginate oligosaccharides	0, 3, 4, 5 or 6 g kg–1 molecular weight of ~10 kDa	*Brown seaweeds, Ascophyllum, Laminaria Macrocystis*	28-d experiment using Duroc × (landrace × Yorkshire) piglets and weaned at 35 days of age	Increase in the average daily gain and weight gain/consumption ratio anti-oxidant capacity of piglets (serum GSH content was increased)	Zhu et al. (2015)
Alginate oligosaccharides	g 0, 50, 100 or 200 mg kg^{-1} molecular weight < 1000 kDa	*Brown seaweeds, Ascophyllum, Laminaria Macrocystis*	Weaned pigs (Duroc × landrace × Yorkshire, weaned at 21 days)	Increased average daily body weight gain anti-oxidant defense properties by enhancing serum catalase activity and GSH content; improved intestinal enzyme activity and digestibility of nutrients.	Wan et al. (2017, 2018a, b, c)

(continued)

Table 18.1 (continued)

Compound/format	Effective concentrations	Seaweed	Experimental model	Significant findings seaweed supplementation	References
Alginate oligosaccharides	100 mg kg^{-1} molecular weight < 1000 kDa	Brown seaweeds, Ascophyllum, Laminaria Macrocystis	24 weaned pigs (Duroc × landrace × Yorkshire, weaned at 21 days	Increase intestinal closure protein abundance and inhibiting the secretion of inflammatory factors. Inhibited the production of pro-inflammatory cytokines and protected the cell integrity of the intestinal wall by inhibiting TLR4/NF-κB and NOD1/NF-κB signaling pathways	Wan et al. (2018b)
Alginate oligosaccharides	100 mg kg^{-1} molecular weight < 1000 kDa	Brown seaweeds, Ascophyllum, Laminaria Macrocystis	Twenty-four weaned pigs (Duroc × landrace × Yorkshire, weaned at 21 days	Restrained enterocyte death, by reducing both mitochondria-dependent and TNFR1-dependent apoptosis and the accelerated enterocyte proliferation, via enhancing the cyclin E-CDK2 complex formation villus height, content of sIgA and goblet cell counts	Wan et al. (2018a, c)
Fish and shrimp					
Hot water extract	100–500 mg/kg in feed, 12 week	*Sargassum oligocystum*	Fish, Pangasius *(Pangasianodon hypophthalmus)*	Increased weight, daily growth rate, feed conversion ratio Increase in WBC, RBC, Hb, Hc, platelets	Baleta and Bolanos (2019)

Compound/format	Effective concentrations	Seaweed	Experimental model	Significant findings seaweed supplementation	References
Aqueous extract	400 mg/kg, 8 week	*Sargassum angustifolium*	Fish, rainbow trout *(Oncohrynchus mykiss)*	Increased weight gain Increase in Hb, Hc, RBC, WBC, total protein, albumin increase survival and immune response to *Yersinia ruckeri* infection	Zeraatpisheh et al. (2018)
Hot water extract, and HCl-EtOH extract	100 mg/kg/day (heat extract), 10 mg/kg/day (HCl-EtOH extract), for 3–7 day	*Laminaria japonica*	Kuruma shrimp (Marsupenaeus japonicas) in vivo WSSV infection in vitro hemocyte analysis	Increased survival upon WSSV infection hemocyte fMet-Leu-Phe stimulation: Increased chemotaxis, superoxide production, phenol oxidase activity and phagocytosis	Imai and Takahashi (2020)
Fucoidan	6 g/kg feed for 21 days	*Laminaria Japonica*	Immunosuppressed African catfish	Macrophages: decrease oxidative burst, increase in phagocytic activity Lymphocytes: Increased transformation index serum: Increased lysozyme, NO and bactericidal activity Increased survival rate in challenge test	El-Boshy et al. (2014)

(continued)

Table 18.1 (continued)

Compound/format	Effective concentrations	Seaweed	Experimental model	Significant findings seaweed supplementation	References
Fucoidan-rich extract	2% inclusion, 45 days feeding trial	*Sargassum wightii*	Sutchi Catfish	Macrophages: Increased oxidative burst, increased phagocytic activity Increase in lymphocyte count Decrease in albumin/globulin ratio Increase in IFN-γ Increased survival rate in challenge test	Prabu et al. (2016)
Laminarin	0.2 g/kg/day for 21 days	*Laminaria digitata*	Rainbow trout	Increased TNF-α, IL-8	Morales-Lange et al. (2015)
Fucoidan	0.1%	*Sargassum horneri*	Yellow catfish	Macrophages: Increased oxidative burst, increased phagocytic activity Serum, lysozyme, catalase, SOD, decreases MDA, increases survival rate in challenge test	Yang et al. (2014)
Fucoidan-rich seaweed extract (FRSE) + Methionine	2% FRSE + 0.3 methionine	*Sargassum wightii*	Carp (Labeo rohita)	Increased respiratory burst activity, phagocytic activity, Increased MPO activity, lysozyme activity, total immunoglobulin and TLC	Mir et al. (2017)
Laminarin	0.5–1.0% inclusion in diet for 48 days	Commercially sourced	Grouper (Epinephelus coioides)	Increased IL-1β, IL-8, and TLR2 Increased LZM, CAT and SOD Increased growth rate and the feed efficiency	Yin et al. (2014)

Compound/format	Effective concentrations	Seaweed	Experimental model	Significant findings seaweed supplementation	References
Fucoidan	250–500 mg/kg feed	*Sargassum, Padina, and Turbinaria*	White shrimp (Litopenaeus vannamei)	Increased THC, PA, and SOD Increased LGBP, toll, and lectin	Setyawan et al. (2018)
Fucoidan extract	0% fucoidan-rich seaweed extract (FRSE), 1% FRSE, 2% FRSE, 3% seaweed powder and 6% seaweed powder for 60 days.	*Sargassum wightii*	Labeo rohita	Increased mRNA expression of antimicrobial peptides in liver, skin and intestine tissues. Increase in survival in bacterial challenge study using Aeromonas hydrophila, non-specific immune response (NBT reduction, serum lysozyme activity, serum albumin: globulin ratio and phagocytic activity) in pre-challenge and post-challenge periods. No difference in growth performance	Gora et al. (2018)
Fucoidan extract	100 mg/kg body weight per day	*Sargassum polycystum*	Black tiger shrimp *Penaeus monodon*	Increase in survival rate after white spot disease, inhibits growth *V. harvey*, *S. aureus* and *E. coli*	Chotigeat et al. (2004), Takahasi et al. (1998), and Baba et al. (1988)

Abbreviations: *FRSE*, fucoidan-rich seaweed extract, *IFN-γ* interferon gamma, *IL-10* interleukin-10, *IL-1β* interleukin-1β, *GSH* glutathione, *IgM* immunoglobulin M, *MDA* malondialdehyde, *ALP* alkaline phosphatase, *ALT* alanine aminotransferase, *GGT* gamma-glutamyl transferase, *VLDL-C* very low-density lipoprotein cholesterol, *IgA* immunoglobulin-A, *IgG* immunoglobulin-G, *TLR4* toll-like receptor-4, *NF-kB* nuclear factor-kappa B, *RBC* red bloood cell, *Hb* hemoglobin, *hc* hematocrit, *WBC* white blood cell, *WSSV* white spot syndrome virus, *NO* nitric oxide, *MPO* myeloperoxidase, *MDA* malondialdehyde, *IL-8* interleukin-8, *LZM* lysozyme, *CAT* catalase. *THC* total hematocytes count, *PA* phagocytic activity, *NBT* nitroblue tetrazolium, *LGBP* lipopolysaccharide- and ß-glucan-binding protein

References

Abbaspour B, Davood SS, Mohammadi-Sangcheshmeh A (2015) Dietary supplementation of *Gracilariopsis persica* is associated with some quality related sera and egg yolk parameters in laying quails. J Sci Food Agric 95:643–648

Abudabos AM, Samara EM, Hussein EOS, Al-Ghadi MQ, Al-Atiyat RM (2013) Impacts of stocking density on the performance and welfare of broiler chickens. Ital J Anim Sci 12(1):66–71. https://doi.org/10.4081/ijas.2013.e11

Aguilar-Briseño JA, Cruz-Suarez LE, Sassi J-F, Ricque-Marie D, Zapata-Benavides P, Mendoza-Gamboa E, Rodríguez-Padilla C, Trejo-Avila LM (2015) Sulphated Polysaccharides from *Ulva clathrata* and *Cladosiphon okamuranus* seaweeds both inhibit viral attachment/entry and cell-cell fusion, in NDV infection. Mar Drugs 13:697–712

Akira S, Uematsu S, Takeuchi O (2006) Pathogen recognition and innate immunity. Cell 124:783–801

Araya N, Takahashi K, Sato T, Nakamura T, Sawa C, Hasegawa D, Ando H, Aratani S, Yagishita N, Fujii R, Oka H, Nishioka K, Nakajima T, Mori N, Yamano Y (2011) Fucoidan therapy decreases the proviral load in patients with human T-lymphotropic virus type-1-associated neurological disease. Antivir Ther 16(1):89–98

Arieli A, Sklan D, Kissil G (1993) A note on the nutritive value of *Ulva lactuca* for ruminants. Anim Prod 57:329–331

Asar M (1972) The use of some weeds in poultry nutrition, Degree Diss. University of Alexandria, Egypt

Azizi AFN, Miyazaki R, Yumito T, Ohashi Y, Uno S, Miyajima U, Kumamoto M, Uchiyama S, Yasuda M (2018) Effect of maternal supplementation with seaweed powder on immune status of liver and lymphoid organs of piglets. J Vet Med Sci 80:8–12

Baba M, Snoeck R, Pauwels R, Clerq ED (1988) Sulfated polysaccharides are potent and selective inhibitors of various enveloped viruses, including herpes simplex virus, cytomegalovirus, vesicular stomatitis virus, and human immunodeficiency virus. Antimicrob Agents Chemother 32(11):1742–1745

Bai J, Wang R, Yan L, Feng J (2019) Co-supplementation of dietary seaweed powder and antibacterial peptides improves broiler growth performance and immune function. Braz J Poultry Sci 21. https://doi.org/10.1590/1806-9061-2018-0826

Baleta FN, Bolanos JM (2019) Growth and immune response of Pangasius hypophthalmus fed diets containing seaweed extracts as immunostimulant. Braz Arch Biol Technol 62

Bendary MM, Bassiouni MI, Ali MF, Gaafar HM, Shamas A (2013) Effect of premix and seaweed additives on productive performance of lactating friesian cows. Int Res J Agric Sci 1(1):174–181

Berteau O, Mulloy B (2003) Sulfated fucans, fresh perspectives: structures, functions, and biological properties of sulfated fucans and an overview of enzymes active toward this class of polysaccharide. Glycobiology 13(6):29–40

Besednova NN, Zaporozhets TS, Somova LM, Kuznetsova TA (2015) Review: prospects for the use of extracts and polysacc harides from marine algae to prevent and treat the diseases caused by *Helicobacter pylori*. Helicobacter 20:89–97

Borzouie S, Rathgeber BM, Stupart CM, MacIsaac J, MacLaren LA (2020) Effects of dietary inclusion of seaweed, heat stress and genetic strain on performance, plasma biochemical and hematological parameters in laying hens. Animals 10:1570

Casas-Valdez M, Hernandez-Contreras H, Marin-Alvarez A, Aguila-Ramirez RN, Hernandez-Guerrero CJ, Sanchez-Rodriguez I, Carrillo-Dominguez S (2006) The seaweed *Sargassum* (*Sargassaceae*) as tropical alternative for goats' feeding. Rev Biol Trop 54:83–92

Cherry P, O'Hara C, Magee PJ, McSorley EM, Allsopp PJ (2019) Risks and benefits of consuming edible seaweeds. Nutr Rev 77(5):307–329

Choi YJ, Lee SR, Oh JW (2014) Effects of dietary fermented seaweed and seaweed fusiforme on growth performance, carcass parameters and immunoglobulin concentration in broiler chicks. Asian Australas J Anim Sci 27(6):862–870

Choi Y, Hosseindoust A, Goel A, Lee S, Jha PK, Kwon IK, Chae BJ (2017) Effects of Ecklonia cava as fucoidan–rich algae on growth performance, nutrient digestibility, intestinal morphology and caecal microflora in weanling pigs. Asian–Australas J Anim Sci 30:64–70

Chotigeat W, Tongsupa S, Supamataya K, Phongdara A (2004) Effect of fucoidan on disease resistance of black tiger shrimp. Aquaculture 233:23–30

Corino C, Clotilde Modina S, Di Giancamillo A, Chiapparini S, Rossi R (2019) Seaweeds in pig nutrition. Animals 9:1126

Cruz-Suarez LE, Ricque-Marie D, Tapia-Salazar M, Guajardo-Barbosa C (2000) Uso de harina de kelp (Macrocystis pyrifera) en alimentos para camaron. In: Cruz-Suarez LE, Ricque-Marie D, Tapia-Salazar M, Olvera-Novoa MA, Cerecedo-Olvera R (eds) Avances en Nutricion Acuıcola V – Memorias del Quinto Simposium Internacional de Nutricion Acuıcola, Merida, Mexico, 19–22 Noviembre 2000, vol 5. Universidad Autonoma de Nuevo Leon, Monterrey, pp 227–266

Cumashi A, Ushakova NA, Preobrazhenskaya ME (2007) A comparative study of the anti-inflammatory, anticoagulant, antiangiogenic, and antiadhesive activities of nine different fucoidans from brown seaweeds. Glycobiology 17(5):541–552

Dantagnan P, Hernandez A, Borquez A, Mansilla A (2009) Inclusion of macroalgae meal (*Macrocystis pyrifera*) as feed ingredient for rainbow trout (*Oncorhynchus mykiss*): effect on flesh fatty acid composition. Aquac Res 41:87–94

Dierick N, Ovyn A, De Smet S (2009) Effect of feeding intact brown seaweed *Ascophyllum nodosum* on some digestive parameters and on iodine content in edible tissues in pigs. J Sci Food Agric 89:584–594

Doh-Ura K, Kuge T, Uomoto M, Nishizawa K, Kawasaki Y, Iha M (2007) Prophylactic effect of dietary seaweed Fucoidan against enteral prion infection. Antimicrob Agents Chemother 51(6):2274–2277

El-Boshy M, El-Ashram A, Risha E, Abdelhamid F, Zahran E, Gab-Alla A (2014) Dietary fucoidan enhance the non-specific immune response and disease resistance in African catfish, *Clarias gariepinus*, immunosuppressed by cadmium chloride. Vet Immunol Immunopathol 162(3–4):168–173

Fitton JH, Stringer DN, Karpiniec SS (2015) Therapies from Fucoidan: an update. Mar Drugs 13(9):5920–5946

Gabrielsen BO, Austreng E (1998) Growth, product quality and immune status of Atlantic Salmon, *Salmo salar* L. fed wet feed with alginate. Aquac Res 29:397–401

Gora AH, Sahu N, Sahoo P, Rehman S (2018) Effect of dietary *Sargassum wightii* and its fucoidan-rich extract on growth, immunity, disease resistance and antimicrobial peptide gene expression in *Labeo rohita*. Showkat Int Aquat Res 10:115–131

Gu HY, Liu YG, Shu ZZ (1988) Nutrient composition of marine algae and their feeding on broilers. Chin J Anim Sci 3:12–14

Halling BP, Vetvicka V, Blakemore WR (2015) Evaluation of the immunomodulatory *in vivo* activity of *Laminaria hyperborea* fucoidan relative to commercial (1,3/1,6) Β D glucans from yeast and mushrooms. J Nutr Health Sci 2(2):205

Hayashi S, Itoh A, Isoda K, Kondoh M, Kawase M, Yagi K (2007) Fucoidan partly prevents CCl4-induced liver fibrosis. Eur J Pharmacol 580(3):380–384

Heim G, O'Doherty JV, O'Shea CJ, Doyle DN, Egan AM, Thornton K, Sweeney T (2015) Maternal supplementation of seaweed derived polysaccharides improves intestinal health and immune status of suckling piglets. J Nutr Sci 4:e27

Higa T, Kuniyoshi M (2000) Toxins associated with medicinal and edible seaweeds. J Toxicol Tox Rev 19:119–137

Holdt S, Kraan S (2011) Bioactive compounds in seaweed; functional food applications and legislation. Review article. J Appl Phycol 23:543–597

Hong ZS, Kim EJ, Jin YC, Lee JS, Choi YJ (2015) Effects of supplementing brown seaweed by-products in the diet of Holstein cows during transition on ruminal fermentation, growth performance and endocrine responses. Asian Australas J Anim Sci 28(9):1296–1302

Hwang JA, Islam MM, Ahmed ST, Mun HS, Kim GM, Kim YJ, Yang CJ (2014) Seamustard (*Undaria pinnatifida*) improves growth immunity fatty acid profile and reduces cholesterol in hanwoo steers Asian–Australasian. J Animal Sci 27(8):1114–1123

Ibraheem IBM, Abdel-Raouf N, Abdel-Hameed MS, Eil-Yamany K (2012) Antimicrobial and antiviral activities against Newcastle disease virus (NDV) from marine algae isolated from Qusier and Marsa-Alam Seashore (Red Sea), Egypt. Afr J Biotechnol 11:8332–8340

Imai T, Takahashi Y (2020) Chemotaxis assay for *Marsupenaeus japonicas* hemocytes and application for the development of an oral immunostimulant against white spot syndrome virus. Front Cell Dev Biol 8(46):1–10

Indergaard M, Minsaas J (1991) Animal and human nutrition 21-64. In: Guiry MD, Blunden G (eds) Seaweed resources in Europe: uses and potential. John Wiley & Sons, Chichester, p 432

Iqbal M, Flick-Smith H, McCauley JW (2000) Interactions of bovine viral diarrhoea virus glycoprotein E(rns) with cell surface glycosaminoglycans. J Gen Virol 81(Pt 2):451–459

Islam MM, Ahmed ST, Mun H-S, Kim Y-B, Yang C-J (2016) Effect of fermented seaweed by-product supplementation on reproduction of Hanwoo cows and growth and immunity of their calves. Anim Prod Sci 56(11):1828–1833

Jayawardena TU, Fernando IPS, Lee WW, Sanjeewa KKA, Kim HS, Lee DS, Jeon YJ (2019) Isolation and purification of fucoidan fraction in *Turbinaria ornata* from the Maldives; inflammation inhibitory potential under LPS stimulated conditions in in-vitro and in-vivo models. Int J Biol Macromol 131:614–623

Kannan G, Saker K, Terrill T, Kouakou B, Galipalli S, Gelaye S (2007) Effect of seaweed extract supplementation in goats exposed to simulated preslaughter stress. Small Rumin Res 73:221–227

Kar S, Sharma G, Das PK (2011) Fucoidan cures infection with both antimony-susceptible and -resistant strains of *Leishmania donovani* through Th1 response and macrophage-derived oxidants. J Antimicrob Chemother 66(3):618–625

Katayama M, Fukuda T, Okamura T, Suzuki E, Tamura K, Shimizu Y, Suda Y, Suzuki K (2011) Effect of dietary addition of seaweed and licorice on the immune performance of pigs. Anim Sci J 82:274–281

Kraan S, Guiry MD (2006) The seaweed resources of Ireland. In: Seaweed resources of the world, A CD-rom project. Alan Critchley, Masao Ohno and Danilo Largo, ETI, Amsterdam

Ktita SR, Chermiti A, Mahouachi M (2010) The use of seaweeds (*Ruppia maritima* and *Chaetomorpha linum*) for lamb fattening during drought periods. Small Rumin Res 91:116–119

Kulshreshtha G, Rathgeber B, Stratton G, Thomas N, Evans F, Critchley A, Hafting J, Prithiviraj B (2014) Immunology, health, and disease: feed supplementation with red seaweeds, *Chondrus crispus* and *Sarcodiotheca gaudichaudii*, affects performance, egg quality, and gut microbiota of layer hens. Poult Sci 93(12):2991–3001

Laurienzo P (2010) Marine polysaccharides in pharmaceutical applications: an overview. Mar Drugs 8(9):2435–2465

Lee HH, Cho Y, Kim GH, Cho H (2020) *Undaria pinnatifida* Fucoidan-Rich extract recovers immunity of immunosuppressed mice. J Microbiol Biotechnol 30(3):439–447

Leonard SG, Sweeney T, Bahar B, O'Doherty JV (2012) Effect of maternal seaweed extract supplementation on suckling piglet growth, humoral immunity, selected microflora, and immune response after an *ex vivo* lipopolysaccharide challenge. J Anim Sci 90:505–514

Li N, Zhang QB, Song JM (2005) Toxicological evaluation of fucoidan extracted from *Laminaria japonica* in Wistar rats. Food Chem Toxicol 43:421–426

Mac Monagail M, Morrison L (2020) The seaweed resources of Ireland: a twenty-first century perspective. J Appl Phycol 32(2):1287–1300

Marín A, Casas M, Carrillo S, Hernandez H, Monroy A (2003) Performance of sheep fed rations with Sargassum spp. sea algae. Cuban J Agric Sci 37:119–123

Marín A, Casas-Valdez M, Carrillo S, Hernandez H, Monroy A, Sangines L, Perez-Gil F (2009) The marine algae *Sargassum* spp. (Sargassaceae) as feed for sheep in tropical and subtropical regions. Rev Biol Trop 57:1271–1281

McDonnell P, Figat S, O'Doherty JV (2010) The effect of dietary laminarin and fucoidan in the diet of the weanling piglet on performance, selected fecal microbial populations and volatile fatty acid concentrations. Animal 4:579–585

McDonnell RP, O'Doherty JV, Earley B, Clarke AM, Kenny DA (2019) Effect of supplementation with n-3 polyunsaturated fatty acids and/or β-glucans on performance, feeding behaviour and immune status of Holstein Friesian bull calves during the pre- and post-weaning periods. J Anim Sci Biotechnol 10:7

Michalak I, Chojnacka K, Dobrański Z, Górecki H, Zielińska A, Korczyński M, Opaliński S (2011) Effect of macroalgae enriched with microelements on egg quality parameters and mineral content of eggs, eggshell, blood, feathers and droppings. J Anim Physiol Anim Nutr 95:374–387

Mir IN, Sahu NP, Pal AK, Makesh M (2017) Synergistic effect of l-methionine and fucoidan rich extract in eliciting growth and non-specific immune response of *Labeo rohita* fingerlings against *Aeromonas hydrophila*. Aquaculture 479:396–403

Morales-Lange B, Bethke J, Schmitt P, Mercado L (2015) Phenotypical parameters as a tool to evaluate the immunostimulatory effects of laminarin in *Oncorhynchus mykiss*. Aquac Res 46:2707–2715

Mustafa MG, Wakamatsu S, Takeda TA, Umino T, Nakagawa H (1995) Effects of algae meal as feed additive on growth, feed efficiency, and body composition in Red Sea Bream. Fisheries Sci 61:25–28

Myers SP, O'Connor J, Fitton JH et al (2011) A combined phase I and II open-label study on the immunomodulatory effects of seaweed extract nutrient complex. Biologics 5:45–60

Nagamine T, Nakazato K, Tomioka S, Iha M, Nakajima K (2014) Intestinal absorption of fucoidan extracted from the brown seaweed, *Cladosiphon okamuranus*. Mar Drugs 13(1):48–64

Ni L, Wang L, Fu X, Duan D, Jeon YJ, Xu J, Gao X (2020) *In vitro* and *in vivo* anti-inflammatory activities of a fucose-rich fucoidan isolated from *Saccharina japonica*. Int J Biol Macromol 156:717–729

O'Doherty JV, Dillon S, Figat S, Callan JJ, Sweeney T (2010) The effects of lactose inclusion and seaweed extract derived from *Laminaria* spp. on performance, digestibility of diet components and microbial populations in newly weaned pigs. Anim Feed Sci Tech 157:173–180

O'Sullivan L, Murphy B, McLoughlin P, Duggan P, Lawlor PG, Hughes H, Gardiner GE (2010) Prebiotics from marine macroalgae for human and animal health applications. Mar Drugs 8:2038–2064

Prabu DL, Sahu NP, Pal AK, Dasgupta S, Narendra A (2016) Immunomodulation and interferon gamma gene expression in sutchi cat fish, *Pangasianodon hypophthalmus*: effect of dietary fucoidan rich seaweed extract (FRSE) on pre and post challenge period. Aquac Res 2016(47):199–218

Ramadan G, Fouda WA, Ellamie AM et al (2020) Dietary supplementation of *Sargassum latifolium* modulates thermo-respiratory response, inflammation, and oxidative stress in bacterial endotoxin-challenged male Barki sheep. Environ Sci Pollut Res 27:33863–33871

Reilly P, O'Doherty JV, Pierce KM et al (2008) The effects of seaweed extract inclusion on gut morphology, selected intestinal microbiota, nutrient digestibility, volatile fatty acid concentrations and the immune status of the weaned pig. Animal 2:1465–1473

Rivera GF, Yoong G, Riofrio B, Reinoso F, Hurtado, Massuh P (2002) Inclusion de harina de kelp (*Macrocystis pyrifera*) en alimento balanceado para el camaron. Congreso Iberoamerica no Virtual de Acuicultura. CIVA, pp 244–252. http://www.civa2002.org

Roque BM, Salwen JK, Kinley R, Kebreab H (2019) Inclusion of *Asparagopsis armata* in lactating dairy cows' diet reduces enteric methane emission by over 50 percent. J Clean Prod 234:132–138

Rupérez P, Ahrazem O, Leal JA (2002) Potential antioxidant capacity of sulphated polysaccharides from the edible marine brown seaweed *Fucus vesiculosus*. J Agric Food Chem 50(4):840–845

Saker KE, Fike JH, Veit H, Ward DL (2004) Brown seaweed- (Tasco™) treated conserved forage enhances antioxidant status and immune function in heat-stressed wether lambs. J Anim Physiol Anim Nutr 88(3/4):122–130

Sanjeewa KK, Jayawardena TU, Kim HS, Kim SY, Shanura Fernando IP, Wang L, Abetunga DTU, Kim WS, Lee DS, Jeon YJ (2019) Fucoidan isolated from *Padina commersonii* inhibit LPS-induced inflammation in macrophages blocking TLR/NF-κB signal pathway. Carbohydr Polym 15(224):115195

Santoso SI, Suprijatna E, Setiadi A, Susanti S (2016) Effect of duck diet supplemented with fermented seaweed wastes on carcass characteristics and production efficiency of indigenous Indonesian ducks. Indian J Anim Res 50(5):699–704

Setyawan A, Isnansetyo A, Murwantoko M, Indarjulianto S, Handayani CR (2018) Comparative immune response of dietary fucoidan from three Indonesian brown algae in white shrimp *Litopenaeus vannamei*. AACL Bioflux 11(6):1707–1723

Shama A, Joyce SG, Mari FD, Ranga Rao A, Ravishankar GA, Hudaa N (2019) Macroalgae and microalgae: novel sources of functional food and feed. In: Ravishankar GA, Ranga Rao A (eds) Handbook of algal technologies and phytochemicals, food, health and nutraceutical applications, vol. I. CRC Press, USA, pp 207–219

Shokaiyan M, Ashayerizadeh O, Shams Shargh M, Dastar B (2019) Algal crude fucoidan alone or with *Bacillus subtilis* DSM 17299 in broiler chickens diet: growth performance, carcass characteristics, blood metabolites, and morphology of intestine. Poult Sci J 7(1):87–94

Soler-Vila A, Coughlan S, Guiry MD, Kraan S (2009) The red alga *Porphyra dioica* as a fish-feed ingredient for rainbow trout (*Oncorhynchus mykiss*): effects on growth, feed efficiency, and carcass composition. J Appl Phycol 21(5):617–624

Sweeney EA, Lortat-Jacob H, Priestley GV, Nakamoto B, Papayannopoulou T (2002) Sulphated polysaccharides increase plasma levels of SDF-1 in monkeys and mice: involvement in mobilization of stem/progenitor cells. Blood 99(1):44–51

Takahasi Y, Uehara K, Watanabe R, Okumura T, Yamashita T, Omura H, Yomo T, Kawano T, Kanemitsu A, Narasaka H, Suzuki N, Itami T (1998) Efficacy of oral administration of fucoidan, a sulphated polysaccharide, in controlling white spot syndrome in Kumura Shrimp in Japan. In: Flegel TW (ed) Advances in shrimp biotechnology. National Center for Genetic Engineering and Biotechnology, Bangkok, pp 171–173

Takeda K, Tomimori K, Kimura R, Ishikawa C, Nowling TK, Mori N (2012) Anti-tumor activity of fucoidan is mediated by nitric oxide released from macrophages. Int J Oncol 40(1):251–260

Tolokonnikov YA, Podokopnaya IS, Pisarskaya TN (1992) *Phyllophora* in the diet for cows. Zootekhniya 5(6):15–17

Torres MD, Kraan S, Dominguez H (2020) Sustainable seaweed technologies; advances in green and sustainable chemistry. Elsevier, Dordrecht, p 330

Trejo-Avila LM, Elizondo-Gonzalez R, Rodriguez-Santillan P, Aguilar-Briseño PA, Ricque-Marie D, Rodriguez-Padilla C, Cruz-Suarez E (2016) Innocuity and anti-Newcastle-virus-activity of *Cladosiphon okamuranus* fucoidan in chicken embryos. Poult Sci 95(12):2795–2802

Usov AI, Smirnova GP, Klochkova NG (2001) Polysaccharides of algae: 55. Polysaccharide composition of several brown algae from Kamchatka. Russ J Bioorg Chem 27:395–399

Valente LMP, Gouveia A, Rema P, Matos J, Gomes EF, Pinto IS (2006) Evaluation of three seaweeds *Gracilaria bursa-pastoris*, *Ulva rigida* and *Gracilaria cornea* as dietary ingredients in European sea bass (*Dicentrarchus labrax*) juveniles. Aquaculture 252(1):85–96

Van Weelden G, Bobiński M, Okła K, van Weelden WJ, Romano A, Pijnenborg J (2019) Fucoidan structure and activity in relation to anti-cancer mechanisms. Mar Drugs 17(1):32

Ventura MR, Castañon JIR, McNab JM (1994) Nutritional value of seaweed (*Ulva rigida*) for poultry. Anim Feed Sci Technol 49:87–92

Verdrengh M, Erlandsson-Harris H, Tarkowski A (2000) Role of selectins in experimental *Staphylococcus aureus*-induced arthritis. Eur J Immunol 30(6):1606–1613

Vetvicka V, Vetvickova J (2017) Fucoidans stimulate immune reaction and suppress cancer growth. Anticancer Res 37:6041–6046

Wan J, Zhang J, Chen D, Yu B, He J (2017) Effects of alginate oligosaccharide on the growth performance, antioxidant capacity and intestinal digestion-absorption function in weaned pigs. Anim Feed Sci Technol 234:118–127

Wan J, Zhang J, Chen D, Yu B, He J (2018a) Alginate oligosaccharide alleviates enterotoxigenic Escherichia coli-induced intestinal mucosal disruption in weaned pigs. Food Funct 9:6401–6413

Wan J, Zhang J, Chen D, Yu B, Huang Z, Mao X, Ping Z, Yu J, He J (2018b) Alginate oligosaccharide enhances intestinal integrity of weaned pigs through altering intestinal inflammatory responses and antioxidant status. RSC Adv 8:13482–13492

Wan J, Zhang J, Chen D, Yu B, Mao X, Zheng P, Yu J, Luo J, He J (2018c) Alginate oligosaccharide-induced intestinal morphology, barrier function and epithelium apoptosis modifications have beneficial effects on the growth performance of weaned pigs. J Anim Sci Biotechnol 9:937–948

Wang WT, Zhou JH, Xing ST (1994) Immunomodulating action of marine algae sulfated polysaccharides on normal and immunosuppressed mice. Chin J Pharm Toxicol 8:199–202

Wang Y, Xing M, Cao Q, Ji A, Liang H, Song S (2019) Biological activities of fucoidan and the factors mediating its therapeutic effects: a review of recent studies. Mar Drugs 17(3):183

Wilke T, Faulkner S, Kealy L, Murphy L, Kraan S, Brouns F (2015) Dietary inclusion of seaweed is associated with increased total fatty acid and n-3 PUFA concentrations in Atlantic salmon. Eur J Lipid Sci Technol 2015(116):1–6

Williams BA, Verstegen MWA, Tamminga S (2001) Fermentation in the large intestine of single-stomached animals and its relationship to animal health. Nutr Res Rev14:207–227

Wu XW, Yang ML, Huang XL, Yan J, Luo Q (2003) Effect of fucoidan on splenic lymphocyte apoptosis induced by radiation. Chin J Radiol Med Prot 23:430–432

Xu HL, Kitajima C, Ito H et al (2001) Antidiabetic effect of polyphenols from brown alga *Ecklonia kurome* in genetically diabetic KK-A(y) mice. Pharm Biol 50(3):393–400

Yan GL, Guo YM, Yuan JM, Liu D, Zhang BK (2011) Sodium alginate oligosaccharides from brown algae inhibit *Salmonella Enteritidis* colonization in broiler chickens. Poult Sci 90(7):1441–1448

Yang XL, Sun JY, Xu HN (1995) An experimental study on immunoregulatory effect of fucoidan. Chin J Marine Drugs (6):9–13

Yang Q, Yang R, Li M, Zhou Q, Liang X, Elmada ZC (2014) Effects of dietary fucoidan on the blood constituents, anti-oxidation and innate immunity of juvenile yellow catfish (*Pelteobagrus fulvidraco*). Fish Shellfish Immunol 41(2):264–270

Yeganeh S, Adel M (2019) Effects of dietary algae (Sargassum ilicifolium) as immunomodulator and growth promoter of juvenile great sturgeon (Huso huso Linnaeus, 1758). J Appl Phycol 31:2093–2102

Yin G, Li W, Lin Q et al (2014) Dietary administration of laminarin improves the growth performance and immune responses in *Epinephelus coioides*. Fish Shellfish Immunol 41:402–406

Zaragozá MC, López D, Sáiz M, Poquet M, Pérez J, Puig-Parellada P, Màrmol F, Simonetti P, Gardana C, Lerat Y, Burtin P, Inisan C, Rousseau I, Besnard M, Mitjavila MT (2008) Toxicity and antioxidant activity in vitro and in vivo of two *Fucus vesiculosus* extracts. J Agric Food Chem 56:7773–7780

Zeraatpisheh F, Firouzbakhsh F, Khalili KJ (2018) Effects of the macroalga Sargassum angustifolium hot water extract on hematological parameters and immune responses in rainbow trout (Oncohrynchus mykiss) infected with Yersinia rukeri. J Appl Phycol 30:2029–2037

Zeynali M, Nafisi Bahabadi M, Morshedi V, Ghasemi A, Torfi Mozanzadeh M (2020) Replacement of dietary fishmeal with *Sargassum ilicifolium* meal on growth, innate immunity and immune gene mRNA transcript abundance in Lates calcarifer juveniles. Aquac Nutr 26:1657–1668

Zhang W, Oda T, Yu Q, Jin JO (2015) Fucoidan from *Macrocystis pyrifera* has powerful immunemodulatory effects compared to three other fucoidans. Mar Drugs 13(3):1084–1104

Zhao X, Xue C-H, Li J-Z, Cai Y-P, Liu H-Y, Qi H-T (2004) Antioxidant and hepatoprotective activities of low molecular weight sulfated polysaccharide from Laminaria japonica. J Appl Phycol 16:111–115. https://doi.org/10.1023/B:JAPH.0000044822.10744.59
Zhao X, Xue CH, Li BF (2008) Study of antioxidant activities of sulphated polysaccharides from *Laminaria japonica*. J Appl Phycol 20:431–436
Zhu W, LiI D, Wang J, Wu H, Xia X, BiI W, Guan H, Zhang L (2015) Effects of polymannuronate on performance, antioxidant capacity, immune status, cecal microflora, and volatile fatty acids in broiler chickens. Poult Sci 94:345–343

Chapter 19
Seaweeds for Animal Feed, Current Status, Challenges, and Opportunities

Khalid M. Mahrose and Izabela Michalak

19.1 Introduction

In the present chapter, the current trends concerning the application of seaweeds in animal feeding are presented. The literature review focuses on publications mainly from the Web of Science database published in the last decade. The interest in seaweeds in livestock nutrition stems from many factors. Nowadays, there is a need for inexpensive and readily available raw materials that can be used as feed additives (Nhlane et al. 2020). For this purpose, the resources occurring in the marine environment – seaweeds can be used to address this challenge (Bonos et al. 2016; Erum et al. 2017; Nhlane et al. 2020). This raw biomass has advantages over land crops, commonly used in animal feeding because its production is independent of arable land and weather (Halmemies-Beauchet-Filleau et al. 2018). Very often it is considered to partially replace common ingredients in the feed with the seaweed biomass (Rjiba et al. 2010). Seaweeds can constitute a substitute for a protein source in the feed instead of soybean or canola meal (Angell et al. 2016; Bay-Larsen et al. 2018; Haberecht et al. 2018; Øverland et al. 2019; Shama et al. 2019) or other nutrients, especially minerals (Michalak et al. 2011; Rey-Crespo et al. 2014; Machado et al. 2015; Cabrita et al. 2016; Momeneh et al. 2018). Angell et al. (2016) indicated that the quality of seaweed protein is similar to, if not better than, traditional protein sources. A novel approach in animal feeding was proposed by El-Maaty et al. (2021).

K. M. Mahrose (✉)
Animal and Poultry Production Department, Faculty of Technology and Development, Zagazig University, Zagazig, Egypt
e-mail: ostrichkhalid@zu.edu.eg

I. Michalak (✉)
Wrocław University of Science and Technology, Faculty of Chemistry, Department of Advanced Material Technologies, Wrocław, Poland
e-mail: izabela.michalak@pwr.edu.pl

A. Ranga Rao, G. A. Ravishankar (eds.), *Sustainable Global Resources of Seaweeds Volume 1*, https://doi.org/10.1007/978-3-030-91955-9_19

Table 19.1 Review and research articles on chemical composition of seaweeds potentially used in animal feeding

Component	References
Micro- and macroelements	Machado et al. (2015), Cabrita et al. (2016), Makkar et al. (2016), Corino et al. (2019), Michalak and Marycz (2019), Abbott et al. (2020), and Morais et al. (2020)
Ash	Makkar et al. (2016), Øverland et al. (2019), Corino et al. (2019), Abbott et al. (2020), Bikker et al. (2020), and Morais et al. (2020)
Protein	Angell et al. (2016), Makkar et al. (2016), Øverland et al. (2019), Corino et al. (2019), Michalak and Marycz (2019), Abbott et al. (2020), and Morais et al. (2020)
Amino acids	Machado et al. (2015), Angell et al. (2016), Makkar et al. (2016), Corino et al. (2019), and Bikker et al. (2020)
Carbohydrate	Øverland et al. (2019), Michalak and Marycz (2019), Bikker et al. (2020), and Morais et al. (2020)
Dietary fiber	Makkar et al. (2016), Corino et al. (2019), Abbott et al. (2020), Bikker et al. (2020), and Morais et al. (2020)
Lipid	Machado et al. (2015), Øverland et al. (2019), Michalak and Marycz (2019), Bikker et al. (2020), and Morais et al. (2020)
Vitamins	Corino et al. (2019), and Morais et al. (2020)

Brown seaweed – *Sargassum latifolium* converted into extract was used for the production of calcium nanoparticles supplemented to the diet of laying hens, which can improve egg quality, including eggshell thickness and shell weight.

In the literature, there are many reviews and research articles, presenting the detailed chemical characteristics of seaweeds, comparing the composition of different species of algae belonging to green (*Chlorophyceae*), red (*Rhodophyceae*), and brown algae (*Phaeophyceae*). Table 19.1 presents a list of publications, which described the most important nutrients potentially used in animal nutrition and Fig. 19.1 shows the examples of seaweed species used in *in vivo* studies on animals.

In general, seaweeds besides the rich content of minerals and proteins with a relatively well-balanced amino acid profile can be utilized in animal feed as a source of carbohydrates, dietary fibers, vitamins and other bioactive compounds (Kulshreshtha et al. 2020). The inclusion level of seaweeds to animal diets usually depends on their mineral profile – the content of toxic metals and iodine, which can limit the use of marine algae in the diet (Cabrita et al. 2016). When utilizing seaweeds in animal nutrition, seasonal variations in the nutrient composition should be considered (Abudabos et al. 2013; Michalak and Mahrose 2020). Additionally, the chemical composition of seaweeds can vary with species, maturity, habitat, geographical location etc. (Erum et al. 2017; Michalak and Mahrose 2020).

Seaweeds can provide animals with nutrients and bioactive compounds, which exhibit nutraceutical properties (Shama et al. 2019; Nhlane et al. 2020) and may have a positive and beneficial effect on animal health (Erum et al. 2017). Seaweeds are known to improve gastrointestinal microflora, thereby immunity and overall health of animals (Kulshreshtha et al. 2020). Especially seaweed polysaccharides are considered as an ideal prebiotic for animals – they stimulate the growth of beneficial bacteria, but inhibit the pathogenic strains (Kulshreshtha et al. 2014).

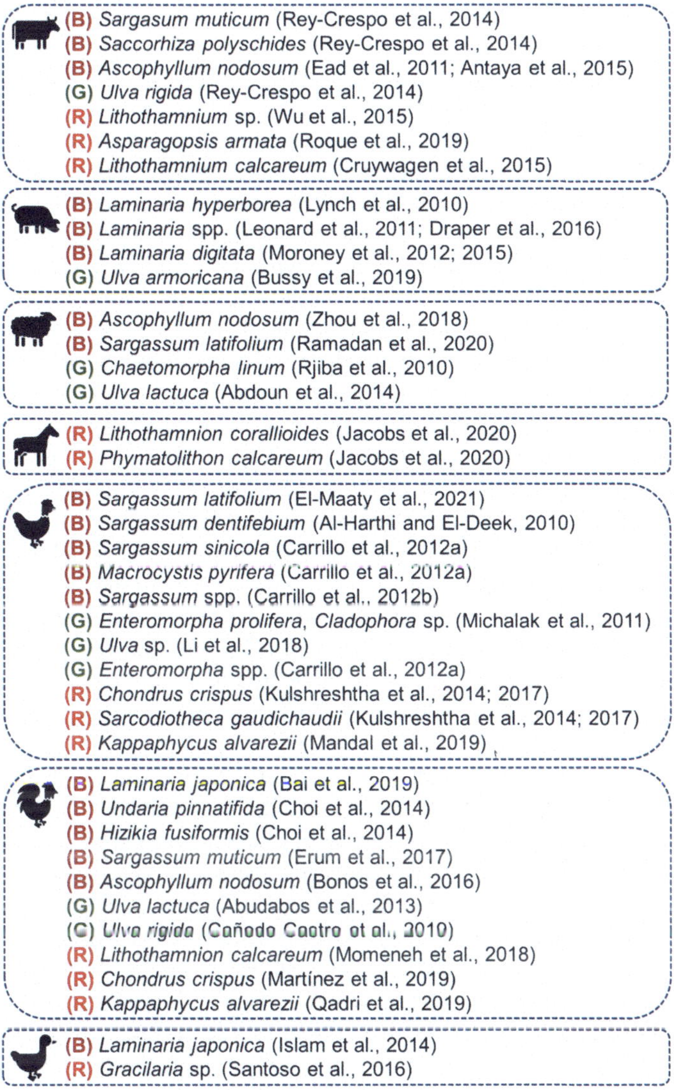

Fig. 19.1 The examples of seaweed species used in *in vivo* studies on animals (own source; (B) – brown seaweeds, (G) – green seaweeds, (R) – red seaweeds)

Seaweeds, in the future, may be an alternative to antibiotics. Nowadays, there is a lot of interest in the mitigation of antibiotics use in animal production by the application of unconventional feed resources such as seaweeds (Islam et al. 2014; Kulshreshtha et al. 2014). Due to the content of bioactive compounds and dietary fiber, seaweeds could improve intestinal integrity and reduce the concentration of lipids in serum (Cañedo-Castro et al. 2019). Some of the animal diseases caused by pathogens, e.g. pig respiratory disease (Bussy et al. 2019), endotoxemia in sheep (Ramadan et al. 2020), the Newcastle disease in poultry (Bai et al. 2019) can be treated with seaweeds, which stimulate immune responses and limit infections in farm animals.

Seaweeds can also be used to stabilize pH in the digestive system and provide favorable environment for microbial activity. For this reason, seaweeds rich in Ca and Mg are chosen – for example calcareous marine algae. These seaweeds contain high levels of calcium (300 g/kg) and magnesium (55 g/kg) (Cruywagen et al. 2015; celticseaminerals.com). Cruywagen et al. (2015) used red seaweed – *Lithothamnium calcareum* – for stabilization of pH in rumen of Holstein cows. Gastric juice pH in horses was stabilized with seaweeds – *Lithothamnion corallioides* and *Phymatolithon calcareum* (Jacobs et al. 2020). These seaweeds can keep long-term the optimum pH in the digestive system as opposite to sodium bicarbonate, which is commonly used as a dietary buffer but is short-lived in the rumen (Cruywagen et al. 2015). Calcified seaweeds can constitute a source of dietary calcium, which is commonly provided in animal diet by limestone. However, high content of limestone in the feed can reduce phosphorus digestibility and can lead to a reduction in the integrity of the skeleton. This novel algal additive can improve broiler leg health – their strength and bone mineral content (Bradbury et al. 2012).

Seaweeds can also be beneficial to cope with environmental problems related to livestock production. The reduction of the carbon footprint of the livestock industry is one of the sustainability goals (Zhou et al. 2018; Maia et al. 2019). Livestock production is known to contribute to greenhouse gas emissions, which are responsible for the global warming (Lee et al. 2018; Roque et al. 2019). Therefore, the development of feed additives, which may reduce the anthropogenic greenhouse gas emissions (particularly enteric methane production in ruminants) is of special interest (Belanche et al. 2016; Lee et al. 2018; Zhou et al. 2018; Maia et al. 2019; Roque et al. 2019; Abbott et al. 2020). Algae added to the feed can solve this problem and additionally can improve the rumen microbiome and reduce the prevalence of food-borne pathogens (Zhou et al. 2018). Seaweeds bioactive compounds like carbohydrates, proteins, to a lesser extent lipids, saponins, and alkaloids could reduce enteric CH_4 production from livestock (Abbott et al. 2020). Abbott et al. (2020) identified in the red seaweed *Asparagopsis taxiformis* a compound bromoform acting as an agent that can reduce CH_4 production. However, most of these studies are performed *in vitro* (e.g. Belanche et al. 2016; Lee et al. 2018; Maia et al. 2019; Abbott et al. 2020). *In vivo* studies are necessary to confirm the effectiveness of algae in reduction of enteric CH_4 (e.g. Roque et al. 2019).

Finally, seaweeds or seaweeds extracts incorporated into animal feed can find an application in the development of novel functional animal-derived products (e.g. meat, milk, eggs) enriched with algal compounds (Michalak et al. 2011; Moroney et al. 2012). Extracted from brown seaweeds polysaccharides (e.g. laminarin and fucoidan) acting as natural antioxidants can enhance not only animal health, but also increase the quality of fresh meat (decrease in lipid oxidation) (Moroney et al. 2012).

In the present chapter we focus on seaweed species that have the potential to be used as common feed additives, current situation in animal feeding with seaweeds based on literature review from the last decade, challenges that arise and possible solutions.

19.2 Preparation of Seaweeds for Animal Feeding

Seaweeds, intact and processed, can constitute a valuable component of animal feed (Michalak and Mahrose 2020). In many regions, locally available seaweeds are used as a feed rich in nutrients – especially by farmers living in the coastal areas (Bay-Larsen et al. 2018). Other option, the more commonly used is to apply the seaweed biomass harvested from the coast, washed with water to remove salt and impurities, dried and ground as a feed additive (e.g. Al-Harthi and El-Deek 2012; Abudabos et al. 2013; Abdoun et al. 2014; Santoso et al. 2016).

Currently, seaweeds are more and more often used in animal feeding in the form of extracts or extracted specific compounds – especially polysaccharides (laminarin, fucoidan) (e.g. Lynch et al. 2010; Moroney et al. 2012, 2015). This approach requires the use of extraction techniques tailored to the isolation of a given component from seaweeds. Red seaweeds – *Grateloupia lanceolata, Hypnea japonica, Pterocladia capillacea, Chondria crassicaulis, Gelidium amansii* after washing, cutting into small pieces, freeze-drying and grinding were extracted with 70 or 80% methyl alcohol, using ultrasounds at room temperature. After extraction, eluates were filtered, solvent was evaporated and the obtained dry mass was re-dissolved in dimethyl sulfoxide (Lee et al. 2018). Wan et al. (2017) proposed the application of alginate oligosaccharide (non-toxic, biodegradable polymer) as a novel feed supplement in pig production obtained by depolymerization of seaweed alginic acid polysaccharide by alginate lyases. Li et al. (2018) used extracted from green seaweeds ulvan (sulphated polysaccharide) as a new feed additive in a diet of laying hens. For this purpose, maceration of dry algal powder in water with the enzyme – cellulase was performed. Before being used in the feed, seaweeds can also be fermented. This approach can facilitate the digestibility of this biomass and its use by animals and additionally can improve animals growth performance and immune responses (Choi et al. 2014). For fermentation, microorganisms like *Bacillus subtilis* and *Aspergillus oryzae* (Choi et al. 2014) or *Aspergillus niger* (Santoso et al. 2016) can be used.

The use of seaweeds as feed additives in *in vivo* experiments, requires detailed research related to the analysis of the seaweeds chemical composition (especially the content of toxic metal ions, which can disqualify them from further use), a standard procedure of *in vitro* seaweeds digestibility and additionally *in vitro* methane production under conditions simulating the rumen fermentation (Bikker et al. 2020).

19.3 Trends in Animal Feeding with Seaweeds

The listed above current trends in animal feeding correspond to the composition and intended action of the commercial seaweed-based preparations available on the market, examples of which are summarized in the Table 19.2. There is a whole range of products based on seaweeds for various species of animals (dairy cows, beef cattle, dairy and meat goats, sheep, swine, horses – mature horses, foals,

Table 19.2 Examples of commercial seaweed-based products for animals

Preparation/Producer	Seaweed species	Animal species	Composition	Intended effect on animals
OceanFeed™ Bovine/Ocean Harvest Technology Limited, Milltown, Ireland; https://oceanharvesttechnology.com	Mixture of green, brown and red seaweeds	Cows	Fat 0.5%, ash 36.5%, protein 7.9%, carbohydrates 42%, fibre 4.8%, amino acids, essential fatty acids (n-3, n-6), vit. B_1, B_3, B_7, β-carotene, minerals	Improved nutrition and milk yields
OceanFeed™ Equine/Ocean Harvest Technology Limited, Milltown, Ireland; https://oceanharvesttechnology.com	Mixture of green, brown and red seaweeds	All horse breeds, ages and workloads	Protein 7%, fat 0.5%, ash 34%, fibre 5%, NaCl 2.5%, K 2.5%, S 4%, I 130 mg/kg	Equine nutrition, sporting performance, prebiotics for digestive health, enhanced coat shine
OceanFeed™ Poultry/Ocean Harvest Technology Limited, Milltown, Ireland; https://oceanharvesttechnology.com	Mixture of green, brown and red seaweeds	Poultry	Fat 0.6%, ash 30.3%, protein 10.2%, carbohydrates 44.1%, fibre 4.1%, amino acids, essential fatty acids (n-3, n-6), vit. B_1, B_3, B_7, β-carotene, minerals	Improved feed intake, increased nutrient uptake, improved digestive balance, body weight gain, feed conversion, carcass yield and livability, support gut microbiome and efficiency, improved growth performance
OceanFeed™ Swine/Ocean Harvest Technology Limited, Milltown, Ireland; https://oceanharvesttechnology.com	Mixture of green, brown and red seaweeds	Gestating and lactating sows, post-wean pigs and growing-finishing pigs	Fat 0.7%, ash 32.4%, protein 9%, carbohydrates 44.5%, fibre 4.5%, amino acids, essential fatty acids (n-3, n-6), minerals, vit. B_1, B_3, B_7, β-carotene	Improved feed intake, weight gain, feed conversion, digestive balance, performance, piglet viability, reduced diarrhea
Seaweed & Multi-Mite Blend/Multi Mite, Wootton Bassett, Wiltshire, UK; https://multi-mite.co.uk	Seaweeds (not specified)	Poultry, cows, horses, sheep, goats, pigs, rabbits, poultry (chicken, turkey, duck)	Full range of amino acids (e.g. L-alanine, methionine, histidine, glycin, arginine, glutamic acid), trace elements (e.g. Se, I), vit. (e.g. E), natural hormones, pigments (e.g. β-carotene), polysaccharide – fucoidan	Boosting of animals digestion capacity, promotion of the growth of animals, enhancement of animals health, improvement of the quality of animal products (e.g. eggs, milk)

Preparation/Producer	Seaweed species	Animal species	Composition	Intended effect on animals
Thorvin for Animals/Thorvin Inc., New Castle, Virginia, USA; http://thorvin.com/	*Ascophyllum nodosum* (B)	Dairy cows, beef cattle, dairy and meat goats, sheep, swine, horses (mature horses, foals, performance / race horse, work horse), poultry (chicken, laying hen, broiler, turkey)	Protein 8.5%, crude fiber 4%, total ash (minerals) 29%, fat 2%, carbohydrates (nitrogen-free extracts) 47.5%, more than 60 minerals (750 mg/kg of I, Zn, Co, Cu and Se), vitamins, amino acids, phytonutrients, polysaccharides	Prebiotic properties, proper immune and thyroid function
Thorvin 1000/Thorvin Inc., New Castle, Virginia, USA; http://thorvin.com/	*Ascophyllum nodosum* (B) and *Laminaria digitata* (B)	Animal feed formulation with a targeted amount of I	Protein 8.5%, crude fiber 3.9%, fat 1.9%, carbohydrates 48.1%, total ash (minerals) 29%, 60 minerals, especially 1000–1400 mg/kg of I (100% natural source of I), amino acids, vit. (B_2, C), phytonutrients	Promotion of mineral balancing, prebiotic properties
Icelandic Geothermal Kelp/ Thorvin Inc., New Castle, Virginia, USA; http://thorvin.com/	*Ascophyllum nodosum* (B)		Protein 8.5%, crude fiber 4%, total ash (minerals) 29%, fat 2%, carbohydrates 47.5%, 600–1200 mg/kg of I (100% natural source), vit. A, B_2, B_3, C, D, E, amino acids	Promotion of mineral balancing and immune response
Vitec Kelp Livestock Supplement/Vitec Organics Pty Ltd, Melbourne, Victoria, Australia; http://www.vitec.com.au	Kelp – *Durvillaea potatorum* (B)	Poultry, cows, horses, pigs	Sources of micro-nutrients, minerals, essential amino acids, alginic acid, complex carbohydrates	Improved overall absorption of ingredients, digestion and metabolism due to the liquid form, promotion of healthy growth, reproduction, milk yield, improved vitality, better immune health, quick recovery from stress or illness

(continued)

Table 19.2 (continued)

Preparation/Producer	Seaweed species	Animal species	Composition	Intended effect on animals
Vitec King Kelp Seaweed Meal/ Vitec Organics Pty Ltd, Melbourne, Victoria, Australia; http://www.vitec.com.au	Kelp – *Durvillaea potatorum* (B)	All livestock and animals	Sources of trace elements (especially I) and complex carbohydrates	Improvement of stress tolerance and hydration in extreme heat, resistance to parasites and diseases, improved health, improvement of fertility and milk production, stimulation of the digestive system, improved absorption of nutrients
Acid Bufis (CalMin/ ProMin); Celtic Sea Minerals, Cork, Ireland; https:// celticseaminerals.com	*Lithothamnium calcareum* (R)	Cattle (beef cattle, dairy cows, goat and sheep)	Mineral supplement – Ca, Mg and 74 trace minerals	Efficient and effective rumen buffer – stabilizes rumen pH, reduction in the risk for rumen acidosis, improvement of rumen fermentation, feed efficiency, higher daily weight gain, increase in milk yield, milk butterfat, milk solids, better meat parameters
CeltiCal; Celtic Sea Minerals, Cork, Ireland; https:// celticseaminerals.com	*Lithothamnium calcareum* (R)	Swine	Mineral supplement – Ca, Mg and 74 trace minerals (ash 95%, Ca 30%, Mg 5.5%, Na 0.6%, K 0.1%, Cl 1%, P 0.05%, S 0.7%)	Healthy gut and digestive function, less stomach gastric ulcers, improvement of feed conversion ratio, sow farrowing, milk and colostrum production, higher piglet birth weights, better survival rate for large litter size
CeltiCal; Celtic Sea Minerals, Cork, Ireland; https:// celticseaminerals.com	*Lithothamnium calcareum* (R)	Broiler	Mineral supplement – Ca, Mg and 74 trace minerals	Maintenance of a healthy digestive function, long term optimum pH in the poultry digestive system, improvement of feed utilization, feed conversion ratio, meat quality parameters and bone mineralization, stronger bones and better bone integrity

Preparation/Producer	Seaweed species	Animal species	Composition	Intended effect on animals
CeltiMin; Celtic Sea Minerals, Cork, Ireland; https://celticseaminerals.com	*Lithothamnium calcareum* (R)	Laying hen	Mineral supplement – Ca and 74 trace minerals	Healthy digestive function, long term optimum pH, improvement of feed utilization, enhancement of egg quality and the egg laying cycle, 'slow release' of Ca, improvement of egg shell formation, Ca metabolism support and bone mineralization
Tasco® /Acadian, Nova Scotia, Canada; https://tasco.ca	*Ascophyllum nodosum* (B)	Horse	Source of prebiotic fibers	Equine health and wellness, prebiotic functions, support immune function, enhanced resistance to environmental stress (e.g. transport, heat), promotion of a healthy skin and shiny coat
		Dairy	Source of prebiotic fibers	Maintenance of optimal rumen function, coping with heat stress, support gastrointestinal tract function and immune system, milk production, reproduction functions
		Swine	Source of prebiotic fibers	Prebiotic functions, maintenance of healthy gut flora, support gastrointestinal tract function and the immune system, improvement of feed utilization, support growth performance
		Poultry	Source of prebiotic fibers	Prebiotic functions, maintenance of healthy gut flora, support gastrointestinal tract function and the immune system, cope with stresses (heat and transport), support growth performance, feed utilization

(continued)

Table 19.2 (continued)

Preparation/Producer	Seaweed species	Animal species	Composition	Intended effect on animals
Algimun®, Olmix SA, Brehan, France; https://www.olmix.com	Marine seaweeds	Livestock	Sulfated polysaccharides extracted from marine algae	Strengthening animal's defense, optimized performance, gut health and immunity
DigestSea®, Olmix SA, Brehan, France; https://www.olmix.com	Marine seaweeds	Livestock	Sulfated polysaccharides extracted from marine algae	Improvement of animals' digestion, liver protection, improve lipid metabolism, prevention of heat and oxidative stress
Searup®, Olmix SA, Brehan, France; https://www.olmix.com	Marine seaweeds	Livestock	Algal sulphated polysaccharides, amino acids, vitamins	Increasing the immunity of animals, strengthening their natural defenses

performance/race horse, work horse, poultry – chicken, laying hen, broiler, turkey) with a wide range of applications. Many products use the prebiotic properties of algae as well as their rich mineral composition. These products were tested in many animal studies, cited in this chapter.

Table 19.3 presents the examples from the recent literature of examined effects of seaweeds on different animal species. In the majority of studies, the effect of seaweeds on animal performance (e.g. body weight gain, feed intake, feed conversion ratio) and production efficiency was examined. Animal health (intestinal microflora – e.g. *Escherichia coli*, *Lactobacillus* spp. and immune status) was also an important issue, especially in the feeding of pigs and poultry. Several Authors examined the influence of seaweeds on meat/eggs/milk quality – their chemical composition. The cow and sheep studies focused mainly on the effect of seaweeds on rumen fermentation and CH_4 production. Seaweeds were also tested for the digestibility of their nutrients and stabilization of rumen pH or gastric juice pH in horses.

19.3.1 Seaweeds Effect on Animal Health

Seaweeds have antimicrobial and antiviral characteristics along with immunomodulatory impacts (Michalak and Mahrose 2020). Marine macroalgae are known to be a rich source of unique polysaccharides, which can be potentially exploited as prebiotics in animal feed, stimulates the growth of intestinal beneficial bacteria. The positive effect of seaweeds on the gastrointestinal flora translates into increased immunity and overall animal health improvement (Moroney et al. 2015; Kulshreshtha et al. 2020). Kulshreshtha et al. (2014) showed that dietary supplementation of laying hens with red seaweeds – *Chondrus crispus* and *Sarcodiotheca gaudichaudii*, acting as potential prebiotics, improved overall gut health, performance, and egg quality. It resulted from the increased population of beneficial bacteria – e.g. *Bifidobacterium longum*, *Streptococcus salivarius* and reduced prevalence of *Clostridium perfringens* in the gut. *Laminaria japonica* powder inhibited the growth of *Escherichia coli*, increased *Lactobacillus* growth in the cecum of broilers, and enhanced their immune function (Bai et al. 2019). *Ulva rigida* also could be considered as a prebiotic that can enhance the health of broilers due to improved growth of intestinal villi and reduced concentration of total cholesterol and triglycerides in serum (Cañedo-Castro et al. 2019). Improved immune response in broilers was also achieved after the application of fermented brown seaweeds – *Undaria pinnatifida* and *Hizikia fusiformis* in the diet (Choi et al. 2014). Polysaccharides – laminarin and fucoidan extracted from *Laminaria* sp. improved gut health of post-weaned pigs and exhibited antimicrobial properties (Leonard et al. 2011). These two polysaccharides, used in the diet of pigs reduced the intestinal population of *Enterobacterium* spp., increased the population of *Lactobacilli* spp. and improved gut health (Lynch et al. 2010). Faeces of fattening pigs fed with the diet supplemented with OceanFeed Swine® (Table 19.2) had reduced *Escherichia coli* CFU (A

Table 19.3 The examples of the tested effects of seaweeds on different species of animals

Animal	Seaweed species	Examined effect (influence of seaweeds) on:
Pigs	(B)	Growth performance (body weight, average daily gain, feed conversion ratio) and health status of the post-weaned pigs (Draper et al. 2016); Quality and shelf-life of fresh pork, susceptibility of liver, heart, kidney and lung tissue to lipid oxidation, plasma antioxidant status (Moroney et al. 2012); Quality indices of fresh pork (Moroney et al. 2015); Growth performance (average daily gain, average daily feed intake), intestinal morphology (villous height, crypt depth), intestinal microflora (*Escherichia coli*, *Lactobacillus* spp.) and immune status in sows and post-weaned pigs (Leonard et al. 2011); Nutrient digestibility (neutral detergent fibre nitrogen, dry matter, organic matter), nitrogen utilization (faecal, urinary and total nitrogen excretion and nitrogen retention), intestinal microflora and concentration of caecal and colonic volatile fatty acid (Lynch et al. 2010)
	(G)	Immunomodulating effect – specific immunoglobulin IgG and total IgA in colostrum, milk, blood (Bussy et al. 2019)
	Mixture	Growth performance (body weight, average daily weight gain, feed efficiency) in nursery and fattening pigs, slaughter weight (Ruiz et al. 2018)
Cows	(B)	Milk production and composition (protein, lactose, fat, fatty acids and I), blood metabolites (triiodothyronine and thyroxine), nutrient intake, digestibility in early lactation dairy cows (Antaya et al. 2015)
	(R)	Dry matter intake, blood (glucose, urea N, total protein, albumin, creatinine, total bilirubin, creatine kinase, aspartate aminotransferase, gamma-glutamyltransferase, Ca, P, Mg, Na, K, Cl) and urine (Na, K, Cl concentration) metabolites, milk yield and composition (fat, protein, lactose) (Wu et al. 2015); Methane production, yield (feed intake), milk yield (Roque et al. 2019); Ruminal pH profile, rumen fermentation, production responses, feed intake, milk production and composition (Cruywagen et al. 2015)
	(B)(G)	Multielemental composition of milk from dairy cattle (Rey-Crespo et al. 2014)
Sheep	(B)	Rumen microbiota, fermentation profile and rumen/fecal *E. coli* O serogroups (Zhou et al. 2018); Thermo-respiratory response, inflammation, oxidative stress in bacterial endotoxin-challenged sheep (Ramadan et al. 2020)
	(G)	Growth performance (initial and final body weight, average daily gain, feed conversion ratio), dry matter intake, water intake, nutrient digestibility, nitrogen balance (N intake, faecal N, urinary N, N retention) (Rjiba et al. 2010); Growth performance (feed intake, feed conversion ratio), blood constituents (total protein, total lipids, albumin, globulin, glucose, cholesterol, Na, K, osmolality), antioxidant capacity, alleviation the impact of heat stress in growing lambs (Abdoun et al. 2014)
Horses	(R)	Gastric juice pH (Jacobs et al. 2020)

(continued)

Table 19.3 (continued)

Animal	Seaweed species	Examined effect (influence of seaweeds) on:
Poultry – ducks	(B)	Growth performance (body weight, feed intake), meat quality (moisture, crude protein, crude fat, cholesterol, ash, fatty acid composition), serum parameters (HDL, LDL, total cholesterol) (Islam et al. 2014);
	(R)	Carcass characteristics (meat antioxidant status, fat content) and production efficiency (Santoso et al. 2016)
Poultry – laying hens	(B)	Exterior egg quality traits, electronic microscopic view of eggshells, Ca and P retention, serum Ca and P concentrations, histology of the uterus (El-Maaty et al. 2021); Plasma (LDL, HDL, total cholesterol, triglycerides, total lipids, Ca, alkaline phosphatase) and yolk lipid profiles (n-3: linolenic, n-6: linoleic, n-9: oleic; stearic acid, palmitic acid, triglycerides, total cholesterol), yolk total carotene and lutein plus zeaxanthin (Al-Harthi and El-Deek 2012); The content of n-3 PUFA in eggs (Carrillo et al. 2012a); Growth performance (feed intake, feed conversion), egg production, egg weight, quality of eggs (yolk color, albumin height, Haugh unit), egg cholesterol content (Carrillo et al. 2012b)
	(G)	Egg quality parameters, mineral content of eggs, eggshell, blood, feathers and droppings, body weight (Michalak et al. 2011); Laying performance (average daily feed intake, average egg weight, feed conversion ratio), egg quality (egg-shaped index, eggshell breaking strength, albumen height, Haugh unit, egg yolk color, egg cholesterol content), immunity function (interleukin-6 (IL-6), interferon-γ (IFN-γ), IgG), antioxidant capacity in blood serum (total antioxidant capacity, malondialdehyde (MDA), superoxide dismutase (SOD)) (Li et al. 2018); The content of n-3 PUFA in eggs (Carrillo et al. 2012a)
	(R)	Growth performance (feed intake, egg production, feed conversion ratio, body weight), egg quality (egg and yolk weight), gut microbiota (Kulshreshtha et al. 2014); Productive performance – egg production, immune responses, egg quality traits (egg weight, shape index, yolk and albumen index, Haugh unit, albumin and yolk pH, shell thickness), eggs composition (free fatty acid, peroxide value, cholesterol) (Mandal et al. 2019); Growth performance (feed intake, body weight), reduction in *Salmonella* Enteritidis in hens, maintenance of body weight, egg production, cecal microbiota, short-chain fatty acids, serum immunoglobulin IgA production (Kulshreshtha et al. 2017)

(continued)

Table 19.3 (continued)

Animal	Seaweed species	Examined effect (influence of seaweeds) on:
Poultry – broilers	(B)	Growth performance (body weight gain, average daily gain, average daily feed intake, feed conversion ratio), immune function (serum Newcastle disease antibody), *E. coli* and *Lactobacillus* spp. count (Bai et al. 2019); Growth performance (body weight, body weight gain, ratio of gain:feed), weights of organs (spleen, bursa of fabricius, abdominal fat, breast muscle), carcass parameters, serum profile (HDL, total cholesterol, triglycerides), immunoglobulin concentration (IgA, IgG, IgM) (Choi et al. 2014); Growth performance (final weight, gain in weight, feed consumption), carcass quality (fat pads and meat color) (Erum et al. 2017); Growth performance (average body weight, feed consumption and feed conversion ratio), meat fatty acid composition, meat resistance to oxidation during refrigerated storage (Bonos et al. 2016)
	(G)	Growth performance (body weight gain, feed intake, feed conversion ratio), carcass characteristics (abdominal fat, breast color), serum constituents (aspartate aminotransferase, creatine kinase, alanine transaminase, alkaline phosphatase, gamma-glutamyltransferase, lactate dehydrogenase) (Abudabos et al. 2013); Growth performance (body weight gain, feed intake, feed conversion ratio), carcass weight and yield, prebiotic effect, intestinal integrity, serum cholesterol and triglyceride content (Cañedo-Castro et al. 2019)
	(R)	Calcium source and phytase on growth performance (body weight, weight gain), serum metabolites (Ca, Fe, P, alkaline phosphatase), ileum mineral contents, bone mineralization (Momeneh et al. 2018); Growth performance (body weight, feed intake, feed conversion ratio), carcass traits, lymphoid organ weight (Bursa of Fabricius, spleen, thymus), intestinal pH (Martínez et al. 2019); Growth performance (body weight gain, feed intake, feed conversion ratio), production performance, immune response and carcass traits (Qadri et al. 2019)
Poultry – chickens	(G)	Growth performance (average weekly body weight gain, feed intake, average weekly feed conversion efficiency), nutrient digestibility, blood indices – haematological parameters and serum biochemical parameters (Nhlane et al. 2020)

colony-forming unit) and increased *Lactobacillus* sp. CFU (Ruiz et al. 2018). Zhou et al. (2018) applied Tasco – a product obtained from *Ascophyllum nodosum* – in rams feeding. As a result, a reduction in the fecal population of *Escherichia coli* and the prevalence of Shiga toxin-producing *E. coli* O45, O103, O111, and O121 was observed, without disturbing the metabolism of the rumen (rumen fermentation).

Seaweeds are also considered as an alternative to antibiotics used in animal nutrition to control bacterial diseases. Unfortunately, the common use of antibiotics led to the increased microbial resistance against antibiotics and showed negative effects on microorganisms in the treated birds (Kulshreshtha et al. 2014; Bai et al. 2019). Dietary prebiotics of seaweed origin can constitute their alternatives (Kulshreshtha et al. 2017; Haberecht et al. 2018). Islam et al. (2014) proposed brown seaweed – *Laminaria japonica* as an alternative to antibiotics in ducks. Red seaweeds – *Chondrus crispus* and *Sarcodiotheca gaudichaudii* supplemented to the diet of laying hens were effective in providing protection against *Salmonella* Enteritidis colonization, which ensured safer egg production and minimization the spread of salmonellosis to humans (Kulshreshtha et al. 2017). Wan et al. (2017) suggested the feasibility of the application of alginate oligosaccharide extracted from brown seaweeds as an alternative to antibiotics in swine production. In a recent study conducted by Kulshreshtha et al. (2020), the Authors concluded that red seaweeds *Chondrus crispus* and *Sarcodiotheca gaudichaudii* and their purified compounds can be exploited to boost the lifespan of prevailing, patented antibiotics and can help to reduce the expensive medicinal and prophylactic usage of antibiotics in poultry. The same Authors also stated that using of a water extract of *Chondrus crispus* at a level of 200 μg/mL plus tetracycline improved the antibacterial activity. *Sarcodiotheca gaudichaudii* water extract, applied at 400 and 800 μg/mL, also in a group with tetracycline, exhibited full suppression of bacterial growth.

19.3.2 Treatment of Animal Diseases

Seaweeds, due to their composition could be useful in the treatment of animal diseases. For example, horses suffer from Equine Gastric Ulcer Syndrome. Jacobs et al. (2020) proposed to use a seaweed-derived calcium supplement (Calmin being the blend of red seaweeds –*Lithothamnion corallioides* and *Phymatolithon calcareum*) to buffer the equine gastric environment.

Seaweed extract from *Ulva armoricana*, containing sulfated polysaccharide, could be used for the treatment of pig respiratory disease caused by pathogen – *Bordetella bronchiseptica*. Bussy et al. (2019) analyzed the immuno-stimulating effect of this extract on sows and piglets – the level of anti-*Bordetella* immunoglobulin G antibodies in milk, blood and colostrum and on total immunoglobulin A in colostrum and milk of sow. It was shown that the examined extract stimulated the immune responses, what in the future may limit infections in animals and consequently reduce the amount of antibiotics used (Bussy et al. 2019).

Seaweeds are also able to promote antioxidant defense in animals under stressful conditions. For example, in pigs weaning stress causes the generation of excessive reactive oxygen species (ROS). Wan et al. (2017) found that extracted from brown seaweeds alginate oligosaccharide increased antioxidant defense in weaned pigs by increasing the content of glutathione in serum and enhancing the catalase activity. The application of ulvan extracted from *Ulva* sp. also increased the total antioxidative capacity, superoxide dismutase and malondialdehyde (blood marker of tissue damage) levels of laying hens blood serum. Additionally, ulvan had a positive effect on interferon-γ and interleukin-6 (Li et al. 2018). Endotoxemia, which is mainly caused by bacterial lipopolysaccharides (LPS) translocated into the bloodstream, enhances inflammation and production of ROS (Ramadan et al. 2020). Ramadan et al. (2020) examined the effect of a diet containing *Sargassum latifolium* on LPS-challenged sheep. This seaweed improved sheeps' antioxidant defence system and regulated inflammatory (increase in the concentration of serum proinflammatory cytokines heat shock protein-70) and thermo-respiratory responses (skin and rectal temperatures, respiration rate). Additionally, an increase in the total antioxidant capacity of the blood and activity of the catalase and superoxide dismutase and decrease in the malondialdehyde level was observed.

The antiviral properties of seaweeds originate from the existence of such bioactive complexes as carrageenan, alginate, fucan and laminaran (Ahmadi et al. 2015). Moreover, these bioactive complexes can stop the connection of the virus into the host cells or manage DNA replication and protein construction (Elizondo-Gonzalez et al. 2012; Ahmadi et al. 2015). Elizondo-Gonzalez et al. (2012) confirmed that fucoidan (a sulfated polysaccharide present in the cell wall of brown seaweed – *Cladosiphon okamuranus*) exhibited an activity against Newcastle disease virus La Sota. Fucoidan operates in the early stages of viral infectivity so as to stop viral–induced syncytia formation, perhaps by hindering the F protein, which is accountable for the fusion of cell membrane and the viral cover and over-conformational alterations (Elizondo-Gonzalez et al. 2012).

19.3.3 Effect of Seaweeds in Animal Diet on Food Quality

In many animal studies it was shown that seaweeds had a positive effect on the quality of animal-derived products. The application of polysaccharides (laminarin and fucoidan) extracted from brown seaweed (*Laminaria digitata*) enhanced pork quality – increased visual sensory descriptors, lowered lipid oxidation in *longissimus thoracis et lumborum* muscle, decreased the content of saturated fatty acids, had no impact on flavour (Moroney et al. 2012, 2015).

Dietary inclusion of seaweeds into poultry diet also has the potential to enhance the product quality – eggs and meat (Nhlane et al. 2020). Michalak et al. (2011) showed that the supplementation of the mixture of green seaweeds (*Enteromorpha prolifera* and *Cladophora* sp.) to the diet of laying hens increased the content of micro- and macroelements in eggs and enhanced the color of yolk. Ulvan extracted from *Ulva* sp. applied in the diet of laying hens led the yolk turning red and significantly decreased yolk cholesterol level (Li et al. 2018). Supplementation of *Sargassum dentifebium* to the diet of laying hen resulted in a significant decrease in plasma and yolk cholesterol and triglycerides and enhancement of carotene and lutein plus zeaxanthin level in eggs (Al-Harthi and El-Deek 2012). These findings were also confirmed in the study of Carrillo et al. (2012b) – *Sargassum* spp. reduced significantly the content of cholesterol in eggs and affected favorably the yolk color. The inclusion of red seaweed – *Kappaphycus alvarezii* to the diet of laying hens also caused the reduction in cholesterol content in eggs and lipid oxidation (Mandal et al. 2019). Kulshreshtha et al. (2014) showed that dietary inclusion of 1% *Sarcodiotheca gaudichaudii* increased egg yolk weight, egg weight; however, egg and eggshell weights were greater in hens consumed 1% *Chondrus crispus* than in 0.5% and 2% *Chondrus crispus* groups. Abu Hafsa et al. (2019) found that dietary green (*Ulva fasciata*) and brown (*Sargassum cinereum*) seaweeds enhanced egg yolk weight, index and yolk color, eggshell thickness and reduced the value of Haugh unit, total lipids and total cholesterol content in egg yolk. The improvement in eggshell thickness may be attributed to the diverse mineral content of seaweeds. The later researchers attributed the lessening in egg yolk cholesterol to the decrease in cholesterol production in the liver. Thus, the decrease in total lipids and total cholesterol can be accredited to the lessening impact of seaweeds on hepatic 3-hydroxy-3-methylglutaryl coenzyme A reductase that is required to produce cholesterol in the liver.

Seaweeds can enrich eggs with polyunsaturated fatty acids (including n-3 PUFA). Carrillo et al. (2012a) showed that *Enteromorpha* sp. increased the content of docosahexaenoic acid in eggs, whereas *Macrocystis pyrifera* and *Sargassum sinicola* had a better effect on the level of eicosapentaenoic acid.

Seaweeds included in poultry diet may also have a positive effect on meat quality. *Laminaria japonica* used in the diet of ducks, reduced the content of cholesterol in meat and increased the content of n-3 fatty acids and reduced the ratio of n-6/n-3 PUFA (Islam et al. 2014). With the increase of *Sargassum muticum* dose used as a feed substitute for broiler (5%, 10%, 15%), meat color changed from yellowish, slightly reddish to reddish. Additionally, for the highest seaweed inclusion level, no fat pads were observed (Erum et al. 2017). Nhlane et al. (2021) indicated that meat from females consumed diet containing green seaweed meal (*Ulva* spp.) at 2.5% had lesser drip loss than meat from females consumed diet containing green seaweed meal at 3.5%.

According to the literature data, algae also affect the composition and quality of milk. Singh et al. (2014) showed that calcium concentration was greater in milk of Sahiwal cows fed with the diet supplemented with *Sargassum wightii* as compared to the control group. Seaweeds can be an alternative to mineral mixtures, which are the main source of calcium in the diet of animals. Cruywagen et al. (2015) compared the effect of limestone (control), calcareous marine algae – *Lithothamnium calcareum* and sodium bicarbonate added to the diet of dairy cows on milk composition. The examined treatments had no effect on the content of milk lactose. Milk fat in the algal group was slightly higher when compared to the group with sodium bicarbonate but statistically significant when compared with the control. Crude protein was comparable in all examined groups. Calcareous algae are known to contain mainly macroelements like Ca – 30% (dry mass, d.m.), Mg – 6%, P – 0.5% and K – 0.1% and numerous trace minerals which are readily available for animals (Wu et al. 2015).

Antaya et al. (2015) demonstrated that the supplementation of *Ascophyllum nodosum* to the diet of early lactation dairy cows had no effect on milk components such as protein, lactose, fat, milk urea N, but increased significantly milk iodine concentration with the increase in seaweed dose (control group – 178 μg/L, 57 g – 602 μg/L, 113 g – 1015 μg/L, 170 g – 1370 μg/L). This seaweed is a rich source of iodine – 820 mg/kg of d.m. and a high caution should be maintained when supplementing the feed with these algae, as high iodine concentrations in milk may be toxic to humans (Rey-Crespo et al. 2014; Antaya et al. 2015). Rey-Crespo et al. (2014) compared the concentration of trace (Co, Cr, Cu, Fe, I, Mn, Mo, Ni, Se, Zn) and toxic elements (As, Cd, Pb) in milk in the algae-supplemented and control Holstein Friesian lactating cows. The addition of the mixture of seaweeds – *Ulva rigida* (80%), *Sargasum muticum* (17.5%) and *Saccorhiza polyschides* (2.5%) resulted in the increase in the milk concentration of Co, Cr, Fe, I, Se and Zn, as well as As and Cd. It was concluded that coastal seaweeds can serve as a source of microelements in dairy cattle.

Lactose and fat percentages, whey protein nitrogen and whey protein concentrations of milk were increased when feeding lactating Friesian cows with the diet supplemented with seaweed meal produced from *Ascophyllum nodosum* (Ead et al. 2011). Hassanien et al. (2015) found that milk composition (fat, protein, lactose and total solid) of lactating Damascus goats consumed diet supplemented with algae (*Ulva rigida*) was greater than that of the control. On the other hand, Hong et al. (2015) concluded that daily milk yield and composition were not impacted in Holstein cows fed diet supplemented with brown seaweed by-products.

19.3.4 Effect of Seaweeds on Rumen Function (Rumen Microbial Production of Methane)

One of the problems with animal husbandry is methane production. Many Authors propose to use seaweeds to reduce the microbial production of methane in rumen. There are many examples of studies performed *in vitro* in a rumen simulation fermenter. For example, Belanche et al. (2016) evaluated brown seaweeds – *Ascophyllum nodosum* and *Laminaria digitata* as an alternative feed for ruminants. The *in vitro* experiments showed that the tested seaweeds had no significant effect on feed degradability, rumen fermentation and methane emissions. Similarly, in the work of Maia et al. (2019) it was shown that seaweeds – green *Ulva rigida*, red *Gracilaria vermiculophylla* and brown *Saccharina latissima* had no effect on the methane production in the *in vitro* experiments. Lee et al. (2018) examined *in vitro* the effect of extracts produced from red seaweeds – *Grateloupia lanceolata*, *Hypnea japonica*, *Pterocladia capillacea*, *Chondria crassicaulis* and *Gelidium amansii* on bovine rumen fermentation and rumen microbial diversity (*Ruminococcus albus*, *Ruminococcus flavefaciens*, *Fibrobacter succinogenes*). These extracts reduced the emission of methane – less methanogens *R. albus* and *R. flavefaciens*. But in the *in vivo* studies, it was shown that seaweed – *Asparagopsis armata* significantly decreased methane production by cows – by 26% for the low (0.5%) seaweed supplementation and by 67% for the high (1%) level of supplementation (Roque et al. 2019). The bioactive compound – bromoform – found in the red seaweed – *Asparagopsis taxiformis* has been recognized as a strategy that may decline enteric methane formation in ruminants (Abbott et al. 2020). The same researchers demonstrated that the ability of seaweeds to decrease methane emissions from ruminants depends on many factors, such as the concentration of bioactive compounds in seaweeds, methods of their collection, transportation, packing and handling methods used to prepare seaweeds as diet components.

19.4 Conclusions

In this chapter we presented the current trends in animal feeding with seaweeds, challenges and possible solutions. Among different species of seaweeds, representatives of brown algae predominate – especially *Ascophyllum nodosum* and *Laminaria* sp. Animal studies using algal feed additives focused mainly on increasing their performance and production efficiency. Many Authors emphasized the positive effect of seaweeds on animal health. Due to prebiotic properties, seaweeds are able to maintain healthy gut flora and support gastrointestinal tract function, which in turn translates into strong immune system. It is underlined that the improvement in growth performance and production may result from the enhanced gut microbiome and efficiency, improved intestinal digestion-absorption function and better immune health. Red, calcareous seaweeds, rich in calcium are used to stabilize pH in the digestive system and provide a favorable environment for microbial activity. The use of seaweeds as antibiotics, prebiotics and heat stress modulators can be of excessive worth for animal farming as they can augment their health. According to this literature review, it could also be established that seaweeds at various doses had an encouraging impact on the blood profile and product quality. Seaweed constituents, due to their bioavailability to animals, can be incorporated into animal-derived products such as meat, eggs and milk. This phenomenon was observed mainly in the case of micro-and macroelements, but special attention should be paid to iodine, which is present in high amounts in brown seaweeds. An important issue, discussed in this chapter is the reduction in enteric methane emission by cattle. Some compounds derived from seaweeds have the potential of enteric methane mitigation. However, most of the performed experiments concern the *in vitro* conditions and they should be verified in real, *in vivo* conditions. This chapter highlighted the beneficial effects of seaweeds and their extracted compounds (especially polysaccharides) used as feed additives on animal health, production and environmental issues.

Acknowledgments Izabela Michalak was funded by a grant entitled: "The effect of bioactive algae enriched by biosorption in the certain minerals such as Cr(III), Mg(II) and Mn(II) on the status of glucose in the course of metabolic syndrome horses. Evaluation *in vitro* and *in vivo*" (No 2015/18/E/NZ9/00607) from the National Science Centre in Poland.

References

Abbott DW, Aasen IM, Beauchemin KA, Grondahl F, Gruninger R, Hayes M, Huws S, Kenny DA, Krizsan SJ, Kirwan SE, Lind V, Meyer U, Ramin M, Theodoridou K, von Soosten D, Walsh PJ, Waters S, Xing XH (2020) Seaweed and seaweed bioactives for mitigation of enteric methane: challenges and opportunities. Animals 10(12):2432. https://doi.org/10.3390/ani10122432

Abdoun KA, Okab AB, El-Waziry AM, Samara EM, Al-Haidary AA (2014) Dietary supplementation of seaweed (*Ulva lactuca*) to alleviate the impact of heat stress in growing lambs. Pak Vet J 34(1):108–111

Abu Hafsa SH, Zeweil HS, Zahran SM, Ahmed MH, Dosoky W, Rwif NA (2019) Effect of dietary supplementation with green and brown seaweeds on laying performance, egg quality, blood lipid profile and antioxidant capacity in Japanese quail. Egypt Poult Sci 39:41–59

Abudabos AM, Okab AB, Aljumaah RS, Samara EM, Abdoun KA, Al-Haidary AA (2013) Nutritional value of green seaweed (*Ulva lactuca*) for broiler chickens. Ital J Anim Sci 12(2):e28. https://doi.org/10.4081/ijas.2013.e28

Ahmadi A, Moghadamtousi SZ, Abubakar S, Zandi K (2015) Antiviral potential of algae polysaccharides isolated from marine sources: a review. Biomed Res Int:1–10. https://doi.org/10.1155/2015/825203

Al-Harthi MA, El-Deek AA (2012) Effect of different dietary concentrations of brown marine algae (*Sargassum dentifebium*) prepared by different methods on plasma and yolk lipid profiles, yolk total carotene and lutein plus zeaxanthin of laying hens. Ital J Anim Sci 11(64):347–353

Angell AR, Angell SF, de Nys R, Paul NA (2016) Seaweed as a protein source for mono-gastric livestock. Trends Food Sci Technol 54:74–84. https://doi.org/10.1016/j.tifs.2016.05.014

Antaya NT, Soder KJ, Kraft J, Whitehouse NL, Guindon NE, Erickson PS, Çonroy AB, Brito AF (2015) Incremental amounts of *Ascophyllum nodosum* meal do not improve animal performance but do increase milk iodine output in early lactation dairy cows fed high-forage diets. J Dairy Sci 98:1991–2004. https://doi.org/10.3168/jds.2014-8851

Bai J, Wang R, Yan L, Feng J (2019) Co-supplementation of dietary seaweed powder and antibacterial peptides improves broiler growth performance and immune function. Braz J Poult Sci 21(2):1–9. https://doi.org/10.1590/1806-9061-2018-0826

Bay-Larsen I, Risvoll C, Vestrum I, Bjorkhaug H (2018) Local protein sources in animal feed – perceptions among arctic sheep farmers. J Rural Stud 59:98–110. https://doi.org/10.1016/j.jrurstud.2018.02.004

Belanche A, Jones E, Parveen I, Newbold CJ (2016) A metagenomics approach to evaluate the impact of dietary supplementation with *Ascophyllum nodosum* or *Laminaria digitata* on rumen function in Rusitec fermenters. Front Microbiol 7:299. https://doi.org/10.3389/fmicb.2016.00299

Bikker P, Stokvis L, van Krimpen MM, van Wikselaar PG, Cone JW (2020) Evaluation of seaweeds from marine waters in Northwestern Europe for application in animal nutrition. Anim Feed Sci Technol 263:114460. https://doi.org/10.1016/j.anifeedsci.2020.114460

Bonos E, Kargopoulos A, Nikolakakis I, Florou-Paneri P, Christaki E (2016) The seaweed *Ascophyllum nodosum* as a potential functional ingredient in chicken nutrition. J Oceanogr Mar Res 4(1). https://doi.org/10.4172/2572-3103.1000140

Bradbury EJ, Wilkinson SJ, Cronin GM, Walk CL, Cowieson AJ (2012) The effect of marine calcium source on broiler leg integrity. In: Proc 23rd Annual Australian Poultry Science Symposium. Sydney NSW. 19–22 Feb, pp 85–88

Bussy F, Le Goff M, Salmon H, Delaval J, Berri M, Pi NC (2019) Immunomodulating effect of a seaweed extract from *Ulva armoricana* in pig: specific IgG and total IgA in colostrum, milk, and blood. Vet Anim Sci 7:100051

Cabrita ARJ, Maia MRG, Oliveira HM, Sousa-Pinto I, Almeida AA, Pinto E, Fonseca AJM (2016) Tracing seaweeds as mineral sources for farm-animals. J Appl Phycol 28(5):3135–3150. https://doi.org/10.1007/s10811-016-0839-y

Cañedo-Castro B, Piñón-Gimate A, Carrillo S, Ramos D, Casas-Valdez M (2019) Prebiotic effect of *Ulva rigida* meal on the intestinal integrity and serum cholesterol and triglyceride content in broilers. J Appl Phycol 31:3265–3273. https://doi.org/10.1007/s10811-019-01785-x

Carrillo S, Ríos VH, Calvo C, Carranco ME, Casas M, Pérez-Gil F (2012a) n-3 fatty acid content in eggs laid by hens fed with marine algae and sardine oil and stored at different times and temperatures. J Appl Phycol 24:593–599. https://doi.org/10.1007/s10811-011-9777-x

Carrillo S, Bahena A, Casas M, Carranco ME, Calvo CC, Ávila E, Pérez-Gi F (2012b) The alga *Sargassum* spp. as alternative to reduce egg cholesterol content. Cuban J Agric Sci 46(2):181–186

Choi YJ, Lee SR, Oh J-W (2014) Effects of dietary fermented seaweed and seaweed fusiforme on growth performance, carcass parameters and immunoglobulin concentration in broiler chicks. Asian Australas J Anim Sci 27(6):862–870

Corino C, Modina SC, Di Giancamillo A, Chiapparini S, Rossi R (2019) Seaweeds in pig nutrition. Animals 9:1126. https://doi.org/10.3390/ani9121126

Cruywagen CW, Taylor S, Beya MM, Calitz T (2015) The effect of buffering dairy cow diets with limestone, calcareous marine algae, or sodium bicarbonate on ruminal pH profiles, production responses, and rumen fermentation. J Dairy Sci 98:5506–5514. https://doi.org/10.3168/jds.2014-8875

Draper J, Walsh AM, McDonnell M, O'Doherty JV (2016) Maternally offered seaweed extracts improves the performance and health status of the postweaned pig. J Anim Sci 94:391–394. https://doi.org/10.2527/jas2015-9776

Ead HME, Maklad EHM, El-Shinnawy MM, Hamza AS, Ibrahim KE (2011) Effects of seaweed supplementation to dairy Friesian cows' rations on: 2- Milk yield, blood parameters and feed efficiency. J Anim Poult Prod Mansoura Univ 2:231–242

Elizondo-Gonzalez R, Cruz-Suárez LE, Marie DR, Mendoza-Gamboa E, Rodriguez-Padilla C, Trejo-Ávila LM (2012) *In vitro* characterization of the antiviral activity of fucoidan from *Cladosiphon okamuranus* against Newcastle disease virus. Virol J 9:307. https://doi.org/10.1186/1743-422X-9-307

El-Maaty HAA, EL-Khateeb AY, Al-Khalaifah H, El Hamed EA, Hamed S, El-Said EA, Mahrose KM, Metwally K, Mansour AM (2021) Effects of ecofriendly synthesized calcium nanoparticles with biocompatible *Sargassum latifolium* algae extract supplementation on egg quality and scanning electron microscopy images of the eggshell of aged laying hens. Poult Sci 100:675–684. https://doi.org/10.1016/j.psj.2020.10.043

Erum T, Frias GG, Cocal CJ (2017) *Sargassum muticum* as feed substitute for broiler. Asia Pac J Educ Arts Sci 4(4):6–9

Haberecht S, Wilkinson S, Roberts J, Wu S, Swick R (2018) Unlocking the potential health and growth benefits of macroscopic algae for poultry. Worlds Poult Sci J 74(1):5–20. https://doi.org/10.1017/S0043933917001052

Halmemies-Beauchet-Filleau A, Rinne M, Lamminen M, Mapato C, Ampapon T, Wanapat M, Vanhatalo A (2018) Review: alternative and novel feeds for ruminants: nutritive value, product quality and environmental aspects. Animal 12(S2):295–309. https://doi.org/10.1017/S1751731118002252

Hassanien HAM, Youssef HFH, Mahrous AA, Hafez YH, Phillip YL, Gaballah MM (2015) Effect feeding of supplemented ration with algae on milk yield and its composition for Damascus goats. Egypt J Nutr Feeds 18:65–75

Hong ZS, Kim EJ, Jin YC, Lee JS, Choi YJ, Lee HG (2015) Effects of supplementing brown seaweed by products in the diet of Holstein cows during transition on ruminal fermentation, growth performance and endocrine responses. Asian-Australas J Anim Sci 28:1296–1302

Islam MM, Ahmed ST, Kim YJ, Mun HS, Kim YJ, Yang CJ (2014) Effect of Sea Tangle (*Laminaria japonica*) and charcoal supplementation as alternatives to antibiotics on growth performance and meat quality of ducks. Asian Australas J Anim Sci 27(2):217–224

Jacobs RD, Gordon MBE, Vineyard KR, Keowen ML, Garza F Jr, Andrews FM (2020) The effect of a seaweed-derived calcium supplement on gastric juice pH in the horse. J Equine Vet Sci 95:103265. https://doi.org/10.1016/j.jevs.2020.103265

Kulshreshtha G, Rathgeber B, Stratton G, Thomas N, Evans F, Critchley A, Hafting J, Prithiviraj B (2014) Feed supplementation with red seaweeds, *Chondrus crispus* and *Sarcodiotheca gaudichaudii*, affects performance, egg quality, and gut microbiota of layer hens. Poult Sci 93:2991–3001

Kulshreshtha G, Rathgeber B, MacIsaac J, Boulianne M, Brigitte L, Stratton G, Thomas NA, Critchley AT, Hafting J, Prithiviraj B (2017) Feed supplementation with red seaweeds, *Chondrus crispus* and *Sarcodiotheca gaudichaudii*, reduce *Salmonella* Enteritidis in laying hens. Front Microbiol 8:567. https://doi.org/10.3389/fmicb.2017.00567

Kulshreshtha G, Hincke MT, Prithiviraj B, Critchley A (2020) A review of the varied uses of macroalgae as dietary supplements in selected poultry with special reference to laying hen and broiler chickens. J Mar Sci Eng 8(7):536. https://doi.org/10.3390/jmse8070536

Lee SJ, Shin NH, Jeong JS, Kim ET, Lee SK, Lee SS (2018) Effect of *Rhodophyta* extracts on *in vitro* ruminal fermentation characteristics, methanogenesis and microbial populations. Asian-Australas J Anim Sci 31(1):54–62. https://doi.org/10.5713/ajas.17.0620

Leonard SG, Sweeney T, Bahar B, Lynch BP, O'Doherty JV (2011) Effects of dietary seaweed extract supplementation in sows and post-weaned pigs on performance, intestinal morphology, intestinal microflora and immune status. Br J Nutr 106(5):688–699

Li Q, Luo J, Wang C, Tai W, Wang H, Zhang X, Liu K, Jia Y, Lyv X, Wang L, He H (2018) Ulvan extracted from green seaweeds as new natural additives in diets for laying hens. J Appl Phycol 30:2017–2027

Lynch MB, Sweeny T, Callan JJ, O'Sullivan JT, O'Doherty JV (2010) The effect of dietary *Laminaria* derived laminarin and fucoidan on nutrient digestibility, nitrogen utilisation, intestinal mircoflora and volatile fatty acid concentration in pigs. J Sci Food Agric 90(3):430–437

Machado L, Kinley RD, Magnusson M, de Nys R, Tomkins NW (2015) The potential of macroalgae for beef production systems in Northern Australia. J Appl Phycol 27(5):2001–2005. https://doi.org/10.1007/s10811-014-0439-7

Maia MRG, Fonseca AJM, Cortez PP, Cabrita ARJ (2019) *In vitro* evaluation of macroalgae as unconventional ingredients in ruminant animal feeds. Algal Res 40:101481. https://doi.org/10.1016/j.algal.2019.101481

Makkar HPS, Tran G, Heuzé V, Giger-Reverdin S, Lessire M, Lebas F, Ankers P (2016) Seaweeds for livestock diets: a review. Anim Feed Sci Technol 212:1–17

Mandal AB, Biswas A, Mir NA, Tyagi PK, Kapil D, Biswas AK (2019) Effects of dietary supplementation of *Kappaphycus alvarezii* on productive performance and egg quality traits of laying hens. J Appl Phycol 31:2065–2072. https://doi.org/10.1007/s10811-018-1707-8

Martínez Y, Ayala L, Hurtado C, Más D, Rodríguez R (2019) Effects of dietary supplementation with red algae powder (*Chondrus crispus*) on growth performance, carcass traits, lymphoid organ weights and intestinal pH in broilers. Braz J Poult Sci 21(4):1–7. https://doi.org/10.1590/1806-9061-2019-1015

Michalak I, Mahrose K (2020) Seaweeds, intact and processed, as a valuable component of poultry feeds. J Mar Sci Eng 8:620. https://doi.org/10.3390/jmse8080620

Michalak I, Marycz K (2019) Algae as a promising feed additive for horses. In: Pereira L, Bahcevandziev K, Joshi NH (eds) Seaweeds as plant fertilizer, agricultural biostimulants and animal fodder, vol 7. CRC Press, Taylor & Francis Group, pp 128–142

Michalak I, Chojnacka K, Dobrzański Z, Górecki H, Zielińska A, Korczyński M, Opaliński S (2011) Effect of macroalgae enriched with microelements on egg quality parameters and mineral content of eggs, eggshell, blood, feathers and droppings. J Anim Physiol Anim Nutr 95:374–387

Momeneh T, Karimi A, Sadeghi G, Vaziry A, Bedford MR (2018) Evaluation of dietary calcium level and source and phytase on growth performance, serum metabolites, and ileum mineral contents in broiler chicks fed adequate phosphorus diets from one to 28 days of age. Poult Sci 97:1283–1289

Morais T, Inácio A, Coutinho T, Ministro M, Cotas J, Pereira L, Bahcevandziev K (2020) Seaweed potential in the animal feed: a review. J Mar Sci Eng 8:559. https://doi.org/10.3390/jmse8080559

Moroney NC, O'Grady MN, O'Doherty JV, Kerry JP (2012) Addition of seaweed (*Laminaria digitata*) extracts containing laminarin and fucoidan to porcine diets: influence on the quality and shelf-life of fresh pork. Meat Sci 92:423–429. https://doi.org/10.1016/j.meatsci.2012.05.005

Moroney NC, O'Grady MN, Robertson RC, Stanton C, O'Doherty JV, Kerry JP (2015) Influence of level and duration of feeding polysaccharide (laminarin and fucoidan) extracts from brown seaweed (*Laminaria digitata*) on quality indices of fresh pork. Meat Sci 99:132–141

Nhlane LT, Mnisi CM, Mlambo V, Madibana M (2020) Nutrient digestibility, growth performance, and blood indices of Boschveld chickens fed seaweed-containing diets. Animals 10(8):1296. https://doi.org/10.3390/ani10081296

Nhlane LT, Mnisi CM, Mlambo V, Madibana MJ (2021) Effect of seaweed-containing diets on visceral organ sizes, carcass characteristics, and meat quality and stability of Boschveld indigenous hens. Poult Sci 100(2):949–956

Øverland M, Mydland LT, Skrede A (2019) Marine macroalgae as sources of protein and bioactive compounds in feed for monogastric animals. J Sci Food Agric 99:13–24. https://doi.org/10.1002/jsfa.9143

Qadri SSN, Biswas A, Mandal AB, Kumawat M, Saxena R, Nasir AM (2019) Production performance, immune response and carcass traits of broiler chickens fed diet incorporated with *Kappaphycus alvarezii*. J Appl Phycol 31:753–760. https://doi.org/10.1007/s10811-018-1498-y

Ramadan G, Fouda WA, Ellamie AM, Ibrahim WM (2020) Dietary supplementation of *Sargassum latifolium* modulates thermo-respiratory response, inflammation, and oxidative stress in bacterial endotoxin-challenged male Barki sheep. Environ Sci Pollut Res 27:33863–33871

Rey-Crespo F, López-Alonso M, Miranda M (2014) The use of seaweed from the Galician coast as a mineral supplement in organic dairy cattle. Animal 8(4):580–586. https://doi.org/10.1017/S1751731113002474

Rjiba SK, Chermiti A, Mahouachi M (2010) The use of seaweeds (*Ruppia maritima* and *Chaetomorpha linum*) for lamb fattening during drought periods. Small Rumin Res 91:116–119. https://doi.org/10.1016/j.smallrumres.2010.01.012

Roque BM, Salwen JK, Kinley R, Kebreab E (2019) Inclusion of *Asparagopsis armata* in lactating dairy cows' diet reduces enteric methane emission by over 50 percent. J Clean Prod 234:132–138. https://doi.org/10.1016/j.jclepro.2019.06.193

Ruiz ÁR, Gadicke P, Andrades SM, Cubillos R (2018) Supplementing nursery pig feed with seaweed extracts increases final body weight of pigs. Austral J Vet Sci 50:83–87

Santoso SI, Suprijatna E, Setiadi A, Susanti S (2016) Effect of duck diet supplemented with fermented seaweed wastes on carcass characteristics and production efficiency of indigenous Indonesian ducks. Indian J Anim Res 50(5):699–704. https://doi.org/10.18805/ijar.11160

Shama A, Joyce SG, Mari FD, Ranga Rao A, Ravishankar GA, Hudaa N (2019) Macroalgae and microalgae: novel sources of functional food and feed. In: Ravishankar GA, Ranga Rao A (eds) Handbook of algal technologies and phytochemicals. Food, Health and Nutraceutical Applications, vol I, CRC Press, USA, pp 207–219

Singh BK, Chopra RC, Rai SN, Verma MP, Mohanta RK (2014) Effect of feeding seaweed as mineral source on mineral metabolism, blood and milk mineral profile in cows. Proc Natl Acad Sci, India, Sect B Biol Sci 86:89–95. https://doi.org/10.1007/s40011-014-0413-9

Wan J, Zhang J, Chen DW, Yu B, He J (2017) Effects of alginate oligosaccharide on the growth performance, antioxidant capacity and intestinal digestion–absorption function in weaned pigs. Anim Feed Sci Technol 234:118–127

Wu Z, Bernard JK, Taylor SJ (2015) Effect of feeding calcareous marine algae to Holstein cows prepartum or postpartum on serum metabolites and performance. J Dairy Sci 98(7):4629–4639. https://doi.org/10.3168/jds.2014-8711

Zhou M, Hünerberg M, Chen Y, Reuter T, McAllister TA, Evans F, Critchley AT, Guan LL (2018) Air-dried brown seaweed, *Ascophyllum nodosum*, alters the rumen microbiome in a manner that changes rumen fermentation profiles and lowers the prevalence of foodborne pathogens. mSphere 3(1). https://doi.org/10.1128/mSphere.00017-18

Chapter 20
Seaweed in Aquaculture: An Overview

Anong Chirapart and Rapeeporn Ruangchuay

20.1 Introduction

Aquaculture refers to the farming of aquatic organisms, and it is one of the fastest-growing food-producing industries in the world. Farming (fish, mollusks, crustaceans, and aquatic plants) may be carried out in various aquatic environments, e.g., ponds, lagoons, oceans, and closed man-made farming systems on land. Over the past decade, aquaculture has played an important role due to the global decline in aquatic resources from overexploitation. Aquaculture is used as a tool to supplement the seafood supply, which is widely recognized as an effective means of meeting the seafood needs of the population. Aquaculture is also a more sustainable option for consumers compared to other farm proteins. In addition, aquaculture emits less greenhouse gas compared to other farming practices. In recent decades, the increased global demand for seafood has been fulfilled by increasing aquaculture, which provided 178.5% of the world fisheries and aquaculture production in 2018 (FAO 2020). With the rapid increase in production, aquaculture activities have affected the environment in a variety of ways. In particular, fed aquaculture includes fish and shrimp, which need to be supplemented with food as an exogenous source of energy (Beveridge 1996; Troell et al. 2004). Unlike unfed aquaculture, this type of system includes seaweed and shellfish (mussels and oysters). Seaweed requires

A. Chirapart (✉)
Department of Fishery Biology, Faculty of Fisheries, Kasetsart University, Chatuchak, Bangkok, Thailand
e-mail: ffisanc@ku.ac.th

R. Ruangchuay
Department of Technology and Industry, Faculty of Science and Technology, Prince of Songkla University, Meaung, Pattani, Thailand

A. Ranga Rao, G. A. Ravishankar (eds.), *Sustainable Global Resources of Seaweeds Volume 1*, https://doi.org/10.1007/978-3-030-91955-9_20

sunlight for use as an energy source to turn nutrient-rich effluents into profitable resources, and shellfish feed on microscopic plankton.

According to the Food and Agriculture Organization (FAO 2014), the global aquaculture production was 90.4 million tons (live weight equivalent) in 2012, with a value of US$ 144 billion. This included 23.8 million tons of aquatic algae and 66.6 million tons of food fish and was estimated at 26.1 million and 70.5 million tons in 2013, respectively. Production increased in 2016 and included 30.1 million tons of aquatic plants and 80.0 million tons of food fish (FAO 2018). Aquaculture has tended to increase its contribution to global seafood production every year. A production volume of 90% was produced in Asia, and China alone produced 13.5 million tons of aquatic algae and 43.5 million tons of food fish, or a 1:3 aquatic algae and food fish ratio. The farmed aquatic plants included mostly seaweeds, and in 2016, China and Indonesia were by far the major producers of aquatic plants (FAO 2018).

Several techniques of seaweed cultivation in aquaculture systems have been continuously developed, but there are still numerous challenges to overcome. In this article, we address the role and application of seaweed in aquaculture and integrated multitrophic aquaculture systems (IMTAs) and their prospects in the future.

20.2 Role of Seaweed Utilization in Aquaculture

The use of seaweeds in aquaculture has been increasingly recognized due to its positive benefits and numerous feedstocks (Shama et al. 2019). In recent years, seaweed aquaculture has received increased attention for promotion and monitoring climate and environmentally friendly bioeconomic development (Chung et al. 2017). Over the decades, seaweed has been utilized in aquaculture in a variety of ways. It has been used for the feeding and treatment of wastewater discharged from shrimp and fish farming. Since seaweed is rich in minerals, vitamins and fatty acids (Ratana-arporn and Chirapart 2006), it has preferentially been used in the farming of aquatic animals, especially as an ingredient in animal diets. In addition, the use of seaweed in aquaculture systems can improve the water quality, and farmers expect seaweed to help increase the rate of growth, survival rate, and disease tolerance to help them make a profit from aquaculture crops (Buschmann 1996).

In the feed industry, seaweed and its colloidal substances (agar, carrageenan, and alginate) are utilized as mineral carbohydrate sources and binders. Seaweed colloids are also used as emulsifiers, thickeners, and sticky substances. Additionally, seaweed is a source of many important nutrients of amino acids, vitamins such as B6, K, A, and C and minerals such as iodine, iron, magnesium, sodium, and manganese. The seaweed mineral content is the most important because minerals are essential for animal cells; thus, seaweed can be used as a substitute for a proportion of raw materials used in food production (Takeshi et al. 2005). However, seaweed is viewed as a complete feed and/or supplementary feed for supplementation to improve color and stimulate immunity. Also, polysaccharides are considered dietary fibers and binders of feed ingredients.

In farming aquatic plants, seaweed is a major dominant practice in over 50 countries. Seaweed farming has expanded by 8% per year in the past, increasing from 6.2% in the previous decade (FAO 2016) and more than doubling in recent years. The major production of seaweed in the world is the farming of *Kappaphycus alvarezii* and *Eucheuma* spp. as raw material for carrageenan extraction, especially in Indonesia (FAO 2018, 2020). According to the FAO (2020), seaweed production in Indonesia increased 15.2-fold, from 1.9% in 2000 to 28.8% in 2018, while the world aquaculture production of aquatic algae increased approximately 205.7%, from 10.6 billion tons in 2000 to 32.4 billion tons in 2018.

For a long time, several cultivation methods have been used to integrate seaweed with aquaculture systems. Seaweed polyculture is one of the known methods that involves the production of two or more cultured species in the same physical space at the same time. The objective is to produce multiple products (e.g., seaweed, fish, and shrimp) that have economic value. Seaweed farming through polyculture systems is prevalent in Southeast Asian countries due to its feasibility of providing additional income to fish farmers. Polyculture systems integrating seaweed farming with other marine fauna, such as clams, abalone, and fish, has been rapidly developing in the region.

In recent years, in Indonesia, the stratified double floating net cage (SDFNC) technique has been developed for seaweed polyculture with fish and shrimp to increase productivity and maintain a healthy water ecosystem for sustainable aquaculture (Putro et al. 2016). This culture system is a combination of red seaweed (*Eucheuma cottonii*), rabbitfish (*Siganus* sp.), black tiger shrimp (*Penaeus monodon*), and Pacific white shrimp (*Litopenaeus vannamei*). *E. cottonii* (3 g) had the highest absolute mean weight (282.35 g) when cocultured with *Gracilaria verrucosa* (3 g), shrimp *Penaeus vannamei* (100 juveniles, ~2.1 g), and milkfish *Chanos chanos* (25 juveniles, ~3.1 g) (Yala et al. 2017). Polyculture conditions could enhance the growth of *E. cottonii* but not *G. verrucosa* due to competition between seaweed species. However, both shrimp and fish species had good growth and were not influenced by the culture system.

Also, integrated intensive aquaculture has been developed from traditional extensive polyculture over the past two decades by integrating the culturing of fish or shrimp with shellfish and/or seaweed (Neori et al. 2004). Integrating seaweed into the farming of fish and shrimp can balance the nutrient input and metabolize dissolved oxygen, acidity and carbon dioxide levels, in one step, in the culture systems. It has been proven that using seaweed in integrated aquaculture provides an excellent system for sustainable environments (Neori et al. 2004; Hernández et al. 2005). Several works have revealed different species of seaweed that are effective in removing nutrients from water bodies. Some species show high potential for use in IMTA in recirculating aquaculture systems (RAS) (Hayashi et al. 2008; Neori 2008; Marinho et al. 2013). In recent years, Kang et al. (2013) developed a seaweed species-selection index for selecting suitable species in seaweed-based integrated aquaculture systems. The available literature-based information, reference data, and physiological seaweed experiments were used to synthesize the index and to identify and prioritize the desired species. The species with the highest scores are considered in the integrated aquaculture suitability index.

20.2.1 Seaweed in Open-Sea and Land-Based Aquaculture

The seaweed types chosen for open-sea aquaculture are typically those with value either as a foodstuff or in industrial applications. In general, open-sea culture systems lack control over the dilution of waste caused by natural seawater movement. In contrast, land-based systems are mostly closed-circulation systems, thus allowing control of nutrient-rich waste. The use of seaweed integrated with fish cultures in open-water and land-based aquaculture systems has been well documented in several regions (Troell et al. 1997; Marinho-Soriano et al. 2002; Chirapart and Lewmanomont 2004; Hernández et al. 2005; Pellizzari and Reis 2011). The red seaweed *Gracilaria chilensis* was cocultured in salmon cage farms in open-sea systems in southern Chilean waters (Troell et al. 1997), and Grote (2019) reviewed the open-sea and land-based cultivation of *Palmaria palmata* in Europe. Seaweed produces cheap nutritious biomass at an average market value of $30–40 per ton -fresh weight basis (Noeri and Guttman 2017). Production mostly takes place in coastal waters in several Southeast Asian countries.

Land-based aquaculture covers the farming systems of aquatic organisms on land, which usually occurs in pond cultures, rather than in the ocean. Several works involving integrated land-based culture have been performed using an integrated design (Neori et al. 2000; Marinho-Soriano et al. 2002; Chirapart and Lewmanomont 2004; Hernández et al. 2005). The intensive land-based culture of abalone (*Haliotis discus hannai*), seaweed (*Ulva lactuca* or *Gracilaria conferta*) and fish (*Sparus aurata*) could achieve nutrient recycling, reduced water use, reduced nutrient discharge and high yields (Neori et al. 2000). Likewise, the cultivation of *Ulva rotundata* and *Gracilariopsis longissima* in effluents from intensive marine culturing (grow-out phase) of gilthead seabream *Sparus aurata* revealed that the dissolved nutrient content was reduced in effluents from the fish tank (Hernández et al. 2005).

20.2.2 Seaweed in Integrated Multitrophic Aquaculture (IMTA)

Currently, seaweed has been widely discussed for its benefits in cultivation in IMTA systems. IMTA is a type of culture system that incorporates marine plants and aquatic animals, and it has the potential to decrease costs and improve efficiency and productivity for a number of species and systems. IMTA is an idea that has been developing over the last 20 years and is becoming increasingly well known.

Barrington et al. (2009) described IMTA as a cultivation practice that combines in appropriate proportions of fed aquaculture species (e.g., finfish/shrimp) with organic extractive aquaculture species (e.g., shellfish/herbivorous fish) and inorganic extractive aquaculture species (e.g., seaweed); the purpose is to create balanced systems for environmental sustainability (biomitigation), economic stability (product diversification and risk reduction) and social acceptability (better management practices). According to Troell (2009), IMTA is defined as fed aquaculture (e.g., fish) combined with inorganic extractive species (e.g., seaweed) and organic extractive species (e.g., shellfish). It also refers to more intensive cultivation of the different species in proximity to each other, connected by nutrient and energy transfers through water. Recently, the production of biomass of *Eucheuma cottonii* has been observed in the Multi Trophic Sea Farming system in the marine aquaculture area of Gerupuk Bay, Central Lombok, Indonesia. The highest biomass production of seaweed was obtained when cocultured with abalone grouper compared to when cocultured with lobster abalone, abalone red carp, and abalone pomfret fish (Sukiman et al. 2014). A meta-analysis of integrated multitrophic aquaculture has been reported, which showed the most successful growth of commercial extractive species within close proximity to open-water fish farms (Kerrigan and Suckling 2016).

20.3 Seaweed Use for Sustainable Aquaculture

According to the FAO report, seaweed in aquaculture systems is mainly produced in the Asian region. Several seaweed species, e.g., *Caulerpa lentillifera*, *Gracilaria fisheri*, *G. tenuistipitata*, *Ulva intestinalis*, and *Ulva rigida*, have been used for the treatment of effluent from farms (Chirapart and Lewmanomont 2004; Chaitanawisuti et al. 2011; Aníbal et al. 2014; Kunawongdet 2020). These seaweed species are raised in effluent ponds due to their ability to effectively absorb and reduce wastewater from brackish aquaculture. Seaweed species commonly grow profusely throughout the year in tropical water ponds and irrigation canals, such in Thailand. They have been used in intensive prawn culture systems for nutrient removal from culture systems before recycling the water or discharging it into the environment (Baliao and Tookwinas 2002; Chaitanawisuti et al. 2011).

20.3.1 Seaweed in Shrimp Culture

Seaweed has been used in shrimp aquaculture as an ingredient in the diet, as a biofilter, and in nursing areas. Using seaweed in white shrimp farming in several countries has led to the highest annual yields and value of aquaculture compared to other marine species.

20.3.1.1 Seaweed in the Shrimp Diet

According to FAO reports, the production of aquaculture worldwide has increased annually, constituting an important human protein source. Thus, to sustain a high rate of growth, it is imperative to have viable production and development of economical aquaculture systems, low costs, and higher-quality feeds because feed is the major operational cost for most aquaculture enterprises (Rana et al. 2009). The rising cost of commercial aquafeed is therefore inducing some farmers to opt for alternative feeds.

In 2013, the shrimp farming business in Thailand collapsed due to the disease outbreak caused by *Vibrio* bacteria (*Vibrio* sp.) and viruses (Sitthimong 2020), resulting in a decline in white shrimp production. Several researchers have attempted to use seaweed species to increase resistance to pathogenesis in white shrimp and have provided associated guidelines (Sirirustananun et al. 2011; Kanjana et al. 2011; Wongprasert et al. 2014). Seaweed has been applied for the prevention of disease through enhancing immune function (Takeshi et al. 2005). It has been used as a dietary supplement, providing nutrients such as vitamin C and polysaccharide components. Among the seaweed species, *Gracilaria fisheri* is mainly applied for diet supplementation in Thailand (Rudtanatip et al. 2014). Following green seaweed, *Ulva rigida* has also been used, as potent protein supplementation in the feed of the white shrimp *Litopeneaus vannamei*, and feed meal is a food source that provides minerals and vitamins for shrimp (Tamtin et al. 2016). This algal species was used as a supplemental protein source at proportions ranging from 6% to 12% in the shrimp diet with a reduction in fishmeal. Generally, the reduction of fishmeal and its replacement with alternative protein sources mostly involve soybeans. Many experiments have reported that seaweed stimulates the growth, immunity, and pathogen resistance of the white shrimp *L. vannamei* (Wongprasert et al. 2014; Zahra et al. 2017; Niu et al. 2019), as well as the taste and color of shrimp meat when compared to the use of soybean meal instead of fishmeal. The white shrimp *L. vannamei* is both omnivorous and herbivorous (Martínez-Córdova and Peña-Messina 2005), and it can use plant nutrients, which are starchy protein-based ingredients. The use of seaweed in the diet could provide effective protein digestion in the diet, which is not different from using fishmeal or soy meal as the main protein source. The protein level found in sea lettuce varied from 10.37% to 11.60%, and the corresponding protein digestibility efficiency of the shrimp was approximately 84.01–84.32% (Tamtin et al. 2016). Furthermore, other work reported that protein in the white prawn diet was reduced by 25–30% (Cuzon et al. 2004). Nevertheless, there is a need for more research to clarify the seaweed state, such as its role in nutrition and disease prevention throughout all the production phases of the feed industry.

Table 20.1 Some research cases showing nutrient removal rates of some seaweed species in wastewater from shrimp ponds in Thailand

Seaweed species	Removal rate (%)				References
	NH_4^+-N	NO_2-N	NO_3-N	PO_4^{3-}-P	
Caulerpa lentillifera	99.76	5.66	–	–	Chokwiwattanawanit (2000)
Caulerpa macrophysa	50.00	16.56	38.30	73.94	Limhang et al. (2014)
Caulerpa sertularioides	49.90	254.19	103.11	41.36	Suthiniam et al. (2009)
Gracilaria fisheri	33.23	–	16.10	32.18	Kaewmesri (2000)
Gracilaria edulis	97.76	92.97	86.74	–	Sriveerachi and Pankdee (2005)
Acanthophora spicifera	98.51	86.89	98.77		Sriveerachi and Pankdee (2005)
Sargassum polycystum	84.04	–	67.25	–	Khidprasert (1995)

20.3.1.2 Seaweed as a Biofilter in Shrimp Culture

Several seaweed species, e.g., *Caulerpa lentillifera, Gracilaria fisheri, Solieria robusta, Halymenia* sp. *Acanthophora spicifera* and *Grateloupia indica*, have been evaluated for their potential use as biofilters, especially *Gracilaria fisheri*, due to its ability to adapt and tolerate nutrient-enriched seawater (Teckayu and Khidprasert 1996; Chokwiwattanawanit 2000). Seaweed has been used as a biofilter in shrimp aquaculture systems for nutrient absorption, especially that of NH_4^+-N, NO_2-N, NO_3-N, and PO_4^{3-}-P (Table 20.1). Chokwiwattanawanit (2000) described the process of water treatment in which shrimp farm effluent circulated to oyster (*Crassostrea lugubris*) or green mussel (*Perna viridis*) ponds to precipitate suspended sediment for 1–2 weeks, after which clear water flowed into the seaweed ponds (*Gracilaria fisheri*), while nutrient-enriched seawater was absorbed to yield improved water quality. The duration of the treatment process was about 2 weeks to allow nitrogen and phosphorous absorption. Cultivation of the *Gracilaria* sp. has been developed in shrimp pond effluents in Brazil, and the cultivation success is related to the techniques used and the environmental conditions of the effluent (Marinho-Soriano et al. 2002). It has been proven that the use of *Gracilaria verrucosa* as a biofilter in traditional aquaculture ponds could increase the production of Vannamei shrimp from 1 ton per ha to 3.5 ton per ha while gaining a seaweed production of 5 ton per ha after 3 months of culture (Andayani et al. 2016). Also, the algal species *Ulva clathrata* cultured in outdoor tanks showed efficiency in removing the main inorganic nutrients from the effluent water of shrimp (*Litopenaeus vannamei*) aquaculture, with uptake rates of 70–82% of the total ammonium nitrogen and 50% of the phosphate within 15 h (Copertino et al. 2009).

Caulerpa lentillifera has shown the potential ability to remove nutrients from aquaculture effluents, especially NO_3 N (Guo et al. 2015; Manori Bambaranda et al. 2019a, b). The efficiency of nutrient uptake has shown a relationship between the nutrient concentration and density of *C. lentillifera*. The highest reduction in the concentration of the nutrient components (NO_2^-, NO_3^-, and PO_4^{3-}) occurred at 30 g/L algal density, while the highest concentration reduction for NH_3-N occurred at 40 g/L algal density (Manori Bambaranda et al. 2019a). In recirculating aquaculture systems, utilizing *C. lentillifera* for biofiltration has shown potential for use in the effective treatment of aquaculture effluent integrating fish and the production of seaweed (Manori Bambaranda et al. 2019b).

In the case of a water treatment system with separate units, the use of shellfish causes sludge to be suspended well, but there is a disadvantage in that sediment filtering causes shellfish to excrete waste in the organic sediment form of nitrogen and phosphorus compounds, while in another unit, seaweed is suitable for the treatment of nitrogen and phosphorus solutions (Tookwinas et al. 2001; Pariyawatee et al. 2003). Therefore, a system that combines the hallmarks of the treatment system in each subsystem is an ideal form for effectively treating wastewater from shrimp culture. In a closed system, the shrimp farm effluent will be drained from the culture pond into a treatment unit consisting of biological systems (shellfish and algae) and physical systems (filtration) by pumping water from the shrimp pond to the biological treatment pond containing mollusks and subsequently pumping it into a treatment pond containing seaweed (*Gracilaria fisheri*). After 3 days of treatment, the treated water will be recycled back through the filter and into the pond (Tookwinas et al. 2001). It has been reported that the removal efficiency of nitrate-N and phosphate-P was decreased by approximately 60% after treatment with this type of system (Pariyawatee et al. 2003). Some species of seaweed, e.g., *Caulerpa lentillifera*, *Ulva* sp. and *Gracilaria fisheri*, had high rates of removing ammonium, with a 92–97% absorption efficiency, and *Caulerpa lentillifera* showed the lowest removal rates, at 90% overall (Kaewmesri 2000; Pariyawatee et al. 2003).

Seaweed is preferably used in treatment systems for wastewater from shrimp farming in Thailand (Fig. 20.1), which can help reduce the amount of hardness and suspended solids in the water. The shrimp farm effluent circulates to the seaweed pond to absorb nutrients in the water, after which the water in the pond is filled with oxygen, and microbes are provided in another unit and recycled to the culture units. The seaweed will absorb the total ammonia–nitrogen, nitrite–nitrogen, nitrate–nitrogen and orthophosphate throughout the biofilter recirculating systems and improve the seawater quality, which is within the safety conditions required for circulation and periodicity.

20.3.1.3 Seaweed for Balancing the Food Chain in Shrimp Culture

Seaweed has been used for cocultivation with tiger prawns (*Penaeus monodon*) in tanks or earthen ponds, which is prevalent in Southeast Asian countries (Songsangjinda and Inek 2003; Tsutsui et al. 2010, 2015; Anh et al. 2018). Another green seaweed, the gut weed *Ulva intestinalis*, has been used in tiger prawn farming in Thailand. Algal species are used as sources for gathering zooplankton, and algal spores are used for feeding zooplankton to increase post larval production, during which *Ulva* thalli are harvested from shrimp ponds as frequently as possible to balance the food chain in shrimp culture ponds (Anonymous 2010). Coculturing tiger prawns with the filamentous green seaweed *Chaetomorpha ligustica* has a dietary advantage, especially in early-age juveniles (Tsutsui et al. 2010), and it reduces feed

Fig. 20.1 Using seaweed for effluent treatment in shrimp farming in Thailand: (**a**) *Ulva intestinalis* pond unit; (**b**) *U. intestinalis* in effluent way; (**c**) *Gracilaria fisheri* pond unit; (**d**) *G. fisheri* in effluent

costs in intensive prawn aquaculture (Tsutsui et al. 2015). Likewise, the coculture of tiger prawns and *G. tenuistipitata* with a reduction in the feeding ratio of up to 50% satiation resulted in improved growth, feed cost, and shrimp color, as well as better water quality in the culture tanks (Anh et al. 2018).

20.3.2 Seaweed in Fish Culture

20.3.2.1 Seaweed in Fish Feed

According to the high cost, fluctuating quality, and uncertain availability of fish meal in fish diets, several sources of conventional plant oilseed meals are used, such as soybean (Voorhees et al. 2019; Abdel-Warith et al. 2020) and cottonseed (El-Sayed 1999; Bu et al. 2017) meal. There have been a few species of seaweed used to feed fish directly. The species of *Ulva* green seaweed are important for bio-filtering fish pond effluents. This alga has been harvested in ponds as a feed additive for red tilapia in the coastal areas of Thailand. The alga has also been used as a replacement raw material for food production or optimization to improve color.

It improves the growth rate, survival rate, efficiency, and feed conversion ratio (FCR.), in addition to enhancing immunity (Prud'homme van Reine and Trono 2001).

Several studies have reported the use of seaweed as an ingredient in tilapia fish feed (Azaza et al. 2008; Siddik and Anh 2015; Putri et al. 2017; Yangthong and Ruensirikul 2020). The green seaweed *Ulva rigida* was used as a replacement for soybean meal in the practical tilapia fish diet formulated to contain 28% crude protein, 7.5% lipids and 15 kJ gross energy (Azaza et al. 2008). They stated that replacement with seaweed species decreased the apparent protein digestibility (APD) with increasing inclusion levels of ulva meal. The use of seaweed as an ingredient causes a reduction in protein and fat contents because of the replacement of feedstock with vegetables in the diet ingredients. In addition, saponins, which are found in several seaweed species (Feroz 2018), interrupt the absorption of dietary fat and inhibit the breakdown of fat in the gallbladder, so the fish cannot fully use fat from the diet. The increased level of algae replacement leads to a reduction in body fat storage (Tacon 1997; Guillaume and Choubert 2001; Azaza et al. 2008).

20.3.2.2 Seaweed as a Biofilter in Fish Culture

Similar to the wastewater treatment system in shrimp culture, seaweed is also used as a biofilter in fish culture. The red seaweed *Gracilaria* and the green seaweed *Caulerpa* and *Ulva* are commonly applied for water treatment in aquaculture systems. *Gracilaria fisheri* is the most widely applied for nutrient absorption in fish farming in Thailand. These seaweed species have high efficiencies in the removal of the total ammonia from fish farm effluent. *G. fisheri* had the best efficiency in the removal of ammonia, nitrite and phosphate (99.6%, 70.9% and 98.4%, respectively), while *C. lentillifera* showed the best efficiency in the removal of nitrate (92.0%) (Thongcanarak and Predalumpaburt 2008). However, in the case of *Ulva intestinalis*, this alga is not suitable for water treatment in standing-water conditions, since the seaweed cannot grow and it dies under nutrient depletion – conditions. Inorganic nitrogen and phosphate are released again into water when seaweed dies and decomposes. Although seaweed species should be used for water treatment, their efficiency and ability to adapt to survive under the specific conditions of effluent from aquaculture ponds should be considered. Fish effluent was treated with seaweed according to the removal periodicity of the system. The total periodicity took approximately 1 month to compensate for inorganic carbon depletion and to reach low levels of nitrate and phosphate. Throughout the farming period, the ammonia, nitrite, nitrate and BOD values in the fish ponds increased, while in the treatment unit, seaweed decreased the ammonia, nitrite, nitrate and BOD (Thongcanarak and Predalumpaburt 2008).

The process of wastewater treatment in fish farming was divided into 4 units, as in the shrimp culture treatment. The effluent from the fish culture unit flowed to the settling unit and then to the seaweed pond for nutrient absorption from the effluent

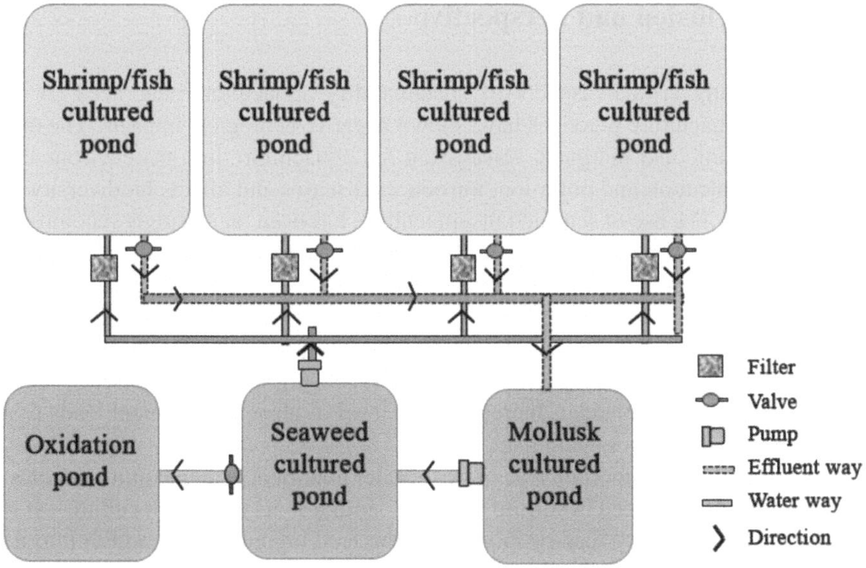

Fig. 20.2 Diagram of wastewater treatment in shrimp or fish culture

seawater. Likewise, in the removal timeline of the process, the total periodicity took a month to complete the treatment. The schematic diagram of the system is shown in Fig. 20.2.

20.3.3 Seaweed in Abalone Culture

Several species of seaweed have been used in abalone aquaculture on several continents, mainly in Asian countries, e.g., China, Japan, Taiwan and Thailand. Abalone farming has been attempted for at least two decades in Thailand; however, farmed abalone has been categorized as a prime cultured species of aquatic animals due to its high value. This was an incentive that made private entrepreneurs raising black tiger prawns or white shrimp initially become interested in abalone instead of sea shrimp. Due to pricing problems and shrimp disease outbreaks in Thailand, the Department of Fisheries has turned to developing abalone farming and has been successful in producing young abalone. For feeding abalone, red seaweed was used as a supplementary feed in parallel with complete feed in pellet form. However, to increase immunity, abalone feed is mixed with seaweed at a proportion of 5%, and the most popular species are *Gracilaria tenuistipitata* and *G. fisheri* (Prud'homme van Reine and Trono 2001).

20.4 Conclusion and Perspective

There are many critical aspects of sustainability aquaculture that need to be addressed. Aquaculture practices have shown negative ecological impacts. The discharge of organic and inorganic waste from fed aquaculture into the environment causes eutrophication and pollution, introduces diseases and affects biodiversity in the ecosystem. The use of seaweed in aquaculture has been increasingly recognized due to its positive benefits and numerous supplies. Seaweed is used for diet supplementation and balancing the food chain in aquaculture (shrimp, fish, and abalone), and it can prevent disease by enhancing immune function. However, more research is needed to clarify its role in nutrition and disease prevention. In particular, the saponins in some seaweeds may affect dietary fat absorption and inhibit the breakdown of fat in the gallbladder. Increasing the level of algae replacement leads to a reduction in body fat storage in fish.

Seaweed plays an important role in wastewater treatment in aquaculture systems. The utilization of seaweed in shrimp culture is similar to its use in fish culture—it is used as a supplementary feed or mixed into the feed ingredients, in addition to its use in the treatment of effluent from aquaculture units. The nutrient removal system consists of the combination of treatment unit using mollusks, such as oysters (*Crassostrea lugubris*) or mussels (*Perna viridis*), with treatment unit using seaweed. This combination effectively improves the effluent quality from shrimp ponds. Treatment using mollusks is a disadvantage that causes organic sludge, but nitrogen and phosphorus are increased and can be absorbed by seaweed in the next treatment unit.

However, seaweed can also be used for many purposes other than absorbing nutrients in seawater. There may be other harmful substances, including heavy metals, that algae take up at the same time as nitrogen and phosphorus adsorption. Therefore, the objectives of its use should be considered. In the case of using seaweed for treating or reducing heavy metals or negative elements, the harvesting or utilization of seaweed after treatment should be considered environmentally friendly. In addition, towards the end of the treatment period, a problem of phosphate release from dead seaweed would occur during the life stage of the algae and in the period of depleted nutrient concentration. Therefore, the ability of seaweed to adapt and the effect of the aquaculture effluent must also be considered. During the treatment period, algae should be left in the system for at least 1 week to treat nitrate nitrogen and 2–3 weeks to treat phosphate phosphorous (Suthiniam et al. 2009). Furthermore, the efficiency of seaweed can provide successful treatment of wastewater in the early stages of the system. However, the nitrogen and phosphate compounds released from the dead seaweed may be troublesome in the maintenance of the seaweed unit, since it may take a period of time to regrow the seaweed in the system. The use of seaweed also has to take into account the adaptability of seaweed in aquaculture. Therefore, monocultivation of seaweed is needed to support the water treatment unit, which requires a support area.

References

Abdel-Warith AA, Younis EM, Al-Asgah NA, Mahboob S (2020) Effect of replacing fish meal by full fat soybean meal on growth performance, feed utilization and gastrointestinal enzymes in diets for African catfish *Clarias gariepinus*. Braz J Biol 80(3):535–543

Andayani S, Yuwanita R, Izzah N (2016) Biofilter application using seaweed (*Gracilaria verrucosa*) to increase production of Vannamei shrimp in traditional pond district Bangil-Pasuruan. Res J Life Sci 3(1):16–22

Anh NTN, Ngan LTH, Vinh NH, Hai TN (2018) Co-culture of red seaweed (*Gracilaria tenuistipitata*) and black tiger shrimp (*Penaeus monodon*) with different feeding rations. Int J Sci Res Publ 8(9):269–277

Aníbal J, Madeira HT, Carvalho LF, Esteves E, Veiga-Pires C, Rocha C (2014) Macroalgae mitigation potential for fish aquaculture effluents: an approach coupling nitrogen uptake and metabolic pathways using *Ulva rigida* and *Enteromorpha clathrata*. Environ Sci Pollut Res 21:13324–13334

Anonymous (2010) Role of gut weeds (*Ulva intestinalis* Linnaeus) on improving water and sediment qualities and producing natural foods for commercial culture of banana shrimp (*Penaeus merguiensis* de Man, 1988). Ministry of Agriculture and Cooperatives, Department of Fisheries, Freshwater Aquaculture Research and Development Division. (in Thai)

Azaza MS, Mensi F, Ksouri J, Dhraief MN, Brini B, Abdelmouleh A, Kraïem MM (2008) Growth of Nile tilapia (*Oreochromis niloticus* L.) fed with diets containing graded levels of green algae *Ulva* meal (*Ulva rigida*) reared in geothermal waters of southern Tunisia. J Appl Ichthyol 24:202–207

Baliao DD, Tookwinas S (2002) Best management practices for a mangrove-friendly shrimp farming. SEAFDEC Aquaculture Department/Association of Southeast Asian Nations, Bangkok/Tigbauan

Barrington K, Chopin T, Robinson S (2009) Integrated multi-trophic aquaculture (IMTA) in marine temperature waters. In: Soto D (ed) Integrated mariculture: a global review. FAO Fisheries and Aquaculture Technical Paper, vol 529, pp 7–46

Beveridge MCM (1996) Cage aquaculture, 2nd edn. Fishing News Books Ltd., Cambridge, p 346

Bu X, Chen A, Lian X, Chen F, Zhang Y, Muhammad I, Ge X, Yang Y (2017) An evaluation of replacing fish meal with cottonseed meal in the diet of juvenile Ussuri catfish *Pseudobagrus ussuriensis*: growth, antioxidant capacity, nonspecific immunity and resistance to *Aeromonas hydrophila*. Aquaculture 479:829–837

Buschmann AH (1996) An introduction to integrated farming and the use of seaweeds as biofilters. Hydrobiologia 326(327):59–60

Chaitanawisuti N, Santhaweesuk W, Kritsanapuntu S (2011) Performance of the seaweeds *Gracilaria salicornia* and *Caulerpa lentillifera* as biofilters in a hatchery scale recirculating aquaculture system for juvenile spotted babylons (*Babylonia areolata*). Aquac Int 19:1139–1150

Chirapart A, Lewmanomont K (2004) Growth and production of Thai agarophyte cultured in natural pond using the effluent seawater from shrimp culture. Hydrobiologia 512:117–126

Chokwiwattanawanit A (2000) Efficiency of the macroalgae *Caulerpa lentillifera* and *Acanthophora spicifera* for the treatment of nitrogen compound from shrimp pond effluent. M Sc Thesis, Environmental Science, Graduate School, Chulalongkorn University. (in Thai)

Chung IK, Sondak CFA, Beardall J (2017) The future of seaweed aquaculture in a rapidly changing world. Eur J Phycol 52(4):495–505. https://doi.org/10.1080/09670262.2017.1359678

Copertino MS, Tormena T, Seeliger U (2009) Biofiltering efficiency, uptake and assimilation rates of *Ulva clathrata* (Roth) J. Agardh (Chlorophyceae) cultivated in shrimp aquaculture waste water. J Appl Phycol 21:31–45

Cuzon G, Brito A, Jiménez-Yan L, Brito R, Tomás G, Taboada G, Gaxiola G (2004) The effects of animal or plant-based diets on energy partitioning in selected ontogenetic stages of the shrimp *Litopenaeus vannamei*. In: Cruz Suárez LE, Ricque Marie D, Nieto Lopez MG, Villarreal D,

Scholz UY, Gonzalez M (eds) Avances en nutricion acuicola VII. Memorias del VII Simposium internacional de nutricion acuicola, Hermosillo, Sonora, Mexico. 16–19 Nov 2004

El-Saidy DMSD (1999) Evaluation of cottonseed meal as partial and complete replacement of fishmeal in practical diets of Nile tilapia, *Oreochromis niloticus* (L.) fingerlings. Egypt J Aquat Biol Fish 3:1–18

FAO (2014) The state of world fisheries and aquaculture. Opportunities and challenges. Rome, 233 p

FAO (2016) The state of world fisheries and aquaculture 2016. Contributing to food security and nutrition for all. Rome, 200 p

FAO (2018) The state of world fisheries and aquaculture 2018 – meeting the sustainable development goals. Rome, 227 p

FAO (2020) The state of world fisheries and aquaculture 2020 – sustainability in action Rome, 224 p

Feroz B (2018) Saponins from marine macroalgae: a review. J Mar Sci Res Dev 8:255. https://doi.org/10.4172/2155-9910.1000255

Grote B (2019) Recent developments in aquaculture of *Palmaria palmata* (Linnaeus) (Weber & Mohr 1805): cultivation and uses. Rev Aquac 11:25–41

Guillaume J, Choubert G (2001) Digestive physiology and nutrient digestibility in fishes. In: Guillaume J, Kaushik S, Bergot P, Métailler R (eds) Nutrition and feeding of fish and crustaceans. Springer-Praxis Publishing, Chichester, pp 27–56

Guo H, Yao JT, Sun ZM, Duan DL (2015) Effects of salinity and nutrients on the growth and chlorophyll fluorescence of *Caulerpa lentillifera*. Chin J Oceanol Limn 33:410–418

Hayashi L, Yokoya NS, Ostini S, Pereira RTL, Braga ES, Oliveira EC (2008) Nutrients removed by *Kappaphycus alvarezii* (Rhodophyta, Solieriaceae) in integrated cultivation with fishes in recirculating water. Aquaculture 277(3–4):185–191

Hernández I, Fernández-Engo MA, Pérez-Lloréns JL, Vergara JJ (2005) Integrated outdoor culture of two estuarine macroalgae as biofilters for dissolved nutrients from *Sparus aurata* waste waters. J Appl Phycol 17:557–567

Kaewmesri A (2000) The treatment of wastewater from *Penaeus monodon* farming by the agarophyte, *Gracilaria fisheri* (Xia & Abbott) Abbott, Zhang & Xia. M Sc Thesis, Graduate School, Mahidol University, Bangkok (in Thai)

Kang YH, Hwang JR, Chung IK, Park SR (2013) Development of a seaweed species-selection index for successful culture in a seaweed-based integrated aquaculture system. J Ocean Univ China (Ocean Coast Sea Res) 12(1):125–133

Kanjana K, Radtanatip T, Asuvapongpatana S, Withyachumnarnkul B, Wongprasert K (2011) Solvent extracts of the red seaweed *Gracilaria fisheri* prevent *Vibrio harveyi* infections in the black tiger shrimp *Penaeus monodon*. Fish Shellfish Immunol 30:389–396

Kerrigan D, Suckling CC (2016) A meta-analysis of integrated multitrophic aquaculture: extractive species growth is most successful within close proximity to open-water fish farms. Rev Aquac 0:1–13

Khidprasert S (1995) Absorption of nitrogen compounds from shrimp pond effluence by marine algae. M Sc Thesis, Graduate School, Kasetsart University, Bangkok (in Thai)

Kunawongdet P (2020) Biological water treatment in recirculating aquaculture system by using *Caulerpa lentillifera* J Agardh. Burapha Sci J 25(2):509–523. (in Thai)

Limhang K, Saeheng L, Khunpol K (2014) Efficiency for treatment of nitrogen compound of the water from rearing pacific white shrimp (*Litopenaeus vannamei*) using sea grapes (*Caulerpa macrophysa*). Proceedings of 52nd Kasetsart University annual conference: fisheries. Agricultural Extension and Home Economics, Bangkok, pp 226–233. (in Thai)

Manori Bambaranda BVAS, Sasaki N, Chirapart A, Salin KR, Tsusaka TW (2019a) Optimization of macroalgal density and salinity for nutrient removal by *Caulerpa lentillifera* from aquaculture effluent. PRO 7(303):1–16. https://doi.org/10.3390/pr7050303

Manori Bambaranda BVAS, Tsusaka TW, Chirapart A, Salin KR, Sasaki N (2019b) Capacity of *Caulerpa lentillifera* in the removal of fish culture effluent in a recirculating aquaculture system. PRO 7(440):1–15. https://doi.org/10.3390/pr7070440

Marinho G, Nunes C, Sousa-Pinto I, Pereira R, Rema P, Valente LMP (2013) The IMTA-cultivated Chlorophyta *Ulva* spp. as a sustainable ingredient in Nile tilapia (*Oreochromis niloticus*) diets. J Appl Phycol 25(5):1359–1367

Marinho-Soriano E, Morales C, Moreira WSC (2002) Cultivation of *Gracilaria* (Rhodophyta) in shrimp pond effluents in Brazil. Aquac Res 33:1081–1086

Martínez-Córdova LR, Peña-Messina E (2005) Biotic communities and feeding habits of *Litopenaeus vannamei* (Boone 1931) and *Litopenaeus stylirostris* (Stimpson 1974) in monoculture and polyculture semi-intensive ponds. Aquac Res 36:1075–1084

Neori A (2008) Essential role of seaweed cultivation in integrated multi-trophic aquaculture farms for global expansion of mariculture: an analysis. J Appl Phycol 20:567–570

Neori A, Guttman L (2017) Thoughts on algae cultivation toward an expansion of aquaculture to the scale of agriculture. In: Proceedings of 7th international conference on innovation in chemical, agricultural, biological and environmental sciences (ICABES-2017), London (UK). 4–6 Dec 2017. https://doi.org/10.15242/HEAIG.H1217234. Accessed 21 Nov 2020

Neori A, Shpigel M, Ben-Ezra D (2000) A sustainable integrated system for culture of fish, seaweed and abalone. Aquaculture 186:279–291

Neori A, Chopin T, Troell M, Buschmann AH, Kraemer GP, Halling C, Shpigel M, Yarish C (2004) Integrated aquaculture: rationale, evolution and state of the art emphasizing seaweed biofiltration in modern mariculture. Aquaculture 231:361–391

Niu J, Xie J-J, Guo T-Y, Fang H-H, Zhang Y-M, Liao S-Y et al (2019) Comparison and evaluation of four species of macro-algae as dietary ingredients in *Litopenaeus vannamei* under normal rearing and WSSV challenge conditions: effect on growth, immune response, and intestinal microbiota. Front Physiol 9(1880):1–15

Pariyawathee S, Songsangjinda P, Deraxbudsarakom S, Tuntichodok P, Chiayvareesajja S (2003) Optimum condition of environmental factors for growth of sea grape (*Caulerpa lentillifera* J. Agardh). Thai Fish Gaz 56:443–448. (in Thai)

Pellizzari F, Reis RP (2011) Seaweed cultivation on the southern and southeastern Brazilian coast. Rev Bras Farmacogn Braz J Pharmacogn 21(2):305–312

Prud'homme van Reine WF, Trono Jr GC (2001) Plant resources of South-East Asia 15(1), cryptogams: algae. Backhuys Publishers, Leiden

Putri NT, Jusadi D, Setiawati M, Sunarno MTD (2017) Potential use of green algae *Caulerpa lentillifera* as feed ingredient in the diet of Nile tilapia *Oreochromis niloticus*. Jurnal Akuakultur Indonesia 16(2):195–203

Putro SP, Widowati W, Rahmansyah R, Suminto S (2016) Application of polyculture using stratified double net cage: a case study at Awerange Gulf, Barru, South Sulawesi, Indonesia. AACL Bioflux 9(4):815–822

Rana KJ, Siriwardena S, Hasan MR (2009) Impact of rising feed ingredient prices on aquafeeds and aquaculture production. FAO Fisheries and Aquaculture Technical Paper. No. 541. Rome, 63 p

Ratana-arporn P, Chirapart A (2006) Nutritional evaluation of tropical green seaweeds *Caulerpa lentillifera* and *Ulva reticulata*. Kasetsart J (Nat Sci) 40(Suppl):75–83

Rudtanatip T, Asuvapongpatana S, Withyachumnarnkul B, Wongprasert K (2014) Sulfated galactans isolated from the red seaweed *Gracilaria fisheri* target the envelope proteins of white spot syndrome virus and protect against viral infection in shrimp haemocytes. J Gen Virol 95:1126–1134

Shama A, Joyce SG, Mari FD, Ranga Rao A, Ravishankar GA, Hudaa N (2019) Macroalgae and microalgae: novel sources of functional food and feed. In: Ravishankar GA, Ranga Rao A (eds) Handbook of algal technologies and phytochemicals. Food, health and nutraceutical applications, vol-I. CRC Press, USA, pp 207–219

Siddik MAB, Anh NTN (2015) Preliminary assessment of the gut weed *Ulva intestinalis* as food for herbivorous fish. Int Aquat Res 7:41–46

Sirirustananun N, Chen JC, Lin YC, Yeh ST, Liou CH, Chen LL, Sim SS, Chiew SL (2011) Dietary administration of a *Gracilaria tenuistipitata* extract enhances the immune response

and resistance against *Vibrio alginolyticus* and white spot syndrome virus in the white shrimp *Litopenaeus vannamei*. Fish Shellfish Immunol 31:848–855
Sitthimong C (2020) The situation of white shrimp farming. https://www.fisheries.go.th/extension/whitesh.htm. Accessed 2 Nov 2020
Songsangjinda P, Inek S (2003) Efficiency of oyster (*Crassostrea lugubris*) green mussel (*Perna viridis*) and seaweed (*Gracilaria fisheri*) on quality improvement of effluent from shrimp pond. Technical Paper No. 2/2003, Coastal Fisheries Research and Development Bureau, Department of Fisheries. Ministry of Agriculture and Cooperatives. 15 p. (in Thai)
Sriveerachi T, Pankdee S (2005) Culture on seaweed *Gracilaria edulis* (Gmelin) Silva and *Acanthophora spicifera* (Vahl) Børgesen in biologically-treated water pond. Technical Paper No. 17/2005, Coastal Fisheries Research and Development Bureau, Department of Fisheries. Ministry of Agriculture and Cooperatives. (in Thai)
Sukiman F, Rohyani IS, Ahyadi H (2014) Growth of seaweed *Eucheuma cottonii* in multi trophic sea farming systems at Gerupuk Bay, Central Lombok, Indonesia. Nus Biosci 6:82–85
Suthiniam S, Chiewwarisajja S, Chiewwarisajja J, Silapaachan K (2009) Efficiency of nitrogen and phosphorus treatment of feather seaweed in effluents from white shrimp farming. Thai Fish Gaz 62(3):215–220. (in Thai)
Tacon AGJ (1997) Fishmeal replacers: review of antinutrients within oilseeds and pulses – a limiting factor for the aquafeed green revolution? In: AGJ T, Basurco B (eds) Feeding tomorrow's fish. CIHEAM, Zaragoza, pp 153–182. (Cahiers Options Méditerranéennes; n. 22)
Takeshi S, Yumiko Y, Joko S (2005) Mineral components and antioxidant activities of tropical seaweeds. J Ocean Univ China 4:205–208
Tamtin M, Chaisri C, Kosawadpat P, Kaukaew J, Lia-Thongkhom N (2016) Nutritional value of sea lettuce (*Ulva rigida*) and application as feed ingredients in Pacific white shrimp (*Litopenaeus vannamei* Boone, 1931) diets. Technical Paper No. 12/2016, Coastal Fisheries Research and Development Bureau, Department of Fisheries. Ministry of Agriculture and Cooperatives. (in Thai)
Teekayu K, Khidprasert S (1996) Wastewater treatment from shrimp farming using seaweed, *Gracilaria*. Reduce the amount of nitrogen compounds. Faculty of Agriculture, Bang Phra Rajamangala Institute of Technology (in Thai)
Thongcanarak P, Predalumpaburt Y (2008) Comparison the efficiency of 4 seaweeds in the removal of nitrogen and phosphate in effluent from orange spotted grouper (*Epinephelus coioides* Hamilton, 1822) culture. Technical Paper No. 40/2008, Coastal Fisheries Research and Development Bureau, Department of Fisheries. Ministry of Agriculture and Cooperatives. (in Thai)
Tookwinas S, Songsangjinda P, Matsuda O, Na-anan P, Ekpanithanpong U (2001) Experiment on integrated physical and biological treatment processes for treatment of intensive marine shrimp farming's effluent. In: Proceedings of the JSPS-NRCT seminar on sustainable shrimp culture and health management diseases and environment. Tokyo University of Fisheries, Tokyo
Troell M (2009) Integrated marine and brackish water aquaculture in tropical regions: research, implementation and prospects. In: Soto D (ed) A global review of integrated marine aquaculture. FAO Fisheries and Aquaculture Technical Paper. No. 529, Rome
Troell M, Halling C, Nilsson A, Buschmann AH, Kautsky N, Kautsky L (1997) Integrated marine cultivation of *Gracilaria chilensis* (Gracilariales, Rhodophyta) and salmon cages for reduced environmental impact and increased economic output. Aquaculture 156:45–61
Troell M, Tyedmers P, Kautsky N, Rönnbäck P (2004) Aquaculture and energy use. In: Encyclopedia of energy, vol 1, pp 97–108. Amsterdam, Boston. Elsevier Academic Press.
Tsutsui I, Kanjanaworakul P, Srisapoome P, Aue-umneoy D, Hamano K (2010) Growth of giant tiger prawn, *Penaeus monodon* Fabricius, under co-culture with a discarded filamentous seaweed, *Chaetomorpha ligustica* (Kützing) Kützing, at an aquarium-scale. Aquacult Int 18:545–553

Tsutsui I, Songphatkaew J, Meeanan C, Aue-umneoy D, Sukchai H, Pinphoo P et al (2015) Co-culture with *Chaetomorpha* sp. enhanced growth performance and reduced feed conversion ratio of the giant tiger prawn, *Penaeus monodon*. Int Aquat Res 7:193–199

Voorhees JM, Barnes ME, Chipps SR, Brown ML (2019) Bioprocessed soybean meal replacement of fish meal in rainbow trout (*Oncorhynchus mykiss*) diets. Cogent Food Agric 5(1):1579482. https://doi.org/10.1080/23311932.2019.1579482. Accessed 9 Dec 2020

Wongprasert K, Rudtanatip T, Praiboon J (2014) Immunostimulatory activity of sulfated galactans isolated from the red seaweed *Gracilaria fisheri* and development of resistance against white spot syndrome virus (WSSV) in shrimp. Fish Shellfish Immunol 36:52–60

Yala ZR, Sulistiawati D, Masyahoro N (2017) Seaweed (*Eucheuma cottonii*) growth in polyculture application. AACL Bioflux 10(5):1064–1073

Yangthong M, Ruensirikul J (2020) Feed intake stimulation of juvenile spotted scat (*Scatophagus argus* Linnaeus, 1766) using dietary seaweed supplementation (*Ulva* sp.). Aquaculture 529:735626. https://doi.org/10.1016/j.aquaculture.2020.735626. Accessed 8 Dec 2020

Zahra A, Sukenda S, Wahjuningrum D (2017) Extract of seaweed *Gracilaria verrucosa* as immunostimulant to controlling white spot disease in Pacific white shrimp *Litopenaeus vannamei*. Jurnal Akuakultur Indonesia 16(2):174–183

Chapter 21
Antimicrobial Potential of Seaweeds: Critical Review

A. Noorjahan, S. Mahesh, P. Anantharaman, and B. Aiyamperumal

Abbreviations

AHPND	Acute hepatopancreatic necrosis disease
CAA	Coastal Aquaculture Authority
DHA	Docosahexaenoic acid
EPA	Eicosapentaenoic acid
MAPs	Marine algae polysaccharides
PLs	Post larvae
PUFA	Polyunsaturated fatty acid
SFA	Saturated Fatty acids
WSSV	White spot syndrome virus

21.1 Introduction

In the past two decades, aquaculture production has been greatly affected by several diseases (Flegel 2006; Lightner 1993, 2011; FAO 2018). The intensity of aquaculture has led to increasing problems caused by viruses, bacteria, fungi, and other pathogens (Planas et al. 2005; Dhaneesh et al. 2012). Shell and finfishes are considered to be important sources protein for human consumption. The threatening of diseases in aquaculture sector decline not only by national economy but also globally. In addition, many researchers reported the presence of gram-negative enteric organisms in edible fish meat can transmit diseases to humans through improper cooked fish and its products (Sarmasik 2000). Infectious diseases of cultured fish are among the most notable constraints on the expansion of aquaculture and the realization of its full potential (Plumb 1999; Klesius et al. 2000). Chemicals

A. Noorjahan (✉) · S. Mahesh · P. Anantharaman · B. Aiyamperumal
Centre of Advanced Study in Marine Biology, Faculty of Marine Sciences, Annamalai University, Parangipettai, Chidambaram, India
e-mail: noorbiotek@gmail.com

A. Ranga Rao, G. A. Ravishankar (eds.), *Sustainable Global Resources of Seaweeds Volume 1*, https://doi.org/10.1007/978-3-030-91955-9_21

and antibiotic treatments are employed to prevent and diagnose diseases which are quite expensive and consequently prompted the advent of drug resistance among the pathogens. Indiscriminate use of chemicals and antibiotics has stimulated intense debate among environmentalists and government agencies as whether to ban these products entirely or to permit them to have sustainable production. Hence it is necessary to search for novel and safe antibacterial agents. These antimicrobial agents from marine habitats are excellent sources in producing several antibiotics.

Algae based value added feed resulted in both growth and reduced mortality at PLs stage for shrimp, better growth performance, carcass quality for Rohu and Tilapia and vibrant skin colour for Ornamental Fishes. Incorporating algae based feed for reducing production cost of fish based aquafeed. Fermented Seaweed incorporated feed improves the growth performance, nutrient digestibility, and microbial flora in *Oreochromis niloticus* (Yuan et al. 2017; Akhilamole et al. 2019a). The effects of spent seaweed *Turbinaria conoids,* at gradient incorporation in feed for Nile tilapia (*Oreochromis niloticus*). As a result better growth performance with no adverse effect or abnormalities in fish was observed (Akhilamole et al. 2019a, b; Noorjahan et al. 2021).

Many researchers also reported the inhibiting potential of seaweed extracts against aquatic pathogenic bacteria which ubiquitous in infecting both shrimp and fish (Vairappan and Suzuki 2000; Bansemir et al. 2004, 2006; Dubber and Harder 2008; Ganeshamurthy et al. 2012; Genovese et al. 2012; Rebecca et al. 2012; Singh et al. 2012; Cavallo et al. 2013; Maheswaran et al. 2013; Mata et al. 2013; Radhika et al. 2014).

Kanjana et al. 2011 studied the protective role in both *in vitro* and *in vivo* of red seaweed *Gracilaria fisheri* solvent extracts against *Vibrio harveyi*. Thanigaivel et al. 2014 demonstrated the therapeutic effect in short term administration of using seaweed extracts as alternatives to antibiotics.

We suggest that seaweed are efficient to maintain healthy status of both shell and finfishes in aquaculture. In the present communication an attempt has been made to review the applications of few seaweeds in rearing and disease prevention of aquaculture animals.

21.1.1 Disease Causing Agents in Aquaculture

Fish bacterial pathogens predominantly found are *Streptococcus sp., Pseudomonas sp., Bacillus sp., Staphylococcus sp., Enterobacter sp. and Escherichia coli*. The slime layer that covers the outer surface of fish has been found to contain bacteriae viz., *Pseudomonas sp., Acinetobacter sp., Moraxella sp., Alkaligens sp., Micrococcus sp., Serratia sp., Vibrio sp.* and *Bacillus sp* (Marudhupandi et al. 2013). Most bacteria that infect fishes are Gram-negative, including *Aeromonas hydrophila, Aeromonas salmonicida, Flavobacterium columnare, Vibrio* sp., and *Pseudomonas sp.*, the major groups of Gram-positive bacteria that cause disease in fish are Streptococcus (Genovese et al. 2012; Cavallo et al. 2013). Some of these pathogenic bacteria, for example *Vibrio* species, can infect both fish and shrimp and in many cases the manifestation and the progress of the associated diseases are affected by the presence of various stressful conditions of crowding, handling, spawning, or holding fish at above-normal temperatures, as well as the stress of external injury, facilitates the transmission of fish diseases. Some of the most obvious signs of bacterial dermatitis include the development of reddened lesions, sores, or ulcers on the body, reddening of the base of the fins, and dulling or darkening of skin color. The distribution of the skin lesions is quite variable and may include the head, face, operculum, mouth, back, trunk, or body wall including the lateral line and caudal peduncle (Noga 1996). *Aeromonas hydrophila* are common inhabitants of soil and water and commonly are found on the surface of fishes, particularly on the gills. It is one among the opportunistic pathogen in fishes and humans. It causes meningitis, bacteremia, bronchopulmonary infection, food-borne gastroenteritis, wound infection, endocarditis, and osteomyelitis in an immune-compromised person (Yousr et al. 2007; Rama Devi et al. 2016). On the other hand, *Aeromonas septicemia* and hemorrhagic septicemia are the major issues in freshwater and marine fish giving rise to serious problems in fish farming industry in developing as well as developed countries (Rama Devi et al. 2016; Azad et al. 2001; Noga 2010). Vibriosis is another major disease in shellfish and finfish responsible for mortality in aquaculture caused by *Vibrio sp* (Chen et al. 2018). *Vibrio parahaemolyticus* was identified as the causative pathogen and caused the pandemic outbreak of AHPND (Acute hepatopancreatic necrosis disease). Virulence profile in AHPND-related *V. parahaemolyticus* (Li et al. 2017; Kongrueng et al. 2015) was unclear. LSS (Loose shell syndrome), WGD (White gut disease), and red disease are bacterial associated and the shrimp industry was still struggling with MBV, IHHNV, YHV and TSV outbreaks. It was hit by an even bigger disaster with the arrival of white spot syndrome virus (WSSV) caused mass mortalities in shrimp culture ponds.

21.1.2 Preventive Measures in Aquaculture

Nearly 300 products of antibiotics and chemicals are marketed globally including USA, Thailand, Malaysia, Belgium and China etc. (Chowdhury et al. 2015) which are commonly in practice for the treatment of bacterial diseases in the aquaculture system. The aquaculture farmers employing various antibiotics like ampicillin, tetracycline and chloroamphinicol (Kamaraj et al. 2018) and chemicals such as Potassium permanganate, Lime, Salt, Virex, Timsen, Aquakleen, Germnil, Pond Safe, Deletrix, Spa, Albez, Ablez are used regularly for disease treatment. Mostly used antibiotics are Renamycin, Oxysentin 20% Chlorsteclin Oxy-D Vet, Aquamycin, Orgamycin 15%, Orgacycline-15% etc. Major active ingredients of these antibiotics are oxytetracycline, chlorotetracycline, amoxicillin, doxycycline etc. Simiarly, Mohammad Ali et al. 2014 also reported the use of a wide variety of antibiotics like Acimox (vet) Powder, Bactitab, Chlorsteclin, Cotrim-Vet, Fish cure, Orgacycline-15%, Otetra vet power 50, Oxin WS, Oxysentin 20%, Ranamox,

Renamycin and Sulfatrim for health and disease management. However, their continuous use in the aquaculture system has led to the development of antibiotic-resistance to the pathogen. Indiscriminate practice of using antibiotics can lead to the advance of antibiotic-resistant pathogens; the presence of these chemicals, left as residues in the meat, dangerous for human consumption also causing harmful bioaccumulation impacts (Hidayat et al. 2018). The results of early investigations (Gunasekaran and Poorniammal 2008; Jia et al. 2000) revealed a potential for antimicrobial peptides to protect against infections. Similarly, the results of the preliminary characterization of the antibacterial component revealed that it also has to be a heat-stable compound. Antibiotics are widely used to prevent and control the outbreaks caused by pathogenic bacteria especially *Vibrio sp* in aquaculture, especially in shrimp and salmonid farming (Holmström et al. 2003). The development of various solutions against diseases, other than the use of antibiotics, was thus initiated to ensure the profitability of aquaculture, for the production of safe and healthy products. For example, using probiotics in aquaculture to limit or inhibit the growth of the pathogenic species clade *Vibrio harveyi* was reported in several studies (Kesarcodi-Watson et al. 2008; Wang et al. 2008; Prado et al. 2010; Jiang et al. 2013).

Hence, application of probiotics and chemicals are in regular practice (Sandeepa and Ammani 2015; Grenni et al. 2018) though banned by Coastal Aquaculture Authority of India, (CAA 2014). The concept of pathogen control in aquaculture, especially disease prevention using herbal and phytochemicals, have received widespread attention over the last decade (Bulfon et al. 2015; Reverter et al. 2017). Nowadays, seaweeds are considered as medicinal plants. They contain numerous health-promoting molecules and materials such as dietary fiber, ω-3 fatty acids, essential amino acids, and vitamins A, B, C, and E (Rajapakse and Kim 2011).

21.1.3 Seaweeds as a source of nutrients

Among the marine algal forms, the red and green seaweeds are rich in carbohydrates, whereas the brown seaweeds are rich in soluble fiber and iodine (Gupta and Abu-Ghannam 2011a). Seaweeds are to be considered as miracle species with high nutritive value with low in calories, with high dietry fibres, good source of polyunsaturated fatty acid DHA and EPA and contain proteins upto 44% with amino acid (Holdt and Kraan 2011) like taurine, high in red algae (Dawczynski et al. 2007). In addition to their nutritional value, seaweeds exhibit pharmacological properties, such as antioxidant, anti-inflammatory, antimicrobial and even anticancer properties (El Gamal 2010; Gupta and Abu-Ghannam 2011a, b; Holdt and Kraan 2011; Mohamed et al. 2012). Within marine macroalgae metabolites, polysaccharides are polymers of simple sugars (monosaccharides) linked together by glycosidic bonds (Praiboon et al. 2017).

Marine algae polysaccharides (MAPs) are the most important type of biological molecules contained in large amounts in marine extracts (Shi et al. 2017), and MAPs with high sulfate functionalization are generally known as sulfate polysaccharides. The type of sulfate polysaccharides in marine algae, such as alginate, fucoidan, agar, carrageenan, porphyran, laminarin, galactan, and ulvan, differs depending on the taxonomic group. The active compounds include polysaccharides (e.g. fucoidan), various phytochemicals (e.g. phlorotannins), carotenoids, minerals, peptides and lipids (Gupta and Abu-Ghannam 2011b; Holdt and Kraan 2011). Some of these are signature compounds, particularly found in seaweeds for example phlorotannins. Additionally, the role of seasonal, environmental and geographical variation in the nutritional properties of seaweeds needs to be better documented to support nutritional claims (Wells et al. 2016). Seaweed biomolecular composition can change markedly in response to seasonal variation in environmental factors such as salinity, nitrogen content and water temperature (Marinho-Soriano et al. 2006; Zhang and Thomsen 2019).

21.2 Seaweed as a Source of Bioactive Compounds

Marine algae are among the richest sources of known and novel bioactive compounds (Blunt et al. 2006). More than 1,50,000 macroalgae or seaweed species are found in oceans around the globe, but only a few of them are identified (Bansemir et al. 2006). Potential activity of some marine plants like mangroves, seaweeds, seagrasses, and lichen have been reported from both India and elsewhere (Naqvi et al. 1980; Bernard and Clement 1983; Premnathan et al. 1992). Marine algae are not only the primary and major producers of organic matter in the Ocean but also exert profound effects on the density and distribution of other inhabitants of the marine environment. Nearly 50 million (50,000,00) species in the ocean are virtually untapped resources of secondary metabolites.

Seaweeds belong to a group of plants known as algae. Seaweeds are classified as Rhodophyta (red algae), Phaeophyta (brown algae), Chlorophyta (green algae) based on the pigment present. The red algae (Rhodophyta) have their typical red coloration due to the pigments phycocyanin and phycoerythrin, in addition to chlorophyll (Kadam et al. 2013; Knowler et al. 2020) having anti-oxidative, anti-inflammatory, anti-viral, anti-tumor, neuroprotective and hepatoprotective activities (Sekar and Chandramohan 2008). The fucoxanthin characterizes the brown seaweeds (Phaeophyceae), (Pangestuti and Kim 2011; Kadam et al. 2013; Knowler et al. 2020). Fucoxanthin contains an anallenic bond and a 5, 6-monoepoxide. Fucoxanthin have anti-tumoral, antioxidant and anti-obesity properties (Mise et al. 2011; Pigmen et al. 2014). The green seaweeds (Chlorophyta) having chlorophyll pigments of both chlorophylls (a and b) and carotenoids (β-carotene and xanthophylls) (Kadam et al. 2013). Ulvans are characterized as a sulfated single polydisperse heteropolysaccharide These compounds have antiviral, antioxidant, anticoagulant, antihyperlipidemic and anticancer activity, in addition to immuno-stimulatory effects (Kraan 2013; Alves et al. 2013).

Seaweeds are considered a pool of bioactive compounds as they produce a great variety of secondary metabolites characterized by a broad spectrum of biological activities such as antibiotics, antioxidants, and anti-inflammatory (Tuney 2006; Patra et al. 2009) antiviral, antifungal, and antimicrobial activities have been determined and reported (Yuan et al. 2005; Bansemir et al. 2006; Chew et al. 2008). Seaweed extracts are considered to be a rich source of phenolic compounds (Athukorala et al. 2003; Heo et al. 2005; Noorjahan et al. 2019) among 60% are terpenes and 20% are fatty acids 20% nitrogenous and biosynthetic compounds comprises the total metabolites. The major metabolites identified as antimicrobial agents are chlorellin derivatives, acrylic acid, halogenated aliphatic compounds, terpenes, sulphur containing heterocyclic compounds, phenolic inhibitors, isoprenoid metabolites and hydrogen peroxide etc., (Mohamed et al. 2012, Saleh and Al-Mariri 2017).

Phenolic compounds mainly involved in plant defence mechanisms against invading microorganisms, environmental stress, such as wounding and excessive light or ultraviolet (UV) radiation (Harbourne 1994, Wallace and Fry 1994). These phenolic compounds are also found in seaweeds like that of terrestrial plants (Duan et al. 2006). These compounds possess antimicrobial activity and are found abundant in brown seaweed than in green and red seaweeds (Noorjahan et al. 2019). The phenolic compounds most present in brown algae are meroditerpenoids (plastoquinones, chromanolsand chromenes), which are found almost exclusively in the Sargassaceae (Reddy and Urban 2009). These phenolic compounds can interfere in the amino acid bioavailability when the seaweed consumed, although these compounds are considered seaweed-flavors, due to the impact in flavors of the seaweeds and in the fish (Whitfield et al. 1999, Tibbetts et al. 2016). Thus, there are aquatic feeds with seaweeds' phenolic to provide "oceanic flavor" to the farmed animals.

Tannins are naturally occurring polyphenolic compounds widespread among terrestrial and marine plants (Seaweed and seagrass) (Haslam 1989; Waterman and Mole 1994). The metabolite are subdivided into condensed and hydrolyzable compounds. Hydrolyzable tannins are gallic and/or egallic acid which easily hydrolyzes in acidic media, and condensed tannins are polymeric flavonoids (Huang et al. 2008). In contrast to terrestrial tannins, phlorotannins are compounds that have been found only in marine algae. Phlorotannins are formed by the polymerization of phloroglucinol (1, 3, 5-trihydroxy benzene) monomer units and are synthesized in the acetate-malonate pathway in marine alga (Waterman and Mole 1994; Arnold and Targett 1998). Phlorotannins purified from several brown algae have been reported to possess strong antioxidant activity which may be associated with their unique molecular skeleton up to eight interconnected rings. They are therefore more potent free radical scavenger than other polyphenols derived from terrestrial plants, including green tea catechins, which only have three to four rings (Hemat 2007).

Terpenes are produced by all divisions of seaweeds, most of the halogenated compounds, are isolated from red alga *Laurencia* (Ceramiales, Rhodomelaceae). The genus *Laurencia* are found in tropical and sub-tropical regions around the world and is an extremely rich source of secondary metabolites, mainly sesquiterpenes and C15- acetogenins (Blunt et al. 2007; Souto et al. 2002). The genus Stypopodium is characterized by its ability to produce diterpenes and prenylated hydroquinones (Gerwick et al. 1985). Wessels et al. 1999 isolated diterpene and various other sesquiterpenes from *S. zonale*. According to Van Heemst et al. 1996 Aquil, phenols have also been found in *S. muticum*.

Flavonoids, second largest group of phenolic compounds known to contain a broad spectrum of chemical and biological activities including antioxidant and free radical scavenging properties (Kahkonen et al. 1999). Flavonoids include flavonols, flavones, catechins, proanthocyanidins, anthocyanidins, and isoflavonoids (Ndhlala et al. 2007). In red seaweeds, phenolic compounds as flavonoids and phlorotannins are abundant; having flavonoids three interconnected rings and phlorotannins above eight, doing more potent and stable antioxidants Namvar et al. 2012. These phenolic compounds of red seaweed are being investigated for various industrial sectors, for example, pharmaceutical and cosmetic, due to their high antioxidant power (Guihéneuf et al. 2018; Torres et al. 2018)

Steroids and fatty acid esters of *Acanthophora spicifera* were reported to exhibit potent antitumor and antibacterial activity against human cancer cell lines and microorganisms (Morales et al. 2006). Cholesterol is one of the major sterols presented in all groups of seaweed (Lopes et al. 2011) Besides that, brown and green algae are rich in other C29 sterols, particularly fucosterol and isofucosterol, respectively (Lopes et al. 2013).

The fatty acids, most commonly abundant SFA are myristic (C14:0) and palmitic (C16:0) acids (Pereira et al. 2012). The red seaweeds contain significant quantities of PUFA, mainly AA and EPA (Kendel et al. 2015; Kumari et al. 2013).

In general, these substances can

(a) Infect the bacterial cell walls and the cell membranes, which results in an extensive release of intracellular substances or/and disruption of the uptake and transportation of substances, as for example various phlorotannins (Hierholtzer et al. 2014)
(b) Reduce the protein and nucleic acid synthesis in the bacterial cells (Cai et al. 2014) and
(c) Inhibit respiration (Cai et al. 2014). Phlorotannins, as many other terrestrial tannins do, may also form complexes with some extracellular bacterial enzymes (Stern et al. 1996) thus reducing their effects. In most cases, the effects are dose dependent.

21.2.1 Exploring Seaweed as a Potent Antibacterial Agent

Secondary or primary metabolites from these organisms may be potential bioactive compounds of interest for the pharmacological industry (Attaway and Zaborsky 1993), especially concerning antiviral, antibacterial, and antifungal activities related to marine algae against several pathogens (Borowitzka and Borowitzka 1992). The antimicrobial compounds derived from these marine florae consist of diverse groups of chemical compounds (Ely et al. 2004). The cell extracts and active constituents of these various algae have been shown to have antibacterial activity against Gram-positive and Gram-negative bacteria (Lima et al. 2002). Hornsey and Hide 1985 found that many species of marine algal crude extracts have inhibition activity against pathogenic bacteria. Seaweeds contain different substances which are incorporated in medicine and pharmacotherapy, whereas some of the isolated substances have bacteriostatic and bactericidal properties (Gorban et al. 2003).

Antibacterial activity has been detected in several seaweeds collected from the coast of Mandapam to Kanyakumari, Tamil Nadu State of South India. Many workers revealed that the crude extracts of these Indian seaweeds are effective against Gram-positive bacteria (Srinivasa and Parekh 1981). The antimicrobial activities of the selected species of marine algae were not uniform. The highest antibacterial activities were found in the class Rhodophyceae (80%) followed by the Chlorophyceae (62.5%) and the Phaeophyceae (61.9%). The maximum antifungal activities were observed in the red algae 37%, brown algae 33.3%, and green algae 8.3% activity. *Staphylococcus aureus* was the most susceptible bacterial pathogen followed by *Vibrio sp* (Padmakumar and Ayyakannu 1997). There are several reports with references to broad spectrum of pathogen inhibitory compounds from marine macroalgae against fish and shrimp viral, bacterial, fungal, and yeast pathogens. The extracts of various marine algae have been shown to exhibit antibacterial activity against Gram-positive and Gram-negative bacteria (Lima et al. 2002).

The red macroalgae (Rhodophyta) stands out as the major producer of halogenated compounds above the green and brown macroalgae groups (Pereira and Teixeira 1999). Important examples of bioactive halogenated terpenes are (i) laurenditerpenol, isolated from *Laurencia intricata*, which inhibits angiogenesis process (Mohammed et al. 2004), and (ii) 7-ethyl desoxiparguerol, isolated from *Jania rubens*, which shows anthelmintic activity and is effective against *Ehrlich carcinoma* (Awad 2004).

Brown algae are a large group, multicellular algae possessing a wide range of bioactive secondary metabolites. They play an important role in marine environments, both as food and the habitats they form. Various bioactivities of compounds from brown algae have been extensively reported (Fang et al. 2015; Manilal et al. 2016; Gupta and Abu-Ghannam 2011a). Among all the seaweeds, the highest phytochemical content has been reported from brown seaweeds (Seafood plus 2008). However, the potent antimicrobial effect of seaweeds resides in the efficiency of the extraction methods (Tuney 2006), the algal species (Valchos et al. 1997), and the solvents being used (Cox et al. 2010). Studies also have shown that higher medicinal effects were obtained from dry seaweeds samples than from fresh samples as indicated by many studies which reported that extracts prepared from fresh seaweeds showed negligible antimicrobial activity compared to that obtained from dried seaweeds (Manivannan et al. 2011).

Many researchers have reported on the antibacterial activity of seaweeds from different geographical patterns globally, that seaweed extracts exhibit antimicrobial activity against various pathogens (Ballesteros et al. 1992; Gonzalez del val et al. 2001a; Kandhasamy and Arunachalam 2008; Karthikaidevi et al. 2009; Kolanjinathan and Stella 2009; Lavanya and Veerappan 2011; Osman et al. 2010; Sreenivasa-Rao

1991, 1995; Seenivasan et al. 2010; Tuney 2006; Vallinayagam et al. 2009, Freile-Pelegrin and Morales 2004; Ibtissam et al. 2009) Active compounds isolated from seaweeds were found to be active against human bacterial pathogens (Kolanjinathan and Stella 2009), fish bacterial pathogens (Bansemir et al. 2006; Kolanjinathan et al. 2009), and marine pathogenic microorganisms (Engel et al. 2006). The antimicrobial activities of the macroalgae have been attributed to the presence of biologically active compounds with antibacterial potential such as Cycloeudesmol, Lyengaroside A, merodíterpenoid, neoirietetraol, diterpene-benzoate, polybrominated indoles, halogenated sesquiterepene alcohol, Lanosol enol ether, diterpene benzoic acids, callophycoic acids, halogenated diterpene-phenols, callophycols and eicosanoids (El Gamal 2010).

Three extracts of two brown seaweeds *Padina tetrastromatica and Sargassum ilicifolium* were investigated as growth promoters and to provide immune protection against *Vibrio parahaemolyticus* in *P.monodon* culture, as result methanol and ethanol extracts of *P. tetrastromatica* showed improved growth performance and the immune capacity of *P. monodon* by increasing their phenoloxidase activity, superoxide anion concentration, and resistance against *V. parahaemolyticus* (Aftab Uddin et al. 2021).

21.2.2 Antifungal Agent

The only report about cartilagineol antimicrobial activity is described against the pathogen *Mycobacterium bovis* (Machado et al. 2011). The Diterpene neophytadiene is rare in seaweed, having been previously isolated from the ethanolic extract of *Himanthalia elongate* and showed antimicrobial activity against the fungi *Aspergillus niger, Cladosporium cladosporioides* and *C. sphaerospermum* (De Felicio 2010).

Prabha et al. 2013, revealed that *Kappaphycus alvarezii* has active secondary metabolites and also exhibited antimicrobial activity against *Aspergillus flavus, Aspergillus fumigates* and *Candida albicans* mainly in the methanolic extract of *K. alvarezii* and this may be mainly due to the presence of phenolic lipids, terepenes and phlorotannins. Ambika et al. 2014 reported that *G. edulis*, *C. racemosa* and *S. myricocystum* reduced the fungal mycelial growth of *Alternaria porri* at increased concentrations of 30%. Pandithurai and Murugesan 2014 reported that extract of *Spatoglossum asperum* showed inhibition of 100% on mycelial growth of *Aspergillus flavus*, 57.14% of *Candiada albicans* and 54.75% of *C. tropicalis*. Renuka et al. 2014 reported that 1% extract of *Chaetomorpha crassa* had inhibitory effect on *Macrophomina phaseolina*, *Sclerotium rolfsi,* and *Pyricularia oryzae*. Khallil et al. 2015 reported that chloroform, ethanol, and cyclohexane extracts had antifungal potential; whereas acetone and ethyl acetate extracts exhibited the lowest antifungal activity.

The seaweed species *Sargassum wightii, Padina gymnospora, Gracilarila edulis, Acanthophora spicifera, Halimeda opuntia* collected from the southeast coast of

Tamilnadu, India, screened for its inhibitory potential against the aquatic pathogen *Aeromonas hydrophila* by poison food technique. No growth was observed in the *Acanthapora spicefera extract* amended plates followed by *Halimeda opuntia* (Noorjahan et al., "unpublished").

Black gill disease was first reported in Japanese Kuruma prawn, *Penaeus japonicus* (*P. japonicus*). *Fusarium* species was the causative agent for this black gill disease, and later *Fusarium* species were considered as the most detrimental pathogens for Kuruma prawn in Japan Khoa et al. 2004.

Black gill disease has been reported in shrimps like *P. japonicas*, rock lobster *Panulirus ornatus*, and mantis shrimp *O. oratoria* induced by *Fusarium* species and anamorphic fungi. However, not even a single report is available on this most economic and nutritional shrimp *L. vannamei.* In this context, black gills of *L. vannamei* which is a serious problem in shrimps growing in the ponds on the east coast of India. *Sargassum wightii, Padina boergesenii, Caulerapa scalpelliformis Ulva lactuca, Acanthophora spicifera, Gracilaria corticata,* against the pathogen *Fusarium sp.* isolated from the black gill. Methanolic extracts from the seaweeds showed maximum inhibitory growth in the Fig. 21.1 (Noorjahan et al., "unpublished") the result of the investigation in agreement with Thinakaran and Sivakumar 2013 who recorded that fungal mycelial growth was strongly inhibited by methanol and ethyl acetate extracts. Reports on the most effective solvent for the extraction of antimicrobials have various results. González del Val et al.,2001 selected

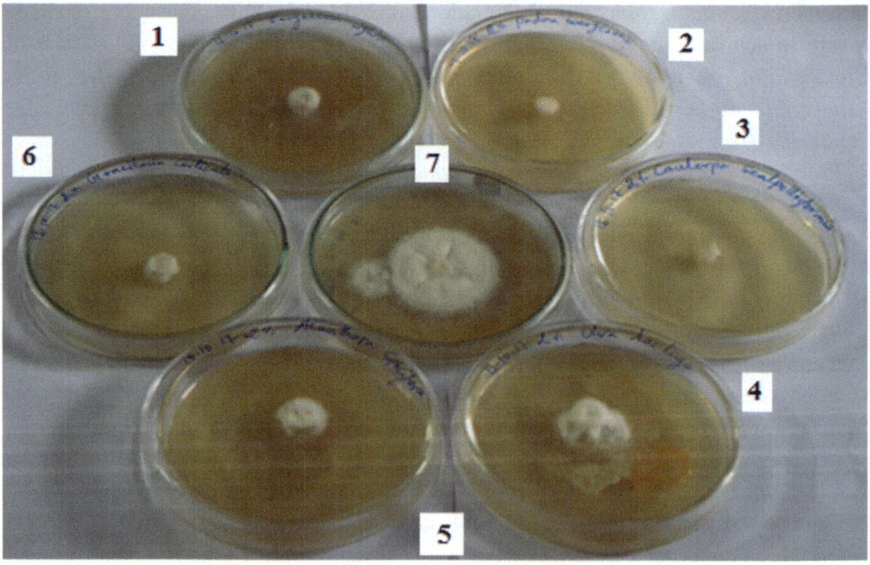

Fig. 21.1 . Inhibitory effect of the seaweed extract against *Fusarium sp*; 1.*Sargassum wightii,* 2.*Padina boergesenii,3. Caulerapa scalpelliformis, 4. Ulva lactuca, 5.Acanthophora spicifera,6. Gracilaria corticata and 7.Fusarium* sp. (Control)

methanol as solvent for extracting antimicrobial compounds for red, green, and brown seaweeds. Salem et al. 2011 also reported that ethyl acetate was the best solvent for the isolation of antimicrobial compounds from the tested marine algae followed by methanol. The steroids, fatty acids, esters of fatty acids and other hydrocarbons were recorded in polar solvents like methanol and ethyl acetate rather than less polar solvents like hexane. This indicates that the polarity of the solvent used for extraction is responsible for antifungal activity.

21.2.3 Antiviral Agent

Currently there is no antiviral drugs in aquaculture industry. The strategies in aquaculture are to control the viral diseases by effective vaccines (farmed fish) and the development of lines of animals resistant to certain diseases through selective breeding (Kibenge et al. 2012). In shrimp farming, oral administration of immunostimulants has been suggested against viral pathogens (Sivagnanavelmurugan et al. 2012), as vaccine is a rather experimental control method (Sudheer et al. 2012).

The major pathogen threatening global shrimp production by 100% mortality within a few days of infection, particularly at larval and juvenile stages is White Spot Syndrome Virus (WSSV - family Nimaviridae). Several reachers have reported that Seaweed have antiviral properties, Seaweed extracts administrated to WSSV infected shrimp either via enriched *Artemia nauplii* (Immanuel et al. 2010, 2012; Sivagnanavelmurugan et al. 2012), or through medicated feeds (Chotigeat et al. 2004; Manilal et al. 2009). Chotigeat et al. 2004, examined in particular, the prophylactic and therapeutic effect of crude fucoidan extracted from *Sargassum polycystum* against WSSV. Takahashi et al. 1998 study the growth rearing in *Penaeus japonicus* by incorporating fucouidan extracted from the brown seaweed *Cladosiphon okamuranus*. Balasubramanian et al. 2006 examined the aqueous extracts of *Sargassum wightii* with mixed suspensions of WSSV, The treated viral preparations were challenged against marine shrimp (*Penaeus indicus*) and freshwater crab (*Paratelphusa hydrodomous*) injected intramuscularly at a concentration of 3 mg per animal and it was observed to have significant less mortality in the infected animals. Recently, Nawanith et al. 2020 determined that the antimicrobial efficacy of the hot water crude extracts (HWCEs) of three species of local Thai green macroalgae *Ulva intestinalis* (Ui), *U. rigida* (Ur), and *Caulopa lentillifera* (Cl) and a commercial ulvan from *U. armoricana* (Ua). HWCEs showed the growth suppression against some pathogenic *Vibrio sp.* interestingly, the HWCEs from Ui at concentrations of 5 and 10 mg/mL completely inhibited white spot syndrome virus (WSSV) in shrimp injected with HWCE–WSSV pre-incubated solutions.

21.2.4 Antifouling Agents

Seaweeds having ubiquitous metabolites on quorum sensing mechanism, by which bacteria attract one another and forming a network by realising substances like peptides or lactones. When the concentration of these substances increased beyond a certain level they are then detected by specific receptors, located in the bacterial cell membranes, or cytoplasms. This in turn regulates the expression of certain genes and many processes, like bioluminescence, formation of biofilms and the production of various virulence factors (Manefield et al. 2001; Rutherford and Bussler 2014). Seaweeds release alleochemicals, such as furanones, which inhibit the biofilm formation, thus by affecting the virulence of many pathogenic bacteria (Defoirdt et al. 2006). Aquaculture industries are gaining interest nowadays in developing new antifouling agents from seaweed (Jha et al. 2013).

In a marine environment, where all surfaces are constantly exposed to the threat of surface colonisation, sessile organisms remain relatively free from biofouling. These sedentary organisms control epibionts in particular marine bacteria by effective antifouling mechanisms (Hellio et al. 2001). Though macroalgae are rich source of bioactive products, lack of knowledge of reports to define the ecological role for these compounds (De Nys et al. 1995; Shanmugam and Mody 2000; Suzuki et al. 2001; Vairappan et al. 2001b). Macroalgae, therefore possess chemical defences to prevent the colonization on their surface. The use of marine natural products capable of inhibiting bacteria development offers rich pharmacological potential (Kornprobst 2005). Many reports showed macroalgae to have a broad range of biological activities such as antibacterial activities (Fenical and Paul 1984; Mtolera and Semesi 1996; Gonzalez del val et al. 2001a, b; Selvin and Lipton 2004; Karabay-yavasoglu et al. 2007; Salvador et al. 2007), antifungal (Moreau et al. 1984; Tariq 1991; Mayer 2002; Mayer et al. 2007, 2009) and antiviral (Bourgougnon et al. 1993, 1994; Hudson et al. 1999; Serkedjieva 2000, 2003; Mayer 2002; Ghosh et al. 2004; Zandi et al. 2007).

The antifouling activities of extracts from nine macroalgae against bacteria, fungi, diatom, have been investigated in bimonthly sampling. Of the extracts tested 48.2% were active against at least one of the fouling organisms and these extracts 31.2% were seasonally active with a peak of activity (Hellio et al. 2004). Most of the bioactive substances isolated from marine algae are chemically classified as brominated, aromatics, nitrogen-heterocyclic, nitrosulphuric-heterocyclic, sterols, dibutanoids, proteins, peptides, and sulphated polysaccharides. The crude extract thus obtained were subjected to broad-based biological screening for antifungal, antiviral, antibacterial, antimalarial, antifilarial, hypoglycaemic and antifertility activity (Garg 1993). The green algae *Calorpha peltada* contains 1-4 diacetoxy butadiene and fatty esters possess antibacterial, anti-ichthyotoxic and anti-hypertensive properties.

The seaweed species collected from Tuticorin coast of Tamilnadu India such as *Halimeda opuntia, Portieria hornemannii, Padina gymnospora, Gracilaria edulis, Gelidiella acerosa* were extracted with methanol against the fouling bacteria isolated from the fibre boat. *Gracilaria edulis* showed maximum inhibition followed by *Portieria hornemannii and Halimeda opuntia* (Noorjahan et al., "unpublished data").

21.3 Recommendation

The current Review highlighted the importance of seaweed extracts and it's bioactivities of diverse secondary metabolites and other compounds.

- Studying different physiological and ecological conditions that have led to the activation of secondary metabolism gene clusters might also be useful in optimizing environmental conditions for maximum production of metabolites.
- Performing detailed research are required to evaluate spectral composition, effectiveness, and potential use of seaweed compounds especially for the development of new antibiotics.
- Algal products can be used to enhance the nutritional value of food and fish feed owing to their chemical compositions and play a crucial role in aquaculture.
- Screening, extracting, and refining seaweed bioactive compounds in formulating aqua feed that could have positive influences on the growth and health of farmed shell and finfish species.
- Associating "Genetic engineering for identifying and developing silent" gene clusters for exploring more "silent" secondary metabolites under laboratory conditions. To develop new antifungal, antibacterial and antiviral agents for application in aquaculture.
- Integrated farming practice to be introduced near aquaculture sites may represent an opportunity to increase the economic and environmental sustainability of the production of all the involved cultures and a mitigation approach to reduce the nutritional and organic inputs.

21.4 Conclusion

It is clearly evident, that algae as dietary additives contribute to an increase in growth and feed utilization of farmed fish and shrimp due to efficacious assimilation of dietary protein, improvement in physiological activity, stress response, starvation tolerance, disease resistance, and carcass quality. Considering its rich diversities of secondary metabolites, drugs formulated from seaweeds might be a promising source to overcome the existing practices of chemicals and drugs in the aquaculture industry. The current review highlighted the importance of some seaweeds as a good antibacterial, antifungal, and antiviral agents. Intensive research and standardization of seaweeds associate in the aquaculture sector, which results not only in the economic upliftment of the coastal and inland dwellers through sustainable culture, but also may open up an avenue of alternative livelihood.

Acknowledgement The author's would like to thank authorities of Annamalai university and our special thanks to Dean and Director Centre of Advanced study in Marine Biology, Faculty of Marine sciences, Annamalai University, Chidambaram for his support and encouragement.

References

Aftab Uddin S, Siddique MAM, Habib A, Akter S, Hossen S, Tanchangya P, Mamun Abdullah A (2021) Effects of seaweeds extract on growth, survival, antibacterial activities, and immune responses of *Penaeus monodon* against *Vibrio parahaemolyticus*. Ital J Anim Sci 20(1):243–255

Akhilamole MA, Noorjahan A, Raghuraman R, Aiyamperumal B, Anantharaman P (2019a) Alternate supplemented diets as spent seaweed on Nile tilapia, *Oreochromis niloticus* performance. Indian Hydrobiol 18

Akhilamole MA, Noorjahan A, Mahesh S, Aiyamperumal B, Anantharaman P (2019b) Nutritive assessment of spending seaweed and fermented spent seaweed *Sargassum wightii*. Adv Biores 10(4/5)

Alves A, Sousa RA, Reis RL (2013) A practical perspective on ulvan extracted from green algae. J Appl Phycol 25:407–424

Ambika S, Sujatha K, Balakrishnan K (2014) Antifungal activity of seaweed extract against *Alternaria porri* in onion. In: National Seminar on Algae for sustainable agricultural production. Madurai, p 87

Arnold TM, Targett NM (1998) Quantifying in situ rates of phlorotannin synthesis and polymerization in marine brown algae. J Chem Ecol 24:577–595

Athukorala LKW, Shahidi MS, Heu HT, Kim JS, Jeon YJ (2003) Antioxidant efficacy of extracts of an edible red alga (*Grateloupia wlicina*) in linoleic acid and fish oil. J Food Lipid 10:313–327

Attaway DH, Zaborsky OR (1993) Marine biotechnology. Pharm Bioac Nat Prod 1(11):63–67.

Awad NE (2004) Bioactive brominated diterpenes from the marine red alga *Jania rubens* (L.) Lamx. Phytother Res 18:275–279

Azad IS, Rajendran KV, Rajan JJS, Vijayan KK, Santiago TC (2001) Virulence and histopathology of *Aeromonas hydrophila* (Sah 93) in experimentally infected tilapia, *Oreochromis mossambicus*. J Aquac Tropics 16:265–275

Balasubramanian G, Sudhakaran R, Syed Musthaq S, Sarathi M, Sahul Hameed AS (2006) Studies on the inactivation of white spot syndrome virus of shrimp by physical and chemical treatments, and seaweed extracts tested in marine and freshwater animal 640 models. J Fish Dis 29:569–572

Ballesteros E, Martin D, Uriz MJ (1992) Biological activity of extracts from some Mediterranean macrophytes. Bot Mar 35:481–485

Bansemir A, Just N, Michalik M, Lindequist U, Lalk M (2004) Extracts and sesquiterpene derivatives from the red alga *Laurencia chondrioides* with antibacterial activity against fish and human pathogenic bacteria. Chem Biodivers 1:463–467

Bansemir A, Blume M, Schröder S, Lindequist U (2006) Screening of cultivated seaweeds for antibacterial activity against fish pathogenic bacteria. Aquaculture 252:79–84

Bernard P, Clement R (1983) Bringing to evidence anti-biotic substances from *Posidonia oceanica*. Rev Int Ocean ogr Mkd 70(71):33–37

Blunt JW, Copp BR, Munro MHG, Northcote PT, Prinsep MR (2006) Marine natural products. Nat Prod Rep 23:26–78

Blunt JW, Copp BR, Hu WP, Munro MHG, Northcote PT, Prinsep MR (2007) Marine natural products. Nat Prod Rep 24:31–86

Borowitzka MA, Borowitzka LJ (1992) Vitamins and fine chemicals from micro algae. In: Micro algal biotechnology. Cambridge University Press, Great Britain, p 179

Bourgougnon N, Lahaye M, Chermann JC, Kornprobst JM (1993) Composition and antiviral activities of a sulphated polysaccharide from *Schizymenia dubyi* (Rhodophyta, Gigartinales). Bioorgan Med Chem Lett 3:1141–1146

Bourgougnon N, Roussakis C, Kornprobst JM, Lahaye M (1994) Effects in vitro of sulphated polysaccharide from *Schizymenia dubyi* (Rhodophyta, Gigartinales) on a non-small-cell bronchopulmonary carcinoma line (NSCLC-N6). Cancer Lett 85:87–92

Bulfon C, Volpatti D, Galeotti M (2015) Current research on the use of plant derived products in farmed fish. Aquac Res 46(3):513–551

Cai J, Feng J, Xie S, Wang F, Xu Q (2014) *Laminaria japonica* extract, an inhibitor of *Clavibater michiganense* subsp. sepedonicum. PLoS One 9(4):e94329

Cavallo RA, Acquaviva M, Stabili L, Cecere E, Petrocelli A, Narracci M (2013) Antibacterial activity of marine macroalgae against fish pathogenic vibrio species. Cent Eur J Biol 8:646–653

Chen WY, Ng TH, Wu JH, Chen JW, Wang HC (2018) Microbiome dynamics in a shrimp grow-out pond with possible outbreak of acute hepatopancreatic necrosis disease. Sci Rep 7:9395

Chew YL, Lim YY, Omar M, Khoo KS (2008) Antioxidant activity of three edible seaweeds from two areas in South East Asia. LWT 41:1067–1072

Chotigeat W, Tongsupa S, Supamataya K, Phongdara A (2004) Effect of fucoidan on disease resistance of black tiger shrimp. Aquaculture 233:23–30

Chowdhury AA, Uddin MS, Vaumi S, Asif AA (2015) Aqua drugs and chemicals used in aquaculture of *Zakigonj upazilla*, Sylhet. Asian J Med Biol Res 1:336–349

Coastal Aquaculture Authority of India (2014) List of antibiotics and other pharmacologically active substances banned for using in shrimp aquaculture. Coastal Aquaculture Authority of India, New Delhi

Cox S, Abu-Ghannam N, Gupta S (2010) An assessment of the antioxidant and antimicrobial activity of six species of edible Irish seaweeds. Int Food Res 17:205–220

Dawczynski C, Schubert R, Jahreis G (2007) Amino acids, fatty acids, and dietary fibre in edible seaweed products. Food Chem 103:891–899

De Felicio R (2010) Trypanocidal, leishmanicidal and antifungal potential from marine red alga *Bostrychia tenella* J. Agardh (Rhodomelaceae, Ceramiales). J Pharm Biomed Anal 52:763–769

De Nys R, Steinberg PD, Willemsen P, Dworjanyn SA, Gabelish CI, King RJ (1995) Broad spectrum effects of secondary metabolites from the red alga *Delisea pulchra* in antifouling assays. Biofouling 8:259–271

Defoirdt T, Crab R, Wood TK, Sorgeloos P, Verstraete W, Bossier P (2006) Quorum sensing disrupting brominated furanones protect the gnotobiotic brine shrimp *Artemia franciscana* from pathogenic Vibrio harveyi, Vibrio campbellii, and *Vibrio parahaemolyticus* isolates. Appl Environ Microbiol 72:6419–6423

Dhaneesh KV, Gopi M, Noushad KM, Ganeshamurthy R, Kumar TTA, Balasubramanian T (2012) Determination of metal levels in thirteen fish species from Lakshadweep Sea. Bull Environ Contam Toxicol 88:69–73

Duan XJ, Zhang WW, Li XM, Wang BG (2006) Evaluation of antioxidant property of extract and fractions obtained from a red alga, *Polysiphonia urceolata*. Food Chem 95:37–43

Dubber D, Harder T (2008) Extracts of *Ceramium rubrum*, *Mastocarpus stellatus* and *Laminaria digitata* inhibit growth of marine and fish pathogenic bacteria at ecologically realistic concentrations. Aquaculture 274:196–200

El Gamal AA (2010) Biological importance of marine algae. Saudi Pharm J 18:1–25

Ely R, Supriya T, Naik CG (2004) Antimicrobial activity of marine organisms collected off the coast of South East India. J Exp Biol Ecol 309 121–127

Engel S, Puglisi MP, Jensen PR, Fenical W (2006) Antimicrobial activities of extracts from tropical Atlantic marine plants against marine pathogens and saprophytes. Mar Biol 149:991–1002

Fang HY, Chokkalingam U, Chio SF, Hwang TL, Chen SL, Wang WL, Sheu JH (2015) Bioactive chemical constituents from the brown alga *Homoeostrichus formosana*. Int J Mol Sci 16:736–746

FAO (2018) The state of world fisheries and aquaculture. Available online: http://203.187.160.133:9011/www.fao.org/c3pr90ntc0td/3/i9540en/i9540en.pdf. Accessed on 27 Feb 2020

Fenical W, Paul VJ (1984) Algae in medicine and pharmacology. Hydrobiologia 116(117):135–170
Flegel TW (2006) Detection of major Penaeid shrimp viruses in Asia, a historical perspective with emphasis on Thailand. Aquaculture 258:1–33
Freile-Pelegrin Y, Morales JL (2004) Antibacterial activity in marine algae from the coast of Yucatan, Mexico. Bot Mar 47(2):140–146
Ganeshamurthy R, Kumar TTA, Dhayanithi NB (2012) Effect of secondary metabolites of the seaweed (*Halimeda micronesia*) at Lakshadweep islands against aquatic pathogens. Int J Pharma Bio Sci 3:B213–B220
Garg HS (1993) Bioactive substance in marine algae, marine biotechnology. Plenum press, New York, pp 1–8
Genovese G, Faggio C, Gugliandolo C, Torre A, Spanò A, Morabito M, Maugeri TL (2012) In vitro evaluation of antibacterial activity of *Asparagopsis taxiformis* from the straits of Messina against pathogens relevant in aquaculture. Mar Environ Res 73:1–6
Gerwick WH, Fenical W, Norris JN (1985) Chemical variation in the tropical seaweed *Stypopodium zonale* (Dictyotaceae). Phytochemistry, Londres 24:1279–1283
Ghosh P, Adhikari U, Ghosal PK, Pujol CA, Carlucci MJ, Damonte EB, Ray B (2004) In vitro anti-herpetic activity of sulphated polysaccharide fractions from *Caulerpa racemosa*. Phytochemistry 65:3151–3157
Gonzalez del val A, Platas G, Basilio A (2001a) Screening of antimicrobial activities of red, green and brown macro algae from *Gran Canaria* (Canary Islands, Spain). Int Microbiol 4:35–40
Gonzalez del Val A, Platas G, Basilo A, Cabello A, Gorrochategui J, Suay I, Vicente F, Portillo E, Jimenez del Rico M, Garcia Reina G, Pelez F (2001b) Screening of antimicrobial activities in red, green and brown macroalgae from *Gran Canaria* (Canary Islands, Spain). Int J Microbiol 4:35–40
Gorban E, Kuprash L, Gorban N (2003) *Spirulina*: perspectives of the application in medicine. LikSprava 7:100–110
Grenni P, Ancona V, Caracciolo AB (2018) Ecological effects of antibiotics on natural ecosystems: a review. Microchem J 136:25–39
Guihéneuf F, Gietl A, Stengel DB (2018) Temporal and spatial variability of mycosporine-like amino acids and pigments in three edible red seaweeds from western Ireland. J Appl Phycol 30:2573–2586
Gunasekaran S, Poorniammal R (2008) Optimization of fermentation conditions for red pigment production from *Penicillium* sp. under submerged cultivation. Afr J Biotechnol 7:1894–1898
Gupta S, Abu-Ghannam N (2011a) Bioactive potential and possible health effects of edible brown seaweeds. Trends Food Sci Tech 22:315–326
Gupta S, Abu-Ghannam N (2011b) Recent developments in the application of seaweeds or seaweed extracts as a means for enhancing the safety and quality of foods. Innov Food Sci Emerg 12:600–609
Harbourne JB (1994) The flavonoids: advances in research since 1986. Chapman and Hall, London
Haslam E (1989) Plant polyphenols: vegetable tannins revisited. Cambridge University Press, Cambridge
Hellio C, De La Broise D, Dufosse L, Le Gal Y, Bourgougnon N (2001) Inhibition of marine bacteria by extracts of macroalgae: potential use for environmentally friendly antifouling paints. Mar Environ Res 52(3):231–247
Hellio C, Marechal JP, Veron B, Bremer G, Clare AS, Le Gal Y (2004) Seasonal variation of antifouling activity of marine from the Brittany coast (France). Mar Biotechnol 6:67–82
Hemat RAS (2007) Fat and muscle dysfunction. In: Hemat RAS (ed) Andropathy. Urotext, Dublin, pp 83–85
Heo SJ, Park EJ, Lee KW, Jeon YJ (2005) Antioxidant activities of enzymatic extracts from brown seaweeds. Bioresour Technol 96:1613–1623
Hidayat Y, Fuad F, Nurhidayati M (2018) Implementation of economic democracy principle in Islamic banking policies through Financial Services Authority (FSA) in Indonesia. At-Taradhi J Stu Eko 8(2):132–154

Hierholtzer A, Chatellard L, Kierans M, Akunna JC, Collier PJ (2014) The impact and mode of action of phenolic compounds extracted from brown seaweed on mixed anaerobic microbial cultures. J Appl Microbiol 114:964–973

Holdt SL, Kraan S (2011) Bioactive compounds in seaweed: functional food applications and legislation. J Appl Phycol 23:543–597

Holmström K, Gräslund S, Wahlström A, Poungshompoo S, Bengtsson BE, Kautsky N (2003) Antibiotic use in shrimp farming and implications for environmental impacts and human health. Int J Food Sci Technol 38:255–266

Hornsey I, Hide D (1985) The production of antimicrobial compounds by British Marine Algae & variation of antimicrobial activity with algal generation. Br Phycol J 20:21–25

Huang J, Liu Y, Wang X (2008) Selective adsorption of tannin from flavonoids by organically modified attapulgite clay. J Hazard Mater 160:382–387

Hudson JB, Kim JH, Lee MK, De Wreede RE, Hong YK (1999) Antiviral compounds in extracts of Korean seaweeds: evidence for multiple activities. J Appl Phycol 10:427–434

Ibtissam C, Hassane R, Martinez-Lopez J, Dominguez Seglar JF, Gomez Vidal JA, Hassan B, Mohamed K (2009) Screening of antibacterial activity in marine green and brown macroalgae from the coast of Morocco. Afr J Biotechnol 8(7):1258–1262

Immanuel G, Sivagnanavelmurugan M, Balasubramanian V, Palavesam A (2010) Effect of hot water extracts of brown seaweeds *Sargassum sp.* on growth and resistance to white spot syndrome virus in shrimp *Penaeus monodon* postlarvae. Aquac Res 41:545–553

Immanuel G, Sivagnanavelmurugan M, Balasubramanian V, Palavesam A (2012) Sodium alginate from *Sargassum wightii* retards mortalities in *Penaeus monodon* postlarvae challenged with white spot syndrome virus. Dis Aquat Org 99:187–196

Jha B, Kavita K, Westphal J, Hartmann A, Schmitt-Kopplin P (2013) Quorum sensing inhibition by Asparagopsis taxiformis, a marine macro alga: separation of the compound that interrupts bacterial communication. Mar Drugs 11:253–265

Jia FY, Greenfield MD, Collins RD (2000) Genetic variance of sexually selected traits in axmoths: maintenance by genotype × environment interaction. Evolution 54:953–967

Jiang H-F, Liu X-L, Chang Y-Q, Liu M-T, Wang G-X (2013) Effects of dietary supplementation of probiotic *Shewanella colwelliana* WA64, *Shewanella olleyana* WA65 on the innate immunity and disease resistance of abalone, *Haliotis discus* hannai Ino. Fish Shellfish Immunol 35:86–91

Kadam SU, Tiwari BK, O'Donnell CP (2013) Application of novel extraction technologies for bioactives from marine algae. J Agric Food Chem 2013(61):4667–4675

Kahkonen MP, Hopia AI, Vuorela HJ, Rauha JP, Pihlaja K, Kujala TS et al (1999) Antioxidant activity of plant extracts containing phenolic compounds. J Agric Food Chem 47:3954–3962

Kamaraj C, Deepak P, Balasubramani G, Karthi S, Arul D, Aiswarya D, Amutha V, Vimalkumar E, Mathivanan D, Suseem SR, Chanthini KMP, Senthil-Nathan S (2018) Target and non-target toxicity of fern extracts against mosquito vectors and beneficial aquatic organisms. Ecotoxicol Environ Saf 161:221–230

Kandhasamy M, Arunachalam KD (2008) Evaluation of in vitro antibacterial property of seaweeds of southeast coast of India. Afr J Biotechnol 7:1958–1961

Kanjana K, Radtanatip T, Asuvapongpatana S, Withyachumnarnkul B, Wongprasert K (2011) Solvent extracts of the red seaweed *Gracilaria fisheri* prevent *Vibrio harveyi* infections in the black tiger shrimp Penaeus monodon. Fish Shellfish Immunol 30:389–396

Karabay-Yavasoglu NU, Sukatar A, Ozdemir G, Horzum Z (2007) Antimicrobial activity of volatile components and various extracts of the red alga *Jania rubens*. Phytother Res 21:153–156

Karthikaidevi G, Manivannan K, Thirumaran G, Anantharaman P, Balasubaramanian T (2009) Antibacterial properties of selected green seaweeds from Vedalai Coastal Waters; Gulf of Mannar marine biosphere reserve. Glob J Pharmacol 3(2):107–112

Kendel M, Wielgosz-Collin G, Bertrand S, Roussakis C, Bourgougnon N, Bedoux G (2015) Lipid composition, fatty acids and sterols in the seaweeds *Ulva armoricana*, and *Solieria chordalis* from Brittany (France): an analysis from nutritional, chemotaxonomic, and antiproliferative activity perspectives. Mar Drugs 13:5606–5628

Kesarcodi-Watson A, Kaspar H, Lategan MJ, Gibson L (2008) Probiotics in aquaculture: the need, principles and mechanisms of action and screening processes. Aquaculture 274:1–14

Khallil AM, Daghman IM, Fady AA (2015) Antifungal potential in crude extracts of five selected brown seaweeds collected from the Western Libya Coast. J Microbiol Creat 1(1):103

Khoa LV, Hatai K, Aoki T (2004) *Fusarium incarnatum* isolated from black tiger shrimp, Penaeus monodon Fabricius, with black gill disease cultured in Vietnam. J Fish Dis 27(9):507–515

Kibenge FSB, Godoy MG, Fast M, Workenhe S, Kibenge MJT (2012) Counter measures against viral diseases of farmed fish. Antivir Res 95:257–281

Klesius PH, Shoemaker CA, Evans JJ (2000) Vaccination. A health management practice for preventing diseases in tilapia and other cultured fish. 5th Int symposium on tilapia aquaculture in the 21st century, Brazil 2:558–564

Klongklaew N, Praiboon J, Tamtin M, Srisapoome P (2020) Antibacterial and antiviral activities of local thai green macroalgae crude extracts in pacific white shrimp (*Litopenaeus vannamei*) Mar. Drugs 18:140

Knowler D, Chopin T, Martínez-Espiñeira R, Neori A, Nobre A, Noce A, Reid G (2020) The economics of integrated multi-trophic aquaculture: where are we now and where do we need to go? Rev Aquac 8. 12(3):1579–1594.

Kolanjinathan K, Stella D (2009) Antibacterial activity of ethanol extracts of seaweeds against human bacterial pathogens. Recent Res Sci Technol 1(1):20–22

Kolanjinathan K, Ganesh P, Govindarajan M (2009) Antibacterial activity of ethanol extracts of seaweeds against fish bacterial pathogens. Eur Rev Med Pharmacol Sci 13:173–177

Kongrueng J, Yingkajorn M, Bunpa S, Sermwittayawong N, Singkhamanan K, Vuddhakul V (2015) Characterization of *Vibrio parahaemolyticus* causing acute hepatopancreatic necrosis disease in southern Thailand. J Fish Dis 38:957–966

Kornprobst JM (2005) Substances naturelles d'origine marine, Tome 1: généralités, micro-organismes, algues. Editions Tec & Doc, Paris, p 598

Kraan S (2013) Mass-cultivation of carbohydrate rich macroalgae, a possible solution for sustainable biofuel production. Mitig Adapt Strateg Glob Chang 18:27–46

Kumari P, Bijo AJ, Mantri VA, Reddy CRK, Jha B (2013) Fatty acid profiling of tropical marine macroalgae: an analysis from chemotaxonomic and nutritional perspectives. Phytochemistry 86:44–56

Lavanya R, Veerappan N (2011) Antibacterial potential of six seaweeds collected from Gulf of Mannar of Southeast Coast of India. Adv Biol Res 5(1):38–44

Li P, Kinch LN, Ray A, Dalia AB, Cong Q, Nunan LM, Camilli A, Grishin NV, Salomon D, Orth K (2017) Acute hepatopancreatic necrosis disease-causing *Vibrio parahaemolyticus* strains maintain an antibacterial type VI secretion system with versatile effector repertoires. Appl Environ Microbiol 83:e00737–e00717

Lightner DV (1993) Diseases of cultured Penaeid shrimps. In: McVey JP (ed) CRC Handbook of mariculture. CRC Press, Boca Raton, pp 393–486

Lightner DV (2011) Virus diseases of farmed shrimp in the Western Hemisphere (the Americas). J Invertebr Pathol 106:110–130

Lima-Filho JVM, Carvalho AFFU, Freitas SM, Melo VMM (2002) Antibacterial activity of extracts of six macroalgae from the northeastern Brazilian coast. Braz J Microbiol 33:311–313

Lopes G, Sousa C, Bernardo J, Andrade PB, Valentão P, Ferreres F, Mouga T (2011) Sterol profiles in 18 macroalgae of the portuguese coast. J Phycol 47:1210–1218

Lopes G, Sousa C, Valentão P, Andrade PB (2013) Sterols in algae and health. Bioactive compounds from marine foods. In: Hernández-Ledesma B, Herrero M (eds) Wiley, Hoboken 173–191

Machado FLS et al (2011) Antileishmanial sesquiterpenes from the Brazilian red algae *Laurencia dendroidea*. Planta Méd Berlin 77:733–735

Maheswaran ML, Padmavathy S, Gunalan B (2013) Screening and characterization of marine seaweeds and its antimicrobial potential against fish pathogens. Int J Fish Aquat Stud 1(788):1–13

Manefield M, Welch M, Givskov M, Salmond GPC, Kjelleberg S (2001) Halogenated furanones from the red alga, *Delisea pulchra*, inhibit carbapenem antibiotic synthesis and exoenzyme virulence factor production in the phytopathogen *Erwinia carotovora*. FEMS Microbiol Lett 205:131–138

Manilal A, Sujith S, Selvin J, Seghal Kiran G, Shakir C (2009) In vivo antiviral activity of polysaccharide from the Indian green alga, *Acrosiphonia orientalis* (J. Agardh): potential implication in shrimp disease management. World J Fish Mar Sci 1:278–282

Manilal A, Gezmu T, Merdekios B (2016) Evaluating the in vitro antagonism of secondary metabolites fractionated from the brown algae, *Sargassum swartzii* against Human Candida spp. Transl Biomed 7:1

Manivannan K, Anantharaman P, Balasubramanian T (2011) Antimicrobial potential of selected brown seaweeds from Vedalai coastal waters, Gulf of Mannar. Asian Pac J Trop Biomed 1(2):114–120

Marinho-Soriano E, Fonseca PC, Carneiro MAA, Moreira WSC (2006) Seasonal variation in the chemical composition of two tropical seaweeds. Bioresour Technol 97:2402–2406

Marudhupandi T, Thankappan T, Kumar A (2013) Effect of fucoidan from *Turbinaria ornata* against marine ornamental fish pathogens. J Coast Life Med 1:282–286

Mata L, Wright E, Owens L, Paul N, de Nys R (2013) Water-soluble natural products from seaweed have limited potential in controlling bacterial pathogens in fish aquaculture. J Appl Phycol 25:1963–1973

Mayer AMS (2002) Current marine pharmacology contributions to new drug development in the biopharmaceutical industry. Pharmaceut News 9:479–482

Mayer AMS, Rodríguez AD, Berlinck RGS, Hamann MT (2007) Marine pharmacology in 2003–2004: marine compounds with anthelmintic antibacterial, anticoagulant, antifungal, anti- inflammatory, antimalarial, antiplatelet, antiprotozoal, antituberculosis, and antiviral activities; affecting the cardiovascular, immune and nervous systems, and other miscellaneous mechanisms of action. Comp Biochem Physiol Part C: Toxicol Pharmacol 145:553–581

Mayer AMS, Rodríguez AD, Roberto BGS, Hamann MT (2009) Marine pharmacology in 2005–2006: marine compounds with anthelmintic, antibacterial, anticoagulant, antifungal, anti-inflammatory, antimalarial, antiprotozoal, antituberculosis, and antiviral activities; affecting the cardiovascular, immune and nervous systems, and other miscellaneous mechanisms of action. Biochim Biophys Acta (BBA)-General Subjects 1790:283–308

Mohamed S, Hashim SN, Rahman HA (2012) Seaweeds: a sustainable functional food for complementary and alternative therapy. Trends Food Sci Technol 23:83–96

Mohammed KA et al (2004) Laurenditerpenol, a new diterpene from the tropical marine alga *Laurencia intricata* that potently inhibits HIF-1 mediated hypoxic hignaling in breast tumor cells. J Nat Prod Washington 67:2002–2007

Morales JL, Cantillo-Ciau ZO, Sanchez-Molina I, Mena-Rejon GJ (2006) Screening of antibacterial and antifungal activities of six marine macroalgae from coasts of Yucatan peninsula. Pharm Biol 44:632–635

Moreau J, Pesando D, Caram B (1984) Antifungal and antibacterial screening of Dictyotales from the French Mediterranean coast. Hydrobiologia 116(117):521–524

Mtolera MSP, Semesi AK (1996) Antimicrobial activity of extracts from six green algae from Tanzania. Curr Trends Mar Bot Res East Afr Region:211–217

Mise T, Ueda M, Yasumoto T (2011) Production of fucoxanthin rich powder from cladosiphon okamuranus. Advance Journal of Food Science and Technology 3(1):73–76.

Namvar F, Mohamed S, Fard SG, Behravan J, Mustapha NM, Alitheen NBM, Othman F (2012) Polyphenol-rich seaweed (*Eucheuma cottonii*) extract suppresses breast tumour via hormone modulation and apoptosis induction. Food Chem 130:376–382

Naqvi SWA, Kamat SSY, Fernandes L, Reddy CVG, Bhakuni DS, Dhawan BN (1980) Screening of some marine plant from the Indian coast for biological activities. Bot Mar 24:51–55

Ndhlala AR, Kasiyamhuru A, Mupure C, Chitindingu K, Benhura MA, Muchuweti M (2007) Phenolic composition of *Flacourtia indica, Opuntia megacantha* and *Sclerocarya birrea*. Food Chem 103:82–87

Noga EJ (1996) Fish diseases diagnosis and treatment. Mosby, New York, pp 152–153

Noga EJ (2010) Text book of fish disease: diagnosis and treatment, 2nd edn. Wiley and Blackwell, USA, p 519

Noorjahan A, Aiyamperumal B, Anantharaman P (2019) Characterization and biochemical properties of Brown seaweed *Sargassum tenerrimum* (J.Aardh). Int J Pharm Biol Sci 9(2):350–357. https://doi.org/10.21276/ijpbs.2019.9.2.34

Noorjahan A, Mahesh S, Manupoori S, Aiyamperumal B, Anatharaman P, Muthukumaran M (2021) Efficacy of incorporating spent seaweeds in aqua feeds as a nutritional source. J Surv Fish Sci. 8(1) 31–45
Osman MEH, Abushady AM, Elshobary ME (2010) In vitro screening of antimicrobial activity of extracts of some macroalgae collected from Abu-Qir bay Alexandria, Egypt. Afr J Biotechnol 9(12):7203–7208
Padmakumar K, Ayyakannu K (1997) Seasonal variation of antibacterial and antifungal activities of the extracts of marine from Southern Coast of India. Bot Mar 40:507–515
Pandithurai M, Murugesan S (2014) Evaluation of antifungal activity of seaweed extracts from marine brown alga *Spatoglossum asperum*. In: National Seminar on Algae for sustainable agricultural production. Madurai, Tamil Nadu, India, p 89
Pangestuti R, Kim SK (2011) Neuroprotective effects of marine algae. Mar Drugs 9:803–818
Patra JK, Patra AP, Mahapatra NK, Thatoi HN, Das S, Sahu RK, Swain GC (2009) Antimicrobial activity of organic solvent extracts of three marine macroalgae from Chilika Lake, Orissa, India. Malays J Microbiol 5(2):128–131
Pereira RC, Teixeira VL (1999) Sesquiterpenos da alga marinha Laurencia Lamouroux (Ceramiales, Rhodophyta). I. Significado ecológico, vol 22. Química Nova, São Paulo, pp 369–373, mai./jun
Pereira H, Barreira L, Figueiredo F, Custódio L, Vizetto-Duarte C, Polo C, Rešek E, Aschwin E, Varela J (2012) Polyunsaturated fatty acids of marine macroalgae: potential for nutritional and pharmaceutical applications. Mar Drugs 10:1920–1935
Pigmen K, Yip WH, Lim SJ, Mustapha WAW, Maskat MY, Said M, Pigmen K (2014) Characterisation and stability of pigments extracted from *Sargassum binderi* obtained from Semporna, Sabah. Sains Malays 2014(43):1345–1354
Planas M, Pérez-Lorenzo M, Vázquez JA, Pintado J (2005) A model for experimental infections with *Vibrio (Listonella) anguillarum* in first feeding turbot (*Scophthalmus maximus* L.) larvae under hatchery conditions. Aquaculture 250:232–243
Plumb JA (1999) Health maintenance and principal microbial diseases of cultured fishes. Iowa State University Press, Ames Iowa, p 344
Prabha V, Prakash DJ, Sudha PN (2013) Analysis of bioactive compounds an antimicrobial activity of marine algae *Kappaphycus alvarezii*. Int J Pharm Sci Res 4(1):306–310
Prado S, Romalde JL, Barja JL (2010) Review of probiotics for use in bivalve hatcheries. Vet Microbiol 145:187–197
Praiboon J, Chirapart A, Soisarp N (2017) Principle and biological properties of sulfate polysaccharides from seaweed. In: Se-Kwon K (ed) Marine glycobiology principles and applications. CRC Press Taylor & Francis Group, Boca Raton 85–120
Premnathan M, Chandra K, Bajpai SK, Kathiresan K (1992) A survey of some Indian marine plants for antiviral activity. Bot Mar 35:321–324
Radhika D, Veerabahu C, Priya R, Mohaideen A (2014) A comparative study of biopotential of crude and fractionated extracts of some seaweeds from Tuticorin coast. Int J Phytopharmacol 5:27–30
Rajapakse N, Kim SK (2011) Nutritional and digestive health benefits of seaweed, Academic Press; San Diego, CA, USA. Adv Food Nutr Res 64:17–28
Rama Devi K, Srinivasan R, Kannappan A, Santhakumari S, Bhuvaneswari M, Rajasekar P, Prabhu NM, Veera Ravi A (2016) *In vitro* and *in vivo* efficacy of rosmarinic acid on quorum sensing mediated biofilm formation and virulence factor production in *Aeromonas hydrophila*. Biofouling 32(10):1171–1183
Rebecca LJ, Dhanalakshmi V, Sharmila S (2012) Effect of the extract of *Ulva sp* on pathogenic microorganisms. J Chem Pharm Res 4:4875–4878
Reddy P, Urban S (2009) Meroditerpenoids from the southern Australian marine brown alga Sargassum fallax. Phytochemistry 70:250–255
Renuka R, Jeyalakshmi C, Rettinassababathy (2014) Studies on antifungal and growth promoting potentialities of algal extracts under in vitro conditions. In: National seminar on algae for sustainable agricultural production. Madurai, Tamil Nadu, India 91–92
Reverter M, Bontemps NT, Sasal P, Saulnier D (2017) Use of medicinal plants in aquaculture. In: Austin B, Newaj-Fyzul A (eds) Diagnosis and control of diseases of fish and shellfish. Wiley, Hoboken, pp 223–261

Rutherford ST, Bassler BL (2014) Bacterial quorum sensing: its role in virulence and possibilities for its control. Cold Spring Harb Perspect Med 2:a012427

Saleh B, Al-Mariri A (2017) Antimicrobial activity of the marine algal extracts against selected pathogens. J Agric Sci Technol 19:1067–1077

Salem WM, Galal H, Nasr El-deen F (2011) Screening for antibacterial activities in some marine algae from the red sea (Hurghada, Egypt). Afr J Microbiol Res 5:2160–2167

Salvador N, Gomez-Garreta A, Lavelli L, Ribera MA (2007) Antimicrobial activity of Iberian macroalgae. Sci Mar 71:101–114

Sandeepa MG, Ammani K (2015) Effect of probiotic bacterium on growth and biochemical parameters of shrimp *Litopenaeus vannamei*. Int J Recent Sci Res 6(2):2871–2875

Sarmasik A (2000) Antimicrobial peptides a potential therapeutic alternative for the treatment of fish diseases. Turk J Biol 26:201–207

Seafoodplus (2008). www.seafoodplus.org/fileadmin/files/news/2004-01-22SFRTD1launchBrussels.pdf. Accessed 25 Mar 10

Seenivasan R, Indu H, Archana G, Geetha S (2010) The antibacterial activity of some marine algae from South East Coast of India. Am-Eurasian J Agric Environ Sci 9(5):480–489

Sekar S, Chandramohan M (2008) Phycobiliproteins as a commodity: trends in applied research, patents and commercialization. J Appl Phycol 2008(20):113–136

Selvin J, Lipton AP (2004) Biopotentials of *Ulva fasciata* and *Hypnea musciformis* collected from the peninsular coast of India. J Mar Sci Technol 12:1–6

Serkedjieva J (2000) Antiherpes virus effect of the red marine alga *Polysiphonia denudata*. Z Naturforsch 55c:830 835

Serkedjieva J (2003) Antiviral activity of the red marine alga *Ceramium rubrum*. Phytother Res 18:480–483

Shanmugam M, Mody KH (2000) Heparinoid-active sulphated polysaccharides from marine algae as potential blood anticoagulant agents. Curr Sci 79(12):1579–1594

Shi Q, Wang A, Lu Z, Qin C, Hu J, Yin J (2017) Overview on the antiviral activities and mechanisms of marine polysaccharides from seaweeds. Carbohydr Res 1(9):453–454

Singh M, Manikandan S, Kumaraguru AK (2012) In vitro antibacterial activity of selected brown marine macroalgae extracts collected from the Pudumadam Coast of "Gulf of Mannar" region against fish pathogens. Int J Human Genet Med Biotechnol Microbiol Stud. ISSN (Online) 2319–1732

Sivagnanavelmurugan M, Marudhupandi T, Palavesam A, Immanuel G (2012) Antiviral effect of fucoidan extracted from the brown seaweed, *Sargassum wightii*, on shrimp *Penaeus monodon* postlarvae against White Spot Syndrome Virus. J World Aquacult Soc 43:697–706

Souto ML et al (2002) Novel marine poly-ethers, vol 58. Tetrahedron, Londres, pp 8119–8125, set

Sreenivasa-Rao PP (1991) Biological investigation of Indian marine algae and screening of some green, red and brown seaweeds for their antimicrobial activity. Seaweed Res Util 14(1):37–43

Sreenivasa-Rao PP (1995) Biological investigation of Indian Phaeophyceae XII, antimicrobial activity of frozen samples of genus *Sargassum* collected from OKHA, west coast of India. Seaweed Res Util 17:105–109

Srinivasa R, Parekh KS (1981) Antibacterial activity of Indian seaweed extracts. Bot Mar 24:577–582

Stern J, Hagerman A, Steinberg P, Magon P (1996) Phlorotannin–protein interactions. J Chem Ecol 22:1877–1899

Sudheer NS, Philip R, Singh ISB (2012) Anti–white spot syndrome virus activity of Ceriops tagal aqueous extract in giant tiger shrimp *Penaeus monodon*. Arch Virol 157:1665–1675

Suzuki M, Daitoh M, Vairappan CS, Abe T, Masuda M (2001) Novel halogenated metabolites from the Malaysian *Laurencia pannosa*. J Nat Prod 64:597–602

Takahashi Y, Uehara K, Watanabe R, Okumura T, Yamashita T, Omura H, Yomo T, Kawano T, Kanemitsu A, Narasaka H, Suzuki N, Itami T (1998) Efficacy of oral administration of fucoidan, a sulfated polysaccharide, in controlling white spot syndrome in kuruma shrimp in Japan. In: Flegel TW (ed) Advances in shrimp biotechnology. National Center for 896 Genetic Engineering and Biotechnology, Bangkok, pp 171–173

Tariq VN (1991) Antifungal activity in crude extracts of marine red algae. Mycol Res 95:1433–1440

Thanigaivel S, Vijayakumar S, Mukherjee A, Chandrasekaran N, Thomas J (2014) Antioxidant and antibacterial activity of *Chaetomorpha antennina* against shrimp pathogen *Vibrio parahaemolyticus*. Aquaculture 433:467–475

Thinakaran T, Sivakumar K (2013) Antifungal activity of certain seaweeds from Puthumadam coast. Int J Res Rev Pharm Appl Sci 3(3):341–350

Tibbetts SM, Milley JE, Lall SP (2016) Nutritional quality of some wild and cultivated seaweeds: nutrient composition, total phenolic content and in vitro digestibility. J Appl Phycol 28:3575–3585

Torres P, Santos JP, Chow F, Pena Ferreira MJ, dos Santos DYAC (2018) Comparative analysis of in vitro antioxidant capacities of mycosporine-like amino acids (MAAs). Algal Res 34:57–67

Tuney İ (2006) Antimicrobial activities of the extracts of marine algae from the coast of Urla (Izmir, Turkey). Turk J Biol 30(3):171–175

Vairappan CS, Suzuki M (2000) Dynamics of total surface bacteria and bacterial species counts during desiccation in the Malaysian sea lettuce, *Ulva reticulata* (Ulvales, Chlorophyta). Phycol Res 48:55–61

Vairappan CS, Suzuki M, Abe T, Masuda M (2001b) Halogenated metabolites with antibacterial activity from the Okinawan *Laurencia* species. Phytochemistry 58:517–523

Vallinayagam K, Arumugam R, Ragupathi Raja Kannan R, Thirumaran G, Anantharaman P (2009) Antibacterial activity of some selected seaweeds from Pudumadam coastal regions. Glob J Pharmacol 3(1):50–52

Van Heemst JDH, Peulvis S, De Leeuw JW (1996) Novel algal polyphenolic biomacromolecules as significant contributors to resistant fractions of marine dissolved and particulate organic matter. Org Geochem, Londres 24(6/7):629–640

Vlachos V, Cristchley AT, von Holy A (1997) Antimicrobial activity of extracts from selected southern Africanmarine *Macroalgae*. South Afr J Sci 93:328–332

Wallace G, Fry SC (1994) Phenolic components of the plant cell wall. Int Rev Cytol 151:229–267

Wang Y-B, Li J-R, Lin J (2008) Probiotics in aquaculture: challenges and outlook. Aquaculture 281:1–4

Waterman PG, Mole S (1994) Analysis of phenolic plant metabolites. In: Methods in ecology. Blackwell Scientific Publications, Oxford

Wells ML, Potin P, Craigie JS, Raven JA, Merchant SS (2016) Algae as nutritional and functional food sources: revisiting our understanding. J Appl Phycol 29:949–982

Wessels M, Konig GM, Wright AD (1999) A new tyrosine kinase inhibitor from the marine brown algae *Stypopodium zonale*. J Nat Prod Washington 62:927–930

Whitfield FB, Helidoniotis F, Shaw KJ, Svoronos D (1999) Distribution of bromophenols in species of marine algae from eastern Australia. J Agric Food Chem 1999(47):2367–2373

Yousr AH, Napis S, Rusul GRA, Son R (2007) Detection of aerolysin and hemolysin genes in Aeromonas spp. isolated from environmental and shellfish sources by polymerase chain reaction. ASEAN Food J 14:115–122

Yuan YV, Carrington MF, Walsh NA (2005) Extracts from dulse (*Palmaria palmata*) are effective antioxidants and inhibitors of cell proliferation in vitro. Food Chem Toxicol 43:1073–1081

Yuan L, Chang J, Yin Q, Lu M, Di Y, Wang P, Wang Z, Wang E, Lu F (2017) Fermented soybean meal improves the growth performance, nutrient digestibility, and microbial flora in piglets. Anim Nutr 3(1):19–24

Zandi K, Fouladvand M, Pakdel P, Sartavi K (2007) Evaluation in vitro antiviral activity of a brown alga (*Cystoseira myrica*) from the Persian Gulf against Herpes simplex virus type 1. Afr J Biotechnol 6:2511–2514

Zhang X, Thomsen M (2019) Biomolecular composition and revenue explained by interactions between extrinsic factors and endogenous rhythms of *Saccharina latissima*. Mar Drugs 17:107

Chapter 22
Can Seaweeds Be Used as Immunity Boosters?

Tiago Sousa, João Cotas, Diana Pacheco, Kiril Bahcevandziev, Ana M. M. Gonçalves, and Leonel Pereira

22.1 Introduction

Several efforts have been done to reduce synthetic chemotherapeutics use (antibiotics, antiviral) in animals and plants. For this reason, novel approaches have been developed for the exploitation of natural immunostimulants that boost animal or plant immune systems (Thépot et al. 2021). In resemblance, the same efforts have been done regarding the development of antibiotics and other drugs against human pathogens.

An immunomodulator is normally an immunostimulant (synthetic or natural) compound that boosts the immune system of an organism, promoting the welfare and enhancing the resistance to pathogens in animals and in plants. Thus, immunomodulators can also protect and ameliorate the plant crops from abiotic stress factors, such as salinity, drought and floods (Cabello 2006; Cook et al. 2018; Thépot et al. 2021).

Thus, consumer awareness regarding the appearance of drug-resistant pathogenic strains intends that there is a need for novel and natural strategies. Thus immunomodulator agents applied in plants and animals, can in fact help to prevent

T. Sousa · J. Cotas · D. Pacheco · L. Pereira (✉)
University of Coimbra, Marine and Environmental Sciences Centre (MARE), Department of Life Sciences, Coimbra, Portugal
e-mail: leonel.pereira@uc.pt

K. Bahcevandziev
Agricultural College of Coimbra (ESAC/IPC), Research Centre for Natural Resources Environment and Society (CERNAS), Institute of Applied Research (IIA), Coimbra, Portugal

A. M. M. Gonçalves
University of Coimbra, Marine and Environmental Sciences Centre (MARE), Department of Life Sciences, Coimbra, Portugal

Department of Biology and CESAM, University of Aveiro, Aveiro, Portugal

A. Ranga Rao, G. A. Ravishankar (eds.), *Sustainable Global Resources of Seaweeds Volume 1*, https://doi.org/10.1007/978-3-030-91955-9_22

and reduce the impacts of the chemotherapeutics in the human health, due the major use of cultivars and farm animals in a human diet (Devendra 2007; Craigie 2011; Carbone and Faggio 2016; Martin and Li 2017). These immunomodulators can elicit an immune response in the host, improve the overall health, increasing the animal and plant resistance to biotic or abiotic aggressive factors (Van Loon et al. 1998; Heil 2002; Akhter et al. 2015; Vallejos-Vidal et al. 2016).

As a source of undiscovered compounds, the marine environment has been one of the major resources in the quest for compounds with immunomodulator activity. In this context, algae present a significant food support with several interesting compounds, that can be exploited as animal and plant immune booster, as well as for humans (Gomez-Zavaglia et al. 2019; Cook et al. 2018; Palstra et al. 2018; Shama et al., 2019). However, only a few of the seaweed compounds have been assessed for this bioactivity. Molecules such as, algal polysaccharides and seaweed extracts, or dried, are the most exploited in several studies for their immunomodulatory properties (Palstra et al. 2018; Vidhya Hindu et al. 2019; Melo et al. 2020).

Thus, seaweeds not only appear to be a natural alternative as food for human consumption, but also as a new hypothesis to enhance animal and plant welfare against pathogens and abiotic stresses (Heil 2002; Khan et al. 2009; Leandro et al. 2020; Morais et al. 2020). This chapter highlights the seaweeds and their compounds as immune boosters in the plant and animal farming regarding their welfare.

22.2 Seaweeds Compounds as Immunity Boosters

Commercially seaweed extracts are used as plant fertilizer and animal feed by different companies (Makkar et al. 2016; Soares et al. 2020). These extracts include many seaweed bioactive molecules (primarily polysaccharides, phenolics, flavonoids, pigments, and carotenoids), as well as, essential minerals that help in disease tolerance. An existing information regarding their immunomodulatory/immunostimulatory activity is scarce (Chatzissavvidis and Therios 2014; Cook et al. 2018; Thépot et al. 2021). Consequently, there is a general lack of knowledge of the mechanism of action of seaweed compounds when activating the plant or animal immune response (e.g., activation of signal pathways controlling immune defense responses) in response to any pathogen attack (Saker et al. 2001; Spiers et al. 2004; Hernández-Herrera et al. 2014). Despite this problem, several studies have already shown that crude or refined seaweed extracts can protect plants against pathogens (Cluzet et al. 2004; Hernández-Herrera et al. 2014; Esserti et al. 2017). Moreover, seaweed extracts can enhance animal resistance to the pathogens (Saker et al. 2001; Spiers et al. 2004).

22.2.1 Plants

Food protection and environmental security are becoming very important around the world, and many countries are limiting the use of chemical pesticides. Plant immunity inducers are safe to use on fruit and vegetable crops, and they work by activating the plant's natural defense mechanism to prevent or control diseases (Mukherjee and Patel 2020). Thus, plant immune inducers have a high level of acceptance among farmers and consumers because they are natural products, which are an important factor in developing and increasing this industry (Dewen et al. 2017).

Plants' latent defense mechanisms can be activated or interfered by seaweeds and/or their molecules, but these mechanisms are only triggered when there is a direct plant–pathogen interaction. These defense mechanisms can result in an induced systemic resistance (ISR) or systemic acquired resistance (SAR), rendering the plant less susceptible to subsequent pathogen attack (Heil 2002), including elicitor synthesis, oxidative bursts, and antimicrobial compound synthesis (Van Loon et al. 1998; Heil 2002; Melo et al. 2020).

Plant immune systems are not the same as mammalian immune systems (Ronald and Beutler 2010); plants have only an innate immune system, while humans have both, an innate and adaptive immune system (Ronald and Beutler 2010). Before getting access to the host, pathogens must conquer several plant obstacles, including a waxy cuticle, robust cell walls, antimicrobial compounds, and secondary metabolites (Muthamilarasan and Prasad 2013). Even if the pathogen reaches the host plant, identification of highly conserved microbe or pathogen-associated molecular pattern (MAMP/ PAMP), molecules such as flagellin, peptidoglycan, lipopolysaccharides, cold shock proteins, or chitin activate innate immunity (Jones and Dangl 2006; Muthamilarasan and Prasad 2013). MAMP/PAMP-induced immunity activates mitogen-activated protein kinases (MAPKs) and various hormone signaling pathways, triggering a series of defensive responses that include alkalization of the growth medium, deposition of callose, development of reactive nitrogen/oxygen species (RNS/ROS), closure of the stomata, production of antimicrobial compounds, and a variety of secondary metabolites (Jones and Dangl 2006; Muthamilarasan and Prasad 2013). Plants can resist to pathogenic attacks by using these defense responses (Jones and Dangl 2006).

If the pathogen is faster than the induced response, elicitors are formed are not effective to exhibit the plant defense reactions, resulting in manifestation of the disease (Sticher et al. 1997).

Elicitors are plant molecules/compounds that induce or activate plant's defense mechanisms. A metabolic stimulus, known as a "signal," is produced when an elicitor interacts with a cell receptor (Aziz et al. 2003; Huffaker et al. 2013). Calcium flux, mitogen-activated protein (MAP) kinase activation, and the production of secondary signals such as reactive oxygen species, nitric oxide, jasmonic acid, ethylene, and salicylic acid can be triggered by the elicitor (Huffaker et al. 2013). Salicylic acid is a phenolic compound found in plants that acts as an elicitor by activating genes that code for pathogenesis-related (PR) proteins and enzymes involved in phytoalexin and lignin formation (Cole 1999). The resistance induction is linked to an increase in the activity of many enzymes in the plant tissues, such as peroxidases, polyphenoloxidases, phenylalanine ammonia lyases, lipoxygenases, 1,3-glucanases, and chitinase (Melo et al. 2020).

22.2.2 Animals

The addition of seaweed as an immunostimulant (such as *Laminaria digitata*, Fig. 22.1b) to a fish's diet improved their immune system and resistance to pathogens (Makkar et al. 2016; Palstra et al. 2018). In general, enhanced innate immune functions in fish are associated with better survival against pathogens or other stressors. Palstra et al. (2018) tested two seaweed-based diets (10% of *Laminaria digitata* (Phaeophyceae) and 10% of a commercial blend of seaweeds) and did not observed differences in fish growth, still the expression of different genes involved in the immune response was recorded. Thus, the correlation between growth and innate immunity remained inconclusive. In addition, the concentration in which the seaweed is supplemented represents a significant factor. In the reviewed studies, improvements in innate immunity following supplementation with seaweed or its derivatives is most common in refined seaweed products, followed by seaweed extract and whole seaweed (Makkar et al. 2016; Gora et al. 2018). This shows that different seaweed and other pre-probiotic ingredients may have synergistic effects (Prabu et al. 2016). As a prophylactic therapy, the combination of seaweed or their extracts with a proven immunostimulant could play a major role in reducing the

Fig. 22.1 *Chondrus crispus* (Rhodophyta) (**a**); *Laminaria digitata* (Ochrophyta, Phaeophyceae) (**b**); *Ascophyllum nodosum* (Ochrophyta, Phaeophyceae) (**c**)

effect of diseases in aquaculture and may eventually replace the industry's use of antibiotics (Thépot et al. 2021).

Chondrus crispus (Fig. 22.1a), a red seaweed with a long history of use as food and medicine, is notable for its bioactive properties (Craigie et al. 2019). Components of this seaweed have been shown to improve host immunity against the human bacteria *Pseudomonas aeruginosa* by suppressing *quorum sensing* and virulence factors, as well as enriching probiotic levels in the host (Liu et al. 2013, 2015). Only 2% of *C. crispus* enriched the short-chain fatty acid concentration, which is thought to serve as an energy source for intestinal epithelial cells, stimulating cell growth, and greatly increased the beneficial (probiotic) bacteria in the guts of layer hens as a supplementary feed ingredient (Kulshreshtha et al. 2014). Sulphated polysaccharides (SP) extracted from *C. crispus* samples harvested of the Irish coast were used to see whether they had any effect on wild mussels (*Mytilus* spp.). These findings showed that the SP from *C. crispus* induced health-promoting indicators (such as, cell viability, lysozyme activity and the expression of immune-related mRNA) in *Mytilus* spp. on a cellular, humoral, and molecular level, with a 10-day effect (Rudtanatip et al. 2018).

MSPs (algal sulfated polysaccharide extracts) can be used as an alternative prophylactic strategy that stimulates innate immune responses and limits infections in farm animals, thereby reducing antibiotic use (Berri et al. 2016). Furthermore, MSP extracts from seaweeds have been shown to have immunomodulatory properties by altering the function of cytokines and macrophages, an important component of the innate immune system. This has been found in murine macrophages *in vitro* (Karnjanapratum et al. 2012; Fang et al. 2015; Jeong et al. 2015), but also *in vivo* in mice (Kim et al. 2011; Liu et al. 2017), as well as *in vitro* system of porcine intestinal epithelial (IPEC-1) cells (Berri et al. 2017). Guriec et al. (2018), recently showed that this was also the case in chicken when the extract was administered orally, as well as *in vitro* on heterophils and monocytes and *in vivo*. In different models, such as in murine macrophages *in vitro*, MSP derived from green algae has demonstrated immunomodulating effects, both as crude algal extracts and as highly purified fractions (Karnjanapratum et al. 2012; Tabarsa et al. 2012; Tabarsa et al. 2018), in mice and in chicken *in vivo* (Song et al. 2015; Guriec et al. 2018), as well as *in vivo* assays with immunoglobulin G anti-Bordetella (Bussy et al. 2019).

22.3 Commercial Exploitation of the Seaweed Compounds as Immunity Boosters

22.3.1 Seaweed-Based Commercially Available Products for Plants Immunity Stimulation

Stella Maris® is a commercially available *Ascophyllum* (Fig. 22.1c) extract that can promote plant growth (Cook et al. 2018). However, there is not much evidence that this extract can increase a plant immunity against pathogens. Biologically active compounds such as ulvans and oligo-ulvans, oligo-alginate, fucans and oligo-fucans, laminarin, and carrageenans and oligo-carrageenans have been fractionated from seaweed extracts (Vera et al. 2011). However, studies showed that an individual biological component of the extracts have activated only one plant immune response pathway in many cases, namely the systemic acquired resistance (SAR) (Klarzynski et al. 2003; Chandía et al. 2004; Cook et al. 2018). It is hypothesized that treating plants with the seaweed extract is more advantageous because it activates many pathways within the innate immune response (Cook et al. 2018). The extract triggered a strong innate immune response in *Arabidopsis thaliana*, as shown by well-characterized high-throughput assays.

The *Ascophyllum* extract triggers a rapid (9 min post-treatment) immune response in the form of hydrogen peroxide production, according to the reactive oxygen species (ROS) chemiluminescence assay. Plants treated with lower concentrations (0.05%) of this extract produced more ROS than plants treated with higher concentrations (0.2% and 0.5%). Despite this, the extract was shown to elicit a strong immune response in the form of hydrogen peroxide production. *Arabidopsis* seedlings treated with both, 100 nM of flg22 and with *Ascophyllum* extract (0.05%), developed 0.8-fold more hydrogen peroxide (Cook et al. 2018). A flg22 is a synthetic flagellar peptide that has been shown to elicit a powerful immune response in *Arabidopsis* (Danna et al. 2011).

Natural compounds for pathogen control are appealing, and the availability of innovative applications and molecular techniques opens new approaches for plant defense. Many organic compounds have already been commercialized and are available as plant growth biostimulants or biofertilizers. These, among others, include Biosept 33SL (grapefruit extract), Bio-Algeen S90 Plus, Labimar 10S, Kelpak SL, Lysodin Alga-Fert, and Vaxiplant SL (all marine algae extracts) (Jamiołkowska 2020). These preparations are generally biodegradable, non-toxic, non-polluting, and non-hazardous to different species. Many of them also explicitly restrict the production of phytopathogens by mitigating stress-induced limitations and regulate/modify physiological processes in plants to promote the plant growth and increase a productivity (Yakhin et al. 2017).

22.3.2 Seaweed-Based Commercially Available Products for Animals' Immunity Stimulation

The magnitude of fescue toxicosis symptoms (caused by ergot alkaloids, produced by the endophyte fungus *Neotyphodium coenophialum* in tall fescue grass) vary depending on the animal's level of environmental stress and thermal pressure. There are currently no cost-effective products available that can effectively mitigate the effects of fescue toxicosis. One product being tested is Tasco (Tasco-EX®; Acadian Seaplants Limited; Nova Scotia, Canada), which is based on a brown seaweed (*Ascophyllum nodosum*) (Spiers et al. 2004). Evans et al. (2002) treated cows with

Tasco, as part of a mineral mix, and found that this treatment decreased rectal temperature during summer heat stress.

Researchers continue to look into the benefits of adding seaweeds to livestock diets, and while protein replacement is one of the most common reasons for developing alternative feeds, other important applications for the meat and dairy industries are emerging (Cornish et al. 2020). For 21 days, young rams were fed with a sun-dried carefully handled, granular extract of the fucoidan (extracted from *A. nodosum*), at rates of 1%, 3%, or 5%. Rumen fermentation was not observed but increasing levels of Tasco administration, decreased the total bacteria and archaea, and increased protozoa, in the rumen. Moreover, introducing seaweed as a feed, decreased the overall population of *E. coli*, a common and widespread foodborne pathogen (Zhou et al. 2018). Benefits to cattle is another example of the beneficial impact that seaweeds can have, when steers were fed with 20 g Tasco/kg diet for 7 days showed similar pathogen mitigation effects. The length and severity of *E. coli* O157:H7 stool shedding were significantly decreased, suggesting an inhibitory effect on the virulent bacteria's growth and its proliferation (Bach et al. 2008). As repeated meat biological samples attests positive for the presence of *E. coli*, this is an example of how a contamination originating from livestock can adversely affect humans (Cornish et al. 2020).

22.4 Conclusions and Future Perspectives

Despite the beneficial effects that seaweed supplementation has had on plants and animals, there is little knowledge on how seaweeds and their compounds influence their immune systems. As a result, there is still a long way to go to fully understand the immunomodulatory pathways in which seaweed compounds affect the immune systems of plants and animals.

Processing seaweed-based extracts is difficult and requires a significant amount of raw material, and the results can be inconsistent (depending on the biologically active compounds) and inaccurate during the processing phase. More research is required to improve the quality of seaweed extract so that scientists can better understand how seaweed compounds function as immune boosters in plants and animals.

Acknowledgments This work is financed by national funds through FCT – Foundation for Science and Technology, I.P., within the scope of the projects UIDB/04292/2020 granted to MARE – Marine and Environmental Sciences Centre and UIDP/50017/2020+UIDB/50017/2020 (by FCT/MTCES) granted to CESAM – Centre for Environmental and Marine Studies. João Cotas thanks to the European Regional Development Fund through the Interreg Atlantic Area Program, under the project NASPA (EAPA_451/2016). Diana Pacheco thanks to PTDC/BIA-CBI/31144/2017-POCI-01 project -0145-FEDER-031144-MARINE INVADERS, co-financed by the ERDF through POCI (Operational Program Competitiveness and Internationalization) and by the Foundation for Science and Technology (FCT, IP). Ana M.M. Gonçalves acknowledges University of Coimbra for the contract IT057-18-7253.

References

Akhter N, Wu B, Mahmood Memon A, Mohsin M (2015) Probiotics and prebiotics associated with aquaculture: a review. Fish Shellfish Immunol 45:733–741. https://doi.org/10.1016/j.fsi.2015.05.038

Aziz A, Poinssot B, Daire X, Adrian M, Bézier A, Lambert B, Joubert JM, Pugin A (2003) Laminarin elicits defense responses in grapevine and induces protection against *Botrytis cinerea* and *Plasmopara viticola*. Mol Plant-Microbe Interact 16:1118–1128. https://doi.org/10.1094/MPMI.2003.16.12.1118

Bach SJ, Wang Y, McAllister TA (2008) Effect of feeding sun-dried seaweed (*Ascophyllum nodosum*) on fecal shedding of *Escherichia coli* O157:H7 by feedlot cattle and on growth performance of lambs. Anim Feed Sci Technol. https://doi.org/10.1016/j.anifeedsci.2007.05.033

Berri M, Slugocki C, Olivier M, Helloin E, Jacques I, Salmon H, Demais H, Le Goff M, Collen PN (2016) Marine-sulfated polysaccharides extract of *Ulva armoricana* green algae exhibits an antimicrobial activity and stimulates cytokine expression by intestinal epithelial cells. J Appl Phycol. https://doi.org/10.1007/s10811-016-0822-7

Berri M, Olivier M, Holbert S, Dupont J, Demais H, Le Goff M, Collen PN (2017) Ulvan from *Ulva armoricana* (Chlorophyta) activates the PI3K/Akt signalling pathway via TLR4 to induce intestinal cytokine production. Algal Res 28:39–47. https://doi.org/10.1016/j.algal.2017.10.008

Bussy F, Matthieu LG, Salmon H, Delaval J, Berri M, Pi NC (2019) Immunomodulating effect of a seaweed extract from *Ulva armoricana* in pig: specific IgG and total IgA in colostrum, milk, and blood. Vet Anim Sci 7:100051. https://doi.org/10.1016/j.vas.2019.100051

Cabello FC (2006) Heavy use of prophylactic antibiotics in aquaculture: a growing problem for human and animal health and for the environment. Environ Microbiol 8:1137–1144. https://doi.org/10.1111/j.1462-2920.2006.01054.x

Carbone D, Faggio C (2016) Importance of prebiotics in aquaculture as immunostimulants. Effects on immune system of *Sparus aurata* and *Dicentrarchus labrax*. Fish Shellfish Immunol 54:172–178. https://doi.org/10.1016/j.fsi.2016.04.011

Chandía NP, Matsuhiro B, Mejías E, Moenne A (2004) Alginic acids in *Lessonia vadosa*: partial hydrolysis and elicitor properties of the polymannuronic acid fraction. J Appl Phycol 16:127–133. https://doi.org/10.1023/B:JAPH.0000044778.44193.a8

Chatzissavvidis C, Therios I (2014) Role of algae in agriculture. In: Seaweeds. Nova Science Publishers, Inc., New York, USA. pp 1–37

Cluzet S, Torregrosa C, Jacquet C, Lafitte C, Fournier J, Mercier L, Salamagne S, Briand X, Esquerre-Tugaye M-T, Dumas B (2004) Gene expression profiling and protection of *Medicago truncatula* against a fungal infection in response to an elicitor from green algae *Ulva* spp. Plant Cell Environ 27:917–928. https://doi.org/10.1111/j.1365-3040.2004.01197.x

Cole DL (1999) The efficacy of acibenzolar-S-methyl, an inducer of systemic acquired resistance, against bacterial and fungal diseases of tobacco. Crop Prot 18:267–273. https://doi.org/10.1016/S0261-2194(99)00026-5

Cook J, Zhang J, Norrie J, Blal B, Cheng Z (2018) Seaweed extract (Stella Maris®) activates innate immune responses in *Arabidopsis thaliana* and protects host against bacterial pathogens. Mar Drugs 16:221. https://doi.org/10.3390/md16070221

Cornish ML, Mac Monagail M, Critchley AT (2020) The animal kingdom, agriculture and seaweeds. J Mar Sci Eng 8:574. https://doi.org/10.3390/jmse8080574

Craigie JS (2011) Seaweed extract stimuli in plant science and agriculture. J Appl Phycol 23:371–393. https://doi.org/10.1007/s10811-010-9560-4

Craigie JS, Cornish ML, Deveau LE (2019) Commercialization of Irish moss aquaculture: the Canadian experience. Bot Mar. https://doi.org/10.1515/bot-2019-0017

Danna CH, Millet YA, Koller T, Han S-W, Bent AF, Ronald PC, Ausubel FM (2011) The *Arabidopsis* flagellin receptor FLS2 mediates the perception of *Xanthomonas* Ax21 secreted peptides. Proc Natl Acad Sci 108:9286–9291. https://doi.org/10.1073/pnas.1106366108

Devendra C (2007) Small farm systems to feed hungry Asia. Outlook Agric 36:7–20. https://doi.org/10.5367/000000007780223641
Dewen Q, Yijie D, Yi Z, Shupeng L, Fachao S (2017) Plant immunity inducer development and application. Mol Plant-Microbe Interact 30:355–360. https://doi.org/10.1094/MPMI-11-16-0231-CR
Esserti S, Smaili A, Rifai LA, Koussa T, Makroum K, Belfaiza M, Kabil EM et al (2017) Protective effect of three brown seaweed extracts against fungal and bacterial diseases of tomato. J Appl Phycol 29:1081–1093. https://doi.org/10.1007/s10811-016-0996-z
Evans RR, Allen VG, Ivy RL, Best TG, Houston JE (2002) Animal response to the ingestion of tasco seaweed meal. J Anim Sci 81:21
Fang Q, Wang J-F, Zha X-Q, Cui S-H, Cao L, Luo J-P (2015) Immunomodulatory activity on macrophage of a purified polysaccharide extracted from *Laminaria japonica*. Carbohydr Polym 134:66–73. https://doi.org/10.1016/j.carbpol.2015.07.070
Gomez-Zavaglia A, Lage MAP, Jimenez-Lopez C, Mejuto JC, Simal-Gandara J (2019) The potential of seaweeds as a source of functional ingredients of prebiotic and antioxidant value. Antioxidants 8:406. https://doi.org/10.3390/antiox8090406
Gora AH, Sahu NP, Sahoo S, Rehman S, Ahmad I, Agarwal D, Dar SA, Rasool SI (2018) Metabolic and haematological responses of *Labeo rohita* to dietary fucoidan. J Appl Anim Res 46:1042–1050. https://doi.org/10.1080/09712119.2018.1456442
Guriec N, Bussy F, Gouin C, Mathiaud O, Quero B, Le Goff M, Collén PN (2018) Ulvan activates chicken heterophils and monocytes through toll-like receptor 2 and toll-like receptor 4. Front Immunol 9. https://doi.org/10.3389/fimmu.2018.02725
Heil M (2002) Induced systemic resistance (ISR) against pathogens in the context of induced plant defences. Ann Bot 89:503–512. https://doi.org/10.1093/aob/mcf076
Hernández-Herrera RM, Virgen-Calleros G, Ruiz-López M, Zañudo-Hernández J, Délano-Frier JP, Sánchez-Hernández C (2014) Extracts from green and brown seaweeds protect tomato (*Solanum lycopersicum*) against the necrotrophic fungus *Alternaria solani*. J Appl Phycol 26:1607–1614. https://doi.org/10.1007/s10811-013-0193-2
Huffaker A, Pearce G, Veyrat N, Erb M, Turlings TCJ, Sartor R, Shen Z et al (2013) Plant elicitor peptides are conserved signals regulating direct and indirect antiherbivore defense. Proc Natl Acad Sci 110:5707–5712. https://doi.org/10.1073/pnas.1214668110
Jamiołkowska A (2020) Natural compounds as elicitors of plant resistance against diseases and new biocontrol strategies. Agronomy 10:173. https://doi.org/10.3390/agronomy10020173
Jeong SC, Jeong YT, Lee SM, Kim JH (2015) Immune-modulating activities of polysaccharides extracted from brown algae *Hizikia fusiforme*. Biosci Biotechnol Biochem 79:1362–1365. https://doi.org/10.1080/09168451.2015.1018121
Jones JDG, Dangl JL (2006) The plant immune system. Nature 444:323–329. https://doi.org/10.1038/nature05286
Karnjanapratum S, Tabarsa M, Cho M, You S (2012) Characterization and immunomodulatory activities of sulfated polysaccharides from *Capsosiphon fulvescens*. Int J Biol Macromol 51:720–729. https://doi.org/10.1016/j.ijbiomac.2012.07.006
Khan W, Rayirath UP, Subramanian S, Jithesh MN, Rayorath P, Hodges DM, Critchley AT, Craigie JS, Norrie J, Prithiviraj B (2009) Seaweed extracts as biostimulants of plant growth and development. J Plant Growth Regul 28:386–399. https://doi.org/10.1007/s00344-009-9103-x
Kim J-K, Cho ML, Karnjanapratum S, Shin I-S, You SG (2011) In vitro and in vivo immunomodulatory activity of sulfated polysaccharides from *Enteromorpha prolifera*. Int J Biol Macromol 49:1051–1058. https://doi.org/10.1016/j.ijbiomac.2011.08.032
Klarzynski O, Descamps V, Plesse B, Yvin J-C, Kloareg B, Fritig B (2003) Sulfated fucan oligosaccharides elicit defense responses in tobacco and local and systemic resistance against tobacco mosaic virus. Mol Plant-Microbe Interact 16:115–122. https://doi.org/10.1094/MPMI.2003.16.2.115
Kulshreshtha G, Rathgeber B, Stratton G, Thomas N, Evans F, Critchley AT, Hafting J, Prithiviraj B (2014) Feed supplementation with red seaweeds, *Chondrus crispus* and *Sarcodiotheca*

gaudichaudii, affects performance, egg quality, and gut microbiota of layer hens. Poult Sci 93:2991–3001. https://doi.org/10.3382/ps.2014-04200

Leandro A, Pacheco D, Cotas J, Marques JC, Pereira L, Gonçalves AMM (2020) Seaweed's bioactive candidate compounds to food industry and global food security. Life 10:140. https://doi.org/10.3390/life10080140

Liu J, Hafting J, Critchley AT, Banskota AH, Prithiviraj B (2013) Components of the cultivated red seaweed *Chondrus crispus* enhance the immune response of *Caenorhabditis elegans* to *Pseudomonas aeruginosa* through the pmk-1, daf-2/daf-16, and skn-1 pathways. Appl Environ Microbiol 79:7343–7350. https://doi.org/10.1128/AEM.01927-13

Liu J, Kandasamy S, Zhang J, Kirby CW, Karakach T, Hafting J, Critchley AT, Evans F, Prithiviraj B (2015) Prebiotic effects of diet supplemented with the cultivated red seaweed *Chondrus crispus* or with fructo-oligo-saccharide on host immunity, colonic microbiota and gut microbial metabolites. BMC Complement Altern Med 15:279. https://doi.org/10.1186/s12906-015-0802-5

Liu Q, Xu S, Li L, Pan T, Shi C, Liu H, Cao M, Su W, Liu G (2017) *In vitro* and *in vivo* immunomodulatory activity of sulfated polysaccharide from *Porphyra haitanensis*. Carbohydr Polym 165:189–196. https://doi.org/10.1016/j.carbpol.2017.02.032

Makkar HPS, Tran G, Heuzé V, Giger-Reverdin S, Lessire M, Lebas F, Ankers P (2016) Seaweeds for livestock diets: a review. Anim Feed Sci Technol 212:1–17. https://doi.org/10.1016/j.anifeedsci.2015.09.018

Martin C, Li J (2017) Medicine is not health care, food is health care: plant metabolic engineering, diet and human health. New Phytol 216:699–719. https://doi.org/10.1111/nph.14730

Melo PC, Collela CF, Sousa T, Pacheco D, Cotas J, Gonçalves AMM, Bahcevandziev K, Pereira L (2020) Seaweed-based products and mushroom β-glucan as tomato plant immunological inducers. Vaccine 8:524. https://doi.org/10.3390/vaccines8030524

Morais T, Inácio A, Coutinho T, Ministro M, Cotas J, Pereira L, Bahcevandziev K (2020) Seaweed potential in the animal feed: a review. J Mar Sci Eng 8:559. https://doi.org/10.3390/jmse8080559

Mukherjee A, Patel JS (2020) Seaweed extract: biostimulator of plant defense and plant productivity. Int J Environ Sci Technol 17:553–558. https://doi.org/10.1007/s13762-019-02442-z

Muthamilarasan M, Prasad M (2013) Plant innate immunity: an updated insight into defense mechanism. J Biosci 38:433–449. https://doi.org/10.1007/s12038-013-9302-2

Palstra AP, Kals J, Garcia AB, Dirks RP, Poelman M (2018) Immunomodulatory effects of dietary seaweeds in LPS challenged Atlantic Salmon *Salmo salar* as determined by deep RNA sequencing of the head kidney transcriptome. Front Physiol:9. https://doi.org/10.3389/fphys.2018.00625

Prabu DL, Sahu NP, Pal AK, Dasgupta S, Narendra A (2016) Immunomodulation and interferon gamma gene expression in sutchi cat fish, *Pangasianodon hypophthalmus*: effect of dietary fucoidan rich seaweed extract (FRSE) on pre and post challenge period. Aquac Res 47:199–218. https://doi.org/10.1111/are.12482

Ronald PC, Beutler B (2010) Plant and animal sensors of conserved microbial signatures. Science 330:1061–1064. https://doi.org/10.1126/science.1189468

Rudtanatip T, Lynch SA, Wongprasert K, Culloty SC (2018) Assessment of the effects of sulfated polysaccharides extracted from the red seaweed Irish moss *Chondrus crispus* on the immune-stimulant activity in mussels *Mytilus* spp. Fish Shellfish Immunol 75:284–290. https://doi.org/10.1016/j.fsi.2018.02.014

Saker KE, Allen VG, Fontenot JP, Bagley CP, Ivy RL, Evans RR, Wester DB (2001) Tasco-Forage: II. Monocyte immune cell response and performance of beef steers grazing tall fescue treated with a seaweed extract. J Anim Sci 79:1022. https://doi.org/10.2527/2001.7941022x

Shama A, Joyce SG, Mari FD, Ranga Rao A, Ravishankar GA, Hudaa N (2019) Macroalgae and microalgae: novel sources of functional food and feed. In: Ravishankar GA, Ranga Rao A (eds) Handbook of algal technologies and phytochemicals. Food, health and nutraceutical applications, vol I. CRC Press, USA, pp 207–219

Soares C, Švarc-Gajić J, Oliva-Teles MT, Pinto E, Nastić N, Savić S, Almeida A, Delerue-Matos C (2020) Mineral composition of subcritical water extracts of *Saccorhiza Polyschides*, a brown seaweed used as fertilizer in the North of Portugal. J Mar Sci Eng 8:244. https://doi.org/10.3390/jmse8040244

Song L, Chen X, Liu X, Zhang F, Hu L, Yue Y, Li K, Li P (2015) Characterization and comparison of the structural features, immune-modulatory and anti-avian influenza virus activities conferred by three algal sulfated polysaccharides. Mar Drugs 14:4. https://doi.org/10.3390/md14010004

Spiers DE, Eichen PA, Leonard MJ, Wax LE, Rottinghaus GE, Williams JE, Colling DP (2004) Benefit of dietary seaweed (*Ascophyllum nodosum*) extract in reducing heat strain and fescue toxicosis: a comparative evaluation. J Therm Biol. https://doi.org/10.1016/j.jtherbio.2004.08.059

Sticher L, Mauch-Mani B, Métraux JP (1997) Systemic acquired resistance. Annu Rev Phytopathol 35:235–270. https://doi.org/10.1146/annurev.phyto.35.1.235

Tabarsa M, Han JH, Kim CY, You SG (2012) Molecular characteristics and immunomodulatory activities of water-soluble sulfated polysaccharides from *Ulva pertusa*. J Med Food 15:135–144. https://doi.org/10.1089/jmf.2011.1716

Tabarsa M, You S, Dabaghian EH, Surayot U (2018) Water-soluble polysaccharides from *Ulva intestinalis* : molecular properties, structural elucidation and immunomodulatory activities. J Food Drug Anal 26:599–608. https://doi.org/10.1016/j.jfda.2017.07.016

Thépot V, Campbell AH, Rimmer MA, Paul NA (2021) Meta-analysis of the use of seaweeds and their extracts as immunostimulants for fish: a systematic review. Rev Aquac 13:907–933. https://doi.org/10.1111/raq.12504

Vallejos-Vidal E, Reyes-López F, Teles M, MacKenzie S (2016) The response of fish to immunostimulant diets. Fish Shellfish Immunol 56:34–69. https://doi.org/10.1016/j.fsi.2016.06.028

van Loon LC, Bakker PAHM, Pieterse CMJ (1998) Systemic resistance induced by rhizosphere bacteria. Annu Rev Phytopathol 36:453–483. https://doi.org/10.1146/annurev.phyto.36.1.453

Vera J, Castro J, Gonzalez A, Moenne A (2011) Seaweed polysaccharides and derived oligosaccharides stimulate defense responses and protection against pathogens in plants. Mar Drugs 9:2514–2525. https://doi.org/10.3390/md9122514

Vidhya Hindu S, Chandrasekaran N, Mukherjee A, Thomas J (2019) A review on the impact of seaweed polysaccharide on the growth of probiotic bacteria and its application in aquaculture. Aquac Int 27:227–238. https://doi.org/10.1007/s10499-018-0318-3

Yakhin OI, Lubyanov AA, Yakhin IA, Brown PH (2017) Biostimulants in plant science: a global perspective. Front Plant Sci 7. https://doi.org/10.3389/fpls.2016.02049

Zhou M, Hünerberg M, Chen Y, Reuter T, McAllister TA, Evans F, Critchley AT, Guan LL (2018) Air-dried brown seaweed, *Ascophyllum nodosum*, alters the rumen microbiome in a manner that changes rumen fermentation profiles and lowers the prevalence of foodborne pathogens. mSphere 3. https://doi.org/10.1128/mSphere.00017-18

Part III
WasteWater Treatment, Bioremediation, Biofuel, Biofertilizer and Miscelaneous Applications of Seaweeds

Chapter 23
Wastewater Cultivated Macroalgae as a Bio-resource in Agriculture

Maja Berden Zrimec, Erik Malta, Martha Bonnet Dunbar, Ana Cerar, Robert Reinhardt, and Rok Mihelič

Abbreviations

DW	Dry Weight
IMTA	Integrated-Multi-Trophic-Aquaculture
TN	Total Nitrogen

23.1 Introduction

Macroalgae and their extracts have a long tradition of being used in the coastal agriculture as the soil conditioners and enhancers of crop productivity (Nabti et al. 2016). Traditionally, seaweeds have been collected from the beach or harvested from the sea. The raising demand for their use for food (Shama et al. 2019) or interesting extracts (agar, alginate, carrageenin), however, resulted in their controlled production, mainly in the coastal seas and in lesser extent in the land-based systems.

Algae cultivation in the wastewater as the parallel (1) bioremediation and (2) biomass production presents an innovative industrial ecology model (Lawton et al. 2017). Nutrients, organic carbon and minerals that would otherwise be lost by the

M. B. Zrimec (✉) · A. Cerar · R. Reinhardt
Algen, Algal Technology Centre, LLC, Ljubljana, Slovenia
e-mail: maja@algen.si

E. Malta
CTAQUA, Aquaculture Technology Centre, El Puerto de Santa María (Cádiz), Spain

M. B. Dunbar
CTAQUA, Aquaculture Technology Centre, El Puerto de Santa María (Cádiz), Spain

ICMAN-CSIC, Institute of Marine Sciences of Andalusia, Spanish National Research Council, Puerto Real (Cádiz), Spain

R. Mihelič
University of Ljubljana, Biotechnical Faculty, Agricultural Department, Ljubljana, Slovenia

A. Ranga Rao, G. A. Ravishankar (eds.), *Sustainable Global Resources of Seaweeds Volume 1*, https://doi.org/10.1007/978-3-030-91955-9_23

discharge into environment, are recovered by algae for their growth. In wastewater cultivation, the large-scale production can be done without consuming large volumes of quality water and expensive commercial growth media. The produced biomass can't be used for human consumption because of the health regulations, except for algae grown in the Integrated Multi-Trophic Aquacultures (IMTA). It can still be exploited for a variety of products, from the low-added-value biofuels, organic fertilizers or biomaterials to the high-added-value compounds for pharmacy, cosmetics, and agriculture.

Macroalgae are an attractive opportunity for multiple industries because they can grow in various wastewaters and their production can be a source of additional jobs and income. In comparison to microalgae, they have an advantage of lower separation and dry mass preparation costs (Lawton et al. 2013; Ge and Champagne 2017). Harvesting is still one of the major setbacks for the large-scale microalgae production due to the costly separation of microscopic cells from the fairly dilute substrate, which usually requires expensive equipment and high energy consumption.

23.2 Wastewater Treatment with Macroalgae

The beginnings of algal cultivation in the wastewater can be traced back to the middle of the last century when W.J. Oswald in H.G. Gotass (1957) suggested that wastewater could be used for a large-scale algae production in the raceway ponds. Indeed, algae can recycle many chemical substances that could cause eutrophication or toxic effects if released into the environment. In the wastewater treatment plants, algal ponds can be used as a final polishing step in the third treatment stage or even as a combination of second and third stage due to the accompanying aerobic bacterial community.

Algae-bacteria community that establishes itself in the wastewater has a symbiotic relationship: algae produce oxygen that aerobic bacteria use for the degradation, while bacteria provide the nutrients and organic carbon for the algal growth. The produced oxygen considerably reduces costs of the energy-demanding technological oxygenation in the wastewater treatment process. Carbon dioxide is consumed in the photosynthesis, decreasing green-house-gas emissions. The odors are significantly reduced as well.

The removal efficiencies of nitrogen and phosphorus with macroalgae can reach levels higher than 90% (Neori et al. 1991; Mulbry et al. 2008; Ge and Champagne 2017; Ross 2017; Ge et al. 2018). The nutrient removal from the environment with macroalgae is most widespread in China where approximately 9,500 tons of phosphorus and 75,000 tons of nitrogen are removed annually by the coastal seaweed aquacultures (Xiao et al. 2017). Nevertheless, their large-scale cultivation still needs optimization. Wastewater is a highly variable medium and variations in the nutrient composition strongly influence the effectiveness of bioremediation.

Nutrient uptake depends most notably on the nitrogen form (NO_3^-, NO_2^-, NH_4^+, urea) and NO_3^-/NH_4^+ and N/P relative molar ratios that can limit the primary production. Ammonium is usually preferred source over the nitrate (Wallentinus 1984; Pedersen and Borum 1997; Abreu et al. 2011; Fan et al. 2014). At the initial NO_3^- and NH_4^+ concentration of 50 μM, *Graciliaria vermiculophylla* removed app. 40% of NO_3^- and 100% of NH_4^+ in just 4 hours (Abreu et al. 2011). Ammonium was removed preferentially also to urea, but the presence of urea enhanced uptake of other co-existing N-forms in the study by Ross (2017). Fan and co-workers (2014) observed that although *Ulva prolifera* preferred NH_4^+-N to NO_3^--N when the NO_3^--N/NH_4^+-N ratio was less than 2.2, the uptake of NO_3^--N was higher at the ratios between 2.2 and 12.9. N-uptake rate (33.9 ± 0.8 μmol·g^{-1} DW h^{-1}) was maximal at N/P ratio 7.5, while P-uptake rate (11.1 ± 4.7 μmol·g^{-1} DW h^{-1}) at N/P ratio 2.2 (Fan et al. 2014). NO_3^- uptake was faster than NH_4^+ at higher initial concentrations (450 μM NO_3^-, 150 μM NH_4^+) by *G. vermiculophylla* (Abreu et al. 2011).

Uptake efficiency was shown to be much higher at lower nutrient concentrations (Abreu et al. 2011). The nutrients' uptake rate can thus be substantially improved by adjusting the protocol of waste stream inflow dynamics, for example, by periodically applying lower nutrient concentrations. With step feeding *Chaetomorpha linum* with 10% centrate wastewater, Ge and Champagne (2017) increased nitrogen and phosphorus removal efficiencies from $72.3 \pm 0.4\%$ and $80.0 + 0.3\%$ to $86.8 \pm 1.1\%$ and $92.6 \pm 0.2\%$, respectively.

The removal rates depend also on the algal species: the uptake of dissolved inorganic nitrogen from the aquaculture effluents by different species can vary as much as 16.9–96.6% (Ross 2017). Considerable research effort has therefore been put into the identification of most promising species for the wastewater treatment. The freshwater genera *Rhizoclonium*, *Cladophora* and *Oedogonium* (Cole et al. 2015, 2016b; Roberts et al. 2015a, b), and marine *Ulva*, *Cladocera*, *Gracilaria*, *Caulerpa* and *Sargassum* (Neori et al. 1991; Ross 2017; Arumugam et al. 2018) are few examples of the most promising macroalgae for bioremediation.

The performance of macroalgae in the wastewater is influenced by several biological factors, for example thallus morphology. Wallentinus and co-workers (1984) found that the species with filamentous, delicately branched, or monostromatic phenotypes had the highest rates of nutrient uptake because of the greater surface/volume ratio. These were short-lived, opportunistic algae like *Cladophora glomerata, Enteromorpha ahlneriana, Scytosiphon lomentaria, Dictyosiphon foeniculaceus* and *Ceramium tenuicorne*. Opportunistic macroalgae exhibit more rapid N uptake to maximally exploit the pulses of nutrient availability, while slower growing, persistent species can have greater N-storage capacity (Pedersen and Borum 1997). The lowest uptake rates thus occurred among the late successional, long-lived, coarse species with a low surface/volume ratio (*Fucus vesiculosus, Furcellaria lumbricalis* and *Phyllophora truncata*). This might be especially relevant for the wastewater input regime and retention times. Uptake of nitrogen and growth rates can be higher following a period of N limitation in opportunistic species as algae strive to replenish their internal N pools ("surge uptake") (Pedersen and Borum 1997; Luo et al. 2012).

Macroalgae have the capacity to accumulate metals from the environment and can be employed in the wastewater treatment as a live (bioaccumulation) or dead (adsorption) biomass (Ross 2017; Michalak 2020). Good biosorption properties result from the macromolecules in the cell wall (e.g., polysaccharides, proteins) offering functional groups for binding metal ions (Michalak 2020). Biochar from macroalgae can also be used for metal and dye removal from the wastewater. These techniques can be used for the wastewater pre-treatment to enable more consistent and controlled influent for the biomass production.

23.2.1 Integrated Multi-trophic Aquaculture

Integrated Multi-Trophic Aquaculture (IMTA) is defined as »Enhanced production of aquatic organisms (with or without terrestrial organisms) of two or more functional groups, that are trophically connected by demonstrated nutrient flows and whose biomass is fully or partially removed by harvesting to facilitate ecological balance« (Dunbar et al. 2020). Primary producers play a key role as they "biofiltrate" inorganic nutrients from the waste of the primary culture, mitigating its impact on the environment and providing a potentially valuable crop. In general, seaweeds are favored over microalgae as the main primary producers of saltwater IMTA for the reasons already mentioned in the introduction (Chopin et al. 2001).

IMTA can be regarded as wastewater treatment integrated into aquaculture; however, it is far more than that. IMTA encompasses a more holistic, ecosystemic approach to aquaculture. Compared to other wastewater treatment systems, algae grown in IMTA are considered safe for human consumption as they are growing in effluent coming from the animal (usually fish) cultivation systems and not from systems containing human waste. It can be viewed as analogous to fertilization of terrestrial crops with manure.

The concept of IMTA has been known and practiced for centuries, albeit using different terminology until the first decade of this century (Neori et al. 2007). In many Asian countries, traditional practices show many examples of integrated aquaculture, such as the integration of carp culture in rice fields. In more recent years, seaweed and mollusk cultures in coastal areas have been integrated into the existing shrimp aquaculture (Edwards 2009; Soto 2009).

IMTA systems can be divided into two main groups. The first group is on-land IMTA, consisting of often compartmentalized systems that can be completely artificial (tanks) but more often comprise earthen ponds through which the water flow is led from the higher trophic level culture (e.g., fed fish) to the lower trophic levels (e.g., mollusks, seaweeds) (Fig. 23.1). The other group is at-sea IMTA (near-shore and off-shore), where the lower trophic level cultures are grown in proximity to and downstream from the higher trophic level cultures. Generally, algae associated with on-land cultures are relatively small, filamentous or foliose seaweeds, such as *Ulva, Codium, Gracilaria, Porphyra, Asparagopsis*; whereas seaweeds associated to at-sea cultivation are more often kelp species, such as *Saccharina latissima* (for instance with salmon farms; Chopin et al. 2001).

Fig. 23.1 Schematic representation of CTAQUA's pilot IMTA installation operating during the INTEGRATE project (2018–2019). Water flows are driven by tidal oscillations, arrows indicate direction. Fish (*Sparus aurata*, sea bream) and Pacific oysters (*Magallana gigas*) were grown in the same pond, which was hydrologically connected to the seaweed pond. Photos clockwise from top left: green seaweed *Ulva* sp., floating seaweed cages, red seaweed *Gracilaria gracilis*, Pacific oyster, sea bream, oyster bag, floating oyster bags. (Photos courtesy of CTAQUA)

23.3 Macroalgae Biomass production in Wastewater

The large-scale cultivation of macroalgae monocultures in the wastewater predominantly occurs in the open pond systems similar to the microalgae cultivation (Fig. 23.2): circular raceways with the paddlewheels for water circulation, maintaining the macroalgae in constant suspension (Lawton et al. 2017). The other predominant culturing system is algal turf scrubber (Fig. 23.3), which has a mixed algal community that is mostly self-seeded and uncontrolled (Mulbry et al. 2008; Lawton et al. 2017).

Wastewater promotes growth of a mixed culture with alternating strain composition, but for the bioproducts a more consistent material is needed. Some control over the culture composition can be provided by the wastewater pre-treatment, like biochar, filtration, dilution and macroalgae species with a tendency for dominant growth in a dynamic substrate, resistance to herbivory and infections, possibly exhibiting stable production rate and high nutrient uptake (Lawton et al. 2013; Ross 2017; Valero-Rodriguez et al. 2020).

Fig. 23.2 Wastewater treatment pond for macroalgae cultivation in Townsville, Australia. (Photo courtesy of Andrew J. Cole): *Oedogonium* sp. was cultivated in the treated municipal wastewater containing mean concentrations of approximately 4 mg· l^{-1} N and 0.8 mg·l^{-1} P (Cole et al. 2016a). Over a 12-months period, a tertiary treatment by *Oedogonium* reduced the concentrations of total N and P by 36% and 68%, resulting in 491 kg DW which recovered 24.4 kg N and 4.8 kg P. The algae were used as the feedstock to produce compost and biochar (Cole et al. 2016b)

The most commonly cultivated freshwater genera are currently *Rhizoclonium*, *Cladophora* and *Oedogonium*, the latest being most beneficial for the wastewater bioremediation and valorization (Cole et al. 2015, 2016b; Roberts et al. 2015a, 2015b). *Oedogonium* can have high biomass productivities (up to 35.7 g·$m^{-2}d^{-1}$ DW) when cultured in a variety of wastewater sources in Australia (Fig. 23.2) (Kidgell et al. 2014; Cole et al. 2015, 2016a; Lawton et al. 2017). The commonly researched seaweeds are from the genera *Gracillaria*, *Ulva* and *Sargassum* (Neori et al. 1991; Arumugam et al. 2018). Peak production for the seaweeds under optimal conditions can exceed 50 g DW $m^{-2}d^{-1}$ as was found for *Ulva lactuca* growing in the fishpond effluent in Eilat, Israel (Neori et al. 1991).

23.3.1 Physical Parameters

Light limitation is one of the main controllers of algal performance in the open systems, as the high concentrations of particulate matter in the wastewater effect the intensity of photosynthetically active radiation reaching the algae (Fig. 23.2). High

Fig. 23.3 First large-scale algal turf scrubber (ATS) system at the Patterson municipal wastewater treatment plant, California. (Photo courtesy Rupert J. Craggs). ATS improves water quality by passing a shallow stream of wastewater over the surface of a gently sloped flow-way (Craggs 2001). The system was 154.2 m long and 6.5 m wide, producing up to 62 g/m per day. Mat forming species were mainly cyanobacteria (mostly *Oscillatoria*), with canopy of filamentous algae (*Ulotrix* sp. and *Stigeoclonium* sp. prevailing in the summer) and diatoms (Craggs 2001)

concentrations of algae have a similar shading effect. This is why the inoculation density, culturing and harvesting regimes all contribute to the consistent biomass production. Lower inoculation density should result in a higher relative growth due to the higher availability of light and nutrients. In the study by Neori and co-workers (1991), the daily growth rates of *Ulva lactuca* ranged from 8.7% to 9.9% at a stocking density of 2 kg·m^{-2}, while they were considerably lower (0.6–4%) at higher densities. Favot and co-workers observed best specific growth rates and biomass production at *Ulva* sp. stocking density of 30 g·m^{-2} when studying a range between 15 and 60 g·m^{-2} (2019).

Light reaching the surface of the pond varies diurnally and seasonally. In the temperate regions, illumination periods are significantly shorter in winter, while summer months are critical due to the periods of high illumination causing photoinhibition and photodamage to the Photosystem II. A seasonal growth rates' variation of 1.78–6.23% was observed in *Graciliaria vermiculophylla* by Abreu and co-workers (2011). Temperature is another important environmental variable. Fan and co-workers (2014) found a considerable effect of temperature on the photosynthetic efficiency and N uptake rates in *Ulva prolifera*: in the temperature range of 5–30 °C, N uptake ranged from 177 to 543 mg·kg^{-1} DW h^{-1}.

The substrate depth in the pond is thus a balance among thermal stability, maximum illumination and photo-damaging effects. Although shallower ponds allow better light availability through the water column, the resulting temperature variation and photoinhibition could reverse the resulting positive effects.

23.3.2 Chemical Parameters

Growth rate and algal biomass composition highly depend on the wastewater chemistry. Ross (2017) demonstrated that various nutrient regimes, characteristic of the wastewaters, resulted in different daily growth rates, i.e., 4.75–11.2% in *Cladophora parriaudii* and 3.98–7.37% in *C. coelothrix*. The presence of urea in the medium enhanced growth and yielded a carbohydrate-rich biomass (38–54% DW) (Ross 2017). N/P relative molar ratios are important for the primary production although Liu and Vyverman (2015) found no marked change in the growth under eight different N/P ratios (ranging from 1 to 20): biomass productivities varied between 52.6 and 56.7 $mg \cdot L^{-1}d^{-1}$ DW in *Cladophora* sp. and 29.6–34.1 $mg \cdot L^{-1}d^{-1}$ DW in *Klebsormidium* sp.

Algal productivity depends also on the wastewater nutrients' loading rate. When growing a consortium of freshwater algae, dominated by *Microspora willeana*, *Ulothrix ozonata*, *Rhizoclonium hieroglyphicum* and *Oedogonium* sp. (Fig. 23.3), Mulbry and co-workers found that the mean algal productivity values increased from approximately 2.5 $g \cdot m^{-2}d^{-1}$ DW at the lowest loading rate (0.3 $g \cdot m^{-2}d^{-1}$ TN) to 25 $g \cdot m^{-2}d^{-1}$ DW at the highest loading rate (2.5 $g \cdot m^{-2}d^{-1}$ TN) (2008). Mean nutrient contents in the dried biomass increased 1.5–2-fold with increasing loading rate to a maximum of 7% N and 1% P (Mulbry et al. 2008). Step-feeding of *Chaetomorpha linum* with 10% centrate wastewater increased the biomass productivity by 26.5% compared to the single feeding of the total load (Ge and Champagne 2017).

pH and CO_2 application are significant as well. In *Oedogonium* cultures, maintained at a pH of 7.5 through the addition of CO_2, the biomass productivity was 8.33 ± 0.51 $g \cdot m^{-2}d^{-1}$ DW, which was 2.5 times higher than in control cultures not supplemented by CO_2 (3.37 ± 0.75 $g \cdot m^{-2}d^{-1}$ DW) (Cole et al. 2014). The rate of carbon fixation was 1380 $g \cdot m^{-2}$ $year^{-1}$ C and 1073.1 $g \cdot m^{-2}$ $year^{-1}$ C for cultures maintained at pH 7.5 and 8.5, respectively, and 481 $g \cdot m^{-2}$ $year^{-1}$ C for the control (Cole et al. 2014).

23.3.3 Culture Rotation

A year-round production is imperative for the industrial applications. A combination of species and culture rotation was proposed and successfully applied to mitigate the seasonal changes in light and temperature (Valero-Rodriguez et al. 2020). Valero-Rodriguez and co-workers found tropical *Oedogonium* sp. had highest specific growth rate in the summer conditions (36–40%), but the temperate *Stigeoclonium* sp. and *Hyalotheca* sp. had higher growth in the winter conditions (2020). When mixed, *Oedogonium* was dominant (>90%) in the warmer conditions, while *Stigeoclonium* and *Hyalotheca* prevailed in the colder ones. Their calculations suggest that a monoculture of *Oedogonium* and *Stigeoclonium* would produce 14.1 and 23.9 $t{\cdot}ha^{-1}year^{-1}$, respectively, while a mixed culture would reach 24.4 $t{\cdot}ha^{-1}year^{-1}$, showing a substantial improvement.

23.4 Macroalgae valorization for Agriculture

The utilization of algal biomass depends highly on its composition and active compounds. Both can be regulated by the cultivation conditions (Lawton et al. 2017; Ross 2017). Although wastewater's chemical composition is very variable, it can be to some point regulated by the pre-treatment or addition of inadequate chemicals. In any case, seaweed biomass cultivated in the wastewater should be examined for the multielemental composition before the further utilization (Michalak 2020). Potentially toxic elements are typical heavy metals such as As, Cd, Hg, Pb, but some of them are microelements necessary for the proper growth and development, e.g., Zn, Cu, Mn, Co, (Tuhy et al. 2014; Michalak 2020).

23.4.1 Biomass Composition

The biomass composition varies extensively among the species as found by Atkinson and Smith (1983) analysis of the C:N:P ratio in 92 macroalgae which varied from 183:9:1 to 3550:61:1. Smaller variations can be found also in the same species during different seasons or depending on the N source (Abreu et al. 2011; Ross 2017). The N and P content in biomass can also reflect the N/P ratio in the growth substrate (Liu and Vyverman 2015).

The most obvious difference arises in different N-regimes: macroalgae generally synthesize proteins and pigments when N is sufficient, and accumulate storage polysaccharides, such as starch, when they are under N-limitation (Smit et al. 1997; Cole et al. 2015). Freshwater macroalgae, growing in the nutrient replete media, have high rates of biomass production (often exceeding 15 $g{\cdot}m^{-2}d^{-1}$ DW) and nutrient uptake (Mulbry et al. 2008; Cole et al. 2015). At high productivities, 50–85% of the supplied nitrogen is incorporated into the algal biomass (Cole et al. 2015). Such algae may be suitable for the food and fertilizer, whereas N-starved algae may be better for the conversion into biofuels via digestion or fermentation processes (Ross 2017).

Macroalgae cultivated in the wastewater from animal production can provide a high-quality source of protein (Cole et al. 2015). *Oedogonium* biomass had an equivalent or higher protein quantity and quality than many terrestrial crops currently used as a source of protein in the animal feeds in Cole et al. study (2015). Additionally, *Oedogonium* accumulated calcium, potassium, magnesium and phosphorous.

23.4.2 Fertilizers

Macroalgae used as fertilizers can improve soil water-holding capacity, reduce erosion and nutrient leaching, increase soil organic matter, carbon, nitrogen, phosphorus and minerals, provide a substrate for the soil microbes, resulting in the increased growth and resilience of crops (Lawton et al. 2013; Sharma et al. 2014; Cole et al. 2015; Roberts et al. 2015a, 2015b; Nabti et al. 2016). They have an advantage of being biodegradable, non-toxic, non-polluting and non-hazardous to human, farm animals and birds (Tuhy et al. 2014; Nabti et al. 2016; Badescu et al. 2017).

Macroalgae cultivated in the wastewater are an effective slow-release fertilizer when applied as the untreated dried and milled biomass (Mulbry et al. 2008). Their effects on the plant mass and nutrient content can be equivalent to the effects of commercial fertilizers (Lawton et al. 2017) as they can have relatively high N and P biomass content. *Oedogonium intermedium* cultivated in the wastewater treatment plants, for example, recovered and concentrated in the biomass up to 5.4% N and 1.1% P (Cole et al. 2016a; Neveux et al. 2018).

Algal biomass with the recovered nutrients can be stabilized for the agricultural use by composting or pyrolysis (Cole et al. 2016b). Slow pyrolysis transforms the biomass into biochar. Biochar produced from *Oedogonium* biomass improved the retention of nutrients from fertilizer (N, P, Ca, Mg, K and Mo) in the low-quality soils and enhanced plant growth and the nutrient uptake (Roberts et al. 2015a, 2015b; Lawton et al. 2017). Radishes grown in the low quality, sandy loam soils with added biochar had 35–40% higher growth rates and 10–50% higher concentrations of the essential trace elements (Ca, Mg, K and Mo) and macronutrients compared to the radishes grown without biochar (Roberts et al. 2015a, 2015b; Lawton et al. 2017).

Algal biomass that was used as a biosorbent can be loaded with metal ions and thus an excellent addition to microelements-depleted soils (Tuhy et al. 2014; Badescu et al. 2017). Such biosorbents have additionally a high content of nitrogen and phosphorus and are readily biodegradable material with a high content of organic matter and other macronutrients (Ca, K) (Badescu et al. 2017). The bioavailability of metals is higher than in the traditional organic fertilizers (Tuhy et al. 2014). When *Ulva* sp. with bound Zn(II) ions (29.6 $mg \cdot g^{-1}$ of biomass) was used as a fertilizer, the content of zinc in the soil increased four-times in 8 weeks (Badescu et al. 2017). Similarly, Tuhy and co-workers prepared micronutrient fertilizer from the Baltic seaweeds and post-extraction residues, previously used as biosorbents of Zn(II) ions. Enriched biomass caused the biofortification of zinc and weight increase of garden cress (*Lepidium sativum*).

When the mature compost from the *Oedogonium intermedium*, cultivated in the municipal wastewater, was added to a low fertility soil, it significantly increased the production of sweet corn (*Zea mays*) (Cole et al. 2016b). Treatments receiving half nutrients with compost and half with mineral fertilizers as well as 100% compost treatment produced 4–9 times more corn biomass than when mineral fertilizer alone was added to the low fertility soil. Additional 15% corn productivity was achieved by addition of biochar, most likely due to its ability to bind labile N and P and prevent its loss from the soil (Cole et al. 2016b).

23.4.3 Bioactive Compounds

Plant biostimulants can be found in the macroalgae extracts and are any products that improve (a) nutrient use efficiency, (b) tolerance to abiotic stress, (c) quality traits or/and (d) availability of soil or rhizosphere confined nutrients (EU 2019). Additionally, several algae have been found to exhibit pesticidal activity (Nabti et al. 2016; Hamed et al. 2018). Bioactive compounds are usually obtained by the different methods of extraction and homogenization, which should be preceded by the biomass pre-treatments like washing to remove particles and impurities, drying, shredding, milling to get homogenous sample and sieving (Michalak and Chojnacka 2016). If enriched subfractions or purified preparations of seaweed extracts are used instead of the crude extracts, the problem of accumulation of salts or metals can be minimized (Nabti et al. 2016).

Macroalgae contain plant growth regulators including auxins, gibberellins and cytokinin, the latter being regarded as the most important in marine algae (Sharma et al. 2014; Michalak and Chojnacka 2016; Hamed et al. 2018). The extracts can promote shoot and root elongation, stimulate seed germination and root development, enhancement of frost and draught resistance, increased nutrient uptake and control of phytopathogenic fungi, bacteria, viruses, insects or other pests and restoration of the plant growth under high salinity stress (for more details see Sharma et al. 2014; Nabti et al. 2016; Hamed et al. 2018).

23.5 Conclusions and Future Perspectives

The on-land cultivation of macroalgae, especially in the wastewater, is still in its infancy. The scarce efforts nevertheless show a great promise. Macroalgae can easily grow in various wastewaters by recycling the nutrients and their productivity can be fairly impressive. Biomass composition is suitable for a range of products and can be further adjusted by the cultivation parameters.

Identifying the algae with best performance in a certain wastewater and environment is the first critical step in the wastewater cultivation, which usually starts with a laboratory screening and should be finally tested in the outdoor conditions, preferably over a whole annual cycle (Borowitzka 2013; Fort et al. 2019). First successful attempts of the annual macroalgae production in the seasonal climate areas have already been made by the culture rotation and utilization of the dominant species isolated from the local environment and should be further explored.

Using wastewater as the cultivating media is a sustainable way for nutrients' recycling and could pave a way to marketable prices. The issue is its very variable chemical composition. Although many strains of macroalgae can function well in such an environment, it is difficult to provide the biomass of constant quality and quantity, necessary for the marketable applications. Different pre-treatments can enable more controlled biomass composition and yield, including a multistep-algal-cultivation system where the biomass production ponds can be preceded by the wastewater treatment ponds.

The IMTA concept encompasses innovative and sustainable idea of utilizing trophically connected organisms to remove excess nutrients and waste from the aquatic environment by valorizing their biomass for different products. IMTA validity and great promise has already been thoroughly demonstrated. Its commercial implementation, especially in the western world, is nevertheless lagging. Improvements in the up-scaling of tested systems, stimulation of innovations in the methods of the different cultures, public outreach to improve general acceptance of aquaculture and a specific eco-label certifying the sustainability of IMTA are the main measures that could drive the industrial-scale adoption (Dunbar et al. 2020).

Macroalgae effectively uptake macro- and microelements from the wastewater, which makes them a rich-nutrition substrate for agricultural production or consumption, but such cultivation can also result in the biomass with toxic substances. Regular chemical analysis of the produced biomass is thus necessary to ensure a quality product. In the case of bioactive compounds, extraction and purification can eliminate the unwanted compounds.

For the algal system optimization, continuous monitoring of the parameters like oxidation-reduction potential, electric conductivity and oxygen concentration can enable optimal application of wastewater. Physiology measurements can be used to monitor algal performance, together with the nutrient concentrations, pH regulation with CO_2 etc. to fine-tune the bioremediation and biomass production.

More research is needed to further understand the roles of influence parameters. Recently developed kinetic models of algal and algal bacterial processes have high predictive power and provide deep insight into the actual processes. Model based control algorithms together with the information on environmental conditions enable significant increase in the productivity of algal ponds (Casagli et al. 2021) and can be used to further develop the large-scale high-production systems fed by the waste streams.

Although there is still a lot of work before the viable and marketable large-scale macroalgae cultivation in the wastewater is achieved, it is worthwhile goal to pursue. The same approach is already widely researched and developed with the microalgae, rapidly gaining in importance as it is supporting the care for our health and the environment.

Acknowledgements The Authors acknowledge projects: INTEGRATE - *Integrate Aquaculture: an eco-innovative solution to foster sustainability in the Atlantic Area*, funded by the ERDF through the INTERREG Atlantic Area 2014-2020 Programme (project grant number EAPA_232/2016); Water2Return - *Recovery and recycling of nutrients: turning waste water into added-value products for a circular economy in agriculture* (H2020 2017-2022) and LIFE AlgaeCan - Adding sustainability to the fruit and vegetable processing industry through solar-powered algal wastewater treatment (LIFE 16/ENV/EC 2017-2021).

References

Abreu MH, Pereira R, Buschmann A, Sousa-Pinto I, Yarish C (2011) Nitrogen uptake responses of *Gracilaria vermiculophylla* (Ohmi) Papenfuss under combined and single addition of nitrate and ammonium. J Exp Mar Biol Ecol 407:190–199

Arumugam N, Chelliapan S, Kamyab H, Thirugnana S, Othman N, Nasri NS (2018) Treatment of wastewater using seaweed: a review. Int J Environ Res Public Health 15:2851

Atkinson M, Smith S (1983) C: N: P ratios of benthic marine plants. Limnol Oceanogr 28:568–574

Badescu IS, Bulgariu D, Bulgariu L (2017) Alternative utilization of algal biomass (*Ulva* sp.) loaded with Zn(II) ions for improving of soil quality. J Appl Phycol 29:1069–1079

Borowitzka MA (2013) Species and strain selection. In: Borowitzka MA, Moheimani NR (eds) Algae for biofuels and energy. Developments in applied phycology, vol 5, pp 77–89. Springer Dordrecht Heidelberg New York London

Casagli F, Zuccaro G, Bernard O, Steyer JP, Ficara E (2021) ALBA: a comprehensive growth model to optimize algae-bacteria wastewater treatment in raceway ponds. Water Res 190:116734

Chopin T, Buschmann AH, Halling C, Troell M, Kautsky N, Neori A, Kraemer GP, Zertuche-González JA, Yarish C, Neefus C (2001) Integrating seaweeds into marine aquaculture systems: a key toward sustainability. J Phycol 37:975–986

Cole AJ, Mata L, Paul NA, de Nys R (2014) Using CO_2 to enhance carbon capture and biomass applications of freshwater macroalgae. GCB Bioenergy 6:637–645

Cole AJ, de Nys R, Paul NA (2015) Biorecovery of nutrient waste as protein in freshwater macroalgae. Algal Res 7:58–65

Cole AJ, Neveux N, Whelan A, Morton J, Vis M, de Nys R, Paul NA (2016a) Adding value to the treatment of municipal wastewater through the intensive production of freshwater macroalgae. Algal Res 20:100–109

Cole AJ, Paul NA, de Nys R, Roberts DA (2016b) Good for sewage treatment and good for agriculture: algal based compost and biochar. J Environ Manag 200:105–113

Craggs RJ (2001) Wastewater treatment by algal turf scrubbing. Water Sci Technol 44(11–12):427–433

Dunbar MB, Malta E-J, Agraso MM, Brunner L, Hughes A, Ratcliff J, Johnson M, Jacquemin B, Michel R, Cunha ME, Oliveira G, Ferreira H, Lesueur M, Lebris H, Luthringer R, Soler A, Edwards M, Pereira R, Abreu H (2020) Defining integrated multi-trophic aquaculture: a consensus. Aquac Europe 45(1):22–27

Edwards P (2009) Traditional Asian aquaculture. In: Burnell G, Allan G (eds) New technologies in aquaculture. Woodhead Publishing Series in Food Science, Technology and Nutrition, Woodhead Publishing, pp 1029–1063

EU Regulation 2019/1009 of the European Parliament and of the Council of 5 June 2019 laying down rules on the making available on the market of EU fertilising products and amending Regulations (EC), No1069/2009 and (EC) No1107/2009 and repealing Regulation (EC) No 2003/2003

Fan X, Xu D, Wang Y, Zhang X, Cao S, Mou S, Ye N (2014) The effect of nutrient concentrations, nutrient ratios and temperature on photosynthesis and nutrient uptake by *Ulva prolifera*: implications for the explosion in green tides. J Appl Phycol 26:537–544

Favot G, Cunha ME, Quental-Ferreira H, Álvares Serrão ME (2019) Production of *Ulva* sp. in multitrophic aquaculture in earth ponds. Aquac Fish Stud 1(1):1–8
Fort A, Lebrault M, Allaire M, Esteves-Ferreira A, McHale M, Lopez F, Fariñas-Franco JM, Alseekh S, Fernie A, Sulpice R (2019) Extensive variations in diurnal growth patterns and metabolism amongst *Ulva* spp strains. Plant Physiol 180:109–123
Ge S, Champagne P (2017) Cultivation of the marine macroalgae *Chaetomorpha linum* in municipal wastewater for nutrient recovery and biomass production. Environ Sci Technol 51(6):3558–3566
Ge S, Madill M, Champagne P (2018) Use of freshwater macroalgae *Spirogyra* sp. for the treatment of municipal wastewaters and biomass production for biofuel applications. Biomass Bioenergy 111:213–223
Hamed SM, Abd El-Rhman AA, Abdel-Raouf N, Ibraheem IBM (2018) Role of marine macroalgae in plant protection & improvement for sustainable agriculture technology. Beni-Suef Univ J Basic Appl Sci 7:104–110
Kidgell JT, de Nys R, Hu Y, Paul NA, Roberts DA (2014) Bioremediation of a complex industrial effluent by biosorbents derived from freshwater macroalgae. PLoS One 9(6):e94706
Lawton RJ, de Nys R, Paul NA (2013) Selecting reliable and robust freshwater macroalgae for biomass applications. PLoS One 8(5):e64168
Lawton RJ, Cole AJ, Roberts DA, Paul NA, de Nys R (2017) The industrial ecology of freshwater macroalgae for biomass applications. Algal Res 24(Part B):486–491
Liu J, Vyverman W (2015) Differences in nutrient uptake capacity of the benthic filamentous algae *Cladophora* sp., *Klebsormidium* sp. and *Pseudanabaena* sp. under varying N/P conditions. Bioresour Technol 179:234–242
Luo MB, Liu F, Xu ZL (2012) Growth and nutrient uptake capacity of two co-occurring species, *Ulva prolifera* and *Ulva linza*. Aquat Bot 100:18–24
Michalak I (2020) The application of seaweeds in environmental biotechnology. In: Bourgougnon N (ed) Seaweeds around the world: state of art and perspectives. Advances in botanical research, vol 95. Elsevier Academic Press, Elsevier Ltd., 85–111
Michalak I, Chojnacka K (2016) The potential usefulness of a new generation of agro-products based on raw materials of biological origin. Acta Sci Pol Hortorum Cultus 15:97–120
Mulbry W, Kondrad S, Pizarro C, Kebede-Westhead E (2008) Treatment of dairy manure effluent using freshwater algae: algal productivity and recovery of manure nutrients using pilot-scale algal turf scrubbers. Bioresour Technol 99:8137–8142
Nabti E, Jha B, Hartmann A (2016) Impact of seaweeds on agricultural crop production as biofertilizer. Int J Environ Sci Technol 14:1119–1134
Neori A, Cohen I, Gordin H (1991) *Ulva lactuca* biofilters for marine fishpond effluents. II. Growth rate, yield and C: N ratio. Bot Mar 34:483–490
Neori A, Troell M, Chopin T, Yarish C, Critchley A, Buschmann AH (2007) The need for a balanced ecosystem approach to blue revolution aquaculture. Environment 49:37–43
Neveux N, Bolton JJ, Bruhn A, Roberts DA, Ras M (2018) The bioremediation potential of seaweeds: recycling nitrogen, phosphorus, and other waste products. In: La Barre S, Bates SS (eds) Blue biotechnology: production and use of marine molecules, vol 1. Wiley Online Library 217–239
Oswald WJ, Gotass HG (1957) Photosynthesis in sewage treatment. Trans Am Soc Civ Eng 122:73–105
Pedersen MF, Borum J (1997) Nutrient control of estuarine macroalgae: growth strategy and the balance between nitrogen requirements and uptake. Mar Ecol-Prog Ser 161:155–163
Roberts DA, Paul NA, Cole AJ, de Nys R (2015a) From waste water treatment to land management: conversion of aquatic biomass to biochar for soil amelioration and the fortification of crops with essential trace elements. J Environ Manag 157:60–68
Roberts DA, Paul NA, Dworjanyn SA, Bird MI, de Nys R (2015b) Biochar from commercially cultivated seaweed for soil amelioration. Sci Rep 5:9665

Ross ME (2017) Wastewater treatment by filamentous macro-algae. PhD, University of Edinburgh & SAMS
Shama A, Joyce SG, Mari FD, Ranga Rao A, Ravishankar GA, Hudaa N (2019) Macroalgae and microalgae: novel sources of functional food and feed. In: Ravishankar GA, Ranga Rao A (eds) Handbook of algal technologies and phytochemicals. Food, health and nutraceutical applications, vol I. CRC Press, USA 207–219
Sharma HSS, Fleming C, Selby C, Rao JR, Martin T (2014) Plant biostimulants: a review on the processing of macroalgae and use of extracts for crop management to reduce abiotic and biotic stresses. J Appl Phycol 26:465–490
Smit AJ, Robertson BL, du Preez DR (1997) Influence of ammonium-N pulse concentrations and frequency, tank condition and nitrogen starvation on growth rate and biochemical composition of *Gracilaria gracilis*. J Appl Phycol 8:473–481
Soto D (2009) Integrated mariculture: a global review. FAO Fisheries and Aquaculture Technical Paper. No. 529. FAO, Rome 183
Tuhy L, Samoraj M, Michalak I, Chojnacka K (2014) The application of biosorption for production of micronutrient fertilizers based on waste biomass. Appl Biochem Biotechnol 174:1376–1392
Valero-Rodriguez JM, Swearer SE, Dempster T, de Nys R, Cole AJ (2020) Evaluating the performance of freshwater macroalgae in the bioremediation of nutrient-enriched water in temperate environments. J Appl Phycol 32:641–652
Wallentinus I (1984) Comparisons of nutrient uptake rates for Baltic macroalgae with different thallus morphologies. Mar Biol 80:215–225
Xiao X, Agusti S, Lin F, Li K, Pan Y, Yu Y, Zheng Y, Wu J, Duarte CM (2017) Nutrient removal from Chinese coastal waters by large scale seaweed aquaculture. Sci Rep 7:46613

Chapter 24
Production and Evaluation of Seaweed-Containing Plant Growth Adjuvant Formulation

Marta Illera-Vives, Adolfo López-Fabal, Felicia Fonseca, and M. Elvira López-Mosquera

Abbreviations

AC	Aeration capacity
AGR	Germination rate
ARLP	Average root length per plant
AWP	Average weight per plant
C. baccata	*Cystoseira baccata*
EAW	Easily available water
EC	Electrical conductivity
GeI	Germination inhibition
GrI	Growth inhibition
MLV	Munoo–Liisa vitality
PGAF	Plant Growth Adjuvant Formulation
S	Shrinkage
sps.	species
UW	Unavailable water
v:v	volume:volume = volumetric fraction
WBC	Water buffer capacity

M. Illera-Vives (✉) · M. E. López-Mosquera
Instituto de Biodiversidad Agraria y Desarrollo Rural (IBADER), Universidad de Santiago de Compostela, Lugo, Spain
e-mail: marta.illera.vives@usc.es

A. López-Fabal
Departamento de Producción Vegetal y Proyectos de Ingeniería, Escuela Politécnica Superior de Ingeniería, Universidad de Santiago de Compostela, Lugo, Spain

F. Fonseca
Centro de Investigação de Montanha (CIMO), Instituto Politécnico de Bragança, Bragança, Portugal

A. Ranga Rao, G. A. Ravishankar (eds.), *Sustainable Global Resources of Seaweeds Volume 1*, https://doi.org/10.1007/978-3-030-91955-9_24

24.1 Introduction

Seaweed has long been known to be an effective fertilizer improving soil quality and also a biostimulant of plant growth. Seaweed contains all the macro and micronutrients needed for plant growth (Verkleij 1992). Also, it has nitrogen (N) content similar to that of most animal manures, high content in potassium (K) (especially brown seaweed) and a modest amount of phosphorus (P) (Stephenson 1974; Senn and Kingman 1978). Seaweed improves soil quality, mainly as a result of its high content in phycocolloids resulting in increased structure, water retention and exchange capacity (Metting et al. 1990; Lynn 1972; Khan et al. 2009). By virtue of its high Calcium (Ca) content and the presence of alginates capable of binding aluminium, seaweeds additionally have liming effects (Crouch et al. 1990; López-Mosquera and Pazos 1997; Eyras et al. 1998). Also, it facilitates the growth of beneficial microbiota, thereby increasing biological activity in soil (Kuwada et al. 2000, 2006; Chen et al. 2003; Khan et al. 2009).

In addition to the previously mentioned benefits, seaweeds have bio-stimulating effects owing to their high contents of vitamins (Bourgougnon et al. 2011), plant hormones, quaternary ammonium compounds (e.g., proline, betaines), trace elements, and lipid-base molecules, having a favourable impact on seed germination (Hernández-Herrera et al. 2019), aerial growth (Basak 2008), root development (Finnie and Van Staden 1985; Crouch and Van Staden 1991), nutrient uptake (Castaings et al. 2011; Di Stasio et al. 2017), flowering (Basak 2008), and tolerance of abiotic stress such as that caused by salinity (Yildiztekin et al. 2018) or drought (Xu and Leskovar 2015). These effects influence the improvement of plant yield and quality of crops (Chouliaras et al. 2009; Rouphael et al. 2018), the extension of their lifespan and the increaseof their nutrient contents(Kamel 2014),especially in crops growing in below optimal conditions(Crouch and Van Staden 1994; Craigie 2011).

In agriculture, seaweed has traditionally been used as an amendment or fertilizer. At present, a number of commercial fertilizers include seaweed flour, compost or extract as a biostimulant in their formulations. Although soil-less cropping has grown steadily in recent decades, few studies have addressed the impact of the presence of biostimulants in growing media (Rady and Rehman 2016), even though they improve crop growth while minimizing the needed for fertilizers and phytosanitary products. Seaweed can in fact be an interesting component of plant growing substrates since it is a natural, biodegradable, nutrient- and biostimulant-rich, and pathogen- and weed-free material that is toxic to neither humans nor animals, so it can facilitate soil-less cropping while helping fight pests and prevent plant diseases. Also, the rheological properties of seaweed could increase the water retention capacity of substrates. Finally, seaweed occasionally accumulates in unwanted zones as waste that could be valorized as a growth media.

However, using seaweed as a substrate component can pose problems arising from its high content in salts and, in polluted areas, also in heavy metals (García and

Martel 2000; Sudharsan et al. 2012). Growth hormones, they possess, can also produce phytotoxic effects if they are present in high amounts. Besides, untreated seaweed is highly biologically unstable and can raise problems such as poor nitrogen fixation, anoxia and the release of phytotoxic substances such as organic acids (Zucconi et al. 1981a, 1981b) or hydrogen sulphide (Craigie 2011).

The aim of this work was to produce and evaluate the agricultural performance of seaweed-based growing substrates. The substrates were made with fresh brown seaweed of *Cystoseira baccata*, a species widely distributed in the NW Atlantic region (García-Fernández and Bárbara 2016), and beach wrack collected from shellfish cultivation areas on the coast of Galicia (NW Spain) consisting mainly of two *Ulva* sps.

24.2 Details of Experimentation

24.2.1 Preparation of Substrates

The raw material used to prepare the substrates was coir consisting of coconut peat and fibre (PeleMix ¾"), which was inert enough to clearly expose the chemical and biostimulating properties of seaweed. The material was supplied with two types of seaweed, namely:

- *Cystoseira baccata* (S.G. Gmelin) P.C. Silva, a brown seaweed species-rich in growth hormones (especially cytokinins) that was obtained in fresh form from a coastal area in A Coruña, NW Spain (43° 21′ 47.5″ N, 8° 20′ 46.5″ W).
- *Ulva* sps. Obtained from shellfish cultivation areas in the Ria of Pontevedra, NW Spain (42° 25′ 32″ N, 8° 41′ 07″ W), and consisting largely of *Ulva rigida* C. and *U. intestinalis* L. This beach wrack was partly stable as it had been air-dried and stored for about 6 months after have been removed from the beach and prior to be used.

The high salinity of *C. baccata* required washing the seaweed before use. Both materials were finely chopped and mixed with PeleMix ¾", which was also used, unmixed, as a control substrate. The seven treatments used were as follows:

- CC. 100% coconut coir (PeleMix ¾").
- 10C. 10% *C. baccata* / 90% coir (v:v).
- 25C. 25% *C. baccata* / 75% coir (v:v)
- 50C. 50% *C. baccata* / 50% coir (v:v)
- 10 U. 10% *Ulva* / 90% coir (v:v)
- 25 U. 25% *Ulva* / 75% coir (v:v)
- 50 U. 50% *Ulva* / 50% coir (v:v)

After mixing, the substrates were supplied with water to the container's capacity and allowed to stand for 1 month, moisture levels being maintained by adding water as required, the mass being turned over at 2-day intervals to facilitate aeration and the temperature being measured periodically. Then, the substrates were assessed for stability by using the self-heating test of Brinton et al. (1995).

24.2.2 Chemical characterization of Substrates

Once the substrates were checked to be stable, they were characterized in chemical terms for pH, electrical conductivity (EC), and contents in organic matter (OM) and elements soluble in water and $CaCl_2$ + DTPA (N-NH_4^+, K^+, Ca^{2+}, Mg^{2+}, Na^+, Cl^- and PO_4^{3-}) according to European standards (viz. EN 13037 2011; EN 13038 2011; EN 13039 2011; EN 13652 2001). Total C and N were determined with a Lecco 2000 autoanalyzer following attack with H_2SO_4 (Thomas et al. 1967); total Ca, Mg, Na, and K by ICP-OES and P colorimetrically (Chapman and Pratt 1997); and heavy metals (Cd, Cu, Cr, Hg, Mn, Ni, and Zn) determined by ICP-MS after attack with HNO_3.

Dry bulk density, particle density, total porosity, water retention properties, and shrinkage were determined in accordance with standard EN 13041. Water retention measurements were made at a suction pressure of 1, 3, 5, 7.5, and 10 kPa, and moisture contents were expressed a volumetric fractions (v:v). Aeration capacity (AC) was calculated as the difference between total porosity and volumetric moisture content (v:v) at 1 kPa; and easily available water (EAW), water buffer capacity (WBC), and unavailable water (UW) were taken to be the volumetric fractions of moisture retained by each substrate at 1–5 kPa, 5–10 kPa and >10 kPa, respectively (Felipó et al. 1979).

24.2.3 Agronomic Evaluation of Substrates

The agronomic potential of the substrates was assessed, and their interactions with plants were examined, with the tests described below.

24.2.3.1 Germination

Phytotoxicity Test

It was carried out following the methodology described in EN 16086-2 (2011) using cress (*Lepidium sativum*) as an indicator species. Screened, wetted samples of the different substrates were placed together with 10 cress seeds in Petri dishes in triplicate, using peat as control treatment. The dishes were tilted vertically by 70–80° and incubated at 25 ± 5 °C for 72 h. Then, they were used to determine the average germination rate (AGR), average root length per plant (ARLP), and Munoo–Liisa vitality (MLV) index, the last being calculated as follows:

$$MLV(\%) = \frac{(GR_{s1} \times RL_{s1}) + (GR_{s2} \times RL_{s2}) + (GR_{s3} \times RL_{s3})}{3 \times (GR_c \times RL_c)} \times 100$$

where GR_{si} is the germination rate of replicate i in each treatment, GR_c that of the control treatment (peat), RL_{si} the average root length in replicate i in each treatment and RL_c that of the control treatment.

Growth Test

It was performed on Chinese cabbage (*Brassica rapa* spp. pekinensis) following the methodology "pot experiment with direct use of the prepared sample" of the EN 16086-1 (2011) Thus, a total of 20 Chinese cabbage seeds were sown in triplicate in the different substrates and grown under controlled conditions. After 5 and 12 days, each pot was assessed for average germination rate (AGR) or the average number of germinated seeds in the 3 replicates for each treatment. The results were used to calculate germination inhibition as follows:

$$GeI(\%) = \frac{(AGR_{control} - AGR_{sample})}{AGR_{control}} \times 100$$

24.2.3.2 Growth

Plant growth tests were performed on two different crops, namely: Chinese cabbage and cucumber.

Chinese Cabbage (*Brassica rapa spp. pekinensis*)

One month after cabbage seeds were sown in the previous test, the influence of the different substrates on plant growth was assessed in terms of plant survival, number of leaves per plant, aerial and root length, and fresh and dry weight of the aerial and root portions. The average weight per plant (AWP) for the substrate and control treatments were used to calculate growth inhibition as follows:

$$GrI(\%) = \frac{APW_{control} - APW_{sample}}{APW_{control}} \times 100$$

Cucumber (*Cucumis sativus*)

Cucumber plants were used to compare growth in the different substrates with or without added fertilizer in order to isolate the potential fertilizing effect of the seaweed-enriched substrates. The experiment was conducted in two seed trays of 14 × 9 cells each. One tray was filled with seven substrates unfertilized (18 cells per substrate) and the other with the substrates that were supplied with 1.5 g L^{-1} substrate of NPK liquid fertilizer 15-10-20. After 45 days, each treatment was assessed for number of leaves per plant, length of the aerial and root portions, and fertilizer efficiency, the last being calculated from the following equation:

$$Efficiency = \frac{average\ length\ of\ fertilized\ plants - average\ length\ of\ unfertilized\ plants}{Average\ length\ of\ unfertilized\ plants}$$

24.2.4 Statistical Analysis

Raw data were processed with the software SPSS Statistics v. 23.0 from IBM. Means for different treatments were compared via analysis of variance (ANOVA) and significant differences ($p < 0.05$) identified by using Duncan's test after checking for normality with the Kolmogorov–Smirnov test and variance homoscedasticity with Levene's test. Data resulting in non-homoscedastic variances were subjected to the Games–Howell test.

24.3 Experimental Results

24.3.1 Raw Materials

Both types of seaweed contained greater amounts of nutrients than coconut coir. Thus, *C. baccata* was richer in K and P, and so was *Ulva* in Ca and Magnesium (Mg). *Cystoseira baccata* additionally had a high content in Na —much higher than that of *Ulva*, which was even poorer in this element than was coir. Both materials had low contents in heavy metals —by exception *C. baccata* had an increased content in Cd (2.55 mg kg^{-1}). In any case, washing reduced the electrical conductivity (EC) of the two materials below the maximum thresholds for use as growing substrates (Table 24.1).

Table 24.1 Chemical characterization of the raw materials used to prepare the mixed substrates (seaweed washed and chopped)

	Cystoseira baccata	*Ulva* sps.	Coconut coir
CBD (g L^{-1})[1]	423.05	538.33	488.94
EC (dS m^{-1})[2]	0.47	0.25	0.39
C (%)	40.31	37.63	51.09
N (%)	2.12	2.32	0.73
C/N	19.00	16.20	69.99
P (%)	0.14	0.01	0.06
K (%)	1.15	0.41	0.93
Ca (%)	1.64	2.11	0.42
Mg (%)	0.76	0.90	0.16
Na (%)	0.73	0.19	0.28
Cd (mg kg^{-1})	2.55	0.02	0.39
Cu (mg kg^{-1})	0.77	1.30	7.57
Ni (mg kg^{-1})	3.44	2.27	4.01
Pb (mg kg^{-1})	3.07	0.00	10.28
Zn (mg kg^{-1})	11.07	8.67	44.82
Hg (mg kg^{-1})	0.02	0.01	0.06
Cr (mg kg^{-1})	4.37	10.15	20.05

All values except those for CBD and EC are expressed on a dry weight basis
[1]Laboratory compacted bulk density according to UNE 13040
[2]Electrical conductivity

24.3.2 Characterization of Mixed Substrates

As can be seen from Table 24.2, adding either type of seaweed increased the pH of the substrates. This was especially so with *C. baccata*, a 50% proportion of which raised the pH to 7.28. This proportion of seaweed additionally increased electrical conductivity (EC) to 1.16 dS m^{-1}. The amount of organic matter —and hence carbon— in the substrates decreased with increasing proportion of seaweed but remained above the threshold recommended by Raviv et al. (1986), 80%, in all cases. Both types of seaweed facilitated N release to a similar extent, so they decreased the C/N ratio in similar proportions.

As regards nutrients, *C. baccata* increased the contents in P and K, and both seaweed species increased those in Ca, Mg and Na —the last was especially markedly raised by *C. baccata*. Adding either seaweed to coir increased the content in nitrate and decreased that in ammonium ion among species soluble in $CaCl_2$–DTPA. Also, both seaweed species increased the content in soluble K with increasing proportion in the mixed substrates. *Cystoseira baccata* additionally raised the content in Na —and hence EC. As regards heavy metals, *C. baccata* substantially increased Cd levels, which must be borne in mind to avoid its potential hazards, if any.

Table 24.2 Chemical characterization of the substrates

	CC	10C	25C	50C	10 U	25 U	50 U
Moisture (%)	86.91	86.48	85.84	84.76	86.47	85.80	84.68
pH	6.03	6.24	6.84	7.28	6.31	6.69	6.99
EC (dS m^{-1})	0.39	0.37	0.64	1.16	0.30	0.33	0.39
C (% dm)	51.09	51.05	47.92	48.20	51.42	50.02	49.21
OM (% dm)	87.87	87.80	82.42	82.91	88.44	86.03	84.64
N (% dm)	0.73	0.87	1.08	1.43	0.89	1.13	1.53
C/N	69.99	58.68	44.37	33.71	57.77	44.26	32.16
P (% dm)	0.06	0.07	0.08	0.10	0.06	0.05	0.04
K (% dm)	0.93	0.96	0.99	1.04	0.88	0.80	0.67
Ca (% dm)	0.42	0.54	0.73	1.03	0.59	0.84	1.26
Mg (% dm)	0.16	0.22	0.31	0.46	0.23	0.34	0.53
Na (% dm)	0.28	0.32	0.39	0.51	0.27	0.26	0.23
Soluble elements extracted by $CaCl_2$ + DTPA (mg L^{-1} substrate)							
NH_4^+	7.75	3.95	4.30	8.65	10.25	12.10	11.70
NO_3^-	10.55	13.70	7.30	2.20	13.85	4.95	1.05
Mg^{2+}	63.72	125.83	194.62	352.39	114.33	353.30	501.72
K^+	1945.8	2742.8	2267.8	2868.4	3061.8	1527.90	1264.9
Na^+	203.77	396.54	467.23	685.77	357.34	151.31	181.78
PO_4^{3-}	7.58	4.93	2.45	4.75	13.35	33.33	17.98
Heavy metals (mg kg^{-1} dm)							
Cd[a] (mg kg^{-1})		0.61	0.93	1.47			

[a]Calculated from its proportion in the substrate

Table 24.3 Physical characterization of the substrates

	CC	10C	25C	50C	10 U	25 U	50 U
BD (g L^{-1})	86.65	81.95	73.93	76.23	78.45	72.68	74.07
PD (g L^{-1})	1944.9	1945.3	1977.5	1974.6	1941.5	1964.1	1955.8
Porosity (% v/v)	95.54	95.79	96.26	96.14	95.96	96.3	96.21
CA (% v/v)	39.6	37.71	38.83	42.46	39.99	43.19	48.17
EAW (% v/v)	17.88	18.34	18.5	16.6	19.05	17.46	15.11
WBC (% v/v)	1.65	2.96	4.09	3.67	1.99	2.01	1.81
UW (% v/v)	3642	36.77	34.85	33.41	34.93	33.64	31.13
R (kPa)	2.14	2.44	2.31	1.93	2.05	1.73	1.05
S (% v/v)	11.77	15.32	18.36	21.11	16.61	18.43	19.59

BD bulk density, *PD* particle density, *CA* aeration capacity, *EAW* easily available water, *WBC* water buffer capacity, *UW* unavailable water, *R* suction that equalizes the water and air contents, *S* shrinkage

As can be seen from Table 24.3, the physical properties of the substrates were not appreciably altered by the addition of either type of seaweed. Porosity was high (96% on average) in all mixed substrates. *Ulva* decreased the total amount of water retained and, especially, that of unavailable water (UW), which was 5.2% smaller in the substrate containing 50% seaweed than in coir. The presence of either seaweed increased the aeration capacity of the substrates, which was 8.6% higher with a proportion of 50% *Ulva*. *R*, which is a measure of water availability at low pressures, ranged from 1 to 3 kPa in the mixed substrates. *R* values above 3 kPa are suggestive of root anoxia by the effect of excessive moisture, and values below 1 of moisture scarcity (Ansorena Miner 1994). As can be seen in Table 24.3, *R* decreased and aeration increased as a result of increasing proportion of green seaweed. The presence of either seaweed, but particularly *C. baccata*, increased shrinkage (*S*) with increasing proportion in the mixed substrates; in any case, *S* was invariably lower than 30%, the maximum recommended value (Abad et al. 1992).

24.3.3 Agronomic Evaluation

24.3.3.1 Germination

No signs of phytotoxicity were detected with any of the substrates in either experiment. Also, the germination results exhibited no significant differences. Thus, although germination was always greater with the mixed substrates, the differences were not significant enough for a biostimulating effect to be inferred (Tables 24.4 and 24.5).

As can be seen from Table 24.4, the presence of seaweed in the substrates increased root growth and the Munoo–Liisa vitality index in cress relative to both coconut coir and the control treatment (peat).

With Chinese cabbage (Table 24.5), the highest germination rate after 5 days was obtained in comparison with the control (peat); however, differences from the other treatments were not significant. On the other hand, inhibition of germination was very negligible in most cases. This trend ceased after 10 days, where an increased number of seeds had germinated —several close to the value for the control treatment or even greater.

Table 24.4 Phytotoxicity test on cress. Means ± standard deviation for 3 replicates

Substrate	AGR (%)			ARLP (%)			MLV (%)		
Peat	97.7	± 4.2	a	46.0	± 9.8	b	1.01	± 13.6	b
CC	93.3	± 4.4	a	29.9	± 10.0	a	0.63	± 7.4	a
10C	100.0	± 0.0	a	50.9	± 6.3	Bc	1.15	± 6.3	Bc
25C	100.0	± 0.0	a	54.5	± 11.5	Bc	1.23	± 11.5	Bc
50C	97.7	± 4.2	a	45.6	± 3.6	b	0.99	± 5.4	b
10 U	100.0	± 0.0	a	58.7	± 10.3	c	1.32	± 10.3	c
25 U	97.7	± 4.2	a	54.7	± 8.6	Bc	1.20	± 12.2	Bc
50 U	100.0	± 0.0	a	56.5	± 9.6	Bc	1.27	± 9.6	Bc

AGR average germination rate, ARLP average root length per plant, MLV Munoo–Liisa vitality index. Different letters in each column denote significant differences between treatments (rows) as per Duncan's test at $p < 0.05$

Table 24.5 Germination test on Chinese cabbage. Means ± standard deviation for 3 replicates

	GR (%)						GeI (%)					
Substrate	5 days			10 days			5 days			10 days		
Peat	85.0	± 8.7	a	95.0	± 0.0	a	0.0	± 10.2	a	0.0	± 0.0	Abc
CC	81.7	± 5.8	a	90.0	± 0.0	a	3.9	± 6.8	a	5.3	± 0.0	Bc
10C	78.3	± 2.9	a	91.7	± 2.9	a	7.8	± 3.4	a	3.5	± 3.0	Bc
25C	83.3	± 5.8	a	95.0	± 0.0	a	2.0	± 6.8	a	0.0	± 0.0	Abc
50C	81.7	± 2.9	a	98.3	± 2.9	a	3.9	± 3.4	a	−3.5	± 3.0	Ab
10 U	70.0	± 13.2	a	98.3	± 2.9	a	17.6	± 15.6	a	−3.5	± 3.0	a
25 U	78.3	± 5.8	a	93.3	± 7.6	a	7.8	± 6.8	a	1.8	± 8.0	a
50 U	70.0	± 13.2	a	93.3	± 2.9	a	17.6	± 15.6	a	1.8	± 3.0	c

GR germination rate, *GeI* germination inhibition. Different letters in each column denote significant differences between treatments (rows) as per Duncan's test at $p < 0.05$

Fig. 24.1 Chinese cabbage growth 20 days after sowing in various substrates. The electrical conductivity of the substrates is shown in brackets

24.3.3.2 Plant Growth

Chinese Cabbage

As can be seen in Fig. 24.1, there were visible differences in growth between the Chinese cabbage plants sown in the substrate containing 10% *C. baccata* and those sown in substrates containing high proportions of this seaweed, as well as with plants grown in the *Ulva*-containing substrates. These last, where salinity level was no limiting unlike mixtures with more than 10% of *C. baccata,* grew to a similar extent as the plants sown in peat or coir; their aerial portion exhibited an increased fresh and dry weight, but the differences were not significant in any case (Table 24.6).

Table 24.6 Growth test on Chinese cabbage. Mean ± standard deviation

	Plants per pot			Total length (cm)			Root length (cm)			Root dry weight (g)			Number of leaves			Aerial length (cm)			Aerial dry weight (g)		
Peat	19.00	± 0.00	b	23.02	± 5.71	a	8.[illegible]9	± 2.54	bc	0.16	± 0.23	a	5.35	± 0.97	a	14.82	± 3.73	a	0.28	± 0.21	a
CC	16.00	± 1.00	b	22.48	± 6.91	a	9.41	± 3.90	c	0.03	± 0.01	a	5.10	± 1.08	a	13.07	± 3.83	a	0.36	± 0.07	a
10C	18.33	± 0.58	b	22.22	± 6.54	a	8.60	± 2.95	Bc	0.03	± 0.01	a	5.24	± 1.04	a	13.62	± 4.12	a	0.44	± 0.02	a
25C	2.50	± 0.71	a	22.00	± 6.11	a	7.30	± 2.80	Abc	0.03	± 0.02	a	6.20	± 1.10	a	14.20	± 3.55	a	0.36	± 0.30	a
50C	–			–			–			–			–			–			–		
10 U	19.67	± 0.58	b	21.8[illegible]	± 7.96	a	8.34	± 3.26	Bc	0.04	± 0.00	a	5.00	± 1.29	a	13.47	± 5.13	a	0.43	± 0.03	a
25 U	18.67	± 1.53	b	2[illegible].22	± 7.12	a	6.56	± 2.31	a	0.04	± 0.00	a	5.43	± 1.25	a	14.66	± 5.19	a	0.51	± 0.05	a
50 U	17.00	± 2.65	b	20.90	± 8.52	a	6.96	± 2.89	Ab	0.04	± 0.01	a	5.47	± 1.33	a	13.94	± 5.82	a	0.52	± 0.12	a

Different letters in each column denote significant differences between treatments (rows) as per Duncan's test at $p < 0.05$

Cucumber

As can be seen from Table 24.7, all fertilizer treatments substantially enhanced the overall growth in the cucumber plants, which suggests that the fertilizing effect of the seaweed was visible, though limited. The greatest effect was that on the aerial part, where the lowest proportions of the two types of seaweed increased growth by 105 and 104%, respectively (Table 24.8). The increase was much greater than that observed with coconut coir despite the very small amounts of nutrients supplied by the seaweed in such low proportions (Table 24.2). The presence of *C. baccata* in the substrates additionally increased root growth, the substrates containing it exhibiting a fertilizer efficiency of up to 67%.

Table 24.7 Growth-related parameters for cucumber. Mean ± standard deviation

Treatment	Total length (cm)			Leaves per plant			Root length (cm)			Aerial length (cm)		
CC	21.39	± 3.40	bc	2.39	± 0.50	ef	9.58	± 1.64	abc	11.81	± 2.84	b
10C	16.08	± 3.45	a	1.56	± 0.51	a	7.89	± 2.34	a	8.19	± 1.86	a
25C	24.29	± 2.38	Cde	2.29	± 0.47	Def	10.53	± 1.80	Bcd	13.76	± 2.06	Cd
50C	26.81	± 3.39	Ef	2.56	± 0.51	f	10.36	± 2.40	Bcd	16.44	± 1.76	Ef
10 U	18.36	± 4.71	Ab	1.78	± 0.43	Ab	10.53	± 3.85	Bcd	7.83	± 1.42	a
25 U	17.89	± 2.56	a	1.89	± 0.47	Bc	8.83	± 1.62	Ab	9.06	± 1.50	a
50 U	18.47	± 3.50	Ab	2.00	± 0.00	Bcd	9.00	± 2.72	Ab	9.47	± 2.31	a
F-CC	29.03	± 3.60	Fg	2.39	± 0.50	Ef	12.08	± 1.76	Def	16.94	± 2.81	f
F-10C	29.94	± 5.12	g	2.33	± 0.49	Def	13.17	± 2.95	Def	16.78	± 3.12	f
F-25C	33.08	± 6.84	h	2.61	± 0.50	f	13.42	± 2.88	f	19.67	± 4.27	g
F-50C	31.72	± 7.69	Gh	2.61	± 0.50	f	11.50	± 3.65	Cde	20.22	± 4.71	g
F-10 U	26.58	± 4.43	Ef	2.32	± 0.48	Def	10.58	± 2.90	Bcd	16.00	± 2.82	Ef
F-25 U	23.11	± 4.75	Cd	2.17	± 0.38	Cde	11.03	± 2.68	Cd	12.08	± 2.88	Bc
F-50 U	25.08	± 3.69	de	2.33	± 0.49	Def	10.33	± 2.01	Bcd	14.75	± 2.42	de

F- fertilized substrate. Different letters in each column denote significant differences between treatments (rows) as per Duncan's test at $p < 0.05$

Table 24.8 Fertilizer efficiency as relative growth increase by effect of fertilization. Mean ± standard deviation

Treatment	Total length		Root length		Aerial length	
F-CC	0.36	±0.17 ab	0.26	±0.18 b	0.44	±0.24 ab
F-10C	0.86	±0.32 c	0.67	±0.37 c	1.05	±0.38 c
F-25C	0.36	±0.28 ab	0.27	±0.27 b	0.43	±0.31 ab
F-50C	0.18	±0.29 a	0.11	±0.35 ab	0.23	±0.29 a
F-10 V	0.45	±0.24 b	0.00	±0.28 a	1.04	±0.36 c
F-25 V	0.29	±0.27 ab	0.25	±0.30 b	0.33	±0.32 ab
F-50 V	0.36	±0.20 ab	0.15	±0.22 ab	0.56	±0.26 b

F- fertilized substrate. Different letters in each column denote significant differences between treatments (rows) as per Duncan's test at $p < 0.05$

Comparing treatments with or without added fertilizer separately, plants grown on substrates containing *C. baccata* invariably grew more than those sown on coir; however, the mixed substrates unfertilized never reached the effects of the coir + fertilizer combination —not even in treatment 50 °C, which was the one that grew the most. Seaweed appreciably increased root growth relative to the control treatment. The increase occurred mainly in the aerial portion, in which the substrates containing a 25 or 50% proportion of *C. baccata* compared to the control and control + fertilizer treatments, respectively.

24.4 Implications of our Studies

24.4.1 Characterization of Seaweeds and Mixed Substrates

No appreciable change in temperature —and hence no potentially adverse effect on the integrity of bio stimulating substances— was observed during the stabilization period in the mixed substrates. In fact, at the end all mixtures were stable according to the self-heating test. This result excluded the presence of strong biological changes but not necessarily that of a slow degradation process.

Mixed substrates should be able to efficiently deliver plant nutrients. The initial characterization analyses showed that both types of seaweed were suitable for use in plant growing substrates; in fact, both contained substantial amounts of nutrients such as N, K, Ca and Mg. *Cystoseira baccata* additionally contained some P, which is interesting given the increasing scarcity of sustainable phosphorus sources some authors (Cordell et al. 2009) have proposed using seaweed as a potential source of P in the future. One potential chemical constraint on the selection of new materials to be used as growth media component is their content in heavy metals (Chong 2005); in fact, cultivation in a substrate avoids the dilution effects of soil and raises toxic risks for plants and humans. Because seaweed is a bioaccumulator, it can increase such risks to an extent dependent on the degree of pollution at the collection site. The ability of seaweed to accumulate toxic substances depends on the composition of its cell walls and increases with the presence of alginates, which are abundant in brown seaweed such as *C. baccata* (Davis et al. 2003). European legislation in force (Regulation (EU) 2019/1009) has set maximum tolerated levels for heavy metals and As in substrates to be CE marked. Spanish law has additionally set domestic limits for heavy metals in growing media and discriminated between those which can be used with edible horticultural crops and those which cannot. Here, only the *C. baccata* based substrates contained substantial amounts of Cd that increased with increasing proportion in the substrate —so much so that 50C contained 1.5 mg kg^{-1}, which is near the limit for CE marking. Also, only substrate 10C had a Cd content below the Spanish legal threshold for use with edible crops (Class A, Cd <0.7 mg kg^{-1}); by contrast, 25C and 50C should only be used for other purposes (Class B, Cd <2 mg kg^{-1}).

The physical properties of a growing substrate are especially important because, unlike its chemical properties, they are very difficult to alter. Some deficiencies or excesses can be offset by combining two or more materials to improve the structure and ensure an appropriate air-water balance during irrigation for optimal plant growth in the absence of anoxia or drought stress. Phycocolloids, which are matrix polysaccharides exclusively present in seaweed, can form grids capable of retaining large amounts of water (Verkleij 1992; García and Martel 2000). The main phycocolloids in green and brown seaweed are ulvans and alginates, respectively (Lahaye and Robic 2007; Kumar et al. 2008). The latter can alter the distribution of soil moisture (Nabti et al. 2017). However, brown seaweed has very little effect on the physical properties of coconut coir despite its high aeration capacity and low water retention capacity (Abad et al. 2005). As expected, *C. baccata* increased the total water retention capacity (EAW + WBC + UW) of the substrates, mainly by increasing WBC (by 5–10 kPa); on the other hand, *Ulva* decreased the water retention capacity —by up to 8% in the mixture containing 50% of this seaweed— to the benefit of aeration.

24.4.2 Substrate Performance

The Chinese cabbage experiment revealed inhibited germination relative to peat 5 days after sowing. The effect disappeared at the second sampling (10 days), which suggests that some unknown factor may have not completely suppressed germination but rather delayed it. One such factor might be salinity, which is especially important in seedlings (Sharma et al. 2004). However, there was no correlation between substrate EC and germination. The addition of *C. baccata*, which is highly saline, improved germination and root growth. Some authors have found seaweed to induce resistance against abiotic stresses such as salinity by virtue of its containing biostimulating and osmoprotective substances (Neily et al. 2008; Nabti et al. 2017) that additionally improve root development (Finnie and Van Staden 1985; Crouch and Van Staden 1991). In the longer term, the high enzyme contents of seaweed can facilitate enzymatic hydrolysis of insoluble substances, and hence demineralization and desalination of the medium (Canales-López, 1999).

Despite the good results obtained with both cress and Chinese cabbage, the latter developed increasing problems in substrates 25C and 50C and eventually died in most cases. Also, plant development and EC were correlated, the plants grown in the most saline substrates being those surviving the least. However, the facts that the previous treatment had no apparent adverse effect on germination, that the initial salinity was not too high and that daily irrigation facilitated leaching —and hence rapid removal of excess salinity— suggest that the adverse effect observed was not due to salinity. One plausible reason for the low survival rate may have been the presence in the seaweed of excessive amounts of biostimulants, which promote growth at low concentrations but inhibit it at high concentrations (Battacharyya et al. 2015). Also, the organic matter may have been unstable and decomposed during cropping —one of the greatest shortcomings of using fresh seaweed in growing substrates. In fact, the stability of a substrate is governed by the resistance of its organic matter to microbial degradation (Komilis et al. 2011). The decomposition of fresh organic matter produces secondary metabolites such as low-molecular weight organic acids (e.g., phenolic acids), ammonium ion and some salts high concentrations of which can hinder plant growth (Zucconi et al. 1981a; Luo et al. 2018). Also, an immature growing medium can lead to poor N fixation and an oxygen-deficient rhizosphere (Raviv et al. 1986), thereby also adversely affecting crop growth.

The results of the cucumber experiment, which was conducted on robust plants after germination, were rather different. Thus, plant growth was optimal with *C. baccata*, particularly at the highest proportions in the mixed substrates, and worse with *Ulva*, both with and without added fertilizer. As regards fertilization efficiency, the substrate containing 10% of *C. baccata* clearly excelled all others. Several studies have shown substrates containing seaweed extracts to improve nutrient uptake by plants (Battacharyya et al. 2015).

24.5 Conclusions and Future Prospects

Our results confirm the starting hypothesis that seaweed can be a useful component of plant growing substrates. Seaweed is a natural, biodegradable, pathogen- and weed-free, non-toxic material capable of enriching common substrates such as coconut coir with nutrients. Also, the biostimulating properties of seaweed boost crop growth and nutrient uptake. As with extracts, however, it is very important not to add excessive amounts of seaweed to the substrates to avoid toxic effects. Additional studies are needed to determine both the most advantageous types of seaweed and the most appropriate proportions to obtain the best effects on the crop, depending on the intended use of the substrates.

References

Abad M, Fornes F, Carrión C, Noguera V, Noguera P, Maquieira Á, Puchades R (2005) Physical properties of various coconut coir dusts compared to peat. HortScience 40(7):2138–2144. https://doi.org/10.21273/HORTSCI.40.7.2138

Abad M, Martínez PF, Martínez MD, Martínez J (1992) Evaluación agronómica de los sustratos de cultivo. Actas Hortic 11:141–154

Ansorena Miner J (1994) Sustratos. Propiedades y caracterización. Mundi-Prensa, Madrid, Spain

Basak A (2008) Effect of preharvest treatment with seaweed products, Kelpak® and Goëmar BM 86®, on fruit quality in apple. Int J Fruit Sci 8(1):1–14. https://doi.org/10.1080/15538360802365251

Battacharyya D, Babgohari MZ, Rathor P, Prithiviraj B (2015) Seaweed extracts as biostimulants in horticulture. Sci Hortic 196:39–48. https://doi.org/10.1016/j.scienta.2015.09.012

Bourgougnon N, Bedoux G, Sangiardi A, Stiger-Pouvreau V (2011) Las algas: potencial nutritivo y aplicaciones cosméticas. In: Las Algas como Recurso, Valorización, Aplicaciones Industriales y Tendencias, ed. Fundación CETMAR. Centro Tecnológico del Mar, Spain, pp 81–94

Brinton WF, Evans E, Droffner ML, Brinton RB (1995) A standardized Dewar test for evaluation of compost self-heating. Biocycle 36:1–16

Canales-López B (1999) Enzimas-algas: posibilidades de su uso para estimular la producción agríola y mejorar los suelos. Terra Latinoam 17(3):271–276

Castaings L, Marchive C, Meyer C, Krapp A (2011) Nitrogen signalling in Arabidopsis: how to obtain insights into a complex signalling network. J Exp Bot 62:1391–1397. https://doi.org/10.1093/jxb/erq375

Chapman H, Pratt PF (1997) Métodos de análisis para suelos, plantas y aguas. Trillas, Mexico City

Chen SK, Edwards CA, Subler S (2003) The influence of two agricultural biostimulants on nitrogen transformations, microbial activity, and plant growth in soil microcosms. Soil Biol Biochem 35:9–19. https://doi.org/10.1016/S0038-0717(02)00209-2

Chong C (2005) Experiences with wastes and composts in nursery substrates. HortTechnology 15:739–747. https://doi.org/10.21273/HORTTECH.15.4.0739

Chouliaras V, Tasioula M, Chatzissavvidis C, Therios I, Tsabolatidou E (2009) The effects of a seaweed extract in addition to nitrogen and boron fertilization on productivity, fruit maturation, leaf nutritional status and oil quality of the olive (*Olea europaea* L.) cultivar Koroneiki. J Sci Food Agric 89:984–988. https://doi.org/10.1002/jsfa.3543

Cordell D, Drangert JO, White S (2009) The story of phosphorus: global food security and food for thought. Global Environ Change 19(2):292–305. https://doi.org/10.1016/j.gloenvcha.2008.10.009

Craigie JS (2011) Seaweed extract stimuli in plant science and agriculture. J Appl Phycol 23(3):371–393. https://doi.org/10.1007/s10811-010-9560-4

Crouch IJ, Beckett RP, Van Staden J (1990) Effect of seaweed concentrate on the growth and mineral nutrition of nutrient-stressed lettuce. J Appl Phycol 2(3):269–272. https://doi.org/10.1007/BF02179784

Crouch IJ, Van Staden J (1991) Evidence for rooting factors in a seaweed concentrate prepared from *Ecklonia maxima*. J Plant Physiol 137(3):319–322. https://doi.org/10.1016/S0176-1617(11)80138-0

Crouch IJ, Van Staden J (1994) Commercial seaweed products as biostimulants in horticulture. J Home Consumer Horticult 1:19–73. https://doi.org/10.1300/J280v01n01_03

Davis TA, Volesky B, Mucci A (2003) A review of the biochemistry of heavy metal biosorption by brown algae. Water Res 37(18):4311–4330. https://doi.org/10.1016/S0043-1354(03)00293-8

Di Stasio E, Rouphael Y, Colla G, Raimondi G, Giordano M, Pannico A, El-Nakhel C, De Pascale S (2017) The influence of *Ecklonia maxima* seaweed extract on growth, photosynthetic activity and mineral composition of *Brassica rapa* L. subsp *sylvestris* under nutrient stress conditions. Eur J Horti. Sci 82(6):286–293. https://doi.org/10.17660/eJHS.2017/82.6.3

EN 13037 (2011) Soil improvers and growing media. Determination of pH. European Commitee for Standarization. Brussels

EN 13038 (2011) Soil improvers and growing media. Determination of electrical conductivity. European Commitee for Standarization. Brussels

EN 13039 (2011) Soil improvers and growing media. Determination of organic matter content and ash. European Commitee for Standarization. Brussels

EN 13652 (2001) Soil improvers and growing media. Extraction of water soluble nutrients and elements. European Commitee for Standarization. Brussels

EN 16086-1 (2011) Soil improvers and growing media. Determination of plant response. Part 1: Pot growth test with Chenese cabbage. European Commitee for Standarization. Brussels

EN 16086-2 (2011) Soil improvers and growing media. Determination of plant response. Part 2: Petri dish test using cress. European Commitee for Standarization. Brussels

Eyras MC, Rostagno CM, Defosse GE (1998) Biological evaluation of seaweed composting. Compost Sci Util 6:74–81. https://doi.org/10.1080/1065657X.1998.10701943

Felipó M, Verdonck O, Cappaert I, De Boodt M (1979) Estudio de las propiedades físicas de los sustratos hortícolas. Anal Edaf Agrobiol 38:603–611

Finnie JF, Van Staden J (1985) Effect of seaweed concentrate and applied hormones on in vitro cultured tomato roots. J Plant Physiol 120:215–222. https://doi.org/10.1016/S0176-1617(85)80108-5

García G, Martel A (2000) Usos y aplicaciones de macroalgas, microalgas y cianobacterias en agricultura ecológica. Actas IV Congreso SEAE. https://fci.uib.es/Servicios/libros/conferencias/seae/Usos-y-aplicaciones-de-macroalgas-microalgas-y.cid221515 (Accessed 1 June 2020)

García-Fernández A, Bárbara I (2016) Studies of *Cystoseira* assemblages in northern Atlantic Iberia. Anales del Jardín Botánico de Madrid 73(1):e035. https://doi.org/10.3989/ajbm.2403

Hernández-Herrera RM, Santacruz-Ruvalcaba F, Hernández-Carmona G (2019) Germination and seedling growth responses of tomato (*Solanum lycopersicum* L.) to seaweed extracts applied on seeds. Rev Latinoam Biotecnol Amb Algal 10(1):28–44

Kamel HM (2014) Impact of garlic oil, seaweed extract and imazalil on keeping quality of Valencia orange fruits during cold storage. J Hortic Sci Ornament Plants 6:116–125. https://doi.org/10.5829/idosi.jhsop.2014.6.3.1145

Khan W, Rayirath UP, Subramanian S, Jithesh MN, Rayorath P, Hodges DM, Critchley AT, Craigie JS, Norrie J, Prithiviraj B (2009) Seaweed extracts as biostimulants of plant growth and development. J Plant Growth Regul 28:386–399. https://doi.org/10.1007/s00344-009-9103-x

Komilis D, Evangelou A, Voudrias E (2011) Monitoring and optimizing the co-composting of dewatered sludge: a mixture experimental design approach. J Environ Manag 92:2241–2249. https://doi.org/10.1016/j.jenvman.2011.04.012

Kumar CS, Ganesan P, Suresh PV, Bhaskar N (2008) Seaweeds as a source of nutritionally beneficial compounds – a review. J Food Sci Technol 45(1):1–13

Kuwada K, Utamura M, Matsushita I, Ishii T (2000) Effect of tangle stock ground extracts on in vitro hyphal growth of vesicular–arbuscular mycorrhizal fungi and their in vivo infections of citrus roots. In: Proceedings of the 9th International Society of Citriculture Congress, Orlando, pp 1034–1037

Kuwada K, Wamocho LS, Utamura M, Matsushita I, Ishii T (2006) Effect of red and green algal extracts on hyphal growth of arbuscular fungi, and on mycorrhizal development and growth of papaya and passionfruit. Agron J 98:1340–1344

Lahaye M, Robic A (2007) Structure and functional properties of ulvan, a polysaccharide from green seaweeds. Biomacromolecules 8(6):1765–1774. https://doi.org/10.1021/bm061185q

López-Mosquera ME, Pazos P (1997) Effects of seaweed on potato yields and soil chemistry. Biol Agric Hortic 14(3):199–205. https://doi.org/10.1080/01448765.1997.9754810

Luo Y, Liang J, Zeng G, Chen M, Mo D, Li G, Zhang D (2018) Seed germination test for toxicity evaluation of compost: Its roles, problems and prospects. Waste Manage 71:109–114

Lynn LB (1972) The chelating properties of seaweed extract *Ascophyllum nodosum* vs. *Macrocystis perifera* on the mineral nutrition of sweet peppers, *Capsicum annuum*. Clemson University, Clemson, SC

Metting B, Zimmerman WJ, Crouch IJ, Van Staden J (1990) Agronomic uses of seaweed and microalgae. In: Akatsuka I (ed) Introduction to applied phycology. SPB Academic Publishing, The Hague, pp 589–627

Nabti E, Jha B, Hartmann A (2017) Impact of seaweeds on agricultural crop production as biofertilizer. In J Environ Sci Technol 14(5):1119–1134. https://doi.org/10.1007/s13762-016-1202-1

Neily W, Shishkov L, Tse T, Titus D (2008) Acadian LSC helps reduce salinity stress in pepper seedlings cv. California Wonder PGRSA Newslett 1:14

Rady MM, Rehman H (2016) Supplementing organic biostimulants into growing media enhances growth and nutrient uptake of tomato transplants. Sci Horticult 203:192–198. https://doi.org/10.1016/j.scienta.2016.03.037

Raviv M, Chen Y, Inbar Y (1986) Peat and peat substitutes as growth media for container grown plants. In: Chen Y, Avnimelech Y (eds) The Role of Organic Matter in Modern Agriculture. Springer, pp 257–287. https://doi.org/10.1007/978-94-009-4426-8_11

Rouphael Y, Giordano M, Cardarelli M, Cozzolino E, Mori M, Kyriacou M, Bonini P, Colla G (2018) Plant-and seaweed-based extracts increase yield but differentially modulate nutritional quality of greenhouse spinach through biostimulant action. Agronomy 8(7):126. https://doi.org/10.3390/agronomy8070126

Senn TL, Kingman AR (1978) Seaweed research in crop production, 1958–1978. Economic Development Administration of the US Department of Commerce. Department of Horticulture, Clemson University

Sharma AD, Thakur M, Rana M, Singh K (2004) Effect of plant growth hormones and abiotic stresses on germination, growth and phosphatase activities in *Sorghum bicolor* (L.) Moench seeds. Afr J Biotechnol 3:308–312. https://doi.org/10.5897/AJB2004.000-2057

Stephenson WA (1974) Seaweed in Agriculture and Horticulture, 3rd edn, Pauma Valley, CA, USA: B and G Rateaver (Ed.)

Sudharsan S, Seedevi P, Ramasamy P, Subhapradha N, Vairamani S, Shanmugam A (2012) Heavy metal accumulation in seaweeds and sea grasses along southeast coast of India. J Chem Pharm Res 4(9):4240–4244

Thomas R, Sheard R, Moyer J (1967) Comparison of conventional and automated procedures for nitrogen, phosphorus, and potassium analysis of plant material using a single digestion. Agron J 59:240–243. https://doi.org/10.2134/agronj1967.00021962005900030010x

Verkleij FN (1992) Seaweed extracts in agriculture and horticulture: a review. Biol Agric Hortic 8:309–324. https://doi.org/10.1080/01448765.1992.9754608

Xu C, Leskovar D (2015) Effects of *A. nodosum* seaweed extracts on spinach growth, physiology and nutrition valued under drought stress. Sci Hortic 183:39–47. https://doi.org/10.1016/j.scienta.2014.12.004

Yildiztekin M, Tuna AL, Kaya C (2018) Physiological effects of the brown seaweed *Ascophyllum nodosum* and humic substances on plant growth, enzyme activities of certain pepper plants grown under salt stress. Acta Biol Hungarica 69(3):325–335. https://doi.org/10.1556/018.68.2018.3.8

Zucconi F, Forte M, Monaco A, De Bertoldi M (1981a) Biological evaluation of compost maturity. Biocycle 22:27–29

Zucconi F, Pera A, Forte M, De Bertoldi M (1981b) Evaluating toxicity of immature compost. Biocycle 22:54–57

Chapter 25
Leveraging Seaweeds as a Potential Biostimulant for Agriculture Sustainability

Dinesh Chandra and Thiyam General

25.1 Introduction

The pervasive inputs of synthetic chemical fertilizers and other chemicals such as pesticides and herbicides into the agricultural system have reduced productivity and nutritional quality of crops and also caused negative impacts on the environment (Kramer et al. 2006). In recent years, biostimulants are receiving much attention in sustainable agriculture as a means of boosting crop growth and productivity in eco-friendly and cost-effective ways and reducing the negative impacts of chemical synthetic fertilizers and displayed positive influence on plant growth and productivity, and soil sustainability (Craigie 2011; Singh et al. 2011a, b). Seaweeds are plant-like organisms with simple internal structures that largely inhabit coastal areas. Seaweeds include the members of red, brown, and green marine algae. The use of natural seaweed as fertilizers has allowed the partial replacement of synthetic chemical fertilizers and they have great potential in improving soil physiochemical properties and have substantial contribution in increasing crop growth and productivity (Li et al. 2017; Renuka et al. 2018; Thilagar et al. 2016; Zodape et al. 2010). They have the immense ability in the mobilization of inorganic and organic acids, nutrient mineralization, and production of biologically active compound which can be utilized for applications for plant improvement making them as a suitable option for biofertilizers (Gayathri et al. 2015; Hernández-Carlos and Gamboa-Angulo 2011; Jäger

D. Chandra
Department of Biological Sciences, CBS&H, G.B. Pant University of Agriculture and Technology, Pantnagar, Uttarakhand, India

GIC Chamtola, Almora, Uttarakhand, India

T. General (✉)
Department of Biological Sciences, CBS&H, G.B. Pant University of Agriculture and Technology, Pantnagar, Uttarakhand, India
e-mail: generalm48@gmail.com

A. Ranga Rao, G. A. Ravishankar (eds.), *Sustainable Global Resources of Seaweeds Volume 1*, https://doi.org/10.1007/978-3-030-91955-9_25

et al. 2010; Prasanna et al. 2016). Seaweed extracts are non-toxic, non-polluting, biodegradable, and non-hazardous to the environment, humans and animals, and have become popular agents as metabolic enhancers.

The products of seaweed extracts can be used as liquid extracts in the agricultural system and applied in the form of a soil drench, foliar spray, and powder or granular form (Thirumaran et al. 2009). A number of studies revealed that a wide range of seaweed extract derived from several species such as *Ascophyllum nodosum*, *Kappaphycus alvarezii*, *Sargassum muticum*, *Ecklonia maxima*, *Durvillaea pota-torum*, *Ulva lactuca*, *Caulerpa sertularioides*, *Padina gymnospora*, *Sargassum liebmannii*, *Sargassum johnstonii* are useful as biofertilizers (Hernández-Herrera et al. 2014). Among marine algae, the extracts of brown marine algae are mainly used in agriculture practices because of their growth-stimulating effects on plants and for their alleviating the negative impacts of abiotic and biotic stresses (Guinan et al. 2012; Sharma et al. 2014). The chemical constituents of seaweed extract include complex polysaccharide, fatty acids, vitamins, phytohormones, and mineral nutrients (Lafarga et al. 2020). Besides, they have a prominent role in maintaining the productivity of terrestrial and aquatic ecosystems through nitrogen fixation and photosynthesis thereby enhancing the accessibility of nutrients by cycling and transformations (Moroney and Ynalvez 2009). Seaweed can be regarded as a potential source of biofertilizers in fresh or dried form and it is a good source of macro and microelements for plant nutrition and it also enhances the biochemical constituents like lipids, proteins, carbohydrates ash, dietary fiber, phenol *etc* in the plant. This technology can be implemented for organic farming for sustainable agriculture which is a better solution for an eco-friendly approach.

25.2 Utilization of Seaweeds as Metabolic Enhancers in Agriculture Field

Seaweeds are valuable resource for plant development as they have higher contents of amino acids, mineral substances, vitamins, and phytohormones (Stirk and Van Staden 1997). The extracts of brown algae are widely used in agriculture and they have been shown to enhance the productivity of agricultural and horticultural plants including tomato, potato, beet, legumes, citrus, grasses, rice, wheat, maize, sugarcane, broccoli, spinach, cabbage, carrot, pepper, cucumber *etc* (Battacharyya et al. 2015; Renuka et al. 2018). The regulator of the plant growth may vary from the fertilizers in many ways such as they alter and manage the cell division, control of shoot and root elongation and initiation of flowering and other metabolic functions. Whereas, fertilizers supply the nutrients required for normal plant growth and development (Allen et al. 2001). Moreover, cytokinin is considered as the most important plant growth regulator in seaweed. The trace elements present in marine algae act as enzyme activators and play important roles in plant nutrition and physiology (Senn 1987). The application of seaweed extract in plants increased the antioxidant metabolites (for example α-tocopherol, carotene, and ascorbic acid) as well as antioxidant enzymes (superoxide dismutase, catalase, and peroxidase) activities

Table 25.1 Seaweeds extract as plant promoting agents and biofertilizers

Seaweed	Benefitted plant	Mechanism of enhanced plant growth	Reference
Ascophyllum nodosum, Sargassum muticum	Rice plants (*Oryza sativa*) and lettuce (*Lactuca sativa*)	Treatment had a positive effect on seed germination, plant development, and production	Silva et al. (2019)
Ascophyllum nodosum	Grapevine	Inoculation increased the physiological performance, and also improved the photosynthesis efficiency under water stress conditions	Frioni et al. (2021)
Ascophyllum nodosum (L.)	Soyabean	Bio-stimulants application provided higher photosynthetic rates, efficient mechanisms for dissipating excess energy and higher activities of antioxidant enzymes	do Rosário Rosa et al. (2021)
Kappaphycus alvarezii	Sugarcane	Increased sugar yield, lowering GHG's	Singh et al. (2018)
Ascophyllum nodosum	Tomato	Biostimulants application changes of chlorophyll, osmolytes levels, MDA production, dehydrin isoform pattern and dehydrin gene expression levels (expression of *tas14* dehydrin gene)	Goñi et al. (2018)
Ecklonia maxima	Bean (*Phaseolus vulgaris* L.)	Increased productivity and nutraceutical quality	Kocira et al. (2018)
Kappaphycus, Gracilaria	Rice	Increased grain yield	Sharma et al. (2017)
Gracilaria edulis, Sargassum wightii	*Withania somnifera*	Increased withanolides contents, enhanced expression of *SE, SS, HMGR and FPPS* genes	Sivanandhan et al. (2015)
Ascophyllum nodosum	Pepper	Enhanced yield, fruit diameter and chlorophyll content	Manna et al. (2012)
Enteromorpha intestinelis, Gelidium pectinutum, Ecklonia Maxima	Cucumber	Improved vegetative growth and yield	Ahmed and Shalaby (2012)
Ascophyllum nodosum	Mint and basil	Increased antibacterial activity, micro- and macronutrient and carbohydrates	Elansary et al. (2016)
Super fifty® and Ecoelicitor® (commercial extract from *Ascophyllum nodosum*)	Lettuce, oilseed rape	Enhanced plant growth and tolerance to biotic and abiotic stresses	Guinan et al. (2012)

(continued)

Table 25.1 (continued)

Seaweed	Benefitted plant	Mechanism of enhanced plant growth	Reference
Ascophyllum nodosum	*Hordeum vulgare*	Induced gibberellic acid independent amylase activity in Barley and promoted seed germination	Rayorath et al. (2008)
Ascophyllum nodosum	*Arabidopsis thaliana*	Improved plant growth by modulation of concentration and Localization of auxin	Rayorath et al. (2008)
AZAL5® (commercial extract from *A. nodosum*)	*Brassica napus*	Promotes plant growth and higher uptake of nitrate and sulfate	Jannin et al. (2013)
Ascophyllum nodosum	Cucumber	Improved plant defense against *Phytopthora melonis*	Abkhoo and Sabbagh (2016)
Ascophyllum nodosum	Tomato	Induced systemic resistance (ISR) against *Phytophthora capsica*	Panjehkeh and Abkhoo (2016)
Ascophyllum nodosum	*Arabidopsis thaliana*	Induced resistance against *pseudomonas syringae* pv. Tomato DC3000	Subramanian et al. (2011)
Ascophyllum nodosum	Carrot	Reduced the progression of disease caused by *Alternaria radicina* and *Botrytis cinerea*	Jayaraj et al. (2008)

(Allen et al. 2001). The impacts of seaweed extract on plant growth are summarized in Table 25.1.

25.3 Amelioration of Abiotic and Biotic Stresses Through Seaweeds

Abiotic and biotic stresses hamper growth and productivity of crops worldwide. In order to boost the agricultural production for the ever-increasing population chemical synthetic fertilizers are used. These synthetic chemical fertilizers posed a serious threat to the health of animals, plants, humans as well as for the entire biosphere (Damalas and Koutroubas 2016). The arable land is continuously shrinking due to the adverse effects of climate change and urbanization. The effects of climatic changes resulted into increased atmospheric CO_2, temperature, nutrient imbalances (such as deficiency and toxicity of mineral) and soil salinization that affected the plant growth, productivity and quality of crops (Anderson et al. 2011; dos Reis et al. 2012; Matesanz et al. 2010). To meet the food demands for growing population, world food production must double by the year 2050 (Qin et al. 2011; Voss-Fels and Snowdon 2016). Therefore, an alternative approach is required to reduce the input of synthetic chemical fertilizers. In this context, plant biostimulant such as seaweeds are a new class of crop input into agricultural practices can reduce the application rate of synthetic chemical fertilizers and pesticides and thereby enhancing the quality and productivity of crops (Calvo et al. 2014; Van Oosten et al. 2017; Yakhin et al. 2017).

The pivotal role of seaweed extracts in enhancing plant growth as well as in increasing plant tolerance to abiotic (drought, salinity, nutrient deficiency, flooding, temperature stress) and biotic (diseases and insects) stresses has been reported (Jayaraman et al. 2011; Sangha et al. 2010). Among seaweeds, brown seaweeds (*Ascophyllum nodosum, Fucus vesiculosus, F. serratus*) are rich in phenolic compounds (Balboa et al. 2013; Keyrouz et al. 2011). Phenolic compounds are secondary metabolites which defend the cell and cellular components from the stressed induced damages and also chelate the metal ions (Wang et al. 2009). The phenolic compounds such as catechol (dihydroxy benzene) or galloyl (trihydroxy benzene), phloroglucinol, eckol and dieck showed strong chelating activities which make brown algae to effectively scavenge free radical (single oxygen, hydroxyl, superoxide, alkoxyl and peroxy radicals) and antioxidant activity (Andjelković et al. 2006). Study reported by Shibata et al. (2002). demonstrated that marine algae derived phlorotannis was highly efficient antioxidant as compared to the ascorbic acid, phlorofucofuroeckol, catechin, resveratrol, epigallocatechin eallate, α-tocopherol . The extracts of seaweed contain a number of plant hormone such as such as auxins, gibberellins, cytokinin, abscisic acid and brassinosteroids (Stirk et al. 2014). Osmolytes such as mannitol, an important protective compound in response to abiotic stresses are found in many brown algae such as *Laminaria digitata, L. hyperborea, L. saccharina, Halidrys siliquosa, Fucus serratus, F. spiralis, F. vesiculosus, Ascophyllum nodosum, Alaria esculenta, Pilayella littoralis* and *Ectocarpus siliculosus*. The chelating activity of mannitol revealed that seaweed can release the inaccessible elements to the soils (Reed et al. 1985).

Previous studies by several researchers revealed that seaweed extracts such as *Lessonia nigrescens, Grateloupia filicina, Kappaphycus alvarezii, Gracilaria dura* improved salinity and drought tolerance in several crops such as wheat, passion fruit, *Arabidopsis*, avocado and rice (Elansary et al. 2017; Di Stasio et al. 2018; Jithesh et al. 2018; Liu et al. 2019; Zou et al. 2019). Under salinity stress, both Rygex® and Super Fifty® the commercial extracts of *A. nodosum* triggered that accumulation of essential amino acids, antioxidant and minerals (Di Stasio et al. 2018). The extract of *A. nodosum* application mitigates the negative impacts of salinity stress in avocado and turfgrass by enhancing nutrient uptake (Elansary et al. 2017; Bonomelli et al. 2018).

Abiotic stress involves hyperosmotic and ionic imbalances leading to an oxidative stress generated from the overproduction of ROS (reactive oxygen species) and the antioxidant mechanisms (Debnath et al. 2011). ROS [includes the singlet oxygen ($^1O_2$), superoxide (O_2^-), hydroxyl (OH^-), hydrogen peroxide (H_2O_2)] are to known to impair biomolecule like proteins, DNA and lipids (Das and Roychoudhury 2014; Mittler 2002). To mitigate such damaging impacts, plant resort to several natural defensive mechanisms. ROS production controlled by the enzymatic and non-enzymatic antioxidants such as SOD (superoxide dismutase), GPX (guaiacol peroxidase), CAT (catalase), APX (ascorbate peroxidase), DHAR (dehydroascorbate reductase), GR (glutathione reductase), MDHAR (monodehydroascorbate reductase), AsA (ascorbic acid), tocopherol, glutathione and phenolic compounds (Gruszka et al. 2018; Yadav et al. 2019). The study of Zou et al. (Zou et al. 2019) demonstrated that potential of *Lessonia nigrescens* derived polysaccharides in the alleviation of salt stress on wheat and found that polysaccharides application significantly increased the root/shoot length and fresh/dry matters of wheat. They also noticed that polysaccharides application maintained osmotic status of salt stressed wheat seedlings by increasing the proline and sugar content as well as regulating the ratio of Na^+/K^+. A study of Liu et al. (Liu et al. 2019) observed that the impact of polysaccharides from *Grateloupia filicina* augmented rice seed development and mitigate the salinity-induced damage. A study by Jayaraman et al. (Jayaraman et al. 2011) examined the impacts of Stimplex™, a commercial extract from *A. nodosum* on some common cucumber fungal pathogen such as *Alternaria cucumerinum, Fusarium oxysporum, Didymella applanata.* They found that Stimplex™ application resulted in a significant increase in activities of various defense related enzymes including peroxidase, polyphenol oxidase, lipoxygenase, chitinase, phenylalanine ammonia lyase, glucanase and β-1,3-glucanase. In addition, treated plants also displayed a higher level of phenolics content and reduced the disease incidence of all tested pathogens. Similarly, the study of Guinan et al. (2012) revealed that extracts of Super Fifty® and Ecoelicitor® were found to be more effective in imparting tolerance to *Sclerotinia sclerotiorum* and *Alternaria brassicae.* The abiotic and biotic stress-responsive genes are elicited by seaweeds are enumerated in Table 25.2.

25.4 Nutritional and Bioactive Potential of Seaweed

A number of studies revealed that macroalgae have huge potential due to their bioactive compounds endowed with a wide range of activities such as antioxidant, anti-inflammatory, anti-tumor and antimicrobial (Rocha et al. 2018; Trifan et al. 2019). Among macroalgae, phaeophyceae members accumulate a variety of compounds including sterols, lipids, proteins, terpenoids, vitamins, phlorotannins with a range of biological activities (Balboa et al. 2013; Jiménez-Escrig et al. 2012). Seaweeds are the macroscopic, multicellular marine algae that form a vital part of the marine coastal ecosystem. Approximately, there are 9000 species of macrolage classified into three groups based on their pigmentation such as Chlorophyta (greem algae), Phaeophyta (brown algae), and Rhodophyta (red algae). Pheophyta is the second most abundant group comprising about 2000 species. This group of microalgae commonly used in agriculture (Blunden and Gordon 1986; Khan et al. 2009). Brown algae such as *Fucus* spp., *Ascophyllum nodosum*, *Sargassum* spp., *Turbinaria* spp., and *Laminaria* spp. are used as biofertilizers in agricultural practices (Hong et al. 2007). For centuries seaweeds are used as source of organic matter and fertilizer nutrients. Around 15 million metric tonnes of seaweed products are produced

Table 25.2 Effect of seaweed extracts on gene expression of plants under abiotic and biotic stress conditions

Seaweed	Plant	Abiotic stress responsive gene(s)	Reference
Lessonia nigrescens	Wheat	Down-regulation of *TaHKT2; 1*, and up-regulation of *TaSOS1* and *TaNHX2*	Zou et al. (2019)
Ascophyllum nodosum	Soybean	Enhanced expression of *GmRD22, GmDREB, GmFIB1a, GmERD1, GmBIPD GmPIP1b*	Shukla et al. (2018b)
Ascophyllum nodosum	*Arabidopsis thaliana*	Up-regulation of *SAUR33, SAUR59*, and *SAUR71*, down-regulation of *SAUR1* and *SAUR50*	Goñi et al. (2016)
AZAL5® (commercial extract from *A. nodosum*)	Oilseed rape (*Brassica napus*)	Enhanced expression of *BnNRT1.1*; *BnNRT2.1, BnSultr4.1*; *BnSultr4.2*	Jannin et al. (2013)
Ascophyllum nodosum	*Arabidopsis*	Up-regulation of CB73, RD29A, and COR15A.	Rayirath et al. (2009)
Ascophyllum nodosum	*Arabidopsis*	Enhanced expression *RAB18, RD29A, DFR, SOD, APX_2*	Santaniello et al. (2017)
Ascophyllum nodosum	Soybean	Induced the expression level of *CYP707A1a, CYP707A3b, GmDREB1B, GmRD22, fibrillin, ABA-response regulators (ABI1, ABI2), FIB1a*	Shukla et al. (2018b)
Ascophyllum nodosum	*Arabidopsis thaliana*	Up-regulation of *DREB1A/CBF3, COR15A, COR78/RD29A*, digalactosyldiacylglycerol synthaseencoding gene *DGD1*, salicylic acid *(SA) (At1g18870)*, Spermine/spermidine biosynthesis *(At5g15950)*, cytokinin conjugation *(UGT73B2, UGT76C1/2, At2g43820, At1g24100)*, increased expression of the ABA responsive genes *RAB18* (At5g66400), RD29A (At5g52310), *PsbS (At1g44575)* and *VDE (At1g08550)* and lower expression of activating protein 2, betaine aldehyde dehydrogenase (*BADH*), glutathione S-transferase, and fucosyltransferase, reduced expression of *NCED3* (*At3g14440*)	De Saeger et al. (2019)
Ascophyllum nodosum	*Arabidopsis thaliana*	Increased expression of proline synthesis genes *P5CS1* and *P5CS2* and a marginal reduction in the expression of the proline dehydrogenase (*ProDH*) gene	Nair et al. (2012)
Ascophyllum nodosum	*Arabidopsis thaliana*	Elicit the expression of glutathione S transferase	Jithesh et al. (2018)
Ascophyllum nodosum	*Arabidopsis thaliana*	Higher transcript accumulation of *SnRK2*	Coello et al. (2011), Jithesh et al. (2018)
Ascophyllum nodosum	*Arabidopsis thaliana*	Induced expression of *IPT3, IPT4*, and *IPT5*	Wally et al. (2013)

(continued)

Table 25.2 (continued)

Seaweed	Plant	Abiotic stress responsive gene(s)	Reference
Ascophyllum nodosum	*Arabidopsis thaliana*	Down-regulate the expression of miR396a-5p, which resulted in a reduction in the expression of its target gene AtGRF7	Yang et al. (2009); Shukla et al. (2018a)
Ascophyllum nodosum	*Arabidopsis thaliana*	Treatment application induced the ath-miR398 targets superoxide dismutase 1 (AtCSD1), ath-miR168a targets ARGONAUTE 1 and ath-miR399a targets ubiquitin-conjugating E2 enzyme (AtUBC24) and wall-associated kinase 2 (AtWAK2)	Shukla et al. (2018a)
Ascophyllum nodosum	*Arabidopsis thaliana*	Induced the expression of JA-related genes such as PDF1.2,	Subramanian et al. (2011)
Ascophyllum nodosum derived extract (Stella Maris®)	*Arabidopsis thaliana*	Up-regulation of *WRKY30*, *CYP71A12*, and *PR-1*	Cook et al. (2018)
Ascophyllum nodosum	Carrot	Increased transcript accumulation of *PR-1, PR-5, NPR-1, LTP*	Jayaraj et al. (2008)

annually (FAO 2006), a substantial portion of which is used for nutrient supplements and as biostimulants to increased plant growth and productivity.

A number of products such as Agri-Gro Ultra, Tasco®, Acadian®, Alg-A-Mic, Bio-GenesisTM High TideTM, Biovita, Espoma, Soluble Seaweed Extract, Guarantee®, Kelp Meal, Synergy, Kelpro, Kelprosoil, Maxicrop, Nitrozime and Stimplex® are formulated from *Ascophyllum nodosum* seaweed and used in agriculture practices as a plant growth stimulant. Kelpak, Seasol®, Profert® and AgroKelp products are obtained from the *Ecklonia maxima*, *Durvillea potatorum*, *Durvillea antarctica* and *Macrocystis pyrifera* seaweed, respectively also these are used in agriculture as plant growth stimulant (Khan et al. 2009). In addition, application of seaweed extracts on plants promotes the seed germination and its establishment, imparts tolerance to abiotic and biotic stresses, enhanced crop performance and productivity, and also enhanced shelf-life of postharvest products (Norrie and Keathley 2006).

Seaweed extracts has been shown to have a positive impact on plant nutrient uptake (Table 25.3). Turan and Köse (Turan and Köse 2004) found that foliar application of commercial seaweed extract enhanced the micro- and macronutrient content of grapevine leaves. Study of Rathore et al. (Rathore et al. 2009) revealed that application of *Kappaphycus alvarezii* increased N, P, K, and S content of soybean under rainfed conditions. Seaweed extracts improve plant via improvement of soil structure and micronutrient solubility in the soil, as well as influencing the plant physiology through changes in root morphology and increased root colonization by AM fungi (Halpern et al. 2015). The polysaccharides such as fucoidans and alginates are found in brown seaweeds and these polysaccharides are bound with

Table 25.3 Effects of seaweed extracts on plant nutrient uptake

Benefitted plant	Seaweed	Impacts on plant	Reference
Maize	*Kappaphycus alvarezii, Gracilaria edulis*	Increased nitrogen, potassium and phosphorus uptake	Basavaraja et al. (2018)
Oilseed rape	*Ecklonia maxima*	Increased leaf phosphorus and potassium Content	Di Stasio et al. (2017)
Tomato	*Ecklonia maxima*	Enhanced fruit calcium concentration	Colla et al. (2017)
Tomato	*Ascophyllum nodosum*	Enhanced concentration of manganese, cooper, and zinc in root and leaf	Carrasco-Gil et al. (2018)
Oilseed rape	*Ascophyllum nodosum*	Increased relative manganese, cooper, and magnesium concentration in whole plant.	Billard et al. (2014)
Wheat	*Ascophyllum nodosum*	Enhanced grain potassium	Stamatiadis et al. (2015)
Oilseed rape	*Ascophyllum nodosum*	Stimulation of root and shoot nitrogen and sulfur	Jannin et al. (2013)
Soybean	*Kappaphycus alvarezii*	Increased nitrogen, phosphorus, potassium, sulfur grain uptake and nitrogen, phosphorus straw uptake	Rathore et al. (2009)
Tomato	Commercial extracts of *A. nodosum*, Rygex® and super fifty®	Macronutrient (N, P, K, ca, S) and micronutrient (mg, Zn, Mn, Fe) contents	Di Stasio et al. (2018)
Olive (*Olea europaea*)	*Ascophyllum nodosum*	Higher uptake of K, Fe, and cu	Chouliaras et al. (2009)
Brassica napus	AZAL5® (commercial extract from *A. nodosum*)	Higher uptake of nitrate and sulfate	Jannin et al. (2013)
Grapevines (*V. vinifera*)	Commercial Extracts Maxicrop®, proton®, and Algipower®	Increased cooper uptake	Turan and Köse (2004)
Ascophyllum nodosum	Avocado	Higher content of Ca^{2+} and K^{+}	Bonomelli et al. (2018)

metallic ions in soil to produce a gel that helps in the absorption of water and form an aggregate structure which assists the plant to grow a vigorous root architecture which in turn enhance the nutrient availability (Khan et al. 2009). Some of the studies also reported that the organic molecules in seaweed extracts can chelate the micronutrients and make them more accessible to plants. Spinelli et al. (Spinelli et al. 2010) showed that commercial extract of *A. nodosum* improved the solubility of micronutrients and that could be used to replace some of the standard iron chelates such as sequestrene.

25.5 Conclusion

Leveraging seaweeds as potential biostimulants have been shown to increase the plant growth, improved physiological and molecular machinery, and provides tolerance to abiotic and biotic stresses and better nutrient uptake in several crops and suggesting their usefulness in reducing chemical fertilizer application in agricultural practices without negatively affecting crop productivity. Although, commercial seaweeds extracts are readily available in the marketplace, they have not substantially reduced chemical fertilizers use in conventional agriculture. Before biostimulants can effectively use in agriculture system, it is important to find the most promising ones for those particular environmental conditions and the mechanism how they are best applied.

References

Abkhoo J, Sabbagh SK (2016) Control of Phytophthora melonis damping-off, induction of defense responses, and gene expression of cucumber treated with commercial extract from *Ascophyllum nodosum*. J Appl Phycol 28:1333–1342

Ahmed YM, Shalaby EA (2012) Effect of different seaweed extracts and composton vegetative growth, yield and fruit quality of cucumber. J Hortic Sci OrnamPlants 4:235–240

Allen VG, Pond KR, Saker KE, Fontenot JP, Bagley CP, Ivy RL, Evans RR, Schmidt RE, Fike JH, Zhang X, Ayad JY (2001) Tasco: influence of a brown seaweed on antioxidants in forages and livestock—A review. J Anim Sci 79:E21

Anderson JT, Willis JH, Mitchell-Olds T (2011) Evolutionary genetics of plant adaptation. Trends Genet 27:258–266

Andjelković M, Van Camp J, De Meulenaer B, Depaemelaere G, Socaciu C, Verloo M, Verhe R (2006) Iron-chelation properties of phenolic acids bearing catechol and galloyl groups. Food Chem 98:23–31

Balboa EM, Conde E, Moure A, Falqué E, Domínguez H (2013) In vitro antioxidant properties of crude extracts and compounds from brown algae. Food Chem 138:1764–1785

Basavaraja PK, Yogendra ND, Zodape ST, Prakash R, Ghosh A (2018) Effect of seaweed sap as foliar spray on growth and yield of hybrid maize. J Plant Nutr 41:1851–1861

Battacharyya D, Babgohari MZ, Rathor P, Prithiviraj B (2015) Seaweed extracts as biostimulants in horticulture. Sci Hortic 196:39–48

Billard V, Etienne P, Jannin L, Garnica M, Cruz F, Garcia-Mina JM, Yvin JC, Ourry A (2014) Two biostimulants derived from algae or humic acid induce similar responses in the mineral content and gene expression of winter oilseed rape (*Brassica napus* L.). J Plant Growth Regul 33:305–316

Blunden G, Gordon SM (1986) Betaines and their sulphono analogues in marine algae. In: Round FE, Chapman DJ (eds) Progress in phycological research, vol 4. Biopress Ltd, Bristol, pp 39–80

Bonomelli C, Celis V, Lombardi G, Mártiz J (2018) Salt stress effects on avocado (*Persea americana* mill). Plants with and without seaweed extract (*Ascophyllum nodosum*) application. Agronomy 8:64

Calvo P, Nelson L, Kloepper JW (2014) Agricultural uses of plant biostimulants. Plant Soil 383:3–41

Carrasco-Gil S, Hernandez-Apaolaza L, Lucena JJ (2018) Effect of several commercial seaweed extracts in the mitigation of iron chlorosis of tomato plants (*Solanum lycopersicum* L.). Plant Growth Regul 86:401–411

Chouliaras V, Tasioula M, Chatzissavvidis C, Therios I, Tsabolatidou E (2009) The effects of a seaweed extract in addition to nitrogen and boron fertilization on productivity, fruit maturation, leaf nutritional status and oil quality of the olive (*Olea europaea* L). Cultivar Koroneiki. J Sci Food Agric 89:984–988

Coello P, Hey SJ, Halford NG (2011) The sucrose non-fermenting-1-related (SnRK) family of protein kinases: potential for manipulation to improve stress tolerance and increase yield. J Exp Bot 62:883–893

Colla G, Cardarelli M, Bonini P, Rouphael Y (2017) Foliar applications of protein hydrolysate, plant and seaweed extracts increase yield but dierentially modulate fruit quality of greenhouse tomato. HortScience 52:1214–1220

Cook J, Zhang J, Norrie J, Blal B, Cheng Z (2018) Seaweed extract (Stella Maris®) activates innate immune responses in *Arabidopsis thaliana* and protects host against bacterial pathogens. Mar Drugs 16:E221

Craigie JS (2011) Seaweed extract stimuli in plant science and agriculture. J Appl Phycol 23:371–393

Damalas C, Koutroubas S (2016) Farmers' exposure to pesticides: toxicity types and ways of prevention. Toxics 4:1

Das K, Roychoudhury A (2014) Reactive oxygen species (ROS) and response of antioxidants as ROS-scavengers during environmental stress in plants. Front Environ Sci 2:53

De Saeger J, Van Praet S, Vereecke D, Park J, Jacques S, Han T, Depuydt S (2019) Toward the molecular understanding of the action mechanism of *Ascophyllum nodosum* extracts on plants. J Appl Phycol:1–25

Debnath M, Pandey M, Bisen PS (2011) An omics approach to understand the plant abiotic stress. Omi J Integr Biol 15:739–762

Di Stasio E, Rouphael Y, Colla G, Raimondi G, Giordano M, Pannico A, El-Nakhel C, De Pascale S (2017) The influence of *Ecklonia maxima* seaweed extract on growth, photosynthetic activity and mineral composition of *Brassica rapa* L. subsp *sylvestris* under nutrient stress conditions. Eur J Hortic Sci 82:286–293

Di Stasio E, Van Oosten MJ, Silletti S, Raimondi G, dell'Aversana E, Carillo P, Maggio A (2018) *Ascophyllum nodosum*-based algal extracts act as enhancers of growth, fruit quality, and adap tation to stress in salinized tomato plants. J Appl Phycol 30:2675–2686

do Rosário Rosa V, Dos Santos ALF, da Silva AA, Sab MPV, Germino GH, Cardoso FB, de Almeida Silva M (2021) Increased soybean tolerance to water deficiency through biostimulant based on fulvic acids and *Ascophyllum nodosum* (L.) seaweed extract. Plant Physiol Biochem 158:228–243

dos Reis SP, Lima AM, de Souza CRB (2012) Recent molecular advances on downstream plant responses to abiotic stress. Int J Mol Sci 13:8628–8647

Elansary HO, Yessoufou K, Abdel-Hamid AME, El-Esawi MA, Ali HM, Elshikh MS (2016) Seaweed extracts enhance Salam turfgrass performance during prolonged irrigation intervals and saline shock. Front Plant Sci 8:830

Elansary HO, Yessoufou K, Abdel-Hamid AME, El-Esawi MA, Ali HM, Elshikh MS (2016) Seaweed extracts enhance Salam turfgrass performance during prolonged irrigation intervals and saline shock. Front Plant Sci 8. https://doi.org/10.3389/fpls.2017.00830

FAO (2006) Yearbook of fishery statistics, vol 98(1–2). Food and Agricultural Organisation of the United Nations, Rome

Frioni T, VanderWeide J, Palliotti A, Tombesi S, Poni S, Sabbatini P (2021) Foliar vs. soil application of *Ascophyllum nodosum* extracts to improve grapevine water stress tolerance. Sci Hortic 277:109807

Gayathri M, Kumar PS, Prabha AML, Muralitharan G (2015) In vitro regeneration of *Arachis hypogaea* L. and *Moringa oleifera* lam. Using extracellular phytohormones from *Aphanothece* sp. MBDU 515. Algal Res 7:100–105

Goñi O, Fort A, Quille P, McKeown PC, Spillane C, O'Connell S (2016) Comparative transcriptome analysis of two *Ascophyllum nodosum* extract biostimulants: same seaweed but different. J Agric Food Chem 64:2980–2989
Goñi O, Quille P, O'Connell S (2018) *Ascophyllum nodosum* extract biostimulants and their role in enhancing tolerance to drought stress in tomato plants. Plant Physiol Biochem 126:63–73
Gruszka D, Janeczko A, Dziurka M, Pociecha E, Fodor J (2018) Non-enzymatic antioxidant accumulations in BR-deficient and BR-insensitive barley mutants under control and drought conditions. Physiol Plant 163:155–169
Guinan KJ, Sujeeth N, Copeland RB, Jones PW, O'brien NM, Sharma HSS, Prouteau PFJ, O'sullivan JT (2012) Discrete roles for extracts of *Ascophyllum nodosum* in enhancing plant growth and tolerance to abiotic and biotic stresses. In: I World Congress on the Use of Biostimulants in Agriculture 1009, pp 127–135
Halpern M, Bar-Tal A, Ofek M, Minz D, Muller T, Yermiyahu U (2015) The use of biostimulants for enhancing nutrient uptake. Adv Agron 130:141–174
Hernández-Carlos B, Gamboa-Angulo MM (2011) Metabolites from freshwater aquatic microalgae and fungi as potential natural pesticides. Phytochem Rev 10:261–286
Hernández-Herrera RM, Santacruz-Ruvalcaba F, Ruiz-López MA, Norrie J, Hernández-Carmona G (2014) Effect of liquid seaweed extracts on growth of tomato seedlings (Solanum lycopersicum L.). J Appl Phycol 26(1):619–628. https://doi.org/10.1007/s10811-013-0078-4
Hong DD, Hien HM, Son PN (2007) Seaweeds from Vietnam used for functional food, medicine and biofertilizer. J Appl Phycol 19:817–826
Jäger K, Bartók T, Ördög V, Barnabás B (2010) Improvement of maize (*Zea mays* L.) anther culture responses by algae-derived natural substances. S Afr J Bot 76:511–516
Jannin L, Arkoun M, Etienne P, Laîné P, Goux D, Garnica M, Fuentes M, San Francisco S, Baigorri R, Cruz F, Houdusse F (2013) *Brassica napus* growth is promoted by *Ascophyllum nodosum* (L). Le Jol. Seaweed extract: microarray analysis and physiological characterization of N, C, and S metabolisms. J Plant Growth Regul 32:31–52
Jayaraj J, Wan A, Rahman M, Punja ZK (2008) Seaweed extract reduces foliar fungal diseases on carrot. Crop Prot 27:1360–1366
Jayaraman J, Norrie J, Punja ZK (2011) Commercial extract from the brown seaweed *Ascophyllum nodosum* reduces fungal diseases in greenhouse cucumber. J Appl Phycol 23:353–361
Jiménez-Escrig A, Gómez-Ordóñez E, Rupérez P (2012) Brown and red seaweeds as potential sources of antioxidant nutraceuticals. J Appl Phycol 24:1123–1132
Jithesh MN, Shukla PS, Kant P, Joshi J, Critchley AT, Prithiviraj B (2018) Physiological and transcriptomics analyses reveal that *Ascophyllum nodosum* extracts induce salinity tolerance in *Arabidopsis* by regulating the expression of stress responsive genes. Plant Growth Regul 38:463–478
Keyrouz R, Abasq ML, Le Bourvellec C, Blanc N, Audibert L, ArGall E, Hauchard D (2011) Total phenolic contents, radical scavenging and cyclic voltammetry of seaweeds from Brittany. Food Chem 126:831–836
Khan W, Rayirath UP, Subramanian S, Jithesh MN, Rayorath P, Hodges DM, Critchley AT, Craigie JS, Norrie J, Prithiviraj B (2009) Seaweed extracts as biostimulants of plant growth and development. Plant Growth Regul 28:386–399
Kocira A, Świeca M, Kocira S, Złotek U, Jakubczyk A (2018) Enhancement of yield, nutritional and nutraceutical properties of two common bean cultivars following the application of seaweed extract (*Ecklonia maxima*). Saudi J Biol Sci 25:563–571
Kramer SB, Reganold JP, Glover JD, Bohannan BJM, Mooney HA (2006) Reduced nitrate leaching and enhanced denitrifier activity and efficiency in organically fertilized soils. Proc Natl Acad Sci 103(12):4522–4527. https://doi.org/10.1073/pnas.0600359103
Lafarga T, Acién-Fernández FG, Garcia-Vaquero M (2020) Bioactive peptides and carbohydrates from seaweed for food applications: natural occurrence, isolation, purification, and identification. Algal Res 48:101909

Li R, Tao R, Ling N, Chu G (2017) Chemical, organic and bio-fertilizer management practices effect on soil physicochemical property and antagonistic bacteria abundance of a cotton field: implications for soil biological quality. Soil Tillage Res 167:30–38

Liu H, Chen X, Song L, Li K, Zhang X, Liu S, Qin Y, Li P (2019) Polysaccharides from *Grateloupia filicina* enhance tolerance of rice seeds (*Oryza sativa* L.) under salt stress. Int J Biol Macromol 124:1197–1204

Manna D, Sarkar A, Maity TK (2012) Impact of biozyme on growth, yield andquality of chilli (*Capsicum annuum* L.). J Crop Weed 8:40–43

Matesanz S, Gianoli E, Valladares F (2010) Global change and the evolution of phenotypic plasticity in plants. Ann N Y Acad Sci 1206:35–55

Mittler R (2002) Oxidative stress, antioxidants and stress tolerance. Trends Plant Sci 7:405–410

Moroney JV, Ynalvez RA (2009) Algal photosynthesis. In: eLS, September 2009. John Wiley & Sons Ltd, Chichester. http://www.els.net

Nair P, Kandasamy S, Zhang J, Ji X, Kirby C, Benkel B, Hodges MD, Critchley AT, Hiltz D, Prithiviraj B (2012) Transcriptional and metabolomic analysis of *Ascophyllum nodosum* mediated freezing tolerance in *Arabidopsis thaliana*. BMC Genomics 13:1–23

Norrie J, Keathley JP (2006) Benefits of *Ascophyllum nodosum* marine-plant extract applications to 'Thompson seedless' grape production. (proceedings of the Xth international symposium on plant bioregulators in fruit production, 2005). Acta Hortic 727:243–247

Panjehkeh N, Abkhoo J (2016) Influence of marine brown alga extract (Dalgin) on damping-off tolerance of tomato. JMES 7:2369–2374

Prasanna R, Kanchan A, Kaur S, Ramakrishnan B, Ranjan K, Singh MC, Hasan M, Saxena AK, Shivay YS (2016) *Chrysanthemum* growth gains from beneficial microbial interactions and fertility improvements in soil under protected cultivation. Hortic Plant J 2:229–239

Qin F, Shinozaki K, Yamaguchi-Shinozaki K (2011) Achievements and challenges in understanding plant abiotic stress responses and tolerance. Plant Cell Physiol 52:1569–1582

Rathore SS, Chaudhary DR, Boricha GN, Ghosh A, Bhatt BP, Zodape ST, Patolia JS (2009) Effect of seaweed extract on the growth, yield and nutrient uptake of soybean (*Glycine max*) under rainfed conditions. S Afr J Bot 75:351–355

Rayirath P, Benkel B, Hodges DM, Allan-Wojtas P, MacKinnon S, Critchley AT, Prithiviraj B (2009) Lipophilic components of the brown seaweed, *Ascophyllum nodosum*, enhance freezing tolerance in *Arabidopsis thaliana*. Planta 230:135–147

Rayorath P, Jithesh MN, Farid A, Khan W, Palanisamy R, Hankins SD, Critchley AT, Prithiviraj B (2008) Rapid bioassays to evaluate the plant growth promoting activity of *Ascophyllum nodosum* (L.) Le Jol. Using a model plant, *Arabidopsis thaliana* (L.) Heynh. J Appl Phycol 20:423–429

Reed RH, Davison IR, Chudek JA, Foster R (1985) The osmotic role of mannitol in the Phaeophyta: an appraisal. Phycologia 24:35–47

Renuka N, Guldhe A, Prasanna R, Singh P, Bux F (2018) Microalgae as multi-functional options in modern agriculture: current trends, prospects and challenges. Biotechnol Adv 36:1255–1273

Rocha DHA, Seca AML, Pinto DCGA (2018) Seaweed secondary metabolites in vitro and in vivo anticancer activity. Mar Drugs 16:410

Sangha JS, Ravichandran S, Prithiviraj K, Critchley AT, Prithiviraj B (2010) Sulfated macroalgal polysaccharides l-carrageenan and i-carrageenan differentially alter *Arabidopsis thaliana* resistance to *Sclerotinia sclerotiorum*. Physiol Mol Plant Pathol 75:38–45

Santaniello A, Scartazza A, Gresta F, Loreti E, Biasone A, Di Tommaso D, Piaggesi A, Perata P (2017) *Ascophyllum nodosum* seaweed extract alleviates drought stress in *Arabidopsis* by affecting photosynthetic performance and related gene expression. Front Plant Sci 8:1362

Senn LT (1987) Seaweed and plant growth. Clemson Univ., Clemson, SC. Seaweed and plant growth. Clemson Univ, Clemson, SC

Sharma HS, Fleming C, Selby C, Rao JR, Martin T (2014) Plant biostimulants: a review on the processing of macroalgae and use of extracts for crop management to reduce abiotic and biotic stresses. J Appl Phycol 26:465–490

Sharma L, Banerjee M, Malik GC, Gopalakrishnan VAK, Zodape ST, Ghosh A (2017) Sustainable agro-technology for enhancement of rice production in the red and lateritic soils using seaweed based biostimulants. J Clean Prod 149:968–975
Shibata T, Yamaguchi K, Nagayama K, Kawaguchi S, Nakamura T (2002) Inhibitory activity of brown algal phlorotannins against glycosidases from the viscera of the turban shell Turbo cornutus. Eur J Phycol 37:493–500
Shukla PS, Borza T, Critchley AT, Hiltz D, Norrie J, Prithiviraj B (2018a) *Ascophyllum nodosum* extract mitigates salinity stress in *Arabidopsis thaliana* by modulating the expression of miRNA involved in stress tolerance and nutrient acquisition. PLoS One 13:e0206221
Shukla PS, Shotton K, Norman E, Neily W, Critchley AT, Prithiviraj B (2018b) Seaweed extract improve drought tolerance of soybean by regulating stress-response genes. AoB Plants 10:plx051
Silva LD, Bahcevandziev K, Pereira L (2019) Production of bio-fertilizer from *Ascophyllum nodosum* and *Sargassum muticum* (Phaeophyceae). J Oceanol Limnol 37:918–927
Singh DP, Prabha R, Yandigeri MS, Arora DK (2011a) Cyanobacteria-mediated phenylpropanoids and phytohormones in rice (*Oryza sativa*) enhance plant growth and stress tolerance. Anton Leeuw 100:557–568
Singh I, Anand KV, Solomon S, Shukla SK, Rai R, Zodape ST, Ghosh A (2018) Can we not mitigate climate change using seaweed based biostimulant: A case study with sugarcane cultivation in India. J Clean Prod 204:992–1003
Singh JS, Pandey VC, Singh DP (2011b) Efficient soil microorganisms: a new dimension for sustainable agriculture and environmental development. Agric Ecosyst Environ 140:339–353
Sivanandhan G, Arunachalam C, Selvaraj N, Sulaiman AA, Lim YP, Ganapathi A (2015) Expression of important pathway genes involved in withanolides biosynthesis in hairy root culture of *Withania somnifera* upon treatment with *Gracilaria edulis* and *Sargassum wightii*. Plant Physiol Biochem 91:61–64
Spinelli F, Fiori G, Noferini M, Sprocatti M, andCosta, G. (2010) A novel type of seaweed extract as a natural alternative to the use of iron chelates in strawberry production. Sci Hortic 125:263–269
Stamatiadis S, Evangelou L, Yvin JC, Tsadilas C, Mina JMG, Cruz F (2015) Responses of winter wheat to Ascophyllum nodosum (L.) Le Jol. Extract application under the effect of N fertilization and water supply. J Appl Phycol 27:589–600
Stirk WA, Tarkowská D, Turečová V, Strnad M, van Staden J (2014) Abscisic acid, gibberellins and brassinosteroids in Kelpak®, a commercial seaweed extract made from *Ecklonia maxima*. J Appl Phycol 26:561–567
Stirk WA, Van Staden J (1997) Isolation and identification of cytokinins in a new commercial seaweed product made from *Fucus serratus* L. J Appl Phycol 9:327–330
Subramanian S, Sangha JS, Gray BA, Singh RP, Hiltz D, Critchley AT, Prithiviraj B (2011) Extracts of the marine brown macroalga, *Ascophyllum nodosum*, induce jasmonic acid dependent systemic resistance in *Arabidopsis thaliana* against *pseudomonas syringae* pv. Tomato DC3000 and *Sclerotinia sclerotiorum*. Eur J Plant Pathol 131:237–248
Thilagar G, Bagyaraja DJ, Rao MS (2016) Selected microbial consortia developed for chilly reduces application of chemical fertilizers by 50% under field conditions. Sci Hortic 198:27–35
Thirumaran G, Arumugam M, Arumugam R, Anantharaman P (2009) Effect of seaweed liquid fertilizer on growth and pigment concentration of *Cyamopsis tetrogonolaba* (L) Taub. Am Eurasian J Agric Environ Sci 2:50–56
Trifan A, Vasincu A, Luca SV, Neophytou C, Wolfram E, Opitz SE, Sava D, Bucur L, Cioroiu BI, Miron A, Constantinou AI (2019) Unravelling the potential of seaweeds from the Black Sea coast of Romania as bioactive compounds sources. Part I: *Cystoseira barbata* (Stackhouse) C. Agardh. Food Chem Toxicol 134:110820
Turan M, Köse C (2004) Seaweed extracts improve copper uptake of grapevine. Acta Agric Scand B Soil Plant Sci 54:213–220

Van Oosten MJ, Pepe O, De Pascale S, Silletti S, Maggio A (2017) The role of biostimulants and bioeffectors as alleviators of abiotic stress in crop plants. Chem Biol Technol Agric 4:5

Voss-Fels K, Snowdon RJ (2016) Understanding and utilizing crop genome diversity via high-resolution genotyping. Plant Biotechnol J 14:1086–1094

Wally OSD, Critchley AT, Hiltz D, Craigie JS, Han X, Zaharia LI, Abrams SR, Prithiviraj B (2013) Regulation of phytohormone biosynthesis and accumulation in *Arabidopsis* following treatment with commercial extract from the marine macroalga *Ascophyllum nodosum*. Plant Growth Regul 32:324–339

Wang T, Jonsdottir R, Ólafsdóttir G (2009) Total phenolic compounds, radical scavenging and metal chelation of extracts from Icelandic seaweeds. Food Chem 116:240–248

Yadav DS, Rai R, Mishra AK, Chaudhary N, Mukherjee A, Agrawal SB, Agrawal M (2019) ROS production and its detoxification in early and late sown cultivars of wheat under future O_3 concentration. Sci Total Environ 659:200–210

Yakhin OI, Lubyanov AA, Yakhin IA, Brown PH (2017) Biostimulants in plant science: a global perspective. Front Plant Sci 7:671

Yang F, Liang G, Liu D, Yu D (2009) *Arabidopsis* MiR396 mediates the development of leaves and flowers in transgenic tobacco. J Plant Biol 52:475–481

Zodape ST, Mukhopadhyay S, Eswaran K, Reddy MP, Chikara J (2010) Enhanced yield and nutritional quality in green gram (*Phaseolus radiata* L) treated with seaweed (*Kappaphycus alvarezii*) extract. J Sci Ind Res 69:468–471

Zou P, Lu X, Zhao H, Yuan Y, Meng L, Zhang C, Li Y (2019) Polysaccharides derived from the brown algae *Lessonia nigrescens* enhance salt stress tolerance to wheat seedlings by enhancing the antioxidant system and modulating intracellular ion concentration. Front Plant Sci 10:48

Chapter 26
Liquid Biofertilizers from Seaweeds: A Critical Review

Debasish Panda, Sananda Mondal, and Ankita Mishra

Abbreviations

CAT	Catalase
NUE	Nutrient Use Efficiency
POD	Peroxidase
RDF	Recommended Dose of Fertilizer
ROS	Reactive Oxygen Species
SLF	Seaweed liquid fertilizer
SLE	Seaweed Liquid Extracts
SOD	Superoxide Dismutase

26.1 Introduction

Seaweeds are the macroscopic and multicellular sea algae belonging to three major groups based on their pigmentation, viz. brown algae or Phaeophyta, red algae or Rhodophyta, and green algae or Chlorophyta (Shama et al. 2019). About 9000 macroalgae species belong to these three classes are the potential sources of organic matter such as biostimulants and biofertilizers in agriculture. According to FAO (2018), the global annual production of seaweeds during 2015 was 30.4 million tonnes and a significant portion of which was utilized as nutrient supplements to improve plant growth and yield as biofertilizers and biostimulants. The liquid biofertilizers extracted from the seaweeds have been used in sustainable crop production practices around the coastal regions of the world. The seaweed extracts are a

D. Panda (✉) · S. Mondal · A. Mishra
Department of Crop Physiology, Institute of Agriculture, Visva-Bharati, Sriniketan, West Bengal, India
e-mail: debasish.panda@visva-bharati.ac.in

A. Ranga Rao, G. A. Ravishankar (eds.), *Sustainable Global Resources of Seaweeds Volume 1*, https://doi.org/10.1007/978-3-030-91955-9_26

rich source of macro and micro-nutrients, vitamins, phytohormones, and other substances such as amino acids and antibiotics. Seaweed extracts and their derivatives can be used for several eco-friendly agricultural practices. The seaweed liquid biofertilizers can either be applied to soil as conditioners and manure or applied as a foliar spray on crop plants (Thirumaran et al. 2009). Since modern agricultural practices involving chemical fertilizers are not eco-friendly, the application of seaweed extracts improves the health of the soil and reduces environmental pollution. Several studies on the use of seaweed extracts in agriculture have proved the beneficial effects of these marine algal extracts on germination, root growth, crop yield, biotic and abiotic stress tolerance, and post-harvest shelf-life fruits, vegetables, and flowers. Seaweed extracts are a rich source of phytohormones such as auxins, gibberellins, cytokinins, abscisic acid, ethylene, betaine and polyamines, and other growth-promoting substances (Panda et al. 2012). The application of different seaweed extract has been found to enhance the tolerance of plants to a wide range of abiotic stresses such as salinity, drought, and temperature extremes. However, the macronutrients such as N, P, K found in seaweed extracts are insufficient to generate any physiological responses at the concentrations that the seaweed extracts are usually applied to the crops in the field (Battacharyya et al. 2015; Blunden 1991; Khan et al. 2009). Seaweeds and their products such seaweed liquid fertilizers and seaweed manures are integral part of sustainable agriculture due to their utilization in crop and nutrient management, plant protection and amelioration of various abiotic stresses in plants (Panda et al. 2012). Frequent use of inorganic fertilizers, pesticides, and insecticides has damaged the soil ecosystem extensively and adversely affected soil fertility, making the soil unsuitable for future use. In this scenario, natural sources of biostimulants and biofertilizers such as seaweeds can be an innovative solution to address sustainable agriculture's challenges to ensure optimal nutrient uptake, crop yield, and tolerance to abiotic stress (Povero et al. 2016). The liquid biofertilizers derived from the seaweed extracts can be a viable alternative to chemical fertilizer as it contains a higher level of organic matter, micro, and macronutrients. The seaweed liquid biofertilizers are known to exert many beneficial effects such as faster germination, enhancement of yield, and quality in many agricultural and horticultural crops (Blunden 1991; Crouch and van Staden 1993; Ganapathy et al. 2013). The seaweed manures are used as soil amendment to improve soil physical properties (Eyras et al. 1998) and are the cheap and popular local resource in coastal areas. The efficiency of seaweed liquid fertilizers obtained from the extracts of several marine algae has been studied and evaluated in the cereals, pulses, and vegetable crops (Ramya et al. 2015).

26.2 Seaweed Extracts as a Source of Bio-Fertilizer

The seaweed liquid fertilizers -SLFs are excellent sources of natural bio-fertilizers rich in macro and micro-nutrients. Several species of marine algae are being utilized for the preparation of SLF. The nutrient content of the SLF varies from species to species. Some of the commonly found minerals elements in SLF are K, Mg, Ca, and Na (Mondal and Panda 2019). The seaweed extracts also contain trace elements such as Fe, Cu, Zn, and Mn (Chojnacka et al. 2012; Sivasankari et al. 2006). Seaweed liquid fertilizer can be applied to the crops as foliar spray or applied to the soil or the seeds (Table 26.1). The seaweeds, particularly kelp and high nitrogen-rich green seaweed (*Ulva ohnoi*), can be either used to the soil as mulch or added to the compost (Cole et al. 2015). SLF applied as soil amendment improves soil health by enhancing the nutrient status of the soil and favouring microbial growth in the rhizosphere. However, the quantity of macronutrients such as N, P, K found in the seaweed extracts is inadequate to induce any physiological responses at the concentrations at which seaweed extracts are usually applied to the crops (Battacharyya et al. 2015; Khan et al. 2009; Blunden 1991).

26.3 Effect of Seaweed Extracts on Crop Plants

The application of seaweed extracts affects the growth and development, and physiology of treated plants affecting the overall health of the plants. A discussion has been made on the effect of seaweed extracts on various morpho-physiological aspects of the plant, such as germination, root development, mineral absorption, shoot growth and photosynthesis, crop yield, and vegetative propagation.

26.4 Effect on Seed Germination

Seed priming or pre-sowing seed treatment with seaweed extracts improves germination and seedling growth in many crops. The germination of tomato seeds was enhanced by the alkaline section of *Ascophyllum esculentum* (Ali et al. 2016). A further study on the effect of germination physiology of tomato seeds using liquid seaweed extracts of *Ulva lactuca* and *Padina gymnospora* (2%) revealed a decrease in mean germination time and increase in germination energy, germination index, and seedling vigour resulting in overall improvement of germination (Hernández-Herrera et al. 2014). The germination rate of fenugreek (*Trigonella foenum-graecum* L.) was increased by the seaweed liquid extracts of brown algae *Sargassum vulgare, Colpomenia sinuosa*, and *Padina pavonica* at a lower concentration. A significant increase in the seed germination rate was observed by applying the extracts of *Sargassum vulgare* (5%SLE) (El-Sheekh et al. 2016). The effect of extracts of the brown algae *Ascophyllum nodosum* on the seed germination and seedling growth of ornamental marigold (*Tagetes erecta*) was studied by Tavares et al. (2020). Regular spraying of the seaweed extracts on the marigold seeds, improved the germination

Table 26.1 Use of Seaweeds as Biofertilizers and Their Effect on Mineral Nutrition of Plants

Seaweed species	Mode of application	Crop Plant	Effect on mineral nutrition (uptake and translocation)	References
Ascophyllum nodosum	Foliar spray in pot experiment	Wheat	Enhanced grain K content	Stamatiadis et al. (2015)
Ascophyllum nodosum	Not mentioned	Tomato	Enhanced Mn, Cu, and Zn content in root and leaf.	Carrasco-Gil et al. (2018)
Ascophyllum nodosum	Nutrient solution in hydroponic system	Rapeseed	Increased Mn, Cu, and Mg concentration in the plant body	Billard et al. (2014)
Kappaphycus alvarezii	Foliar spray in field experiment	Maize	Enhanced N, P and K uptake (grain + Stover)	Basavaraja et al. (2018)
Gracilaria edulis	Foliar spray in field experiment	Maize	Enhanced N, P and K uptake (grain + Stover)	Basavaraja et al. (2018)
Ascophyllum nodosum	Nutrient solution in potculture experiment in greenhouse condition	Rapeseed	Stimulation of root and shoot N and S	Jannin et al. (2013)
Ecklonia maxima	Foliar spray under greenhouse condition	Tomato	Enhanced fruit Ca concentration	Colla et al. (2017)
Kappaphycus alvarezii	Foliar spray in field experiment	Soybean	Enhanced N, P, K and S uptake of grain and N, P uptake of straw	Rathore et al. (2009)
Ecklonia maxima	Root application in pot experiment	Rapeseed	Enhanced Leaf P and K concentration	Di Stasio et al. (2017)
Ecklonia maxima	Liquid feed with Hoaglands solutions	Lettuce	Enhanced Ca, K and Mg concentrations in produce	Crouch et al. (1990)
Ecklonia maxima	Liquid feed with Hoaglands solutions	Bean	Enhanced Ca, K and Mg concentrations	Beckett et al. (1994)
Laminaria spp.	Nutrient solution in potculture experiment	Maize	Enhanced Ca, Mg and Zn concentrations	Ertani et al. (2018)
Ascophyllum nodosum	Nutrient solution in potculture experiment	Maize	Enhanced the ability of the plant to absorb Ca, Mg, S, Fe, Cu, Mn, Mo, Zn, and B	Ertani et al. (2018)
Kappaphycus alvarezii	Foliar spray in field experiments	Rice	Enhanced the uptake of N and P	Layek et al. (2018)
Gracilaria edulis	Foliar spray in field experiments	Rice	Enhanced the uptake of N and P	Layek et al. (2018)

(continued)

Table 26.1 (continued)

Seaweed species	Mode of application	Crop Plant	Effect on mineral nutrition (uptake and translocation)	References
Ascophyllum nodosum and other commercial Extracts	Foliar application in pot experiment	Grapevine	Improved uptake of N, P, K, Ca, Mg, Fe, Zn, Mn and Cu	Turan and Kose (2004)
Ecklonia maxima	Root flush in pot experiment	Wheat	Improved uptake of K	Beckett and van Staden (1989a, b)

percentage, germination index and average time of seed germination, seedling height, fresh, dry mass of shoots. Liquid seaweed fertilizer made from *Sargassum plagiophyllum* and other commercial seaweed extract promoted seed germination and seedling growth of black gram and green gram seeds (Venkataraman and Mohan 1997). The crude extracts of brown seaweeds *Sargassum* and *Padina* promoted germination of red gram (*Cajanus cajan*) when seeds were soaked for 24 hours in the seaweed extracts (Mohan et al. 1994).

26.5 Root and Shoot Growth

Seaweed extracts play a crucial role in promoting root growth and development (Metting et al. 1990; Jeannin et al. 1991). The application of seaweed extracts not only improves root growth but also the overall vegetative growth of the plant (Fig. 26.1). The stimulatory effect of the seaweed extracts on root growth is more pronounced when applied at an early growth stage of maize. The promotion of root growth by application of seaweed extracts similar to that of auxin may be attributed to the presence of the phytohormone in the marine algal extracts (Jeannin et al. 1991). The application of seaweed extracts increases root size and vigor, which helps in reducing the transplant shock in marigold, cabbage, and tomato seedlings (Aldworthm and van Staden 1987; Crouch and van Staden 1992). Crouch et al. 1990 found that application of seaweed extracts greatly influenced growth and mineral nutrition of lettuce by increasing the concentration of minerals such as Ca, K and Mg in the leaf tissues as well as yield. Crouch and van Staden (1992) found that the application of seaweed concentrates stimulated root growth, increased root: shoot ratio and promoted biomass accumulation of tomato. The foliar spray of seaweed extracts as well as root application promoted root growth (Finnie and van Staden

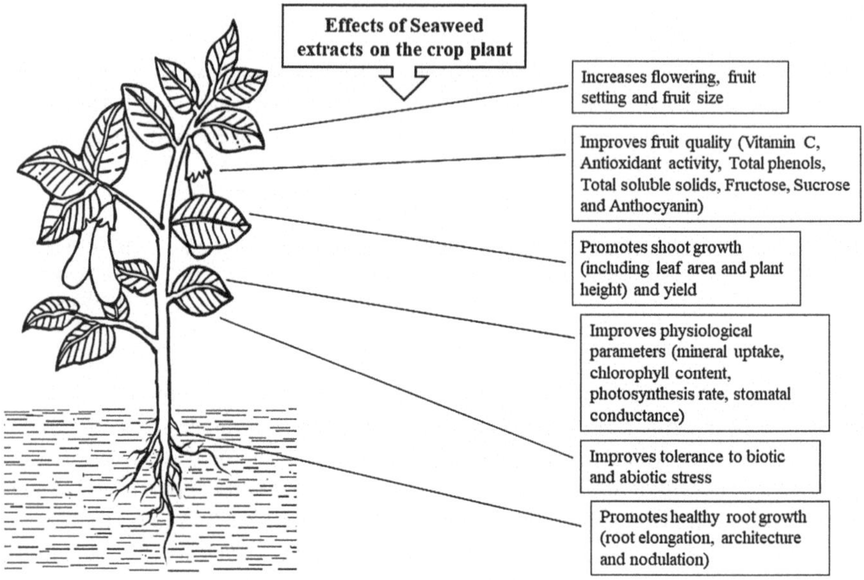

Fig. 26.1 Effects of the seaweed extracts on the crop plan

1985). However, the efficacy of the seaweed extracts in the promotion of root growth was found to be influenced by the concentration of the algal extract as seen in tomato plants. The higher concentrations (1:100 of seaweed extract: water) impaired root growth but a lower concentration (1:600) stimulated root growth. The seaweed extract-based biostimulants are found to influence root growth both by strengthening lateral root structure and by increasing the overall root system volume (Vernieri et al. 2005; Mancuso et al. 2006). According to Tavares et al. (2020), the application of extracts of the brown algae *Ascophyllum nodosum* improved the roots and root system morphology of ornamental marigold (*Tagetes erecta*). The bio-stimulants present in the marine algal extracts can promote root development by enhancing formation of lateral roots and increasing the total root volume (Vernieri et al. 2005; Khan et al. 2009). It has been reported that the application of seaweed extracts not only improved crop growth but also increased the number of functional nodules in treated plants (Begum et al. 2018), which might be attributed to the presence of cytokinins including trans-zeatin riboside and its dihydro derivatives in the extracts of several brown algae (Khan et al. 2009). It has also been documented that the bioactive compounds and their organic sub-fractions in *Ascophyllum nodosum* extract have impaired signaling processes of rhizobia-legume symbiosis, resulting in more stable nodules and an overall increase in plant growth (Khan et al. 2013; Begum et al. 2018).

Seaweed products are also used in conventional vegetative propagation (Crouch and van Staden 1991; Atzmon and van Staden 1994; Kowalski et al. 1999) in many crop species. When treated with 10% seaweed concentrate Kelpak (*Ecklonia maxima*) for about 18 h, enhanced rooting has been observed in marigold (*Tagetes patula*) (Crouch and van Staden 1991). Application of Kelpak (*Ecklonia maxima*) at a dilution of 1: 100, was found to increase rooting and improve the vigor of the roots in stem cuttings of *Pinus pinea,* which was otherwise difficult to root (Atzmon and van Staden 1994). Foliar spraying of the extracts of *Ascophyllum nodosum*, supplemented with BA and IBA was found to increase the number of propagules per plant in the ornamental herbaceous perennial *Hemerocallis sp.* (Leclerc et al. 2006).

26.6 Effect on Mineral Absorption

Application of various seaweed products such as seaweed liquid fertilizer and seaweed manure improve soil texture and structure, which facilitates better aeration of the soil, enhanced nitrogen fixation, and improved proliferation of beneficial soil microorganisms. The capillary property of the soil is also improved by applying seaweed extracts, which favours the growth of the root system. Seaweed extracts also improved mobility of some trace elements, which otherwise remain in unavailable form (Zodape 2001). The nutrient use efficiency (NUE)which is the measure of nutrient uptake of a plant can be improved by the application of seaweed extracts. Many experiments were conducted to study the effect of different marine algal extracts on the nutrient uptake in a number of crop plants (Table 26.1). The application of the extracts of the seaweeds *Laminaria* spp. and *Ascophyllum nodosum* significantly improved absorption of a number of mineral elements such as Ca, Mg, S, Fe, Cu, Mn, Mo, Zn, and B by the maize plants (El Boukhari et al. 2020; Ertani et al. 2018). According to Di Stasio et al. (2017) extracts of the marine alga *Ecklonia maxima* increased P and K composition of leaves of *Brassica rapa* L. under nutrient stress conditions (El Boukhari et al. 2020). The seaweed extracts improved water and nutrient use efficiency, leading to increased growth and vigor in plants (Khan et al. 2009; Begum et al. 2018) as well as improved nutrient uptake by the roots (Crouch et al. 1990). The foliar application of the sap of *Kappaphyccus* spp. and *Gracilaria* spp. increased the absorption of N and P by the sesame grain (Pramanick et al. 2014). The role of seaweed extracts in the improvement of nutrient use efficiency of plants might be attributed to their role in the upregulation of some genes encoding nutrients transporters of the root (El Boukhari et al. 2020).

26.7 Effect on Photosynthesis

The application of seaweed extracts is also known to improve photosynthesis by enhancing that the chlorophyll content of plants (Blunden et al. 1997). Soil or foliage application of extracts of the brown algae *Ascophyllum nodosum* at a low concentration was found to increase the chlorophyll content of tomato plant leaves.

Betaine compounds present in the seaweed extracts retarded chlorophyll degradation, which may be attributed to an increase in chlorophyll content of leaves of treated plants (Fig. 26.1). The inhibition of chlorophyll degradation by glycine betain during storage conditions in isolated chloroplast might be responsible for the maintenance of photosynthesis without any loss of photosynthetic activity. Different concentrations of *Ascophyllum nodosum* extracts at very low concentration (0.1 g l^{-1}) was found to affect the root growth of *Arabidopsis*, whereas a higher concentration of the seaweed extract (1 g l^{-1}) improved plant height and number of leaves greatly (Rayorath et al. 2008). The application of a novel seaweed extract obtained from *Sargassum horneri*, was found to improve chlorophyll content and photosynthetic capacity of tomato leaves (Yao et al. 2020). Krajnc et al. (2012) studied the effect of the extracts of the brown algae *Ecklonia maxima* on the photosynthetic pigment content of pelagornium (*Pelargonium peltatum*) leaf tissue. The chlorophyll a, chlorophyll b and carotenoid content of the leaves of treated pelagornium cuttings was found to be significantly higher than the control. Application of *Ascophyllum nodosum* extracts in hydroponic solution influenced the levels of expression of genes involved in ABA-responsive and antioxidative system pathways and induced partial stomatal closure in *Arabidopsis* plants grown under drought stress. The *Ascophyllum nodosum* extracts-treated plants show better photosynthetic performance compared to control throughout the dehydration period and higher capacity to dissipate the excess of energy as heat in the reaction centers of photosystem II may be attributed to the pre-activation of these pathways by the application of the seaweed extract (Santaniello et al. 2017).

26.8 Effect on Crop Yield

Effects of seaweed extracts on crop yield may be attributed to various hormonal substances present, especially cytokinins (Featonby-Smith and van Staden 1984a, b). Cytokinin in seaweed concentrate helps to shift photosynthate distribution from vegetative parts (roots, stem, and young leaves) to the developing fruit and promotes fruit development (Panda et al. 2012).

The application of seaweed extract induces early flowering and increases fruit yield when sprayed on tomato plants during the vegetative stage. It led to large-sized fruits with superior quality (Crouch and van Staden 1992). Seaweed concentrates also trigger early flowering and fruit set in many crop plants. Seaweed extract of *Ecklonia maxima* increased the number of flowers and seeds flowerhead in marigold when applied immediately after transplanting (Aldworthm and van Staden 1987). Application of *Ascophyllum nodosum* increased the yield of cauliflower, lettuce, and maize (Abetz and Young 1983; Jeannin et al. 1991). Foliar application of *Ecklonia maxima* extracts is known to enhance yield in beans, wheat, barley, and peppers (Temple and Bomke 1989; Nelson and van Staden 1984; Beckett and van Staden 1989a, b; Arthur et al. 2003). Application of

Ascophyllum nodosum extract has also been shown to have positive effects on the yield of Thompson seedless grape (Norrie and Keathley 2006). Seaweed extracts also encourage flowering by initiating robust plant growth (Abetz and Young 1983). Yao et al. (2020) studied the effect of novel seaweed extract obtained from the brown algae *Sargassum horneri* on the yield, quality, ripening time of tomatoes. The results revealed that the application of seaweed extracts shortened the ripening time of tomato and significantly increased tomato yield which can be attributed to the improved photosynthetic capacity of the treated plant. The seaweed extracts of *Sargassum horneri* also increased the hardness of tomato which can help to minimize losses during transportation and storage. Foliar application of *Kappaphyccus* extract along with the required dose of fertilizers improved yield attributes such as the number of panicles m^{-2}, filled grain panicle^{-1}, panicle length, and 1000 grain weight and thus improved grain yield and straw yield (Begum et al. 2018). Foliar spray of *Gracilaria* sap along with recommended dose of fertilizers also increased the yield attributes of rice. (Pramanick et al. 2014). The yield attributes rice such as the number of panicles hill^{-1} and the number of productive grains panicle^{-1} also increased significantly with application of higher concentration (15 percent) of *Kappaphyccus* sap (Layek et al. 2018). The foliar application of seaweed extracts of Kappaphycus *alvarezii* and *Gracilaria edulis* at different concentrations twice during the crop growth period enhanced the yield and improved the quality of the black gram [*Vigna mungo* L.) Hepper]. The seed yield, as well as other yield characteristics such as number of pods per plant; pod weight; seed weight per plant; and test weight of seed of black gram; were found to be increased with the application of 10% concentration of and *Kappaphycus alvarezii*, *Gracilaria edulis* respectively (Jadhao et al. 2015). The foliar application of seaweed extracts also significantly increased uptake of most of the mineral nutrients. The field studies on the effects of the extracts of *Kappaphycus alvarezii* (K sap) and *Gracilaria edulis* (G sap) on potato revealed that application of 10% both K and G sap along with a recommended dose of fertilizer (RDF) significantly improved the growth and yield attributes like plant height, number of stems and tubers per hill. The marketable and total tuber yield of potato was found to increase markedly by application of 10% sap of both *Kappaphycus alvarezii* and *Gracilaria edulis* along with RDF (Prajapati et al. 2016). According to Sethi and Adhikary (2008), foliar application of 1% of seaweed extracts enhanced the yield parameters such as fruit length and fruit weight of vegetables such as brinjal and tomato.

26.9 Role in Abiotic Stresses Tolerance in Crop Plants

Different abiotic stresses such as drought, salinity, high temperature, and nutrient deficiency not only affects the growth and physiology but also the productivity of the crops adversely. Many of the recent studies conducted to study the effect of seaweed extracts on the mitigation of abiotic stresses experienced by crop plants showed prominent results. The application of various seaweed extracts was found to improve the stress tolerance ability of many crops (Fig. 26.1). It has been reported that seaweed extracts from *Ascophyllum nodosum* contain different betaine compounds such as glycine betaine, γ-aminobutyric acid betaine, and δ-aminovaleric acid betaine which have antioxidant activity (Mondal and Panda 2019). According to Mancuso et al. (2006), foliar application of seaweed extracts enhanced salt and freezing tolerance in potted grape (*Vitis vinifera*) plants. Commercial formulations of *Ascophyllum* extracts improve freezing tolerance in grapes. The substantial amounts of cytokinins present in the seaweed extracts help to mitigate stress-induced free radicals. It may be due to direct scavenging and restricting ROS formation by inhibition of xanthine oxidation (Fike et al. 2001). Cytokinin components in seaweed extracts may help heat tolerance in seaweed extract-induced plants (Zhang and Ervin 2008). Commercial seaweed extracts like Kelpak mediate stress tolerance by enhancing K+ uptake in plants. Exogenous application of the extracts of the brown algae *Ascophyllum nodosum* and *Ecklonia maxima* also helped in mitigation of a number of abiotic stresses in crop plants (Panda et al. 2012). Zou et al. (2019) studied the potential of polysaccharides extracted from the brown algae *Lessonia nigrescens* on the salinity tolerance ability of wheat seedlings. It has been reported that the application of this seaweed extract enhanced lengths of shoot and root and dry and fresh matter accumulation of wheat significantly under salinity stress, which may be attributed to the positive role of the seaweed extracts in attenuation of the salt stress-induced oxidative damage of plants by decreasing the electrolyte leakage and malondialdehyde content and by increasing the activity of antioxidant enzymes such as SOD, POD and CAT involved in ROS scavenging (El Boukhari et al. 2020). The application of the polysaccharides extracted from *Lessonia nigrescens* also helped in maintaining the osmotic status of wheat seedlings under stress by increasing their sugar and proline content and regulating the Na+/K+ ratio (El Boukhari et al. 2020). Liu et al. (2019) investigated on the effect of polysaccharides extracted from *Grateloupia filicina* on the alleviation of salt stress in rice at seed germination stage and reported that the application of this algal extract stimulated rice seed development under salt stress. According to Elansary et al. (2017), application of the extract of the brown algae, *Ascophyllum nodosum* reduced H_2O_2 content and increased CAT and APX activity and ascorbate content of *Paspalum vaginatum* plants grown under drought and salinity stress (El Boukhari et al. 2020). It has been reported that the application of *Kappaphycus alvarezii* sap alleviated the effects of salt and drought stress in wheat varieties during vegetative and reproductive stages through enhancing the morphological parameters such as shoot and root length and weight (Patel et al. 2018). Application of *Kappaphycus alvarezii* extract also increased chlorophyll and carotenoids content, tissue water content, and reduced electrolyte leakage and lipid peroxidation in treated plants. K sap treatment also increased the concentration of phytohormones like Abscisic acid (ABA) and zeatin under stress conditions. This algal extract also decreased the ionic imbalance of the plant under abiotic stress by reducing the Na+/K+ ratio and improving calcium, proline, total protein, and amino acids content, and enhancing the accumulation of osmoprotectants (El Boukhari et al. 2020). Sharma et al. (2019) studied the effect of *Gracilaria dura* extract on the wheat plant and found that the seaweed extracts increased biomass and yield of wheat grown under drought stress by 57% and 70%, respectively, by facilitating the mechanisms involved in water-saving strategies.

26.10 Commercial Seaweed Products Used in Agriculture

A number of commercial seaweed products are available in the market for their use in agriculture and horticulture. Different companies across the world have developed several commercial seaweed products (Table 26.1). These seaweed products are either used as plant biostimulants or liquid biofertilizers. The use of seaweed products in agriculture was mainly confined to the coastal area of the world. But with the advent of commercial production of seaweed extracts, seaweed products are now available across the globe for sustainable crop production practices. Among various marine macroalgae commercially used to produce seaweed products such as plant growth regulators and biofertilizers, *Ascophyllum nodosum* is the most commonly used species (Khan et al. 2009). However, *Kappaphycus alvarezii* is also used In India for production of seaweed products by various companies. The seaweed products have huge market potential considering the proportion of seaweeds utilized at present for production of biofertilizers and biostimulants. More production of commercial seaweed products will not only promote affordable use of seaweed products in crop production practices but also help effective utilization of marine macro algae for sustainable agriculture (Table 26.2).

26.11 Conclusion

The seaweeds are a source of biofertilizers and bio-stimulants substances including phytohormones. The micronutrients present in seaweed extracts favour growth and yield in many crops. The positive effects of these extracts on various aspects of crop growth and development must be exploited more for their proper utilization in agriculture. The exhaustive use of seaweeds in agriculture can further help in the improvement of crop productivity. It can be achieved by further extensive research about the biochemical nature and mechanism of actions of the extracts' components. The application of SLFs can be an effective solution for crop production under various adverse conditions. It will further help reduce the use of harmful agrochemicals, leading to a sustainable way of crop production. This huge mass of renewable resources can be exploited to achieve food security through sustainable agriculture with minimal use of inorganic fertilizers.

Table 26.2 Commercial products of seaweeds used in the agriculture and horticulture

Name of the seaweed product	Company	Name of the algal species	Application
Algae green	Ocean knowledge, Ireland	*Ascophyllum nodosum*	Plant biostimulant
Seamac lion	Headland agrochemicals limited, UK	*Ascophyllum nodosum*	Liquid biofertilizer
Opteine	United potash limited, India	*Ascophyllum nodosum*	Plant biostimulant
Acadian	Acadian Agritech, Dartmouth, Canada	*Ascophyllum nodosum*	Plant growth stimulant
Agri-Gro ultra	Agri Gro marketing Inc. Doniphan, United States	*Ascophyllum nodosum*	Plant growth stimulant
Biovita granules Biovita liquid	PI industries, Udaipur, India	*Ascophyllum nodosum*	Plant growth regulators
Algaetech-10	Divine laboratory, Vadodora, India	Ascophyllum nodosum	Liquid biofertilizer
AgroKelp	Algas y Bioderivados Marinos, S.A. de C.V., Mexico	*Macrocystis pyrifera*	Plant growth stimulant
Organic dews premium liquid seaweed concentrate	OrganicDews, India	*Ascophyllum nodosum*	Liquid biofertilizer
Alg-A-mic	BioBizz worldwide N.V., Spain	*Ascophyllum nodosum*	Plant growth stimulant
Aquasap sx	AquAgri processing, Delhi, India	*Kappaphycus alvarezii*	Liquid biofertilizer
Espoma	The Espoma company, New Jersey, USA	*Ascophyllum nodosum*	Plant growth stimulant
Guarantee	Ocean organics, USA	*Ascophyllum nodosum*	Plant growth stimulant
Kelp meal	Acadian Seaplants ltd., Canada	*Ascophyllum nodosum*	Plant growth stimulant
Kelpak	BASF, Germany	*Ecklonia maxima*	Plant growth stimulant
Kelpro	Tecniprocesos Biologicos, S.A. de C.V., Mexico	*Ascophyllum nodosum*	Plant growth stimulant
Superfifty	BioAtlantis, Ireland	*Ascophyllum nodosum*	Liquid biofertilizer
Kelprosoil	Productos del Pacifico, S.A. de C.V., Mexico	*Ascophyllum nodosum*	Plant growth stimulant
Maxicrop	Maxicrop USA, Inc.	*Ascophyllum nodosum*	Plant growth stimulant
Nitrozime	Hydrodynamics international Inc., USA	*Ascophyllum nodosum*	Plant growth stimulant
Profert	BASF, Germany	*Durvillea antartica*	Plant biostimulant

(continued)

Table 26.2 (continued)

Name of the seaweed product	Company	Name of the algal species	Application
Sagarika	IFFCO, New Delhi, India	*Kappaphycus alvarezii*	Plant biostimulant
Seasol	Seasol international Pty ltd., Victoria, Australia	*Durvillea potatorum*	Plant growth stimulant
Alagaezyme	The unique Enterprise, Vadodara, Gujarat	*Ascophyllum nodosum*	Seaweed liquid biofertilizer
Soluble seaweed extract	Technaflora plant products, LTD	*Ascophyllum nodosum*	Plant growth stimulant
Stimplex	Acadian Agritech, Canada	*Ascophyllum nodosum*	Plant growth stimulant
Basfoliar algae SL	Agricare, New Delhi	Durvillea antarctic	Plant biostimulant
Synergy	Green air products, Inc., USA	*Ascophyllum nodosum*	Plant growth stimulant
Supernova	Noble crop science, Rajkot, India	*Ascophyllum nodosum*	Seaweed liquid fertilizer

Acknowledgement It is a great pleasure to thank all who have contributed to this review. The authors acknowledge the assistance from Visva-Bharati Library Network, West Bengal, India. Authors are also thankful to the editor(s) of this book for their suggestions on the manuscript of this chapter.

References

Abetz P, Young CL (1983) The effect of seaweed extract sprays derived from *Ascophyllum nodosum* on lettuce and cauliflower crops. Bot Mar 26:487–492

Aldworthm SJ, van Staden J (1987) The effect of seaweed concentrate on seedling transplants. S Afr J Bot 53:187–189

Ali N, Farrell A, Ramsubhag A, Jayaraman J (2016) The effect of *Ascophyllum nodosum* extract on the growth, yield and fruit quality of tomato grown under tropical conditions. J Appl Phycol 28:1353–1362. https://doi.org/10.1007/s10811-015-0608-3

Arthur GD, Stirk WA, van Staden J (2003) Effect of a seaweed concentrate on the growth and yield of three varieties of *Capsicum annuum*. S Afr J Bot 69:207–211

Atzmon N, van Staden J (1994) The effect of seaweed concentrate on the growth of *Pinus pinea* seedlings. New For 8:279–288

Basavaraja PK, Yogendra ND, Zodape ST, Prakash R, Ghosh A (2018) Effect of seaweed sap as foliar spray on growth and yield of hybrid maize. J Plant Nutr 41:1851–1861

Battacharyya D, Babgohari MZ, Rathor P, Prithiviraj B (2015) Seaweed extracts as biostimulants in horticulture. Sci Hortic 196:39–48

Beckett RP, Mathegka ADM, van Staden J (1994) Effect of seaweed concentrate on yield of nutrient-stressed tepary bean (*Phaseolus acutifolius* gray). J Appl Phycol 6:429–430

Beckett RP, van Staden J (1989a) The effect of seaweed concentrate on the growth and yield of potassium stressed wheat. Plant Soil 116:29–36

Beckett RP, van Staden J (1989b) The effect of seaweed concentrate on the growth and yield of potassium stressed wheat. Plant Soil 116:29–36

Begum M, Bijnan CB, Dhiman DS, Nayan JO (2018) Role of seaweed extract on growth, yield and quality of some agricultural crops: a review. Agric Rev 39(4):321–326

Billard V, Etienne P, Jannin L, Garnica M, Cruz F, Garcia-Mina JM, Yvin JC, Ourry A (2014) Two biostimulants derived from algae or humic acid induce similar responses in the mineral content and gene expression of winter oilseed rape (*Brassica napus* L.). J Plant Growth Regul 33:305–316

Blunden G (1991) Agricultural uses of seaweeds and seaweed extracts. In: Guiry MD et al (eds) Seaweed resources in Europe: uses and potential. John Wiley & Sons, Chichester, pp 65–81

Blunden G, Jenkins T, Liu Y (1997) Enhanced leaf chlorophyll levels in plants treated with seaweed extract. J Appl Phycol 8:535–543

Carrasco-Gil S, Hernandez-Apaolaza L, Lucena JJ (2018) Effect of several commercial seaweed extracts in the mitigation of iron chlorosis of tomato plants (*Solanum lycopersicum* L.). Plant Growth Regul 86:401–411

Chojnacka K, Saeid A, Witkowska Z, Tuhy L (2012) Biologically active compounds in seaweed extracts-the prospects for the application. Open Conf Proc J 3:20–28

Cole AJ, Roberts DA, Garside AL, De Nys R, Paul NA (2015) Seaweed compost for agricultural crop production. J Appl Phycol 28:629–642. https://doi.org/10.1007/s10811-015-0544-2

Colla G, Cardarelli M, Bonini P, Rouphael Y (2017) Foliar applications of protein hydrolysate, plant and seaweed extracts increase yield but differentially modulate fruit quality of greenhouse tomato. HortScience 52:1214–1220

Crouch IJ, Beckett RP, van Staden J (1990) Effect of seaweed concentrate on growth and mineral nutrition of nutrient stressed lettuce. J Appl Phycol 2:269–272

Crouch IJ, van Staden J (1991) Evidence for rooting factors in a seaweed concentrate prepared from *Ecklonia maxima*. J Plant Physiol 137:319–322

Crouch IJ, van Staden J (1992) Effect of seaweed concentrate on the establishment and yield of greenhouse tomato plants. J Appl Phycol 4:291–296

Crouch IJ, van Staden J (1993) Evidence for the presence of plant growth regulators in commercial seaweed products. Plant Growth Regul 13:21–29

Di Stasio E, Rouphael Y, Colla G, Raimondi G, Giordano M, Pannico A, El-Nakhel C, De Pascale S (2017) The influence of *Ecklonia maxima* seaweed extract on growth, photosynthetic activity and mineral composition of *Brassica rapa* L. subsp sylvestris under nutrient stress conditions. Eur J Hortic Sci 82:286–293

El Boukhari MEM, Barakate M, Bouhia Y, Lyamlouli K (2020) Trends in seaweed extract based biostimulants: manufacturing process and beneficial effect on soil-plant systems. Plants (Basel) 9:359. https://doi.org/10.3390/plants9030359

Elansary HO, Yessoufou K, Abdel-Hamid AME, El-Esawi MA, Ali HM, Elshikh MS (2017) Seaweed extracts enhance Salam Turfgrass performance during prolonged irrigation intervals and saline shock. Front Plant Sci 8:830. https://doi.org/10.3389/fpls.2017.00830

El-Sheekh MM, Ismail MM, Hamouda MM (2016) Influence of some Brown seaweed extracts on germination and cytological responses of *Trigonella foenum-graecum* L. Biotechnol Ind J 12:104

Ertani A, Francioso O, Tinti A, Schiavon M, Pizzeghello D, Nardi S (2018) Evaluation of seaweed extracts from *Laminaria* and *Ascophyllum nodosum* spp. as biostimulants in *Zea mays* L. using a combination of chemical, biochemical and morphological approaches. Front Plant Sci 9:428. https://doi.org/10.3389/fpls.2018.00428

Eyras MC, Rostagno CM, Defosse GE (1998) Biological evaluation of seaweed composting. Comp Sci Util 6:74–81

FAO (2018) In: Ferdouse F, Yang Z, Holdt SL, Murúa P, Smith R (eds) The global status of seaweed production, trade and utilization, volume 124. FAO Consultants. Globefish Research Programme, Rome, Italy

Featonby-Smith BC, van Staden J (1984a) The effect of seaweed concentrate and fertilizer on growth and the endogenous cytokinin content of *Phaseolus vulgaris*. S Afr J Bot 3:375–379

Featonby-Smith BC, van Staden J (1984b) The effect of seaweed concentrate and fertilizer on growth and the endogenous cytokinin content of *Phaseolus vulgaris*. S Afr J Bot 3:375–379

Featonby-Smith BC, van Staden J (1987) Effects of seaweed concentrate on grain yield in barley. S Afr J Bot 53:125–128

Fike JH, Allen VG, Schmidt RE, Zhang X, Fontenot JP, Bagley CP, Ivy RL, Evans RR, Coelho RW, Wester DB (2001) Tasco-forage I: influence of a seaweed extract on antioxidant activity in tall fescue and in ruminants. J Anim Sci 79:1011–1021

Finnie JF, van Staden J (1985) Effect of seaweed concentrate and applied hormones on in vitro cultured tomato roots. J Plant Physiol 120:215–222

Ganapathy SG, Balamurugan M, Thinakaran T, Sivakumar K (2013) Developmental changes in the germination, growth and chlorophyllase activity of *Vigna mungo* L. using seaweed extract of *Ulva reticulata* Forsskål. Int Res J Pharma 4:252–254

Hernández-Herrera RM, Santacruz-Ruvalcaba F, Ruiz-López MA, Jeffrey N, Hernández-Carmona G (2014) Effect of liquid seaweed extracts on growth of tomato seedlings (*Solanum lycopersicum* L.). J Appl Phycol 26:619–628. https://doi.org/10.1007/s10811-013-0078-4

Jadhao GR, Chaudhary D, Khadse VA, Zodape ST (2015) Utilization of seaweeds in enhancing productivity and quality of black gram (*Vigna mungo* L.) Hepper for sustainable agriculture. Indian J Nat Prod Resour 6(1):16–22

Jannin L, Arkoun M, Etienne P, Laîné P, Goux D, Garnica M, Fuentes M, Francisco SS, Baigorri R, Cruz F et al (2013) *Brassica napus* growth is promoted by *Ascophyllum nodosum* (L.) Le Jol. Seaweed extract: microarray analysis and physiological characterization of N, C, and S metabolisms. J Plant Growth Regul 32:31–52

Jeannin I, Lescure JC, Morot-Gaudry JF (1991) The effects of aqueous seaweed sprays on the growth of maize. Bot Mar 34:469–473

Khan W, Palanisamy R, Critchley AT, Smith DL, Papadopoulos Y, Prithiviraj B (2013) *Ascophyllum nodosum* extract and its organic fractions stimulate rhizobium root nodulation and growth of *Medicago sativa* (alfalfa). Commun Soil Sci Plant Anal 44:900–908

Khan W, Rayirath UP, Subramanian S, Jithesh MN, Rayorath P, Hodges DM, Critchley AT, Craigie JS, Norrie J, Prithiviraj B (2009) Seaweed extracts as biostimulants of plant growth and development. J Plant Growth Regul 28:386–399

Kowalski B, Jager AK, van Staden J (1999) The effect of a seaweed concentrate on the in vitro growth and acclimatization of potato plantlets. Potato Res 42:131–139

Krajnc AU, Ivanus A, Kristl J, Susek A (2012) Seaweed extract elicits the metabolic responses in leaves and enhances growth of Pelagornium cuttings. Eur J Hortic Sci 77:170–181

Layek J, Das A, Idapuganti RG, Sarkar D, Ghosh A, Zodape ST, Lal R, Yadav GS, Panwar AS, Ngachan S, Meena RS (2018) Seaweed extract as organic bio-stimulant improves productivity and quality of rice in eastern Himalayas. J Appl Phycol 30:547–558

Leclerc M, Caldwell CD, Lada RR, Norrie J (2006) Effect of plant growth regulators on propagule formation in *Hemerocallis* spp. and *Hosta* spp. Hortic Sci 41:651–653

Liu H, Chen X, Song L, Li K, Zhang X, Liu S, Qin Y, Li P (2019) Polysaccharides from Grateloupia filicina enhance tolerance of rice seeds (*Oryza sativa* L.) under salt stress. Int J Biol Macromol 124:1197–1204. https://doi.org/10.1016/j.ijbiomac.2018.11.270

Mancuso SF, Azzarello E, Mugnai S, Briand X (2006) Marine bioactive substances (IPA extract) improve ion fluxes and water stress tolerance in potted *Vitis vinifera* plants. Adv Hortic Sci 20:156–161

Mettling B, Zimmerman WJ, Crouch IJ, van Staden J (1990) Agronomic uses of seaweed and microalgae. In: Akatsuka I (ed) Introduction to applied phycology. SPB Academic Publishing, The Hague, Netherlands, pp 269–627

Mohan VR, Venkataraman K, Murugeswari R, Muthusamy S (1994) Effect of crude and commercial seaweed extract on seed germination and seedling growth in *Cajanus cajan* L. Phykos 33:47–51

Mondal S, Panda D (2019) Seaweed as source of plant growth promoters and bio-fertilizers: an overview. In: Ravishankar GA, Ranga Rao A (eds) Handbook of algal technologies and phytochemicals: volume II phytoremediation, biofuels and global biomass production, vol 2. Boca Raton, CRC Press, pp 111–122. https://doi.org/10.1201/9780429057892

Nelson WR, van Staden J (1984) The effect of seaweed concentrate on wheat culms. J Plant Physiol 115:433–437

Norrie J, Keathley JP (2006) Benefits of Ascophyllum nodosum marine-plant extract applications to 'Thompson seedless' grape production. Acta Hortic 727:243–247

Panda D, Pramanik K, Nayak BR (2012) Use of seaweed extracts as plant growth regulators for sustainable agriculture. Int J Bio-resource Stress Manag 3:404–411

Patel K, Agarwal P, Agarwal PK (2018) *Kappaphycus alvarezii* sap mitigates abiotic induced stress in *Triticum durum* by modulating metabolic coordination and improves growth and yield. J Appl Phycol 30:2659–2673

Povero G, Mejia JF, Tommaso DD, Piaggesi A, Warrior P (2016) A systematic approach to discover and characterize natural plant biostimulants front plant Scil. https://doi.org/10.3389/fpls.2016.00435

Prajapati A, Patel CK, Singh N, Jain SK, Chongtham SK, Maheshwari MN, Patel CR, Patel RN (2016) Evaluation of seaweed extract on growth and yield of potato. Environ Ecol 34(2):605–608

Pramanick B, Brahmachari K, Ghosh A (2014) Efficacy of Kappaphycus and Gracilaria sap on growth and yield improvement of sesame in new alluvial soil. J Crop and Weed 10(1):77–81

Ramya SS, Vijayanand N, Rathinavel S (2015) Foliar application of liquid biofertilizer of brown alga *Stoechospermum marginatum* on growth, biochemical and yield of Solanum melongena. Int J Recycl Org Waste Agricult 4:167–173. https://doi.org/10.1007/s40093-015-0096-0

Rathore SS, Chaudhary DR, Boricha GN, Ghosh A, Bhatt BP, Zodape ST, Patolia JS (2009) Effect of seaweed extract on the growth, yield and nutrient uptake of soybean (*Glycine max*) under rainfed conditions. S Afr J Bot 75:351–355

Rayorath P, Narayanan JM, Farid A, Khan W, Palanisamy R, Hankins S, Critchley AT, Prithiviraj B (2008) Rapid bioassays to evaluate the plant growth promoting activity of *Ascophyllum nodosum* (L.) Le Jol. Using a model plant, *Arabidopsis thaliana* (L.) Heynh. J Appl Phycol 20:423–429

Santaniello A, Scartazza A, Gresta F, Loreti E, Biasone A, Tommaso DD, Piaggesi A, Perata P (2017) *Ascophyllum nodosum* seaweed extract alleviates drought stress in *Arabidopsis* by affecting photosynthetic performance and related gene expression. Front Plant Sci. https://doi.org/10.3389/fpls.2017.01362

Sethi SK, Adhikary SP (2008) Effect of seaweed liquid fertilizer on vegetative growth and yield of black gram, brinjal and tomato. Seaweed Res Utiln 30:241–248

Shama A, Joyce SG, Mari FD, Ranga Rao A, Ravishankar GA, Hudaa N (2019) Macroalgae and microalgae: novel sources of functional food and feed. In: Ravishankar GA, Rao AR (eds) Handbook of Algal Technologies and Phytochemicals: Volume-I: Food, Health and Nutraceutical Applications. CRC Press, USA, pp 207–219

Sharma S, Chen C, Khatri K, Rathore MS, Pandey SP (2019) *Gracilaria dura* extract confers drought tolerance in wheat by modulating abscisic acid homeostasis. Plant Physiol Biochem 136:143–154. https://doi.org/10.1016/j.plaphy.2019.01.015

Sivasankari S, Venkatesalu V, Anantharaj M, Chandrasekaran M (2006) Effect of seaweed extracts on the growth and biochemical constituents of *Vigna sinensis*. Bioresour Technol 97:1745–1751

Stamatiadis S, Evangelou L, Yvin JC, Tsadilas C, Mina JMG, Cruz F (2015) Responses of winter wheat to *Ascophyllum nodosum* (L.) Le Jol. Extract application under the effect of N fertilization and water supply. J Appl Phycol 27:589–600

Tavares AR, dos Santos PLF, Zabotto AR et al (2020) Seaweed extract to enhance marigold seed germination and seedling establishment. SN Appl Sci 2:1792. https://doi.org/10.1007/s42452-020-03603-3

Temple WD, Bomke AA (1989) Effects of kelp (*Macrocystis integrifolia* and *Ecklonia maxima*) foliar applications on bean crop growth. Plant Soil 117:85–92. https://doi.org/10.1007/BF02206260

Thirumaran G, Arumugam M, Arumugam R, Anantharaman P (2009) Effect of seaweed liquid fertilizer on growth and pigment concentration of *Abelmoschus esculentus* (I) medikus. Am-Euras J Agron 2:57–66

Turan M, Kose C (2004) Seaweed extracts improve copper uptake of grapevine. Acta Agric Scand B 54:213–220

Venkataraman K, Mohan VR (1997) The effect of liquid seaweed fertilizer on blackgram. Phykos 36:43–47

Vernieri P, Borghesi E, Ferrante A, Magnani G (2005) Application of biostimulants in floating system for improving rocket quality. J Food Agric Environ 3:86–88

Yao Y, Wang X, Chen B, Zhang M, Ma J (2020) Seaweed extract improved yields, leaf photosynthesis, ripening time, and net returns of tomato (*Solanum lycopersicum* mill.). ACS. Omega 5(8):4242–4249. https://doi.org/10.1021/acsomega.9b04155

Zhang X, Ervin EH (2008) Impact of seaweed extract-based cytokinins and zeatin riboside on creeping bent grass heat tolerance. Crop Sci 48:364–370

Zodape ST (2001) Seaweeds as biofertilizer. J Sci Ind Res 60:378–382

Zou P, Lu X, Zhao H, Yuan Y, Meng L, Zhang C, Li Y (2019) Polysaccharides derived from the brown algae *Lessonia nigrescens* enhance salt stress tolerance to wheat seedlings by enhancing the antioxidant system and modulating intracellular ion concentration. Front Plant Sci 10:1–15

Chapter 27
Energy and Economic Nexus of Seaweeds

T. V. Ramachandra and Deepthi Hebbale

27.1 Introduction

27.1.1 Macroalgae

Macroalgae, commonly known as seaweeds are multicellular, photosynthetic plants exhibiting wide range of variations in their morphology and reproduction. Multicellular algae consist of haptera (root like), stipes (stem) and blades (leaves). They are devoid of specialized conducting tissues and their reproductive organs are simple cells or masses of cells producing gametes. They lack the embryo and multicellular envelope around sporangia and gametangia, the freshwater charophytes being some exception. They are comparable to higher plants in their biochemical and metabolic pathways, especially in those where chlorophylls constitute as the main photosynthetic pigments (Guiry and Guiry 2008; Pereira and Neto 2014). Algae encompass several phyla and a vast array of forms, ranging from benthic to free-floating forms, from microscopic phytoplankton to minute cells and colonies or filaments epiphytic on larger plants to larger macroalgae or seaweeds. Over the last thirty years', classification of algae has witnessed major transitions while yet to

T. V. Ramachandra (✉)
Energy & Wetlands Research Group, Centre for Ecological Sciences, Bangalore, India

Centre for Sustainable Technologies, Bangalore, India

Centre for infrastructure, Sustainable Transportation and Urban Planning [CiSTUP], Indian Institute of Science, Bangalore, India
e-mail: tvr@iisc.ac.in

D. Hebbale
Energy & Wetlands Research Group, Centre for Ecological Sciences, Bangalore, India

Centre for Sustainable Technologies, Bangalore, India

A. Ranga Rao, G. A. Ravishankar (eds.), *Sustainable Global Resources of Seaweeds Volume 1*, https://doi.org/10.1007/978-3-030-91955-9_27

Table 27.1 Classification scheme of different algal groups

Kingdom	Phylum	Subphylum	Class
Prokaryota	Cyanophyta		Cyanophyceae
Eukaryota	Glaucophyta		Glaucophyceae
	Rhodophyta	Cyanidiophytina	Cyanidiophyceae
		Eurhophytina	Compsopogonophyceae Poryphyridophyceae Rhodellophyceae Stylonematophyceae Florideophyceae
	Cryptophyta		Cryptophyceae
	Dinotophyta		Dinophyceae
	Haptophyte		Haptophyceae
	Ochrophyta	Khakista	Bacillariophyceae Bolidophyceae
		Phaeista	Chryosophyceae Synurophyceae Eustigmatophyceae Raphidophyceae Dictyochophyceae Pelagophyceae Pinguiophyceae Phaeothamniophyceae Chrysomerophyceae Xanthophyceae Phaeophyceae
	Euglenophyta		Euglenophyceae
	Chlorarachinophyta		Chlorarachinophyceae
	Chlorophyta	Prasinophytina	Prasinophyceae
		Tetraphytina	Chlorophyceae Chlorodendrophyceae Trebouxyophyceae Ulvophyceae Dasycladophyceae
	Charophyta		Coleochaetophyceae Conjugatophyceae Mesotigmatophyceae Klebsormidiophyceae Charophyceae

Source: (Pereira and Neto 2014; Smith 1938)

arrive consensus of an acceptable general scheme. Table 27.1 summarizes a system as per Yoon et al. (2006) for red algae, Leliaert et al. (2012) for green algae, Riisberg et al. (2009) and Yoon et al. (2009) for Ochrophytes, as available from Algaebase (Guiry and Guiry 2008. Seaweeds grow predominantly in marine environment and to lesser extent in brackish waters. Their classification is primarily based on their photosynthetic pigments, which impart them characteristic ranges of colors as well. The Chlorophyta group are green algae, while Phaeophyta are brown and Rhodophyta are red.

27.1.2 Seaweed Structure

Seaweeds consist of thallus which are devoid of roots, stem and leaf. Thallus include superficial leaflike blades, stem-like stipes and often having attaching organs called holdfast or haptera. The ***blades*** have varied shapes ranging from flat, tubular or round depending on taxa. The ***stipe*** is elongated, often-thick stalk of seaweed, superficially resembling stem, keeping the plant erect, while properly exposing the blades to incident light. The ***holdfast*** and ***haptera*** are structures attaching the seaweeds to the substratum. Not all seaweeds (Fig. 27.1) have definite stipe and holdfast, which are characteristic of Laminaria species. Holdfast is a specialized structure, an attaching organ on the base of a seaweed, which anchors the thallus firmly to stable surfaces like rocks. These rhizoidal haptera secrete mucilaginous substance to adhere to the substratum (Hardy and Moss 1979; Moss 1975; Tovey

Fig. 27.1 Different structures represented in all Red, Green and Brown seaweed of fragmentation. Fragments after settlement forms one or more adhesive rhizoids and subsequently new filamentous shoots

and Moss 1978). ***Floats*** are inflated vesicles or balloon like structures or parts of the blade itself helping to keep buoyancy in water. The seaweeds absorb mineral nutrients directly from the seawater through their entire surface. The photosynthetic pigments of these algae are located along a membrane system, forming flat vesicles called thylakoids. These thylakoids are free in stroma, the main body of the plastid, remaining in singles or two or more thylakoids remain in groups called lamella. In green algae, thylakoids are in clusters interconnected by other thylakoids forming a compact stack known as grana. In red algae, thylakoids are not grouped and are associated with granules, the Phycobilisomes, where Phycobiliproteins (mainly phycoerythrin and phycocyanin) are contained whereas in brown algae, thylakoids form packs of three surrounded by a band of three thylakoids or a girdle lamella (Guiry 2014; Pereira and Neto 2014).

27.1.3 Seaweed Reproduction

Seaweeds reproduction are either vegetative, asexual or sexual mode (Fig. 27.2). Stoloniferous outgrowths of creeping axes on the substratum, if break up, the separated thalli can live as independent plants. Such fragmentation is a form of vegetative reproduction. Physical forces like wave action, or chemical damages caused by insolation can result in fragmentation. *Gelidium* and *Caulerpa* propagate on hard substrata (Santelices 1990) while *Gracilaria* propagates when thallus fragments are partially buried in soft bottoms (Barnes 1999). *Enteromorpha* and *Sargassum* propagate by various forms. Asexual reproduction is a common mode of reproduction in seaweeds. This is accomplished by special cells called spores produced in special chambers called sporangia. The spore producing parent plant is called sporophyte. After their release from the sporophyte, the spores settle down on the substratum and into male and female plants called gametophytes. The gametophytes reproduce sexually by producing gametes (sperm or eggs). The sperm and eggs are either retained within the gametophyte plant body or released into the water. Egg is fertilized on the sperm and forms a zygote. Zygotes develops and into sporophytes, and the life cycle continues in alternation, the diploid sporophyte (formed by fusion of haploid gametes) produce spores after meiosis; the haploid spores germinate into haploid gametophyte plants resulting in alternation of generations, typical of several algae. This is only a generalized pattern of life cycle in marine algae. In a few species, there is an alternating sexual and asexual reproductive process with every generation. All offspring resulting from vegetative reproduction are called clones. They are genetically identical to each other and the parent seaweed. Vegetative and sexual reproduction, however, involves larger expenditures of resources than asexual reproduction, with greater risks of reproductive failures. Therefore, asexual reproduction, by small-sized, multi or unicellular propagules is an economic way by which population can be increased (Fenner 2012).

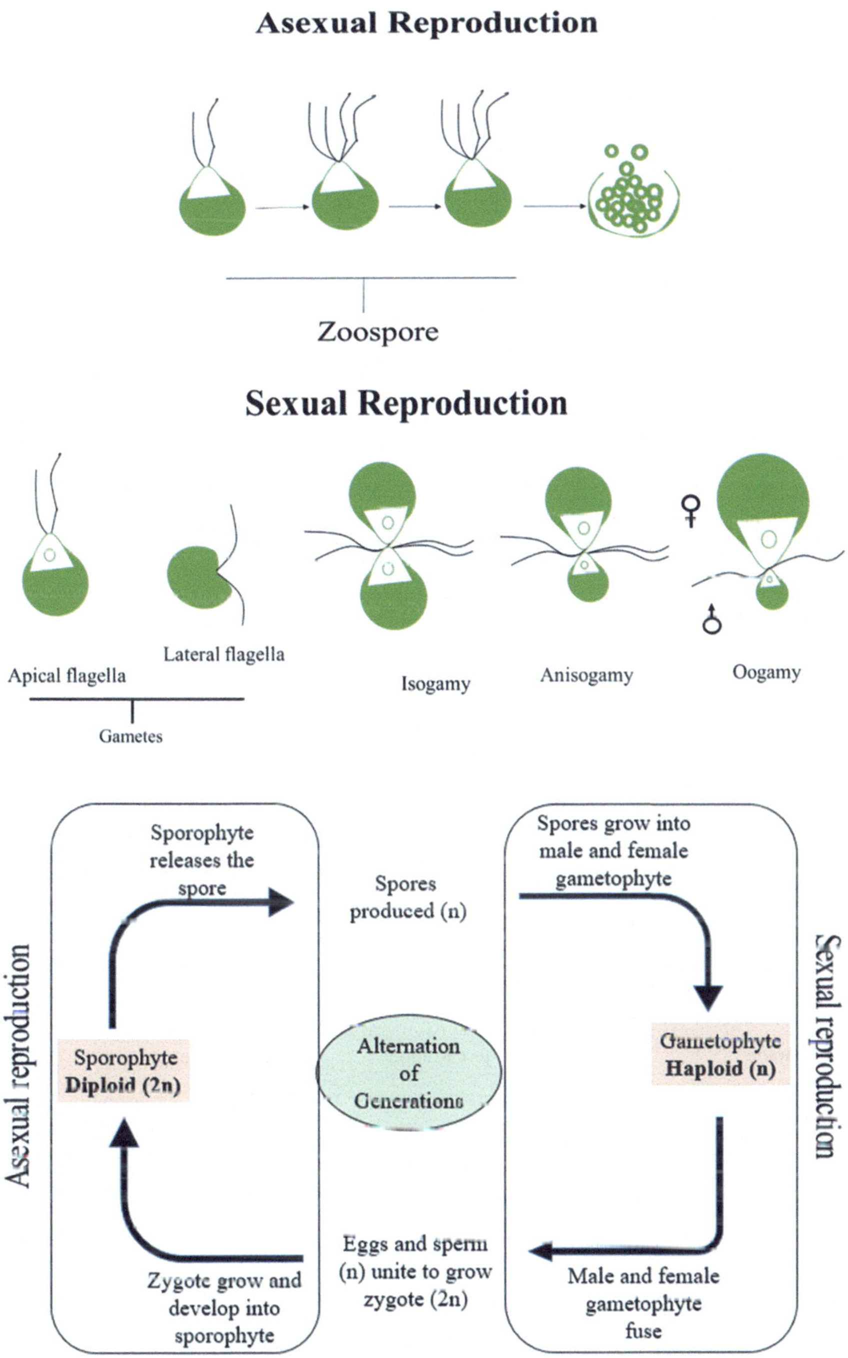

Fig. 27.2 Schematic representation of Alternation of generations observed in Seaweed reproduction

27.1.4 Types of Seaweeds

Seaweeds based on the presence of pigments are categorised into three broad types namely chlorophyta (green), rhodophyta (red) and phaeophyta (brown) and main characteristics of are summarised in Table 27.2 (Borines et al. 2013; Jung et al. 2013; Kjellman 1891; Kumar et al. 2013; Pascher 1914; Percival and McDowell 1967; Smith 1938; Yanagisawa et al. 2013). Chlorophyta have definite geographical

Table 27.2 Detailed characteristics of different types of Seaweed

Characteristics	Green seaweed	Red seaweed	Brown seaweed
Number of species recorded	6032[a]	7105[b]	2039[c]
Habitat	Freshwater and Marine	Strictly marine	Strictly marine
Photosynthetic pigment present	Chlorophyll *a, b,* carotene and Xanthophyll	Phycoerythrin	Fucoxanthin
Photosynthetic rate (μmol CO_2 /h) g/dry	30 to 1786	20–1808.7	100–500
Productivity [dry g/(m^2 year)]	7100	3300–11,300	3300–11,300
Nature of cell wall	Cellulose, pectin rarely hemi-cellulose	Cellulose and pectic material with polysulphate esters	Cellulose with alginic acid and fucocinic acid
Sexuality	Isogamy to oogamy	Advanced and complex (oogamous)	Isogamy to oogamy
No. of flagella and their insertion	2 or 4, equal anterior, whiplash	Absent	Only in reproductive cells, 2 unequal, lateral whiplash and tinsel
Phycolibins	Absent	Allophycocyanin,r-Phycoerythrin,r-Phycocyanin	Absent
Carotenoids	α-, β-, γ- carotene	α-, β- carotene	α-, β-, ε- carotene
Xanthophylls	Lutein Prasinoxanthin	Lutein	Fucoxanthin,Violaxanthin, Diadinoxanthin,Heteroxanthin, Vacheriaxanthin
Carbohydrate (%)	30–60	30–50	20–30
Protein (%)	10–20	6–15	10–15
Lipid (%)	1–3	0.5–1.5	1–2
Ash (%)	13–22	5–15	14–28
Photosynthetic reserve* (Stored food)	Starch	Floridean starch (intermediate between true starch and dextrin)	Laminarin and mannitol (hexahydride alcohol)

(*Stored food)

distribution, primarily dependent upon temperature of the water. The plant body (thallus) may be unicellular or multicellular and have either a definite or an indefinite number of cells. In multicellular forms, cells may be arranged in irregular masses, in filaments, as expanded sheets or as solid or hollow cylinders. Cell wall is composed of two concentric portions, the innermost portion wholly of cellulose (Tiffany 1924), and the outer mainly of pectose. The pigmentation of chloroplast is extremely variable and ranges all the way from a quantity sufficient to colour the plastid a brilliant green to an amount so small that there is only a tinge of color. Old cells of many species have the chlorophyll diffused throughout the cytoplasm, but young cells have chloroplast (Smith 1938).

Phaeophyta or brown algae range from microscopic ones to giant kelps, which attain lengths of 50 meters or more. Pigments in the brown algae are similar in chemical composition to those of green plants, but the proportion of chlorophyll *b* is lesser. The unique pigment of the Phaeophyta, fucoxanthin masks the other pigments in the chromatophores. Phaeophyta are known as algae of cold waters, however certain brown algae of orders *Dictyotales* and *Fucales* are distinctly warm-water plants. Cells of the Phaeophyta have a distinct wall and differentiated into an inner firm portion and outer gelatinous portion constituting cellulose and algin respectively. These algin or alginate are extracted from brown algae due to their adhesive property which has many uses in dairy, textile, adhesives, rubber, pharmaceutical, paper industries, etc. Rhodophyta or Red algae are multicellular plants, containing red pigment-*phycoerythrin* in the plastids in addition to chlorophyll, these pigments are present in such quantities as to mask the other pigments and give the plant a distinctive red color. Variation in the proportion of chlorophyll, phycoerythrin and phycocyanin accounts for the diversity of shades and color among Rhodophyceae (Smith 1938).

Algae in Rhodophyta are placed in a single class, the Rhodophyceae. Majority of the red algae are strictly marine, under normal conditions all marine species are sessile and, in most cases, dry up easily if the thallus is detached and free-floating. Rhodophyceae are confined to zones of amplitude of approximately 5 °C of the summer temperature, but certain species extend over zones representing 10 °C amplitude, and a few are known in zones with an amplitude of 20 °C. Cells of Rhodophyceae lack central vacuole, but those of a majority of species have a large central vacuole and the cytoplasm restricted to a thin peripheral layer next to the cell wall. The cell wall contains cellulose and various other pectic compounds.

27.1.5 Seaweeds Habitat, Ecology and Distribution

Seaweeds occur in a wide range of environment, including fresh, brackish, and marine waters. They require an aquatic environment for reproduction. Seaweeds grow by attaching to a substrate (natural or artificial); the need for a stable anchorage restricts the development of large seaweed beds on rocky substrate (Speight and Henderson 2013). Seaweeds grow well in photic zone at a depth where light reaches.

Benthic seaweeds mostly occupy continental shelf, whereas in much clearer waters they reach up to great depths, like in Caribbean Sea where red algae have been found at depths of over 260 m (Littler et al. 1986). The ability of seaweeds to occur at different depths is closely related to the composition of their photosynthetic pigments. Guiry (2014) observed that in macroalgae, accessory pigments absorb light with specific wavelengths selectively. Light intensity of aquatic ecosystems corresponds to wavelengths of blue-green region of the spectrum. They cover rocky shores in horizontal bands, with the green seaweeds growing along the high-tide line (Supra littoral zone), brown seaweeds found between the high-and low tide lines (littoral zone), and red seaweeds in the waters below the low tide line (subtidal zone).

Green seaweeds have chlorophyll as their light absorbing pigment and are typically found in intertidal or shallow water zone. Red seaweeds have phycoerythrin and phycocyanin pigments, which can efficiently absorb light with wavelengths of photosynthetically active radiation (PAR) that can penetrate seawater to the deep zone, some red algae inhabit the deep sea (over 25 m below the surface) where sunlight availability is limited (Santelices 1991). Whereas, brown seaweeds have fucoxanthin pigment which are efficient in absorbing wavelengths of light not filtered by the water column and can grow in deeper seas. Red and brown seaweeds also contain chlorophyll but are masked by the accessory pigments like phycoerythrin, phycocyanin and fucoxanthin. Substrate, topography, temperature, salinity, humidity, tides, waves, wind, and pollutants can all affect the growth and distribution of the algae, similar to light (Fig. 27.3). Temperature is an important factor for seaweeds, as maximum and minimum temperatures allow seaweeds to survive or complete their reproductive life cycle. Higher biomass and predominance of brown algae is observed in cold and temperate seas (except for few *Sargassum* species) than in warmer seas, where red and green algae are predominant. Some algae prefer the

Fig. 27.3 Vertical zonation of seaweeds covering the rocky shores in horizontal bands

areas with strong waves; others prefer calm water or sea bays. Certain seaweed species tolerate drier conditions, hence can grow in the supralittoral fringe; others survive in the extra ordinal environmental changes (dry or wet) in the littoral zone, in which low tide occurs twice daily (Figs. 27.4, 27.5 and 27.6).

27.1.5.1 Worldwide Seaweed Resources

Tropical seas are characterized by warm waters (>22 °C), having low concentration of inorganic nutrients (i.e. oligotrophic), etc. Seaweeds in these waters are found up to a depth of 268 m (Littler et al. 1986) due to the high incident photon flux density and deeper penetration of light rays (Lobban et al. 1994). Temperate seas are characterized by progressive seasonal cycles of light and temperature resulting in seasonal stratification of water column. Canopy forming large brown seaweed of the orders Fucales and/or Lamninariales dominate the intertidal and subtidal regions. About 33% of seaweeds recorded in this region are endemic communities restricted to sheltered habitats, including crevices. *Ulothrix* spp., *Enteromorpha bulbosa, Acrosiphonia* spp., *Pyropia ediviifolia,* and *Prasiola crispa* dominate this region. The physical conditions of tidal pools differ from those of the adjacent seawater

Fig. 27.4 Seaweeds attached to substratum using rhizoidal roots (**a**) *Ulva lactuca,* (**b**) *Enteromorpha intestinalis* (**c**) *Sargassum ilicifolium,* (**d**) *Gracilaria corticata* and *Coralline algae* forming mat on the substratum

Fig. 27.5 (**a**) *Sargassum sp.* flourishing in Intertidal zone, (**b**) *Enteromorpha sp.* flourishing in supratidal zone (splash zone), (**c**) *Enteromorpha intestinalis.* in tidal pools. (**d**) *Enteromorpha sp.* buried in sediments in estuary

Fig. 27.6 Profuse growth of Ulva and Sargassum species in calm waters, Gulf of Kutch, India

depending on the size of the pool, height on the shore and atmospheric conditions. Tolerant seaweed species like *Enteromoprha (Ulva) intestinalis, Chaetomorpha aerea, Ralfasia verrucosa* (Fig. 27.5) populate these tidal pools which face rapid changes in temperature, salinity, pH, nutrients and oxygen concentrations (Lobban et al. 1994). Seaweed communities in estuary is richest towards the mouth of the estuary, they become progressively poorer with more stress tolerant species towards the head (Lobban et al. 1994; Martins et al. 1999). Seaweeds such as *Enteromorpha* can withstand burial within the sediments and spread on the surface of estuarine mudflats (Martins et al. 1999). Seaweed communities for deep water (>30 m) are well documented for clear tropical waters (Spalding et al. 2003). Mostly coralline algae, and few foliose red seaweeds and geniculate coralline algae are seen populating these waters at depth of 40–55 m (Lobban et al. 1994). Seaweeds that detach from the shore during storms, raft and travel long distance colonizing; mostly brown seaweeds such as *Fucales* for e.g. *Ascophyllum* with buoyant structures and *Druvillaea antarctica* having uniquely inflated medulla (Rothäusler et al. 2012). The Sargasso Sea is famous for its floating population of *Sargassum.* Seaweeds cover rocky shores in horizontal bands, with the green seaweeds growing along the high-tide line (supra-littoral zone), brown seaweeds between the high-and low tide lines (littoral zone), and red seaweeds living in the waters below the low tide line (sub-tidal zone). This zonation is regulated by the light availability and level of sunlight penetration in the seawater. Accessory pigments in macroalgae absorb light with specific wavelengths selectively (Guiry 2014). The Northwest Atlantic Provinces of Canada have well established *Chondrus* spp. One of the richest seaweed resources in the world is believed to be in the Nova Scotia/Gulf of St. Lawrence area. Seaweed resources and their uses are very well established in regions of The Northeast Atlantic region including Norway, Scotland, Iceland, Ireland, France, Spain, Portugal and Denmark. Brown rockweeds particularly abundant on the Iceland's southern coasts, where broad belts of *Ascophyllum* dominates the vast areas of the littoral slopes (Naylor 1976). The coastal seaweed resources of West Central Atlantic are believed to be only moderate; Gulf coast, especially the offshore waters, is found out to be one of the richest centers for seaweeds. Rich algal resources in the Caribbean Sea occurs at a depth of over 260 m (Littler et al. 1986), although possibly the commercially attractive species do not occur in economically adequate quantities. This region includes the floating seaweed resource "Sargasso Sea".

Northeast and Northwest Pacific regions with cold and temperate water seas, show predominance of brown algae (large kelps), in particular Macrocystis and Nereocystis, and the algal biomass in this region is higher than in warmer seas (Pereira and Neto 2014). Seaweeds like *Gracilaria* and the bull-kelp, *Durvillea* are mainly recorded from Southwest Pacific, New Zealand and the southeastern coasts of Australia whereas rich *Gracilaria* and *Macrocystis* occur at Southeast Pacific coastlines of Chile and Peru. Mediterranean seaweed resources are of moderate quantities, Laminaria off the south coast of Spain, *Cladophora* and *Fucus* off coast of southern Italy and *Sargassum* and *Hypnea* off Yugoslavia. Black sea is rich in red algae resources. *Ulva, Eucheuma, Sargassum, Hypnea, Gracillaria, Gelidium, Turbinaria* are spread out along the coasts of Pakistan, India, Sri Lanka, and parts of Australian coastline. Luxurious growth of seaweeds mainly of Bryopsidales, Laminaria, Ulvales, and Siphonales are observed in Antarctic regions, (Fig. 27.7 and Table 27.3).

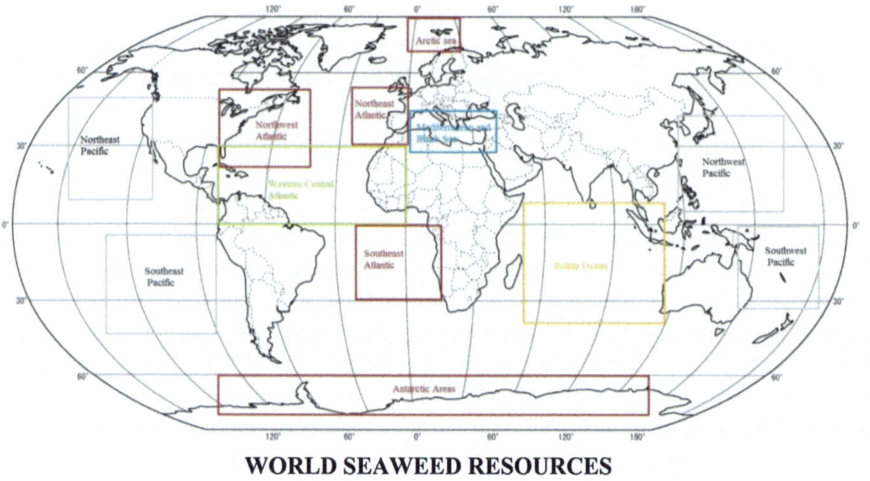

Fig. 27.7 Prominent coastal regions of the world rich in seaweed resources

The Indian coast has an amazing diversity of about 1153 marine algal species, out of which 60 species are commercially important (Chennubhotla et al. 1990; Kaliaperumal and Chennubhotla 1997; Kaliaperumal et al. 1995, 2004). Indian west coast is endowed with the higher seaweed resources, compared to the east coast and the islands due to the availability of suitable substratum in the rocky shore (beach rocks) (Wagle 1990). Rocky beaches, estuaries, mudflats, lagoons and coral reefs are the habitats preferred by macroalgae. Indian coast harbors predominantly intertidal and sub-tidal algal communities (Subba Rao and Mantri 2006). Tamil Nadu and Gujarat coasts are richest in seaweeds in the country, while rocky shores along Mumbai, Goa, Karnataka, and Kerala have moderate flora of seaweeds.

27.1.6 Biochemical Composition of Seaweeds

Macroalgae and terrestrial plants have different carbohydrate profile, both the group contains hemicelluloses-heterogeneous polysaccharide composed of pentoses, mainly xylose (Daroch et al. 2013) associated with the cell wall and intercellular spaces (Pereira and Neto 2014) (Figs. 27.8 and 27.9). Seaweed polysaccharides show range of structures and fulfil a variety of functions, most such constituents are also present in neutral sugars and sugar acids of land plants (Table 27.4). Certain seaweeds also contain acidic half ester sulphate groups attached to hydroxyl group of sugars. Hexose sugars such as, glucose, galactose and mannose found in these polysaccharides have identical chemical formula but their constituent atoms have different spatial arrangements and linkages, resulting in vast array of polysaccharides in different shapes and having diverse properties.

Table 27.3 Worldwide distribution of seaweed resources

Red Algae	
Gelidium	Japan, Spain, Portugal, Morocco, Algeria, Senegal, U.S.A., Mexico, Ireland, Chile, India, Philippines, Madagascar
Gracilaria	South Africa, Japan, Philippines, coastal areas of South China Sea, India, Sri Lanka, Australia, Chile, Peru, Brazil, Argentina, Adriatic, U.S.A. (Florida), Canada (British Columbia)
Chondrus	Canada (Nova Scotia, Newfoundland), Portugal, France (Brittany), U.K. (Scotland), Republic of Korea, Japan
Gigartina	South Africa, New Zealand, Portugal
Hypnea	U.S.A. (Florida), north Brazil, South Africa, Gulf of Oman
Eucheuma	Indonesia, Philippines, Malaysia, East Africa
Iridea	U.S.A. (California), Japan, Chile, South Africa
Furcellaria	Denmark, Baltic, Canada
Brown Algae	
Macrocystis	Northeast Pacific, California, Mexico, Peru, Chile, Argentina, South Africa, New Zealand, Tasmania
Alaria	Alaska, Japan
Laminaria	Northwest Atlantic, Greenland, Iceland, Norway, Ireland, Scotland, France, Spain, Morocco, Japan, U.S.S.R. (White Sea, Murmansk, Kamchatka, Okhotsk Sea)
Nereocystis	Northeast Pacific
Ecklonia	South Africa, Japan, Australia, New Zealand
Eisenia	Japan
Fucales order	Northeast and Northwest Pacific, Northeast and Northwest Atlantic, Chile, Murmansk, White Sea, New Zealand, Australia, Gulf of Oman
Green algae	
Bryopsidales	Eastern Atlantic (Africa canaries), Western Atlantic, Indo-Pacific, North Pacific Ocean, Caribbean, Gulf of Mexico, Federal Republic of Somalia, Gulf of Mexico, Indian Ocean, Kenya, Madagascar, North Atlantic Ocean, Republic of Mauritius, Seychelles, Tanzania, Venezuela.
Cladophorales	Gulf of Mexico, Caribbean Sea, Puerto Rico, Australia
Ulvales	Australia, North Pacific Ocean, Belgium, Federal Republic of Somalia, France, Gulf of Mexico, Indian Ocean, Ireland, Kenya, Madagascar, Mediterranean Sea, North Atlantic Ocean, Seychelles, South Africa
Siphonales	Western Atlantic, Indo-Pacific, North Pacific Ocean, Caribbean, Brazil, Gulf of Mexico, Gulf of Mexico, Indian Ocean, Kenya, Madagascar, North Atlantic Ocean, Southeast pacific

Carbohydrate formation in brown algae is comparable to that in sugar storing vascular plants rather than to the starch storing ones. These sugars are stored in dissolved form; very small amount of simple sugars is found in brown algae as these simple sugars are converted immediately to complex carbohydrates. One of the widely distributed of these is the dextrin-like polysaccharides known as ***laminarin.*** Laminarin can accumulate sufficient quantity constituting 7–35% of the dry weight of the plant. Gradual increase in the amount of laminarin throughout the growing season and the diminution at the time of reproduction or when new parts are

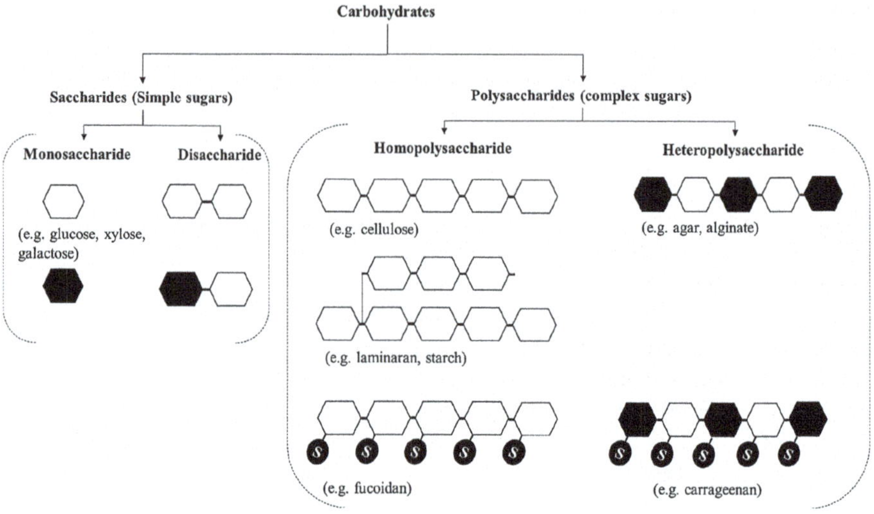

Fig. 27.8 Different types of sugars occurring in macroalgae

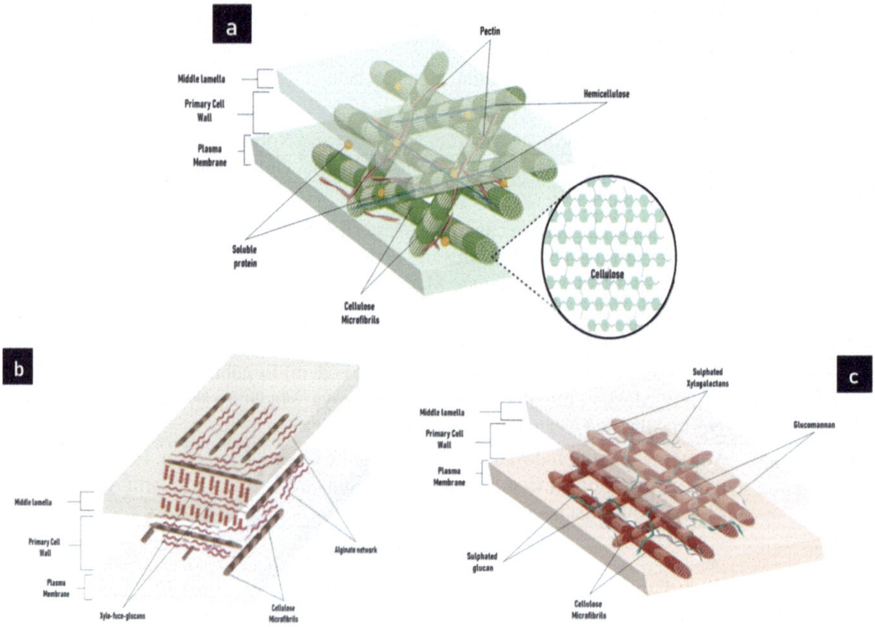

Fig. 27.9 Schematic representation of cell wall of green (**a**), brown (**b**) and red (**c**) macroalgae

regenerated shows that it serves as a reserve food (Adams et al. 2011; Guiry 2014; Mautner 1954). ***Mannitol*** a hexahydride alcohol is another widely distributed carbohydrate of Phaeophyta (Smith 1938). Major polysaccharide of brown algae is alginic acid (i.e., alginate), which accounts for up to 40% dry wt. as a principal

Table 27.4 Carbohydrate profile of seaweeds

Seaweeds	Carbohydrate reserve	Polysaccharides	Monosaccharides	Sugar alcohols	Sugar acids
Chlorophyta	Starch	Starch, Cellulose, Ulvan, Mannan	Glucose, Mannose, Rhamnose, Xylose	–	Uronic acid, Glucuronic acid
Rhodophyta	Floridean starch	Agar, Carrageenan, Cellulose, lignin	Glucose, Galactose, Agarose	–	
Phaeophyta	Laminarin	Laminarin, Alginate, Fucoidan, Cellulose	Glucose, Galactose, Fucose, Xylose	Mannitol	Uronic acid, mannuronic acid, Glucuronic acid. Alginic acid

Source: (Percival and McDowell 1967; Sudhakar et al. 2016; Yanagisawa et al. 2013)

material of the cell wall (Draget et al. 2005). Alginate is composed of uronic acids: mannuronic acid blocks, glucuronic acid blocks, and alternative blocks of mannuronic and glucuronic units (Lobban and Wynne 1981). Alginate tends to become gel due to its high affinity for divalent cations such as calcium, strontium, barium, and magnesium (Draget et al. 2005). Various brown algae tend to have their respective alginate structure (Fig. 27.9) and proportions of mannuronic and glucuronic acids in alginate (Lobban and Wynne 1981). Brown algae also have fucoidan, which is a sulphated complex polysaccharide consisting of fucose and a linear backbone of sulfate monosaccharides (Holtkamp et al. 2009). *Fucus vesiculosus* is used to produce relatively pure fucoidan (Duarte et al. 2001). In addition, brown algae have glucose and glyoxylic acid in small amounts (Mautner 1954).

Carbohydrate reserves of red algae are usually stored in the form of small grains that lie in the cytoplasm outside the algal plastids, the chromatophores. These plastids turn light brown to wine red, when treated with iodine, instead of taking the deep-blue colour characteristic of the iodine-starch reaction. On this account, the insoluble carbohydrate reserve of red algae has been called floridean starch (intermediate between true starch and dextrin). Floridean starch is an α-1,4-glucosidic linked glucose homopolymer and accounts for up to 80% of the cell volume (Yu et al. 2002). During the production of agar and carrageenan by using thermophilic α-amylase treatment, floridean starch is removed as an impurity, as it has a property to weaken gel strength (Yu et al. 2002). The major polysaccharide constituents of red algae are galactans such as carrageenan (up to 75% dry wt.) and agar (up to 52%), which are the most commercially important polysaccharides from red algae (Lobban and Wynne 1981; McHugh 2003). Carrageenan consists of repeating D-galactose unit and an-hydrogalactose, which may or may not be sulfated (Lobban and Wynne 1981). By dissolving red seaweeds into an aqueous solution carrageenan can be readily extracted (McHugh 2003), which are later purified and used for forming thick solution or gel (Lobban and Wynne 1981). Commercial carrageenan have been extracted from *Chondrus, Gigartina, and Eucheuma sp.* (Vera et al. 2011).

Agar, another major constituent, is made up of alternating β-D-galactose and α-L-galactose with scarce sulfations (Lobban and Wynne 1981). Agar cannot make a gel structure (Lobban and Wynne 1981), when these galactose compounds are highly sulfated.

Agar is produced from *Gracilaria*, *Gelidium,* and *Pterocladia sp.* by treating them with acid/heat or alkali (McHugh 2003). In Green algae polysaccharides proportions are small (1–4% for starch; and 0–6% for lipids) (Burton et al. 2009). Lahaye and Robic (2007) reported the insoluble cellulose (38–52% dry wt.) and water-soluble ulvan in the cell walls of *Ulva* and *Enteromorpha sp.* Ulvan, a distinctive carbohydrate of green algae, is composed mainly of D-glucuronic acid, D-xylose, L-rhamnose, and sulfate (Lobban and Wynne 1981).

27.1.7 Economic Utilization of Seaweed Resources

27.1.7.1 Food

Seaweeds are being used as sources of food, which can be traced back to the fourth century AD in Japan and the sixth century AD in China. These two countries and the Republic of Korea are the largest consumers of seaweed as food. However, as nationals from these countries have migrated to other parts of the world, the demand for seaweed for food has followed them, for example, in some parts of the United States of America and South America, *Ulva* (Chlorophyta), *Poryphyra* (Rhodophyta), *Undaria, Laminaria, Himanthalia* and *Saccharina* (Phaeophyta) are direct sources of food (Pereira and Neto 2014; Shama et al. 2019) Food products for human consumption, mainly associated with the Asian market, account for 83 to 95% of the total value of macroalgae. China is leading country in world market in seaweed production, followed by North Korea, South Korea, Japan, Phillippines, Chile, Norway, Indonesia, USA and India, altogether contributing upto 95% of world's commercial seaweed volume (Khan and Satam 2003; Roesijadi et al. 2010) (Table 27.5). The cultivation industries are producing more than 90 percent of the market's demand with increase in research into the life cycles of these seaweeds. Fresh seaweeds as vegetables and in salads are being used traditionally in an informal market that exists among coastal dwellers in some developing countries (McHugh 2003).

27.1.7.2 Hydrocolloid

Seaweeds are the only source of hydrocolloids, viz., agar-agar, algin and carrageenan. Hydrocolloids extraction from seaweeds dates to 1658, when the gelling properties of agar, extracted with hot water from a red seaweed, were first discovered in Japan. Extracts of Irish moss, another red seaweed, contain carrageenan and

Table 27.5 Different edible seaweed with their common names

Species	Type	Country	Local name/ product
Laminaria	Brown	Japan	Kombu
		China	Hai Dai
Porphyra	Red	Japan	Nori/amanori/hoshi-nori / Yaki-nori
		China	Zicai
		Korea	Kim
		UK (wales)	Purple laver/Laver bread
Hizikia fusiforme	Brown	Republic of Korea	Hoshi hiziki
Cladosiphon okamuranus	Brown	Japan	Mozuku
Alaria esculenta	Brown	Ireland	Winged kelp
		Scotland	
		Iceland	
		Hawaii	Limu
Undaria stipes *Undaria pinnatifida*	Brown	Japan	Wakame
		China	Quindai cai
Rhodymeni palmata *Palmaria palmate*	Red	Scotland	Dulse
		Ireland	Dillisk
		Iceland	Sol
		Canada	Sea parsley
Chondrus crispus	Red	Europe	Irish Moss/Carragenan
Callophyllis variegate	Red	Chile	Carola
Gracillaria spp.	Red		Ogo,ogonori or sea moss
Asparogopsis taxiformis	Red	Hawaii	Limu kohu
Caulerpa lentillifera, Caulerpa. Racemosa	Green	Phillipines	Sea grapes or green caviar
Monostroma spp and Enteromorpha spp.	Green	Japan	Aonori or green laver
Ulva spp.	Green	Japan	Sea lettuce

Source: (Chapman and Chapman 1980; Khan and Satam 2003; McHugh 2003)

were prevalent as thickening agents in the nineteenth century. Alginates were produced commercially in late 1930s, extracted from brown seaweeds, containing alginate and sold as thickening and gelling agents. Industrial uses of seaweed extracts extended rapidly after the Second World War, but were sometimes restricted by the availability of raw materials (McHugh 2003) (Table 27.8). These phytochemicals are widely used in various industries like food, confectionary, textile, cosmetics, paper, pharmaceutical, dairy, paint etc., as gelling, stabilizing and thickening agents (Table 27.6) (Roesijadi et al. 2010).

Table 27.6 Estimated global value of seaweed products per annum as reported in 2003 by McHugh

Product	Value
Human Food (Nori, aonori, kombu, wakame, etc.)	$5 billion
Algal hydrocolloids	
Agar (Food ingredient, pharmaceutical, biological/ microbiological)	$132 million
Alginate (Textile printing, food additive, pharmaceutical, medical)	$213 million
Carrageenan (Food additive, pet food, toothpaste)	$240 million

27.1.7.3 Bioactive Compounds

Seaweeds are well known for their broad spectrum of bioactive compounds that are useful for formulation of cosmetics, function food and pharmaceuticals (such as antimicrobial, antiviral, anti-allergic, anticoagulant, anticancer, antifouling and antioxidant activities) (Table 27.7) (Pereira and Neto 2014). Chemically active metabolites such as alkaloids, polyketides, sterols, quinones, lipid and glycerol are released by macroalgae as an aid to protect themselves against other organisms, these metabolites having a broad range of biological activities and have found place in pharmaceutical industries as well. Macroalgae *Ulva lactuca* and *Enteromorpha intestinalis* have been studied for their antioxidant and antimicrobial activities, showcasing very promising antimicrobial activity against numerous bacterial, fungal, human, animal and plant pathogens, mycotoxin producers, and food spoilage agents (Kosanić et al. 2015). Sulphated polysaccharides, sodium alginate, laminarin have been explored for medical and pharmaceutical uses (Torres et al. 2019).

27.1.7.4 Biofertilizers

Seaweeds are rich in minerals and trace elements due to which they are preferred as fertilizers for plants (Table 27.7). High fiber content of macroalgae aids in retaining moisture in the soil and improve soil conditioning. Fresh seaweeds are mechanically pressed to extract sap, which is used as liquid fertilizers, sprayed to the plants as growth stimulants. Dried seaweeds are powdered and used as micronutrients for plants. Fertilizers from macroalgae act as biostimulant and biofungicide also by improving the plant yield and quality. It is seen to affect composition of essential oils i.e. rosemary oil, α- phellandrene, β-pinene, α-thujene etc. in medicinal plants (Tawfeeq et al. 2016; Torres et al. 2019).

27.1.7.5 Feed

Macroalgae rich in essential nutrients, minerals, soluble and insoluble fiber, vitamins and trace elements are utilized as feed for farm animals, poultry and aquaculture. Enormous amounts of brown seaweeds are washed ashore along the coasts of

Table 27.7 Seaweeds species used for extraction of various bioactive compounds

Seaweeds	Antimicrobial	Antifungal	Antiviral	Antitumor
Turbinaria conoides	+			
Padina gymnospora	+			
Sargassum tenerrimum	+			
Codium decorticatum		+		
Caulerpa scalpelliformis		+		
Sargassum wightii		+		
Acanthophora spicifera		+		
Dictyota mertensii			+	
Lobophora variegate			+	
Spatoglossum schroederi			+	
Fucus vesiculosus			+	
Grateloupia sp.			+	
Undaria pinnatifida			+	
Sargassum muticum				+
Sargassum vulgare				+

Source*:* Modified from Pereira and Neto, 2015

Table 27.8 Biofuel potential of seaweeds

Seaweeds	Fermentation yield (%)	Methane yield (m^3 kg^{-1} VS)	H2 yield (L kg^{-1} TS)	HV (MJ kg^{-1})	Biochar (%, at 500 °C)	Bio-oil (%, at 500 °C)
Alaria crassifolia	38					
Ascophyllum nodusum		0.18		21.2	21.4	
Chaetomorpha linum	39					
Cladophora glomerata					40	30
Codium fragile			49			
Cystoseira barbata					0.21	0.32
Ecklonia stolonifera			43			
Enteromorpha clathrata				12–12.1		41.2–45
Eucheuma denticulatum	47					
Eucheuma spinosum	40					
Fucus vesiculosus		0.12				
Fucuss serratus						11

(continued)

Table 27.8 (continued)

Seaweeds	Fermentation yield (%)	Methane yield (m^3 kg^{-1} VS)	H2 yield (L kg^{-1} TS)	HV (MJ kg^{-1})	Biochar (%, at 500 °C)	Bio-oil (%, at 500 °C)
Gelidia dura	46					
Gelidiella acerosa	47					
Gelidium amansii	50		34–53			
Gelidium elegans	38					
Gelidium pusillum	47					
Gracilaria gracilis					28	65
Gracilaria verrucosa	48		46			
Hizikia fusiforme			10			
Kappaphycus alvarezii	47					
Laminaria digitata	38	0.36	26	23.1		17
Laminaria hyperborea	29					
Laminaria japonica	41	0.17	28–110	10.3–33.5		29–37.5
Padina tetrastromatica			16–1000			
Porphyra tenera			16	29.7		47.4
Saccharina latissima	13	0.22				
Sargassum fusiforme					38.3	
Sargassum muticum		0.11				
Ulva fasciata	45					
Ulva lactuca		0.25	10			
Ulva ohnoi					50	
Ulva pertusa	38					
Ulva rigida	50	0.63				
Undaria pinnatifida			13–23		67.7	39.5

UK, which are utilized as fertilizers. Seaweed incorporated meal in feed has resulted in several benefits such as increased milk production in cows, increased iodine content in eggs in poultry and increased wool production in lambs (Kraan and Guiry 2006). Sodium alginate fed poultry saw reduction in ceacal *salmonella* spp., that infected the birds (Yan et al. 2011) (Table 27.8).

27.1.7.6 Biofuels

Seaweed species are explored for their potential as third generation feedstock for biofuel production (Table 27.8), due to the limitations of land availability encountered by the first- and second-generation feedstocks. Several macroalgal biomass have been explored for biogas, biodiesel, bioethanol and biobutanol productions (Mohammad et al. 2019; Hessami et al. 2019). Macroalgal biomass are potential candidate for bioethanol production, achieved by conversion of their polysaccharides by fermentation process. The conversion process of macroalgal biomass to bioethanol involves (i) pretreatment or hydrolysis of complex sugars to simpler forms and (ii) fermentation of these simple sugars (Fig. 27.10). Pretreatment

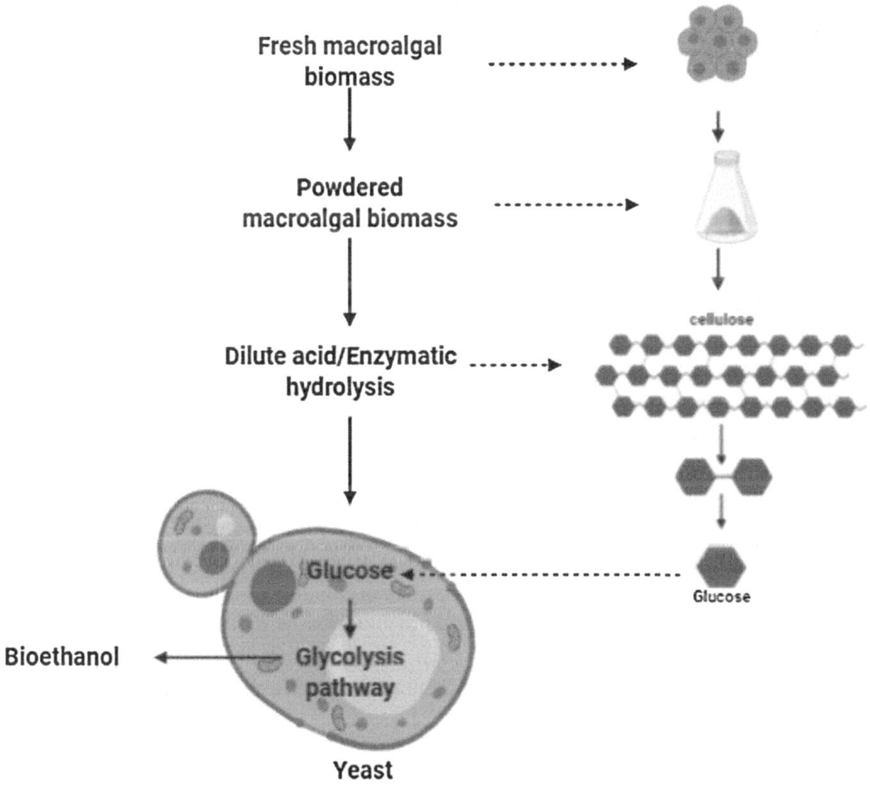

Fig. 27.10 Production of bioethanol from macroalgal biomass

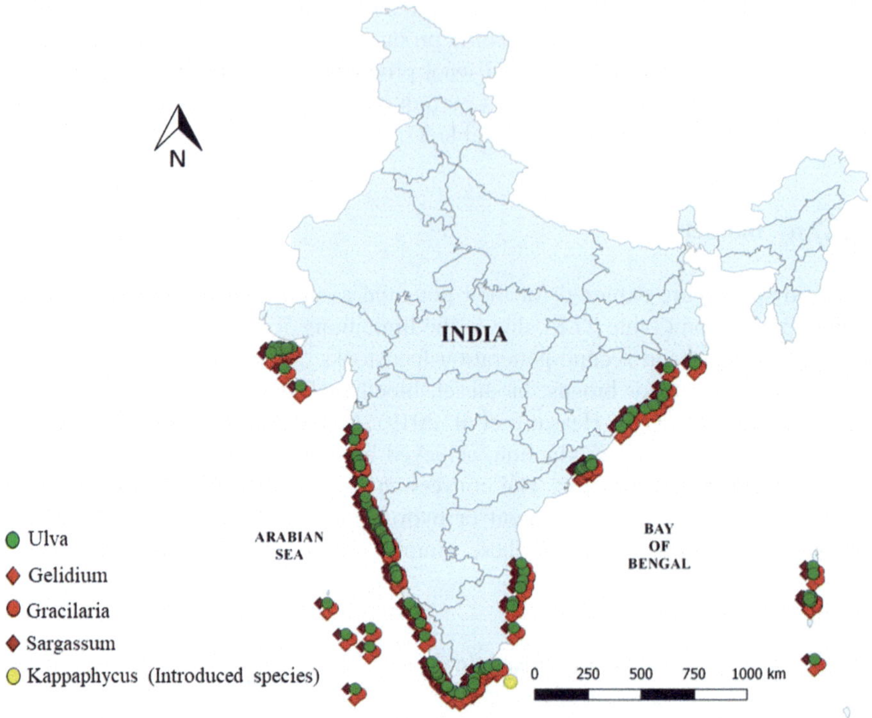

Fig. 27.11 Distribution of potential macroalgal feedstock of Indian coast

involves breaking down of complex polysaccharides to simple fermentable sugars by employing physical, chemical or biological agents. Pre-treatment with H_2SO_4 at different temperature and incubation period, is one of the most widely used procedures for seaweed cell wall depolymerization. Apart from cellulose, there are certain polymers such as alginate, mannitol and fucoidan present in cell wall of macroalgae, which requires additional processing like pre-treatment and enzyme hydrolysis before fermentation. Enzymatic hydrolysis involves dilute-acid pre-treatment followed by enzyme for further saccharification of seaweed polysaccharide. Enzymes are mostly extracted from terrestrial fungi sources such as *Trichoderma, Penicillum, and Aspergillus,* bacterial sources such as *Bacilus subtilis, Vibrio paraheamolyticusi* etc. (Hebbale et al. 2019; Trivedi et al. 2011). Fermentation is carried out using yeast microorganisms. Majority of macroalgal species utilized for production of bioethanol belongs to *Kappaphycus, Gelidium, Gracilaria, Sargassum, Laminaria* and *Ulva* genera, which are recorded abundantly from Indian east and west coast (Ramachandra and Hebbale 2020; Mohammad et al. 2019; Hessami et al. 2019) in Fig. 27.11.

27.1.8 Bioethanol Prospects of Seaweed in the West Coast: Energy and Economic Nexus

Twenty-five macroalgal species were recorded from West coast of Karnataka, among which eight species were available in extractable amounts from the rocky shores and were selected as potential source for bioethanol production. Based on the biochemical compositions, seasonality studies and euryhaline nature of the species two green macroalgae; *Enteromorpha intestinalis* and *Ulva lactuca* were prioritized as suitable feedstock for bioethanol production. The production of bioethanol from seaweeds (Fig. 27.12) involves two major steps, which are described below:

27.1.8.1 Hydrolysis (Acid or Enzyme) of the Biomass to Release Fermentable Sugars

Reducing sugar (RS) from dilute acid hydrolysis (DAH) of *E. intestinalis* (**DAH:** 5% w/w, 0.7 N H_2SO_4 at 121 °C for 45 min, **RS:** 239.94 ± 1.3 mg/g) and *U. lactuca* (**DAH:** 5% w/v, 0.5 N H_2SO_4 at 121 °C for 45 min, RS: 214.67 ± 0.9 mg/g). Cellulose degrading bacteria were isolated from wide-ranging sources including marine habitats, herbivore residues and gastrointestinal region. *Vibrio parahaemolyticus* (Hebbale et al. 2019) isolated from marine environment is capable of hydrolysing CMC as well as *E. intestinalis* and *U. lactuca* pretreated biomass, highlighting cellulolytic activity of enzyme capable of hydrolysing structural polysaccharide (cellulose) in green seaweed. Pre-treated macroalgal biomass produced one-fold higher reducing sugar in enzymatic hydrolysis (EH) *E. intestinalis* (**EH:** pH 6 at

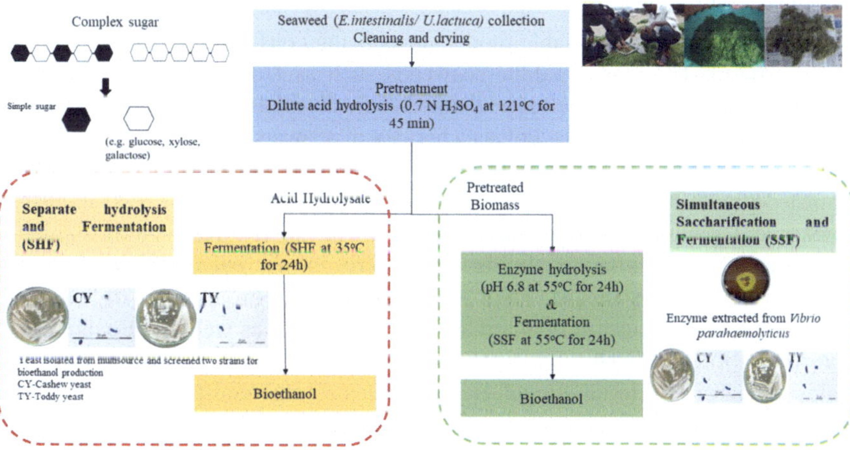

Fig. 27.12 Bioethanol production from macroalgal biomass

50 °C for 24 h, **RS:** 289.89 ± 2.4 mg/g) and *U.lactuca* (**EH:** pH 6 at 50 °C for 24 h, **RS:** 261.76 ± 0.9 mg/g) as compared to the dilute acid hydrolysis.

27.1.8.2 Fermentation by Microorganisms

Marine macroalgae contain low amounts of polysaccharides composed of glucose. Production of ethanol, therefore, needs to be from carbohydrates including sulphated polysaccharides, sugar acids and sugar alcohols. However, the major hurdle is the inability of microorganisms in fermenting all sugars present in seaweeds. Therefore, exploration of yeast strains that can ferment both pentose (C5) and hexose (C6) sugars is crucial for higher ethanol yield. Yeast strains were isolated from various fruit sources and fermented products and 19 strains were prioritized based on the performance in glucose and xylose media, with the carbohydrate fermentation capabilities. Yeast strains Meyerozyma *caribbica* (Cashew yeast: CY) and *Pichia kudriavzevii* (toddy yeast: TY) were chosen based on longer exponential growth, maximum conversion efficiency with respect to glucose and macroalgal sugar fermentation (**F**) apart from being tolerant to temperature and ethanol (ETOH). Separate hydrolysis and fermentation (SHF) (**DAH:** 0.7 N H_2SO_4 121 °C for 45 min, **F**: 35 °C, 100 rpm for 24 h, **ETOH**: 0.16 g) yielded higher ethanol conversion efficiency of 51% using *Pichia kudriavzevii* for *Enteromorpha intestinalis.* Simultaneous Saccharification and Fermentation (SSF) (**DAH:** 0.5 N H_2SO_4 121 °C for 45 min, **F**: 35 °C, 100 rpm for 24 h, **ETOH**: 0.14 g) yielded higher ethanol conversion efficiency of 80.9% using enzyme extracted from *Vibrio parahaemolyticus* and *Pichia kudriavzevii* for *U. lactuca* biomass. Wild yeast strain *Pichia kudriavzevii* exhibited higher fermentation capabilities using macroalgal biomass.

27.1.8.3 Economic Nexus

Macroalgae are readily available food sources being consumed by coastal communities, particularly in Asia. Edible seaweeds such as *Ulva* species are popularly known as "Sea lettuce", which is consumed as fresh salads or cooked as vegetables along with rice (WHO 2003). Polysaccharides of seaweeds are used as thickening agents in sweet and savoury sauces and condiments, stabilizing food products against degradation, staling and heating, and for replacing fat in food industries (Forster and Radulovich 2015). Chemically active metabolites such as alkaloids, polyketides, sterols, quinones, lipids and glycerol in macroalgae aid in protecting them against other organisms. These metabolites with a broad range of biological properties have been useful in pharmaceutical industries as well. Macroalgae *Ulva lactuca* and *Enteromorpha intestinalis* with the promising antimicrobial activities against numerous bacterial, fungal, etc. have been commercially used as mycotoxin producers and food spoilage agents (Kosanić et al. 2015; Pereira and Neto 2014). Carrageenans – the water-soluble polysaccharide present in red algae are extracted using water, alkali or acid treatment. Major portion of carrageenan and agar

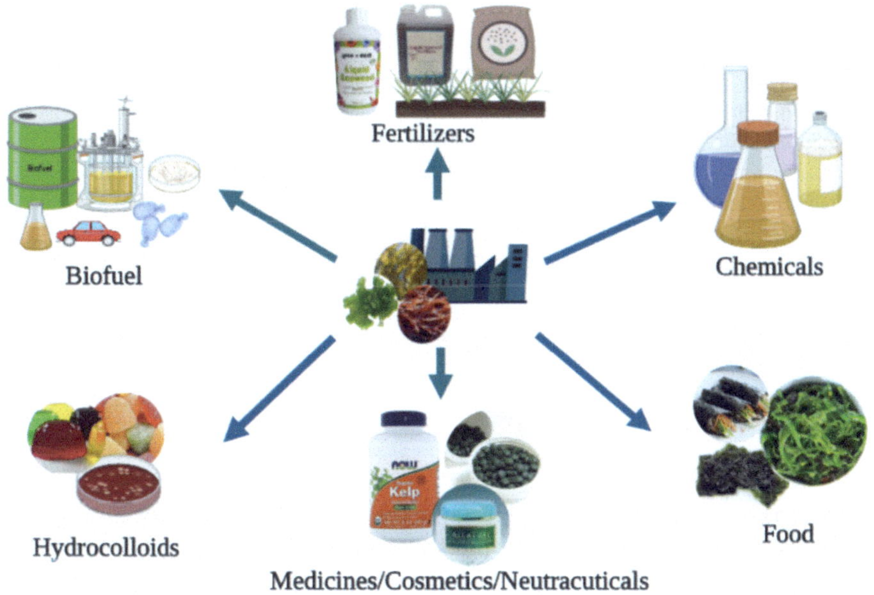

Fig. 27.13 Economic nexus of macroalgal biomass

extraction is done using cultivated red algae, whereas alginate (from brown algae) is mostly obtained from natural populations. These have been used to make jellies and milk puddings (Blanc mange) (from *C. crispus* and *G. stellata*) as well as utilization of carrageenan granules (from *K. alvarezii*) as raw material for bioethanol production (Khambhaty et al. 2012; Stanley 1987; Tuvikene et al. 2006). These phytochemical compounds constitute a part of healthy balanced diet, as protein content in seaweeds is higher than other food materials such as cereals, eggs and fish they are also excellent sources of vitamins A, B1, B12, C, D and E, riboflavin, niacin, pantothenic acid and folic acid as well as minerals such as Ca, P, Na, K. Apart from production of bioethanol, fermentation of algal hydrolysate also produces many by-products, such as glycerol, organic acids (e.g., acetate, succinate), biomass protein, and other minor products. Industries using seaweed as feedstock for production of bioethanol will be economical with the simultaneous utilization of the fermentation by-products, similar to the petroleum industry where many such products besides gasoline are profitable (Fig. 27.13). Cultivation prospects for economically important macroalgal biomass is discussed in the next section.

27.1.9 Macroalgal Cultivation in India

The seaweed industry in India (Table 27.9) is still at infant stage, functioning more as a cottage industry. Agar and algin yielding seaweeds are collected from the from the natural stock along the coast (Khan and Satam 2003). However, this type of

Table 27.9 Distribution of Agar, Carrageen yielding and edible seaweeds in India

Seaweed species	Culture Location
Agar yielding	
Gelideilla acerosa	Gulf of Manner and Palk Bay near Mandapam
Gracilaria edulis	Gulf of Mannar and Palk Bay
	Krusadai Island (Mandapam) and Karvaratti Island (Lakshadweep)
Carrageen yielding	
Hypnea musciformis	Krusadai island, Lakshadweep
Acanthophora spicifera	Hare Island near Mandapam
Kappaphycus alvarezii	Saurastra and Mandapam region, Narakkal (Cochi) and Calicut
Edible seaweed	
Enteromorpha Flexuosa	Okha (Gujarat)

Source: (Subba Rao and Mantri 2006)

collection is not sustainable and not favoring the stability of an industry. More attention must be paid for bridging the gap between demand and supply that is possible only through cultivation of various seaweeds as raw materials for food production and for other seaweed-based industries. Cultivation can ensure improved yield, continuous supply, and conserve the natural seaweed beds.

Few genera namely, *Laminaria, Undaria, Sargassum, Poryphyra, Eucheuma* (*Kappaphycus*), *Gracilaria, Gelidium* and *Ulva* are mostly focused on the aquaculture production as they are potential source for production of various raw materials. Red seaweeds such as *Kappaphycus, Gelidium, Gracilaria,* and brown seaweeds including *Sargassum,* and *Laminaria* are mostly utilized for extraction of hydrocolloid (agar, algin and carrageenan). Green seaweed (such as *Ulva*) serves as direct source of food (e.g. Salads) (Pereira and Neto 2014), and support a well-established multi-billion-dollar industry in Asia (Milledge et al. 2014). Currently, these seaweeds are also the leading feedstock for bioethanol production (Trivedi et al. 2013, 2015).

India's mainland coastline of 5422.6 km endows several major and minor estuaries along east and west coast. Indian estuarine area of 3.9 million ha, of which 1.2 million ha are salt-affected lands dedicated for brackish water shrimp cultivation, which covers about 15% for aquaculture purpose. as potential sites the rest 85% of the land serves as potential sites for cultivation of macroalgae. These salt-affected aquaculture ponds are known as *gazni* (Karnataka), *pokkali land* (Kerala), *kharland* (Maharashtra), *bheri* (West Bengal), *gheri* (Odisha) across Indian states. The stagnant water in these ponds are conducive for macroalgal cultivation as it prevents the algae from drying out.

Seaweed cultivation along the coast is a challenging task due to wave action, epiphyte fouling, ice-ice disease and algae feeding fishes.Off-shore macroalgal cultivation has been explored in India along 10 km stretch of Palk Bay, Mandapam, where red seaweed *Kappaphycus alvarezii* and *Hypnea musciformis* were cultivated in an area of 100 ha through contract farming system involving the local

Fig. 27.14 Estuaries and Lagoons in Indian

communities. These two species other than serving as raw materials for industries are also explored for bioethanol production potential. However, drifting of broken algal fragments from the rafts were a hindrance as they attached to the nearby hard corals (in Mandapam Marine National Park), growing profusely by vegetative propagation, thereby affecting the growth of the coral. Such difficulties can be overcome by cultivating seaweeds in enclosed ponds, lagoon systems, or in coastal brackish water *gazni* rice fields, which experience sufficient inflow of seawater to keep the seaweeds from drying out. Figure 27.14 represents estuaries and lagoons of India, which serves as potential sites for macroalgal cultivation.

Coastal and marine livelihoods include a wide range of stakeholders, who are dependent completely or partially on the direct use of the goods and services generated from coastal and marine areas. Most of these coastal resources are utilized for "self- subsistence", although market forces have strongly come into the fishery sector. The entire economy of coastal areas is intimately linked to the earnings generated from the use of those resources (Townsley 2004). Land based macroalgal cultivation integrated with shrimp cultivation is a beneficial process as it caters to employment of coastal women, who, with little effort can contribute significantly to the household income and supports the livelihood of the fisherman's family during any failure in shrimp cultivation. Labour from the fishing community and skill from estuarine farming community will provide a perfect platform for launch of seaweed cultivation in estuaries. Estimated production potential of one million tons of dried seaweeds from India can generate employment to 200 thousand families with annual earnings of around 0.1 million per family (Radhika et al. 2014). Setting up of a small-scale viable seaweed bio refinery, brings economy and employment to the coastal communities residing in the region. Seaweed cultivation can be taken up in these estuaries where suitable sea conditions prevail, advocating full-time or part-time large-scale job opportunities in coastal rural sector. Therefore, such estuarine backwaters can serve as suitable site for Seaweed cultivation at large scale.

27.1.10 Conclusion

Macroalgae are multicellular, photosynthetic algae occurring in marine and brackish environment. Broadly classified into three types as chlorophyta, rhodophyta and phaeophyta based on the presence of pigment. Seaweeds are rich in carbohydrates. Polysaccharides from red and brown algae are natural source of hydrocolloids such as agar, algin and carrageenan. Seaweeds are utilized as food source and for production of biofertilizers, feed, biofuel, cosmetics etc. Production of bioethanol from seaweeds is regarded as promising and sustainable option. Major unit operations involved in bioethanol productions are pretreatment or hydrolysis of macroalgal biomass and fermentation of sugars released. Cultivation of economically important native species of macroalgae in abandoned aquaculture ponds along estuaries is a sustainable and income generating option. The prospects visualized here is for fetching a value for these seaweeds which can become feedstock for biofuel production, so much so such 'sea-weeds' of today could be easily transformed into 'sea-wealth'. Bioethanol production not only ensures the strategic energy security of the nation, but also helps in mitigating GHG footprint, judicious use of feedstock, lowering import burdens, empowers rural women with the sustainable livelihood through integrated approaches in fishery, etc.

27.1.11 Prospects of Macroalgae in India

Seaweed biomass are promising resources for production of various products catering to industries related to food, fertilizers, cosmetics, biofuel, pharmaceuticals. Future direction towards economic utilization of macroalgal biomass should focus mainly on:

(i) biorefinery approach which is profitable with production of multi products including bioethanol.
(ii) Optimization of process parameters to obtain high yield of bioethanol production.
(iii) Innovative macroalgal cultivation technologies on land-based aquaculture ponds for achieving higher biomass productivity.
(iv) Development of decentralised seaweed-based industries in coastal regions of India where resources are abundantly available.
(v) To contribute to the growth of maritime sectors by improving existing infrastructures for fisherman and training the fisher folks in large scale ocean farming of seaweeds.
(vi) Seaweed cultivation as a notable future enterprise can open up platform for establishing seed hatcheries, seeding units and processing units and enhance employment opportunities in rural coastal area.

References

Adams JMM, Toop TA, Donnison IS, Gallagher JA (2011) Seasonal variation in laminaria digitata and its impact on biochemical conversion routes to biofuels. Bioresource Technol 102(21):9976–9984. https://www.sciencedirect.com/science/article/pii/S0960852411011205 (January 3, 2019)

Barnes H (1999) Oceanography and marine biology. CRC Press

Borines MG, de Leon RL, Cuello JL (2013) Bioethanol production from the macroalgae sargassum spp. Bioresour Technol 138(1):22–29. https://doi.org/10.1016/j.biortech.2013.03.108

Burton, T., Lyons, H., Lerat, Y., Stanley, M., & Rasmussen, M. B. 2009. "A review of the potential of marine algae as a source of biofuel in Ireland"

Chapman VJ, Chapman DJ (1980) Sea Vegetables (Algae as Food for Man). In: Seaweeds and Their Uses. Springer, pp 62–97

Chennubhotla VS, Najmuddin M, Ramalingam JR, Kaliaperumal N (1990) Bulletin 44. In: CMFRI Bulletin-national symposium on research and development in marine fisheries sessions III & IV, vol 1987, pp 442–446

Daroch M, Geng S, Wang G (2013) Recent advances in liquid biofuel production from algal feedstocks. Appl Energy 102:1371–1381. https://doi.org/10.1016/j.apenergy.2012.07.031

Draget KI, Smidsrød O, Skjåk-Bræk G (2005) Alginates from Algae. Biopolymers Online: Biol Chem Biotechnol Applicat 6

Duarte MER, Cardoso MA, Noseda MD, Cerezo AS (2001) Structural studies on fucoidans from the brown seaweed *sargassum stenophyllum*. Carbohydr Res 333(4):281–293

Fenner MW (2012) Seed ecology. Springer Science & Business Media

Forster J, Radulovich R (2015) Seaweed and food security. In: Seaweed sustainability: food and non-food applications (October), pp 289–313

Guiry MD (2014) The seaweed site: information on marine algae. Seaweed. ie

Guiry MD, G M 2013 Guiry (2008) AlgaeBase. AlgaeBase

Hardy FG, Moss BL (1979) Attachment and development of the zygotes of pelvetia canaliculata (L.) Dcne. et Thur.(Phaeophyceae, Fucales). Phycologia 18(3):203–212

Hebbale D, Bhargavi R, Ramachandra TV (2019) Saccharification of macroalgal polysaccharides through prioritized cellulase producing bacteria. Heliyon 5(3):e01372. https://linkinghub.elsevier.com/retrieve/pii/S2405844018374954

Hessami MJ, Cheng SF, Ranga Rao A, Yin YH, Phang SM (2019) Bioethanol production from agarophyte red seaweed, *Gelidium elegans* using a novel sample preparation method for analysing bioethanol content by gas chromatography. Biotech J 9(1):25

Holtkamp AD, Kelly S, Ulber R, Lang S (2009) Fucoidans and fucoidanases—focus on techniques for molecular structure elucidation and modification of marine polysaccharides. Appl Microbiol Biotechnol 82(1):1

Jung KA, Lim SR, Kim Y, Park JM (2013) Potentials of macroalgae as feedstocks for biorefinery. Bioresour Technol 135:182–190. https://doi.org/10.1016/j.biortech.2012.10.025

Kaliaperumal N, Chennubhotla VSK (1997) Sea weed distribution and resources in Kerala coast. Seaweed Res Uliln 19(1):29–32

Kaliaperumal N, Kalimuthu S, Ramalingam JR (2004) Kaliaperumal, kalimuthu, ramalingam - 2004 - present scenario of seaweed exploitation and industry in India. Seaweed Res Utiln 26(1 & 2):47–53

Kaliaperumal N, S Kalimuthu, Ramalingam J (1995) Economically important seaweeds 1–44

Khambhaty Y et al (2012) *Kappaphycus alvarezii* as a source of bioethanol. Bioresour Technol 103(1):180–185

Khan SI, Satam SB (2003) Seaweed mariculture: scope and potential in India. Aquaculture Asia VIII (4):26–29. http://aquaticcommons.org/205/1/AquacultureAsia_8_4_2003.pdf#page=28

Kjellman FR (1891) Phaeophyceae (Fucoideae). Die Naturlichen Pflanzenfamilien I Teil 2:176–192

Kosanić M, Ranković B, Stanojković T (2015) Biological activities of two macroalgae from adriatic coast of montenegro. Saudi J Biol Sci 22(4):390–397. http://www.sciencedirect.com/science/article/pii/S1319562X14001417

Kraan S, Guiry MD (2006) The seaweed resources of Ireland. ETI, Amsterdam, Seaweed resources of the worldm, A CD-rom project

Kumar S et al (2013) Bioethanol production from gracilaria verrucosa, a red alga, in a biorefinery approach. Bioresour Technol 135:150–156. https://doi.org/10.1016/j.biortech.2012.10.120

Lahaye M, Robic A (2007) Structure and function properties of ulvan, a polysaccharide from green seaweeds. Biomacromolecules 8(6):1765–1774

Leliaert F et al (2012) Phylogeny and molecular evolution of the green algae. Crit Rev Plant Sci 31(1):1–46

Littler MM, Littler DS, Blair SM, Norris JN (1986) Deep-water plant communities from an uncharted seamount off san salvador Island, Bahamas: distribution, abundance, and primary productivity. Deep Sea Research Part A Oceanographic Research Papers 33(7):881–892

Lobban CS, Harrison PJ, Harrison PJ (1994) Seaweed ecology and physiology. Cambridge University Press

Lobban CS, Wynne MJ (1981) The biology of seaweeds, vol 17. Univ of California Press

Martins I, Oliveira JM, Flindt MR, Marques JC (1999) The Effect of Salinity on the Growth Rate of the Macroalgae *Enteromorpha Intestinales* (Chlorophyta) in the Mondega Estuary (West Portugal). Acta Oecol 20(4):259–265

Mautner HG (1954) The chemistry of brown algae. Econ Bot 128:174–175

McHugh DJ (2003) FAO. Fisheries technical paper A guide to the seaweed industry. Food and Agriculture Organization of the United Nations Rome. http://www.fao.org/documents/card/en/c/24efc5c0-d2fe-55b3-ade9-6bdffee4e66e/

Milledge JJ, Smith B, Dyer PW, Harvey P (2014) Macroalgae-derived biofuel: a review of methods of energy extraction from seaweed biomass. Energies 7(11):7194–7222

Mohammad JH, Ranga Rao A, Ravishankar GA (2019) Opportunities and challenges in seaweeds as feed stock for biofuel production. In: Ravishankar GA, Rao AR (eds) Handbook of algal technologies and phytochemicals: volume-II phycoremediation, biofuels and global biomass production. CRC Press, USA, pp 39–50

Moss B (1975) Attachment of zygotes and germlings of *Ascophyllum nodosum* (L.) Le Jol. (Phaeophyceae, Fucales). Phycologia 14(2):75–80

Naylor J (1976) Production, trade and utilization of seaweeds and seaweed products. In: FAO Fisheries Technical Papers (FAO). Documents Techniques FAO sur les Peches (FAO)-Documentos Tecnicos de la FAO sobre la Pesca (FAO), p 159

Organization, World Health (2003) 916 Diet, nutrition, and the prevention of chronic diseases: report of a joint WHO/FAO expert consultation. World Health Organization

Pascher A (1914) Uber flagellaten und algen. Ber Deutschen Bot Ges 32:136–160

Percival E, McDowell RH (1967) Chemistry and enzymology of marine algal polysaccharides

Pereira L, Neto JM (2014) Marine algae: biodiversity, taxonomy, environmental assessment, and biotechnology. CRC Press

Ramachandra TV, Hebbale D (2020) Bioethanol from macroalgae: prospects and challenges. Renew Sustainable Energy Rev 117:109479. https://linkinghub.elsevier.com/retrieve/pii/S1364032119306872 (October 18, 2019)

Riisberg I et al (2009) Seven gene phylogeny of heterokonts. Protist 160(2):191–204

Roesijadi G, Jones SB, Zhu Y (2010) Macroalgae as a biomass feedstock: a preliminary analysis. Analysis:1–50. http://www.pnl.gov/main/publications/external/technical_reports/PNNL-19944.pdf

Rothäusler E, Gutow L, Thiel M (2012) Floating seaweeds and their communities. In: Seaweed Biology. Springer, pp 359–380

Santelices B (1991) Production ecology of gelidium. In: International workshop on gelidium. Springer, pp 31–44

Santelices B (1990) Patterns of reproduction, dispersal and recruitment inseaweeds. Ocearrogr A Mar Biol 471h Rez 28:177–276

Shama A, Joyce SG, Mari FD, Ranga Rao A, Ravishankar GA, Hudaa N (2019) Macroalgae and microalgae: novel sources of functional food and feed. In: Ravishankar GA, Rao AR (eds) Handbook of algal technologies and phytochemicals: volume-I: food, health and nutraceutical applications. CRC Press, USA, pp 207–219

Smith GM (1938) Cryptogamic botany. volume I. In: Algae and fungi

Spalding H, Foster MS, Heine JN (2003) Composition, distribution and abundance of deep-water (>30m) macroalgae in central California. J Phycol 39(2):273–284

Speight MR, Henderson PA (2013) Marine ecology: concepts and applications. John Wiley & Sons

Stanley N (1987) Production, properties and uses of carrageenan. production and utilization of products from commercial seaweeds. In: FAO fisheries technical paper, vol 288, pp 116–146. http://www.fao.org/3/x5822e/x5822e05.htm (September 16, 2019)

Subba Rao PV, Mantri VA (2006) Indian seaweed resources and sustainable utilization: scenario at the dawn of a new century. Curr Sci 91(2):164–174

Sudhakar MP, Ravel Merlyn K, Arunkumar, and K. Perumal. (2016) Characterization, pretreatment and saccharification of spent seaweed biomass for bioethanol production using baker's yeast. Biomass Bioenergy 90:148–154. https://doi.org/10.1016/j.biombioe.2016.03.031

Tawfeeq A, Culham A, Davis F, Reeves M (2016) Does fertilizer type and method of application cause significant differences in essential oil yield and composition in rosemary (*Rosmarinus Officinalis L.*)? Ind Crop Prod 88:17–22

Tiffany, Lewis Hanford. 1924. "A physiological study of growth and reproduction among certain green algae"

Torres MD, Kraan S, Domínguez H (2019) 18 Reviews in environmental science and biotechnology. Seaweed Biorefinery

Tovey DJ, Moss BL (1978) Attachment of the haptera of *laminaria digitata* (Huds.) lamour. Phycologia 17(1):17–22

Townsley P (2004) Review of coastal and marine livelihoods and food security in the Bay of Bengal large marine ecosystem region. Bay of Bengal Large Marine Ecosystem Programme

Trivedi N, Gupta V, Reddy CRK, Jha B (2013) Detection of ionic liquid stable cellulase produced by the marine bacterium *pseudoalteromonas* Sp. Isolated from brown alga sargassum polycystum C. Agardh. Bioresour Technol 132:313–319. https://doi.org/10.1016/j.biortech.2013.01.040

Trivedi N, Reddy CRK, Radulovich R, Jha B (2015) Solid State Fermentation (SSF)-derived cellulase for saccharification of the green seaweed ulva for bioethanol production. Algal Res 9:48–54. https://doi.org/10.1016/j.algal.2015.02.025

Trivedi N et al (2011) An alkali-halotolerant cellulase from bacillus flexus isolated from green seaweed ulva lactuca. Carbohydr Polym 83(2):891–897. https://doi.org/10.1016/j.carbpol.2010.08.069

Tuvikene R, Truus K, Vaher M, Kailas T (2006) Extraction and quantification of hybrid carrageenans from the biomass of the red algae furcellaria lumbricalis and coccotylus truncatus. Proc Estonian Acad Sci Chem 55(1):40–53

Vera J, Castro J, Gonzalez A, Moenne A (2011) Seaweed polysaccharides and derived oligosaccharides stimulate defense responses and protection against pathogens in plants. Mar Drugs 9(12):2514–2525

Wagle BG (1990) 10 Beach rocks of the central west coast of India. https://link.springer.com/content/pdf/10.1007%2FBF02431028.pdf (December 26, 2018)

Yan GL et al (2011) Sodium alginate oligosaccharides from brown algae inhibit salmonella enteritidis colonization in broiler chickens. Poult Sci 90(7):1441–1448

Yanagisawa M, Kawai S, Murata K (2013) Strategies for the production of high concentrations of bioethanol from seaweeds. Bioengineered 4(4):224–235

Yoon HS, Andersen RA, Boo SM, Bhattacharya D (2009) Stramenopiles. In: Encyclopedia of microbiology. Elsevier Inc., pp 721–731

Yoon HS et al (2006) Defining the major lineages of red algae (RHODOPHYTA) 1. J Phycol 42(2):482–492

Yu S et al (2002) Physico-chemical characterization of floridean starch of red algae. Starch-Stärke 54(2):66–74

Chapter 28
Gracilaria Cultivation and the Role of Its Associated Bacteria for Biomass Production

V. R. Umashree and K. Arunkumar

28.1 Introduction

The importance of ocean as a source of protein is well known. Since the ocean covers ¾th of the earth surface and 2 to 3 times more productive than landmasses it gained much attraction recent years. Recently the fishing activity is increased largely all over the world (Raghu Prasad 1964). Together with this, the utilization of algae as a source of food and as a source of other secondary metabolites has also increased tremendously. Long-years back countries like China, Japan, and South Korea used algae as food. But now a days many other countries involved in the cultivation and utilization of algae when the researches showed their benefits to mankind.

One of the most important problems faced by our generation is global warming and climate change. Global warming is the term used to describe a gradual increase in the average temperature of the Earth's atmosphere and its oceans. According to the scientific consensus on climatic changes, the average temperature of the Earth has risen between 0.4 and 0.8 °C over the past 100 years (https://www.livescience.com/topics/global-warming). Research has shown that the carbon dioxide (CO_2) released into the atmosphere has increased significantly since the beginning of the industrial era (https://www.globalccsinstitute.com/institute). The increased volumes of carbon dioxide and other greenhouse gases released by the burning of fossil fuels, land clearing, agriculture, and other human activities, are believed to be the main sources of the global warming (https://www.livescience.com/topics/global-warming).

Carbon sequestration is the process of capturing atmospheric carbon and depositing it in reservoirs, a process to mitigate global warming and climate change

V. R. Umashree · K. Arunkumar (✉)
Department of Plant Science, Central University of Kerala, Kasaragod, Kerala, India
e-mail: arunkumark@cukerala.ac.in

A. Ranga Rao, G. A. Ravishankar (eds.), *Sustainable Global Resources of Seaweeds Volume 1*, https://doi.org/10.1007/978-3-030-91955-9_28

(Dhanwantri et al. 2014). Various reports suggested that harvested seaweeds can be processed as biofuel feedstock which shall offer additional carbon sequestration potential (Zacharia et al. 2015; Mohammad et al. 2019; Hessami et al. 2019). They also suggested the conversion of the large seaweed biomass into biological charcoal known as biochar through pyrolysis which has agricultural applications (Zacharia et al. 2015).

Haoyang (2018) suggested that algae cultivation would reduce the greenhouse effect in the atmosphere. Algae can absorb carbon dioxide. Carbon, together with the remnants of algae, would be stored in deep oceans, on the seafloor, for several centuries. Since the algae have high metabolism rates and can also provide shellfish with abundant food that contains carbon. Shellfish's shells, which are difficult to be decomposed, are reliable storage of carbon, compared to dead organisms like trees and algae. They suggest that since the algae are fast-growing species, the effort we would need to spend to cultivate them is very little (Haoyang 2018).

Recently the bacteria associated with algae have gained much attraction due to their ability to produce a variety of secondary metabolites. Studies have also shown that these bacteria help in enhancing the growth of algae. Apart from these they also protect the algae from other pathogens and heavy metals.

Gracilaria is a group of warm water seaweeds with a great range of temperature and latitudinal tolerance. Some species were used as food (Shama et al. 2019) and as binding material in the preparation of lime for painting walls in China. On the discovery of agar content, its use is expanded to several Asiatic countries. From then diversity of cultivation methods have been developed in other different places (FAO 2014–2021).

With all these backgrounds, this review focuses on the different methods of cultivating the red macroalgae *Gracilaria* species, the significance of bacterial role in growth promotion and the importance of improving cultivation to gain multiple benefits including blue carbon sequestration.

28.2 *Gracilaria* Stock for Agar Production

According to the statistics reported by the Food and Agriculture Organization of the United Nations (FAO 2014–2021), there is about 2,257,919 tonnes of *Gracilaria* were reported by the year 2011. About 94.2 percent (some 697,240 tonnes) is produced by the cultivation while the remaining (42,224 tonnes) are gathered from wild stocks. The reports also indicate that the most productive countries in America are Chile, Peru and Argentina. In Asia, there is productive cultivation in Indonesia,Vietnam, the Philippines, and Korea while in Africa only Namibia reports *Gracilaria* production (FAO 2014–2021).

According to the FAO (2014), *Gracilaria* is one of the world's most cultivated seaweeds with over 0.8 million tons of annual production and nearly $160 million annual values (Kim et al. 2014). According to the Food and Agriculture Organization of the United Nations (2017), it is the world's most cultivated seaweeds with over 3.8 million tons of annual production and worth annually about US $1 billion (Kim et al. 2017). Most of the biomass of *Gracilaria* is used in the phycocolloid industry as the main source of food-grade agar (Pereira and Yarish 2008; Kim et al. 2017) and as an animal feed (Johnson et al. 2014; Kim et al. 2017; Shama et al. 2019). *Gracilaria* contributes about 66% of the total agar production (Pereira and Yarish 2008; Kim et al. 2017).

It is expected that the Asia Pacific is the largest producer of agar due to the easily available raw material – the red algal species in eastern and southwestern Asian countries. North America is expected to be the second-largest one. It is widely used in the food consumed by the North American population, which plays a crucial role in increasing the overall demand in the region (http://www.algaeindustrymagazine.com/new-report-explores-agar-market/).

Global agar production increased from 7500 to 9600 tonnes, with sale prices of USD 17/kg in 1999, increasing to USD 18/kg in 2009. The world agar sale value increased from USD 128 million in 1999 to USD173 million in 2009. In 1999, about 63 percent of the total agar production was produced by *Gracilaria* which have been increased to 80 percent in 2009 (FAO 2014–2021). Among the seaweed cultivation around the world, the genus *Gracilaria* ranked highest with over 3.8 million tons of annual production. It contributes about 91% of total agar production (Kavale et al. 2018).

In 2016, Agar for Food Industry occupied more than 57% of the total production of agar world wide. Since the agar is widely used in Food Industry, Pharmaceutic, Cosmetics, Daily Chemical, and Scientific Research, the demand for agar is expected to increase during the remaining years of the forecast period of 2017–2022. The global Agar market is valued at 280 million US$ in 2017. With a growth at a CAGR of 4.0% during 2018–2025, it is expected to reach 380 million US$ by the end of 2025 (http://www.abnewswire.com/pressreleases/global-agar-market-2018-industry-key-players-trends-sales-supply-demand-analysis-forecast-to-2025_234917.html).

There is a high pressure on wild populations of *Gracilaria* because of the high harvesting of wild *Gracilaria* crops and an increase in demand for agarophytes (Guanzon and de Castro 1992; Wilson and Critchley 1997). To reduce the over-exploitation of wild crops and to meet the ever-increasing demands for agarophytes and their products, there have been rapid developments in *Gracilaria* mariculture during recent years (Hansen et al. 1981; Wilson and Critchley 1997).

However, information about the influence of environmental parameters such as temperature, irradiance, salinity, and nutrients on the growth of the alga is required to evaluate commercial utilization of *Gracilaria* spp., (Rebello et al. 1996; Wilson and Critchley 1997).

28.3 Agar Yielding Potential Red Algal Species

The genus *Gracilaria* belongs to the class *Rhodophyceae* (Red algae). Most of them are marine species except a few ones. They are having well-developed branched thalli and multicelluar form. They vary in size and shape. They grow in crust on the rocks or shells as a large fleshy, branched, or blade like thalli. The thallus is basically filamentous. It may be simple or branched, free or compacted, and thus forming pseudoparenchyma with uni or multi axial construction (Fig. 28.1). They occupy intertidal to subtidal zones of coastal areas (Jones 1959; Kolanjinathan and Stella 2011).

Members of *Gracilaria* are among the most economically important seaweeds because of their ability to achieve high yields and to produce commercially valuable extracts (Cynthia et al. 2011). They are having a variety of uses, ranging from traditional foods and medicines to biological and industrial applications (Gressler et al. 2010; Kim et al. 2017). They are important for the industrial and biotechnological uses because of the presence of phycocolloids in their cellwalls mainly the agar (Cynthia et al. 2011).

The important and commonly occurring agar yielding seaweeds in different localities of Indian coast are species of *Gelidiella, Gracilaria, Gelidium and Pterocladia*. Among these red algae, only *Gelidiella acerosa, Gracilaria edulis,*

Fig. 28.1 *Gracilaria edulis* (S.G. Gmelin) P.C. Silva fresh thallus found along the coast of Tondi (9° 44′ 30“ N; 79° 1’ 3” E), South India

G. corticota var. corticata G. foliifera and G. verrucosa are available in exploitable quantities (Kaliaperumal and Kalimuthu 1997).

There are three quality grade agars are produced namely, sugar reactive agar, standard agar and food-grade agar. In the sugar reactive agar, the gels are stronger as a function of sugar concentration. It is obtained largely from *Gracilariopsis lemaneiformis*, the most important species under cultivation in China at present (FAO 2014–2021). In the standard agar the gel has the temperature, consistency and structure for microbiological purposes. It is produced largely by *Gelidium, Pterocladia* or *Pterocladiella*. Any kinds of agar that are not meeting the requirements for sugar-reactive or bacteriological agar are designated as the food-grade agar. It is extracted from a wide variety of *Gracilaria* species (FAO 2014–2021).

28.4 Methods of Cultivation

Gracilaria species are mainly cultivated using vegetative fragments. For its cultivation, sustainable seedstock is very important. Seedstock has been supplied from the wild; either the healthy branches of *Gracilaria* from natural stock were collected or reproductive plants were selected to collect spores (either carpospores or tetraspores) for seeding (Buschmann et al. 2008; Kim et al. 2017).

The tropical species of agarophytic seaweeds can be obtained either from gathering natural stocks which is greatly influenced by seasonal changes in the weather (monsoons) or from farming or culturing of these species which is more predictable and stable, and targeted outputs are easily attained (Gavino 1989).

According to some reports,*Gracilaria* cultivation is done mainly in three different ways, including open water, pond, or tank cultures (Pereira and Yarish 2008; Kim et al. 2014). But some other reports say that it is being cultivated in four ways including open water rope cultivation, nearshore bottom cultivation, pond culture, and tank cultures (Kim et al. 2017). Though there are many more methods are practiced for the cultivation.

Open water cultivation is practiced in estuaries, bays, and upwelling areas. A nursery (tank culture) system provides sufficient seedstock (Pereira and Yarish 2008).

In the **pond culture** of *Gracilaria*, the water is introduced into the pond that is dried for several days. Seaweed cuttings are directly staked onto the bottom (Castanos and Buendia 1998).The non-intensive ponds are usually made of an uncovered earthen construction and are lack an artificial water agitation system, while the intensive cultivation ponds are made of a concrete or plastic structure with a water agitation system (Friedlander and Levy 1995).

Tank cultivation has the advantage in its simplicity of controlling the culture system (Pereira et al. 2013). This ensures that production meets high quality standards and biosafety for human consumptions, as well as for other high-value applications such as cosmetics or pharmaceutical products.

In the **rope cultivation** either vegetative materials are tied or inserted within a rope or spores are left to settle on the surface of the ropes. The ropes or lines used here can be monofilament, nylon or other suitable lines. Suspended rope culture is a relatively simple fixed grow-out system (Yarish et al. 2012). Seeded ropes are suspended, stretched between stakes buried in the sediment, or supported at different levels by buoys or rafts (FAO 2014–2021).

In **near shore bottom cultivation,** fronds are either 'seeded' onto rocks and spread on the bottom in a shallow area, or attached to lines that are strung on stakes and suspended just above the bottom (Chen 1989). Here the spore collection and farm sites are usually the same (Veeragurunathan et al. 2015).

In the **floating culture** method, seaweed stocks are inserted in seedling rope made up of palm thread or artificial fiber which isthen fixed to a floating raft (Chen 1989).

In **tube- net method**, tube-net modules (polypropylene commercial fishnet) are used (Pereira and Yarish 2008). After seeding directly lengthwise into the tube they are tied to floating bamboo raft (Ganesan et al. 2017; Mantri et al. 2017).

In the **net bag method,** about 200–300-g seedlings are seeded onto a 75-cm-long bag prepared from commercial fishnet which is covered with agro net. Then the bag is tied onto an 8-mm polypropylene rope. The rope is then tied on both sides to the vertically erected bamboo poles (Ganesan et al. 2017).

In **net pouch method**, net pouch is made with a 3-m-long tube. The bag is then made in to five equal compartments by a hand stitch up to the nylon thread in which *Gracilaria* fragments are seeded (Ganesan et al. 2017).

28.5 Bacterial Association in the Algal Growth

Microorganisms are found everywhere, in all ecosystems around the globe. They can survive even in extreme conditions. It is suggested that at least ten million microbial species remaining unidentified in nature. Only less than 1% of all bacterial species and less than 5% of all fungi species are described (Berdy 2012). Goecke and his colleagues mentioned that seawater containsup to 107 viruses, 106 bacteria, 103 fungi per ml (Goecke et al. 2010). Macroalgae are highly susceptible to epibiosis. Because of the rich content of organic material, a number of bacterial species get associated with macroalgae (Goecke et al. 2010).

Previous studies reported that the number of seaweed-associated bacteria is 100–10,000 times more than those from the surrounding seawater (Chan and McManus 1969; Weinberger et al. 1994). Sutha et al. (2011) enumerated 15 bacterial isolates from six seaweeds (*Gracilaria edulis, Hypnea valentiae, Acanthophora spicifera, Enteromorpha intestinalis, E. flexuosa, Ulva lactuca*) and identified the presence of *Bacillus licheniformis, B. subtilis, B. pumilus, B.marinus, Staphylococcus aureus* and *Streptomyces coelicolar* (Sutha et al. 2011).

Seaweed associated bacteria so far isolated belong to the (super) phyla Proteobacteria, Actinobacteria, Bacteroidetes (CFB group), Cyanobacteria, Firmicutes, Planctomycetes, Verrucomicrobia, Chloroflexi, Deinococcus-Thermus, Fusobacteria, Tenericutes, and the candidate division OP11 (Hollants et al. 2012).

In the review of Hollants and collegues (2012), they mentioned that, in all studies reviewed, the most common bacterial clade associated with seaweeds were Gammaproteobacteria with 37% relative abundance (percentage of published records), followed by the CFB group (20%), Alphaproteobacteria (13%), Firmicutes (10%), and Actinobacteria (9%). On a lower taxonomic level, the orders Flavobacteriales (14%), Alteromonadales (12%), Vibrionales (10%), Pseudomonadales (9%), Bacillales (9%), Actinomycetales (8%), and Rhodobacterales (7%) were most abundant in seaweed-associated bacterial communities (Hollants et al. 2012).

Suvega and Arunkumar (2014), isolated 673 bacteria from different algae *Caulerpa scalpelliformis, Ulva lactuca, U. fasciata, Chaetomorpha linum, Gracilaria edulis, G.corticata* var. *corticata, Hypnea valentiae, Grateloupia filicina, Kappaphycus alvarezii and Sargassum wightii* as well as from sediments and seawater. They reported the presence of 26 bacterial genera with with species of *Bacillus* recording a maximum of 40.2%. They indicated that bacterial populations were considerably higher in seaweeds as compared to seawater and sediments (Suvega and Arunkumar 2014).

Algal–Bacterial Interactions Marine algae and bacteria have come a long way since algal plastids originated from Endosymbiotic Cyanobacteria (Margulis and Schwartz 1998; Ramanan et al. 2016). They developed a diversified bacterial association like beneficial (mutualistic), harmful (parasitic), neutral (commensal), etc. Bacteria either live on the surface (epiphytes) or in the cytoplasm and/or vacuolar systems of the cells (endophytes) (Hollants et al. 2012; Friedrich 2012).

Beneficial Bacterial–Macroalgal Interactions Seaweed-associated bacteria gained a lot of importance because of theirability to produce a variety of secondary metabolites. A large number of endophytic, epiphytic and epibiotic bacteria are associated with macroalgae, which constantly interact with their host positively or negatively.

Studies by Sturz et al. (2000) have shown that bacterial isolates can contribute to the growth, health, and development of seaweeds by the direct production of the plant growth regulators and nitrogen fixation (Sturz et al. 2000). These plant growth regulators have been reported to play an important role in the morphogenesis of callus from Rhodophyta members (Yokoya 2000; Reddy et al. 2003).

Bacterial effects on morphogenesis have been reported in foliaceous green macroalgae such as *Ulva* and *Monostroma*. It has been reported that this morphogenesis controlled by bacteria belonging to the genera *Cytophaga, Pseudomonas, Staphylococcus, Vibrio, Bacillus*, and *Flavobacterium* (Marshall et al. 2006; Singh et al. 2011a).

The plant growth-promoting nature of bacteria associated with *Laminaria japonica* was reported by Dimitrieva et al. (2006). In addition to these, studies showed that microorganisms play a role in the protection of macroalga against toxic heavy metals (Goecke et al. 2010) or crude oil (Goecke et al. 2010). Microorganisms are able to detoxify, for example, heavy metals by precipitation, adsorption, or transformation to fewer toxic forms (Yurkov and Beatty 1998; Goecke et al. 2010).

Bacteria produce morphogenic factors, fixed nitrogen, enzymes, and vitamins which promote algal growth. Morphogenesis is controlled by a highly potent differentiation inducer, thallusin, isolated from well-defined associated bacteria (Hollants et al. 2012).

Seaweed-associated bacteria were isolated and studied for their morphogenesis capability in axenic cultures of the green alga *Ulva fasciata* (Singh et al. 2011a). This was the first study providing evidence of the effect of bacterial isolates on zoospore induction. Later the role of bacterial isolates in enhancing the bud induction in the industrially important red alga *Gracilaria dura* was studied. The findings revealed for the first time that IAA coupled with nitrogen fixation induces and regenerates new buds in *G. Dura* (Singh et al. 2011b). They suggested that bud-inducing bacterial isolates are important for nitrogen supply to the *G. dura*. These associated nitrogen-fixing bacteria will fix atmospheric nitrogen and supply nitrogen supplement sources.

Studies by Weinberger et al. (2007) on *Acrochaetium* sp., showed that the spore release is controlled by bacteria. He reported that suppression of bacterial epiphytes with antibiotics prevented spore release of *Acrochaetium* sp. completely, that demonstrated that this alga strongly depends on bacterial AHL. They suggested that the life cycle completion in *Acrochaetium* sp. strongly depends on bacteria (Weinberger et al. 2007).

The studies done by Shin (2008) showed that enhanced settlement of spores on mixed microbial biofilm of *Ulva fasciata* (Shin 2008). The studies done by Imchen (2012) on the influence of biofilms on the zoospore settlement of *Enteromorpha flexousa* showed that the biofilms provide a favorable substratum for settlement (Imchen 2012).

Studies to evaluate the effect of quorum sensing (QS) molecules produced by the epiphytic and endophytic bacteria associated with green macroalgae *Ulva* (*U.fasciata* and *U.lactuca*) and red macroalgae *Gracilaria* (*G.corticata* and *G.dura*) on the carpospores liberation from *G.dura* has shown that carpospore liberation is positively induced by the isolated bacteria (Singh et al. 2015). Studies also demonstrated that some Gram-negative epiphytic and endophytic seaweed-associated bacteria produce different types of AHLs. They reported that these bacterial isolates can effectively be used for mass carpospore liberation.

These studies show the significance of associated bacteria in the growth of macro algae. It is obvious that there are bacteria that contribute to bud induction, spore release, spore settlement, morphogenesis, nitrogen fixation, etc. Thus, isolation and identification of such bacteria and their mechanism of interaction will be very helpful for improving the cultural practices of *Gracilaria* and other algae.

28.6 Potential of *Gracilaria* Mariculture in India

Within the red macroalgae, the genus *Gracilaria*, represented by more than 100 species (Oliveira and Plastino 1994) living in warm water areas (Steentoft and Farham 1997), is notable for its economic importance as a source of agar. Culturing of seaweeds in Indian coastal waters started with cultivation of *Gracilaria edulis* due to its high regenerative capacity (Zacharia et al. 2015). Currently, 185 *Gracilaria* species are taxonomically accepted (Guiry and Guiry 2016).

The surveys carried out by CSMCRI, CMFRI and other research organizations have revealed vast seaweed resources along the coastal belts of South India. On the West Coast, especially in the state of Gujarat, abundant seaweed resources are present on the intertidal and sub tidal regions (Govt. of Gujarath 2017). These resources have great potential for the development of seaweed-based industries in India.

There are about 25 agar and alginate industries situated in states of Tamil Nadu, Karnataka, Kerala Pondicherry, Andhra Pradesh and Gujarat. Among these, more than 20 industries produce food grade agar by using *Gracilaria edulis* (Kaliaperumal et al. 2004; Kaliaperumal and Ramalingam 2005).

Thirty-one species of *Gracilaria* are found on the Indian coastline of the Bay of Bengal. The estimated biomass of *Gracilaria* from Indian waters is 1700 tonnes (Krishnamurthy 1989). India produces 110–132 tons of dry agar annually utilizing about 880–1100 tons of dry agarophytes (Govt. of Gujarath 2017).

Central Salt and Marine Chemical Research Institute (CSMCRI), Marine Algal Research Station (MARS), Mandapam, Tamil Nadu a CSIR Institute developed viable and commercially sustainable methods for cultivating *Gracilaria edulis*. CSMCRI developed single Rope Floating Raft (SRFR) method which is suitable for culturing seaweeds in wide area and greater depth (Kaliaperumal and Kalimuthu 1997; Gulshad 2016).

Floating raft technology has been recommended to be used on the coasts of Kerala for agarophytic algal cultivation. Certain areas in the Gulf of Kutch have been suggested as suitable for deep-water seaweed cultivation. CMFRI has developed techniques for culturing *Gelidiella acerosa, Gracilaria edulis, Hypnea musciformis* and *Acanthophora spicifera*, and to find improved techniques for propagation and large-scale culture of other economically important seaweeds, attempts are being made (Gulshad 2016; Govt. of Gujarat 2017).

In different localities of Indian coast, *Gracilaria edulis, G. corticola var. corticata, G. foliifera* and *G.verrucosa* are available in exploitable abundant quantities. *G.arcuala* and *G. verrucosa* occur in harvestable quantities in some estuaries and backwaters of Tamil Nadu and Pondicherry (Kaliaperumal and Kalimuthu 1997). A 17-fold increase in yield of *G. edulis* was obtained in 76 days in the first harvest at Minicoy Lagoon, Minicoy Island, of Lakshadweep, India during south west monsoon season by adopting single bottom coir rope method (Krishnamurthy 1989; Kaliaperumal and Kalimuthu 1997).

Single bottom coir rope method and single bottom nylon rope method are best method for culture during the southwest monsoon season at Islands of U.T. of

Lakshadweep. This is cost effective seaweed farming technology and practiced by fisherman of Lakshadweep to generate income during lean fishing season of south-west monsoon (May to September) because this time site becomes enriched with nutrients, which help to get good growth of seaweeds (Kaliaperumal and Kalimuthu 1997; Gulshad 2016).

According to the project done by Government of Gujarat (2017), India can grow more than one million tonnes of seaweeds in six states-Gujarat, Tamil Nadu, Kerala, Andhra Pradesh, Maharashtra and Andaman & Nicobar Islands. In the global markets each tons of average quality agar-agar is sold for more than US$2000 (INR120,000) and the country has the potential to generate more than INR200 crore in foreign exchange annually apart from providing additional income and gainful employment to thousands of people on the coastline (Govt. of Gujarat 2017; Krishnamurthy 1989). This shows the importance of *Gracilaria* culturing in India.

28.7 Conclusions and Future Perspectives

The increasing population pressure and energy needs have resulted in issues related to global warming. To avoid the dangerous after-effects of this, there is need for ecofriendly solution. It also demands multi-pronged approaches. One of them being seaweed farming to achieve sustainable climate resilient strategy with multiple benefits as means of carbon sequestration and mitigation of ocean acidification, also as feedstock for bioenergy production with an underpinning of economic gains for the dependent populations. The red macro algae *Gracilaria* are economically very important seaweeds for their agar yielding properties and other biotechnological and industrial applications. A number of cultivation techniques are being developed for the *Gracilaria* as a sustainable alternative to natural stock which is already over expoloited. Knowledge on the life cycle and environmental factors that affect growth of *Gracilaria* will help to improve the techniques for cultivation. One of the ways to improve the algal growth is isolating and identifying growth promoting seaweed associated bacteria. The recent studies showed that the seaweed associated bacteria help in improving the growth of algae thus improving the biomass productivity. When compared to other countries, India is still improving methodologies for cultivation of *Gracilaria*. However, India, which has a large coastline area, has a high possibility for developing cultivation techniques and improving algae-based industries. It will offer jobs for millions of people and provide a large income to the nation. It will also address the issues of global warming and facilitate a move towards carbon-neutral economy.

Acknowledgements The authors gratefully acknowledge the Department of Plant Science, Central University of Kerala. India. The authors are thankful to CSIR- India, for the financial support.

References

Berdy J (2012) Thoughts and facts about antibiotics: where we are now and wherewe are heading. J Antibiot 65:385–395

Buschmann AH, Varela DA, Hernández-González MC, Huovinen P (2008) Opportunities and challenges for the development of an integrated seaweed-based aqua culture activity in Chile: determining the physiological capabilities of *Macrocystis*and*Gracilaria*as biofilters. J Appl Phycol 20:571–577

Castanos M, Buendia R (1998) Farming techniques for seaweeds. SEAFDEC Asian. Aquaculture XX No. 1:14–25

Chan ECS, McManus EA (1969) Distribution, characterization, and nutrition of marine micro-organisms from the algae *Polysiphonia lanosa* and *Ascophyllum nodosum*. Can J Microbiol 15:409–420

Chen JX (1989) Gracilaria culture in China. 18 pp. NACA-SF/WP/89/13. Sea farming project, RAS/86/024. Gracilaria production and utilization in the bay of Bengal. BOBP/REP/45

Cynthia Layse F, Almeida D, De Heloina S, FalcãoGedson R, Lima DM, Montenegro CDA, Lira NS, De Athayde-Filho PF, Rodrigues LC, De Souza MDFV, Barbosa-Filho JM, Batista LM (2011) Bioactivities from marine algae of the genus *Gracilaria*. Int J Mol Sci 12:4550–4573. https://doi.org/10.3390/ijms12074550

Dhanwantri K, Sharma P, Mehta S, Prakash P (2014) Carbon sequestration, its methods and significance, environmental sustainability: concepts. Principles, Evidences and Innovations. ISBN: 978–93–83083-75-6

Dimitrieva GY, Crawford RL, Yuksel GU (2006) The nature of plant growth-promoting effect of a pseudomonad associated with the marine algae Laminaria japonica and linked tocatalase extraction. J ApplMicrobiol 100:1159–1169

FAO (Food and Agriculture Organization of the United Nations) (2014–2021) Cultured aquatic species information programme. *Gracilaria* spp. Cultured aquatic species information programme. Text by Santelices, B. In: FAO Fisheries Division [online]

Friedlander M, Levy I (1995) Cultivation of *Gracilaria* in outdoor tanks and ponds. J Appl Phycol 7:315–324

Friedrich MW (2012) Bacterial communities on macroalgae. In: Seaweed biology. Springer, Berlin, Germany, pp 189–201

Ganesan M, Eswaran K, Reddy CK (2017) Farming of agarophytes in India—a long-time sustainability for the industry and preserving wild stocks. J Appl Phycol 2017(29):2239–2248. https://doi.org/10.1007/s10811-017-1128-0

Gavino C, Trono Jr (1989) Present status of *Gracilaria* culture. pp. 5–12. Gracilaria production and utilization in the Bay of Bengal. BOBP/REP/45

Goecke F, Labes A, Wiese J, Imhoff JF (2010) Chemical interactions between marine macroalgae and bacteria. Mar Ecol Prog Ser 409:267–299. https://doi.org/10.3354/meps08607

Govt. of Gujarath (2017) Development of seaweed culture. Agro and Food Processing https://gaic.gujarat.gov.in/writereaddata/images/pdf/26-Seaweed-Culture

Gressler V, Yokoya NS, Fujii MT, Colepicolo P, Filho JM, Torres RP, Pinto E (2010) Lipid, fatty acid, protein, amino acid and ash contents in four Brazilian red algae species. Food Chem 120:585–590

Guanzon NGJ, De Castro T (1992) The effects of different stocking densities and some abiotic factors on cage culture of *Gracilariasp.* (Rhodophyta. Gigartinales), Bot Mar 35:239–243

Guiry MD, Guiry GM (2016) AlgaeBase. World-wide electronic publication, National University of Ireland, Galway. Available from: http://www.algaebase.org. Accessed June 21, 2016

Gulshad M (2016) Current trends and prospects of seaweed farming in India. In: Joseph I, Boby I (eds) Winter school on technological advances in mariculture for production enhancement and sustainability. Course Manual, Central Marine Fisheries Research Institute, Kochi, pp 78–84

Hansen JE, Packard JE, Doyle WT (1981) Mariculture of red seaweeds. Cited in: Haoyang C (2018) Algae-based carbon sequestration, IOP Conf. Series: Earth and Environmental Science 120, 012011. https://doi.org/10.1088/1755-1315/120/1/012011

Haoyang C (2018) Algae-Based Carbon Sequestration. IOP Conference Series: Earth and Environmental Science 120 012011. https://doi.org/10.1088/1755-1315/120/1/012011

Hessami MJ, Cheng SF, Ranga Rao A, Yin YH, Phang SM (2019) Bioethanol production from agarophyte red seaweed, *Gelidium elegans* using a novel sample preparation method for analysing bioethanol content by gas chromatography. Biotech J 9(1):25

Hollants J, Leliaert F, De Clerck O, Willems A (2012) What we can learn from sushi: a review on seaweed– bacterial associations. FEMS Microbiol Ecol 83:1–16. https://doi.org/10.1111/j.1574-6941.2012.01446x

http://www.abnewswire.com/pressreleases/global-agar-market-2018-industry-key-players-trends-sales-supply-demand-analysis-forecast-to-2025_234917.html

http://www.algaeindustrymagazine.com/new-report-explores-agar-market/

https://www.globalccsinstitute.com/institute

https://www.livescience.com/topics/global-warming

Imchen T (2012) Effect of temperature, salinity and biofilm on the zoospores settlement of *Enteromorphaflexuosa (*Wulfen) J. Agardh. Indian J Geo- Marine Sci 41(4):355–358

Johnson RB, Kim JK, Armbruster LC, Yarish C (2014) Nitrogen allocation of *Gracilariatikvahiae*grown in urbanized estuaries of Long Island sound and new York City, USA: a preliminary evaluation of ocean farmed *Gracilaria*for alternative fish feeds. Algae 29:227–235

Jones WE (1959) The growth and fruiting of *Gracilariaverrucosa*(Hudson) Papenfuss. J Mar boil Ass 38:47–56

Kaliaperumal N, Kalimuthu S (1997) Seaweed potential and its exploitation in India. Seaweed Res Utiln 19(1 & 2):33–40

Kaliaperumal N, Kalimuthu S, Ramalingam JR (2004) Present scenario of seaweed exploitation and industry in India. Seaweed Res Utilization 27(1 & 2) (in press)

Kaliaperumal N, Ramalingam JR (2005) Effect of different fertilizers on the growth of *Gracilaria*edulis?(Gmelin) silva in onshore cultivation. Indian Hydrobiologia 7:63–67

Kavale MG, Veeragurunathan V, Mantri VA (2018) Handbook on farming of *Gracilaria* dura. Govt. of India, National Fisheries Development Board. https://www.researchgate.net/publication/327418918

Kim JK, Kraemer GP, Yarish C (2014) Field scale evaluation of seaweed aquaculture as a nutrient bioextraction strategy in Long Island sound and the Bronx River estuary. Aquaculture 433:148–156

Kim JK, Yarish C, Hwang EK, Park M, Kim Y (2017) Seaweed aquaculture: cultivation technologies, challenges and its ecosystem services. Algae 32(1):1–13

Kolanjinathan K, Stella D (2011) Pharmacological effect of *Gracilaria corticata* solvent extracts against human pathogenic bacteria and fungi. *I*. J Pharmaceutic Biol Arch 2(6):1722–1728

Krishnamurthy V (1989) *Gracilaria* resources of India with particular reference to the bay of Bengal. Gracilaria production and utilization in the Bay of Bengal, BOBP/REP/45, pp 97–98

Mantri VA, Ashok KS, Musamil TM, Gobalakrishnan M, Saminathan KR, Behera DP, Veeragurunathan V, Eswaran K, Thiruppathi S, Pothal JK, Ghosh PK (2017) Tube-net farming and device for efficient tissue segregation for industrially important agarophyte*Gracilariaedulis* (Rhodophyta). Aquac Eng 77(2017):132–135

Margulis L, Schwartz KV (1998) Five kingdoms: an illustrated guide to the phyla of life on earth, 3rd edn. Freeman, New York, NY, USA. ISBN 0-7167-3027-8. xx, p 520

Marshall K, Joint I, Callow ME, Callow JA (2006) Effect of marine bacterial isolates on the growth and morphology of axenic plantlets of the green alga *Ulvalinza*. MicrobEcol 52:302–310

Mohammad JH, Ranga Rao A, Ravishankar GA (2019) Opportunities and challenges in seaweeds as feed stock for biofuel production. In: Ravishankar GA, Rao AR (eds) Handbook of algal technologies and phytochemicals: volume-II Phycoremediation, biofuels and global biomass production. CRC Press, USA, pp 39–50

Oliveira EC, Plastino EM (1994) *Gracilaria*ceae. In: Akatsuka I (ed) Biology of economic algae. SSB Academic, The Hague, pp 185–226

Pereira R, Yarish C (2008) Mass production of marine macroalgae. In: Jørgensen SE, Fath BD (eds) Encyclopedia of ecology. Ecological engineering, vol 3. Elsevier, Oxford, pp 2236–2247

Pereira R, Yarish C, Critchley AT (2013) Seaweed aquaculture for human foods in land based and IMTA systems. In: Meyers RA (ed) Encyclopedia of sustainability science and technology. Springer, New York, pp 9109–9128

Raghu Prasad R (1964) Study of primary production and its importance in an integra ted fisheries research programme. Fishery Technology I(I)

Ramanan R, Kim BH, Cho DH, Oh HM, Kim HS (2016) Algae–bacteria interactions: evolution, ecology and emerging applications. Biotechnol Adv 34:14–12

Rebello J, Ohno M, Critchley AT, Sawamura M (1996) Growth rates and agar quality of *Gracilariagracilis*(stack house) Steentoft from Namibia, southern Africa. Bot Mar 39:273–279

Reddy CRK, Kumar GRK, Siddhanta AK, Tewari A (2003) Invitro somatic embryogenesis and regeneration of somatic embryos from pigmented callus of *Kappaphycus alvarezii*(DOTY) DOTY (Rhodophyta, Gigartinales). J Phycol 39:610–616

Shama A, Joyce SG, Mari FD, Ranga Rao A, Ravishankar GA, Hudaa N (2019) Macroalgae and microalgae: novel sources of functional food and feed. In: Ravishankar GA, Rao AR (eds) Handbook of algal technologies and phytochemicals, volume-I: food, health and nutraceutical applications. CRC Press, USA, pp 207 219

Shin HW (2008) Rapid attachment of spores of the fouling alga *Ulva fasciata* on biofilms. J Environ Biol 29:613–619

Singh RP, Baghel RS, Reddy CRK, Jha B (2015) Effect of quorum sensing signals produced by seaweed associated bacteria on carpospores liberation from *Gracilaria dura*. FrontiersinPlantScience, Plant- Microbe Interaction 6:117

Singh RP, Bijo AJ, Baghel RS, Reddy CRK, Jha B (2011b) Role of bacterial isolates in enhancing the bud induction in the industrially important red alga *Gracilaria dura*. FEMS Microbiol Ecol 76:381–392. https://doi.org/10.1111/j.1574-6941.2011.01057.x

Singh RP, Mantri VA, Reddy CRK, Jha B (2011a) Isolation of seaweed-associated bacteria and their morphogenesis-inducing capability in axenic cultures of the green alga *Ulva fasciata*. Aquat Biol 12:13–21. https://doi.org/10.3354/ab00312

Steentoft M, Farham WF (1997) Northern distribution boundaries and thermal requirements of *Gracilaria* and Gracilariopsis (*Gracilaria*les, Rhodophyta) in Atlantic Europe and Scandinavia. Nord J Bot 5:87–93

Sturz AV, Christie BR, Nowak J (2000) Bacterial endophytes:potential role in developing sustainable systems of cropproduction. Crit Rev Plant Sci 19:1–30

Sutha SP, Venkatesan M, Arunkumar K (2011) Endobiotic bacteria in some seaweeds of Thondi coastal region in Palk Bay, Tamil Nadu. India J Mar Biol Ass India 53(2):251–256

Suvega T, Arunkumar K (2014) Antimicrobial activity of bacteria associated with seaweeds against plant pathogens on par with bacteria found in seawater and sediments. Br Microbiol Res J 4(8):841–855

Veeragurunathan V, Eswaran K, Malarvizhi J, Gobalakrishnan M (2015) Cultivation of *Gracilaria dura* in the open sea along the southeast coast of India. J Appl Phycol 27:2353. https://doi.org/10.1007/s10811-014-0514-0

Weinberger F, Beltran J, Correa JA, Lion U, Pohnert G, Kumar N, Steinberg P, Kloareg B, Potin P (2007) Spore release in acrochaetium sp. (rhodophyta) is bacterially controlled. J Phycol 43:235–241. https://doi.org/10.1111/j.1529-8817.2007.00329.x

Weinberger F, Friedlander M, Gunkel W (1994) A bacterial facultative parasite of *Gracilariaconferta*. Dis Aquat Org 18:135–141

Wilson AJ, Critchley AT (1997) Studies on *Gracilariagracilis*(Stackhouse) Steen toft, Farnham and Irvine and *Gracilariaaculeata*(Hering) Papenfuss from southern Africa. I. the influence of temperature, irradiance, salinity and nitrogen-nutrition on growth. S Afr J Bot 63(6):465–471

Yarish C, Redmond S, Kim JK (2012) *Gracilaria* culture handbook for New England. Wrack Lines 72. http://digitalcommons.uconn.edu/wracklines/72

Yokoya NS (2000) Apical callus formation and plant regenerationcontrolled by plant growth regulators on axenic culture of thered alga *Gracilariopsis tenuifrons* (*Gracilaria*les, Rhodophyta). Phycol Res 48:133–142

Yurkov VV, Beatty JT (1998) Aerobic anoxygenic phototrophic bacteria. Microbiol Mol Biol Rev 62:695–724

Zacharia PU, Kaladharan P, Rojith G (2015) Seaweed farming as a climate resilient strategy for Indian Coastal Waters. In: The international conference on integrating climate, crop, ecology - the emerging areas of agriculture, horticulture, livestock, fishery, forestry, biodiversity and policy issues, pp 59–62. ISBN:978-8-930585-9-6

Chapter 29
Biosorption of Heavy Metals by Seaweed Biomass

Josefina Plaza Cazón and Edgardo R. Donati

Abbreviations

$CaCl_2$	Calcium chloride
Hg(II)	Mercury
Zn(II)	Zinc
Cd(II)	Cadmium
Ni(II)	Nickel
Cr(III)	Chromium
Ca(II)	Calcium
Na(I)	Sodium
$Cr(OH)_3$	Chromium(III) hydroxide
$Cr(OH)^{2+}$	Chromium(II) hydroxide
$HgCl_2$	Mercury(II) chloride
Hg(OH)Cl	Mercury(II) hydroxide chloride
$Hg(OH)_2$	Mercury(II) hydroxide
M	Molar
min	Minutes
mm	millimetre or, millimetres
rpm	revolutions per minute
µm	Micrometer
$mg.L^{-1}$	Milligram or, milligrams per litre
$mmol.g^{-1}$	Millimole per gram
$mol.J^{-1}$	Mole per Joule
$kJ.mol^{-1}$	KiloJoule per mole
$mol.g^{-1}$	Mole per gram

J. P. Cazón (✉) · E. R. Donati
Centro de Investigación y Desarrollo en Fermentaciones Industriales (CINDEFI-CONICET-La Plata). Facultad de Ciencias Exactas. Universidad Nacional de La Plata (UNLP), La Plata, Argentina
e-mail: donati@quimica.unlp.edu.ar

A. Ranga Rao, G. A. Ravishankar (eds.), *Sustainable Global Resources of Seaweeds Volume 1*, https://doi.org/10.1007/978-3-030-91955-9_29

R Gas Constant
T Temperature in Kelvin degrees

29.1 Introduction

Various industrial processes generate heavy metals that in many countries are partially or totally released into the environment. Heavy metals do not degrade and can accumulate in different compartments of the food chain affecting plants, animals and even human health (Vijayaraghavan and Balasubramanian 2015). For this reason, it is essential to treat industrial effluents before they are discharged into the environment. There are several methodologies for the removal of metal ions from wastewater, such as precipitation, evaporation, ion exchange, membrane processes, etc. Many of these conventional technologies are expensive and applicable only when metal ion concentrations are high and the volumes to be treated are small (Beni and Esmaeili 2020). Biosorption could be a suitable alternative for moderate concentrations of metal ions and large volumes to be treated. The process consists of passive capture of metal ions by metabolically inactive biological materials while alternatives such as bioprecipitation and bioaccumulation involve the use of active biomasses. Biosorption, in general, is fast, efficient and highly specific for the removal of heavy metals. Another aspect in biosorption field is the desorption process. For this purpose, it is desirable to desorb metals from the biosorbent and to regenerate the biomaterial for more than one cycle of adsorption-desorption. When the biomaterial loses the capacity for adsorb heavy metals, it is incinerated to remove metals from the ash (Volesky 2003).

The materials of biological origin that have been used include mainly biomasses from different microorganisms and different agro-industrial residues, among others (Vijayaraghavan and Balasubramanian 2015); also other materials have also been used, such as chitosan, which is a biopolymer extracted from the shells of crustaceans (Kyzas and Bikiaris 2015). Among these biological materials, algae biomasses and in particular brown algae stand out, not only because of their abundance and availability in nature but also because of the chemical composition of their cell wall (Anastopoulos and Kyzas 2015; Redha 2020). Alginate (from alginic acid) constitutes 40% of their cell wall, which is mainly responsible for the adsorption of heavy metals. *Macrocystis pyrifera* (Linnaeus) C. Agardh is a brown macroalgae belonging to the class Phaeophyceae. In Argentina, it is found in the coastal waters from Peninsula de Valdez (Chubut Province) to the Beagle Channel (Tierra del Fuego). *M. pyrifera* reaches the coasts through the arrivals, depositing itself and causing a bad smell, provoking a negative impact for the tourist activity. Here we have presented our finding on the use of *M. pyrifera* waste in biosorption of different heavy metals for applications in pollution abatement.

29.2 Methodologies Adapted by us for Biosorption Studies

29.2.1 Biomass Pretreatment

Macrocystis pyrifera biomass was collected in Bahía Camarones (Chubut Province; 44° 47′24′′S, 65° 43′ 12′′ W). Blades, stipes and floats were separated, the pieces were ground and sieved, and finally 1.18–2 mm size fraction was selected; the particles were washed several times with distilled water and dried at 50 °C. This biomass was treated with 0.2 M $CaCl_2$ at pH 5.0 for 24 h. Figure 29.1 shows the biomass after and before grinding, sieving and pretreatment. Then, it was repeatedly washed with distilled water and dried as mentioned before.

29.2.2 Kinetic Studies

The kinetics of the biosorption process was studied through batch experiments in which 1.0 g of biomass (particle size 1.18–2 mm and pre-treated with 0.2 M $CaCl_2$) was added to 50 mL of individual solution of Hg(II), Zn(II), Cd(II), Ni(II) and Cr(III) with a concentration of 50 mg L^{-1}. The initial pH value of the solutions was adjusted to different values: 3.0, 4.0 and 5.0. The bottles were kept at 160 rpm and 25 °C; samples were taken at different times during 24 hours, except for Hg(II), whose pH solution was adjusted from 3.0 to 6.0 and sampling time was extended to 72 hours. All the experiments were carried out, at least, in duplicate.

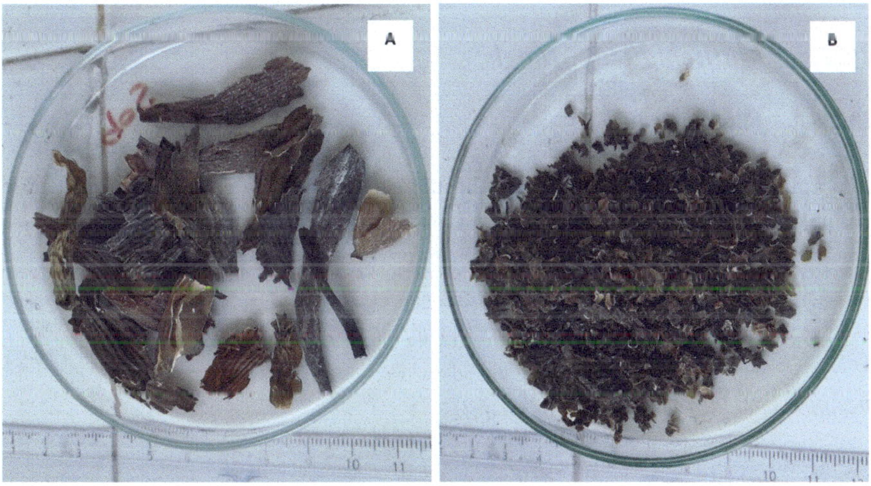

Fig. 29.1 (**a**) *M. pyrifera* biomass; (**b**) pre-treated *M. pyrifera* biomass

29.2.3 Adsorption Isotherms

The equilibrium studies in the biosorption of the selected metals were carried out in the range of metal concentrations between 10 and 400 mg L^{-1}. These experiments were carried out under optimal pH conditions and using the contact time necessary to reach the equilibrium that was obtained in the previous kinetic studies. All the experiments were carried out, at least, in duplicate.

29.2.4 Analytical Methods

Samples taken at certain intervals were filtered through a 0.45 μm membrane. Atomic absorption spectrophotometry (Shimadzu AA6650, Shimadzu Corporation Kyoto, Japan) and Inductively Couple Plasma Atomic Emission Spectroscopy (ICP-AES) using a JY 2000 spectrometer (Jobin–Yvon, Longjumeau, France) were used to measure initial and final metal concentrations after biosorption. The amount of metal adsorbed was calculated by the mass balance equation, according to the equation:

$$q = \frac{V(C_i - C_f)}{m} \quad (29.1)$$

where q is the solute uptake (mg.g^{-1} or mmol.g^{-1}); C_i and C_f the initial and final solute concentrations in solution (mg.L^{-1} or mM), respectively; V solution volume (L) and m the mass of biosorbent (g, dry weight basis).

29.2.5 Mathematical Models

29.2.5.1 Kinetic Studies

Adsorption kinetics were analyzed with the pseudo-first order (PPO) and pseudo-second order (PSO) kinetic models described below.

Pseudo-first order model (Lagergren model)

$$\frac{q_t}{t} = K_1(q_{eq} - q_t) \quad (29.2)$$

Pseudo-second order model (Ho model)

$$\frac{\delta q_t}{\delta_t} = K_2(q_{eq} - q_t)2 \quad (29.3)$$

where q_{eq} and q_t (mmol.g^{-1}) are the sorption capacity at equilibrium and at time t, respectively and K_1 (1.min^{-1}) is the rate constant of pseudo first order and K_2 is the rate constant of pseudo second order [g (mmol.min^{-1})].

The prediction of the rate-limiting step is an important factor to be considered in the adsorption process. In the solid–liquid sorption process, the solute transfer is usually characterized by external mass transfer (boundary layer diffusion), or intra-particle diffusion, or both.

The external mass transfer coefficient (βL) in the liquid film boundary can be evaluated by using the following equation:

$$\ln\left(\frac{C_t}{C_o} - \frac{1}{1+mK_a}\right) = \ln\left(\frac{mK_a}{1+mK_a}\right) - \left(\frac{1+mK_a}{mK_a}\right)\beta L S_s t \tag{29.4}$$

where C_t and C_o (both in mg.L^{-1}) are the concentrations of sorbent at time t and zero, respectively; K_a (L.g^{-1}) is a constant defined as the product of the Langmuir constants ($q_m b$), m (g.L^{-1}) and S_s (m^2) are the adsorbent mass and surface area, respectively. The coefficient βL can be calculated from the slope of the regression line $\ln[(C_t/C_o) = (1/1 + mK_a)]$ vs t.

On the other hand, the intra particle diffusion was explored by using the following simplified equation:

$$q_t = K_{dif} t^{1/2} + C \tag{29.5}$$

where C is the intercept and K_{dif} is the intra-particle diffusion rate constant. The value of q_t correlated linearly with values of $t^{1/2}$ and the rate constant K_{dif} was directly evaluated from the slope of the regression line.

29.2.5.2 Adsorption Isotherms Models

Langmuir and Freundlich isotherms were used to describe the equilibrium state for metal ion adsorption. The Langmuir isotherm is represented by the following equation:

$$q_{eq} = \frac{q_m b C_{eq}}{1 + bC_{eq}} \tag{29.6}$$

where q_{eq} is the metal uptake at the equilibrium (mmol.g^{-1}); q_m is the maximum Langmuir uptake (mmol.g^{-1}) at saturation of the monolayer; C_{eq} is the final concentration at the equilibrium (mmol.L^{-1}); b is the Langmuir affinity constant (L.mmol^{-1}). The maximum adsorption capacity (q_m) and Langmuir constant (b) were obtained after linearization of Eq. (29.6) (C_{eq}/q_{eq} vs. C_{eq}). Langmuir constant b = 1/K is related to the energy of adsorption through the Arrhenius equation. The higher b (and the smaller K), the higher the affinity of the sorbent for the sorbate.

The general form of Freundlich model is:

$$q_{eq} = K_f C_{eq}^{1/n} \tag{29.7}$$

where K_f is a constant related to the biosorption capacity and 1/n is an empirical parameter related to the biosorption intensity, which varies with the heterogeneity of the material. The parameters were obtained by linear regression after linearization (ln q_{eq} vs ln C_{eq}).

The equilibrium data were also analyzed with the Dubinin–Radushkevich (D–R) isotherm model to determine the nature of the biosorption processes (i.e., physical vs. chemical). The linear representation of D–R isotherm equation is given by:

$$\ln q_{eq} = \ln q_m - \beta_\epsilon^2 \tag{29.8}$$

where q_{eq} is the amount of metal ions sorbed per unit weight of biomass (mol.g^{-1}), q_m is the maximum sorption capacity (mol.g^{-1}), β is the activity coefficient related to biosorption mean free energy (mol.J^{-1}) and ε is the Polanyi potential (ε = RT ln(1 + 1/C_{eq}).

The sorption energy E (kJ.mol^{-1}) was calculated using:

$$E = \frac{1}{\sqrt{-2\beta}} \tag{29.9}$$

29.3 Our Findings with Regard to Use of Seaweed in the Removal of Heavy Metals

29.3.1 Biosorption Kinetics

Figure 29.2 shows the adsorption kinetics of Hg(II), Zn(II), Cd(II), Ni(II), and Cr(III), at the different pH values tested. The adsorption process was faster for Ni(II), Zn(II) and Cd(II), with equilibrium times between 60 and 120 min, somewhat slower for Cr(III) where the time needed to reach equilibrium was 360 min and considerably slower for Hg(II) where 1440 min were needed to reach equilibrium.

The times needed to reach equilibrium are similar to those found by other researchers on other biomasses, with the possible exception of Hg. For example, Tabaraki et al. (2013) obtained an optimal contact time of 90 min for the removal of Zn(II) using *Acinetobacter sp.* biomass isolated from an oil spill as a biosorbent; Limcharoensuk et al. (2015) reported that the biosorption of Cd(II) using *P. areuginosa* B237 and the biosorption of Zn(II) using *T. paurometabola* A155 (bacteria isolated from a Thai mine) reached equilibrium within 15 min and 120 min, respectively, while Shanmugaprakash and Sivakumar (2015)) determined that the optimal

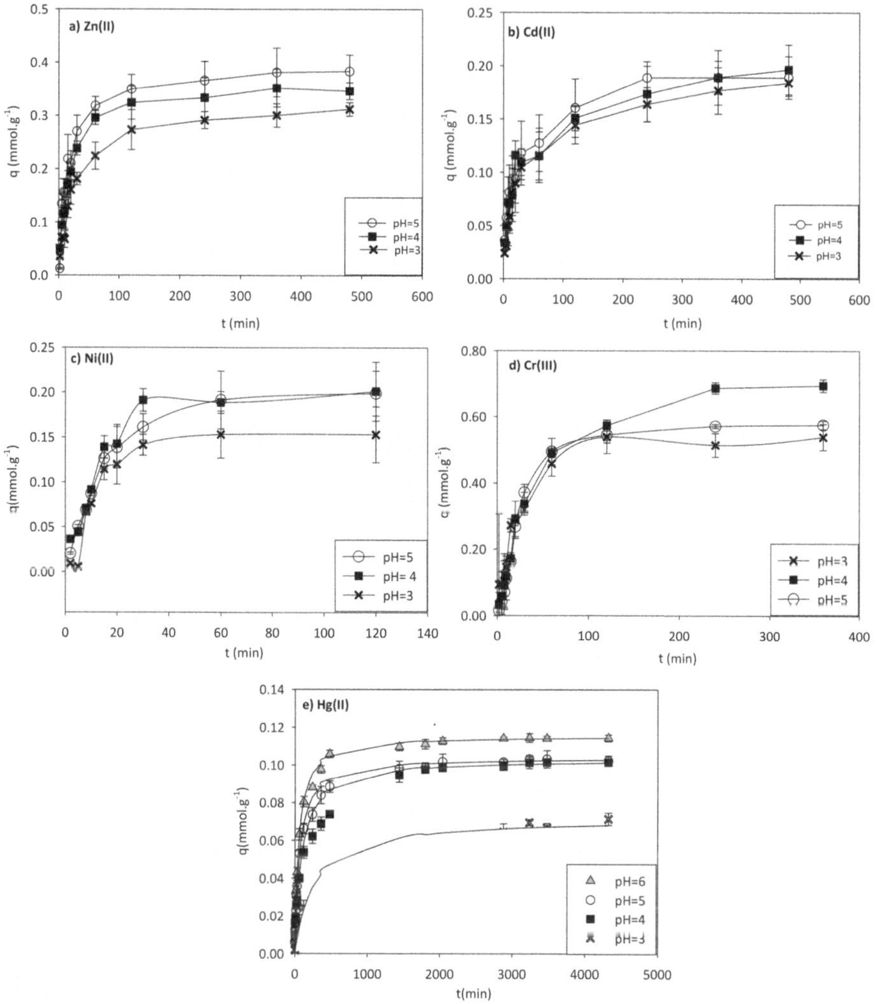

Fig. 29.2 Biosorption kinetics for the removal of **a**) Zn(II), **b**) Cd(II), **c**) Ni(II), **d**) Cr(III) and **e**) Hg(II) on pre-treated biomass of *M. pyrifera*

contact time for the adsorption of Zn(II) using *Pongamia pinnata* as a biosorbent was 60 min. More recently, Barquilha et al. (2019) reported that an equilibrium time of 90 minutes for Ni(II) biosorption on a brown algae *Sargassum sp*, while Ni(II) biosorption by *Streptomyces roseorubens* needed 60 min of contact (Long et al. 2018). Al-Qahtani (2016) used different agro-industrial wastes (tangarine, banana and kiwi shells) for the biosorption of Zn(II), Cd(II), and Cr(III); in all cases, the same optimal equilibrium time was obtained, which was close to 60 min.

In contrast, the equilibrium time for the adsorption of Cr(III) by *M. pyrifera* obtained in this study (360 min) was greater than that determined by Murphy et al.

(2008) who reported that the equilibrium times required for adsorption of Cr(III) by *Polysphonia lanosa*, *Ulva Lactuca* and *Fucus vesiculosus*, were 30 min, 60 min, and 120 min, respectively.

Similarly, the rate of Hg(II) adsorption observed in this study was very low compared to that observed, for example, using microalgae biomass *Chlorella vulgaris* where the time required to reach equilibrium was 90 min (Kumar et al. 2020). In contrast, an optimal time similar to that obtained in this study, was used in the adsorption of Hg(II) using two brown algae, *Lessonia nigrescens* and *Lessonia trabeculata* (Reategui et al. 2009).

Another important parameter that influences the biosorption process is the pH of the solution. It is well known that the pH affects the solubility of metal ions, as well as the availability of functional groups present in the cell wall of the biomass (Plaza et al. 2011). The data represented in Fig. 29.2 allowed the determination of the optimal adsorption pH for all metals tested: 6 for Hg(II), 4 for Zn(II), Ni(II) and Cr(III) and 3 for Cd(II).

The optimum pH values for Cd(II), Zn(II), Ni(II) reported in the literature for other algae are in the pH range of 3 to 4.5 (Barquilha et al. 2019). However, Castro et al. (2017) determined that the optimal operating pH for their Zn(II) biosorption assays was 5 using *Fucus vesiculosus* and sugar pulp as adsorbents. Something similar was obtained by Long et al. (2018) in the adsorption of Ni(II) by *Streptomyces roseorubens*. In the biosorption of Zn(II) and Cd(II) using biomasses of three bacteria isolated from a mine in Thailand, and in the biosorption of Zn(II), Cd(II) and Cr(III) using banana peel, tanjarin and kiwi, the optimum pH was even higher, determining a value equal to 6 (Limcharoensuk et al. 2015; Al-Qahtani 2016.

In most reports, an increase in adsorption has been observed with an increase in pH; this effect is a product of the relationship between biosorption and the amount of negative charges on the surface of the biomass and, in turn, of the nature and behaviour of the functional groups. The ionic state of these groups was determined by the pH value. In the particular case of brown algae the main component is alginate which is formed by blocks of glucuronic acid and manuronic acid. The carboxylic groups of these acids have a pKa of 3.38 and 3.65, respectively. At pH values below pKa, the ligands in the cell wall are mostly covered by protons, while metal ions may be associated with the biomass in increasing numbers above pKa. This means that at higher pH an increase in the absorption of metals by the biomass is expected (Plaza-Cazón et al. 2014). However, an excessive increase in pH would cause hydrolysis and eventual precipitation of transition metal hydroxides that would compete with the uptake (Castro et al. 2017; Barquilha et al. 2019; Beni and Esmaeili 2020).

Within the pH range studied, in solutions of $HgCl_2$ as it was used in our experiments, mercury exists essentially in the form of neutral species such as $HgCl_2$, Hg(OH)Cl and $Hg(OH)_2$, with hydroxide formation being more important as the pH increases. For this reason, in our study it was decided to perform the Hg(II) equilibrium isotherms at pH 5 (Plaza et al. 2011). In other reported studies, such as the study of Hg(II) biosorption by *Chlorella vulgaris*, a pH equal to 6 was used without making a distinction between both possible (biosorption and bioprecipitation) mercury sequestering processes (Kumar et al. 2020).

Chromium as trivalent cation can undergo successive hydrolysis processes, the most significant of which is represented by the following equation:

$$Cr3++H2O \rightleftarrows Cr(OH)2++H+$$

At pH values below 3, the predominant species is Cr^{3+}; at pH values close to 4, the concentrations of Cr^{3+} and $Cr(OH)^{2+}$ are approximately equal, while from pH 4 and up to 6 the predominant species is the cation $Cr(OH)^{2+}$. Finally, if the pH is above 6, chromium is precipitated as $Cr(OH)_3$ (Plaza-Cazón et al. 2012). In our studies of chrome biosorption, a pH equal to 4 was selected as optimal.

In most of the reported biosorption works, a better fit of their experimental kinetic data to the PSO model has been obtained (Al-Qahtani 2016; Castro et al. 2017; Long et al. 2018; Kumar et al. 2020); in this model, it is proposed that the rate-limiting step involves chemical adsorption, where the removal of the metal from the solution is almost exclusively due to physicochemical interactions between the metal and the biosorbent, through ion exchange or covalent bonding. However, the fact that the experimental data can be adjusted to PSO model is not sufficient evidence to ensure the limiting mechanism of the process (Robalds et al. 2016). The values of the specific constants found in our studies when applying the PSO model follow the following order: $K_2Ni(II) > K_2Zn(II) > K_2Cd(II) > K_2Cr(III) > K_2Hg(II)$ (Tables 29.1 and 29.2).

The correlation coefficients found when trying to adjust the equations that use external mass transfer or intra-particle diffusion as limiting steps are significantly lower than those determined for the PSO model, which would indicate that neither process is really the limiting stage (Tables 29.1 and 29.2). In a system with continuous agitation, it is reasonable that the effect of mass transfer on the speed of the biosorption process is not significant. On the other hand, due to the high surface area used, it could justify that the diffusion of the adsorbate into the interior of the adsorbent particle is not the limiting factor either. Similar results and conclusions were obtained by Mata et al. (2008) when they studied the adsorption of Cd(II), Pb(II) and Cu(II) by *Fucus vesiculosus*. In contrast, El-Sikaily et al. (2007) found that intraparticle diffusion was the mechanism that controlled the adsorption of Cr(III) on activated carbon and on the biomass of *Ulva lactuca*. Different models have been proposed to correct for the spherical particle condition included in the intraparticle model (Robalds et al. 2016) but they have not been used in this study.

29.3.2 Biosorption Isotherms

Figure 29.3 shows the adsorption isotherms of Zn(II), Cd(II), Ni(II), Cr(III), and Hg(II) on *M. pyrifera* biomass, under the optimal conditions defined for each metal. In practically all cases the typical plateau is observed, which indicates the saturation of the biosorbent binding sites. The observed saturation values for the different metals are in the range of 0.55 to 0.80 $mmol.g^{-1}$.

Table 29.1 Kinetic parameters for the biosorption of Zn(II), Cd(II), Ni(II) and Cr(III) on pretreated biomass of *M. pyrifera*

	Zn(II)			Cd(II)		
	pH 3	pH 4	pH 5	pH 3	pH 4	pH 5
Pseudo-first order model (PPO)						
q_{exp} (mmol.g^{-1})	0.30	0.34	0.38	0.18	0.21	0.20
q_{eq} (mmol.g^{-1})	0.21	0.21	0.22	0.12	0.14	0.15
K_1(min^{-1})	0.012	0.01	0.012	2.76x10^{-3}	4.41x10^{-3}	5.9x10^{-3}
R^2	0.92	0.92	0.95	0.94	0.94	0.93
Pseudo-second order model (PSO)						
q_{eq}(mmol.g^{-1})	0.32	0.36	0.40	0.18	0.21	0.20
K_2 [g(mmol.min)$^{-1}$]	0.13	0.17	0.11	0.20	0.15	0.23
R^2	0.99	0.99	0.99	0.99	0.99	0.99
External mass transfer						
βL (m.s^{-1})	2.5x10^{-5}	5.5x10^{-5}	5.8x10^{-5}	1.48x10^{-4}	2.4x10^{-4}	2.5x10^{-4}
R^2	0.90	0.92	0.89	0.92	0.97	0.98
Intraparticle diffusion						
$K_{id\,(}$ (mmol.g^{-1}.min$^{-1/2}$)	0.025	0.029	0.031	0.02	0.02	0.01
R^2	0.92	0.91	0.83	0.97	0.90	0.94
	Ni(II)			Cr(III)		
	pH 3	pH 4	pH 5	pH 3	pH 4	pH 5
Pseudo-first order model (PPO)						
q_{exp} (mmol.g^{-1})	0.15	0.19	0.20	0.55	0.78	0.67
q_{eq} (mmol.g^{-1})	0.17	0.20	0.19	0.43	0.59	0.55
K_1(min^{-1})	0.086	0.066	0.057	0.013	7.3x10^{-3}	0.016
R^2	0.96	0.93	0.98	0.95	0.95	0.96
Pseudo-second order model (PSO)						
q_{eq}(mmol.g^{-1})	0.15	0.20	0.19	0.55	0.78	0.66
K_2 [g(mmol.min)$^{-1}$]	0.71	0.47	0.53	0.076	0.019	0.026
R^2	0.99	0.99	0.99	0.99	0.99	0.99
External mass transfer						
βL (m.s^{-1})	2.4x10^{-5}	6.4x10^{-5}	6.1x10^{-5}	7.1x10^{-5}	2.5x10^{-6}	1.2x10^{-4}
R^2	0.94	0.91	0.95	0.96	0.98	0.98
Intraparticle diffusion						
K_{id}(mmol.g^{-1} .min$^{-1/2}$)	0.023	0.028	0.016	0.033	0.049	0.041
R^2	0.85	0.85	0.86	0.94	0.97	0.93

In this work, the most widely used models have been used to adjust the results of heavy metal biosorption: Langmuir, Freundlich and Dubinin-Radushkevich (D- R).

The isotherms, in general, fit satisfactorily to the Langmuir model (R^2 = 0.98–0.99). However, the Hg(II) isotherm fits Freundlich's model better than Langmuir's (R^2 = 0.98 versus R^2 = 0.91). This is probably due to the fact that the concentration range used was not wide enough to reach the saturation plateau. In

Table 29.2 Kinetic parameters for the biosorption of Hg(II) on pre-treated biomass of *M. pyrifera*

	Hg(II)			
	pH 3	pH 4	pH 5	pH 6
Pseudo-first order model (PPO)				
q_{exp} (mmol.g^{-1})	0.071	0.10	0.10	0.11
q_{eq} (mmol.g^{-1})	0.06	0.07	0.06	0.06
K_1(min^{-1})	9.2x10^{-4}	1.6x10^{-3}	1.6x10^{-3}	1.8x10^{-3}
R^2	0.95	0.96	0.90	0.93
Pseudo-second order model (PSO)				
q_{eq}(mmol.g^{-1})	0.07	0.10	0.10	0.23
K_2 [g(mmol.min)$^{-1}$]	0.06	0.11	0.16	0.16
R^2	0.99	0.99	0.99	0.97
External mass transfer				
βL (m.s^{-1})	2.0x10^{-6}	5.1x10^{-6}	1.6x10^{-5}	2.8x10^{-5}
R^2	0.91	0.91	0.96	0.97
Intraparticle diffusion				
K_{id} (mmol.g^{-1}.min$^{-1/2}$)	0.0016	0.0025	0.0028	0.0031
R^2	0.97	0.92	0.82	0.78

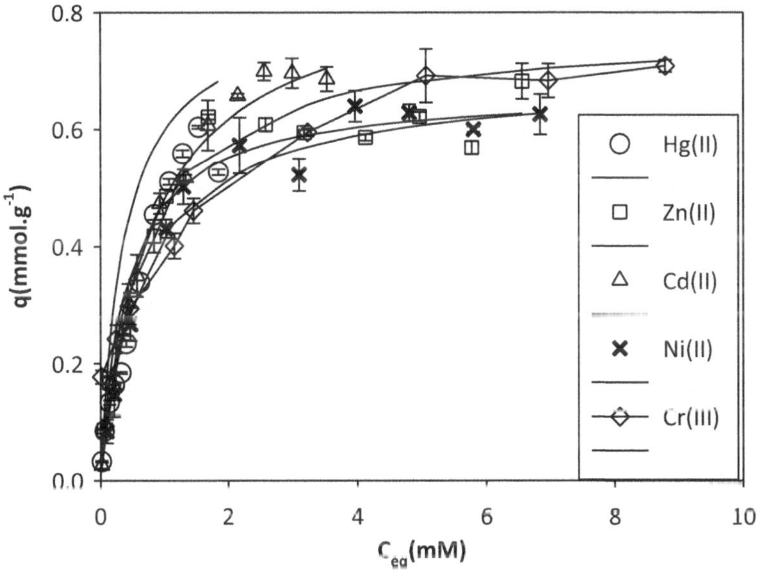

Fig. 29.3 Biosorption isotherms for removal of Zn(II), Cd(II), Ni(II), Cr(III) and Hg(II) on pre-treated biomass of *M. pyrifera*

this case, the adsorption isotherm maintains a tendency to increase consistent with the typical form of the Freundlich equation.

The parameters corresponding to the Langmuir, Freundlich and D-R models obtained can be seen in Table 29.3. According to the maximum capacity (q_m)

Table 29.3 Langmuir, Freundlich and D-R parameters for the biosorption of Zn(II), Cd(II), Ni(II), Cr(III) and Hg(II) on pre-treated biomass of *M. pyrifera*

	Zn(II)	Cd(II)	Ni(II)	Cr(III)	Hg(II)
Langmuir					
q_m (mmol. g^{-1})	0.67	0.87	0.69	0.77	0.82
b (L. $mmol^{-1}$)	2.4	1.25	1.48	1.20	2.7
R^2	0.98	0.99	0.99	0.99	0.91
Freundlich					
K_f (mmol.g^{-1})	0.37	0.41	0.32	0.38	0.44
n	2.5	1.56	1.37	3.22	1.53
R^2	0.91	0.94	0.90	0.94	0.98
D-R					
E (kJ.mol^{-1})	10.50	8.70	9.53	14.14	8.90
R^2	0.95	0.99	0.94	0.88	0.98

Table 29.4 Langmuir parameters for the adsorption of Zn(II) using different bioadsorbents

Bioadsorbents	q_m (mmol.g^{-1})	b (L.$mmol^{-1}$)	References
Acinetobacter sp	0.80	1.17	Tabaraki et al. (2013)
U. pinnatifida	1.53	0.75	Plaza-Cazón et al. (2013)
P. canaliculata	1.20	–	Giardi et al. (2014)
*T.paurometabola*A155	0.27	955.85	Limcharoensuk et al. (2015)
Banana peels	0.14	142.52	Al-Qahtani (2016)
Kiwi peels	0.56	126.18	Al-Qahtani (2016)
Tangarine peels	0.58	90.22	Al-Qahtani (2016)
Sugar beet pulp	0.08	7.69	Castro et al. (2017)
F. vesiculosus	0.94	6.6	Castro et al. (2017)

Note: --- not reported

obtained with the Langmuir model for each metal, the following order (in mmol. g^{-1}) was established Cd(II) ˃ Hg(II) ˃ Cr(III) > Ni(II) > Zn(II). The parameter *b* of the Langmuir equation, related to the affinity of the biosorbent for the adsorbate, obtained for all biosorptions is shown in Table 29.3. The following order of affinity of the metals on the biomass of *M. pyrifera* was established: Hg(II) > Zn(II) > Ni(I I) > Cd(II) > Cr)(III).

Although a comparison of the values of the maximum adsorption capacity of the different metals on different adsorbents would require considering both the operating conditions (agitation, dosage, pH, etc.) and the eventual pre-treatment of the biosorbent, some results found in the literature for the removal of Zn(II), Cd(II), Ni(II), Cr(III) and Hg(II) are shown in Tables 29.4, 29.5, 29.6, 29.7, and 29.8. For this purpose, not only the values of q_m and b were compared (Robalds et al. 2016).

The sorption capacity of Zn(II) exhibited by the pre-treated biomass of *M. pyrifera* was superior to other adsorbents such as sugar beet pulp, banana peels, kiwi and tangarine (Table 29.4) but lower than the biomasses of *F. vesiculosus*, *P. canaliculata*, *Acinetobacter* sp. and *U. pinnatifida*; the affinity obtained in this study was even

Table 29.5 Langmuir parameters for the adsorption of Cd(II) using different bioadsorbents

Bioadsorbents	q_m (mmol.g^{-1})	b (L.mmol^{-1})	References
Lessonia nigrescens	0.99	116.48	Boschi et al. (2011)
Lessonia trabeculata	1.47	9.07	Boschi et al. (2011)
U. pinnatifida	1.08	0.96	Plaza-Cazón et al. 2013
P. Areuginosa B237	0.15	1643.43	Limcharoensuk et al. (2015)
Banana peels	0.24	161.87	Al-Qahtani (2016)
Kiwi peels	0.14	1863.75	Al-Qahtani (2016)
Tangarine peels	0.15	1728.86	Al-Qahtani (2016)

Table 29.6 Langmuir parameters for the adsorption of Ni(II) using different bioadsorbents

Bioadsorbents	q_m (mmol.g^{-1})	b (L.mmol^{-1})	References
S. filipendula	1.07	4.06	Kleinübing et al. (2011)
U. pinnatifida	0.92	1.24	Plaza-Cazón et al. (2012)
Sargassum sp.	0.86	3.89	Barquilha et al. (2019)
S. Roseorubens	3.55	2.75	Long et al. (2018)
Phanerochaete Chrysosporium	0.79	---	Noormohamad et al. (2019)
Aloe barbadensis	0.49	1.11	Gupta et al. (2019)

Note: not reported

Table 29.7 Langmuir parameters for the adsorption of Cr(III) using different bioadsorbents

Bioadsorbents	q_m (mmol.g^{-1})	b (L.mmol^{-1})	References
Fucus vesiculosus	1.21	1.88	Murphy et al. (2008)
Fucus spiralis	1.17	1.77	Murphy et al. (2008)
Ulva lactuca	0.71	1.98	Murphy et al. (2008)
Ulva sp.	1.02	1.38	Murphy et al. (2008)
Palmaria palmata	0.57	4.94	Murphy et al. (2008)
Polysiphonia lanosa	0.65	1.34	Murphy et al. (2008)
U. pinnatifida	0.74	1.06	Plaza-Cazón (2012)
Alginate residue from Sargassum sp.	0.61	106.45	Bertagnolli et al. (2014)
Spirulina sp.	1.74	2.33	Rezaei (2016)
Cladophora glomerata	2.06	2.59	Godlewska et al. (2018)

higher than that obtained with the biomasses in the last two cases (Table 29.4). On the other hand, the adsorption capacity of Cd(II), although lower than that reported for the biomasses of *Lessonia nigrescens*, *Lessonia trabeculata*, *P. canaliculata* and *U. pinnatifida*, was higher than for all the other adsorbents cited in Table 29.5.

In contrast, both the maximum adsorption capacity of Ni(II) and its affinity coefficient (b) were lower than those obtained for most of the bioadsorbents cited in Table 29.6, with the only exceptions of *Aloe barbadensis* and *P. chrysosporium*. In these last cases, pH values equal to or greater than 6 were used in the experiments causing that in addition to metal adsorption there has been metal precipitation.

Table 29.8 Langmuir parameters for the adsorption of Hg(II) using different bioadsorbents

Bioadsorbents	q_m (mmol. g^{-1})	b ($L.mmol^{-1}$)	References
Sargassum muticum (without treatment)	0.90	4.8	Carro et al. (2011)
Sargassum muticum (with treatment)	1.3	4.7	Carro et al. (2011)
U. pinnatifida	0.84	4.4	Plaza-Cazón et al. (2011)
Sargassum fusiforme	0.15	198.17	Huang and Lin (2015)
Pleurotus eryngii	0.17	50.14	Amin et al. (2016)
Raw almond shell	0.018	26.07	Taha et al. (2018)
Activated almond shell	0.18	22.06	Taha et al. (2018)
Untreated mango leaves powder (UMLP)	0.11	5.01	Adelaja et al. (2019)
Treated mango leaves powder (with potassium persulphate) (TMLP)	0.14	15.04	Adelaja et al. (2019)
Chlorella vulgaris	0.20	3.71	Kumar et al. (2020)

Table 29.7. contains values of the maximum adsorption capacity (q_m) of Cr(III) previously reported. The q_m values vary within the range of 0.078 to 2.06 $mmol.g^{-1}$. It can be seen that the pre-treated biomass of *M. pyrifera* is better adsorbent than the biomasses of *Ulva lactuca*, *Palmaria palmata* and *U. pinnatifida* and the alginate residues obtained from *Sargassum sp.* but has lower adsorption capacity than the rest of the bioadsorbents cited in Table 29.7 and, especially, the adsorption on *Cladophora glomerata* biomass (Godlewska et al. 2018).

The affinity coefficient for Cr(III) sorption on pre-treated biomass of *M. pyifera* (1.20 L $mmol^{-1}$) is within the range of *b* values reported for other biomaterials; however, it was much lower than that obtained for *Palmaria palmata*, *Cladophora glomerata*, alginate residue from *Sargassum* sp., *Spirulina* sp., and *Laminaria digitata* (Table 29.7).

The maximum adsorption capacity (q_m) of Hg(II) on the pre-treated biomass of *M. pyrifera* was considerably higher than that found on other bioadsorbents cited in Table 29.8, with the only exceptions of the biomasses of *Sargassum muticum* and *U. pinnatifida*. The affinity coefficient determined on the biomass of *M. pyrifera* (2.7 $L.mmol^{-1}$) was lower than the affinity coefficients for all the other biomaterials cited in Table 29.8.

The D-R model is useful to estimate the value of energy involved in the adsorption process that allows to differentiating whether the sorption process is physical or chemical. In the first case, energy values lower than 8 $kJ.mol^{-1}$ are expected, indicating that the interaction between sorbate and adsorbent is weak; in contrast, energy values in the range of 8–16 $kJ.mol^{-1}$ would indicate that the process is a chemosorption (Beni and Esmaeili 2020). The values for the adsorption of Zn(II), Cd(II), Ni(II), Cr(III) and Hg(II) obtained in this study, determine that these were chemical type adsorption (Table 29.3).

Sorption processes may involve ion exchange between the ions retained with ions previously retained in those active sites (Plaza-Cazón et al. 2014). When biomass is pre-treated with $CaCl_2$, most of the active sites will be occupied by Ca(II)

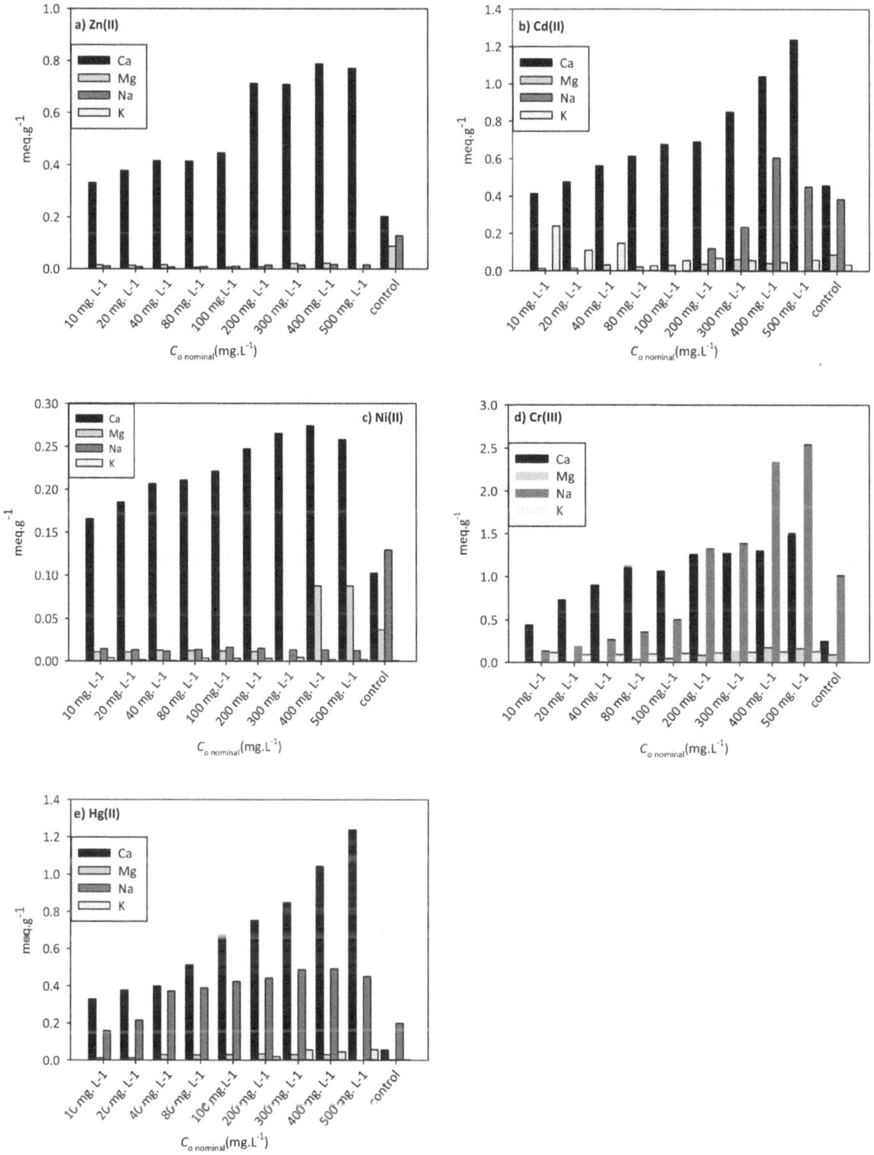

Fig. 29.4 Exchange cations released during the biosorption of Zn(II) (a), Cd(II) (b), Ni(II) (c), Cr(III) (d) and Hg(II) (e) on pre-treated biomass of *M. pyrifera*

ions. However, in the experiences carried out it was observed that in addition to Ca(II), light cations naturally present in the cell wall of *M. pyrifera* (mainly Na(I)) were also exchanged during sorption (Fig. 29.4).

During the studied biosorption there was, in general, an increase of the pH during the adsorption, being higher when the initial concentration of the metal is smaller. This effect is probably due to the fact that the adsorption of the $Cr(OH)^{2+}$ species is favored over Cr^{3+} on the biomass provoking the increase of the hydrolysis of the latter ion and releasing more protons into the medium.

29.4 Conclusions

The results of this work show that the biomass of *M. pyrifera* pretreated with $CaCl_2$ is a good sorbent of the heavy metals tested and, in particular, of Hg(II), presenting a higher adsorption capacity than most of the reported sorbents. The biosorption process adjusted to a pseudo-first order kinetics and was quite fast for Zn(II), Ni(II), and Cd(II) but much slower for Cr(III) and Hg(II). The biosorption isotherms fit quite well with the Langmuir equation and showed similar maximum adsorption capacity values in terms of $mmol.g^{-1}$ (between 0.67 and 0.87) indicating a saturation of the available active sites.

Finally, for real industrial application, it is necessary equilibrium and dynamic biosorption-desorption studies based on the parameters determined in previous kinetics and equilibrium batch studies.

Acknowledgements This work was partially supported by ANPCyT, PICT-2016-2535 and PICT-2015-0463 grants.

References

Adelaja OA, Bankole AC, Oladipo ME, Lene DB (2019) Biosorption of hg(II) ions, Congo red and their binary mixture using raw and chemically activated mango leaves. Inter J Energy Water Resources 3:1–12

Al-Qahtani KM (2016) Water purification using different waste fruit cortexes for the removal of heavy metals. J Taibah University Sci 10:700–708

Amin F, Talpur FN, Balouch A, Chandio ZA, Surhio MA, Afridi HI (2016) Biosorption of mercury(II) from aqueous solution by fungal biomass *Pleurotus eryngii*: isotherm, kinetic, and thermodynamic studies. Environ Prog Sustain Energy 35(5):1274–1282

Anastopoulos I, Kyzas GZ (2015) Progress in batch biosorption of heavy metals onto algae. J Mol Liq 209:77–86

Barquilha CER, Cossich ES, Tavares CRG, Silva EA (2019) Biosorption of nickel(II) and copper(II) ions by *Sargassum* sp. in nature and alginate extraction products. Bioresource Technol Reps 5:43–50

Beni AA, Esmaeili A (2020) Biosorption, an efficient method for removing heavy metals from industrial effluents: A review. Environ Technol Innovat 17:100503

Bertagnolli C, da Silva MGC, Guibal E (2014) Chromium biosorption using the residue of alginate extraction from *Sargassum filipendula*. Chem Eng J 237:362–371

Boschi C, Maldonado H, Ly M, Guibal E (2011) Cd(II) biosorption using *Lessonia* kelps. J Colloid Interface Sci 357:487–496

Carro L, Barriada JL, Herrero R, Sastre de Vicente ME (2011) Adsorptive behaviour of mercury on algal biomass: competition with divalent cations on organic compounds. J Hazard Mater 192:184–291

Castro L, Blázquez ML, González F, Muñoz JA, Ballester A (2017) Biosorption of Zn(II) from industrial effluents using sugar beet pulp and *Fucus vesiculosus*: from laboratory tests to a pilot approach. Sci Total Environ 598:856–866

El-Sikaily A, El Nemr A, Khaled A, Abdelwehab O (2007) Removal of toxic chromium from wastewater using green alga *Ulva lactuca* and its activatd carbon. J Hazard Mater 148(1–2):216–228

Girardi F, Hackbarth FV, De Souza GUSMA, De Souza AA, Boaventura RAR, Vilar VJP (2014) Marine macroalgae *Pelvetia canaliculata* (Linnaeus) as natural cation exchanger for metal ions separation: A case study on copper and zinc ions removal. Chem Eng J 247:320–329

Godlewska K, Marycz K, Michalak I (2018) Freshwater green macroalgae as a biosorbent of Cr(III) ions. Open Chemistry Journal 16:689–701

Gupta W, Sharma SK, Kumar A (2019) Biosorption of Ni(II) ions from aqueous solution using modified *Aloe barbadensis* (miller) leaf powder. Water Science and Engineering 12(1):27–36

Huang S, Lin G (2015) Biosorption of hg(II) and cu(II) by biomass of dried *Sargassum fusiforme* in aquatic solution. J Environ Health Sci Eng 13:21

Kleinübing SJ, Silva EA, da Silva MGC, Guibal E (2011) Equilibrium of cu(II) and Ni(II) biosorption by marine alga *Sargassum filipendula* in a dynamic system: competitiveness and selectivity. Bioresour Technol 102:4610–4617

Kumar M, Kumar SA, Sikandar M (2020) Biosorption of hg (II) from aqueous solution using algal biomass: kinetics and isotherm studies. Heliyon 6:e03321

Kyzas GZ, Bikiaris DN (2015) Recent modifications of chitosan for adsorption applications: a critical and systematic review. Mar Drugs 13:312–337

Limcharoensuk T, Sooksawat N, Sumarnrote A, Awutpet T, Kruatrachue M, Pokethitiyook P, Auesukaree C (2015) Bioaccumulation and biosorption of Cd^{2+} and Zn^{2+} by bacteria isolated from a zinc mine in Thailand. Ecotoxicol Environ Saf 122:322–330

Long J, Gao X, Su M, Li H, Chen D, Zhou S (2018) Performance and mechanism of biosorption of nickel(II) from aqueous solution by non-living *Streptomyces roseorubens* SY. Colloids and Surfaces A 548:125–133

Mata YN, Blázquez ML, Ballester A, González F, Muñoz JA (2008) Characterization of the biosorption of cadmium, lead and copper with the brown alga *Fucus vesiculosus*. J Hazardous Mat 158:316–323

Murphy V, Hugues H, McLoughlin P (2008) Comparative study of chromium biosorption by red, green and brown seaweed biomass. Chemosphere 70(6):1128–1134

Noormohamadi HR, Fathi MR, Ghaedi M, Ghezelbash GR (2019) Potentiality of white-rot fungi in biosorption of nickel and cadmium: Modeling optimization and kinetics study. Chemosphere 216:124–130

Plaza J, Viera M, Donati E, Guibal E (2011) Biosorption of mercury by *Macrocystis pyrifera* and *Undaria pinnatifida*: influence of zinc, cadmium and nickel. J Environ Sci 23(11):1778–1786

Plaza-Cazón J, Viera M, Donati E (2013) Zinc and cadmium removal by biosorption on *Undaria pinnatifida* in batch and continuous processes. J Environ Manag 129:423–434

Plaza-Cazón J, Benitez L, Viera M, Donati E (2012) Equilibrium and kinetic study of the biosorption of chromium (III) by two brown algae: *Macrocystis pyrifera* and *Undaria pinnatifida*. Eng Life Sci 12(1):95–103

Plaza-Cazón J (2012) Tesis doctoral: Remoción de metales pesados empleando algas marinas. Facultad de Ciencias Exactas, Universidad Nacional de La Plata, Argentina

Plaza-Cazón J, Viera M, Sala S, Donati E (2014) Biochemical characterization of *Macrocystis pyrifera* and *Undaria pinnatifida* (Phaeophyceae) in relation to their potentiality as biosorbents. Phycologia Volume 53(1):100–108

Reategui M, Maldonado H, Ly M, Guibal E (2009) Biosorption of mercury(II) with deiffernt marine algae. Advanc Mat Res (71-73):585–588

Redha AA (2020) Removal of heavy metals from aqueous media by biosorption. Arab J Basic Applied Sci 27(1):183–193

Rezaei H (2016) Biosorption of chromium by using *spirulina* sp. Arab J Chem 9:846–853

Robalds A, Naja GM, Klavins M (2016) Highlighting inconsistencies regarding metal biosorption. J Hazard Mater 304:553–556

Shanmugaprakash M, Sivakumar V (2015) Batch and fixed-bed column studies for biosorption of Zn(II) ions onto pongamia oil cake (*Pongamia pinnata*) from biodiesel oil extraction. J Environ Manag 164:161–170

Tabaraki R, Ahmady-Asbchin S, Abdi (2013) Biosorption of Zn(II) from aqueous solutions by *Acinetobacter* sp. isolated from petroleum spilled soil. J Environ Chem Eng 1:604–608

Taha AA, Moustafa AHE, Abdel-Rahman HH, El-Hameed MMAA (2018) Comparative biosorption study of hg (II) using raw and chemically activated almond shell. Adsorpt Sci Technol 36(1–2):521–548

Vijayaraghavan K, Balasubramanian R (2015) Is biosorption suitable for decontamination of metal-bearing wastewaters? A critical review on the state-of-the-art of biosorption processes and future directions. J Environ Manag 160:283–296

Volesky B (2003) Sorption and biosorption. BV Sorbex. Inc. Montreal-St. Lambert, Quebec, Canada, p 336

Chapter 30
Potential of Silver Nanocomposites from Seaweeds for Plant Protection: An Overview

T. Antony Roseline and K. Arunkumar

Abbreviation

Ag	Silver
AgNPs	Silver nanoparticles
Au	Gold
CMV	*Cucumber mosaic virus*)
FIFRA	Federal Insecticide, Fungicide, and Rodenticide Act)
FTIR	Fourier Transform Infrared spectroscopy
MoNPs	Molybdenum nanoparticles
MSNs	Mesoporous silica nanoparticles)
Nm	Nanometer
NP Synthesis	Nanoparticle synthesis
PVOH	polyvinyl alcohol
PVP	polyvinyl pyrrolidone
TiO_2	Titanium dioxide
ZnO_2	Zinc oxide
µg/L	Microgram per Litre

T. A. Roseline
Department of Botany, Alagappa Government Arts College Karaikudi, Tamil Nadu, India

K. Arunkumar (✉)
Phycoscience Laboratory, Department of Plant Science Central University of Kerala, Kasaragod, Kerala, India
e-mail: arunkumark@cukerala.ac.in

A. Ranga Rao, G. A. Ravishankar (eds.), *Sustainable Global Resources of Seaweeds Volume 1*, https://doi.org/10.1007/978-3-030-91955-9_30

30.1 Introduction

The common application of nanotechnology includes nanoformulation of agrochemical for applying pesticides and fertilizers for crop improvement, the application of nanosensors /Nano biosensors for crop protection for the identification of the disease and residues of agrochemicals and Nano devices for the genetic manipulation of plants, plant disease diagnostic and post-harvest management (Sekhon 2014). In the recent past, engineered nanoparticles have received attention as potential candidates for improving crop yield, résistance, and plant disease management technologies (Husen and Siddiqi 2014; Ocsoy et al. 2013). Metal -based/carbon-based nanomaterials were widely used for crop protection (Jeyaseelan et al. 2013) but these methods are expensive and require the use of toxic and aggressive chemicals as reducing and/or capping agents (Arunkumar et al. 2010). But the biological method of nanoparticlesynthesis is receiving attention owing to its unique advantages (Chandran et al. 2006; Gajbhiye et al. 2009).

Silver ions are very reactive, they inhibit microbial respiration and metabolism and they cause physical damage (Bragg and Rannie 1974). Silver has been used to treat medical ailments for over 100 years due to its natural antibacterial and antifungal properties (Morones et al. 2005). Recently, nanotechnology practices have amplified the effectiveness of silver particles as antimicrobial agents (Elechiguerra et al. 2005). Silver nanoparticles have relatively large surface areas which increase their contact with bacteria and fungi, vastly improving its bactericidal and fungicidal effectiveness. When in contact with bacteria and fungus, they will adversely affect cellular metabolism and inhibit cell growth.

30.2 Silver as Antimicrobial Agent

Silver (Ag) is being used as an antimicrobial agent since centuries, in Ayurvedic therapeutics as a broad spectrum and multiple modesof antimicrobial activity (Guggenbichler et al. 1999). Silver has been widely used throughout history for its antibacterial, antiviral,and antifungal properties (Castellano et al. 2007; Fabrega et al. 2011; Aziz et al. 2014). Since then the antimicrobial properties of silver have been investigated and exploited more extensively than any other molecule showing antibacterial potentiality in treatment of diverse diseases caused by bacterial pathogens (Aziz et al. 2015, 2016; Joshi et al. 2018). The Ag ions exhibit higher toxicity to microorganisms and lower toxicity to mammalian cells (Clement and Jarrett 1994). Although different forms of Ag,available commercially in the market, such as silver acetate, silver nitrate and silver sulfadiazine are being used for antimicrobial activity (Jung et al. 2008), silver components have been proven as an effective tool for retarding and preventing the microbial infections. In addition, silver is known to exhibit oligo dynamic effect because of its ability to exert inhibitory effect at minute concentrations (Tien et al. 2009).

30.3 Biological Sources for the Synthesis of Nanoparticles

A wide variety of physical and chemical processes have been developed for the synthesis of metal nanoparticles (Kumar and Yadav 2009), but these methods are expensive and require the use of toxic and aggressive chemicals as reducing and/or capping agents (Li et al. 2009). But biological method of nanoparticles synthesis is a rapidlygrowing technique in the field of nanotechnology. The biological sources offer trouble-free protocols and when applied for the human health associated fields (Rastogi and Arunachalam 2011). Therefore, biological approaches to nanoparticle synthesis have been suggested as valuable alternatives to physical and chemical methods (Mohanpuria et al. 2008; Gnanadesigan et al. 2012). In biological medium,NPs may interact with biomolecules such as proteins, nucleic acid, lipids and even biological metabolites due to their nanosized and large surface to mass ratio.

For the exploitation of the green nanotechnology, a number of plant species and microorganisms including bacteria, algae and fungi are being currently used for NP synthesis. For example, *Medicago sativa* and *Sesbania* plant species are used to formulate gold nanoparticles. Likewise, inorganic nanomaterials, made of silver, nickel, cobalt, zinc and copper, can be synthesized inside live plants, as demonstrated in *Brassica juncea*, *Medicago sativa* and *Helianthus annus* (Panpatte et al. 2016; Ghormade et al. 2011; Kitching et al. 2015; Iravani 2011). Microorganisms, such as diatoms, *Pseudomonas stuzeri*, *Desulfovibrio desulfuricans* NCIMB 8307 *Clostridium thermoaceticum* and *Klebsiella aerogens* are used to synthesize silicon, gold, zinc sulphide and cadmium sulphide nanoparticles, respectively. Although a large number of microorganisms are used to synthesize green NPs, fungi, mainly *Verticillium* sp., *Aspergillus flavus*, *Aspergillus furnigatus*, *Phanerochaete chrysoparium* and *Fusarium oxysporum* are considered to be the most efficient systems for the biosynthesis of metal and metal sulphide containing NPs (Panpatte et al. 2016; Kitching et al. 2015)

30.4 Silver Based Nanoparticles Effective for Antimicrobial Activity

Silver nanoparticles (AgNPs) are one of the most commonly used nanomaterials in various fields, especially in the agricultural sector. Plants are the basic component of the ecosystem and the most important source of food for mankind; therefore, understanding the impacts of AgNPs on plant growth and development is crucial for the evaluation of potential environmental risks on food safety and human health imposed by AgNPs (Yan and Zhong 2019).

AgNPs together with their nontoxicity to living tissues at low concentrations make them more attractive to be utilized as antimicrobial agents in different biological fields particularly plant disease control however, only a few studies were done on microbial plant pathogens. Interestingly, their results showed that silver nanoparticles with the size (20–30 nm) could efficiently penetrate and colonize within the plant tissue than larger ones. The results also indicated that Ag nanoparticles had a great potential for use in controlling spore-producing fungal plant pathogens. They suggested that these nanoparticles might be less toxic than synthetic fungicides. In the other study, Mishra and his co-workers also indicated astrong antifungal activity exhibited against the phytopathogenic fungus *Bipolaris sorokiniana*, causing spot blotch disease of wheat crop, by using small-sized silver nanoparticles (Mishra et al. 2014). The antimicrobial effect of different controlled sized silver nanoparticles to controldifferent soil borne fungi as well as *Bipolaris sorokiniana* and *Magnaporthe grisea* was investigated.

On the other side, other reports assayed the antibacterial activity of a well-formed Silver nanoparticles with 19 nm size against different eight microorganisms using the disk-diffusion method. On the other hand, Roseline and his co-workers revealed the antimicrobial activity of a well forms silver nanoparticles with 19 nm against bacteria and fungi using disk diffusion method as well as broth assay (Roseline et al. 2019).

Recently, the International Center for Technology Assessment (ICTA) has submitted a petition to the Environmental Protection Agency (USA) requesting that it regulates nanosilver used in products as a pesticide under the Federal Insecticide, Fungicide, and Rodenticide Act (FIFRA). Silver is now globallyaccepted as agrochemical replacement. Silver that exhibits excellent qualities should be tasteless, strong nontoxic disinfectant, and good growth stimulator. In 2013, Ocsoy and his co-workers indicated that DNA-directed silver nanoparticles grown on graphene oxide (GO) composites effectively decrease cell viability in culture and on plants of *Xanthomonas perforans* causing bacterial spot of tomatoes (*Solanum lycopersicum*) in Florida while the pathogen has developed resistance to Cu fungicides (Ocsoy et al. 2013). These compounds (Ag@dsDNA@GO) show excellent antibacterial activity in culture at a very low concentration of 16 ppm with higher adsorption rate. Severity of tomato bacterial spot is significantly reduced by application of Ag@dsDNA@GO at 100 ppm in greenhouse when compared to untreated and showed no phytotoxicity.

30.5 Eco-Friendly Approach of Biologically Synthesis of Silver Nanoparticles Used for Plant Disease Management

A variety of physical and chemical methods including high-energy ball milling, arc discharge, laser pyrolysis or ablation, electrochemical and chemical vapour deposition, micro-emulsion sol-gel, and reverse precipitation have been adopted for synthesis of metallic nanoparticles (Sastry et al. 2003). In the biosynthetic process, the biomolecules involved are non-toxic, of low cost, and most importantly environment friendly and the obtained nanoparticles are more biocompatible (Mandal et al.

Table. 30.1 Activities of biological Silver Nanomaterials against Plant Pathogens

Biological source for the synthesis of AgNPs	Application	Reference
Pine cone	Antibacterial activity against Bacillus megaterium, *Pseudomonas syringae,Burkholderiaglumae, Xanthomonas oryzae*, and *Bacillus thuringiensis*	Velmurugan et al. (2013)
Serratia sp	BHUS4 Antifungal activity against *Bipolaris sorokiniana*	Mishra et al. (2014)
Trichoderma viride	Vegetable and fruit preservation	Fayaz et al. (2009)
Piper nigrum	Antifungal activity against phytopathogens	Paulkumar et al. (2014)
Spirulina platensis	Bactericidal activity against phytopathogens	Mala et al. (2009)
Bacillus sp	GP-23 Antifungal activity against *Fusarium oxysporum*	Gopinath and Velusamy (2013)
Acalypha indica	Antifungal activity against plant pathogen	Krishnaraj et al. (2010)
Milk	Antifungal activity against phytopathogens	Lee et al. (2013)

2006; Gericke and Pinches 2006). In this regard, owing to the promising application of AgNPs for plant disease management, researchers have started using biological agents for AgNPs synthesis (Table 30.1).

30.6 Green Synthesis of Silver Nanoparticles Using Seaweeds

The biologically diverse marine environment has a great promise for nano science and nanotechnology. Biomolecules such as polysaccharides present in algal species also play an important role in controlling the size and desired shape of Ag NPs. Among the biological materials, algae are called as "bionanofactories"; because both the live and dead dried biomass was used for synthesis of metallic nanoparticles. It is low cost and environmentally effective, macroscopic structured material, and has the distinct advantage due to its high metal uptake capacity (Davis et al. 2003). Recently many biological sources like green seaweed *Ulva flexuosa Ulva* lactuca brown seaweed *Sargassum weightii* (Thirumalairaj et al. 2014) *S. longifolium* (Rajeshkumar et al. 2014), *Turbinaria conoides* (Rajeshkumar et al. 2012a), red seaweed, *Gracilaria dura, Gracilaria corticata, Gracilaria edulis, Spyridea hypnoides and Hypnea musciformis* (Roseline et al. 2019) crude extracts are used for metallic conversion (Table 30.2).

Table 30.2 Summary of the works related to AgNPs green synthesized using Marine Macroalgae(Seaweeds)

Group/Seaweed		Size (nm)/ Shape	Pathogens tested	Assay	Reference
Red	*Gracilaria birdiae*	20.2–94.9 /Spherical	*Escherichia coli, Staphylococcus aureus*	Broth	Aragao et al. (2016)
	Gracilaria crassa	60–200 Spherical	*Escherichia coli, Proteus mirabilis, Bacillus subtilis, Pseudomonas aeruginosa*	Disc	Lavakumar et al. (2015)
	Gracilaria corticata	51.08 nm Spherical	*Candida albicans* and *C. glabrata*	Disc	Kumar et al. (2012)
	Gelidiella acerosa	22 Spherical	*Mucor indicus* and *Trichoderma reesei*	Well	Vivek et al. (2011)
	Amphiroafragilissima	–	*Escherichia coli, Bacillus subtilis, Klebsiella pneumonia, Staphylococcus aureus, Pseudomonas aeruginosa*	Disc	Sajidha and Lakshmi (2016)
	Kappaphycus alverazii	74 nm	–	–	Ganesan et al. (2013)
	Kappaphycus sps	52–102	–	–	Baskar (2013)
	Laurencia aldingensis and Laurenciella sp	100 nm hexagonal and triangular	–	–	Adriana et al. (2016)
	Halymenia poryphyroides	34–80 spherical	–	–	Kiran and Murugesan (2013)
	Halymenia poryphyroides	34–80 nm	*Staphylococcus aureus, Salmonella typhi, Escherichia coli, Klebsiella pneumoniae and Proteus vulgaris*	Disc	Kiran and Murugesan (2014)
	**Gracilaria corticata*	37/Spherical	*Xanthomonas oryzae* pv. *oryzae* *Xanthomonas axonopodis* pv. *citri* *Ustilaginoidea virens*	Disc and Broth	Roseline et al. (2019)
	**Gracilaria edulis*	54/Spherical			
	**Hypnea musciformis*	53/Spherical			
	**Spyridia hypnoides*	49/Spherical			

Green	*Ulva lactuca*	20/Spherical	*Bacillus* sp., *E.coli,Pseudomonas* sp.	Well	Sangeetha and Saravanan. (2014).
	Ulva lactuca	20–56 nm	–	–	Devi and Bhimba (2012)
	**Ulva fasciata*	28–41 Spherical	*X campestris* pv. *malvacearum*	Well	Rajesh et al. (2012)
	Ulva flexousa	2–32/Circular	–	–	Rahimi et al. (2014)
	Ulva reticulata	–	*Antibacterial activity Staphylococcus aureus, Escherichia coli, Pseudomonas aeruginosa, Bacillus sp., Klebsiella pneumonia Candida albicans, Candida parapsilosis and Aspergillus niger.*	well	Devi and Bhimba (2014)
	Urospora sp.	20 to 30	*Staphylococcus aureus, Escherichia coli and Pseudomonas aeruginosa*	well	Suriya et al. (2012)
	Pithophora oedogonia	34.03 nm	*Escherichia coli, Pseudomonas aeruginosa, Vibrio cholerae, Shigellaflexneri, Bacillus subtilis,Staphylococcus aureus, Micrococcus luteus.*	Well	Sankar et al. (2014)
	Chaetomophora linum	3–44 Cubic	–	–	Kannan et al. (2012)
	Codiumcapitatum	30 nm Cubic 2013	–	–	Kannan et al. (2013b)
	Caulerpa racemosa	5–25 spherical/ triangular	*Staphylococcus aureus Proteus mirabilis*	Well	Kathiraven et al. (2015)

(continued)

Table 30.2 (continued)

Group/Seaweed		Size (nm)/ Shape	Pathogens tested	Assay	Reference
Brown	*Padina gymnospora*	25–40 spherical	*Bacillus cereus and Escherichia coli*	Well	Shiny et al. (2013)
	Padina Tetrastromatica	14 Spherical	*Bacillus sps, Klebsiellaplanticola, Bacillus subtilis and Pseudomonas sps.*	Well	Rajeshkumar et al. (2012b)
	Cystophora moniliformis	75 nm face-centred cubic	–	–	Tollamadugu et al. (2013)
	Sargassum wightii	8–27 Spherical	*S. aureus, B. rhizoids, E. coli* and *P. aeruginosa*	Well	Govindaraju et al. (2009)
	Sargassum wightii	15–20 Sperical	*P. aeruginosa, V. cholerae, K. pneumoniae, S. aureus, E. coli, S. pneumoniae* and *S. typhi*	Disc	Shanmugam et al. (2014)
	Sargassum wightii	8–14	*Bacillus cereus, Bacillus anhtracis, Staphylococcus aureus and Vibrio alginoyticus*	Disc	Vinothkumar et al. (2014)
	Sargassum wightii	35 Spherical	*Staphylococcus aureus, Escherichia coli, Klebsiellapneumoniae and Pseudomonas aeroginosa*	Disc	Sunitha et al. (2015)
	Sargassum cinereum	45–75 triangular	*Enterobactor aerogens, Staphylococcus aureus, Salmonella typhi* and *Proteus vulgaris*	Disc	Mohandass et al. (2013)
	Sargassum polycystum	– –	*Escherichia coli, Streptococcus pyogenes, Pseudomonas aeruginosa, Staphylococcus flexneri* and *Morengillamorrgani*	Disc	Asha et al. (2015)
	Sargassum polyphyllum	37–43 nm	–	–	Arun Kumar et al. (2014)
	Sargassummuticum	5–15 Spherical	–	–	Susan et al. (2013)
	Sargassum plagiophyllum	20 Spherical	–	–	Dhanalakshmi et al. (2012)
	Sargassum longifolium	40–85 Spherical	*Aspergillus fumigatus, Candida albicans, Fusarium* sp.	Well	Rajeshkumar et al. (2014)
	Sargassum tenerrimum	20 Spherical	*Clinical bacteria*	Well	Kumar et al. (2012)
	Turbinaria conoides	96/ Spherical	*Bacillus subtilis,Klebsiella planticola*	Disc	Rajeshkumar et al., (2012a)
	Padina tetrastromatica and*Turbinaria ornata*).	18–90	–	–	Kayalvilzhi et al. (2014)

30.7 Seaweed Polysaccharides with Potential Use in Nanoparticle Synthesis

Seaweeds are rich sources of sulphated polysaccharides, including some that have become valuable additives in the food industry because of their rheological properties as gelling and thickening agents (e.g., alginates, agar, and carrageenan).

Polysaccharides have been used as stabilizing and reducing agents for the synthesis of silver nanoparticlesSulphated polysaccharides are recognized to possess a number of biological activities including anticoagulant, antiviral, antitumor, anti-inflammatory, and immune stimulating activities that might find relevance in nutraceutical/functional food, cosmetic, and pharmaceutical applications. Some seaweeds produce hydrocolloids, associated with the cell wall and intercellular spaces. Members of the red algae (Rhodophyta) produce galactans (e.g., carrageenan and agars) and the brown algae (Heterokontophyta, Phaeophyceae) produce uronates (alginates) and other sulphated polysaccharides (e.g., fucoidans and laminaran).

Polysaccharides present in several seaweeds (*Kappaphycus alvarezii*, *Calliblepharis jubata*, and *Chondrus crispus*—Gigartinales, Rhodophyta; *Gelidium corneum* and *Pterocladiella capillacea*—Gelidiales, Rhodophyta; *Laurencia obtusa* Ceramiales, Rhodophyta; *Himanthalia elongata*, *Undaria pinnatifida*, *Saccorhiza polyschides*, *Sargassum vulgare*, and *Padina pavonica*—Phaeophyceae, Ochrophyta) are analyzed by spectroscopic techniques. The nature of the polysaccharides (with extraction and without any type of extraction) present in these seaweeds was determined with FTIR-ATR and FT-Raman analysis of extracted phycocolloids and ground dry seaweed. Polysaccharides from marine algae such as *Sargassum* can be used as a bio-factory for synthesis of nanoparticles because these are highly stable, safe, not toxic and with known biological activities (Liu et al. 2008; Vasquez and Ramos 2012).

30.8 Silver Nanoparticles in Agriculture for Plant Disease Management

In general, the agricultural production is largely affected by various kinds of plant diseases leading to huge economic losses. Though the idea of using biopesticides helped in modernization of the agricultural sector owing to their eco-friendly and sustainable approach, it is worth mentioning here that the major drawbacks associated with their usage recently has paved way for nanotechnology applications in agriculture (Ghormade et al. 2011) especially for plant disease management (Park et al. 2006; Kim et al. 2012; Mishra et al. 2014). Many workers have advocated the antimicrobial activity of AgNPs against a diverse and broad range of plant pathogens (Table 30.1). Initially, Park et al. (2006) effectively developed a nanosized silica-silver particle formulation demonstrating antimicrobial activity against various phytopathogens, viz. *Pythium* sp., *Colletotrichum* sp., *Pseudomonas syringae*,

Xanthomonas compestris, etc. Interestingly, Kim et al. (2009) investigated the inhibitory effect of AgNPs suspension on the fungal growth and conidial germination of ascomycetous phytopathogen *Raffaelea* sp.causing oak wilt disease. Furthermore, comprehensive studies made by many researchers provided evidence of a wider applicability of AgNPs for controlling a variety of plant pathogens such as *Bipolaris sorokiniana, Magnaporthe grise,,Colletotrichum* sp. (Lamsal et al. 2011), sclerotium-formingfungi (Min et al. 2009), powdery mildew on cucumber and pumpkin (Lamsal et al. 2011).

30.9 Nanoparticle Uptake, Translocation, and Biological Impact in Plants

The small-sized AgNPs can pass through the pores, whereas larger AgNPsare unable to enter into plant cells and are thereby sieved out (Tripathi et al. 2017). Ma et al. (2010) reported that particle size of 20 nm silver nanoparticles may be transported inside the cells through plasmadesmata. Xylem is one of the main passages of uptake and transportations to shoot and leaves of plant. Pores size of cell wall was in the range of 3–8 nm, smaller than engineered nanoparticles. The cell wall is a porous network of polysaccharide fibre matrices and, thus, acts as natural sieve (Navarro et al. 2008). Applications of nanotechnology strategies in plants need a preventive accurate evaluation of nanoparticle-plant interactions, including the comprehension of the mechanisms of their uptake, translocation and accumulation, together with the assessment of potential adverse effects on plant growth and development. Plant uptake of NPs is hardly predictable, depending on multiple factors related to the nanoparticle itself (size, chemical composition, net charge and surface functionalization), but also on the application routes, the interactions with environmental components (soil texture, water availability, microbiota), the constraint due to the presence of a cell wall, the physiology and the multifaceted anatomy of individual plant species.

30.9.1 Merits and Demerits of Nanoparticles in Plant Disease Management

Many gaps need to be filled in our knowledge regarding the toxicity of those nanomaterials on the environment and the ecological systems. Therefore, when talking about the application of nanomaterials in agroecosystems, their major interaction with residing soil biota cannot be ignored. The negative effect of nanoparticles is found to be more pronounced on denitrifying bacteria, disrupting the process of denitrification in soil (VandeVoort and Arai 2012). As a result, nanoparticles in soil have been used as a model system to evaluate the dose-dependent effects of metal itself (Throbäck et al. 2007). In this viewpoint, Yang et al. (2013) studied the interaction of a carbon coated AgNPs with 35 nm in size and Ag + (provided as AgNO3) with *Pseudomonas stutzeri*(denitrifier), *Azotobacter vinelandii* (nitrogen fixer), and *Nitrosomonaseuropaea*(nitrifier). They concluded lower toxicity of AgNPs toward these bacteria compared to 20–48 times higher toxicity exerted by the Ag + ions. Conversely, low and sub lethal concentrations of Ag + and AgNPs (20–25 μg/L) yielded no significant impact on the expression pattern of denitrifying genes and nitrogen-fixing genes but showed 2.1- to 3.3-fold up regulation in nitrifying genes. Another study representing the impact of nanosilver on aerobic denitrification process by Shahrokh et al. (2014) advocated that a low dose of AgNPs had no adverse effect on nitrate reductase activity of *Rhizobium* and *Azotobacter*.

Also and in another study, when nanosized silica-silver particles were applied under *in vivo* condition to control the fungal disease cucurbit powdery mildew, 100% control was achieved after 25 days (Park et al. 2006). These nanoparticles were found to be phytotoxic only at a very high dose of 3200 ppm when tested in cucumber and pansy plants. Taken together, these findings clearly indicate a dose-dependent effect of AgNPs on the microbial process of nitrogen cycle, implying that entering an optimum concentration of AgNPs into the environment could be favourable for microbial processes with no hindrance in beneficial plant-microbe interactions in agroecosystems. On the other hand, size-dependent toxicity of AgNPs has also been evidenced by Choi and Hu (2008), where they found that AgNPs of size less than 5 nm were more toxic to nitrification bacteria. So, beside the need to understand the possible benefits of applying nanotechnology to agriculture, there is also an urgent need to feel secure about nanomaterials phytotoxicity when applied on crop plants. In consequence, the first step should be to analyse penetration of those particles and the pathway of their transport inside plant cells. Formulation stability is also an important aspect of the biosafety of nanomaterials. Liu et al. (2008) successfully formulated a stable nanopesticide (bifenthrin) using polymer stabilizer such as poly (acrylic acid)-b-poly (butyl acrylate) (PAA b PBA), polyvinyl pyrrolidone (PVP), and polyvinyl alcohol (PVOH). A flash nanoprecipitation technique was used to prepare 60–200 nm bifenthrin particles. While using such techniques commercially, the stability profile of the polymers over an extended time period needs to be firstly considered.

On the other hand, Current research work revealed that the uptake, translocation, and accumulation of nanoparticles depend on different factors including the microbial pathogen that needs to be controlled, the plant part which will be sprayed, the species of plant, and the size, chemical composition, functionalization, and stability of the nanoparticles used. Among the carbon-based nanoparticles, only the fullerene C70 and fullerols were shown to get readily accumulated in plants (Rico et al. 2011). Most of the data corresponds to the germination stage and cell culture, because the protocols for quantification of nanoparticles within tissues are not well-defined yet. The discussion of the current research is more oriented to the effect of the nanoparticles on plants. A very few of the nanoparticles to the next generation of plants exposed to nanoparticles are unknown.

Table 30.3 Multifarious factors determining relative advantages of using nanoformulation in agriculture in comparison to Biopesticides

Multifarious factors	Biopesticides	Nanoformulation
Shelf life	Major issue determining efficiency	Not required
Stability on field	Low	High
Effect on fluctuating environment	High effect	Not effect
Surface area	Low	High
Coverage	Moderate	High
Recommended dose	High	Very low

A silver nanoparticle (AgNPs) are now used to enhance seed germination, plant growth, and photosynthetic quantum efficiency and as antimicrobial agents to control plant diseases. There have been relatively few studies on the applicability of silver to control plant diseases; especially for sclerotia-forming species of *Rhizoctonia solani, Sclerotinia sclerotiorum* and *S. minor* (Min et al. 2009) and powdery mildew in cucurbits (Lamsal et al. 2011). Antifungal activity of ionic or nanoparticle silver has a great potential for use in controlling spore-producing fungal plant pathogens,such as *Bipolaris sorokiniana* and Magnaporthe grisea (Young et al. 2009).

30.10 Nanoformulation Pesticide

Today use of chemicals such as pesticides, fungicides and herbicides is the fastest and cheapest way to control pests and diseases. Uncontrolled use of pesticides has caused many problems such as: adverse effects on human health, adverse effects on pollinating insects and domestic animals, and entering this material into the soil and water and its direct and indirect effect on ecosystems. Intelligent use of chemicals on the nano scale can be a suitable solution for this problem (Table 30.3).

30.11 Application of Nanotechnology in Agriculture

Nanotechnology considered as one of the key technologies in the twenty-first century that promises to advance traditional agricultural practices and offer sustainable development by improving the management and conservation tactics with reduced waste of agricultural inputs (Jampilek and Kral'ova 2015; Dubey and Mailapalli 2016). Implementation of new technology like nanoscience in agriculture sector is of extreme importance, particularly in dealing with major problems facing this sector like plant growth, climate change, pest management, and nutrient shortage.

Recent advancements in the fabrication of nanomaterials of different sizes and shapes have yielded their wide array of applications in medicine, environmental

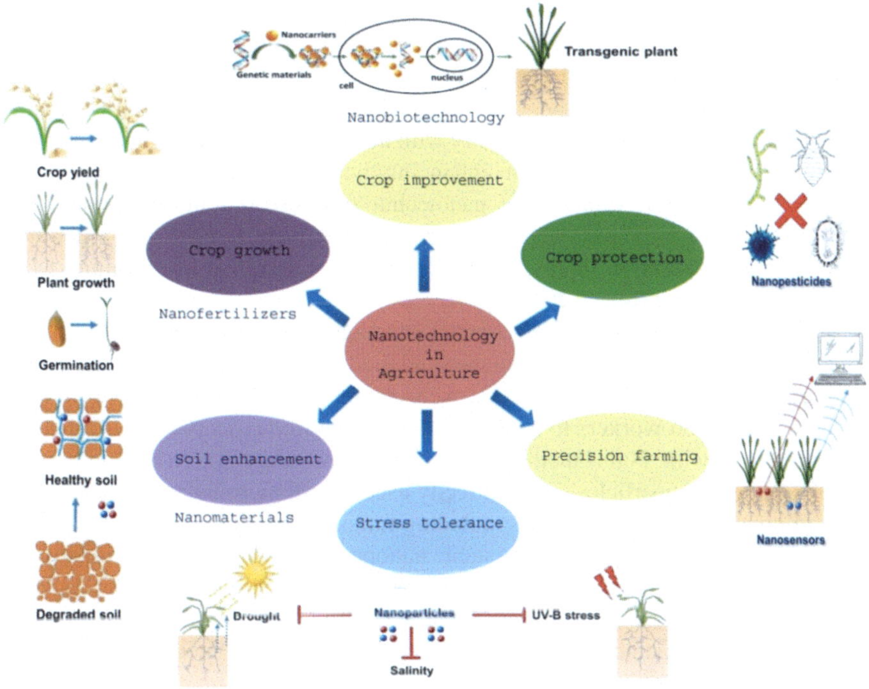

Fig. 30.1 Application of Nanotechnology in Agriculture for various methods

science, agriculture and food processing. Throughout history, agriculture has always benefited from these innovations (Chen et al. 2016). In continuation, as agriculture faces numerous and unprecedented challenges, such as reduced crop yield due to biotic and abiotic stresses, including nutrient deficiency and environmental pollution, the emergence of nanotechnology has offered promising applications for precision agriculture (Fig. 30.1).

30.12 Types of Nanoparticles Used in Plant Disease Management

Nanoparticles alone have the potential to be directly applied to plant seeds, foliage, or roots for protection against pest and pathogens, such as insects, bacteria, fungi, and viruses. Metal nanoparticles such as silver, copper, zinc oxide, and titanium dioxide have been intensively researched for their antibacterial and antifungal properties, and are known for their antiviral properties (Kah and Hofmann 2014; Gogos et al. 2012). This section gives a brief overview and an update on current literature reviews pertaining to the individual nanoparticles that already exist (Kah and Hofmann 2014; Mishra and Singh 2015; Sadeghi et al. 2017; Malerba and Cerana 2016).

Different types of nanomaterials like copper, zinc, titanium, magnesium, gold (Zhao et al. 2010), alginate (Ahmad et al. 2006), and silver (Bhattacharyya et al. 2016) have come up in recent years, and most of them have proven to be effective against diverse microbial pathogens. Nanotechnology has the potential to revolutionize the agricultural and food industry with new tools for the molecular treatment of diseases, rapid disease detection, enhancing the ability of plants to absorb nutrients, etc. In the agricultural sector, nanotechnology research and development is likely to facilitate and frame the next stage of development of genetically modified crops, animal production inputs, chemical pesticides and precision farming techniques (Mousavi and Rezaei 2011).

There has been an increasing amount of attention in the literature regarding effects of TiO2NPs on plant performance and their potential as antimicrobial agents (Roy et al. 2010). Zinc nanoparticles were also evaluated as antiviral agents by El-Sawy and his co-workers to control *Cucumber mosaic virus* (CMV) in eggplant in comparison with 2-nitromethyl phenol and seaweeds extract (El–Sawy et al. 2017). Copper nanoparticles can be directly used as antimicrobial agents against an array of phytopathogenic microbes including fungi and bacteria. Mesoporous silica nanoparticles (MSNs) are thermally stable nanoparticles with tractable porosity and channels, having the tendency to deliver biomolecules likebio-antimicrobial agents, chemicals, pesticides, and nucleic acids into plant tissuesby targeted release (Wang et al. 2010). This system is successfully testedusing different modeling plants like tobacco, Arabidopsis, and maize.

Molybdenum nanoparticles (MoNPs) are important element of nitrogen-fixation system in plants. In this context Taran et al. (2014) reported that Mo nanoparticle with nitrogen-fixing bacteria treatment to chickpea seed showed enhanced growth two to three times in comparison to water, only Mo nanoparticle, indicating that,Mo nanoparticle increases the microbial activity.

Chitosan is a naturally occurring compound that has broad-spectrum antimicrobial and antiviral activities (El Hadrami et al. 2010; Choudhary et al. 2017). Chitosan nanoparticles were shown to inhibit the systemic propagation of viruses and viroids throughout the plant and to enhance the host's hypersensitive response to infection (Chirkov 2002). Different reports also confirmed this issue on different viral pathogens including the potato virus X, tobacco mosaic and necrosis viruses, alfalfa mosaic virus, peanut stunt virus, and cucumber mosaic virus (Chirkov 2002)

30.13 Role of Nanomaterials on Crop Physiology and Plant Protection

Recently, nanotechnology is gaining interest also in plant science, due to the need to develop miniaturized efficient systems to improve seed germination, growth and plant protection to abiotic and biotic stresses (Wang et al. 2016). Metallic nanoparticles (NPs), such as gold (Au), and silver (Ag) NPs, have been widely introduced in plant science for different applications (Table 30.4). The greener approaches

Table 30.4 Various concentration of different nanoparticles influencing plant growth

Nanomaterials	Crop species	Mode of application	Concentration used	Duration of treatments	Responses	Reference
ZnO	Triticum aestivum	Growth substrate	20 mg/L	Growth cycle	Increased grain yield and biomass accumulation	Du et al. (2019)
FeS2	Cicer arietinum; pinacia oleracea; Daucuscarota, Brassica juncea and Sesamumindicum	Seed priming	80100 μg/mL	12–14 h	Increased germination and crop yield	Srivastava et al (2017) Das et al. (2016)
ZnO	Nicotinatabacum	Hydroponics	0.2 μM and 1 μM	21 days	Positively affected growth physiology, increased metabolites, enzymatic activities and anatomical properties of plants	Tirani et al. (2019)
Fe/SiO2	Arachishypogaea,Zea mays	As fertilizers	15 mg/kg	3 days	Enhanced plants growth and biomass accumulation	Disfani et al. (2017)
ZnO Foliar spray 10 mg/L 45 days	Coffeaarabica	Foliar spray	10 mg/L	45 days	Enhanced growth, biomass accumulation and net photosynthesis	Rossi et al. (2019)
CuO	Spinaciaoleraceae	Mixed with soils	200 mg/kg	60 days	Improved photosynthesis and biomass production	Wang et al. (2019)
AgNPs	Triticum aestivum	Mixed with pot soils	50 mg/L and 75 mg/L	Trifoliate stage	Improved growth and tolerance to heat stress	Iqbal et al. (2019)
Ag NPs	Vigna sinensis	Foliar application	50 mg/L 40 days	40 days	Enhanced growth and biomass by stimulating root nodulation and soil bacterial diversity	Pallavi et al. (2016)

(continued)

Table 30.4 (continued)

Nanomaterials	Crop species	Mode of application	Concentration used	Duration of treatments	Responses	Reference
AgNPs	Vigna unguiculata	Foliar application	50100 μg/mL	7 days	Showed no phytotoxicity,but could inhibit growth of Xanthomonas axonopodis pv. malvacearum and Xanthomonas campestris pv. campestris in vitro	Vanti et al. (2019)
CuO	Solanum lycopersicum	Foliat application	150–340 μg/ml	11 days	Effectively controlled late blight disease caused by Phytophthora infestans	Giannousi et al. (2013)
MgO	Solanum lycopersicum	Drenching	7–10 μg/mL	7 days	Controlled bacterial wilt disease by suppressing pathogen Ralstonia solanacearum	Imada et al. (2016)
TiO2 and SiO2 Oryza sativa	Oryza sativa	Foliar application	20 and 30 mg/L	55 days	Mitigated Cd toxicity and improved growth by stimulating antioxidant potential and inhibiting Cd translocation	Rizwan et al. (2019)
MWCNTs	Triticum aestivum, Zea mays, Arachis hypogaea, Allium sativum	Seed priming	50 μg/mL	Over night	Improved and rapid germination, increased biomass accumulation and water absorption potential of seeds	Srivastava and Rao (2014)
ZnO	Cyamopsis tetragonoloba	Foliar spray	10 mg/L	6 weeks	Improved plant growth, biomass accumulation and nutrient content	Raliya et al. (2013)
MWCNTs	Hordeum vulgare, Glycine max, Zea mays	Seed priming	100 μg/mL	24 h	Enhanced germination and growth of seedlings	Lahiani et al. (2013)

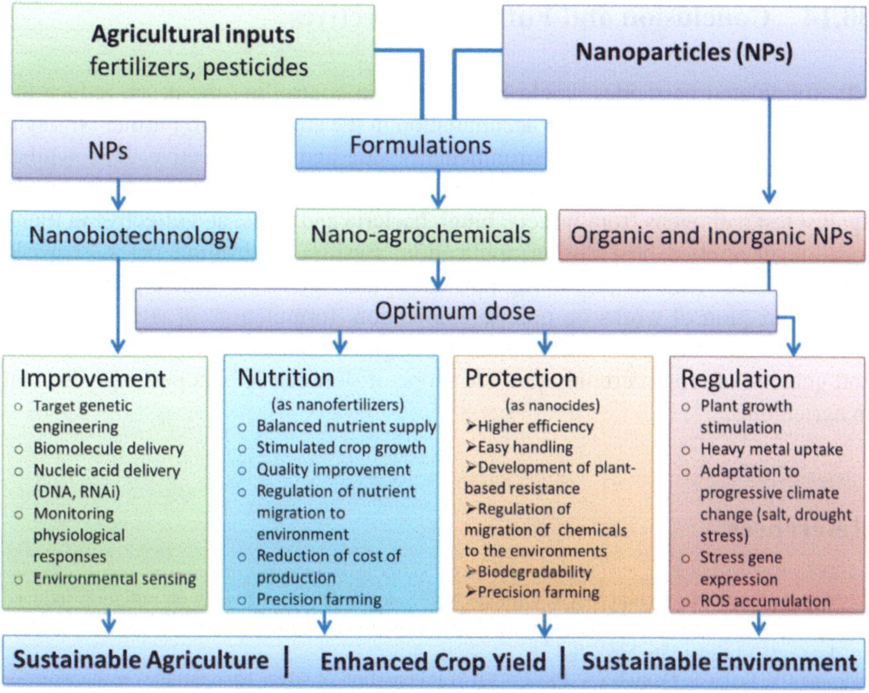

Fig. 30.2 Nanoparticles: A Novel Approach for sustainable Agro productivity

based on the use of plant extract as well as ionizing radiation chemistry in aqueous solutions have been developed (Abedini et al. 2013). This technologyhelps to improve agricultural productionby increasing the efficiency of inputs and minimizing relevant losses. Nanomaterials offer a widerspecific surface area to fertilizers and pesticides. In addition, nanomaterials as unique carriersof agrochemicals facilitate the site-targeted controlled delivery of nutrients with increased cropprotection. Due to their direct and intended applications in the precise management and control ofinputs (fertilizers, pesticides, herbicides), nanotools, such as nanobiosensors, support the developmentof high-tech agricultural farms. The integration of biology and nanotechnology into nanosensors hasgreatly increased their potential to sense and identify the environmental conditions or impairments (Chen et al. 2016).

The potential of nanomaterials encourages a newgreen revolution with reduced farming risks. However, there are still huge gaps in our knowledge ofthe uptake capacity, permissible limit and the ecotoxicity of direntrent nanomaterials (He et al. 2018). Therefore, further research is urgently needed to unravel the behaviour and fate of altered agriculture inputs andtheir interaction with bio macromolecules present in living systems and environment (Fig. 30.2).

30.14 Conclusion and Future Perspectives

NPs formulated pesticides would be effective in controlling the pests, reduce the usage and minimize the residue accumulation in the environment. Further NPs synthesized by green route are environmentally safe and cost effective. NPs synthesized using seaweeds characterized showed varied size, shape and bioactivity against test pathogens from insects, fungi, bacteria and virus; and effective in translocation in applied host during application. Besides controlling pests, seaweed aided NPs are found as promoting nutrient uptake their by enhance plant growth. Future direction of works on precise application, formulation, *in vivo* large scale field trail against various pests in different crops and toxicity study at biochemistry and genome of crops certainly provide scope of seaweed NPs in pest management in agriculture.

References

Abedini A, Daud AR, Hamid MAA, Othman NK, Saion E (2013) A review on radiation-induced nucleation and growth of colloidal metallic nanoparticles. Nanoscale Res Lett 8:1–10. https://doi.org/10.1186/1556-276X-8-474

Adriana PV, Erika S, Daniel XA, Pio C (2016) Preparation of silver nanoparticles using aqueous extracts of the red algae *Laurencia aldingensis* and *Laurenciella* sp. and their cytotoxic activities. J Appl Phycol 28(4):2615–2622. https://doi.org/10.1007/s10811-015-0757-4

Ahmad Z, Pandey R, Sharma S, Khuller G (2006) Alginate nanoparticles as antituberculosis drug carriers: formulation development, pharmacokinetics and therapeutic potential. Indian J Chest Dis Allied Sci 48(3):171

Aragão AP, De Oliveir TA, Quelemes PV, Silva DA (2016) Green synthesis of silver nanoparticles using the seaweed *Gracilaria birdiae* and their antibacterial activity. Arabian J Chem 12:4182–4188

Arunkumar K, Sivakumar SR, Rengasamy R (2010) Review on bioactive potential in seaweeds (marine macroalgae): a special emphasis on bioactivity of seaweeds against plant pathogens. Asian J Plant Sci 9:227–240

Arunkumar M, Suhashini K, Mahesh N, Ravikumar R (2014) Quorum quenching and antibacterial activity of siver nanoparticles synthesized from *Sargassum polyphyllum*. Bangladesh J Pharm 9(1):54–59

Asha KS, Johnson M, Chandra KP, Shibila T, Revathy I (2015) Extracellular synthesis of silver nanoparticles from a marine alga, *Sargassum polycystum* C. Agardh and their biopotentials. World J Pharm Pharm Sci 4(9):1388–1400

Aziz N, Fatma T, Varma A, Prasad R (2014) Biogenic synthesis of silver nanoparticles using *Scenedesmus abundans* and evaluation of their antibacterial activity. J Nanoparticles. https://doi.org/10.1155/2014/689419

Aziz N, Faraz M, Pandey R, Shakir M, Fatma T, Varma A et al (2015) Facile algae-derived route to biogenic silver nanoparticles: synthesis, antibacterial, and photocatalytic properties. Langmuir 16:11605–11612. https://doi.org/10.1021/acs.langmuir.5b03081

Aziz N, Pandey R, Barman I, Prasad R (2016) Leveraging the attributes of *Mucorhiemalis*-derived silver nanoparticles for a synergistic broad-spectrum antimicrobial platform. Front Microbiol 7:1984. https://doi.org/10.3389/fmicb.2016.01984

Baskar BB (2013) Biosynthesis of silver nanoparticles using *Kappa phycus* species. Int J Pharm Sci Rev Res 3:55–63

Bhattacharyya A, Duraisamy P, Govindarajan M, Buhroo AA, Prasad R (2016) Nano-biofungicides: emerging trend in insect pest control. In: Prasad R (ed) Advances and applications through fungal nanobiotechnology. Springer International Publishing, Cham, pp 307–319. https://doi.org/10.1007/978-3-319-42990-8_15

Bragg PD, Rannie DJ (1974) The effect of silver ions on the respiratory chain of *Escherichia coli*. Can J Microbiol 20:883–889

Castellano JJ, Shai SM, Ko F, Donate G, Wright TE, Mannari RJ et al (2007) Comparative evaluation of silver–containing antimicrobial dressings and drugs. Int Wound J 4(2):114–122

Chandran SP et al (2006) Synthesis ofgolnanotriangles and silver nanoparticles using *Aloevera* plant extract. Bio Technol Prog 22(2):577–583

Chen YW, Lee HV, Juan JC, Phang SM (2016) Production of new cellulose nanomaterial from red algae marine biomass *Gelidium elegans*. Carbohydr Polym 151:1210–1219

Chirkov SN (2002) The antiviral activity of chitosan. Appl Biochem Micro 38:1–8

Choi O, Hu Z (2008) Size dependent and reactive oxygen species related Nano silver toxicity to nitrifying bacteria. Environ SciTechnol 42:4583–4588

Choudhary SMJ, Joshi A, Saharan V (2017) Assessment of Cu-chitosan nanoparticles for its antibacterial activity against *Pseudomonas syringae* pv. *Glycinea*. Int J Curr Microbiol App Sci 6:1335–1350

Clement JL, Jarrett PS (1994) Antibacterial silver. Met Based Drugs 1(5–6):467–482

Das CK, Srivastava G, Dubey A, Roy M, Jain S, Sethy NK, Saxena M, Harke S, Sarkar S, Misra K (2016) Nano-iron pyrite seed dressing: a sustainable intervention to reduce fertilizer consumption in vegetable (beetroot, carrot), spice (fenugreek), fodder (alfalfa), and oilseed (mustard, sesamum) crops. Nanotechnol Environ Eng 1:2. https://doi.org/10.1007/s41204-016-0002-7

Davis TA, Volesky B, Mucci A (2003) A review of the biochemistry of heavy metal biosorption by brown algae. Water Res 37(18):4311–4330

Devi JS, Bhimba BV (2014) Antimicrobial potential of silver nanoparticles synthesized using Ulva recticulata. Asian J Pharm Clin Res 7(2):82–85

Dhanalakshmi PK, Riyazulla A, Rekha R, Poonkodi S, Thangaraju N (2012) Synthesis of silver nanoparticles using green and brown seaweeds. Phykos 42(2):39–45

Disfani MN, Mikhak A, Kassaee MZ, Maghari A (2017) Effects of nano Fe/SiO2 fertilizers on germination and growth of barley and maize. Arch Agron Soil Sci 63:817–826

Du W, Yang J, Peng Q, Liang X, Mao H (2019) Comparison study of zinc nanoparticles and zinc sulphate on wheat growth: from toxicity and zinc biofortification. Chemosphere 227:109–116

Dubey A, Mailapalli DR (2016) Nanofertilisers, nanopesticides, nanosensors of pest and nanotoxicity in agriculture. In Sustainable agriculture reviews; Lichtfouse, E., Ed.; Springer, Cham 19:307–330

El-Samy MM, Elsharkavy MM, Abass JM, Kasem MH (2017) Antiviral activity of 2 nitro methyl phenol.Zink nanoparticles and seawee extract against cucumber mosaic virus in egg plant. JVirol Antivir Res 6:2

El Hadrami A, Adam LR, El Hadrami I, Daayf F (2010) Chitosan in plant protection. Mar Drugs 8:968–987

Elechiguerra JL, Burt JL, Morones JR et al (2005) Interaction of silver nanoparticles with HIV. J Nanobiotechnol 3:1–10

Fabrega J, Luoma SN, Tyler CR, Galloway TS, Lead JR (2011) Silver nanoparticles: behavior and effects in the aquatic environment. Environ Int 37:517–531

Fayaz AM, Balaji K, Girilal M, Kalaichelvan PT, Venkatesan R (2009) Mycobased synthesis of silver nanoparticles and their incorporation into sodium alginate films for vegetables and fruit preservation. J Agric Food Chem 57:6246–6252

Gajbhiye MB, Kesharwani JG, Ingle AP et al (2009) Fungus-mediated synthesis of silver nanoparticles and its activity against pathogenic fungi in combination of fluconazole. Nanomed NBM 5:282–286

Ganesan V, Aruna DJ, Astalakshmi A, Nima P, Thangaraja A (2013) Eco-friendly synthesis of silver nanoparticles using a sea weed, *Kappaphycus alvarezii* (Doty) Doty ex PC Silva. Int J Eng Adv Technol 2:559–563

Gericke M, Pinches A (2006) Biological synthesis of metal nanoparticles. Hydrometallurgy 83(1):132–140. https://doi.org/10.1016/j.hydromet.2006.03.019

Ghormade V, Deshpande MV, Paknikar KM (2011) Perspectives for nano-biotechnology enabled protectionand nutrition of plants. Biotechnol Adv 29:792–803

Giannousi K, Avramidis I, Dendrinou-Samara C (2013) Synthesis, characterization and evaluation of copper based nanoparticles as agrochemicals against *Phytophthora infestans*. RCS Adv 3:21743–21752

Gnanadesigan M, Anand M, Ravikumar S, Maruthupandy M, Ali MS, Vijayakumar V, Kumaraguru AK (2012) Antibacterial potential of biosynthesized silver nanoparticles using *Avicennia marina* mangrove plant. Appl Nanosci 2(2):143–147

Gogos A, Knauer K, Bucheli TD (2012) Nanomaterials in plant protection and fertilization: current state, foreseen applications, and research priorities. J Agric Food Chem 60:9781–9792

Gopinath V, Velusamy P (2013) Extracellular biosynthesis of silver nanoparticles using Bacillus sp. GP-23 and evaluation of their antifungal activity towards *Fusarium oxysporum*. Spectrochim Acta A 106:170–174

Govindaraju K, Kiruthiga V, Ganesh Kumar V, Singaravelu G (2009) Extracellular synthesis of silver nanoparticles by a marine alga, *Sargassum wightii,* Grevilli and their antibacterial effects. J Nanosci Nanotechnol 9(9):5497–5501

Guggenbichler JP, Boswald M, Lugauer S, Krall T (1999) A new technology of micro dispersed silver in polyurethane induces antimicrobial activity in central venous catheters. Infection 27:16–23

He X, Deng H, Hwang H-M (2018) The current application of nanotechnology in food and agriculture.J. Food Drug Anal 27:1–21

Husen A, Siddiqi KS (2014) Phytosynthesis of nanoparticles: concept,controversy and application. Nano Res Lett 9:229

Imada K, Sakai S, Kajihara H, Tanaka S, Ito S (2016) Magnesium oxide nanoparticles induce systemic resistance in tomato against bacterial wilt disease. Plant Pathol 65:551–560

Iqbal M, Raja NI, Hussain M, Ejaz M, Yasmeen F (2019) Effect of silver nanoparticles on growth of wheat under heat stress. IJST A Sci 43(387–395):65

Iravani S (2011) Green synthesis of metal nanoparticles using plants. Green Chem 13:2638–2650

Jampilek J, Kral'ova K (2015) Application of nanotechnology in agriculture and food industry, its prospects and risks. Ecol Chem Eng S-Chemia I Inzynieria Ekologiczna S 22(321–361):38

Jeyaseelan C, Abul R, Selvaraj MR,Vishnu K (2013) Biological approach to synthesize TiO2 nanoparticles using Aeromonashydrophila and its antibacterial activity. Spectrochimica Acta Part A Mol Biomol Spectroscopy 107C. https://doi.org/10.1016/j.saa.2012.12.083

Joshi N, Jain N, Pathak A, Singh J, Prasad R, Upadhyaya CP (2018) Biosynthesis of silver nanoparticles using Carissa carandas berries and its potential antibacterial activities. Journal of Sol-Gel Science and Technology 86:682–689

Jung WK, Koo HC, Kim KW, Shin S, Kim SH, Park YH (2008) Antibacterial activity and mechanism of action of the silver ion in *Staphylococcus aureus* and *Escherichia coli*. Appl Environ Microbiol 74:2171–2178

Kah M, Hofmann T (2014) Nanopesticide research: current trends and future priorities. Environ Int 63:224–235

Kannan RRR, Arumugam R, Ramya D, Manivannan K, Anantharaman P (2013a) Greensynthesis of nanoparticles using marine macroalga *Chaetomorpha linum*. Appl Nanosci 3:229–233

Kannan RRR, Stirk WA, Wanstaden J (2013b) Green synthesis of silver nanoparticles using the seaweed *Codiumcapitatum* P.C.Silva (Chlorophyceae). S Afr J Bot. https://doi.org/10.1016/j.sajb.2013.01.003

Kathiraven T, Sundaramanickam A, Shanmugam N, Balasubramanian T (2015) Green synthesis of silver nanoparticles using marine algae *Caulerpa racemosa* and their antibacterial activity against some human pathogens. Appl Nanosci 5:449–504

Kayalvizhi K, Nabikhan A, Vasuki S, Kandasamy K (2014) Purification of silver and gold nanoparticles from two species of brown seaweeds *Padinatetrastromatica* and *Turbinariaornata*. J Med Plants Stud 2(4):32–37

Kim KJ, Sung WS, Suh BK, Moon SK, Choi JS, Kim JG, Lee DG (2009) Antifungalactivity and mode of action of silver nanoparticles on Cnia albicans. Biometals 22:235
Kim SW, Jung JH, Lamsal K et al (2012) Antifungal effects of silver nanoparticles (AgNPs) against various plant pathogenic fungi. Mycobiol 40:53–58. https://doi.org/10.5941/MYCO.2012.40.1.053
Kiran MV, Murugesan S (2013) Biogenic silver nanoparticles by *Halymenia Poryphyroides* and its *in Vitro* anti-diabetic efficacy. J Chem Pharm Res 5(12):1001–1008
Kiran VM, Murugesan S (2014) Biosynthesis of silver nano particles from marine alga *Halymenia poryphyroides* and its antibacterial efficacy. Int J Curr Microbiol App Sci 3(4):96–103
Kitching M, Ramani M, Marsili E (2015) Fungal biosynthesis of goldnanoparticles: mechanism and scale up. J Microbial Biotechnol 8:904–917
Krishnaraj C, Jagan EG, Rajasekar S, Selvakumar O, Kalaichelvan PT, Mohan NJCSBB (2010) Synthesis of silver nanoparticles using *Acalypha indica* leaf extracts and its antibacterial activity against water borne pathogens. Colloids Surf B Biointerfaces 76:50–56
Kumar V, Yadav SK (2009) Plant mediated synthesis of silver and gold nanoparticles and their application. J Chem Technol Biotechnol 84:151–157
Lahiani MH, Dervishi E, Chen JH, Nima Z, Gaume A, Biris AS, Khodakovskaya MV (2013) Impact of carbon nanotube exposure to seeds of valuable crops. ACS Appl Mater Interfaces 5:7965–7973
Lamsal K, Kim S, Jin JH, Kim YS, Kim KS, Lee YS (2011) Inhibition effects of silver nanoparticles against powdery mildews on Cucumber and Pumpkin. Mycobiol 39(1):26–32. https://doi.org/10.4489/MYCO.2011.39.1.026
Lavakumar V, Masilamani K, Ravichandiran V, Venkateshan N, Saigopal DVR, AshokKumar CK, Sowmya C (2015) Promising upshot of silver nanoparticles primed from *Gracilaria crassa* against bacterial pathogens. Chem Cent J 9:42
Lee KJ, Park SH, Govarthanan M, Hwang PH, Seo YS et al (2013) Synthesis of silver nanoparticles using cow milk and their antifungal activity against phytopathogens. Mater Lett 105:128–131. https://doi.org/10.1016/j.matlet.2013.04.076
Li WR, Xie XB, Shi QS, Zeng HY, Ou-Yang YS, Chen YB (2009) Antibacterial activity and mechanism of silver nanoparticles on *Escherichia coli*. Appl Microbiol Biotechnol 85:1115–1122
Liu Z, Jiao Y, Wang Y, Zhou C, Zhang Z (2008) Polysaccharides-based nanoparticles as drug delivery systems. Adv Drug Deliv Rev 60:1650–1662
Ma X, Geisier Lee J, Yang D, Kolmokov A (2010) Interaction between engineered nanoparticles (ENP) and plants,phytotoxicity,uptake and accumulation. Sci Total Environ 108:3053–3061
Mala R, Sarojini M, Saravanababu S, Umadevi G (2009) Screening for antimicrobial activity of crude extracts of *Spirulina platensis*. J Cell Tissue Res 9:1951–1955
Malerba M, Cerana R (2016) Chitosan effects on plant systems. Int J Mol Sci 17:996
Manal D, Mukhopadhyay D, Sarkar G, Polander ME (2006) The use of microorganism for the formation of metal nanoparticles and their application. Appl Microbiol Biotechnol 69(5):485–492. https://doi.org/10.1007/s00253-005-0179-3
Min JS, Kim KS, Kim SW, Jin JH, Lamsal K, Kim SB (2009) Effects of colloidal silver nanoparticles on Sclerotium-forming phytopathogenic fungi. Plant Pathol J 25(4):376–380. https://doi.org/10.5423/ppj.2009.25.4.376
Mishra S, Singh H (2015) Biosynthesized silver nanoparticles as a nanoweapon against phytopathogens: exploring their scope and potential in agriculture. Appl Microbiol Biotechnol 99:1097–1107
Mishra S, Singh BR, Singh A, Keswani C, Naqvi AH, Singh HB (2014) Biofabricated silver nanoparticles act as a strong fungicide against *Bipolaris sorokiniana* causing spot blotch disease in wheat. PLoS One 9(5):e97881. https://doi.org/10.1371/journal.phone.0097881
Mohandass C, Vijayraj R, Satheeshbabu S, Rao SV, Shiva C, De Mello I (2013) Biosynthesis of silver nanoparticles from marine seaweed *Sargassum cinereum* and their antibacterial activity. Pharm Sci 75(5):606–610
Mohanpuria P, Rana NK, Yadav SK (2008) Biosynthesis of nanoparticles: technical concept and future applications. J Nanopart Res 10:507–517

Morones JR, Elechiguerra JL, Camacho A, Holt K, Kouri JB, Ramirez JT, Yacaman MJ (2005) Thebacterial effect of silver nanoparticles. Nanotechnology 16:2346–2353

Mousavi SR, Rezaei M (2011) Nanotechnology in agriculture and food production. J Appl Environ Biol Sci 1(10):414–419

Navarro E, Piccapietra F, Wagner B, Marconi F, Kaegi R, Odzak N, Sigg L, Behra R (2008) Toxicity of silver nanoparticles to *Chlamydomonas reinhardtii*. Environ Sci Technol 42:8959–8964

Ocsoy I, Paret ML, Oscsoy MS, Kunwar S (2013) Nanotechnology in plant disease management: DNA-directed silver nanoparticles on graphene oxide as an antibacterial against *Xanthomonas Perforans*. ACS Nano 7(10):8972–8980. https://doi.org/10.1021/nn4034794

Panpatte DG, Jhala YK, Shelat HN, Vyas RV (2016) Nanoparticles: the next generation technology for sustainable agriculture. In: Microbial inoculants in sustainable agricultural productivity. Springer, New Delhi, pp 289–300

Park HJ, Kim SH, Kim SJ, Choi SH (2006) A new composition of nanosized silica-silver for control of various plant diseases. Plant Pathol J 22:295–302

Paulkumar K, Gnanajobitha G, Vanaja M, Rajeshkumar S, Malarkodi C, Pandian K, Annadurai G (2014) *Piper nigrum* leaf and stem assisted green synthesis of silver nanoparticles and evaluation of its antibacterial activity against agricultural plant pathogens. Research Article | Open Access. https://doi.org/10.1155/2014/829894

Rahimi Z, Yousefzadi M, Noori A, Akbarzadeh A (2014) Green synthesis of silver nanoparticles using *Ulva flexousa* from the Persian Gulf. Iran J Persian Gulf (Mar Sci) 5:9–16

Rajesh S, Raja DP, Rathi JM, Sahayaraj K (2012) Biosynthesis of silver nanoparticles using Ulva fasciata (Delile) ethyl acetate extract and its activity against *Xanthomonas compestris* pv. malvacearum. J Biopest 5:119–128

Rajeshkumar S, Chelladurai M, Gnanadas G, Kanniah P, Mahendran V, Chellapandian K, Gurusamy A (2012a) Seaweed-mediated synthesis of gold nanoparticles using Turbinariaconoides and its characterization. J Nanostruct Chem 3:44

Rajeshkumar S, Kannan CP, Annadurai G (2012b) Synthesis and characterization of antimicrobial silver nanoparticles using marine brown sea weed *Padina tetrastromatica*. Drug Invent Today 4(10):511–513

Rajeshkumar S, Malarkodi C, Paulkumar K, Vanaja M, Gnanajobitha G, Annadurai G (2014) Algae mediated green fabrication of silver nanoparticles and examination of its antifungal activity against clinical pathogens. Int J Met 8. https://doi.org/10.1155/2014/692643

Raliya R, Tarafdar JC (2013) ZnO nanoparticle biosynthesis and its effect on phosphorous-mobilizing enzyme secretion and gum contents in Clusterbean (*Cyamopsistetragonoloba* L.). Agric Res 2:48–57

Rastogi and Arunachalam (2011) Sunlight based irradiation strategy for rapid green synthesis of highly stable silver nanoparticles using aqueous garlic (*Allium sativum*) extract and their antibacterial potential. Mater Chem Phys 129(2011):558–563

Rico CM, Majumdar S, Gardea MD, Videa JP (2011) Interaction of nanoparticles with edible plants and their possible implications in food chain. J Agric Food Chem 59(3):3485–3498. https://doi.org/10.1021/jf104517j

Rizwan M, Ali S, ur Rehman MZ, Malik S, Adrees M, Qayyum MF, Alamri SA, Alyemeni MN, Ahmad P (2019) Effect of foliar applications of silicon and titanium dioxide nanoparticles on growth, oxidative stress, and cadmium accumulation by rice (*Oryza sativa*). Acta Physiol Plant 41:35. https://doi.org/10.1007/s11738-019-2828-7

Roseline TA, Murugan M, Sudhakar MP, Arunkumar K (2019) Nanopesticidal potential of silver nanocomposites synthesized from the aqueous extract of red seaweeds. Environ Technol Innov. https://doi.org/10.1016/j.eti.2018.10.005

Rossi L, Fedenia LN, Sharifan H, Ma X, Lombardini L (2019) Effects of foliar application of zinc sulfate and zinc nanoparticles in coffee (*Coffeaarabica* L.) plants. Plant Physiol Biochem 135:160–166

Roy A, Parveen A, Koppalkar AR, MVN AP (2010) Effect of Nano - titanium dioxide with different antibiotics against methicillin- resistant staphylococcus aureus. J Biomater Nanobiotechnol 1(01):37. https://doi.org/10.4236/jbnb.2010.11005

Sadeghi R, Rodriguez RJ, Yao Y, Kokini JL (2017) Advances in nanotechnology as they pertain to food and agriculture: benefits and risks. Annu Rev Food Sci Technol 8:467–492

Sajidha PK, Lakshmi D (2016) Biosynthesis of silver nanoparticles using red algae, *Amphiroafragilissima* and its antibacterial potential against gram positive and gram negative bacteria. Int J Curr Sci 19:93–100

Sangeetha N, Saravanan K (2014) Biogenic silver nanoparticles using marine seaweed (*Ulva lactuca*) and valuation of its antibacterial activity. J Nanosci Nanotechnol 2:99–102

Sankar NS, Dipak P, Nilu H, Samir KP (2014) Green synthesis of silver nanoparticles using fresh water green alga *Pithophora oedogonia* (Mont.) Wittrock and evaluation of their antibacterial activity. Appl Nanosci 5:703–709

Sastry M, Ahmad A, Khan I, Kumar R (2003) Biosynthesis of metal nanoparticles using fungi andactinomycete. Curr Sci 85:162–170

Sekhon B (2014) Nanotechnology in agri-food production: an overview. Nanotechnol Sci Appl 7(2):31–53. https://doi.org/10.2147/NSA.S39406

Shahrokh S, Hosseinkhani B, Emtiazi G (2014) The impact of nano-silver on bacterial aerobic nitrate reductase. J Bioprocess Biotechnol 4:162. https://doi.org/10.4172/2155-9821.1000162

Shanmugam N, Rajkamal P, Cholan S, Kannadasan N, Sathiskumar K, Viruthagiri G, Sumdramanikam A (2014) Biosynthesis of silver nanoparticles from the marine seaweed Sargassum wightii and their antibacterial activity against some human pathogens. Appl Nano Sci 4:881–888

Shiny PJ, Amirtava M, Chandrasekaran N (2013) Marine algae mediated synthesis of the silver nanoparticles and its antibacterial efficiency. Int J Pharm Sci 5(2):239–241

Srivastava A, Rao D (2014) Enhancement of seed germination and plant growth of wheat, maize, peanut and garlic using multiwalled carbon nanotubes. Eur Chem Bull 3:502–504

Sunitha S, Nageswararao A, Abragam LS, Dhayalan E, Thirunanasambanadam R, Ganeskumar V (2015) Enhanced bacterial effect of silver nanoparticles using marine brown macro algae. J Chem Pharm Res 7(3):191–195

Suriya J, Bharathi Raja S, Sekar V, Rajasekaran R (2012) Biosynthesis of silver nanoparticles and its antibacterial activity using seaweed Urospora sp. Afr J Biotechnol 11:12192–12198

Susan A, Namvar F, Mahdavi M, Ahmad MP, Mohamad R (2013) Biosynthesis of silver nanoparticles using brown marine macroalga, *Sargassum Muticum* aqueous extract. Materials 6:5942–5950

Taran NY, Gonchar OM, Lopatko KG, Batsmanova LM, Patyka MV, Volkogon MV (2014) The effect of colloidal solution of molybdenum nanoparticles on the microbial composition in rhizosphere of *Cicer arietinum* L. Nanoscale Res Lett 9:289

Thirumalairaj V, Vijayan MP, Durairaj G, Shanmugaasokan L (2014) Potential antibacterial activity of crude extracts and silver nanoparticles synthesized from *Sargassum wightii*. Int Curr Pharm J 3(10):322–325. https://doi.org/10.3329/icpj.v3i10.20337

Throback IN, Johansson M, Rosenquist M, Pell M (2007) Silver (Ag+) reduces denitrification and induces enrichment of novel nirK genotypes in soil. FEMS Microbiol Lett 270(2):189–194. https://doi.org/10.1111/j.1574-6968.2007.00632.x

Tien D, Tseng K, Liao C, Tsung T (2009) Identification and quantification of ionic silver from colloidal silver prepared by electric spark discharge system and its antimicrobial potency study. J Alloys Compd 473:298–302

Tirani MM, Haghjou MM, Ismaili A (2019) Hydroponic grown tobacco plants respond to zinc oxide nanoparticles and bulk exposures by morphological, physiological and anatomical adjustments. Funct Plant Biol 46:360–375

Tripathi DK, Singh S, Singh S, Pandey R, Singh VP, Sharma NC, Prasad SM, Dubey NK, Chauhan DK (2017) An overview on manufactured nanoparticles in plants: uptake, translocation, accumulation and phytotoxicity. Plant Physiol Biochem 110:2–12

Vandevoort AR, Arai Y (2012) Effect of silver nanoparticles on soil denitrification kinetics. Industrial Biotechnology 8(6):358–364

Vanti GL, Nargund VB, Basavesha KN, Vanarchi R, Kurjogi M, Mulla SI, Tubaki S, Patil RR (2019) Synthesis of *Gossypiumhirsutum*-derived silver nanoparticles and their antibacterial efficacy against plant pathogens. Appl Organomet Chem 33:e4630. https://doi.org/10.1002/aoc.4630

Vasquez RD, Ramos JDA (2012) Immunomodulatory effects of sulfated polysaccharides from *Sargassum siliquosum* J.G. Agardh on peripheral blood mononuclear cells. Int J Res Phytochem Pharmacol 2(1):55–63

Velmurugan P, Lee SM, Iydrrose M, Lee KJ (2013) Pine cone-mediated green synthesis of silver nanoparticles and their antibacterial activity against agricultural pathogens. Appl Microbiol Biotechnol 97(1):361–368. https://doi.org/10.1007/s00253-012-3892-8

VinothKumar T, Mahitha PV, Geetharamani D, Lakshmanasenthil S, Rincy Y, Suja G (2014) Potential antibacterial activity of crude extracts and silver nanoparticles synthesized from *Sargassum wightii*. Int Currt Pharm J 3(10):322–325

Vivek M, Kumar PS, Steffi S, Sudha S (2011) Biogenic silver nanoparticles by *Gelidiellaacerosa* extract and their antifungal effects. Avicenna J Med Biotechnol 3(3):143–148

Wang LS, Wu LC, Lu SY, Chang LL, Teng IT et al (2010) Biofunctionalized phospholipid-capped mesoporous silica nanoshuttles for targeted drug delivery: improved water suspensibility and decreased nonspecific protein binding. ACS Nano 4:4371–4379

Wang Z, Xu L, Zhao J, Wang X, White JC, Xing B (2016) CuO nanoparticle interaction with *Arabidopsis thaliana*: toxicity, parent-progeny transfer, and gene expression. Environ Sci Technol 50:6008–6016. https://doi.org/10.1021/acs.est.6b01017

Wang Y, Lin Y, Xu Y, Yin Y, Guo H, Du W (2019) Divergence in response of lettuce (var. ramosa Hort.) to copper oxide nanoparticles/microparticles as potential agricultural fertilizer. Environ Pollut Bioavailab 31:80–84

Yan A, Zhong C (2019) Impacts of silver nanoparticles on plants: a focus on the Phytotoxicity and underlying mechanism. Int J Mol Sci 26;20(5):1003. https://doi.org/10.3390/ijms20051003

Yang YU, Wang J, Xiu Z, Pedro JJ, Varez AL (2013) Impact of silver nanoparticles on cellular and transcriptional activity of nitrogen cycling bacteria. Environ Toxicol Chem 32(7):1488–1494

Young KJ, Byung HK, Geunhwa J (2009) Antifungal activity of silver ions and nanoparticles on phytopathogenic fungi. Plant Dis 93:1037–1043. https://doi.org/10.1094/PDIS-93-10-1037

Zhao X, Wang J, Wu F, Wang T, Cai Y, Shi Y, Jiang G (2010) Removal of fluoride from aqueousmeia by Fe3O@AI(OH) 3 magnetic nanoparticles. Jhazar Mater 173:102–109

Chapter 31
Seaweed-Based Polymers from Sustainable Aquaculture to "Greener" Plastic Products

Diana Pacheco, João Cotas, João Carlos Marques, Leonel Pereira, and Ana M. M. Gonçalves

31.1 Introduction

Bio-based biodegradable plastics are gaining popularity to minimize reliance on finite fossil fuels and provide a sustainable alternative. Currently, several plastic items have become an indispensable part of quotidian life. Petrochemical plastics accounted for 99% of all common plastics, whereas fossil fuels are rapidly depleting (Visnji 2019). Moreover, recycled waste has resulted in a variety of environmental concerns, including greenhouse gas emissions, the manufacture of microplastics, and their possible toxic effects (Paula et al. 2018; Prata et al. 2019). As a result, the need to use sustainable and environmentally friendly polymers is extremely pressing to achieve long-term social growth.

Bio-based plastics can be produced as a replacement for petroleum-based plastics (Mekonnen et al. 2013). Lipids, proteins, and carbohydrates are commonly used to make bio-based plastics (Zhang et al. 2019). In addition, short-term biodegradable bio-based plastics have been widely used in packaging, medical/pharmaceutical materials, and agricultural products in the industrial sector. In 2018, the global demand for bio-based plastics was projected to be 2.11 million tons, and by 2023, it is expected to reach 2.62 million tons (European Bioplastics 2021).

D. Pacheco · J. Cotas · J. C. Marques · L. Pereira
University of Coimbra, MARE – Marine and Environmental Sciences Centre, Department of Life Sciences, Coimbra, Portugal

A. M. M. Gonçalves (✉)
University of Coimbra, MARE – Marine and Environmental Sciences Centre, Department of Life Sciences, Coimbra, Portugal

Department of Biology and CESAM, University of Aveiro, Aveiro, Portugal
e-mail: amgoncalves@uc.pt

A. Ranga Rao, G. A. Ravishankar (eds.), *Sustainable Global Resources of Seaweeds Volume 1*, https://doi.org/10.1007/978-3-030-91955-9_31

Polylactide (PLA), polyhydroxyalkanoate (PHA), and starch compost bags outperform conventional petroleum-based polymers in terms of non-renewable energy usage and global warming potential, according to a life cycle assessment (Yates and Barlow 2013). These bio-based plastics can solve the issues of sustainability and waste management thanks to advanced technologies. Other raw bio-based plastics products, such as polysaccharides and proteins, are primarily extracted from agricultural plants (corn, wheat, potato, and soybean), which can pose a risk to food security, fertile land, and irrigation (Fabra et al. 2017). As a result, designing next-generation bio-based materials to replace existing bio-based plastics is crucial.

Due to their rapid growth rate and extensive environmental tolerance, seaweeds have been considered as a possible feedstock for bio-based plastics, greatly alleviating the competition for food and water resources (Venkatesan et al. 2016; Mathiot et al. 2019). Furthermore, algae may effectively reduce carbon dioxide emissions from flue gas while also accumulate significant quantities of polysaccharides (>60%), proteins (60%), and lipids (>50%) that can be used as potential bio-based plastics raw materials (Leandro et al. 2020).

31.2 General Overview in Polymers

Packaging accounts for most of the plastic waste; non-biodegradability and reliance on non-renewable natural resources are two major roadblocks to increased plastic use. The accumulation of plastic products in natural environments has placed immense pressure on the ecosystem, negatively affecting biodiversity, wildlife habitat, and humans. As a result, packaging for novel polymers is needed to resolve the shortcomings of traditional plastics (Abdul Khalil et al. 2017).

However, as opposed to their non-biodegradable counterparts, biopolymers have low mechanical and barrier properties, which limits their use as packaging materials (Saurabh et al. 2013).

Fabricating composite films by combining one polymer with another polymer, a hydrophobic component, and/or nanoparticles is one of the most common techniques for overcoming this downside. This method allows you to take advantage of each aspect of the composite film's unique functional characteristics. As a result, these hybrid films outperform pure polymeric films in terms of mechanical and barrier properties (Saurabh et al. 2015, 2016).

31.3 Seaweed Based Polymers

The polysaccharide content of seaweeds, such as agar, alginate, and carrageenan, is valuable (Masarin et al. 2016; Abdul Khalil et al. 2017). Bioplastics, biomaterials, and bio-scaffolds can all be made from these polysaccharides. Alternatives to petrochemical-based plastics include biodegradable and recycled polymers, such as polysaccharides from seaweeds (Hanani and Husna 2018). Polysaccharides derived from seaweeds are highly colloidal in nature, have a low production cost, and can be used in a variety of industrial applications (Sudhakar et al. 2020).

Seaweeds may provide high biomass productivity, while still allowing productive land to be used for other purposes (Jang et al. 2013). Algal polymers are a natural source with numerous benefits, including excellent heat insulation and heat power, total biodegradability, and heavy carbon dioxide fixation. Fibrils have energy-absorbing properties, resulting in excellent insulation and sound-absorption properties, as well as a flame retardant (Hassan et al. 2008).

31.3.1 Alginate

Alginate is a polysaccharide found in marine brown seaweeds, such as *Macrocystis pyrifera* and *Saccharina japonica* (formerly known as *Laminaria japonica*), accounting for 22–44% of their dry cell weight (Lin et al. 2018; Li et al. 2019; Venkatesan et al. 2016). Alginate is a linear copolymer made up of 1,4-glycosidic linkages connecting mannuronic acid (M) and guluronic acid (G) (Pereira and Cotas 2020). Different physicochemical properties are expressed by the ratios of alternate GG, MM, and MG blocks. The higher the G content, the stronger the moisture barriers and the lower the water vapor permeability (Oms-Oliu et al. 2008). Because of its excellent biocompatibility, alginate has been commonly used in biomedical applications such as drug delivery systems, wound dressings, and tissue regeneration. Furthermore, the calcium alginate-polyacrylamide complex can provide improved rigidity, recoverability, and durability, making it a viable cartilage substitute (Sun et al. 2012).

Alginates may, for example, crosslink with cations to form hydrogels or packaging films, with Ca^{2+} having higher binding power, mechanical strength, and water barrier than Na^{+} (Cazón et al. 2017). After exchanging cations with Ca^{2+}, the strength and broken rate of straw-alginate reached 216 g and 4%, respectively (Xue et al. 2019). The swelling inhibition was found to be over 95% with the addition of Ca^{2+} to the cellulose-alginate complex, and the Young's modulus and tensile strength were estimated to be 135 and 17 MPa, respectively (Benselfelt et al. 2018). As a result, cations-associated with alginate has the potential to be used in the manufacture of several environmentally friendly packaging materials.

31.3.2 Carrageenan

Carrageenan is a polysaccharide constituted by D-galactose and 3,6-anhydo-D-galactose that is specifically isolated from red seaweeds which can reach up to 30–50% of their dry weight (DW) (Rupérez and Saura-Calixto 2001) in species, such as *Chondrus crispus* (Azevedo et al. 2015) and *Mastocarpus stellatus*

Fig. 31.1 Red seaweeds *Chondrus crispus* (**a**) and *Mastocarpus stellatus* (**b**)

(Rhodophyta) (Fig. 31.1) (Torres et al. 2016). Based on the number and location of sulphate ester groups, three forms of carrageenan fractions have been identified: kappa (κ), iota (ι), and lambda (λ) (Pereira et al. 2009). Carrageenan's double helix arrangement, similar to alginate, provides anion-hosting positions for enhanced gel structure. Monovalent ions (such as K^+) are usually accepted by κ-carrageenan, whereas divalent ions (such as Ca^{2+}) are favored by ι-carrageenan. The goal of many carrageenan bioplastics is to produce edible packaging films (Cazón et al. 2017). In this context, ι-carrageenan, for example, may be a good food coating material for ham preservation (Carocho et al. 2019).

Polyvinyl alcohol (PVA) has also been used as a synthetic material to enhance the material's poor mechanical properties. Carrageenan-PVA films were found to have a 439.31% rise in elongation (Meng et al. 2018), while the tensile strength of κ-carrageenan-gelatin composites could also be increased to 10.32 MPa with the addition of 5% SiO_2 nanoparticles (Hashemi Tabatabaei et al. 2018). Currently, the global carrageenan market accounts for the majority of the global hydrocolloids market, which is expected to reach US$ 1 billion by 2024 (Zhang et al. 2019). However, more than 80% of the products are channeled for food applications like processed foods, dairy, sweets, and jellies (Campbell and Hotchkiss 2017).

31.3.3 Agar

Agar is a polysaccharide that is commonly found in the Gracilariaceae and Gelidiaceae families of red seaweed (Rhodophyta). Being agarose and agaropectin the two most important components (Arvizu-Higuera et al. 2007), this is a hydrophilic colloid that can form reversible gels after being cooled from a hot aqueous solution (Guerrero et al. 2014). Before being used as a raw material for the development of bioplastic film formation, agar was first used in the fields of food, biotechnology, and pharmaceutical applications. Agar-based films, according to The et al. (2009), are clear, solid, and versatile. It is also heat-sealable, making it ideal for use in the food packaging industry.

31.3.4 Seaweed Polymers Composites and Quality Enhancement

Several algal polysaccharides can be used to enhance the consistency of these polymers (i.e., mechanical, thermal, and antimicrobial properties). They not only improve hydrophilicity and mechanical properties including tensile strength and elongation, but they also make it possible to use it as an active packaging (Sharma et al. 2021). This is possible due to the naturally occurring antioxidant properties of seaweed, which can minimize lipid oxidation and thereby increase the shelf life and nutritional value of food while also reducing free radicals that could be carcinogenic, mutagenic, or cytotoxic (Dima et al. 2014; Vital et al. 2016).

31.3.4.1 Plasticizer

Plasticizers are an effective tool for increasing the efficiency of most bio-based plastics, aside from sustainability concerns and raw material biodegradation. Plasticizers can minimize brittleness and crystallinity, lower the glass transition temperature, increase durability, and impart versatility in general (Mekonnen et al. 2013). Grafting with raw polymer as part of the plasticization process will weaken dispersion forces and hydrogen bonds (Mekonnen et al. 2013). Biodegradable, non-toxic, stable, or non-volatile properties are often required by an ideal plasticizer, depending on the target. For instance, glycerol is a well-known plasticizer for hydrophilic polymers (Fabra et al. 2017, Shi et al. 2017).

31.3.4.2 Strengthen Solution

When seaweed-based polysaccharides films were reinforced with organically modified or unmodified nano-clay for the production of food packaging, the mechanical strength of the film was improved (Rhim 2011; Martins et al. 2013). Furthermore, natural or synthetic antimicrobial agents such as grapefruit seed extracts and silver nanoparticles are used to develop strong inhibitory activity against foodborne pathogens in seaweed-based food packaging (Rhim et al. 2013; Kanmani and Rhim 2014). According to the comprehensive literature review, additives such as nonmaterial and antimicrobial materials effectively boost different properties of composite films. Seaweed-based composites are also being researched for pharmaceutical applications due to their excellent properties.

Rhim (2013) created a nanocomposite film using PLA and a laminated agar/carrageenan/clay bio-nanocomposite film, which showed improved tensile strength, water vapor permeability, water uptake ratio, and water solubility. Thermal stability and water resistance were also improved in the newly formed composite. Paramita et al. (2015) developed a high-solid matrix consisting of κ-carrageenan and polydextrose that is an amorphous molecule in nature, improving the stability as well as quality control, and exhibits diffusional mobility of α-lionic acid for use in the food and pharmaceutical industries. Furthermore, the film constituted by polylysine, κ -carrageenan, and pectin demonstrated electrostatic attraction between the constituents, resulting in a stronger complex that acted as an antimicrobial delivery mechanism for foods and beverages (Lopez-Pena and McClements 2014).

31.3.4.3 Biodegrability

Despite the addition of plastifiers has added strength to this biopolymer, they need to be biodegradable. Plastics are made up of polymers, which, depending on their constituents, can go through a complex biodegradation process (Siracusa et al. 2008). Polymers decompose due to biotic and abiotic factors, providing microorganisms (such as, bacteria, algae, fungi) with organic compounds (i.e., monosaccharides, essential amino acids) (Song et al. 2011). During chemical reactions, polymers are oxidized, resulting in digestion, in which organic matter is converted to compounds such as carbon dioxide (Li et al. 2017). These changes that occur during deterioration deteriorating the material, leading to its fragmentation that can be caused by the shortening of the polymer chain.

Biopolymers, on the other hand, have a core feature of compostability, which allows packaging to be disposed in the soil. Compostability, on the other hand, is a key function of biopolymers, enabling packaging to be disposed of in the soil. Biopolymers must be biodegradable, particularly in terms of composting, according to the European bioplastics, so that they can be used as fertilizers and soil conditioners. Some bioplastics based on natural monomers, on the other hand, may lose biodegradability due to chemical polymerization modification. Many bioplastics involve a combination of ingredients, such as synthetic polymers and additives, to improve the functional characteristics of the finished products and broaden their application possibilities. It is possible to obtain a polymer with approximately 100% biodegradable compounds by weight as both the additives and pigments are derived from renewable energy sources (Siracusa et al. 2008). Thus, it is also vital to investigate the changes that can occur when bioplastics interact with food products.

Natural polymers like rubber, humus, and lignin must follow the oxo-biodegradation mechanism, whereas synthetic polymers (e.g., polyolefins) must follow the oxo-biodegradation mechanism (Rydz et al. 2014). Peroxidation, which is triggered by heat or light, produces low-molecular-mass aldehydes, ketones, and alcohols during the degradation of oxocarboxylic acid molecules, which is the key cause of the loss of mechanical properties of carbohydrate polymers. Bacteria, fungi, and enzymes then begin bio-assimilation, which increases biomass and CO_2 levels, forming cavities. Antioxidants and stabilizers are widely applied to synthetic polymers to prevent them from oxidation during mechanical processing and to ensure that they have the desired shelf life. On the other hand, antioxidants are needed to improve the performance of materials, but in the case of biodegradability, it is preferable to avoid adding antioxidants during polymer processing (Siracusa et al. 2008). Hydro-biodegradation is a process that converts cellulose, starch, and polyesters into bioassimilable items. In this context, in an aqueous medium, the aliphatic polyester is rapidly hydrolyzed and bioassimilated (Scott and Wiles 2001).

31.4 Seaweed Cultivation for Polymer Production

To meet the rising demand for alternatives to petroleum-derived fuels and products, the future landscape of sustainable biobased polymers and plastic products would likely depend on a portfolio of different feedstock sources (van Hal et al. 2014; Laurens et al. 2017). Biomass derived from seaweed is one of the most promising emerging sources in this field (Shama et al. 2019; Mohammad et al. 2019; Hessami et al. 2019). Seaweeds can produce more biomass per acre in offshore marine farms than terrestrial crops, and they can be harvested and produced sustainably without consuming precious arable land or needing excessive nutrient levels (Hafting et al. 2015; Buschmann et al. 2017). As a result, seaweed polymers are the primary goal of today's aquaculture around the world (Porse and Rudolph 2017).

In Europe, wild population harvesting is the most common method of collecting seaweed biomass, while aquaculture accounts for just 13% of total seaweeds production. The transformation industries' requirements are shifting away from biomass quantity and toward biomass efficiency, traceability, and sustainability. This demand comes in part from the general product protection EU directive (2001/95/EC), which has strict quality management policies. "Sustainable feedstock sourcing is a requirement for more sustainable products" (Unkown 2021). Seaweed cultivation under managed conditions allows for high traceability, biomass composition and properties management, high quality, and long-term sustainability (preservation of natural populations) (Unkown 2021) .

In conclusion, seaweed has many benefits over the raw materials commonly used in biomass-based plastics: reduces carbon dioxide emissions, has no competition for land with other industries (food and biofuel processing, power and heat generation), has higher productivities, reduces nutrient intake in marine coastal areas (released from fed aquaculture or agriculture runoff), has no chance of biodiversity loss (if cultivated), avoids deforestation, has no freshwater consumption, and there are no pesticides or fertilizers used (and no nutrient added when cultivated in integrated multi-trophic aquaculture (IMTA) systems (Unkown 2021).

31.5 Conclusions and Future Perspectives

The modern food industry cannot work without packaging. It is required for food preservation and ensures the food's protection and integrity. The utilization of synthetic plastics for packaging is an integral aspect of our supply chain, but it also has disadvantages, such as contaminant residue migration, cost and energy usage, and sustainability.

Food packaging has a rapid accumulation of plastic in our environments due to its short lifespan, which thus has a direct effect on environmental pollution. Biodegradable and biobased polymers have been developed and are now available on the market to reduce these effects and provide a more sustainable alternative to food packaging. In this context, seaweed polymers present a sustainable and natural source for the development of novel food packaging and greener plastics.

Acknowledgments This work is financed by national funds through FCT - Foundation for Science and Technology, I.P., within the scope of the projects UIDB/04292/2020 granted to MARE - Marine and Environmental Sciences Centre and UIDP/50017/2020 + UIDB/50017/2020 (by FCT/MTCES) granted to CESAM - Centre for Environmental and Marine Studies. João Cotas thanks to the European Regional Development Fund through the Interreg Atlantic Area Program, under the project NASPA (EAPA_451/2016). Diana Pacheco thanks to PTDC/BIA-CBI/31144/2017-POCI-01 project -0145-FEDER-031144-MARINE INVADERS, co-financed by the ERDF through POCI (Operational Program Competitiveness and Internationalization) and by the Foundation for Science and Technology (FCT, IP). Ana M. M. Gonçalves acknowledges University of Coimbra for the contract IT057-18-7253 and also thanks to the project MENU – Marine Macroalgae: Alternative recipes for a daily nutritional diet (FA_05_2017_011), funded by the Blue Fund under Public Notice No. 5 – Blue Biotechnology, which partially financed this research.

References

Abdul Khalil HPS, Tye YY, Saurabh CK, Leh CP, Lai TK, Chong EWN, Nurul Fazita MR, Mohd Hafiidz J, Banerjee A, Syakir MI (2017) Biodegradable polymer films from seaweed polysaccharides: a review on cellulose as a reinforcement material. Express Polym Lett 11:244–265. https://doi.org/10.3144/expresspolymlett.2017.26

Arvizu-Higuera DL, Rodríguez-Montesinos YE, Murillo-Álvarez JI, Muñoz-Ochoa M, Hernández-Carmona G (2007) Effect of alkali treatment time and extraction time on agar from *Gracilaria vermiculophylla*. In: Nineteenth international seaweed symposium. Springer Netherlands, Dordrecht, pp 65–69. https://doi.org/10.1007/978-1-4020-9619-8_9

Azevedo G, Torres MD, Sousa-Pinto I, Hilliou L (2015) Effect of pre-extraction alkali treatment on the chemical structure and gelling properties of extracted hybrid carrageenan from *Chondrus crispus* and *Ahnfeltiopsis devoniensis*. Food Hydrocoll 50:150–158. https://doi.org/10.1016/j.foodhyd.2015.03.029

Benselfelt T, Engström J, Wågberg L (2018) Supramolecular double networks of cellulose nanofibrils and algal polysaccharides with excellent wet mechanical properties. Green Chem 20:2558–2570. https://doi.org/10.1039/C8GC00590G

Buschmann AH, Camus C, Infante J, Neori A, Israel Á, Hernández-González MC, Pereda SV et al (2017) Seaweed production: overview of the global state of exploitation, farming and

emerging research activity. Eur J Phycol. Taylor & Francis. 52:391–406. https://doi.org/10.1080/09670262.2017.1365175
Campbell R, Hotchkiss S (2017) Carrageenan industry market overview. In: Tropical seaweed farming trends, problems and opportunities. Springer International Publishing, Cham, pp 193–205. https://doi.org/10.1007/978-3-319-63498-2_13
Carocho M, Heleno S, Rodrigues P, Filomena Barreiro M, Barros L, Ferreira ICFR (2019) A novel natural coating for food preservation: effectiveness on microbial growth and physicochemical parameters. LWT 104:76–83. https://doi.org/10.1016/j.lwt.2019.01.031
Cazón P, Velazquez G, Ramírez JA, Vázquez M (2017) Polysaccharide-based films and coatings for food packaging: a review. Food Hydrocoll 68:136–148. https://doi.org/10.1016/j.foodhyd.2016.09.009
de Paula FC, de Paula CBC, Contiero J (2018) Prospective biodegradable plastics from biomass conversion processes. In: Biofuels - state of development. InTech. https://doi.org/10.5772/intechopen.75111
Dima C, Cotârlet M, Alexe P, Dima S (2014) Microencapsulation of essential oil of pimento [*Pimenta dioica* (L) Merr.] by chitosan/k-carrageenan complex coacervation method. Innov Food Sci Emerg Technol 22:203–211. https://doi.org/10.1016/j.ifset.2013.12.020
European Bioplastics e.V (2021). https://www.european-bioplastics.org/. Accessed March 12
Fabra MJ, Martínez-Sanz M, Gómez-Mascaraque LG, Coll-Marqués JM, Martínez JC, López-Rubio A (2017) Development and characterization of hybrid corn starch-microalgae films: effect of ultrasound pre-treatment on structural, barrier and mechanical performance. Algal Res 28:80–87. https://doi.org/10.1016/j.algal.2017.10.010
Guerrero P, Etxabide A, Leceta I, Peñalba M, de la Caba K (2014) Extraction of agar from *Gelidium sesquipedale* (Rodhopyta) and surface characterization of agar based films. Carbohydr Polym 99:491–498. https://doi.org/10.1016/j.carbpol.2013.08.049
Hafting JT, Craigie JS, Stengel DB, Loureiro RR, Buschmann AH, Yarish C, Edwards MD, Critchley AT (2015) Prospects and challenges for industrial production of seaweed bioactives. J Phycol 51:821–837. https://doi.org/10.1111/jpy.12326
Hashemi Tabatabaei R, Jafari SM, Mirzaei H, Nafchi AM, Dehnad D (2018) Preparation and characterization of nano-SiO_2 reinforced gelatin-k-carrageenan biocomposites. Int J Biol Macromol 111:1091–1099. https://doi.org/10.1016/j.ijbiomac.2018.01.116
Hassan MM, Mueller M, Wagners MH (2008) Exploratory study on seaweed as novel filler in polypropylene composite. J Appl Polym Sci 109:1242–1247. https://doi.org/10.1002/app.28287
Hessami MJ, Cheng SF, Ranga Rao A, Yin YH, Phang SM (2019) Bioethanol production from agarophyte red seaweed, *Gelidium elegans* using a novel sample preparation method for analysing bioethanol content by gas chromatography. 3 Biotech J 9(1).25
Jang YH, Han SO, Sim IN, Kim H-I (2013) Pretreatment effects of seaweed on the thermal and mechanical properties of seaweed/polypropylene biocomposites. Compos A: Appl Sci Manuf 47:83–90. https://doi.org/10.1016/j.compositesa.2012.11.016
Kanmani P, Rhim J-W (2014) Development and characterization of carrageenan/grapefruit seed extract composite films for active packaging. Int J Biol Macromol 68:258–266. https://doi.org/10.1016/j.ijbiomac.2014.05.011
Laurens LML, Markham J, Templeton DW, Christensen ED, Van Wychen S, Vadelius EW, Chen-Glasser M, Dong T, Davis R, Pienkos PT (2017) Development of algae biorefinery concepts for biofuels and bioproducts; a perspective on process-compatible products and their impact on cost-reduction. Energ Environ Sci 10:1716–1738. https://doi.org/10.1039/C7EE01306J
Leandro A, Pacheco D, Cotas J, Marques JC, Pereira L, Gonçalves AMM (2020) Seaweed's bioactive candidate compounds to food industry and global food security. Life 10:140. https://doi.org/10.3390/life10080140
Li J, Li R, Li J (2017) Current research scenario for microcystins biodegradation – a review on fundamental knowledge, application prospects and challenges. Sci Total Environ 595:615–632. https://doi.org/10.1016/j.scitotenv.2017.03.285

Li S-Y, Wang Z-P, Wang L-N, Peng J-X, Wang Y-N, Han Y-T, Zhao S-F (2019) Combined enzymatic hydrolysis and selective fermentation for green production of alginate oligosaccharides from Laminaria japonica. Bioresour Technol 281:84–89. https://doi.org/10.1016/j.biortech.2019.02.056

Lin JD, Lemay MA, Parfrey LW (2018) Diverse bacteria utilize alginate within the microbiome of the giant kelp *Macrocystis pyrifera*. Front Microbiol 9:1914. https://doi.org/10.3389/fmicb.2018.01914

Lopez-Pena CL, McClements DJ (2014) Optimizing delivery systems for cationic biopolymers: competitive interactions of cationic polylysine with anionic κ-carrageenan and pectin. Food Chem 153:9–14. https://doi.org/10.1016/j.foodchem.2013.12.024

Martins JT, Bourbon AI, Pinheiro AC, Souza BWS, Cerqueira MA, Vicente AA (2013) Biocomposite films based on κ-carrageenan/locust bean gum blends and clays: physical and antimicrobial properties. Food Bioproc Tech 6:2081–2092. https://doi.org/10.1007/s11947-012-0851-4

Masarin F, Cedeno FRP, Chavez EGS, de Oliveira LE, Gelli VC, Monti R (2016) Chemical analysis and biorefinery of red algae *Kappaphycus alvarezii* for efficient production of glucose from residue of carrageenan extraction process. Biotechnol Biofuels 9:122. https://doi.org/10.1186/s13068-016-0535-9

Mathiot C, Ponge P, Gallard B, Sassi J-F, Delrue F, Le Moigne N (2019) Microalgae starch-based bioplastics: screening of ten strains and plasticization of unfractionated microalgae by extrusion. Carbohydr Polym 208:142–151. https://doi.org/10.1016/j.carbpol.2018.12.057

Mekonnen T, Mussone P, Khalil H, Bressler D (2013) Progress in bio-based plastics and plasticizing modifications. J Mater Chem A 1:13379. https://doi.org/10.1039/c3ta12555f

Meng F, Zhang Y, Xiong Z, Wang G, Li F, Zhang L (2018) Mechanical, hydrophobic and thermal properties of an organic-inorganic hybrid carrageenan-polyvinyl alcohol composite film. Compos Part B 143:1–8. https://doi.org/10.1016/j.compositesb.2017.12.009

Mohammad JH, Ranga Rao A, Ravishankar GA (2019) Opportunities and challenges in seaweeds as feed stock for biofuel production. In: Ravishankar GA, Rao AR (eds) Handbook of algal technologies and phytochemicals: volume-II phycoremediation, biofuels and global biomass production. CRC Press, Boca Raton, pp 39–50

Nur Hanani ZA, Aelma Husna AB (2018) Effect of different types and concentrations of emulsifier on the characteristics of kappa-carrageenan films. Int J Biol Macromol 114:710–716. https://doi.org/10.1016/j.ijbiomac.2018.03.163

Oms-Oliu G, Soliva-Fortuny R, Martín-Belloso O (2008) Using polysaccharide-based edible coatings to enhance quality and antioxidant properties of fresh-cut melon. LWT- Food Sci Technol 41:1862–1870. https://doi.org/10.1016/j.lwt.2008.01.007

Paramita VD, Bannikova A, Kasapis S (2015) Release mechanism of omega-3 fatty acid in κ-carrageenan/polydextrose undergoing glass transition. Carbohydr Polym 126:141–149. https://doi.org/10.1016/j.carbpol.2015.03.027

Pereira L, Cotas J (2020) Introductory chapter: alginates - a general overview. In: Alginates - recent uses of this natural polymer. IntechOpen. https://doi.org/10.5772/intechopen.88381

Pereira L, Amado AM, Critchley AT, van de Velde F, Ribeiro-Claro PJA (2009) Ideentification of selected seaweed polysaccharides (phycocolloids) by vibrational spectroscopy (FTIR-ATR and FT-Raman). In Food hydrocolloids 23. Elsevier Ltd: 1903–1909. https://doi.org/10.1016/j.foodhyd.2008.11.014

Porse H, Rudolph B (2017) The seaweed hydrocolloid industry: 2016 updates, requirements, and outlook. J Appl Phycol 29:2187–2200. https://doi.org/10.1007/s10811-017-1144-0

Prata JC, da Costa JP, Lopes I, Duarte AC, Rocha-Santos T (2019) Effects of microplastics on microalgae populations: a critical review. Sci Total Environ 665:400–405. https://doi.org/10.1016/j.scitotenv.2019.02.132

Rhim J-W (2011) Effect of clay contents on mechanical and water vapor barrier properties of agar-based nanocomposite films. Carbohydr Polym 86:691–699. https://doi.org/10.1016/j.carbpol.2011.05.010

Rhim J-W (2013) Effect of PLA lamination on performance characteristics of agar/κ-carrageenan/clay bio-nanocomposite film. Food Res Int 51:714–722. https://doi.org/10.1016/j.foodres.2013.01.050

Rhim J-W, Wang LF, Hong SI (2013) Preparation and characterization of agar/silver nanoparticles composite films with antimicrobial activity. Food Hydrocoll 33:327–335. https://doi.org/10.1016/j.foodhyd.2013.04.002

Rupérez P, Saura-Calixto F (2001) Dietary fibre and physicochemical properties of edible Spanish seaweeds. Eur Food Res Technol 212:349–354. https://doi.org/10.1007/s002170000264

Rydz J, Sikorska W, Kyulavska M, Christova D (2014) Polyester-based (bio)degradable polymers as environmentally friendly materials for sustainable development. Int J Mol Sci 16:564–596. https://doi.org/10.3390/ijms16010564

Saurabh CK, Gupta S, Bahadur J, Mazumder S, Variyar PS, Sharma A (2013) Radiation dose dependent change in physiochemical, mechanical and barrier properties of guar gum based films. Carbohydr Polym 98:1610–1617. https://doi.org/10.1016/j.carbpol.2013.07.041

Saurabh CK, Gupta S, Bahadur J, Mazumder S, Variyar PS, Sharma A (2015) Mechanical and barrier properties of guar gum based nano-composite films. Carbohydr Polym 124:77–84. https://doi.org/10.1016/j.carbpol.2015.02.004

Saurabh CK, Gupta S, Variyar PS, Sharma A (2016) Effect of addition of nanoclay, beeswax, tween-80 and glycerol on physicochemical properties of guar gum films. Ind Crop Prod 89:109–118. https://doi.org/10.1016/j.indcrop.2016.05.003

Scott G, Wiles DM (2001) Programmed-life plastics from polyolefins: a new look at sustainability †. Biomacromolecules 2:615–622. https://doi.org/10.1021/bm010099h

Shama A, Joyce SG, Mari FD, Ranga Rao A, Ravishankar GA, Hudaa N (2019) Macroalgae and microalgae: novel sources of functional food and feed. In: Ravishankar GA, Rao AR (eds) Handbook of algal technologies and phytochemicals: volume-I: food, health and nutraceutical applications. CRC Press, Boca Raton, pp 207–219

Sharma S, Barkauskaite S, Jaiswal AK, Jaiswal S (2021) Essential oils as additives in active food packaging. Food Chem 343:128403. https://doi.org/10.1016/j.foodchem.2020.128403

Shi B, Liang L, Yang H, Zhang L, He F (2017) Glycerol-plasticized spirulina-poly(vinyl alcohol) films with improved mechanical performance. J Appl Polym Sci 134. https://doi.org/10.1002/app.44842

Siracusa V, Rocculi P, Romani S, Dalla Rosa M (2008) Biodegradable polymers for food packaging: a review. Trends Food Sci Technol 19:634–643. https://doi.org/10.1016/j.tifs.2008.07.003

Song F, Tang D-L, Wang X-L, Wang Y-Z (2011) Biodegradable soy protein isolate-based materials: a review. Biomacromolecules 12:3369–3380. https://doi.org/10.1021/bm200904x

Sudhakar MP, Peter DM, Dharani G (2020) Studies on the development and characterization of bioplastic film from the red seaweed (*Kappaphycus alvarezii*). Environ Sci Pollut Res. https://doi.org/10.1007/s11356-020-10010-z

Sun J-Y, Zhao X, Illeperuma WRK, Chaudhuri O, Oh KH, Mooney DJ, Vlassak JJ, Suo Z (2012) Highly stretchable and tough hydrogels. Nature 489:133–136. https://doi.org/10.1038/nature11409

The DP, Debeaufort F, Voilley A, Luu D (2009) Biopolymer interactions affect the functional properties of edible films based on agar, cassava starch and arabinoxylan blends. J Food Eng 90:548–558. https://doi.org/10.1016/j.jfoodeng.2008.07.023

Torres MD, Chenlo F, Moreira R (2016) Rheology of κ/ι-hybrid carrageenan from Mastocarpus stellatus: critical parameters for the gel formation. Int J Biol Macromol 86:418–424. https://doi.org/10.1016/j.ijbiomac.2016.01.096

Unkown (2021) Final report summary - SEABIOPLAS (seaweeds from sustainable aquaculture as feedstock for biodegradable bioplastics) | report summary | seabioplas | fp7 | cordis | European Commission. https://cordis.europa.eu/project/id/606032/reporting. Accessed 13 Mar

van Hal JW, Huijgen WJJ, López-Contreras AM (2014) Opportunities and challenges for seaweed in the biobased economy. Trends Biotechnol 32:231–233. https://doi.org/10.1016/j.tibtech.2014.02.007

Venkatesan J, Anil S, Kim SK, Shim MS (2016) Seaweed polysaccharide-based nanoparticles: preparation and applications for drug delivery. Polymers 8:1–25. https://doi.org/10.3390/polym8020030

Visnji M (2019) The production of plastic and petrochemical feedstocks. http://revenuesandprofits.com/how-exxonmobil-makes-money/. Accessed 12 Mar

Vital ACP, Guerrero A, Monteschio JDO, Valero MV, Carvalho CB, de Abreu Filho BA, Madrona GS, do Prado IN (2016) Effect of edible and active coating (with rosemary and oregano essential oils) on beef characteristics and consumer acceptability. PLoS One 11:e0160535. https://doi.org/10.1371/journal.pone.0160535

Xue J, Wu Y, Shi K, Xiao X, Gao Y, Li L, Qiao Y (2019) Study on the degradation performance and kinetics of immobilized cells in straw-alginate beads in marine environment. Bioresour Technol 280:88–94. https://doi.org/10.1016/j.biortech.2019.02.019

Yates MR, Barlow CY (2013) Life cycle assessments of biodegradable, commercial biopolymers—a critical review. Resour Conserv Recycl 78:54–66. https://doi.org/10.1016/j.resconrec.2013.06.010

Zhang C, Show P-L, Ho S-H (2019) Progress and perspective on algal plastics – a critical review. Bioresour Technol 289:121700. https://doi.org/10.1016/j.biortech.2019.121700

Chapter 32
Valorisation of Macroalgal Biomass for Sustainable Biorefineries

Mukul S. Kareya, Anju Mehra, Asha A. Nesamma, and Pannaga P. Jutur

Abbreviations

CAZymes	Carbohydrate-active enzymes
CCAP	Culture Collection of Algae and Protozoa
COX-2	Cyclooxygenase-2
DNP	Dinitrophenyl
EST	Expressed Sequence Tag
GHG	Greenhouse Gases
HHV	Higher Heating Value
HTL	Hydrothermal Liquefaction
KO	Kyoto Encyclopedia of Genes and Genomes Orthology
KU-MACC	Kobe University Macroalgae Culture Collection
RBL	Rat Basophile Leukemia
RCC	Roscoff Culture Collection
RNA	Ribonucleic acid
ROS	Reactive Oxygen Species

Mukul S. Kareya and Anju Mehra contributed equally with all other contributors.

M. S. Kareya · A. Mehra · A. A. Nesamma · P. P. Jutur (✉)
Omics of Algae Group, Industrial Biotechnology, International Centre for Genetic Engineering and Biotechnology, Aruna Asaf Ali Marg, New Delhi, India
e-mail: jppavan@icgeb.res.in

A. Ranga Rao, G. A. Ravishankar (eds.), *Sustainable Global Resources of Seaweeds Volume 1*, https://doi.org/10.1007/978-3-030-91955-9_32

32.1 Introduction

The demand for food with the ever-growing global population is estimated to increase by 70% of the current production of food. This led to a rise in the agriculture sector and further exacerbate the environment and climate, hence there is a rapid need for sustainable food from alternative natural resources (Leandro et al. 2020a, b). The use of algae as a supplementary feed source has long been discussed and tried to be implemented. Algae are diverse photosynthetic organisms that manifest in both micro and macro forms. Macroalgae taxonomically divided into three divisions, namely, red algae (Rhodophyta), brown algae (Ochrophyta-Phaeophyceae), and green algae (Chlorophyta) (Leandro et al. 2019). These macroalgae, which are usually referred to as seaweeds, long been cultivated by humans and used as food and as fertilizers for soil enrichment (Milledge and Harvey 2016; Øverland et al. 2019; Shama et al. 2019). Seaweeds are distributed along the sea-shore from polar to tropical zones and forms the foundation of the aquatic food chain (García-Poza et al. 2020). Like land plants, seaweeds too play important ecological functions, where some species are reported to serve as bioindicators of water quality while other species could be playing a role in bioremediation (La Barre et al. 2018; Henriques et al. 2017; Leandro et al. 2019). As the primary producers, these edible benthic organisms are reported to be rich in fibres, proteins, and lipids along with different bioactive compounds such as carotenoids, vitamins, minerals, and polyphenols (Dawczynski et al. 2007; MacArtain et al. 2007). As a result of their high nutritional content, seaweeds in course of time emerged as a potential candidate and as an alternative to the vegetable diet, thus sustaining food chain supply. Moreover, seaweeds can be cultivated offshore, thereby eliminating the need to use additional nutrients, agricultural lands which are extremely critical for land crops, and the carbon fixation by these marine cell factories which are capable of assisting and playing a crucial role in bioremediation which comes at a great expense to the environment (Hebbale et al. 2017; Sharpley et al. 2015; Shore et al. 2017). It was estimated to enhance the production of food by 70% by the year 2050 with policies to substantially reduce the emission of greenhouse gases (GHG) by 5.6% by 2012 (UNFCCC 2009). Consequently, it was declared by the world summit to broaden the investigation and research in the field of bio-renewables and biofuels from sustainable sources such as marine macroalgae (Forster and Radulovich 2015; Tester and Langridge 2010). Therefore, cultivation of seaweeds in oceans encouraged and increased simultaneously, as there are no practical implications and added advantage due to the availability of water, space, sunlight and the readily available carbon dioxide in oceans whereas availability of inorganic nutrients such as potassium, nitrate and phosphate remains a bottleneck/constrains (Forster and Radulovich 2015; Kim et al. 2014). Food and fuels from seaweeds as a combined approach, however, received very less attention due to the lack of downstream processing (Chen et al. 2015). Conversely, biofuel production from seaweeds is technically feasible and these marine organisms offers bioethanol, bio-oils, biomethane and

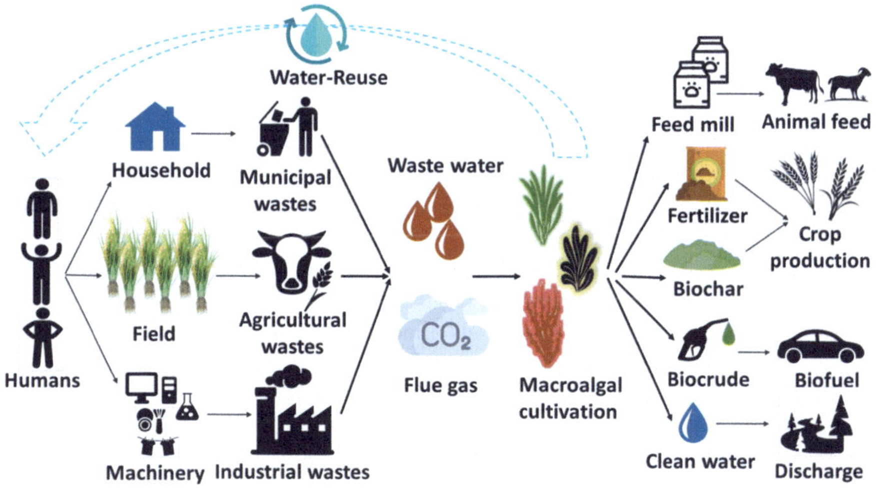

Fig. 32.1 Schematic overview of integration of macroalgae biomass to unveil valorization towards industrial and ecological systems, thus predicting multifaceted applications for sustainable biorefineries

biodiesel by employing various conversion methods (Chen et al. 2015; Demirbas 2010; Mohammad et al. 2019; Hessami et al. 2019).

In a recent report, the US Energy Department reported that seaweed biofuel productivity is two and five times greater than the productivity of ethanol from sugarcane and maize, respectively. (Soliman et al. 2018; Del Río et al. 2020). Presences of higher carbohydrate content and lack of lignin in the cell walls of seaweeds, make them a promising remarkable renewable source of biobutanol and bioethanol production (Dave et al. 2019; Hong et al. 2017). The seaweed industry is a multi-billion dollar industry and perhaps Asia being the major market for the seaweed production with products ranging from fertilizers, human foods, biofuels, phycocolloids and cosmetics (Milledge et al. 2014, Kraan 2013, Smit 2004;). Finally, with these many various factors in combination with the value-added biorenewables concept further demonstrates seaweeds as a potential candidate, which enable the feasibility of a biorefinery-based approach for the valorisation of these macroalgae (Fig. 32.1).

32.2 Genome Annotation

For many species, completely sequenced genomes are available in publicly accessible domains, these genomes offer important insights in particular areas such as in the fields of comparative evolution, biochemistry, developmental biology and physiology. Although, currently a large progression is made in the sequencing of eukaryotic genomes which are remarkably faster and more reliable than before and simultaneously a large number of different eukaryotic organisms are also successfully and completely sequenced which include a variety of model different organisms, algae, land plants and fungi (Pertea et al. 2018; Necsulea 2020). Identification and functionality of gene products are referred as one-dimensional genome annotation whereas two-dimensional genome annotation identifies protein-protein interactions, regulatory interactions, metabolite transformations existing during different cellular mechanisms, giving some chemical and physical interactions that contribute to network reconstruction (Palsson 2004). Finally, such interactions and networks help to provide insights on their genotype-phenotype relationship and more details of the intracellular framework, such as cell packaging and positioning of the genome plays an important role in its function and these come under "three-dimensional annotation" (Woldringh 2002). "Next level dimensional annotation" is the study of changes taking place at the genomic level because of the adaptive evolution (Ibarra et al. 2002).

Research on the model brown algae *Ectocarpus siliculosus*, started during the nineteenth century which includes details of taxonomy and species, accompanied by experiments aimed at unwinding the propagation and life cycle of the species (Peters et al. 2004). A complete reference genome sequence is currently available and the transcribed regions have been demonstrated by the use of Sanger ESTs (Dittami et al. 2009) and RNA-seq datasets (Macaisne et al. 2017; Cormier et al. 2017; Dittami et al. 2020). Collectively, genome and genetic data developed to create a sequence-anchored genetic map, a high-quality assembly, and a Hi-C physical map (Avia et al. 2017; Montecinos et al. 2017). A total of 353 species were updated in the database in the genus *Ectocarpus* (www.algaebase.org, December 2020) and 97 flagged as taxonomically accepted and publicly available on different culture collections (www.algaebase.org, December 2020). Major culture collections include Roscof Culture Collection (RCC), France, the Culture Collection of Algae and Protozoa (CCAP), Scotland, and the Kobe University Macroalgae Culture Collection (KU-MACC), Japan (Coelho et al. 2020). Availability of the *E. siliculosus* genome sequence facilitates the analysis of its metabolic pathways and the identification of Carbohydrate-active enzymes (CAZymes) genes, as these can be used to investigate in several aspects of the different carbohydrate metabolism (Zhang et al. 2018) (Table 32.1).

32.3 Molecular Aspects

Seaweeds are considered an important dietary and feed component throughout the world because of their high protein content and nutritional value (Menon 2008; Haque et al. 2010). Edible marine algae contain large amounts of protein, lipids, minerals, and vitamins (Ravi and Kasi 2014; Hani and Ching 2000), although the nutritional content of seaweeds can vary with species, environmental conditions, seasonal variations, temperature, and humidity (Kaehler and Kennish 1996). Freshwater–grown macroalgae possessed a high amount of protein and trace

Table 32.1 Functional genes of E. siliculosus encoding CAZymes with KO numbers, enzymes names and their molecular function are summarized as potential targets to enhance the biorenewables

Query	KO	Definition	Biological activity
tr|D7FZA8|D7FZA8_ECTSI	K00434	E1.11.1.11; L-ascorbate peroxidase [EC:1.11.1.11]	Peroxidase activity
tr|D7G168|D7G168_ECTSI	K00434	E1.11.1.11; L-ascorbate peroxidase [EC:1.11.1.11]	Peroxidase activity
tr|D8LPB5|D8LPB5_ECTSI	K00698	CHS1; chitin synthase [EC:2.4.1.16]	Chitin synthase activity
tr|D7FQ07|D7FQ07_ECTSI	K00721	DPM1; dolichol-phosphate mannosyltransferase [EC:2.4.1.83]	Protein glycosylation
tr|D7FXV4|D7FXV4_ECTSI	K00726	MGAT1; alpha-1,3-mannosyl-glycoprotein beta-1,2-N-acetylglucosaminyltransferase [EC:2.4.1.101]	Dolichyl-phosphate beta-glucosyltransferase activity
tr|D8LR86|D8LR86_ECTSI	K00729	ALG5; dolichyl-phosphate beta-glucosyltransferase [EC:2.4.1.117]	Dolichyl-phosphate beta-glucosyltransferase activity
tr|D7FTA2|D7FTA2_ECTSI	K01180	E3.2.1.6; endo-1,3(4)-beta-glucanase [EC:3.2.1.6]	Endohydrolysis
tr|D7FGU3|D7FGU3_ECTSI	K01180	E3.2.1.6; endo-1,3(4)-beta-glucanase [EC:3.2.1.6]	Endohydrolysis
tr|D8LU99|D8LU99_ECTSI	K01180	E3.2.1.6; endo-1,3(4)-beta-glucanase [EC:3.2.1.6]	Endohydrolysis
tr|D8LL13|D8LL13_ECTSI	K01180	E3.2.1.6; endo-1,3(4)-beta-glucanase [EC:3.2.1.6]	Endohydrolysis
tr|D8LGS1|D8LGS1_ECTSI	K01188	E3.2.1.21; beta-glucosidase [EC:3.2.1.21]	Beta-glucosidase activity
tr|D7FS01|D7FS01_ECTSI	K01190	lacZ; beta-galactosidase [EC:3.2.1.23]	Beta-glucosidase activity
tr|D7FWP3|D7FWP3_ ECTSI	K01190	lacZ; beta-galactosidase [EC:3.2.1.23]	Beta-glucosidase activity
tr|D7FJT1|D7FJT1_ECTSI	K01191	MAN2C1; alpha-mannosidase [EC:3.2.1.24]	Alpha-mannosidase activity
tr|D7G6T3|D7G6T3_ECTSI	K01194	TREH, treA, treF; alpha,alpha-trehalase [EC:3.2.1.28]	Alpha,alpha-trehalase activity
tr|D7G419|D7G419_ECTSI	K01210	E3.2.1.58; glucan 1,3-beta-glucosidase [EC:3.2.1.58]	Glucan exo-1,3-beta-glucosidase activity
tr|D8LQD7|D8LQD7_ ECTSI	K01227	ENGASE; mannosyl-glycoprotein endo-beta-N-acetylglucosaminidase [EC:3.2.1.96]	Mannosyl-glycoprotein endo-beta-n-acetylglucosaminidase activity
tr|D8LE92|D8LE92_ECTSI	K01228	MOGS; mannosyl-oligosaccharide glucosidase [EC:3.2.1.106]	Mannosyl-oligosaccharide glucosidase activity

(continued)

Table 32.1 (continued)

<table>
<tr><th>Query</th><th>KO</th><th>Definition</th><th>Biological activity</th></tr>
<tr><td>tr|D7FS40|D7FS40_ECTSI</td><td>K01230</td><td>MAN1A_C, MNS1_2; mannosyl-oligosaccharide alpha-1,2-mannosidase [EC:3.2.1.113]</td><td>Mannosyl-oligosaccharide glucosidase activity</td></tr>
<tr><td>tr|D8LGK4|D8LGK4_ECTSI</td><td>K01230</td><td>MAN1A_C, MNS1_2; mannosyl-oligosaccharide alpha-1,2-mannosidase [EC:3.2.1.113]</td><td>Mannosyl-oligosaccharide 1,2-alpha-mannosidase activity</td></tr>
<tr><td>tr|D7FJU5|D7FJU5_ECTSI</td><td>K01230</td><td>MAN1A_C, MNS1_2; mannosyl-oligosaccharide alpha-1,2-mannosidase [EC:3.2.1.113]</td><td>Mannosyl-oligosaccharide 1,2-alpha-mannosidase activity</td></tr>
<tr><td>tr|D7FGZ4|D7FGZ4_ECTSI</td><td>K03715</td><td>MGD; 1,2-diacylglycerol 3-beta-galactosyltransferase [EC:2.4.1.46]</td><td>1,2-diacylglycerol 3-beta-galactosyltransferase activity</td></tr>
<tr><td>tr|D7FH03|D7FH03_ECTSI</td><td>K03715</td><td>MGD; 1,2-diacylglycerol 3-beta-galactosyltransferase [EC:2.4.1.46]</td><td>1,2-diacylglycerol 3-beta-galactosyltransferase activity</td></tr>
<tr><td>tr|D8LHP4|D8LHP4_ECTSI</td><td>K03715</td><td>MGD; 1,2-diacylglycerol 3-beta-galactosyltransferase [EC:2.4.1.46]</td><td>1,2-diacylglycerol 3-beta-galactosyltransferase activity</td></tr>
<tr><td>tr|D8LTD7|D8LTD7_ECTSI</td><td>K03809</td><td>wrbA; NAD(P)H dehydrogenase (quinone) [EC:1.6.5.2]</td><td>NAD(P)H dehydrogenase (quinone) activity</td></tr>
<tr><td>tr|D7FIM2|D7FIM2_ECTSI</td><td>K03842</td><td>ALG1; beta-1,4-mannosyltransferase [EC:2.4.1.142]</td><td>Chitobiosyldiphosphodolichol beta-mannosyltransferase activity</td></tr>
<tr><td>tr|D7G9B9|D7G9B9_ECTSI</td><td>K03843</td><td>ALG2; alpha-1,3/alpha-1,6-mannosyltransferase [EC:2.4.1.132 2.4.1.257]</td><td>GDP-man:man1glcnac2-pp-dol alpha-1,3-mannosyltransferase activity</td></tr>
<tr><td>tr|D8LI23|D8LI23_ECTSI</td><td>K03844</td><td>ALG11; alpha-1,2-mannosyltransferase [EC:2.4.1.131]</td><td>GDP-man:man3glcnac2-pp-dol, alpha-1,2-mannosyltransferase activity,glycosyltransferase</td></tr>
<tr><td>tr|D7FLD1|D7FLD1_ECTSI</td><td>K03845</td><td>ALG3; alpha-1,3-mannosyltransferase [EC:2.4.1.258]</td><td>Mannosyltransferase activity</td></tr>
<tr><td>tr|D7G1E8|D7G1E8_ECTSI</td><td>K03847</td><td>ALG12; alpha-1,6-mannosyltransferase [EC:2.4.1.260]</td><td>Dolichyl-pyrophosphate man7glcnac2 alpha-1,6-mannosyltransferase activity</td></tr>
<tr><td>tr|D7G191|D7G191_ECTSI</td><td>K03848</td><td>ALG6; alpha-1,3-glucosyltransferase [EC:2.4.1.267]</td><td>Dolichyl pyrophosphate man9glcnac2 alpha-1,3-glucosyltransferase activity</td></tr>
<tr><td>tr|D7G161|D7G161_ECTSI</td><td>K03849</td><td>ALG8; alpha-1,3-glucosyltransferase [EC:2.4.1.265]</td><td>Dolichyl pyrophosphate glc1man9glcnac2 alpha-1,3-glucosyltransferase activity</td></tr>
</table>

Query	KO	Definition	Biological activity
tr|D8LPS4|D8LPS4_ECTSI	K03850	ALG10; alpha-1,2-glucosyltransferase [EC:2.4.1.256]	Dolichyl pyrophosphate glc2man9glcnac2 alpha-1,2-glucosyltransferase activity
tr|D8LMM7|D8LMM7_ECTSI	K03857	PIGA, GPI3; phosphatidylinositol N-acetylglucosaminyltransferase subunit A [EC:2.4.1.198]	Transferase activity
tr|D8LC83|D8LC83_ECTSI	K05349	bglX; beta-glucosidase [EC:3.2.1.21]	Beta-glucosidase activity
tr|D7G789|D7G789_ECTSI	K05546	GANAB; mannosyl-oligosaccharide alpha-1,3-glucosidase [EC:3.2.1.207]	Glucan 1,3-alpha-glucosidase activity
tr|D7G3U2|D7G3U2_ECTSI	K07151	STT3; dolichyl-diphosphooligosaccharide---protein glycosyltransferase [EC:2.4.99.18]	Dolichyl-diphosphooligosaccharide-protein glycotransferase activity
tr|D7FQ40|D7FQ40_ECTSI	K09667	OGT; protein O-GlcNAc transferase [EC:2.4.1.255]	Transferase activity
tr|D7FX68|D7FX68_ECTSI	K09667	OGT; protein O-GlcNAc transferase [EC:2.4.1.255]	Protein o-glcnac transferase activity
tr|D8LTW3|D8LTW3_ECTSI	K10085	EDEM2; ER degradation enhancer, mannosidase alpha-like 2	Mannosyl-oligosaccharide 1,2-alpha-mannosidase activity
tr|D7FXN7|D7FXN7_ECTSI	K11000	CALS; callose synthase [EC:2.4.1.-]	1,3-beta-d-glucan synthase activity
tr|D7FY26|D7FY26_ECTSI	K11000	CALS; callose synthase [EC:2.4.1.-]	1,3-beta-d-glucan synthase activity
tr|D7FPK0|D7FPK0_ECTSI	K11000	CALS; callose synthase [EC:2.4.1.-]	1,3-beta-d-glucan synthase activity
tr|D7G654|D7G654_ECTSI	K11718	HUGT; UDP-glucose:glycoprotein glucosyltransferase [EC:2.4.1.-]	UDP-glucose:glycoprotein glucosyltransferase activity
tr|D7FL73|D7FL73_ECTSI	K16055	TPS; trehalose 6-phosphate synthase/phosphatase [EC:2.4.1.15 3.1.3.12]	Alpha,alpha-trehalose-phosphate synthase (udp-forming) activity
tr|D8LRB5|D8LRB5_ECTSI	K16055	TPS; trehalose 6-phosphate synthase/phosphatase [EC:2.4.1.15 3.1.3.12]	Alpha,alpha-trehalose-phosphate synthase (udp-forming) activity
tr|D7FT10|D7FT10_ECTSI	K16055	TPS; trehalose 6-phosphate synthase/phosphatase [EC:2.4.1.15 3.1.3.12]	Alpha,alpha-trehalose-phosphate synthase (udp-forming) activity

(continued)

Table 32.1 (continued)

<table>
<tr><th>Query</th><th>KO</th><th>Definition</th><th>Biological activity</th></tr>
<tr><td>tr|D7G011|D7G011_ECTSI</td><td>K16055</td><td>TPS; trehalose 6-phosphate synthase/phosphatase [EC:2.4.1.15 3.1.3.12]</td><td>Alpha,alpha-trehalose-phosphate synthase (udp-forming) activity</td></tr>
<tr><td>tr|D8LMG5|D8LMG5_ECTSI</td><td>K16055</td><td>TPS; trehalose 6-phosphate synthase/phosphatase [EC:2.4.1.15 3.1.3.12]</td><td>Alpha,alpha-trehalose-phosphate synthase (udp-forming) activity</td></tr>
<tr><td>tr|D7FL31|D7FL31_ECTSI</td><td>K17525</td><td>CHID1; chitinase domain-containing protein 1</td><td>Chitin binding activity</td></tr>
<tr><td>tr|D8LGW4|D8LGW4_ECTSI</td><td>K23741</td><td>MAN1B, MNS3; endoplasmic reticulum Man9GlcNAc2 1,2-alpha-mannosidase [EC:3.2.1.209]</td><td>Mannosyl-oligosaccharide 1,2-alpha-mannosidase activity</td></tr>
</table>

minerals when grown in aquaculture or agriculture wastewater (rich in N and P) and turn into an appropriate animal feed. Macroalgae cultivated in dairy manure and aquaculture reported 31–44% and 21–28% rich in protein content (Wilkie and Mulbry 2002, Lawton et al. 2013).

Seaweeds are multicellular plants that are grouped as Chlorophyta, Rhodophyta, and Phaeophyta based on their pigmentation and possess high polysaccharide chains in their cell wall (Manilal 2012) and basically consist of neutral polysaccharides, cellulose, and hemicelluloses to provide physical protection. According to the algal physiology and taxonomic classification, the cell wall structure and storage polysaccharides are unique. Macroalgal species from Chlorophyta contain sulfuric acid polysaccharides, sulfated galactans, and xylans, while Phaeophyta consist mainly of algin, fucoidans (sulfated fucose), sargassum, and laminarin (β-1,3 glucan) and the last Rhodophyta have mainly Floridian, starch, carrageenans, agars, xylans, water-soluble sulfated galactan, mucopolysaccharides. Some biochemical contents show variation according to the seasons such as mannitol, laminarin, alginic acid (Schiener et al. 2015). Major biochemical components proteins, carbohydrates, lipids, and ash are summarized in Table 32.2.

Table 32.2: Biochemical constituents of different macroalgal genus representing protein, carbohydrate, lipid and ash content.

Constituents	Macroalgae species Chlorophyta	Rhodophyta	Phaeophyta	References
Protein	12–13%	10–16%	7–12%	Jard et al. (2013); Cole et al. (2015); Neveux et al. (2014); Kan et al. (2014)
Lipid	2–3%	0–3%	0–2%	Jard et al. (2013); Neveux et al. (2014); Neveux et al. (2015)
Carbohydrates	25–50%	30–60%	30–50%	Jung et al. (2013), Neveux et al. (2014); Neveux et al. (2015)
Polysaccharides	Ulvan, Alginate, Starch, Cellulose, Mannan	Agar, Alginate, Lignin, Cellulose, Carrageenan	Agar, Alginate, Cellulose, Laminarin, Carrageenan, Mannitol, Fucoidan	Jung et al. (2013); Roesijadi et al. (2010); Jang et al. (2012)
Ash	18–53%	26–48%	33–55%	Jard et al. (2013)

32.4 Genetic Engineering in Macroalgae

Limited genomes information is available for macroalgae, like *Ectocarpus siliculosus* (214 megabase pairs), *Chondrus crispus* (105 megabase pairs), *Pyropia yezoensis*, which are commercially used macroalga, has a genome of 43 megabase pairs, and *Saccharina japonica*, with a 580–720 megabase pairs of genome size (Cock et al. 2010; Collén et al. 2013; Le Gall et al. 1993; Nakamura et al. 2013). Due to the diversity and complexity in the physiological and genetic information existing among macroalgae, there are only a few reported stable and transient genetic transformations achieved in few strains of green, brown, and red macroalgae (Mikami 2013; Qin et al. 2012). Hirata et al. (2011) reported the successful expression of a foreign gene in *P. yezoensis*, moreover, a codon-optimized gene of β-glucuronidase (Gust and Moore 1989) was successfully expressed under the control of an actin promoter in the gametophytic cells. A complete actin1 gene from *Ulva prolifera* recently cloned thus establishing expression machinery in macroalgae. Moreover, genome walking was performed to obtain 5 'flanking sequence from *U. prolifera* (Wu et al. 2018). Moreover, reports suggest the use of synthetic biology for the conservation of the seaweed forests by the manipulation of their genomes using CRISPR-based technologies (Coleman and Goold 2019). Additionally, genetic engineering of *S. japonica* was carried out successfully by incorporating a vaccine gene into the genome of these macroalga (Robinson et al. 2013). Moreover, there is an emphasis on genetically improved-based techniques to enhance the quality of the macroalga-derived nutrients or nutraceuticals (Reddy et al. 2008; Robinson et al. 2013).

32.5 Industrial Applications of Macroalgae

32.5.1 Human Food

Among the Asian countries, especially Japan and China, are the largest consumers of seaweed used for human consumption, such as *Laminaria* sp., *Sargassum fusiforme, Undaria pinnatifida*, etc. (Kılınç et al. 2013). Due to its low calorific and high nutritional content like a high amount of proteins, long-chain polysaccharides, pigments like carotenoids, vitamins, minerals such as micronutrient constituents ie., 5–17 mg/100 g; K, Ca, Na, Mg, Mn, Zn, Fe, Cu (Anyanwu et al. 2018) and certain soluble and insoluble dietary fibers (Yuan et al. 2009), moreover, 600 species of seaweeds are acclaimed edible and used as supplements (Shannon and Abu-Ghannam 2016; Kılınç et al. 2013; Pereira 2016). Also, macroalgae can be used like other vegetables and eaten in different forms, for example-freshly harvested, after dried, and flour or powder form (Kılınç et al. 2013), used in amalgamation with different other food products are been commercialized by multiple food companies. There are healthier and natural constituents of multiple foods, for example-*Himanthalia elongata* as spaghetti, and *Palmaria palmate* as sea bacon, *Porphyra* sp. (nori) in rolls, or some other snacks, wraps of *U. pinnatifida*, etc. Carrageenan is a sulfated gel-forming polysaccharide, extracted from order Gigartinales (includes *Chondrus crispus* the major carrageenan producer) and mostly used as an additive by various food-based industries like dairy e.g. ice creams, jellies, etc. and as a thickening and stabilizing agent in meat products e.g. hams (Vidotti and Maria do Carmo 2004). Products such as Agar, produced by various species of Rhodophyceae such as *Gelidium* sp., *Gracilaria* sp. and *Pterocladiella* sp. etc., is also another widely used phycocolloid, composed of a primary component called agarose and agaropectin. Depending on the level of purity it is employed accordingly, highly pure grade used for gel electrophoresis and chromatography, medium quality agar generally for making medium substrate and lower grade agar is used by food industries in ice-creams, bakery etc. (Table 32.3).

It acts as a thickener in various food products and as a vegetarian replacement to gelatin (Kılınç et al. 2013; Pereira 2016). Contemporarily, it acts as a sustainable biomaterial used to make wrappers instead of plastics (Leandro et al. 2020a; b). Alginate is the world's most abundant marine biopolymer and serves as a gelling agent, acquired and extracted from brown seaweed such as *Ascophyllum* sp., *Laminaria* sp., etc. (Levine 2016), and acts as an additive in many packaged foods, including dairy products, sauces, etc. (Brownlee et al. 2009; Garcia-Ceja et al. 2015).

Table 32.3 Illustration of seaweed applications for different industrial uses and food products

	Macroalgae species			
Constituents	Chlorophyta	Rhodophyta	Phaeophyta	References
Industrial use	Human food, supplements, medicinal use	Human and animal food, as an agricultural, laboratory and cosmetic thickener, emulsifier and gelling agent	Human food, animal feed, textiles, paper fibre, cosmetics, pharmaceutical industry, fermentative production of organic acids	Jung et al. (2013); Jard et al. (2013); Cardoso et al. (2014)
Industrial extracts	Sulfated galactan, vitamin C, antiviral and anticoagulating agents	Sulfated galactan, vitamins, mineral nutrients, agar, phycobiliproteins	Fucoidan, fucan hydrocolloids, polyphenols, mineral nutrients, pigments	
Lipids and lipophilic molecules	Oleic, linoleic, and linolenic acid, Vitamins D2, K1 and E, Antioxidant, anti-inflammatory, skin elasticity, collagen synthesis, anti-wrinkle, emollient, moisturizing	Omega-3 and 6 (Emollient, moisturizing, sheaths, anti-inflammatory) -Linolenic acid Palmitic, linolenic, oleic acids, Vitamins A, D2, K1 and E	Fucoxanthin (Anti-aging, anti-wrinkle and smoothing agent) Fucosterol, polyunsaturated fatty acids (Antioxidant and anticancer activities, UV protection, anti-aging agent, moisturizing properties)	Couteau and Coiffard (2016); Kim et al. (2013); Ravi and Kasi (2014); Debbarma et al. (2016); Joshi et al. (2018)

32.5.2 Livestock and Agriculture

The utilization of brown seaweeds or their extracts for livestock is an advantageous way of contributing to global health. Laminarin is an essential bioactive compound, a source of antioxidants, and a rich source of β-glucan, a natural compound known for its anti-inflammatory properties, boosting immunity hence widely used as a functional food in the diet (Glynn and Martin 2013). Supplementation of pig diets with *Laminaria* sp. or *U. pinnatifida* powder, lead to some bioactivities such as the positive impact on productivity, inhibition towards *Escherichia coli,* some immunomodulatory effects observed, these measures resemble a positive step towards minimizing the widespread use of antibiotics (Kristinsson and Jónsdóttir 2015; Hemmingson et al. 2006). Seaweed meals are produced predominantly from the kelps *Ascophyllum nodosum*, *Fucus* sp., *Laminaria* sp., *Macrocystis* sp., used mostly as the vitamin and mineral supplements (Craigie 2011). Reports on algal extracts supplemented with animal feed can enhance intestinal integrity, the effective immune response of animals, and decrease the release of enteric methane from the fermentation of rumen (Li et al. 2018) and thus control biotic methane (Pereira et al. 2019).

Besides, seaweeds and products derived from seaweeds have been widely used in agriculture to improve the quality and production of crops because of the availability of a number of plant growth-stimulating products (Nabti et al. 2017; Khan et al. 2009; Tuhy et al. 2013). Seaweeds were used to fertilize the fields to improve plant fertility and food production in different regions of the world during ancient times (Craigie 2011). Seaweeds or their compounds affect plant process such as better seed germination, root and shoot growth, promote freezing tolerance, biotic stress resistance, increase the absorption capacity of plant nutrients (Akila and Jeyadoss 2010). Hormones such as auxins, which are essential for vegetative growth are present in seaweed (Crouch and Van Staden 1993). Besides, alginate oligomers have growth-promoting effects on certain higher plant species (Khan et al. 2009). Addition of depleted fucoidan, alginate, etc. to soils enhances the structure and aeration of scraps while promoting microorganisms in root systems to enhance plant development (Zodape et al. 2008). Currently, commercial extracts are derived from brown algae *A. nodosum*, *Ecklonia maxima*, green algae *U. lactuca*, *Codium* sp., and red algae *Gelidium* sp., and *C. crispus* (Craigie 2011). Seaweed biomass cultured in animal-husbandry waste is also used as feedstock for the development of nutrient-rich biochar that enriches the soil quality and can serve as effective slow-release fertilizers. (Mulbry et al. 2007). Another product is Biochar (produced by pyrolysis) which helps to improve the retention of nutrients (N, P, Ca, Mg, K, and Mo) in low-quality soil and acts as the carbon source (Roberts et al. 2015), used in the treatment of wastewater (Lehmann and Joseph 2009).

32.5.3 Pharmaceutics

Seaweeds have adapted to survive under multiple environmental pressures and challenges simultaneously confronting consistently high concentrations of pathogenic and contagious bacteria prevalent in ocean waters (Shannon and Abu-Ghannam 2016) and emerged as a defense strategies with the production of certain bioactive compounds (Table 32.4).

Seaweeds resist bacterial invasion by producing specific substances such as phlorotannins, polysaccharides and peptides (Shannon and Abu-Ghannam 2016), and consequently explored for other possible pharmacological consequences like, antiviral, antitumoral, immunogenic effects. Kahalalide F and its isomer, iso-kahalalide F, are peptides isolated from green macroalga *Bryopsis pennata* exhibits cytotoxic effects and used in clinical anticancer research. Extracts from *Cladophora* sp., *Monostroma* sp., *F. vesiculosus*, *A. nodosum* patented for the treatment of type 2 diabetes and related health problems (Daniels 2004). The methanolic extracts of some specific algae, *Corallina offinalis*, *Cystoseira barbata*, *Dictyota dichotoma*, and *U. rigida* are capable of preventing the growth of some pathogenic bacteria such as *Staphylococcus aureus*, *Enterococcus faecalis* (Taskin et al. 2007). Sulfated polysaccharides seaweed-derived compounds such as carrageenan, alginates, etc. gained interest from pharmaceutical firms, because of their antibacterial, antiviral, antitumoral, and immunomodulatory activities. Brown algae derived polymers are extensively used as bio-adhesives in pharmaceutical formulations, as a traditional excipient, and as tools in polymer-controlled drug delivery (Shannon and Abu-Ghannam 2016; Tønnesen and Karlsen 2002; Guo et al. 1998). Among polysaccharides, fucoidans have been demonstrated for their promising biological properties such as anti-thrombotic, anti-coagulant, anti-cancer, anti-proliferative, and anti-inflammatory properties (Chizhov et al. 1999; Church et al. 1989; Kim et al. 2010; Nakazato et al. 2010). Similarly, MAAs have skin protection and wound healing properties e.g. Porphyra-334 can suppress ROS (reactive oxygen species) produced in human skin fibroblast cells (Choi et al. 2015). The reddish-brown carotenoid mainly found in diatoms, fucoxanthin, acquired from *S. japonica*, can suppress tyrosinase activity in UVB-irradiated guinea pig and melanogenesis in UVB-irradiated mice (Thomas and Kim 2013). Phlorotannin extracts from *A. nodosum* and *E. cava* are antioxidants with anti-diabetic properties (Daniels 2004; Yanagibayashi et al. 2012; Zhang et al. 2008; Lee et al. 2002). Both topical treatment and dietary feeding of brown algal polyphenols suppressed cyclooxygenase-2 (COX-2) expression and cell proliferation in SKH-1 hairless mouse skin model (Hwang et al. 2006; Ragan and Glombitza 1986; Kang et al. 2003). This suggests the chemo-preventive and anti-carcinogenic role of brown algal polyphenols, phlorotannins against adverse effects of UVB exposure, revealing the use of these compounds as active ingredients in drug formulations (Hwang et al. 2006). The *in vitro* studies with methanolic extract from marine brown alga, *Eisenia arborea*

Table 32.4 Pharmaceutical applications of different macroalgal products and their properties.

Compounds	Properties	Applications	References
Mycosporine-like amino acids (MAAs)	Low molecular weight, hydro-soluble and stable molecules	Sunscreen ingredient	Pereira (2018); Alparslan et al. (2018)
Arginine	Precursor of urea, Stable at pH ranges of 5–9, Antioxidant, antidiabetic	Cosmetic formulations	Guillerme et al. (2017)
Agar	Gelling	vegetal jelly	Kılınç et al. (2013); Pereira (2016)
Alginate	Emulsifying, Gelling, Stabilizer	yogurts, ice creams	Pereira et al. (2013)
Carrageenan: [a] Iota [b] Kappa [c] Lambda	[a] and [b]-gelling [c]-thickening/viscosifier	yoghurts, flans, jellies, ice creams, meat products (ham)	Pereira (2016); Pereira and van de Velde (2011); Armisen (1995)
Zeaxanthin	Anti-tyrosinase activity	Prevention of skin spots formation, Whitening agent.	Levine (2016); Ariede et al. (2017)
Phycobiliproteins	Antioxidant, antidiabetic, anti-hypertensive, immune-modulator, anti-inflammatory, anti-aging.	Reduce synthetic colorants. Dyes in cosmetics.	Sonani et al. (2016); Dumay et al. (2014); Arad and Yaron (1992)
Fucoxanthin	Reduction of: ROS, MMPs expression, (MMP13, related to tumor) and tyrosinase activity	Oral or cutaneous administration, Inhibition of melanogenesis	Levine (2016); Wang et al. (2017)
Pheophytin a and pheophorbide a	Antioxidant bioactivity	Cosmetic industry as antioxidants	Chen et al. (2017)
Chromene meroterpenoid and tetraprenyltoluquinol	Reduction of ROS by 21%, Anti-photo-oxidative stress activity	Sunscreen ingredient	Wang et al. (2017)
Spatane diterpenoids	Anti-tumor capacity	Creams, sunscreens and other products	Wang et al. (2017)
Halogenated compounds derived from Iodine	Promoting lipolysis (thyroid metabolism)	Cosmetic industry	Levine (2016)
Jacinto et al. (2013)	Antimicrobial properties	Preservatives	
Eckstolonol	Reduction of ROS. Increment in enzymatic activity, Decrease of protein levels, pro-apoptotic factors caspases 3 and 9, Antioxidant and photo-protective properties	Anticancer therapies, Sunscreens	Balboa et al. (2014); Wang et al. (2017)

(continued)

Table 32.4 (continued)

Compounds	Properties	Applications	References
Phloroeckol and phloroglucinol	Anti-diabetic and antioxidant properties	Anti-skin aging products	Ariede et al. (2017)
Phenolic compounds and other phlorotannins	Inhibiting the MMP overexpression, Photoprotective properties	Prevention of collagen premature degradation, wrinkle formation and the appearance of cancer cells	Anyanwu et al. (2018)
Vitamins A and E	Protective effects against DNA damage. Ability to favour cell renewal.	Sunscreens	Anyanwu et al. (2018)

demonstrated anti-histamine activity from anti-dinitrophenyl (DNP) IgE sensitized and DNP-BSA stimulated rat basophile leukemia cells (RBL-2H3) (Sugiura et al. 2006; Ferreres et al. 2012).

32.5.4 *Cosmetics*

In recent times there is an increasing appetite for natural cosmetics, as they lack many dangerous chemicals found in traditional cosmetic products. Simultaneously with increased demands, the cosmetic industry started using macroalgae because they generate metabolites such as sporopollenin, scytonemin, and mycosporin-like amino acids (Sellimi et al. 2018) that shield skin from UV radiation (Hwang et al. 2006). Owing to its bioavailable nature, seaweed cosmetic products are readily absorbed by the skin, reduced redness effect, discoloration, rejuvenates, moisturizes, re-mineralize, and reduces the appearance of sun damage (Thomas and Kim 2013; Pereira 2018). Seaweeds can be utilized in cosmetics in several ways for the preparation of various substances like in the form of a vehicle, stabilizing agent, emulsifier, or as an active medicinal ingredient in the product. Skincare products such as exfoliating lotions, face masks, etc. use extracts of crushed and grounded bits of dried seaweed (Rupérez 2002; Fabrowska et al. 2015; Kılınç et al. 2013; Sellimi et al. 2018). Alginate or carrageenans act as water-binding agents, which are capable of moisturizing the skin and hair (Buck et al. 2006). Along with bioactive compounds, macroalgae are abundant with saturated and unsaturated fatty acids and are used as emulsifiers in cosmetics e.g. palmitic acid derivative ascorbyl-palmitate that acts as an antioxidant and is effective against persistent complications such as wrinkles and aging (Fabrowska et al. 2015; Yarnpakdee et al. 2018). High purity phlorotannins derived from brown seaweeds are in an anti-aging role in skincare products, due to their reduced free-radicals and hyaluronic acid concentrations (Ferreres et al. 2012).

32.5.5 Biosorbents

Besides chemical precipitation, membrane filtration, etc., sorption is currently employed for removing heavy metals from wastewater, mostly when their concentrations ranges between 1 to 100 mg L^{-1} (Davis et al. 2003). Biochar is used as fertilizer in agriculture capable of effectively binding to metals (Kidgell et al. 2014, Bird et al. 2011, Zhou et al. 2018, De Bhowmick et al. 2018, Cole et al. 2017, De Ramon N'Yeurt and Iese 2015). Biomass treatment with iron (Fe) will convert biochar to biosorbent efficiently binds to metalloids (Kidgell et al. 2014, Johansson et al. 2015) and widely used in the sequential treatment of industrial wastewater by toxic-metal/metalloid removal (Kidgell et al. 2014). Combustion of this metal-loaded biosorbent after sorption called phyto-mining, allows the sale of the retrieved metals (Jiang et al. 2015). Algal biochar effectively eradicates metal ions from industrial effluent of the coal-fired power stations (Kidgell et al. 2014) and some specific metals such as arsenic, molybdenum, selenium, which cannot be removed through conventional techniques, can easily be removed by Fe treated biochar of *Gracilaria* and *Oedogoium* (Johansson et al. 2016).

32.5.6 Bioenergy

Recent decades, algae-produced bioenergy gained broad recognition as a renewable energy alternative because of its various advantages and environmental implementations (Wijffels et al. 2010). Freshwater macroalgae cultured in wastewater exhibit significant potential as a source of biofuels and bioenergy (Yun et al. 2014). Higher heating value (HHV) is referred to as a measure of reserved energy of these freshwater macroalgae, is quite good and comparable to alternative bioenergy sources i.e., terrestrial crop. HHV of the industrial biomass *Miscanthus* sp. is 18.5 MJ kg^{-1} greater than terrestrial crop wood which is 12 MJ kg^{-1} and crop residues varying between 11–28 MJ kg^{-1} (Samolada et al. 2000, Jun et al. 2010, Wang et al. 2010). HHV of freshwater macroalgae biomass range 12–22 MJ kg^{-1} could be increased through hydrothermal liquefaction (HTL) up to 33.7 and 33.5 MJ kg^{-1} for *Oedogonium* sp. and *Cladophora* sp. respectively (Lawton et al. 2013; Yun et al. 2014; Cole et al. 2014). To escalate and increase the energy yield, some other techniques such as combustion, gasification, pyrolysis could be employed for the conversion of biomass into biogas and bio-oil (Cole et al. 2014; Saidur, et al. 2011; Li et al. 2013; Suib 2013; Demirbaş 2001; Roberts et al. 2015).

32.6 Future Perspectives and Conclusions

Nowadays, various difficulties for converting the carbohydrates from macroalgae into bioethanol have changed drastically and overcome due to the advances in the biotechnological processes (Dave et al. 2019; Kawai and Murata 2016). Moreover, because of their remarkable properties and added health benefits to humans, these seaweeds have gained tremendous popularity (Kawai and Murata 2016; Leandro et al. 2020a; b; Reddy et al. 2008). Currently, there is no paucity in the research areas based on macroalgae and the number keeps on increasing. To answer the need for food and its alternative supplements with the growing population new policies that will further help to advance the use of seaweeds for bioproducts and human consumption need to be developed, promoted, and sustained (Leandro et al. 2020a; b). Macroalgae also offer a platform for the production of biofuels, biomass, and bio-renewables with a higher growth rate, no requirement of agricultural land and carbon-neutral processes along with applications in the nutraceutical and pharmaceutical industries (Biris-Dorhoi et al. 2020; García-Poza et al. 2020). These advantages resemble the potential of macroalgae in terms of broad range applications as a prospective candidate for adopting the biorefinery approach (Rajak et al. 2020; Reddy et al. 2008). These ideas are countered by a few problems and bottlenecks, however, with the significant benefits of macroalgae, these cell factories offer an alternative source over traditional crops and with the developing biotechnological advances it will be possible to realize the unseen potential of macroalgae.

Acknowledgments The work was supported by grants from the Department of Biotechnology, Government of India, to PPJ (Sanction No. BT/PB/ Center/03/2011-Phase II). Senior Research Fellowships to MSK and AM from the Department of Biotechnology (DBT) and Council of Scientific and Industrial Research (CSIR), Government of India, India are duly acknowledged.

Conflict of Interest The authors declare that there is no conflict of interest.

References

Akila N, Jeyadoss T (2010) The potential of seaweed liquid fertilizer on the growth and antioxidant enhancement of *Helianthus annuus* L. Orient J Chem 26:1353–1360

Alparslan L, Şekeroglu N, Kijjoa A (2018) The potential of marine resources in cosmetics. Curr Perspect Med Aromatic Plants 1:53–66

Anyanwu R, Rodriguez C, Durrant A, Olabi A-G (2018) Micro-macroalgae properties and applications. In: Reference module in materials science and materials engineering. Elsevier BV

Arad SM, Yaron A (1992) Natural pigments from red microalgae for use in foods and cosmetics. Trends Food Sci Technol 3:92–97

Ariede MB, Candido TM, Jacome ALM, Velasco MVR, de Carvalho JCM, Baby AR (2017) Cosmetic attributes of algae-a review. Algal Res 25:83–487

Armisen R (1995) World-wide use and importance of *Gracilaria*. J Appl Phycol 7:231–243

Avia K, Coelho SM, Montecinos GJ, Cormier A, Lerck F, Mauger S, Faugeron S, Myriam Valero J, Cock M, Boudry P (2017) High-density genetic map and identification of QTLs for responses to temperature and salinity stresses in the model brown alga *Ectocarpus*. Sci Rep 7:1–15

Balboa EM, Soto ML, Nogueira DR, González-López N, Conde E, Moure A, Vinardell MP, Mitjans M, Domínguez H (2014) Potential of antioxidant extracts produced by aqueous processing of renewable resources for the formulation of cosmetics. Ind Crop Prod 58:104–110

Bird MI, Wurster CM, de Paula Silva PH, Bass AM, De Nys R (2011) Algal biochar–production and properties. Bioresour Technol 102:1886–1891

Biris-Dorhoi E-S, Michiu D, Pop CR, Rotar AM, Tofana M, Pop OL, Socaci SA, Farcas AC (2020) Macroalgae—A sustainable source of chemical compounds with biological activities. Nutrients 12:3085–3108

Brownlee IA, Seal CJ, Wilcox M, Dettmar PW, Pearson JP (2009) Applications of alginates in food. In: Alginates: biology and applications. Springer

Buck CB, Thompson CD, Roberts JN, Müller M, Lowy DR, Schiller JT (2006) Carrageenan is a potent inhibitor of papillomavirus infection. PLoS Pathog 2(7):e69

Cardoso SM, Carvalho LG, Silva PJ, Rodrigues MS, Pereira OR, Pereira L (2014) Bioproducts from seaweeds: A review with special focus on the *Iberian Peninsula*. Curr Org Chem 18:896–917

Chen H, Zhou D, Luo G, Zhang S, Chen J (2015) Macroalgae for biofuels production: Progress and perspectives. Renew Sust Energ Rev 47:427–437

Chen K, Ríos JJ, Pérez-Gálvez A, Roca M (2017) Comprehensive chlorophyll composition in the main edible seaweeds. Food Chem 228:625–633

Chizhov AO, Dell A, Morris HR, Haslam SM, McDowell RA, Shashkov AS, Nikolay E, Nifant'ev, Elena A Khatuntseva, and Anatolii I Usov. (1999) A study of fucoidan from the brown seaweed *Chorda filum*. Carbohydr Res 320(1–2):108–119

Choi Y-H, Yang DJ, Kulkarni A, Moh SH, Kim KW (2015) Mycosporine-like amino acids promote wound healing through focal adhesion kinase (FAK) and mitogen-activated protein kinases (MAP kinases) signaling pathways in keratinocytes. Mar Drugs 13(12):7055–7066

Church FC, Meade JB, Treanor RE, Whinna HC (1989) Antithrombin activity of fucoidan. The interaction of fucoidan with heparin cofactor II, antithrombin III, and thrombin. J Biol Chem 264(6):3618–3623

Cock JM, Sterck L, Rouzé P, Scornet D, Allen AE, Amoutzias G, Anthouard V, Artiguenave F, Aury J-M, Badger JH (2010) The Ectocarpus genome and the independent evolution of multicellularity in brown algae. Nature 465:617–621

Coelho S, Peters A, Müller D, Cock J (2020) *Ectocarpus*: an evo-devo model for the brown algae. EvoDevo 11:1–9

Cole AJ, de Nys R, Paul NA (2014) Removing constraints on the biomass production of freshwater macroalgae by manipulating water exchange to manage nutrient flux. PLoS One 9:1–9

Cole AJ, De Nys R, Paul NA (2015) Biorecovery of nutrient waste as protein in freshwater macroalgae. Algal Res 7:58–65

Cole AJ, Paul NA, De Nys R, Roberts DA (2017) Good for sewage treatment and good for agriculture: Algal based compost and biochar. J Environ Manage 200:105–113

Coleman MA, Goold HD (2019) Harnessing synthetic biology for kelp forest conservation. J Phycol 55:745–751

Collén J, Porcel B, Carré W, Ball SG, Chaparro C, Tonon T, Barbeyron T, Michel G, Noel B, Valentin K (2013) Genome structure and metabolic features in the red seaweed Chondrus crispus shed light on .3the evolution of the Archaeplastida. Proc Natl Acad Sci 110:5247–5252

Cormier A, Avia K, Sterck L, Derrien T, Wucher V, Andres G, Monsoor M, Godfroy O, Lipinska A, Perrineau M-M (2017) Re-annotation, improved large-scale assembly and establishment of a catalogue of noncoding loci for the genome of the model brown alga *Ectocarpus*. New Phytol 214:219–232

Couteau C, Coiffard L (2016) Seaweed application in cosmetics. In: Fleurence J, Levine I (eds) Seaweed in health and disease prevention. Academic, San Diego

Craigie JS (2011) Seaweed extract stimuli in plant science and agriculture. J Appl Phycol 23:371–393

Crouch IJ, Van Staden J (1993) Evidence for the presence of plant growth regulators in commercial seaweed products. Plant Growth Regul 13:21–29

Daniels, Bruce. 2004. Seaweed extract composition for treatment of diabetes and diabetic complications. EP20040752007

Dave N, Selvaraj R, Varadavenkatesan T, Vinayagam R (2019) A critical review on production of bioethanol from macroalgal biomass. Algal Res 42:101606

Davis TA, Volesky B, Mucci A (2003) A review of the biochemistry of heavy metal biosorption by brown algae. Water Res 37:4311–4330

Dawczynski C, Schubert R, Jahreis G (2007) Amino acids, fatty acids, and dietary fibre in edible seaweed products. Food Chem 103:891–899

Debbarma J, Rao M, Murthy LN, Mathew S, Gudipati V, Ravishankar C (2016) Nutritional profiling of the edible seaweeds *Gracilaria edulis, Ulva lactuca and Sargassum* sp. Indian J Fish 63:81–87

De Bhowmick G, Sarmah AK, Sen R (2018) Lignocellulosic biorefinery as a model for sustainable development of biofuels and value-added products. Bioresour Technol 247:1144–1154

Del Río PG, Gomes-Dias JS, Rocha CMR, Romaní A, Garrote G, Domingues L (2020) Recent trends on seaweed fractionation for liquid biofuels production. Bioresour Technol 299:122613

De Ramon N'Yeurt A, Iese V (2015) The proliferating brown alga Sargassum polycystum in Tuvalu, South Pacific: assessment of the bloom and applications to local agriculture and sustainable energy. J Appl Phycol 27:2037–2045

Demirbaş A (2001) Relationships between lignin contents and heating values of biomass. Energy Convers Manag 42:183–188

Demirbas A (2010) Use of algae as biofuel sources. Energy Convers Manag 51:2738–2749

Dittami SM, Corre E, Brillet-Guéguen L, Lipinska AP, Pontoizeau N, Aite M, Avia K, Caron C, Cho CH, Collen J (2020) The genome of *Ectocarpus subulatus*–A highly stress-tolerant brown alga. Mar Genomics 52:100740

Dittami SM, Scornet D, Petit J-L, Ségurens B, Da Silva C, Corre E, Dondrup M, Glatting K-H, König R, Sterck L (2009) Global expression analysis of the brown alga *Ectocarpus siliculosus* (Phaeophyceae) reveals large-scale reprogramming of the transcriptome in response to abiotic stress. Genome Biol 10:1–20

Dumay J, Morancais M, Munier M, Le Guillard C, Fleurence J (2014) Phycoerythrins: valuable proteinic pigments in red seaweeds. Adv Bot Res 71:321–343

Fabrowska J, Leska B, Schroeder G (2015) Freshwater *Cladophora glomerata* as a new potential cosmetic raw material. Chemik 69:495–497

Ferreres F, Lopes G, Gil-Izquierdo A, Andrade PB, Sousa C, Mouga T, Valentão P (2012) Phlorotannin extracts from fucales characterized by HPLC-DAD-ESI-MSn: approaches to hyaluronidase inhibitory capacity and antioxidant properties. Mar Drugs 10:2766–2781

Forster, John, and Ricardo Radulovich. 2015. Seaweed and food security. Seaweed Sustainability:289-313

Garcia-Ceja A, Mani-López E, Palou E, López-Malo A (2015) Viability during refrigerated storage in selected food products and during simulated gastrointestinal conditions of individual and combined *lactobacilli* encapsulated in alginate or alginate-chitosan. LWT-Food Sci Technol 63:482–489

García-Poza S, Leandro A, Cotas C, Cotas J, Marques JC, Pereira L, Gonçalves AMM (2020) The evolution road of seaweed aquaculture: cultivation technologies and the industry 4.0. Int J Environ Res Public Health 17:6528

Glynn, Kenneth P, and Alain Martin. 2013. Biodegradable wound care products with biocompatible artificial skin treatment and healing accelerator. USA: US20130018334A1

Guillerme J-B, Couteau C, Coiffard L (2017) Applications for marine resources in cosmetics. Cosmetics 4:35

Guo J-H, Skinner GW, Harcum WW, Barnum PE (1998) Pharmaceutical applications of naturally occurring water-soluble polymers. Pharm Sci Technol Today 1:254–261

Gust D, Moore TA (1989) Mimicking photosynthesis. Science 244(4900):35

Hani N, Ching C (2000) Nutritional composition of edible seaweed Gracilaria changgi. Food Chem 68:69–76

Haque KM, Chy S, Akter S, Nath KK (2010) Collection, identification and biochemical analyses of different sea weeds from Saint Martin's island. Bangladesh J Agric Res 34:59–65

Hebbale D, Subash Chandran MD, Joshi NV, Ramachandra TV (2017) Energy and food security from macroalgae. J Biodiversity 8:1–11

Hemmingson JAH, Falshaw R, Furneaux RH, Thompson K (2006) Structure and antiviral activity of the galactofucan sulfates extracted from *Undaria pinnatifida* (Phaeophyta). J Appl Phycol 18:185

Henriques B, Lopes CB, Figueira P, Rocha LS, Duarte AC, Vale C, Pardal MA, Pereira E (2017) Bioaccumulation of hg, cd and Pb by Fucus vesiculosus in single and multi-metal contamination scenarios and its effect on growth rate. Chemosphere 171:208–222

Hessami MJ, Cheng SF, Ranga Rao A, Yin YH, Phang SM (2019) Bioethanol production from agarophyte red seaweed, *Gelidium elegans* using a novel sample preparation method for analysing bioethanol content by gas chromatography. 3-Biotech J 9(1):25

Hirata R, Takahashi M, Saga N, Mikami K (2011) Transient gene expression system established in Porphyra yezoensis is widely applicable in Bangiophycean algae. Mar Biotechnol 13:1038–1047

Hong Y, Chen W, Luo X, Pang C, Lester E, Wu T (2017) Microwave-enhanced pyrolysis of macroalgae and microalgae for syngas production. Bioresour Technol 237:47–56

Hwang H, Tong Chen R, Nines HS, Stoner G (2006) Photochemoprevention of UVb induced skin carcinogenesis in SKH 1 mice by brown algae polyphenols. Int J Cancer:119

Ibarra RU, Edwards JS, Palsson BO (2002) *Escherichia coli* K-12 undergoes adaptive evolution to achieve in silico predicted optimal growth. Nature 420:186–189

Jacinto, Marta SC, Hugo R Monteiro, and Marco FL Lemos. 2013. Impact of the invasive macroalgae *Asparagopsis armata* on coastal environments: an ecotoxicological assessment. Curr Opin Biotechnol, 75

Jang SS, Shirai Y, Uchida M, Wakisaka M (2012) Production of mono sugar from acid hydrolysis of seaweed. Afr J Biotechnol 11:1953–1963

Jard G, Marfaing H, Carrère H, Delgenes JP, Steyer JP, Dumas C (2013) French Brittany macroalgae screening: composition and methane potential for potential alternative sources of energy and products. Bioresour Technol 144:492–498

Jiang Y, Lei M, Duan L, Longhurst P (2015) Integrating phytoremediation with biomass valorisation and critical element recovery: a UK contaminated land perspective. Biomass Bioenergy 83:328–339

Johansson CL, Paul NA, de Nys R, Roberts DA (2015) The complexity of biosorption treatments for oxyanions in a multi-element mine effluent. J Environ Manage 151:386–392

Johansson CL, Paul NA, de Nys R, Roberts DA (2016) Simultaneous biosorption of selenium, arsenic and molybdenum with modified algal-based biochars. J Environ Manage 165:117–123

Joshi S, Kumari R, Upasani V (2018) Applications of algae in cosmetics: an overview. Int J Innov Sci Eng Technol 7:1269–1278

Jun D, Ping L, Liu Z-H, Sun D-g, Tao C-y (2010) Fast pyrolysis of biomass for bio-oil with ionic liquid and microwave irradiation. J Fuel Chem Technol 38:554–559

Jung KA, Lim S-R, Kim Y, Park JM (2013) Potentials of macroalgae as feedstocks for biorefinery. Bioresour Technol 135:182–190

Kaehler S, Kennish R (1996) Summer and winter comparisons in the nutritional value of marine macroalgae from Hong Kong. Bot Mar 39:11–17

Kan T, Grierson S, de Nys R, Strezov V (2014) Comparative assessment of the thermochemical conversion of freshwater and marine micro- and macroalgae. Energy Fuel 28:104–114

Kang K, Park Y, Hwang H, Kim S, Lee J, Shin H-C (2003) Antioxidative properties of brown algae polyphenolics and their perspectives as chemopreventive agents against vascular risk factors. Arch Pharm Res 26:286–293

Kawai S, Murata K (2016) Biofuel production based on carbohydrates from both brown and red macroalgae: recent developments in key biotechnologies. Int J Mol Sci 17:145

Khan W, Rayirath UP, Subramanian S, Jithesh MN, Prasanth Rayorath D, Hodges M, Critchley AT, Craigie JS, Norrie J, Prithiviraj B (2009) Seaweed extracts as biostimulants of plant growth and development. J Plant Growth Regul 28:386–399

Kidgell JT, de Nys R, Yi H, Paul NA, Roberts DA (2014) Bioremediation of a complex industrial effluent by biosorbents derived from freshwater macroalgae. PLoS One 9(6):1–9

Kılınç B, Cirik S, Turan G, Tekogul H, Koru E (2013) Seaweeds for food and industrial applications. In: Food industry. IntechOpen, Izmir

Kim EJ, Park SY, Lee J-Y, Park JHY (2010) Fucoidan present in brown algae induces apoptosis of human colon cancer cells. BMC Gastroenterol 10(1):1–11

Kim JK, Kraemer GP, Yarish C (2014) Field scale evaluation of seaweed aquaculture as a nutrient bioextraction strategy in Long Island Sound and the Bronx River Estuary. Aquaculture 433:148–156

Kim M-S, Ga-Hui O, Kim M-J, Hwang J-K (2013) Fucosterol inhibits matrix metalloproteinase expression and promotes Type-1 procollagen production in UVB-induced HaCaT cells. Photochem Photobiol 89:911–918

Kraan S (2013) Mass-cultivation of carbohydrate rich macroalgae, a possible solution for sustainable biofuel production. Mitig Adapt Strateg Glob Chang 18:27–46

Kristinsson, G, and Rósa Jónsdóttir. 2015. Novel bioactive seaweed based ingredients and products.72–110

La Barre S, Bates S, Neveux N, Bolton JJ, Bruhn A, Roberts DA, Ras M (2018) The bioremediation potential of seaweeds: recycling nitrogen, phosphorus, and other waste products. In: Blue biotechnology, pp 217–239

Lawton RJ, de Nys R, Paul NA (2013) Selecting reliable and robust freshwater macroalgae for biomass applications. PLoS One 8:1–7

Le Gall Y, Brown S, Marie D, Mejjad M, Kloareg B (1993) Quantification of nuclear DNA and GC content in marine macroalgae by flow cytometry of isolated nuclei. Protoplasma 173(3–4):123–132

Leandro, A., L. Pereira, and A. M. M. Gonçalves. 2019. Diverse applications of marine macroalgae. Mar Drugs. 18

Leandro A, Pacheco D, Cotas J, Marques JC, Pereira L, Gonçalves AMM (2020a) Seaweed's bioactive candidate compounds to food industry and global food security. Life (Basel, Switzerland) 10:140–176

Leandro A, Pereira L, Gonçalves AMM (2020b) Diverse applications of marine macroalgae. Mar Drugs 18(1):1–15

Lee B-H, Choi B-W, Ryu G-S, Kim S-K, Shin H-C (2002) Material separated from *Ecklonia cava*, method for extracting and purifying the same, and use thereof as antioxidants. US6384085B1

Lehmann J, Joseph S (2009) Biochar systems. Environ Manag Sci Technol 10:147–168

Levine, Ira. 2016. Algae: A way of life and health. Seaweed Health Dis Prev. 1–5

Li L, Rowbotham JS, Christopher Greenwell H, Dyer PW (2013) An Introduction to Pyrolysis and Catalytic Pyrolysis: Versatile Techniques for Biomass Conversion. In: Suib SL, editor. New and Future Developments in Catalysis. Amsterdam: Elsevier, pp. 173–208

Li X, Norman HC, Kinley RD, Laurence M, Wilmot M, Bender H, de Nys R, Tomkins N (2018) Asparagopsis taxiformis decreases enteric methane production from sheep. Anim Prod Sci 58:681–688

Macaisne N, Liu F, Scornet D, Peters AF, Lipinska A, Perrineau M-M, Henry A, Strittmatter M, Coelho SM, Mark J, Cock. (2017) The Ectocarpus immediate upright gene encodes a member of a novel family of cysteine-rich proteins with an unusual distribution across the eukaryotes. Development 144:409–418

MacArtain P, Gill CI, Brooks M, Campbell R, Rowland IR (2007) Nutritional value of edible seaweeds. Nutr Rev 65:535–543

Manilal A (2012) Anticoagulant potential of polysaccharide isolated from Indian red alga, Asparagopsis taxiformis (Delile) Trevisan. Thalassas 28:9–15

Menon V (2008) Marine products for healthcare: functional and bioactive nutraceutical compounds from the ocean. Taylor & Francis Ltd., Hoboken, pp 1–527

Mikami, Koji. 2013. Current advances in seaweed transformation. An integrated view of the molecular recognition and toxinology—From analytical procedures to biomedical applications. 1:323–347

Milledge JJ, Harvey PJ (2016) Potential process 'hurdles' in the use of macroalgae as feedstock for biofuel production in the British Isles. J Chem Technol Biotechnol (Oxford, Oxfordshire : 1986) 91:2221–2234

Milledge JJ, Smith B, Dyer PW, Harvey P (2014) Macroalgae-derived biofuel: A review of methods of energy extraction from seaweed biomass. Energies 7:7194–7222

Mohammad JH, Ranga Rao A, Ravishankar GA (2019) Opportunities and challenges in seaweeds as feed stock for biofuel production. In: Ravishankar GA, Rao AR (eds) Handbook of algal technologies and phytochemicals: volume-II phycoremediation, biofuels and global biomass production, vol II. CRC Press, Boca Raton, pp 39–50

Montecinos AE, Guillemin M-L, Couceiro L, Peters AF, Stoeckel S, Valero M (2017) Hybridization between two cryptic filamentous brown seaweeds along the shore: analysing pre-and postzygotic barriers in populations of individuals with varying ploidy levels. Mol Ecol 26:3497–3512

Mulbry W, Kondrad S, Pizarro C (2007) Biofertilizers from algal treatment of dairy and swine manure effluents: characterization of algal biomass as a slow release fertilizer. J Veg Sci 12:107–125

Nabti E, Jha B, Hartmann A (2017) Impact of seaweeds on agricultural crop production as biofertilizer. Intern J Environ Sci Technol 14:1119–1134

Nakamura Y, Sasaki N, Kobayashi M, Ojima N, Yasuike M, Shigenobu Y, Satomi M, Fukuma Y, Shiwaku K, Tsujimoto A (2013) The first symbiont-free genome sequence of marine red alga, Susabi-nori (*Pyropia yezoensis*). PLoS One 8:1–11

Nakazato K, Takada H, Iha M, Nagamine T (2010) Attenuation of N-nitrosodiethylamine-induced liver fibrosis by high-molecular-weight fucoidan derived from *Cladosiphon okamuranus*. J Gastroenterol Hepatol 25:1692–1701

Necsulea A (2020) Phylogenomics and genome annotation. No commercial publisherl Authors open access book

Neveux N, Yuen AKL, Jazrawi C, Magnusson M, Haynes BS, Masters AF, Montoya A, Paul NA, Maschmeyer T, De Nys R (2014) Biocrude yield and productivity from the hydrothermal liquefaction of marine and freshwater green macroalgae. Bioresour Technol 155:334–341

Neveux N, Magnusson M, Maschmeyer T, de Nys R, Paul NA (2015) Comparing the potential production and value of high-energy liquid fuels and protein from marine and freshwater macroalgae. GCB Bioenergy 7:673–689

Øverland M, Mydland LT, Skrede A (2019) Marine macroalgae as sources of protein and bioactive compounds in feed for monogastric animals. J Sci Food Agric 99:13–24

Palsson B (2004) Two-dimensional annotation of genomes. Nat Biotechnol 22:1218–1219

Pereira L (2016) Edible seaweeds of the world, vol 1, 1st edn. CRC Press, Boca Raton

Pereira L (2018) Seaweeds as source of bioactive substances and skin care therapy—cosmeceuticals, algotheraphy, and thalassotherapy. Cosmetics 5:68–108

Pereira L, Bahcevandziev K, Joshi NH (2019) Seaweeds as plant fertilizer, agricultural biostimulants and animal fodder, vol 1. CRC Press, Boca Raton

Pereira L, Gheda SF, Ribeiro-Claro PJA (2013) Analysis by vibrational spectroscopy of seaweed polysaccharides with potential use in food, pharmaceutical, and cosmetic industries. Intern J Carbohydr Chem 2013:1–8

Pereira L, van de Velde F (2011) Portuguese carrageenophytes: carrageenan composition and geographic distribution of eight species (Gigartinales, Rhodophyta). Carbohydr Polym 84:614–623

Pertea M, Shumate A, Pertea G, Varabyou A, Chang Y-C, Madugundu AK, Pandey A, Salzberg SL (2018) Thousands of large-scale RNA sequencing experiments yield a comprehensive new human gene list and reveal extensive transcriptional noise. BioRxiv:1–21

Peters AF, Marie D, Scornet D, Kloareg B, Mark J, Cock. (2004) Proposal of *Ectocarpus siliculosus* (Ectocarpales, phaeophyceae) as a model organism for brown algal genetics and genomics. J Phycol 40(6):1079–1088

Qin S, Lin H, Jiang P (2012) Advances in genetic engineering of marine algae. Biotechnol Adv 30:1602–1613

Ragan MA, Glombitza KW (1986) Phlorotannins, brown algal polyphenols. In: Hellebust JA, Craigie JS (eds) Progress in phycological research. Cambridge University Press, New York, pp 129–241

Rajak RC, Jacob S, Kim BS (2020) A holistic zero waste biorefinery approach for macroalgal biomass utilization: A review. Sci Total Environ 716:1–17

Ravi S, Kasi PD (2014) Evaluation of physicochemical properties, proximate and nutritional composition of *Gracilaria edulis* collected from Palk Bay. Food Chem 174:68–74

Reddy CRK, Gupta MK, Mantri VA, Jha B (2008) Seaweed protoplasts: status; biotechnological perspectives and needs. J Appl Phycol 20:619–632

Roberts DA, Paul NA, Bird MI, de Nys R (2015) Bioremediation for coal-fired power stations using macroalgae. J Environ Manag 153:25–32

Robinson N, Winberg P, Kirkendale L (2013) Genetic improvement of macroalgae: status to date and needs for the future. J Appl Phycol 25:703–716

Roesijadi G, Jones S, Snowden-Swan L, Zhu Y (2010) Macroalgae as a biomass feedstock: A preliminary analysis. Analysis 1:1–50

Rupérez P (2002) Mineral content of edible marine seaweeds. Food Chem 79:23–26

Samolada M, Papafotica A, Vasalos I (2000) Catalyst evaluation for catalytic biomass pyrolysis. Energy Fuels 14:1161–1167

Schiener P, Black K, Stanley M, Green D (2015) The seasonal variation in the chemical composition of the kelp species *Laminaria digitata, Laminaria hyperborea, Saccharina latissima and Alaria esculenta*. J Appl Phycol 27:363–373

Sellimi S, Maalej H, Rekik DM, Benslima A, Ksouda G, Hamdi M, Sahnoun Z, Li S, Nasri M, Hajji M (2018) Antioxidant, antibacterial and in vivo wound healing properties of laminaran purified from *Cystoseira barbata* seaweed. Int J Biol Macromol 119:633–644

Shama A, Joyce SG, Mari FD, Ranga Rao A, Ravishankar GA, Hudaa N (2019) Macroalgae and microalgae: novel sources of functional food and feed. In: Ravishankar GA, Rao AR (eds) Handbook of algal technologies and phytochemicals: Volume-I: food, health and nutraceutical applications. CRC Press, Boca Raton, pp 207–219

Shannon E, Abu-Ghannam N (2016) Antibacterial derivatives of marine algae: an overview of pharmacological mechanisms and applications. Mar Drugs 14:81–103

Sharpley AN, Bergström L, Aronsson H, Bechmann M, Bolster CH, Börling K, Djodjic F, Jarvie HP, Schoumans OF, Stamm C, Tonderski KS, Ulén B, Uusitalo R, Withers PJA (2015) Future agriculture with minimized phosphorus losses to waters: research needs and direction. Ambio 44:163–179

Shore M, Murphy S, Mellander P-E, Shortle G, Melland AR, Crockford L, O'Flaherty V, Williams L, Morgan G, Jordan P (2017) Influence of stormflow and baseflow phosphorus pressures on stream ecology in agricultural catchments. Sci Total Environ 590–591:469–483

Smit AJ (2004) Medicinal and pharmaceutical uses of seaweed natural products: a review. J Appl Phycol 16:245–262

Soliman RM, Younis SA, El-Gendy NS (2018) Batch bioethanol production via the biological and chemical saccharification of some Egyptian marine macroalgae. J Appl Microbiol 125:422–440

Sonani RR, Rastogi RP, Patel R, Madamwar D (2016) Recent advances in production, purification and applications of phycobiliproteins. World J Biol Chem 7:100

Sugiura Y, Takeuchi Y, Kakinuma M, Amano H (2006) Inhibitory effects of seaweeds on histamine release from rat basophile leukemia cells (RBL-2H3). Fish Sci 72:1286–1291

Suib SL (2013) New and future developments in catalysis: catalysis by nanoparticles. Netherlands: Elsevier, pp. 1–512

Taskin E, Ozturk M, Kurt O (2007) Antibacterial activities of some marine algae from the Aegean Sea (Turkey). Afr J Biotechnol 6:2746–2751

Tester M, Langridge P (2010) Breeding technologies to increase crop production in a changing world. Science 327:818–822

Thomas NV, Kim S-K (2013) Beneficial effects of marine algal compounds in cosmeceuticals. Mar Drugs 11:146–164

Tønnesen HH, Karlsen J (2002) Alginate in drug delivery systems. Drug Dev Ind Pharm 28:621–630

Tuhy Ł, Chowańska J, Chojnacka K (2013) Ekstrakty glonowe jako biostymulatory wzrostu roślin: przegląd piśmiennictwa. Chemik 67(7):636–641

UNFCCC, U. N (2009) Kyoto protocol reference manual on accounting of emissions and assigned amount. *esocialsciences.com*, Working Papers 1:1–130

Vidotti EC, Maria do Carmo ER (2004) Algae: from economics in aquatic environments to bioremediation and analytical chemistry. Química nova 27:139–145

Wang P, Zhan S, Yu H, Xue X, Hong N (2010) The effects of temperature and catalysts on the pyrolysis of industrial wastes (herb residue). Bioresour Technol 101:3236–3241

Wang H-MD, Li X-C, Lee D-J, Chang J-S (2017) Potential biomedical applications of marine algae. Bioresour Technol 244:1407–1415

Wijffels RH, Barbosa MJ, Eppink MH (2010) Microalgae for the production of bulk chemicals and biofuels. Biofuel Bioprod Biorefin 4:287–295

Wilkie AC, Mulbry WW (2002) Recovery of dairy manure nutrients by benthic freshwater algae. Bioresour Technol 84:81–91

Woldringh CL (2002) The role of co-transcriptional translation and protein translocation (transertion) in bacterial chromosome segregation. Mol Microbiol 45:17–29

Wu C, Jiang P, Guo Y, Liu J, Zhao J, Huihui F (2018) Isolation and characterization of *Ulva prolifera* actin1 gene and function verification of the 5′ flanking region as a strong promoter. Bioengineered 9:124–133

Yanagibayashi S, Kishimoto S, Ishihara M, Murakami K, Aoki H, Takikawa M, Fujita M, Sekido M, Kiyosawa T (2012) Novel hydrocolloid-sheet as wound dressing to stimulate healing-impaired wound healing in diabetic db/db mice. Biomed Mater Eng 22:301–310

Yarnpakdee S, Benjakul S, Senphan T (2018) Antioxidant activity of the extracts from freshwater macroalgae (*Cladophora glomerata*) Grown in Northern Thailand and its preventive effect against lipid oxidation of refrigerated eastern little tuna slice. Turk J Fish Aquat Sci 19:209–219

Yuan YV, Westcott ND, Chun H, Kitts DD (2009) Mycosporine-like amino acid composition of the edible red alga, *Palmaria palmata* (dulse) harvested from the west and east coasts of Grand Manan Island, New Brunswick. Food Chem 112:321–328

Yun J-H, Smith VH, deNoyelles FJ, Roberts GW, Stagg-Williams SM (2014) Freshwater macroalgae as a biofuels feedstock: mini-review and assessment of their bioenergy potential. Ind Biotechnol 10(3):212–220

Zhang H, Yohe T, Huang L, Entwistle S, Peizhi W, Yang Z, Busk P, Ying X, Yin Y (2018) DbCAN2: A meta server for automated carbohydrate-active enzyme annotation. Nucleic Acids Res 46:95–101

Zhang J, Ewart HS, Jing Kai Shen, and Colin James Barrow (2008) *Ascophyllum* compositions and methods. US20080280994A1

Zhou H, Meng H, Zhao L, Shen Y, Hou Y, Cheng H, Song L (2018) Effect of biochar and humic acid on the copper, lead, and cadmium passivation during composting. Bioresour Technol 258:279–286

Zodape ST, Kawarkhe VJ, Patolia JS, Warade AD (2008) Effect of liquid seaweed fertilizer on yield and quality of okra (*Abelmoschus esculentus* L.). J Sci Ind Res 67:1115–1117

Chapter 33
Recent Advances in Biotechnology of Seaweeds: An Overview

Nitin Trivedi, Ritu Sharma, Arijit Sankar Mondal, and Dhara Dixit

33.1 Introduction

The diversification in the advancement of industrial commercialization in recent decades relies upon the exploitation of marine macroalgae or seaweed and its derived products. The domestication of seaweed cultivation and novel seaweed discovery will lead to the development of various sophisticated products using distinct biotechnological applications. Though most of the marine resources of the microbial kingdom are only accessible to the wealthier countries due to huge economic investment and advanced technological applications but marine seaweed cultivation can be the solution to this problem for developing countries. The increment in seaweed cultivation and biomass production has been established intentionally for food, feedstock, biomedical, and bioenergy applications (Mazarrasa et al. 2014). The global seaweed industrial market has increased to about 6 billion USD per annum (FAO 2018). Moreover, the upbringing of various policies with advanced technological approaches has resulted in the expansion of global industrial markets. Apart from food and feed (Shama et al., 2019), several bioenergy products i.e. biofuel obtained from seaweed biomass will become the alternative strategy for finite fossil fuels with less carbon (C) accumulation (Rajak et al. 2020; Mohammad et al.

N. Trivedi (✉)
DBT-ICT Centre for Energy Biosciences, Institute of Chemical Technology, Matunga, Mumbai, India
e-mail: seaweed204@gmail.com

R. Sharma · A. S. Mondal
Department of Microbiology, Guru Nanak Institute of Pharmaceutical Science and Technology (Life Science), Kolkata, India

D. Dixit
Department of Earth and Environmental Science, Krantiguru Shyamji Krishna Verma Kachchh University, Bhuj, Gujarat, India

A. Ranga Rao, G. A. Ravishankar (eds.), *Sustainable Global Resources of Seaweeds Volume 1*, https://doi.org/10.1007/978-3-030-91955-9_33

2019; Hessami et al. 2019). Therefore, here we discuss different advanced seaweed cultivation techniques and biotechnological options for improvement of seaweed biomass and its utilization.

33.2 Growth Engineering and Cultivation of Seaweeds

Seaweeds or macroalgae are marine organisms that grow in the inter-and sub-tidal regions and are widely distributed throughout the marine coast (García-Poza et al. 2020). Seaweed cultivation has been traditionally started in Asian countries for about centuries (Loureiro et al. 2015). The earliest utilization of seaweeds by humans has been originated in China since 1700 years ago (Buschmann et al. 2017). However, the commercial macroalgal program of more than 20 seaweed varieties with improved physiological and genetic tolerance started in the 1950s. The primary cultivation engineering started with the improvement of the seed quality either by clonal propagation or via propagation-based sexual reproduction (Loureiro et al. 2015). The advancement in seaweed cultivation relies on the interaction between scientific knowledge and the product demand of consumers (Hafting et al. 2015).

33.2.1 Cultivation of Seaweeds/Macroalgae

Aquaculture is the well-established technique for upscaling the cultivation of seaweed biomass which comprises both the monoculture and Integrated Multi-Trophic Aquaculture (IMTA) systems. Several Asian countries such as China, the Philippines, Indonesia, North, and South Korea are engaged in upscaling the seaweed biomass cultivation for both human consumption and also for industrial application (Stévant et al. 2017). Moreover, seaweed cultivation confers several ecological balances not only by contributing to the food chain, oxygen (O_2), carbon (C), and nutrient cycles but also reduces global warming and eutrophication (Buschmann et al. 2017). Several macroalgae such as *Pyropia*, *Glacilaria*, *Geledium*, *Pterocladia*, *Kappaphycus*, and *Eucheuma* are widely cultivated by aquaculture technique. Besides cultivation process improvement discussed below, advancement in multidisciplinary analysis via computational fluid dynamics (CFD) and other technological engineering will uplift the production and economy to the next level (García-Poza et al. 2020).

33.2.2 Offshore Cultivation

Offshore cultivation of seaweeds is the most widely practiced technique for the production of a large number of seaweed biomass harvest for the commercial production of industrial bioactive components (Hafting et al. 2015). The installment and maintenance of in-sea seaweed cultivation practices are cost-effective either by attaching the seaweed seedlings directly to the ropes, nets, and lines in the sea or by cultivating the seedlings in greenhouse tanks indoors followed by transplanting them onto ropes in the ocean or sea (García-Poza et al. 2020). Apart from the minimal cost advantage, there are several drawbacks in open sea cultivation techniques such as unfavorable environmental conditions, fluctuations in biochemical constituents due to physiological plasticity, labor-intensive, limited nutrients availability, and light source manipulation (Hafting et al. 2015). A small advancement in offshore cultivation has been established by replacing the ropes with floating rafts of 100 Km^2 which are now being used as beds for macroalgal growth till harvest thereby producing 10^6 tonnes $year^{-1}$ fresh weight (Buschmann et al. 2017). Nevertheless, offshore cultivation has been identified to expand in recent years to satisfy the industrial demand in recent decades. A mathematical model has been established for analyzing physicochemical characteristics of *Ulva* sp. The model ran annually once a month each on a global 1° grid with a single output file. The algal growth rate (μ) was determined as a function of light intensity, temperature, salinity, nutrients, and respiration rate. This model described the potential for biomass production could be extended 400Km from the shore and classified them into either future deep- or shallow-water provinces. The climatological oceanographic data can be used to analyze the environmental risks which can be controlled to develop futuristic potential offshore biorefineries globally (Lehahn et al. 2016). The problem with the nutrient and light supply can be controlled by establishing huge offshore floating reactors accompanied by external solar energy (sun) which can be used for exploiting photon capture and carbon fixation rates via algal cells to increase biomass yield per unit area. Moreover, strain improvement and various phenotypic imaging techniques can be utilized for improving the properties of seaweeds and their products (Buschmann et al. 2017).

33.2.3 Nearshore Cultivation

The nearshore cultivation, derived from offshore cultivation is the most common seaweed aquaculture technique ubiquitously practiced in near-coast sites. This cultivation strategy is economically feasible compared to in- and onshore cultivation. The major advantage of this technique relies on the bioremediation of polluted water bodies from agricultural fields (García-Poza et al. 2020). Distinct seaweed species such as *Gracilaria*, *Porphyra*, *Laminaria*, and *Sargassum fusiforme*are widely cultivated in the eutrophic water bodies due to their high carbon (C) sequestration, oxygen (O_2) release, nitrogen (N_2), and phosphorus (P) remediation capability (iZheng et al. 2019).

33.2.3.1 Onshore Cultivation

Onshore cultivation, commonly known as land-based cultivation introduced in the 1970s–1980s (Hafting et al. 2015) which is mostly practiced in closed systems i.e. raceways, tanks, ponds, or lagoons where the seaweeds are suspended and exposed to light under controlled agitation (García-Poza et al. 2020). The salinity, photoperiod via artificial light exposure, the addition of nutrients, onshore seawater pumping, in-, and out-flow are practiced manually to improve the production of bioactive compounds with limited discharge (Hafting et al. 2015; Hafting et al. 2012). Moreover, light quantity and quality can be artificially manipulated by using greenhouse coverings along with tight control over CO_2 and pH (García-Poza et al. 2020). The onshore cultivation system has a huge advantage over the offshore cultivation system. Tanks of diverse proportions and different cultivars are easily accessible due to their localization at a single place without any advanced equipment facility. However, the high labour and infrastructure cost along with demand for energy requirement with time-consuming process brings out the major disadvantages of this technique (Hafting et al. 2012).

33.2.3.2 Inland Saline Aquaculture

The inland saline aquaculture is the advanced onshore promising seaweed cultivation technique practiced by using saline groundwater. The major sources of saline groundwater include saline lakes, agricultural sites, saline water obtained from aquifers, and coal steam gas. The economic feasibility and nutrient uptake capability become major advantages of this technique compared to the offshore ocean and sea cultivation (García-Poza et al. 2020).

33.2.3.3 Integrated Multi Trophic Aquaculture (IMTA)

The Integrated Multi Trophic Aquaculture (IMTA) or macroalgal polyculture has been introduced with a dual intention of increasing seaweed biomass with high specific growth rates (SGR) and amelioration of polluted waters via bioremediation. This advanced aquaculture technique was first initiated in the USA in the 1970s and Israel in the 1980s followed by Canada and Asian countries. Distinct species such as *Ulva*, *Gracilaria*, and *Chondracanthus* have been cultivated in IMTA either with aquatic vertebrates and invertebrates (Giangrande et al. 2020). The IMTA technique is mostly used for farming different macroalgal species at various trophic levels. Several countries such as Portugal, Norway, South Africa, and Israel have started pilot-based production of seaweed biomass based on this technique (Buschmann et al. 2017). The IMTA is an interconnected cultivation technique where the co- or by-products of one species serve as valuable nutrients for the other making the production economically feasible. The bioremediation of wastewater is mostly done by absorbing the organic and inorganic (chemical) nutrients from various sources.

Light plays a crucial role in the IMTA technique for increasing the seaweed biomass to speed up the bioremediation followed by pH, nutrient, and CO_2 balance. Marine seaweeds are the most valuable photosynthetic organisms both for their fast ontogeny cycle and essential bioactive compounds composition. In recent decades, the IMTA technique is also used to meet up the macroalgal biofuel production (García-Poza et al. 2020). Moreover, a new scheme was proposed by Gracia-Blairsy Reina is the 'Green Sahara Concept' on several coasts of South Africa, Israel, and Jordan which is based on advanced marine polyculture technique for simultaneous production of several aquatic life forms, seaweeds, biogas, and hydroelectricity in a cost-effective manner (Buschmann et al. 2017). Nevertheless, at an industrial scale, the seaweed IMTA technique is the most efficient way of advancing cultivation involving both economic and ecological stability with high commercial marketing values.

33.2.4 Advancement by Genetic Modulation

The advancement of seaweed cultivation also relies on the manipulation of the seaweed genome and ontogeny cycle. In contrast to microalgae, the genomic information of macroalgae is limited which makes the advancement challenging with wider research scopes (Lin and Qin 2014). The industrial problem in seaweed biomass production can be successfully overcome via several genetic engineering techniques such as trans-conjugation, transformation (natural, induced, and biolistic), electroporation, particle bombardment, glass beads, microinjection, RNA interference induced gene silencing, artificial transposon method, silicon carbon whiskers method, and recombinant algal virus- and *Agrobacterium tumefaciens*-mediated vector transformation (Qin et al. 2012; Lin and Qin 2014). The macroalgal genes are identified, localized, and manipulated via specific reporter and marker genes responsible for protein expression. GUS gene was primarily used as a potential reporter gene that encodes ß-glucuronidase for protein expression in various seaweeds however imparts negative control due to its toxicity hampering the cellular structure. This problem was overcome by the lacZ gene isolated from bacteria which confers antimicrobial activity thus making this reporter gene more potential and effective compared to the GUS gene. Two macroalgae such as *Porphyra haitanensis* and *Laminaria japonica* have been transformed with the lacZ gene. The selection of marker genes is a crucial parameter while preparing transformants which must obey two factors i.e. antibiotic resistance and homologous complementation. Moreover, distinct vectors such as shuttle vectors, higher plant-based vectors, bacterial vectors (*Escherichia coli*), and other vectors from cyanophages and cyanobacterial chromosome segments can be developed for the desired gene expression in the macroalgae which are highly promoter and codon specific (Qin et al. 2012). The introduction of omics (genomics) technology can be a helping hand for uplifting the engineering of seaweeds with the desired gene of interest. The selection of model organism along with its complete genome sequencing can be comparatively established via a cost-efficient method i.e. next-generation sequencing

(NGS) which might open a wider area for precisely targeted genome engineering. Few techniques such as zinc-finger nucleases (ZFN), transcription activators-like effectors (TALs), and clustered regularly interspaced short palindromic repeats (CRISPR) may open a wider knowledge in gene editing. Though promoter selection and trapping in seaweeds are still foggy which makes the manipulation of metabolic pathways, mRNA stability improvement, codon usage, ribosome-binding sites design, and RNase III activity manipulation more difficult. However, few eukaryotic class II viral promoters such as CaMV35S and SV40 with a TATA box have been used in macroalgal genetic engineering with limited scientific knowledge (Lin and Qin 2014). Moreover, transcriptomics, proteomics, metabolomics, and transgenesis technique have also been developed to prepare growth phase-dependent markers for understanding the primary developmental stages and cell-specific expression through RNA amplification and knocking endogenous gene regulation with limited experimental models i.e. *Ulva*, *Ectocarpus*, *Volvox*, *Porphyra*, and *Gonium*(Charrier et al. 2017).

33.2.5 Improving Developmental Cycle of Seaweeds

The bottleneck of seaweed aquaculture can be enhanced by preparing various artificial hybrids via clonal propagation, vegetative propagation, somatic hybridization, and cross-fertilization which potentially manipulate the seaweed ontogeny cycle thereby increasing seaweed diversity with controlled breeding. Large macroalgal biomass production can be produced by controlling the circadian rhythm and abiotic factors which are the key parameters to improve physiologically (growth, structure development, and composition) and physicochemical factors. However, the knowledge gap in understanding their signaling pathways makes it challenging. The reproduction cycle of the seaweed can be manipulated by using specific light, temperature, phytohormones, and tissue ablation which speed up the vegetative transition. Moreover, vegetative propagation can be established by parthenogenesis induced by chemical treatments or hybridization. Several macroalgae such as *Porphyra*, *Codium*, *Polysiphonia*, *Palmaria*, *Ulva*, *Portieria*, *Himanthalia*, and *Ochtodes* have been well cultivated in photobioreactors using parthenogenesis to meet the consumer food demand. *Ulva* and *Monostroma*exhibit good symbiotic interactions with bacteria which not only characterize their interaction but also influence the improvement in macroalgal metabolites production (Charrier et al. 2017). Gupta and his coworkers (2011) studied various plant growth regulators such as gibberellic acid (GA_3), indole-3-acetic acid (IAA), salicylic acid, indole-3-butyric acid (IBA), and kinetin riboside (KR) in *Monostroma* and several *Ulva* sp. which not only stimulates various physiological processes but also resist environmental stress. These phytohormones can be effectively used as plant stimulants for enhancing the physicochemical responses in agricultural fields. Moreover, innovative macroalgal cultivation of *Ulva lactuca* was studied using its protoplast as seeding material. Two distinct acrylic tanks such as horizontal and vertical tanks were constructed for the regeneration of the protoplast. The protoplast is further cultured under controlled temperature and illumination for sporulation development followed by germination via parthenogenesis. The plantlets obtained from protoplast can be further transferred to the open sea as seed stocks for huge biomass cultivation. The major advantage of the study is that 250 g of biomass can be produced from 100 mg of biomass through protoplast at $0.33m^2$ surface areas only in 44 days which could be an efficient integrated model for fulfilling the sustainable production and demand of seaweed-based products (Gupta et al. 2018).

33.2.6 Photobioreactors Advancement

Apart from microalgae, seaweeds are also cultivated in photobioreactors (PBRs) for improving the production of bioactive compounds and biofuel sources. PBRs are broadly classified into two categories: open and closed systems. The open system commonly includes raceways, cascades and sloping ponds with a paddle wheel exhibits limited overheating issues, high dissolved oxygen concentrations, easy construction and economic feasibility make the system attractive. But fluctuations in biomass concentration, nutrient uptake, water loss, light utilization, respiratory losses, and mass transfer ratio with high contamination possibility and large site requirement make this system less advantageous in contrast to the closed system (Qin et al. 2012; Gupta et al. 2015). Moreover, the advancement in open systems was installed by enhancing light utilization, mixing, and limited sedimentation. The closed PBRs were established to overcome the open system cultivation which is broadly classified as stirred tank PBRs (mechanical agitator), vertical column PBRs (airlift or bubble column), horizontal tubular PBRs (pump based recirculation), flat-panel PBRs (airlift or bubble column from side or bottom). The disadvantages of the open system can be overcome via a closed system which makes the cultivation effective. However the efficiency can be improved by installing sensors to regulate the energy demand, modeling PBRs via simulations and phase fluid dynamics, model development to achieve dynamic cells to fasten growth, high biomass by modulating PBRs, improved light utilization strategy, efficient O_2mixing approaches via baffles, impellers, sparger, motor driven and peristaltic pumps (Gupta et al. 2015). Moreover, an offshore flat panel PBR was installed at a pilot scale for analyzing the annual cultivation of *Ulva lactuca* with poultry litter as a feasible N_2 source. The process exhibited substantial growth, biomass accumulation along with the good photosynthetic activity of the seaweed across the panels. However, reduced irradiance was observed only during the monsoon season which slightly affected growth yields with decreased biomass density (Mhatre et al. 2018) (Fig. 33.1).

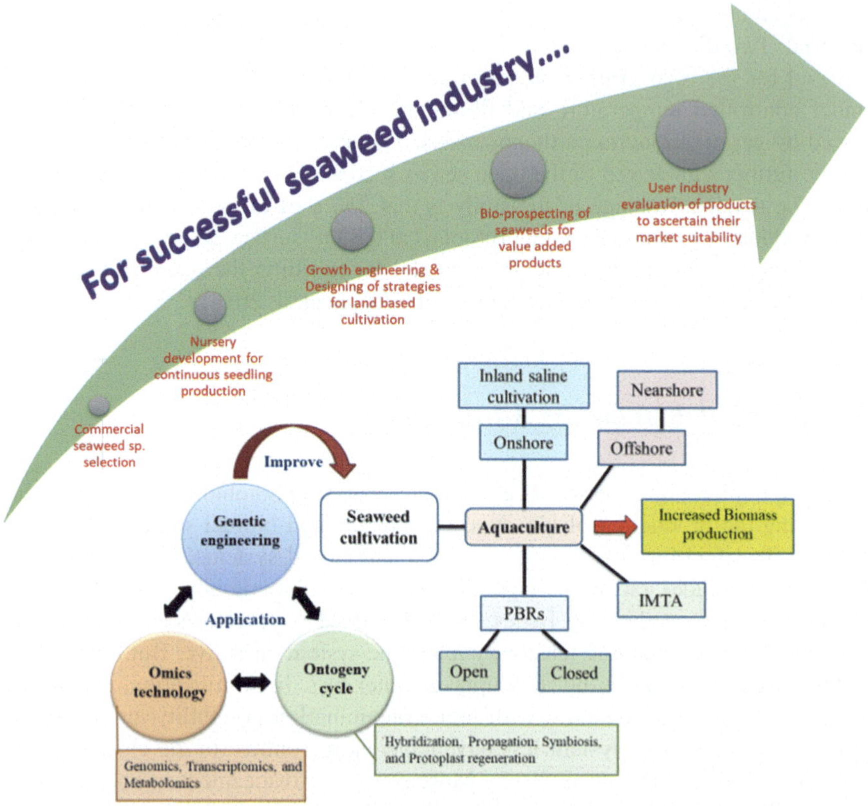

Fig. 33.1 Advancement in the cultivation of seaweed aquaculture to obtain higher biomass

33.3 Seaweeds as Value-Added Products

Since seaweeds possess about 60% of polysaccharides of all bioactive compounds occurring in them (Pereira 2018), most of the value-added products are manufactured using polysaccharides. Seemingly, the enormous application of seaweeds has improved the development of several value-added products of nutraceuticals, functional foods, pharmaceuticals, and cosmeceuticals applications (Lehahn et al. 2016). Moreover, the seaweed-derived biofertilizer approach is a very trending eco-friendly strategy in the field of agriculture (Nabti et al. 2016).

33.3.1 Seaweed Metabolomics

Seaweeds represent a vital coastal community of the marine milieu. They offer the probability of yielding a broad array of natural primary as well as secondary metabolites possessing biochemical, nutritional, or pharmaceutical virtues. The primary metabolites are known to be involved directly in regular growth and functions associated with physiological aspects viz.; reproduction. However, the secondary metabolites are primarily generated as an outcome caused due to various stress viz.; salinity and temperature fluctuations, ultraviolet (UV) radiation, environmental hazards, pollutants, etc. Both the types of metabolites may exhibit properties such as antitumor, anti-inflammatory, neuroprotective, antidiabetic, antioxidant, anticoagulant, antibacterial, anti-hypertensive, etc. The bioactive compounds derived from macroalgae include lipids, proteins, phlorotannins, polysaccharides, sterols, terpenes, vitamins, and minerals. It has been reported that the metabolite concentration profiles exhibit unique make-up that is predominantly based on the taxonomic group as well as the individual species under consideration (Hamid et al. 2019). Earlier studies have revealed that >50,000 metabolites have been deposited to the KNApSAck database (http://kanaya.naist.jp/KNApS AcK/) (Nakamura et al. 2014) that have been reported from terrestrial plants. In contrast to this, the data available for seaweed originated metabolites is quite restricted. The macroalgal metabolite repository (SWMD; http://www.swmd.co.in/) consists of a mere 1109 metabolites, a majority of which have been obtained from *Laurencia* species as reported by Davis and Vasanthi (2011). Nevertheless, the quest for specialized metabolites from seaweeds belonging to Rhodophyta, Chlorophyta, and Ochrophyta has now led to the identification of over 3000 metabolites (Pereira et al. 2016). Table 33.1 represents a few of such bioactive metabolites reported from different macroalgal species globally.

A comprehensive set of primary and secondary metabolites together forms a metabolome (Dixit and Reddy 2017). Metabolomics – a fast emerging integrative branch of 'omics' shapes the modern understanding of systems biology by providing comprehensive biological information, both qualitatively as well as quantitatively (Young and Alfaro 2016). It aids in generating the metabolite profile/s of small bioactive moieties formed as an outcome of complex biochemical responses and are associated with a specific set of amalgams involved in particular pathways (Dixit et al. 2020). Advanced analytical procedures have eased the mammoth task of acquiring elaborate data to a certain extent. These techniques include Gas chromatography coupled with mass spectrometry (GC-MS), nuclear magnetic resonance (NMR), Liquid chromatography coupled with mass spectrometry (LC-MS), High-performance liquid chromatography coupled to mass spectrometry (HPLC-MS), Fourier transform–ion cyclotron resonance–mass spectrometry (FT-ICR-MS or FTMS), Capillary Electrophoresis–Mass Spectrometry (CE-MS), matrix-assisted laser desorption/ionization time-of-flight mass spectroscopy (MALDI-TOF-MS), liquid chromatography (LC) with electrochemical detection, FTIR-MS, Ultra Performance Liquid Chromatography-Tandem Mass Spectrometer

Table 33.1 A list of few bioactive metabolites reported from different macroalgal species globally

Detected Metabolites/Bioactives	Species	Collection area/Season/Time	Analytical Technique Used	References
Phlorotannins	*Eisenia arborea*	Mugizaki, Mie prefecture, Japan	Reversed Phase HPLC NMR MALDI- TOF-MS	Sugiura et al. (2009)
Floridoside D-isofloridoside	*Laurencia undulata*	Cheju Island coast South Korea October 2007.	Column chromatography High-performance liquid chromatography (HPLC) NMR	Li et al. (2010)
(4R,7R,14S)-4α,7α-diacetoxy-14-hydroxydolast-1(15),8-diene	*Canistrocarpus cervicornis*	Ribeira Bay, Rio de Janeiro Brazil, August 2004,	Column chromatography NMR	Bianco et al. (2010)
8-hydroxy-4E,6E-octadien-3-one Loliolide 3b-hydroxy-5a,6a-epoxy-7-megastigmen-9-one N-phenethylacetamide Squamolone 2-ethylidene-4-methylsuccinimide	*Gracilaria lemaneiformis*	South China Sea February 2009	NMR and HRESIMS	Lu et al. (2011)
Diphlorethohydroxycarmalol	*Ishige okamurae*	Kimnyung, Jeju Island, Korea July 2005	Column chromatography and TLC	Ahn et al. (2011)
(1R*,2S*,3R*,5S*,8S*,9R*)-2,3,5,9-tetramethyltricyclo[6.3.0.01,5]undecan-2-ol(1S*,2S*,3S*,5S*,8S*,9S*)-2,3,5,9-tetramethyltricyclo-[6.3.0.01,5]undecan-2-ol	*Laurencia dendroidea*	Parati beach in Anchieta, Espírito Santo State, Brazil, in October, 2006	GCMS NMR	Gressler et al. (2011)

Detected Metabolites/Bioactives	Species	Collection area/Season/Time	Analytical Technique Used	References
6-hydroxy-isololiolide Isololiolide	*Galaxaura filamentosa*	Fiji July 2008	Reverse phase chromatography HR-ESI-MS NMR	Rasher et al. (2011)
(E)-2-((3E,7E)-4,8,12-trimethyltrideca-3,7,11-trienyl) but-2-ene-1,4-diyl diacetate (1E,3E)-2-((3E,7E)-12-formyl-4,8-dimethyl-10-oxotrideca-3,7,12-trienyl)buta-1,3-diene-1,4-diyl diacetate	*Chlorodesmis fastigiata*			
Dictyol E, dictyotadiol, dictyoxide, isopachydictyol A and pachydictyol A	*Dictyota guineensis*	Penha Beach, Bahia, Brazil April 2004, January 2005.	HRGC-MS	De-Paula et al. (2012)
3-(2,3-dibromo-4,5-dihydroxybenzyl) pyrrolidine-2,5-dione methyl 4-(2,3-dibromo-4,5-dihydroxybenzylamino)-4-oxobutanoate 4-(2,3-dibromo-4,5-dihydroxybenzylamino)-4-oxobutanoic acid 3-bromo-5-hydroxy-4-methoxybenzamide 2-(3-bromo-5-hydroxy-4-methoxyphenyl)acetamide	*Rhodomela confervoides*	Dalian coastline of Liaoning Province, People's Republic of China, in April, 2007	HRESIMS NMR	Li et al. (2012)
2,3-dibromo-4,5-dihydroxybenzylaldehyde 2,2′,3-tribromo-3′,4,4′,5-tetrahydroxy-6′-hydroxymethyldiphenylmethane bis(2, 3-dibromo-4,5-dihydroxylbenzyl) ether 5,5″-oxybis(methylene)bis (3-bromo-4-(2′,3′-dibromo-4′,5′-dihydroxylbenzyl)benzene-1,2-diol)	*Vertebrata lanosa*	Oldervik, Ullsfjorden, Norway Spring 2010	HPLC UPLC-ToF-MS NMR	Olsen et al. (2013)
1,1,3,3-tetrabromo-2-heptanone	*Bonnemaisonia hamifera*	Tjarno, Swedish west coast	GCMS	Nylund et al. (2013)

(continued)

Table 33.1 (continued)

Detected Metabolites/Bioactives	Species	Collection area/Season/Time	Analytical Technique Used	References
2-dodecanoyloxyethanesulfonate	*Asparagopsis taxiformis*	Tamil Nadu, India	ICR-FT/MS	Jha et al. (2013)
Phlorotannins	*Eisenia bicyclis*	Ulleung Island, East Sea, Korea	Hydrophilic interaction chromatography (HILIC)	Kim et al. (2013)
1,3,5-trihydroxybenzene (phloroglucinol), dibenzo [1,4] dioxine-2,4,7,9-tetraol hexahydroxyphenoxydibenzo [1,4] dioxine (eckol)	*Ecklonia maxima*	Kommetjie, west coast of South Africa	ESI-TOF-MS 1H and 13C NMR	Kannan et al. (2013)
Eleganolone	*Bifurcaria bifurcate*	Basse-Normandie November 2005 – September 2007	NMR and HREIMS	Galle et al. (2013)
Pheophytin a	*Odonthalia corymbifera*	Hakodate, south-western Hokkaido, Japan	NMR	Kurihara et al. (2014)
Phlorotannins	*Cystoseira abies-marina*	Spanish Bank of Algae (Marine Biotechnology Center, University of Las Palmas de Gran Canaria, Gran Canaria, Spain)	HILIC × RP-DADMS/MS. 2D LC (LC × LC)	Montero et al. (2014)
Pachydictyol A Isopachydictyol A Dichotomanol	*Dictyota menstrualis*	Praia do Forno Rio de Janeiro Brazil	Silica gel-column chromatography Reverse-phase chromatography NMR	Moura et al. (2014)
Stigmast-5,24-dien-3-ol (fucosterol)	*Sargassum tenerrimum*	–	Column chromatography NMR	Majik et al. (2015)
Triquinane sesquiterpene Hexadecanoic acid Cholesterol (10) (E)-2-tridecyl heptadecenal	*Laurencia dendroidea* J. AGARDH	Southeast Brazilian coast.	GCMS NMR	Machado et al. (2016)

Detected Metabolites/Bioactives	Species	Collection area/Season/Time	Analytical Technique Used	References
Dieckol	*Ecklonia cava*	–	Column chromatography HPLC	Kim et al. (2016)
Phloroglucinol, gallic acid, catechin, protocatechuic acid, rutin, chlorogenic acid, caffeic acid, coumaric acid, ferulic acid, myricetin and quercetin	*Himanthalia elongate*	Algamar (Pontevedra, Spain)	HPLC/DAD	Belda et al. (2016)
hydroxybenzaldehyde (meta/para), phloroglucinol, kaempferol, cirsimaritin, gallic acid 4-O-glucoside, carnosic acid and gallic acid.	*Himanthalia elongate*	Ireland autumn (September/October)	LC-DAD–ESI-MS/MS	Rajauria et al. (2016)
β-sitosterol Oleic acid	*Halimeda gracilis*	Teluk Lampung	NMR	Hendri et al. (2017)
(6Z,9Z,12Z,15Z,18Z)-1,6,9,12,15,18-henicosahexaene (6Z,9Z,12Z,15Z)-1,6,9,12,15-henicosapentaene	*Sargassum thunbergii*	Aio bay, Japan December 2015.	GCMS and NMR	Lu et al. (2018)
Fucosterol, mixture of saringosterols	*Cystoseira foeniculacea*	Algeria	Column chromatography NMR	Bouzidi et al. (2019)
Phlorotannin	*Fucus vesiculosus*	Algaplus Lda. Ria de Aveiro coastal lagoon, Northern Portugal	UHPLC-DAD-ESI-MS	Catarino et al. (2019)
fatty acids and phenols	*Laminaria ochroleuca*	Galicia (Spain)	Pressurized liquid extraction (PLE) GCMS	Otero et al. (2019)
N-benzyl cinnamamide α-resorcylic acid.	*Gracilaria fisheri*	Southern coast of Surat Thani province, Thailand	FTIR, NMR, and HR-TOF-MS	Karnjana et al. (2020)
Laminarin	*Saccharina latissima*	Swedish west coast July 2018	Size exclusion chromatography, NMR	Sterner and Gröndahl. (2021)

(UPLC-MS), etc. (Segers et al. 2019). Every technique has its virtues and limitations. The choice of an appropriate analytical technique depends upon the type or class of metabolites intended to be targeted. A combination of protocols may be used to augment the information from the analysis. The collective data is then pooled together and subjected to further analysis and simplification using several software packages applicable to such large data sets.

Metabolite profiling is characteristically performed following three main approaches: 1) Untargeted/Non-targeted approach 2) Semi-targeted approach and 2) Targeted approach. Untargeted/Non-targeted metabolite profiling focuses on the assessment and comparison of detected peaks across a sample set followed by their identification using various metabolomics databases (Bingol 2018). A semi-targeted metabolomics approach aims at an explicit identification as well as quantification of a large number of metabolites (Liu and Locasale 2017). While targeted approach circles around the identification as well as quantification of specific, defined, and limited metabolites of interest. Figure 33.2 represents a general metabolomics workflow used for determining the metabolites present in a biological specimen. Despite the recent technological advancements and scientific know-how, the scientific community still faces the limitation of measuring every single metabolite within the metabolome present in a biological system. No sole evaluation strategy or technique can deliver all the desired information (Doerr 2017). Hence, it becomes

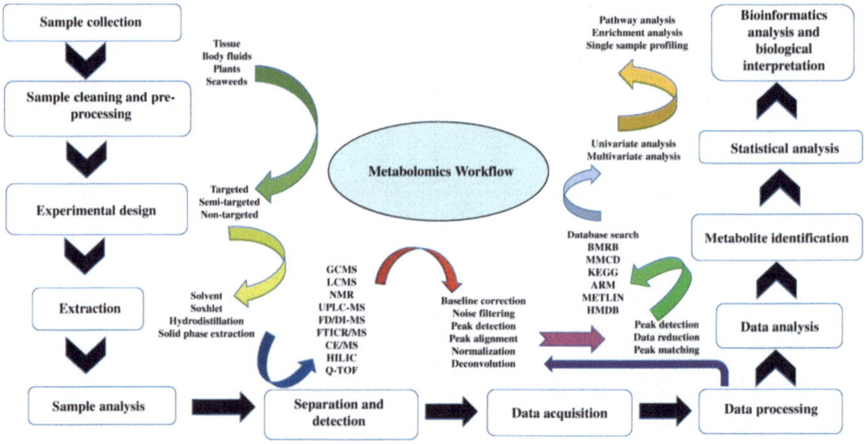

Fig. 33.2 A general summary of the metabolomics workflow. GCMS Gas Chromatography, Mass Spectrometry, LCMS Liquid Chromatography-Mass Spectrometry, NMR Nuclear Magnetic Resonance, UPLC-MS Ultra Performance Liquid Chromatography–Tandem Mass Spectrometry, FD/DI-MS Flow- or Direct-Injection/Infusion Mass Spectrometry, FTICR/MS Fourier Transform–Ion Cyclotron Resonance– Mass Spectrometry, CE/MS Capillary Electrophoresis–Mass Spectrometry, HILIC Hydrophilic Interaction Liquid Chromatography, Q-TOF Quadrupole Time-Of-Flight, BMRB Biological Magnetic Resonance Bank, MMCD Madison-Qingdao Metabolomics Consortium Database, KEGG Kyoto Encyclopedia of Genes and Genomes, ARM Atomic Reconstruction of Metabolism, METLIN METLIN Metabolite and Chemical Entity Database, HMDB Human Metabolome Database

very crucial to make appropriate choices of both, the preferred metabolites of interest as well as the analytical platform. However, marine metabolites possess the potential to emerge as promising entities provided with due research impetus.

33.4 Conclusion and Future Prospectives

The demand for seaweed-derived value-added products is increasing in recent years due to its utilization in various applications. Several nutraceuticals, food, pharmaceutical, cosmeceutical, non-food, and bioenergy products have been extracted from the seaweeds. Aquaculture is the most common technique practiced globally for seaweed cultivation. However, the cultivation of seaweeds still does not satisfy the industries and population demand. To fulfill this demand gap, various modifications and advancements in seaweed cultivation have been optimized. Moreover, 'omics' technology (mostly genomics and metabolomics) along with growth engineering via genetic and developmental cycle modulation has been arbitrarily introduced. Contrast to outdoor seaweed cultivation, PBRs have been identified to overcome the environmental issues regarding seaweed growth and metabolism. Nevertheless, the seaweed-based product markets are taking off with huge profit returns but still, a scientific knowledge gap reflects in the field of cultivation, genetic engineering, and species selection which will provide a huge research scope in the future. The bioenergy production from seaweeds is the most advanced strategy which will be the alternative source of the limited fossil fuel with reduced carbon accumulation in the ecosystem. Advancement in protocols and technologies will reduce macroalgal biomass wastage. Therefore, the sequential extraction process has enlightened the disadvantages of the direct extraction method. Progressive technology development with advanced strategies has exhibited the introduction of seaweed biorefineries which will be the next-generation advancement in the cultivation and utilization of seaweed biomass. The metabolomics approach enables the elucidation of bioactive natural algal metabolites that possess immense and untapped pharmaceutical as well as nutraceutical potential. The advancements in the fields of bioinformatics, as well as computational biology, will boost metabolomics oriented scientific exploration of algal species for human well-being.

Acknowledgments Author (NT) would like to thank the Department of Science & Technology, Delhi, India for the DST INSPIRE faculty award. All the authors have contributed equally to the book chapter with no conflict of interest.

References

Ahn M, Moon C, Yang W et al (2011) Diphlorethohydroxycarmalol, isolated from the brown algae *Ishige okamurae*, protects against radiation-induced cell damage in mice. Food Chem Toxicol 49:864–870

Belda M, Sanchez D, Bover E et al (2016) Extraction of polyphenols in *Himanthalia elongata* and determination by high performance liquid chromatography with diode array detector prior to its potential use against oxidative stress. J Chromatogr B 1033-1034:334–341

Bianco ÉM, Teixeira VL, Pereira RC (2010) Chemical defenses of the tropical marine seaweed *Canistrocarpus cervicornis* against herbivory by sea urchin. Braz J Oceanogr 58:213–218

Bingol K (2018) Recent advances in targeted and untargeted metabolomics by NMR and MS/NMR methods. High-Throughput. https://doi.org/10.3390/ht7020009

Bouzidi N, Viano Y, Ortalo-Magné A et al (2019) Sterols from the brown alga *Cystoseira foeniculacea*: degradation of fucosterol into saringosterol epimers. Arab J Chem 12:1474–1478

Buschmann AH, Camus C, Infante J et al (2017) Seaweed production: overview of the global state of exploitation, farming and emerging research activity. Eur J Phycol 52:391–406

Catarino MD, Silva AMS, Mateus N et al (2019) Optimization of phlorotannins extraction from *Fucus vesiculosus* and evaluation of their potential to prevent metabolic disorders. Mar Drugs. https://doi.org/10.3390/md17030162

Charrier B, Abreu MH, Araujo R et al (2017) Furthering knowledge of seaweed growth and development to facilitate sustainable aquaculture. New Phytol 216:967–975

Davis GDJ, Vasanthi AHR (2011) Seaweed metabolite database (SWMD): a database of natural compounds from marine algae. Bioinformation 5:361–364

De-Paula JC, Cavalcanti DN, Yoneshigue-Valentin Y et al (2012) Diterpenes from marine brown alga *Dictyota guineensis* (Dictyotaceae, Phaeophyceae). Rev Bras Farmacogn 22:736–740

Dixit D, Reddy CRK (2017) Non-targeted secondary metabolite profile study for deciphering the cosmeceutical potential of red marine macro alga *Jania rubens*—an LCMS based approach. Cosmetics. https://doi.org/10.3390/cosmetics4040045

Dixit D, Reddy CRK, Trivedi MH et al (2020) Non-targeted metabolomics approach to assess the brown marine macroalga *Dictyota dichotoma* as a functional food using liquid chromatography with mass spectrometry. Sep Sci Plus 3:140–149

Doerr A (2017) Global metabolomics. Nat Methods. https://doi.org/10.1038/nmeth.4112

Food and Agriculture Organization (FAO) (2018) The global status of seaweed production, trade and utilization. Globefish Research Programme 124:120

Galle JB, Attioua B, Kaiser M et al (2013) Eleganolone, a diterpene from the French marine alga *Bifurcaria bifurcata* inhibits growth of the human pathogens *Trypanosoma brucei* and *plasmodium falciparum*. Mar Drugs 11:599–610

García-Poza S, Leandro A, Cotas C et al (2020) The evolution road of seaweed aquaculture: cultivation technologies and the industry 4.0. Int J Environ Res Public Health. https://doi.org/10.3390/ijerph17186528

Giangrande A, Pierri C, Arduini D et al (2020) An innovative IMTA system: Polychaetes, sponges and macroalgae co-cultured in a southern Italian in-shore mariculture plant (Ionian Sea). J Mar Sci Eng. 10.3390/jmse8100733

Gressler V, Stein ÉM, Dörr F et al (2011) Sesquiterpenes from the essential oil of *Laurencia dendroidea* (Ceramiales, Rhodophyta): isolation, biological activities and distribution among seaweeds. Rev Bras Farmacogn 21:248–254

Gupta V, Kumar M, Brahmbhatt H et al (2011) Simultaneous determination of different endogenetic plant growth regulators in common green seaweeds using dispersive liquid-liquid microextraction method. Plant Physiol Biochem 49:1259–1263

Gupta PL, Lee SM, Choi HJ (2015) A mini review: Photobioreactors for large scale algal cultivation. World J MicrobiolBiotechnol 31:1409–1417

Gupta V, Trivedi N, Simoni S et al (2018) Marine macroalgal nursery: a model for sustainable production of seedlings for large scale farming. Algal Res 31:463–468

Hafting JT, Critchley AT, Cornish ML et al (2012) On-land cultivation of functional seaweed products for human usage. J Appl Phycol 24:385–392
Hafting JT, Craigie JS, Stengel DB et al (2015) Prospects and challenges for industrial production of seaweed bioactives. J Phycol 51:821–837
Hamid SS, Wakayama M, Ichihara K et al (2019) Metabolome profiling of various seaweed species discriminates between brown, red, and green algae. Planta 249:1921–1947
Hendri M, Darmanto JS, Prayitno B et al (2017) The isolation of metabolite compounds from seaweed (*Halimeda gracillis*) in the waters of Teluk Lampung as a source of antibacterial compounds. Int J Mar Sci 7:297–307
Hessami MJ, Cheng SF, Ranga Rao A, Yin YH, Phang SM (2019) Bioethanol production from agarophyte red seaweed, *Gelidium elegans* using a novel sample preparation method for analysing bioethanol content by gas chromatography. 3-Biotech Journal 9(1):25
Jha B, Kavita K, Westphal J et al (2013) Quorum sensing inhibition by *Asparagopsis taxiformis*, a marine macro alga: separation of the compound that interrupts bacterial communication. Mar Drugs 11:253–265
Kannan RRR, Aderogba MA, Ndhlala AR et al (2013) Acetylcholinesterase inhibitory activity of phlorotannins isolated from the brown alga, *Ecklonia maxima* (Osbeck) Papenfuss. Food Res Int 54:1250–1254
Karnjana K, Nobsathian S, Soowannayan C et al (2020) Purification and evaluation of N-benzyl cinnamamide from red seaweed *Gracilaria fisheri* as an inhibitor of *Vibrio harveyi* AI-2 quorum sensing. Mar Drugs. https://doi.org/10.3390/md18020080
Kim SM, Kang SW, Jeon J-S et al (2013) Determination of major phlorotannins in *Eisenia bicyclis* using hydrophilic interaction chromatography: seasonal variation and extraction characteristics. Food Chem 138:2399–2406
Kim J, Um M, Yang H et al (2016) Method development and validation for dieckol in the standardization of phlorotannin preparations. Fish Aquatic Sci. https://doi.org/10.1186/s41240-016-0003-2
Kurihara H, Kagawa Y, Konno R et al (2014) Lipoxygenase inhibitors derived from marine macroalgae. Bioorganic Med Chem Lett 24:1383–1385
Lehahn Y, Ingle KN, Golberg A (2016) Global potential of offshore and shallow waters macroalgal biorefineries to provide for food, chemicals and energy: feasibility and sustainability. Algal Res 17:150–160
Li Y-X, Li Y, Lee S-H et al (2010) Inhibitors of oxidation and matrix metalloproteinases, floridoside, and d-isofloridoside from marine red alga *Laurencia undulata*. J Agric Food Chem 58:578–586
Li K, Li X-M, Gloer JB et al (2012) New nitrogen-containing bromophenols from the marine red alga *Rhodomela confervoides* and their radical scavenging activity. Food Chem 135:868–872
Lin H, Qin S (2014) Tipping points in seaweed genetic engineering: scaling up opportunities in the next decade. Mar Drugs. https://doi.org/10.3390/md12053025
Liu X, Locasale JW (2017) Metabolomics: a primer. Trends Biochem Sci 42:274–284
Loureiro R, Gachon CMM, Rebours C (2015) Seaweed cultivation: potential and challenges of crop domestication at an unprecedented pace. New Phytol 206:489–492
Lu H, Xie H, Gong Y et al (2011) Secondary metabolites from the seaweed *Gracilaria lemaneiformis* and their allelopathic effects on *Skeletonema costatum*. Biochem Syst Ecol 39:397–400
Lu SJ, Yosemoto S, Satomi D et al (2018) Two types of volatile polyenes in the brown alga *Sargassum thunbergii*. J Oleo Sci 67:1463–1471
Machado FLS, Duarte HM, Gestinari LMS et al (2016) Geographic distribution of natural products produced by the red alga *Laurencia dendroidea*. J Agardh Chem Biodivers 13:845–851
Majik MS, Adel H, Shirodkar D et al (2015) Isolation of stigmast-5,24-dien 3 ol from marine brown algae *Sargassum tenerrimum* and its antipredatory activity. RSC Adv 5:51008–51011
Mazarrasa I, Olsen YS, Mayol E et al (2014) Global unbalance in seaweed production, research effort and biotechnology markets. Biotechnol Adv 32:1028–1036

Mhatre A, Navale M, Trivedi N et al (2018) Pilot scale flat panel photobioreactor system for mass production of *Ulvalactuca* (Chlorophyta). BioresourTechnol 249:582–591

Mohammad JH, Ranga Rao A, Ravishankar GA (2019) Opportunities and challenges in seaweeds as feed stock for biofuel production. In: Ravishankar GA, Rao AR (eds) Handbook of algal technologies and phytochemicals: volume-II Phycoremediation, biofuels and global biomass production. CRC Press, Boca Raton, pp 39–50

Montero L, Herrero M, Ibáñez E et al (2014) Separation and characterization of phlorotannins from brown algae *Cystoseira abies-marina* by comprehensive two-dimensional liquid chromatography. Electrophoresis 35:1644–1651

Moura LA, de Almeida ACM, Domingos TFS et al (2014) Antiplatelet and anticoagulant effects of diterpenes isolated from the marine alga, *Dictyota menstrualis*. Mar Drugs 12:2471–2484

Nabti E, Jha B, Hartmann A (2016) Impact of seaweeds on agricultural crop production as biofertilizers. Int J Environ Sci Technol. https://doi.org/10.1007/s13762-016-1202-1

Nakamura Y, Mochamad Afendi F, Kawsar Parvin A et al (2014) KNApSAcK metabolite activity database for retrieving the relationships between metabolites and biological activities. Plant Cell Physiol 55:1–9

Nylund GM, Enge S, Pavia H (2013) Costs and benefits of chemical defence in the red alga *Bonnemaisonia hamifera*. PLoS One. https://doi.org/10.1371/journal.pone.0061291

Olsen EK, Hansen E, Isaksson J et al (2013) Cellular antioxidant effect of four bromophenols from the red algae, *Vertebrata lanosa*. Mar Drugs 11:2769–2784

Otero P, Lopez-Martinez MI, García-Risco MR (2019) Application of pressurized liquid extraction (PLE) to obtain bioactive fatty acids and phenols from *Laminaria ochroleuca* collected in Galicia (NW Spain). J Pharm Biomed Anal 164:86–92

Pereira L (2018) Seaweeds as source of bioactive substances and skin care therapy- cosmeceuticals, algotheraphy, and thalassotherapy. Cosmetics. https://doi.org/10.3390/cosmetics5040068

Pereira RC, da Gama BAP, Sudatti DB (2016) In: Õlafsson E (ed) Marine Macrophytes as foundation species. CRC Press Taylor & Francis Group, Boca Raton, pp 26–42

Qin S, Lin H, Jiang P (2012) Advances in genetic engineering of marine algae. Biotechnol Adv 30:1602–1613

Rajak RC, Jacob S, Kim BS (2020) A holistic zero waste biorefinery approach for macroalgal biomass utilization: a review. Sci Total Environ. https://doi.org/10.1016/j.scitotenv.2020.137067

Rajauria G, Foley B, Abu-Ghannam N (2016) Identification and characterization of phenolic antioxidant compounds from brown Irish seaweed *Himanthalia elongata* using LC-DAD–ESI-MS/MS. Innov Food Sci Emerg Technol 37:261–268

Rasher DB, Stout EP, Engel S et al (2011) Macroalgal terpenes function as allelopathic agents against reef corals. PNAS 108:17726–17731

Segers K, Declerck S, Mangelings D et al (2019) Analytical techniques for metabolomic studies: a review. Bioanalysis 24:2297–2318

Shama A, Joyce SG, Mari FD, Ranga Rao A, Ravishankar GA, Hudaa N (2019) Macroalgae and microalgae: novel sources of functional food and feed. In: Ravishankar GA, Rao AR (eds) Handbook of algal technologies and phytochemicals: Volume-I: food, health and nutraceutical applications. CRC Press, Boca Raton, pp 207–219

Sterner M, Gröndahl F (2021) Extraction of laminarin from *Saccharina latissima* seaweed using cross-flow filtration. J Appl Phycol. https://doi.org/10.1007/s10811-021-02398-z

Stévant P, Rebours C, Chapman A (2017) Seaweed aquaculture in Norway: recent industrial developments and future perspectives. Aquac Int 25:1373–1390

Sugiura Y, Matsuda K, Okamoto T et al (2009) The inhibitory effects of components from a brown alga, *Eisenia arborea*, on degranulation of mast cells and eicosanoid synthesis. J Funct Foods 1:387–393

Young T, Alfaro AC (2016) Metabolomic strategies for aquaculture research: a primer. Rev Aquac 10:26–56

Zheng Y, Jin R, Zhang X et al (2019) The considerable environmental benefits of seaweed aquaculture in China. Stoch Environ Res Risk 33:1203–1221

Index

A. Ranga Rao, G. A. Ravishankar (eds.), *Sustainable Global Resources of Seaweeds Volume 1*, https://doi.org/10.1007/978-3-030-91955-9

B

C

N

O

R

S

The manufacturer's authorised representative in the EU is Springer Nature Customer Service Centre GmbH, Europaplatz 3, 69115 Heidelberg, Germany. If you have any concerns regarding our products, please contact ProductSafety@springernature.com

Printed and bound by CPI Group (UK) Ltd, Croydon, CR0 4YY
01/10/2026
02225893-0017